THE ROUTLEDGE HANDBOOK OF PHILOSOPHY OF RELATIVISM

Relativism can be found in all philosophical traditions and subfields of philosophy. It is also a central idea in the social sciences, the humanities, religion and politics. This is the first volume to map relativistic motifs in all areas of philosophy, synchronically and diachronically. It thereby provides essential intellectual tools for thinking about contemporary issues like cultural diversity, the plurality of the sciences, or the scope of moral values.

The Routledge Handbook of Philosophy of Relativism is an outstanding major reference source on this fundamental topic. The 57 chapters by a team of international contributors are divided into nine parts:

- Relativism in non-Western philosophical traditions
- Relativism in Western philosophical traditions
- Relativism in ethics
- Relativism in political and legal philosophy
- Relativism in epistemology
- Relativism in metaphysics
- Relativism in philosophy of science
- Relativism in philosophy of language and mind
- Relativism in other areas of philosophy

Essential reading for students and researchers in all branches of philosophy, this handbook will also be of interest to those in related subjects such as politics, religion, sociology, cultural studies and literature.

Martin Kusch is professor of philosophy at the University of Vienna, Austria. He was the principal investigator of an ERC Advanced Grant project "The Emergence of Relativism" (2014–2019), and he is co-editor of *The Emergence of Relativism*, also published by Routledge (2019).

Routledge Handbooks in Philosophy

Routledge Handbooks in Philosophy are state-of-the-art surveys of emerging, newly refreshed, and important fields in philosophy, providing accessible yet thorough assessments of key problems, themes, thinkers, and recent developments in research.

All chapters for each volume are specially commissioned, and written by leading scholars in the field. Carefully edited and organized, *Routledge Handbooks in Philosophy* provide indispensable reference tools for students and researchers seeking a comprehensive overview of new and exciting topics in philosophy. They are also valuable teaching resources as accompaniments to textbooks, anthologies, and research-orientated publications.

Also available:

The Routledge Handbook of Virtue Epistemology
Edited by Heather Battaly

The Routledge Handbook of Moral Epistemology
Edited by Karen Jones, Mark Timmons, and Aaron Zimmerman

The Routledge Handbook of Love in Philosophy
Edited by Adrienne M. Martin

The Routledge Handbook of the Philosophy and Psychology of Luck
Edited by Ian M. Church and Robert J. Hartman

The Routledge Handbook of Emergence
Edited by Sophie Gibb, Robin Hendry, and Tom Lancaster

The Routledge Handbook of the Philosophy of Evil
Edited by Thomas Nys and Stephen de Wijze

The Routledge Handbook of Social Epistemology
Edited by Peter Graham, Nikolaj Jang Lee, David Henderson, and Miranda Fricker

The Routledge Handbook of Philosophy of the City
Edited by Sharon M. Meagher, Samantha Noll, and Joseph S. Biehl

The Routledge Handbook of Panpsychism
Edited by William Seager

The Routledge Handbook of Emotion Theory
Edited by Andrea Scarantino

The Routledge Handbook of Philosophy of Relativism
Edited by Martin Kusch

For more information about this series, please visit: www.routledge.com/Routledge-Handbooks-in-Philosophy/book-series/RHP

THE ROUTLEDGE HANDBOOK OF PHILOSOPHY OF RELATIVISM

Edited by Martin Kusch

Routledge
Taylor & Francis Group

LONDON AND NEW YORK

First published 2020
by Routledge
4 Park Square, Milton Park, Abingdon, Oxon OX14 4RN
605 Third Avenue, New York, NY 10017

First issued in paperback 2023

Routledge is an imprint of the Taylor & Francis Group, an informa business

British Library Cataloguing-in-Publication Data
A catalogue record for this book is available from the British Library

Library of Congress Cataloging-in-Publication Data
A catalog record for this book has been requested

ISBN: 978-1-03-257033-4 (pbk)
ISBN: 978-1-138-48428-3 (hbk)
ISBN: 978-1-351-05230-6 (ebk)

DOI: 10.4324/9781351052306

Typeset in Bembo
by Apex CoVantage, LLC

Publisher's Note
The publisher has gone to great lengths to ensure the quality of this reprint but points out that some imperfections in the original copies may be apparent.

CONTENTS

Contents

Contents

ACKNOWLEDGEMENTS

Work on this project was made possible by ERC Advanced Grant Project #339382 "The Emergence of Relativism: Historical, Philosophical and Sociological Issues." I am grateful to all team members – Natalie Ashton, Katherina Kinzel, Robin McKenna, Katharina Sodoma, Johannes Steizinger, and Niels Wildschut – for numerous discussions, many of which have found their way into this volume.

In recent years my Viennese department has become something of a hub for research on relativism. This explains why many of my immediate colleagues were able to contribute. Several of them have also helped by suggesting topics or authors. I am particularly grateful to Delia Belleri, Dirk Kindermann, Anne-Kathrin Koch, Veli Mitova, and Herlinde Pauer-Studer. I also owe a great debt to Karoline Paier and Paul Tucek who acted as my research assistants and helped with the nitty-gritty of the editing process.

I thank the Routledge philosophy editors – Tony Bruce and Adam Johnson – for their wonderful support throughout. Four referees made a number of useful suggestions.

Last but not least I am grateful to my family for tolerating my working through (far too) many weekends. I dedicate this volume to my daughter Annabelle.

NOTES ON CONTRIBUTORS

Natalie Ashton is a postdoc in philosophy at Stirling University, United Kingdom.

Jussi Backman is a research fellow in philosophy at the University of Jyväskylä, Finland.

Bob Beddor is an assistant professor of philosophy at the National University of Singapore.

Delia Belleri is a postdoc in philosophy at the University of Vienna, Austria.

Jessica N. Berry is an associate professor of philosophy at Georgia State University, U.S.A.

David Bloor is a professor emeritus of sociology at the University of Edinburgh, United Kingdom.

Anna Boncompagni is an assistant professor of philosophy at the University of California at Irvine, U.S.A.

J. Adam Carter is a lecturer in philosophy at the University of Glasgow, United Kingdom.

Hasok Chang is a professor of history and philosophy of science at the University of Cambridge, United Kingdom.

Annalisa Coliva is a professor of philosophy at the University of California at Irvine, U.S.A.

George Crowder is a professor of political science at Flinders University, Austria.

Lawrence Dallman is a PhD student in philosophy at the University of Chicago, U.S.A.

E. Díaz-León is an associate professor of philosophy at the University of Barcelona, Spain.

Catarina Dutilh Novaes is a professor of philosophy at the VU Amsterdam, Netherlands.

Andy Egan is a professor of philosophy at Rutgers University, U.S.A.

Nader El-Bizri is a professor of philosophy at the American University of Beirut, Lebanon.

Filippo Ferrari is a postdoc in philosophy at the University of Bonn, Germany.

Vera Flocke is an assistant professor of philosophy at Indiana University at Bloomington, U.S.A.

Manuel García-Carpintero is a professor of philosophy at the University of Barcelona, Spain.

Kristin Gjesdal is a professor of philosophy at Temple University, U.S.A.

Anke Graness is a member of the research project "Histories of Philosophy in a Global Perspective" at the University of Hildesheim, Germany.

Steven D. Hales is a professor of philosophy at Bloomsburg University, U.S.A.

Espen Hammer is a professor of philosophy at Temple University, U.S.A.

Christoph Hanisch is an assistant professor of philosophy at Ohio University, U.S.A.

Eli Hirsch is a professor of philosophy at Brandeis University, U.S.A.

Drew Khlentzos is an adjunct professor of linguistics at Macquarie University, Austria.

Brian Kim is an assistant professor of philosophy at Oklahoma State University, U.S.A.

Dirk Kindermann is a postdoc in philosophy at the University of Vienna, Austria.

Katherina Kinzel is a postdoc in philosophy at the University of Vienna, Austria.

Charlotte Knowles is an assistant professor of philosophy at the University of Groningen, Netherlands.

Max Kölbel is a professor of philosophy at the University of Vienna, Austria.

Daniel Z. Korman is a professor of philosophy at the University of California, Santa Barbara, U.S.A.

Inkeri Koskinen is a postdoc in philosophy at the University of Helsinki, Finland.

Martin Kusch is a professor of philosophy at the University of Vienna, Austria.

John Christian Laursen is a professor of political science at the University of California, Riverside, U.S.A.

Brian Leiter is a professor of jurisprudence at the University of Chicago, U.S.A.

Martin Lenz is a professor of philosophy at the University of Groningen, Netherlands.

Geoffrey Brahm Levey is an associate professor of political science at the University of New South Wales, Australia.

Dustin Locke is an assistant professor of philosophy at Claremont McKenna College, U.S.A.

Edouard Machery is a professor of philosophy at the University of Pittsburgh, U.S.A.

John Marenbon is senior research fellow at Trinity College, Cambridge, United Kingdom.

Teresa Marques is a researcher at the University of Barcelona, Spain.

Robin McKenna is a lecturer in philosophy at the University of Liverpool, United Kingdom.

Matthew J. Moore is a professor of political science at Cal Poly University, U.S.A.

Victor Morales is an assistant professor of philosophy at the University of California, Riverside, U.S.A.

Tamer Nawar is an assistant professor of philosophy at the University of Groningen, Netherlands.

Paul O'Grady is a professor of philosophy at Trinity College Dublin, Ireland.

Timo Pankakoski is a postdoc in the Centre of European Studies at the University of Helsinki, Finland.

Alexandra Plakias is an assistant professor of philosophy at Hamilton College, U.S.A.

Gerald Posselt is a lecturer in philosophy at the University of Vienna, Austria.

Duncan Pritchard is professor of philosophy at the University of California at Irvine, U.S.A. and at the University of Edinburgh, United Kingdom.

Stathis Psillos is a professor of philosophy at the University of Athens, Greece.

Mark R. Reiff is a professor of philosophy at the University of California, Davis, U.S.A.

Sonja Rinofner-Kreidl is a professor of philosophy at the University of Graz, Austria.

David Rose is a professor of philosophy at Florida State University, U.S.A.

Howard Sankey is a professor of philosophy at Melbourne University, Australia.

Peter Seipel is an assistant professor at the University of South Carolina, Lancaster, U.S.A.

Sergej Seitz is a doctoral student in philosophy at the University of Vienna, Austria.

Jamie Shaw is a postdoc in philosophy the University of Western Ontario, Canada.

Torben Spaak is a professor of jurisprudence at the University of Stockholm, Sweden.

Florian Steinberger is a lecturer in philosophy at Birkbeck College, University of London, United Kingdom.

Johannes Steizinger is an assistant professor of philosophy at McMaster University, Hamilton, Canada.

Stephen Stich is a professor of philosophy and cognitive science at Rutgers University, U.S.A.

David J. Stump is a professor of philosophy at the University of San Francisco, U.S.A.

Henry Tam is a member of the Centre for Welfare Reform, Sheffield, United Kingdom.

Sthaneshwar Timalsina is an assistant professor of religious studies, San Diego State University, U.S.A.

Stephen Turner is a professor of philosophy at the University of South Florida, U.S.A.

Jared Warren is an assistant professor of philosophy at Stanford University, U.S.A.

David B. Wong is a professor of philosophy at Duke University, U.S.A.

Dan Zeman is a postdoc in philosophy at the University of Vienna, Austria.

INTRODUCTION

A primer on relativism

Martin Kusch

One could say of relativism what Hermann Ebbinghaus once observed with respect to psychology: to wit, that it has a "long past but a short history" (1908, 3). Although relativistic motifs have always played a significant role in philosophy, their systematic investigation – and thus the explicit formulation of different forms and strengths of relativism – is a child only of the twentieth century. Perhaps one could even maintain that most of the really important, detailed and systematic work on relativism was done by philosophers alive today. This volume documents both the long past and the short history of relativism.

The structure of the volume is straightforward. The first two parts cover relativistic motifs in Indian, Islamic, African and Western traditions. (Unfortunately, it proved impossible to secure chapters on relativistic motifs in Japanese and Chinese philosophy.) The following parts divide by subfields of philosophy: ethics (Part 3), political and legal philosophy (Part 4), epistemology (Part 5), metaphysics (Part 6), philosophy of science (Part 7), and philosophy of language and mind (Part 8). The last part ("Part 9: Relativism in Other Areas of Philosophy") contains chapters on philosophy of religion and experimental philosophy. (An entry on aesthetics fell through too late for it to be re-assigned.)

As is to be expected, the authors of this volume take very different positions concerning the forms of relativism they discuss. Some *support* particular versions of relativism, others *oppose* it vigorously in some or all of its variants. Still, I like to think that all contributions assembled here investigate relativism in a scholarly and respectful manner. This confirms my belief that disagreements over relativism need not have the character of "wars." Remember that, in the U.S., it is customary to speak of "wars" either when intellectual exchanges become acrimonious ("science wars") or when one launches an intellectual campaign ("war on cancer"). I regret that, in the past, disputes over relativism have all too often had the feel of "relativism wars" or "wars on relativism." Of course, in lamenting "relativism wars" I am not advocating a soggy attitude of "anything goes." I am urging that in discussing relativism we rely on the same epistemic virtues of curiosity, open-mindedness, fairness and charity that serve us so well in other – less emotionally charged – areas of philosophy.

Relativism is not easy to define, and no definition has found general approval. Still, readers new to the area might profit from at least a rough characterization of the spectrum of views falling under the term.

To begin with, it is common to capture forms of relativism as different instantiations of the scheme "x is relative to y" (Haack 1998, 149). Here are some examples:

"x" stands for. . .		forms of relativism
objects, properties, facts, worlds	. . .	ontological
truth(s)	. . .	alethic or semantic
classifications, concepts, meanings	. . .	semantic
moral values, norms, commitments, justifications	. . .	moral
knowledge or epistemic justification	. . .	epistemic
tastes	. . .	gustatory
"y" stands for. . .		forms of relativism
individuals	. . .	Protagorean
cultures	. . .	cultural
scientific paradigms	. . .	Kuhnian
classes, religions, genders	. . .	standpoint

A further important divide is between *descriptive, normative,* and *methodological relativisms.* To facilitate the exposition, I shall use "culture" as the relevant "y" and "morals" as the relevant "x." Forms of *descriptive* relativism claim that, as far as moral beliefs or standards are concerned, one finds fundamentally different standards in different cultures. Forms of *methodological* relativism insist that in investigating moralities we had better approach cultural differences in an "impartial" and "symmetrical" way. For instance, we had better be

> . . . impartial with respect to truth and falsity, rationality or irrationality, success or failure. Both sides of these dichotomies . . . require explanation.
>   . . . symmetrical in [the] style of explanation. The same types of cause would explain say, true and false beliefs.
>
> *(Bloor 1991, 7)*

Descriptive and methodological forms of relativism leave open the possibility that there are absolute norms or truths. As far as descriptive or methodological relativisms are concerned, one of the cultures might well be on the (absolutely) right track. *Normative* forms of relativism go further and deny that there are any absolutely true or absolutely correct beliefs or standards.

The last sentence gives only a very minimalist characterization of normative relativism. Although there are authors happy with this definition (Bloor 2011), others – friends and foes of relativism alike – go further and add various additional assumptions. A list of such assumptions follows. To be sure, I am not suggesting that these items constitute necessary and sufficient conditions for relativism. Different authors disagree over their importance and relevance. I am offering them here merely to give the (novice) reader a rough idea of the kinds of theses with which relativism is often associated. I am not addressing the question which combination of these theses leads to the most plausible version (by my lights). Note also that I have not aimed for the smallest possible set of theses; a limited degree of redundancy has been accepted for the sake of greater clarity. I am using *epistemic* relativism as my example. It should be obvious how the key parameters need to be changed to arrive at moral or other forms of relativism.

I have collected these assumptions from both friends and foes of relativism, including Maria Baghramian (2019), Barry Barnes and David Bloor (1982), Simon Blackburn (2005), David Bloor (2011), Paul Boghossian (2006), Adam Carter (2016), Lorraine Code (1995), Annalisa Coliva (2010), Hartry Field (2009), Steven Hales (2014), Gilbert Harman (1996), Christopher Herbert (2001), Barbara Herrnstein Smith (2018), Max Kölbel (2004), John MacFarlane (2014), Duncan Pritchard (2009), Gideon Rosen (2001), Richard Rorty (1979), Carol Rovane (2013), F. F. Schmitt (2007), Markus Seidel (2014), Harvey Siegel (1987), Sharon Street (2011), David Velleman (2015), Bernard Williams (1981), Michael Williams (2007), Timothy Williamson (2015) and Crispin Wright (2008).

(DEPENDENCE) A belief has an epistemic status only relative to either. . .

(a)   system of epistemic principles (REGULARISM), or
(b)   a coherent bundle of precedents (or paradigms) (PARTICULARISM).

The distinction between (a) and (b) is meant to clarify that *what a belief is relative to* is different for different versions of relativism. Some make the "regularist" assumption that *what a belief is relative to* is a set of more or less fundamental rules (e.g. Boghossian 2006); others rely on the "particularist" thought that it is individual and concrete precedents that guide our epistemic life (Kuhn 1962). (I am here borrowing a conceptual distinction from Dancy 2017.)

(PLURALISM) There is (has been, or could be) more than one such system or bundle.

Given PLURALISM, relativism is compatible with the idea that our current system or bundle is without an *existing* alternative. Moreover, PLURALISM permits the relativist to be highly selective in choosing those systems or bundles with respect to which relativism applies. She might for example restrict her relativistic thesis to just two systems or bundles.

(NON-ABSOLUTISM) None of these systems or bundles is absolutely correct.

I already mentioned NON-ABSOLUTISM as the minimal characterization of normative relativism. It can of course be combined with the other assumptions listed here.

(CONFLICT) Some of these systems or bundles are such that their epistemic verdicts on the epistemic status of given beliefs exclude one another. This can happen either. . .

(a)   because the two systems or bundles give incompatible answers to the same question, or
(b)   because the advocates of one system or bundle find the answers suggested by the advocates of another system or bundle unintelligible.

(a) is an "ordinary" disagreement; (b) captures cases of "incommensurability"; that is, cases where the advocates of two different systems or bundles find the categories and values of the other side unintelligible (cf. Kuhn 1962; Feyerabend 1975; van Fraassen 2002).

(SYMMETRY) Different systems or bundles are symmetrical in that they all are. . .

    (a) based on nothing but local causes of credibility (LOCALITY); and/or
    (b) impossible to rank except on the basis of a specific system or bundle (NON-NEUTRALITY); and/or
    (c) equally true or valid (EQUAL VALIDITY); and/or
    (d) impossible to rank since the evaluative terms of one system or bundle seem not applicable to another system or bundle (NON-APPRAISAL).

(a) is central, e.g. in Barnes and Bloor (1982); (b) can be found in Field (2009); and (d) in B. Williams (1981). (c) is routinely attributed to relativism by its critics (e.g. Baghramian 2019; Boghossian 2006; Williamson 2015), but typically rejected by card-carrying relativists (cf. e.g. Bloor 2011; Field 2009; Herbert 2001).

(CONVERSION) For some pairs of systems or bundles it is true that switching from one to the other has the character of a "conversion." "Conversion" stands for a switch (to new rules or precedents) that is not licensed by the rules or precedents of the old system or bundle.

This assumption plays of course a central role in relativism debates in the philosophy of science after Kuhn (Kuhn 1962; Feyerabend 1975; van Fraassen 2002).

(FAULTLESS DISAGREEMENT) If two epistemic subjects, committed to different epistemic systems or bundles, disagree over an epistemic issue, and if their differing views are based on their respective epistemic systems or bundles, then their disagreement is faultless: neither side can be faulted for their positions on the issue.

This assumption has played a crucial role in recent discussions in semantic relativism. A central paradigm has been questions of taste (Kölbel 2002; MacFarlane 2014).

(SEMANTIC RELATIVITY) An utterance of the form "Subject S is epistemically justified (unjustified) to believe that p" expresses a proposition of the following form:

    (a) *According to the epistemic system or bundle that I (the speaker) am committed to, S is epistemically justified (unjustified) to believe that p.* This proposition is absolutely true or false. (SEMANTIC CONTEXTUALISM)
    (b) *S is epistemically justified (unjustified) to believe that p.* This proposition is true or false relative to different systems or bundles. (SEMANTIC RELATIVISM)

The first option is formulated and criticized in Boghossian (2006). Wright (2008) defends (b) as dealing the relativist a better hand. Much contemporary debate concerns these two (as well as other, more complex) semantic options.

(METAPHYSICAL COMMITMENT)

    (a) (FACTUALISM) The property of *being epistemically justified* has as one of its relata (an element of) a system of rules or a bundle of precedents. This relatum is usually overlooked.

(b) (NON-FACTUALISM) Epistemic relativism is not a claim about the property of *being epistemically justified*; it is a claim about the meaning of the term "justified."

Here too Boghossian (2006) is to be credited with first having put this distinction on the table.

(CONTINGENCY) Which epistemic system or bundle a given group is committed to is a question of historical contingency.

If the history of the given group g had been different – for instance, if the members of g had had a different evolutionary or cultural history – g's current system or bundle could be substantially, perhaps even radically, different from what it is now. The contingency might reach deep: even those beliefs that group members deem "self-evident" might be discovered to be contingent. Becoming aware of the contingency of one's views in this sense can, but need not, undermine the strength of one's conviction (cf. Rosen 2001; Street 2011; Kinzel and Kusch 2018).

(GROUNDLESSNESS) A given epistemic system or bundle cannot be justified in anything but a circular fashion.

GROUNDLESSNESS is rarely formulated as a distinct ingredient of epistemic relativism. But it is sometimes invoked in arguments meant to establish the truth of relativism. For instance, it is occasionally put forward that epistemic relativism results from the recognition that all systems or bundles are on a par insofar as none of them is able to justify itself without moving in an (illegitimate) circle (cf. Williams 2007, 95).

(UNDERDETERMINATION) Epistemic systems and practices are not determined by facts of nature or truths that "are there anyway."

UNDERDETERMINATION is not to be confused with the thesis that the world has *no* causal impact on epistemic systems or bundles. Instead the relativist is committed to the view that more than one system or bundle is compatible with the given causal impact of the world (cf. Seidel 2014).

(SELF-VINDICATION) Every system or bundle is such that it vindicates as true or correct all beliefs formed by relying on its norms or precedents.

This view is sometimes attributed to relativists by their absolutist critics (e.g. Baghramian 2019). Relativists might retort that they do not wish to rule out that systems or bundles might be self-correcting, or that advocates of a given system or bundle might recognize – by their own lights – that another system or bundle would serve them better (Kusch 2019).

(ARBITRARY CHOICE) Assume an epistemic subject S, information I, known to S, and a belief B that S would like to hold. S is epistemically blameless if S picks such epistemic norms or precedents (system or bundle) E as make holding B epistemically rational. The choice of E is unconstrained by other epistemic standards.

ARBITRARY CHOICE is in the vicinity of a wide-spread interpretation of Feyerabend's formula "anything goes" (Feyerabend 1975; Boghossian 2001).

(TOLERANCE) Epistemic systems or practices other than one's own, must be tolerated.

Relativist views are often motivated by the wish or demand to be tolerant. But it is an open question whether one needs to be a relativist to be tolerant.

It is easy to appreciate that some of the preceding theses seem more important than others. For instance, to be counted a relativist, a philosopher must surely commit to (some version of) DEPENDENCE, PLURALISM, NON-ABSOLUTISM, CONFLICT or SYMMETRY. It is much less clear whether they would need to also adopt SELF-VINDICATION, FAULTLESS DISAGREEMENT, GROUNDLESSNESS, TOLERANCE or ARBITRARY CHOICE. Still, some relativists start from FAULTLESS DISAGREEMENT, TOLERANCE or GROUND-LESSNESS and then seek to argue in defense of, say, NON-ABSOLUTISM or CONFLICT on this basis.

Note also that arguments over epistemic relativism often take the form of a debate about which of the preceding the relativist is explicitly or implicitly committed to. For instance, few relativists endorse EQUAL VALIDITY or ARBITRARY CHOICE. Their critics aim to show that the counterintuitive assumptions EQUAL VALIDITY or ARBITRARY CHOICE follow from a combination of the other theses listed.

I am delighted to have had the opportunity to edit the to-date most extensive handbook on relativism. Of course, no handbook can pretend to be 100% complete in its coverage of relevant issues, and there are bound to be some omissions that experts will quickly identify. Still, I hope this work will be useful to non-philosophers, philosophy students and professional philosophers alike.

References

Baghramian, M. (2019), "The Virtues of Relativism," *Aristotelian Society Supplementary Volume* 93: 247–269.
Barnes, B. and Bloor, D. (1982), "Relativism, Rationalism and the Sociology of Knowledge," in *Rationality and Relativism*, edited by M. Hollis and S. Lukes, Oxford: Blackwell, 21–47.
Blackburn, S. (2005), *Truth: A Guide to the Perplexed*, Oxford: Oxford University Press.
Bloor, D. (1991), *Knowledge and Social Imagery*, second edition, Chicago: University of Chicago Press.
———. (2011), "Relativism and the Sociology of Scientific Knowledge," in *The Oxford Companion to Relativism*, edited by S. Hales, Oxford: Wiley-Blackwell, 433–455.
Boghossian, P. (2001), "How Are Objective Epistemic Reasons Possible?" *Philosophical Studies* 106: 1–40.
———. (2006), *Fear of Knowledge: Against Relativism and Constructivism*, Oxford: Clarendon Press.
Carter, A. (2016), *Metaepistemology and Relativism*, London: Palgrave Macmillan.
Code, L. (1995), *Rhetorical Spaces: Essays on Gendered Location*, London: Routledge.
Coliva, A. (2010), "Was Wittgenstein an Epistemic Relativist?" *Philosophical Investigations* 33: 1–23.
Dancy, J. (2017), "Moral Particularism," in *Stanford Encyclopedia of Philosophy*, Winter 2017 edition, edited by E. N. Zalta, https://plato.stanford.edu/entries/moral-particularism/.
Ebbinghaus, H. (1908), *Psychology: An Elementary Textbook*, Boston: Heath.
Feyerabend, P. (1975), *Against Method: Outline of an Anarchistic Theory of Knowledge*, London: New Left Books.
Field, H. (2009), "Epistemology Without Metaphysics," *Philosophical Studies* 143: 249–290.
Haack, S. (1998), *Manifesto of a Passionate Moderate: Unfashionable Essays*, Chicago: University of Chicago Press.
Hales, S. (2014), "Motivations for Relativism as a Solution to Disagreements," *Philosophy* 89: 63–82.
Harman, G. (1996), "Moral Relativism," in *Moral Relativism and Moral Objectivity*, edited by G. Harman and J. J. Thomson, Oxford: Blackwell, 1–64.

Herbert, C. (2001), *Victorian Relativity: Radical Thought and Scientific Discovery*, kindle version, Chicago: Chicago University Press.

Herrnstein Smith, B. (2018), *Practicing Relativism in the Anthropocene: On Science, Belief and the Humanities*, London: Open Humanities Press.

Kinzel, K. and M. Kusch (2018), "De-Idealizing Disagreement, Rethinking Relativism," *International Journal of Philosophical Studies* 26: 40–71.

Kölbel, M. (2002), *Truth Without Objectivity*, London: Routledge.

Kölbel, M. (2004), "Faultless Disagreement," *Proceedings of the Aristotelian Society* 104: 53–73.

Kuhn, T. (1962), *The Structure of Scientific Revolutions*, Chicago: University of Chicago Press.

Kusch, M. (2019), "Relativist Stances, Virtues and Vices," *Aristotelian Society Supplementary Volume* 93: 271–291.

MacFarlane, J. (2014), *Assessment Sensitivity: Relative Truth and Its Applications*, New York and Oxford: Oxford University Press.

Pritchard, D. H. (2009), "Defusing Epistemic Relativism," *Synthese* 166: 397–412.

Rorty, R. (1979), *Philosophy and the Mirror of Nature*, Princeton, NJ: Princeton University Press.

Rosen, G. (2001), "Nominalism, Naturalism, Epistemic Relativism," *Philosophical Perspectives* 15: 60–91.

Rovane, C. (2013), *The Metaphysics and Ethics of Relativism*, Boston: Harvard University Press.

Schmitt, F. F. (2007), "Introduction: Epistemic Relativism," *Episteme* 4: 1–9.

Seidel, M. (2014), *Epistemic Relativism: A Constructive Critique*, London: Palgrave Macmillan.

Siegel, H. (1987), *Relativism Refuted*, Dordrecht: Reidel.

Street, S. (2011), "Evolution and the Normativity of Epistemic Reasons," *Canadian Journal of Philosophy* 35: 213–248.

van Fraassen, B. (2002), *The Empirical Stance*, Princeton, NJ: Princeton University Press.

Velleman, D. (2015), *Foundations for Moral Relativism*, second edition, Cambridge: Open Book Publishers.

Williams, B. (1981), "The Truth in Relativism," in his *Moral Luck*, Cambridge: Cambridge University Press, 132–143.

Williams, M. (2007), "Why (Wittgensteinian) Contextualism Is Not Relativism," *Episteme* 4: 93–114.

Williamson, T. (2015), *Tetralogue: I'm Right, You're Wrong*, Oxford: Oxford University Press.

Wright, C. (2008), "Fear of Relativism?" *Philosophical Studies* 141: 379–390.

PART 1

Relativism in non-Western philosophical traditions

1

RELATIVISM IN THE INDIAN TRADITION

Examining the viewpoints (dṛṣṭis)

Sthaneshwar Timalsina

Two birds, paired companions, occupy the same tree.
Of the two, one eats the sweet fig.
The other, not eating, looks on.

(Ṛgveda I.164.20)

1. Introduction

Thinking about classical Indian philosophy in light of relativism is a challenging hermeneutic task. There are no readymade volumes in the classical literature that we can identify under this category. Siderits argues along these lines that "the cultural factors that make relativism a pressing issue for us were largely absent from the classical Indian context, so that the various forms of relativism do not receive philosophical scrutiny in the Indian tradition" (2016, 24). The fundamental problem in thinking about Indian philosophy through relativism is not that there are no readymade texts but that scholars refrain from engaging relativism, as if it is taboo or a disease that philosophers need to stay away from (e.g. Siderits 2016, 31, 35). My own approach to relativism is relativistic, as I believe that endorsing relativism in one respect does not require one to be relativistic in all accounts. Just like any other "ism," relativism should be handled as a device to fathom human nature and to help humanity negotiate a perplexing, complex social reality. When we open ourselves to read classical Indian materials through the lens of relativism, we encounter a wealth of materials. Dialogues recorded in Vedic literature epitomize cultural fluidity, diversity and an openness to perspectives. Traditions have adopted perspectivism to make sense of an otherwise bewildering variety of commentarial literature with conflicting interpretations. The problem then is we encounter a semblance of relativism and can be easily misdirected. Before we assign epistemic relativism in the Jain "multiperspectivism" (*anekāntavāda*) or moral relativism in the *Mahābhārata* or meaning relativism in Bhartṛhari's philosophy of language, we need to carefully define the categories and explore the parameters.

Cultural pluralism was a norm in classical India and every region dealt with religious differences. Everyday society also incorporated linguistic differences and grammarians such as Patañjali were keenly aware of dialectical variations even within a single language. Combined with polytheism and panpsychism, India is founded upon the co-existence of different and

11

at times, conflicting viewpoints. Written in this cultural milieu, texts such as *Bhagavadgītā* endorsed different soteriological approaches by necessity, to combine multiple methods for liberation. It is not possible to address all these issues in a few pages. I therefore limit myself to re-examination of some of Nāgārjuna's claims, keeping in mind both classical and contemporary interpretations. I explore, in particular, the doctrine of "two truths" and Nāgārjuna's interpretation of the "viewpoints" (*dṛṣṭis*). In so doing, I am open to drawing parallels and initiating a cross-cultural dialogue on relativism. In conclusion, this conversation boils down to relativism leading to truth skepticism on the one hand and pluralism and hierarchical truth predications on the other.

2. Nāgārjuna on viewpoints (dṛṣṭis)

Nāgārjuna (150–250) is one of the major Buddhist philosophers and the founder of the Mādhyamika school. Scholars have primarily read his philosophy for its dialectical methods, rejection of substantialism, and interpretation of the doctrine of "emptiness" (*śūnyatā*). Most importantly, he is known for his pioneering doctrine of "two truths" (*dve satye*) and deconstruction of "viewpoints" (*dṛṣṭi*). Nāgārjuna introduces a unique logical method that reduces the opponent's viewpoints to absurdity (*reductio ad absurdum*) to defend his position that there is no inherent nature (*svabhāva*), whether by ontological truth claims regarding substance, or epistemic claims regarding reality – including the limits to human rationality. Nāgārjuna explores any proposition in terms of fourfold possible extremes (*koṭi*), eventually proving it absurd to adopt any one of those extremes.

Regarding the inherent nature (*svabhāva*) of being and things, Nāgārjuna posits and then refutes that:

(1) Things have inherent nature ("is" thesis).
(2) Things do not have inherent nature ("is not" thesis).
(3) Things simultaneously possess and lack inherent nature ("is and is not" thesis).
(4) Things lack both the inherent nature and the lack thereof (not – "is and is not" thesis).

Regarding causality, he likewise proposes as categories that:

(1) Things emerge because of the internal factors ("*svataḥ*" or "from within" thesis).
(2) Things emerge because of the external factors ("*parataḥ*" or "from without" thesis).
(3) Things emerge due both to the internal as well as the external factors ("*dvābhyām*" or "from both" thesis).
(4) Things emerge without any cause ("*ahetutaḥ*" or from "no cause" thesis).

In rendering this thesis of an "intrinsic nature" (*svabhāva*) absurd, Nāgārjuna establishes the doctrine of "emptiness" (*śūnyatā*). Examining this discussion historically, what he says is that, just like aggregates do not have their own intrinsic nature (the position that the Abhidharma school has endorsed), so also do the building blocks of the manifest reality, the so-called *dharmas*, not have any inherent nature. The tricky part is that he is not advocating this last statement as his thesis. The argument is if the emptiness of inherent nature were a thesis, this would be tantamount to endorsing absolutism by another name. Therefore, the negation of intrinsic nature is just a negation. The problem is that this understanding only partially captures the way Nāgārjuna has been historically understood. Reading Nāgārjuna is perplexing for both the classical commentators and contemporary scholars alike. The following verse is ground-zero of our investigation:

> The teaching of the dharma(s) by the Buddha relies on two truths: the limited conventional truth and the truth as it is.
>
> *(MMK XXIV.8)*[1]

There are obviously two different ways to understand this passage. It can mean that phenomenal truth exists and only applies to conventional reality and that absolute truth transcends language and concepts. This understanding of a hierarchy of truth does not reject truth claims, and can be interpreted in two different ways: first, that there are two tiers of truth, or second, that there are different sets of truths. In another possible interpretation of "two truths," this verse can also be explained by truth that is conceived of in the "covered" (*saṃvṛti*) state. For example, a truth such as seeing a sand dune as mirage or a rope as a snake, does not amount to actual truth due to its origination within a state of delusion. As a result, this view asserts that truth only exists corresponding to the way the entities are (*parama-artha-taḥ*). Therefore, a correspondence theory of truth underlies this interpretation. And if this position is followed, Nāgārjuna would not be making any anti-foundational claim in the exalted sense. This reading, however, would contradict Nāgārjuna's own proclamation that there is no "inherent nature" (*svabhāva*), as this would simply be replacing one form of absolutism with another. This would also contradict Nāgārjuna's direct statement that openly rejects absolutism regarding emptiness (*śūnyatā*):

> It is not our fault that you resort to emptiness. No foundation (sa = *adhilaya*) can be established on emptiness.
>
> *(MMK XXIV.13)*

> If what is described in terms of [the entities] lacking their inherent nature is the very being of the lack of the inherent nature, this would negate the lack of inherent nature and only the being of inherent nature would be established.
>
> *(VV 26, see Bhattacharya et al. 1978)*[2]

Keeping these straightforward stanzas in mind, Siderits argues that the term *paramārtha* or "the way the things are" does not confirm any ultimate truth, but on the contrary, "the ultimate truth is that there is no ultimate truth" (Siderits 1989, 231). Garfield confirms this same interpretation:

> Suppose that we take a conventional entity, such as a table. We analyze it to demonstrate its emptiness, finding that there is no table apart from its parts. . . . So, we conclude that it is empty. But now let us analyze that emptiness. . . . What do we find? Nothing at all but the table's lack of inherent existence. . . . To see the table as empty . . . is to see the table as conventional, as dependent.
>
> *(Garfield 2002, 38–39)*

There are two possible responses to the preceding statements, and both were historically applied by Nāgārjuna. One response is to reject such a claim, demonstrating circularity in its logic, arguing that even this amounts to a truth claim. The other is to apply linguistic or conceptual tactics to interpret negation while keeping open the possibility of speaking about the truth. The current conversation on relativism claims a central place in this shift from a correspondence theory of truth. Whether to understand Nāgārjunian claims as metaphysical or semantic is not a new quandary. So far, recent discussions and arguments are a flimsy replica of the debate between the Prāsaṅgika and the Svātantrika readings sustained over millennia.[3] The dilemma though is if this is a rejection of the absolute truth, and the conventional is not the "truth" per se, there is no truth to defend. With this view, the category "truth" would be fictitious, like rabbit-horn. And if this is only the rejection of absolute truth but not of relative truth and therefore interdependent truth, this would mean that truth is always relative, perspectival, and this position is not a rejection of "truth."

3. Truth: metaphysical or semantic

If what Nāgārjuna meant is that there are two truths, this would be a metaphysical theory, a theory about the ultimate nature of reality. The semantic interpretation recognizes this proclamation as not about the nature of reality but about the nature of truth. Siderits explains that, "all things are empty [means] that the ultimate truth [has] no ultimate truth – there is only conventional truth" (Siderits 2003, 11). This would help to separate truth claims from metaphysical reality and we could say, the statement "Rāvaṇa had ten heads" is true based on narratives, irrespective of the possible existence of such a monster. Returning to the position of "two truths," a semantic interpretation claims that no statement can be ultimately true. Siderits argues further, "Given that *dharmas* must be things with intrinsic natures, if nothing can bear an intrinsic nature, then there is nothing for ultimately true statements to be about; hence the very notion of ultimate truth is incoherent" (2003, 11–12). It appears Siderits draws from Hilary Putnam to develop a thesis that requires the rejection of any singular truth regarding the nature of reality that would presuppose a model of metaphysical realism. The target is to reject "emptiness" (*śūnyatā*) itself as a metaphysical claim. And this position omits the demolition of such a premise by the logical fallacy of circularity. To say that "there is no final truth about reality" would also apply to the claim that "all things are empty," which of course one would expect the Mādhyamika philosophers to reject. And historically some have taken this route. Siderits, however, suggests that even the claim "all things are empty" is only conventionally true.

Re-contextualization of the claims is necessary to establish any form of relativism based on the aforementioned position. To say that truth is only conventional, the conclusion derived from Siderits' reading, opens up a potential space for multiple perspectives in which all retain a degree of validity. This, however, is not what Siderits proposes and it deviates from Nāgārjuna's position, as it yet again underlies a supposition on the truth per se; specifically that, in an underlying metaphysical claim, even absolute truth can only be relatively revealed. The rejection of absolute truth does not, however, confirm the validity of viewpoints (*dṛṣṭi*), as has already been argued. To assume all that can be spoken of truth are just viewpoints does not mean the same judgment can't be true in one perspective while false in another. Nāgārjuna is not proposing that the human encounter with reality is mediated by language or culture. But if we were to read that "two truth" theory affirms perspectives, while not discrediting the category truth in the ultimate sense, we can derive that truth is relatively revealed in different modes. We can now engage G. Ferraro's (2013) arguments with this new accommodation to address relativism.

Ferraro argues against this semantic reading, maintaining that Nāgārjuna's doctrine of "two truths" upholds "two visions of reality on which the Buddhas, for soteriological and pedagogical reasons, build teachings of two types" (2013, 563). Emptiness (*śūnyatā*), in this reading, is in fact "equivalent to supreme truth." To make his claim, Ferraro first divides the metaphysical claims into two groups:

(1) a realistic metaphysical reading that considers "supreme truth an existing and somehow characterizable dimension," and
(2) an anti-realistic metaphysical reading that denies the "existence of supreme truth" and affirms "existence exclusively of ordinary reality" (2013, 566).

Now the argument is that whatever applies to our pedagogical approach also applies to the use of language: our use of language or words are relational, and while our objective may be to speak the "truth," given that there are metaphysical truths to be conveyed by language, our approaches can vary. Consequently, we can derive that the conventional is a necessary step, that we can discuss truth only conventionally. And since it is counterintuitive to conceive of the "conventional" as being a single perspective, the discourse on truth automatically becomes perspectival and relational. This claim, therefore, could reject both the metaphysical claim, and the validity of the so-called supreme truth. The fundamental divergence in this interpretation with Siderits and Garfield (2013) arises due to confusion between metaphysical and semantic interpretations. Siderits and Garfield argue that semantic interpretation does not interpret "two truths," but demonstrates that truth is a semantic property. In the Buddhist historical context, if reality is analyzed based on *dharmas* or essential factors, the emptiness doctrine says that even *dharmas* lack inherent characteristics and thus are devoid of intrinsic nature. In this sense, what Siderits and Garfield propose only negates the reality of what is proposed as a higher reality of *dharmas*.

There is not much new to add, except to point out that contemporary conversations are enriched with nuances borrowed from a global philosophical discourse. While we should persist in the hermeneutic task, our first loyalty goes to reading the texts the way that they are. One can be relative about different interpretations but not about the actual words. And when we look back to the texts themselves, we encounter that the term *satya* for example, is not just for the truth but also for reality. Derived from the Sanskrit verbal root *asa*, the term only describes the mode of *sat*, or that which is. Nāgārjunian terms for the so-called two truths are *saṃvṛti* and *paramārtha*, where the first does not translate to "relative" but "covered," and it also means "covering." There is nothing "supreme" in the *paramārtha* either, as the term is a compound of "*parama + artha*" with the first being in the superlative of *para* meaning the other, and thus meaning the last or the final, and *artha* referring to both "meaning" and "reality."

Multiple interpretations of the same text lead to hermeneutic relativism. Even when we ignore the examples where the same commentator derives different meanings from the same passage, texts come with multiple commentaries with contrasting meanings. For a reader, there are always options in determining meaning. A *relativistic* hermeneutic approach, however, does not open a text to anarchy in meaning. Even the skeptics such as Jayarāśi were not skeptical about reason per se. And the openness of interpretation only meant that readers needed to be openminded about perspectives as far as the semantic power of words can accommodate. When we read Sanskrit literature, we not only come across multiple commentaries, we even encounter different interpretations in the commentary written by the same author who composed the original text. Buddhist literature is no exception to the phenomenon of different interpretations for the same passage or the same author composing the text and its commentary. All in all, there is no taboo for a multi-façade-interpretation as far as classical exegesis is concerned. If we give credit to Nāgārjuna for being the philosopher that he is, it is not hard to conceive that he is aware of both possibilities, and is leaving the text open-ended regarding the ways it can be read. The interpretative prowess within the context of MMK is epitomized in the commentarial literature.

For our current purpose, let us say Nāgārjuna makes a realistic metaphysical claim that he considers a two-tier truth theory. Even if this does not directly confront relativism, we can accept that teaching methods and what is described are relative to the audience. In other words, our words can mean what they mean based on external factors. Accordingly, the teaching of the four noble truths (suffering, origination, cessation of suffering, and the means to end suffering) relates to "adopting the limited perspective"; while teaching emptiness relates to "following the supreme truth." On the other hand, if we follow Siderits and Garfield (2013), we are left with just perspectives and can only confirm relativism. However, these are very different types of relativism. The relativism that fits better with perspectivism should not be conflated with relativism regarding rationality. Even following Siderits and Garfield in this regard, there is no need to confirm that all epistemic claims are equal or that knowledge is a norm of assertion governing rational inquiry (see Walsh 2015). Whichever position, reading Nāgārjuna in light of relativism remains valid. However, if we mean "hard" relativism, we can argue along the lines of what Siderits says:

> The Prāsaṅgikas, with their no-theory approach to conventional truth, would be forced to accept the relativism about rationality that such evidence seems to suggest. But the Svātantrikas could, I think, be pluralists without being relativists: pluralists in admitting a plurality of possible canons of rationality, no single one of which is ideally suited to uncover the ultimate nature of reality; but they could not be considered relativists in that one such canon may quite straightforwardly be said to be better than another.
>
> *(Siderits 2016, 35)*

4. Moving beyond Nāgārjuna

Even more important than asking, "why did Nāgārjuna start with causation?" (Garfield 1994) would be to ask, "why did Nāgārjuna end his masterpiece with 'viewpoints' (*dṛṣṭi*)?" Rejection of causality grounds the Mādhyamika philosophy. Deconstruction of the "viewpoints," on the other hand, destabilizes the entire philosophical enterprise. The imprints of Nāgārjuna are visible in the lines of Śrīharṣa, a prominent Advaita philosopher who lived one millennium after Nāgārjuna.[4] By critiquing other viewpoints, Nāgārjuna is not proposing his own thesis, which would be counterintuitive. He himself cautions, "the victorious ones have proclaimed that there is no foundation as there is emptiness of all views. However, to whom emptiness [itself] is a view, they are considered incorrigible" (MMK XIII.8). It is therefore not the case that Nāgārjuna is rejecting the theory of causality; he is rejecting the viewpoints, and the first among them happens to be the theory of causality. For him the fundamental human problem is not the lack of theories but our obsession with them

Another key position to derive relativism comes from Maṇḍana Miśra. For him, our everyday reality is composed of our own ignorance (*avidyā*) and the individual subjects are the locus of this metaphysical ignorance. This position results in saying that all we can encounter by means of our cognitive faculties and semantic analysis are just the perspectives, each conditioned by our own preconceived notions, and filtered by means of the habit patterns (*saṃskāras*). Every individual, in this paradigm, projects his own world. Accordingly, each has his own conceptualized truth, guided by one's own presuppositions and misconceptions.[5] Since all that we can communicate regarding the nature of reality is mediated by our concepts, which in turn are the conditions from our past experiences, this thesis does lead to some form of relativism. Expanding upon the philosophy of Maṇḍana, the non-dualist philosophers (Advaitins) argue that collectively shared experiences are what they are because subjects having homogenous experiences do share a common history. Borrowing their own example, this is similar to multiple subjects having the dream of a snakebite and coming to the conclusion that they all dreamt of the same snake. Just as dream experiences are subjectively circumscribed, so also are other experiences. Even our experiences of pain and pleasure corresponding to certain stimuli are rooted in habitual tendencies that constitute some experiences as painful and others as pleasant. This does involve bodily memory. This is to say that we are not able to escape our corporeal and psychological horizons in our pursuit for grounding our experience. What are we left with then? Just our "viewpoints" or "perspectives" (*dṛṣṭis*). However, this is as far as their agreement goes, as the Advaita philosophers are not relativistic with regard to the absolute reality of the being equated with consciousness (sat-cit). For them, every mode of experience and every perspective underlies the same principle of being and consciousness. For them, being and consciousness are a logical necessity for every is affirmation or negation. They see this as something that cannot be rejected by means of negation, and for them, the foundational being and awareness is not yet another perspective but only the possibility for the perspectives to be, and not the truth of all the truths, but merely the categorical possibility that makes us think about truth in general. Our everyday modes of experience, accordingly, do not negate experience as a category. The argument here is that subjects can bracket the factors that condition experience, including the ego, and enable being in a mode that is not subject to conceptualization. This is not to say that there is nothing real; this is not surrender to any form of nihilism. This is a proclamation that any truth-claim is relative, or perspectival.

One may conflate this position with Kantian transcendental idealism. And some early scholars reading Nāgārjuna such as Tiruppattur R. Venkatachala Murti have found comfort in such a charge. Following this, just as the objects we intuit in space and time are appearances, the mental states that we intuit in introspection are likewise appearances. We can nevertheless think of things in themselves using categories, as they affect our sense faculties. This conflation, however, misses a major distinction whether it be a Nāgārjunian or Advaita position: the entire philosophical endeavor cannot be isolated from the goal of "apprehending the way things are" (*yathābhūtārtha-darśana*), or "direct apprehension" (*sākṣātkāra*). When scholars say that our experiences are shaped by our habit tendencies (*saṃskāra*) and that all we experience, conceptualize, and verbally express are mere copies of the way things are and that what it actually is cannot be expressed; this is never meant to conclude that we are incapable of overcoming our own subjectivity. The resultant position advocates some variations of semantic and epistemic relativisms, while retaining the possibility of different types of metaphysical realism.

In the discourse on relativism, the Jain "multiperspectivalism" (*anekāntavāda*) is sometimes imagined to be relativism itself. This, however, is not the case. In its most systematized form, this doctrine for any given situation consists of sevenfold possibilities:

(1)   It may be.
(2)   It may not be.
(3)   It may and may not be.
(4)   It may be but is not describable.
(5)   It may not be while being indescribable.
(6)   It may both be and not be while being indescribable.
(7)   It may simply be indescribable.

This is not a thesis that truth is relative to individual subjects, or that everyone has her own truth conditioned by her language and culture. Another way this has been confused is by equating it with perspectivism. "May be," to begin with, is not proposed as yet another perspective, and none of these are individually circumscribed to be true. This is rather saying that truth is manifold, or that each of these constitutes a part of the truth that is revealed only globally when all aspects have been analyzed. Another way this has been understood, is as a form of pluralism. It seems appealing to argue that there are multiple perspectives to the truth, but in fact, what the doctrine is saying, is that while different doctrines make different truth-claims, none of these have the total picture of the reality when accepted individually. That is, there is a truth claim when the totality of the possibilities is accepted, but not that truth is only revealed as a perspective and that all of them have some sort of validity if taken individually. What this implies is that one who has all the perspectives has the truth. And this can be better explained as "mosaicism": that each component of a mosaic comprises a necessary element for constituting the truth, but no single piece of the mosaic alone can reveal the truth the way it is.

5. Conclusion

It would be wrong to equate any of the aforementioned positions with relativism. But fortunately, there are many kinds of relativism and when engaging Sanskrit philosophical literature, we may have encountered a different variety, or varieties of relativism that are not just antecedent to contemporary forms of relativism. What applies to most Indian traditions, is that being relative about truth is not to deny the category "truth" but to assert that our rationality and comprehension of what is true is relative, and that there are external factors to underscore our ways of reasoning or our grasping of what we consider to be true. But in all accounts, truth as a category underpins this assumption. Different subjects from varied cultural backgrounds might share different values and different systems of judgment and from a meta-gaze we may see relativism in their perspectives. However, this does not apply, that subjects endorsing such views consider them as relative. Each and every cultural subject has their own unique experiential and epistemic horizon that is for them the only truth. Those who are capable of distinguishing their personal perspective from among other viewpoints, are subjects possessing a "meta-gaze" and in some regards, are the liberated (*mukta*) subject, able to transcend their own subjective horizon.

The preceding discussion provides a framework for re-contextualizing moral relativism in the *Mahābhārata*. Overall the text teaches non-violence (*ahiṃsā*) although every page of it is saturated with the blood of the antagonists and heroes. A small section from it, the *Bhagavadgītā*, epitomizes the tension between relative and absolute perspectives on morality, vividly portrayed as the clash between the individual duty of a warrior (Arjuna) to fight, and the universal *dharma* of non-violence. There is no relativism about non-violence: this is the single most absolute upon which the other absolutes such as truth (*satya*) and "not stealing" (*asteya*) are founded. In this tension between the universal and individual *dharmas*, Arjuna recognizes the necessity to perform his individual *dharma*. Is this a simple justification for a war? If the book is teaching anything, it is that individual perspectives or truths triumph over global perspectives, but this can be allowed when and only when the global perspective is at peril. *Ahiṃsā*, it seems, is not always capable of defending itself. Not by choice but as the final resort when all options have been exhausted, Arjuna is left to decide between a lesser evil of confronting violence with violence, or a greater evil of avoiding it. This isn't because a warrior wants to kill or craves fame but because those being killed and raped are unable to defend themselves with a mere vow of non-violence. A warrior allows himself to act within the universal *dharma* so that the others can uphold it and Arjuna chooses his personal truth: as a warrior he has to fight. The difference in perspective is, prior to the teachings, there is Arjuna a prince, a brother, and a husband deeply wounded by the atrocities of his enemies, while after the teachings, there is just a warrior who recognizes his role, his moral responsibility which makes the global sense of morality possible.

Notes

1 Refercens to Nāgārjuna's *Mūlamādhyamakakārikā* are cited using the abbreviation MMK, number of the chapter and verse or half-verse, e.g. "MMK XXIV.8." Please find the full reference in the bibliography under Kalupahana (ed.) (1986).
2 Refercens to Nāgārjuna's *Vigrahavyāvartanī* are cited using the abbreviation VV and the verse, e.g. "VV 26." Please find the full reference in the bibliography under Bhattacharya et al. (1978).
3 The classical analysis of "two truths" is complex. Candrakīrti, for example, divides *saṃvṛti* as real empirical and unreal empirical in order to make a distinction between the conventional and erroneous objects. Bhāvaviveka makes a distinction between the conceptualized and actual truths when addressing the *paramārtha*. For further analysis of the Prāsaṅgika-Svātantrika distinction, see Dreyfus and McClintock (2003).
4 For the convergence of the philosophy of Nāgārjuna with the Advaita of Śrīharṣa, see Timalsina (2017).
5 For Maṇḍana's philosophy of *avidyā*, see Timalsina (2009).

References

Bhattacharya, K., E. H. Johnson and A. Kunst (1978), *Vigrahavyāvartanī. The Dialectical Method of Nāgārjuna: Vigrahavyāvartanī*, New Delhi: Motilal Banarsidass.
Dreyfus, G. B. J. and S. L. McClintock (2003), *The Svātantrika-Prāsaṅgika Distinction: What Difference Does a Difference Make*, Boston: Wisdom Publications.
Ferraro, G. (2013), "Outlines of a Pedagogical Interpretation of Nāgārjuna's Two Truths Doctrine," *Journal of Indian Philosophy* 41: 563–590.
Garfield, J. (1994), "Dependent Arising and the Emptiness of Emptiness: Why Did Nāgārjuna Start with Causation?" *Philosophy East and West* 44(2): 219–250.
———. (2002), *Empty Words: Buddhist Philosophy and Cross-Cultural Interpretation*, Oxford: Oxford University Press.
Kalupahana, D. (ed.) (1986), *Madhyamakakārikā. Mūlamadhyamakakārikā of Nāgārjuna: The Philosophy of the Middle Way*, New York: State University of New York.
Siderits, M. (1989), "Thinking on Empty: Madhyamaka Anti-Realism and Canons of Rationality," in *Rationality in Question: On Eastern and Western Views of Rationality*, edited by S. Biderman and B-A. Scharfstein, Leiden: Brill, 231–249.
———. (2003), "On the Soteriological Significance of Emptiness," *Contemporary Buddhism* 4(1): 9–23.
———. (2016), *Studies in Buddhist Philosophy*, edited by J. Westerhoff, Oxford: Oxford University Press.
Siderits, M. and J. L. Garfield (2013), "Defending the Semantic Interpretation: A Reply to Ferraro," *Journal of Indian Philosophy* 41: 655–664.
(2009), "Bhartṛhari and Maṇḍana on Avidyā," *Journal of Indian Philosophy* 37: 367–382.
———. (2017), "Śrīharṣa on Knowledge and Justification," *Journal of Indian Philosophy* 45(2): 313–329.
Walsh, E. (2015), "Relativism in Buddhist Philosophy: Candrakīrti on Mutual Dependence and the Basis of Convention," in *The Moon Points Back*, edited by K. Tanaka, Y. Deguchi, J. Garfield and G. Priest, New York: Oxford University Press.

2

RELATIVISM IN THE ISLAMIC TRADITIONS

Nader El-Bizri

1. Sectarian factionalism

The differences in scriptural interpretation in Islam resulted in a relativizing sectarian pluralism despite the Muslim ecumenical monotheistic belief in the existence of a single divine absolute and universal truth (*al-ḥaqq*).[1] The principal doctrinal disputations in religious thinking emerged between *Sunnī* and *Shī'ī* factions (*madhāhib*) over the succession (*khilāfa*) in leadership following the death of the Prophet Muḥammad (632 CE), which continued to split the Muslim communities politically across various dynastic lines, with additional bifurcations arising due to polemics concerning jurisprudence, eschatology, hagiography, and heresiology. These disputes were also impacted by the dialectical interactions of Muslim scholars with their Arab Christian counterparts in the Fertile Crescent, in addition to the adaptive assimilation of ancient Greek knowledge. A relativizing current ran in concealment within the folds of scriptural exegesis and hermeneutics, as furthermore differentiated via individuated approaches in scholarship. Epistemic differences also emerged between the theologians (*mutakallimūn*), mystics (*'ārifūn*), and philosophers (*falāsifa*) in terms of their distinct claims concerning the absolute truth in Islam. They all believed in its uniqueness while quarrelling over its meaning and the methods that disclose the essence of its truism. This translated into relativizing disputations and apologetics, in epistemic, cultural, and moral terms, over the religious law and lifestyles, as concretely influenced by political and societal undercurrents. Moreover, the methods of textual and linguistic interpretation differed in terms of adopting a literal (*ẓāhir*) exegesis (*tafsīr*) versus allegorical hermeneutics (*ta'wīl*) that disclose hidden (*bāṭin*) meanings.

The belief in an absolute (*muṭlaq*), universal (*kullī*), eternal (*azalī*), and unique (*wāḥid*) truth (*ḥaqq*) was relativized in interpretive acculturation and comportment. Relativism arose from polemics over the essence of monotheistic absolutism and monism. Moral factionalism translated into an epistemic relativism, as reflected in emic and etic dispositions. For instance, an emic insider is an agent within a faction who normatively affirms the group's doctrine with absolutism, and may charge co-religious opponents with heresy; while an etic observer pictures the claim concerning the absolute truth in factional disputes as being relativistic in its world view and lifestyle. The belief in an absolute truth in Islam is rendered relative (*nisbī*) and particular (*juz'ī*) via Muslim sectarianism; albeit, without resulting in scepticism. Even if individual Islamic scholars battled with doubt, their sceptical stance is directed towards the closed belief systems of Muslim factions and not Islam altogether. They would move away from philosophy towards mysticism, from theology to Sufism, without renouncing the Islam in the manner of an apostate (*murtadd*). Relativism in Islam is not declared as a doubt concerning the existence of the religious truth, but is instead directed at some Muslim sects. Islam is not relativistic *per se*; rather its competing factions (*firaq*) relativize the claim concerning the truth of the faith. Relativism emerges in such sectarian syncretic fissures, especially in epochs of strife.

Grasping what is alethic in Islam depends on the ontic and epistemic horizons within which it is disclosed. The disputations concerning the truth got exhausted when the orthodoxy closed the gateway to interpretation (*iqfāl bāb al-ijtihād*). How an opinion becomes more entrenched than another is dependent on historical, societal, political, and quotidian factors in praxis, as much as it rests on onto-theology, epistemology, and logic. When a doctrine imposes itself socially and becomes mainstream historically, it henceforth stands as a vocal tradition that ideologically judges other cults of the creed. The incongruent sects managed to coexist in *Realpolitik* pragmatic and apologetic terms, and yet during the episodes of oppression and conflict the divisions erupt with bellicosity and violence. These doctrines do not yield unanimity and consensus (*ijmā'*), even if agreements may have been achievable by the first generation of the Prophet's companions (*ṣaḥāba*) in isolated devotional rites within a narrow societal milieu.

Grosso modo, all Muslims are expected to analogically differentiate their beliefs and comportments from those who are religiously described as being "liars" (*kāzibūn*), "deniers" (*munkirūn*), "deceitful hypocrites" (*munāfiqūn*), "treading the wrong path" (*fī ḍalāl*), "suppressing the truth" (*muktimūn*), or "doubting it" (*mushakkikūn*). An emphasis is placed on emulating (*tamaththul*) the prophetic sayings and doings (*al-qawl wa'l-'amal*) in accordance with the Islamic law (*al-shar'*), which ethically promotes honesty (*ṣidq*) and correctness (*ṣaḥīḥ*).[2]

2. Theology and mysticism

The theological conception of truth is mediated via the elucidation of the question of the divine essence and attributes (*al-dhāt wa'l-ṣifāt*) (El-Bizri 2008) in the disputations of Muslim theologians (*al-mutakallimūn*) amongst themselves, or in their quarrels with the philosophers, or in their dialectical argumentations with Arab Christian scholars. The question arises over the affirmation of divine transcendence (*tanzīh*) and unity (*tawḥīd*), while eschewing analogical anthropomorphism (*tashbīh*). This reduced the divine attributes to the divine essence by asserting absolute unity; even if this undermined the semantic character of the divine names in the ritual of worship. This eventually occasioned a schism within Muslim theology via the emergence of *Ash'arīsm* besides *Mu'tazilīsm*. The paradoxical nature of the question of the divine essence and attributes was veiled by accepting it "without how or why" (*bi-lā kayf wa lā limā*), and without showing how this would not entail pluralism instead of monism.

Some Muslim theologians advocated occasionalism under the influence of al-Ghazālī's (1058–1111 CE) (Al-Ghazālī 2000) doubts concerning the justification of induction, by arguing that *the connection between what is habitually taken to be a cause and what is taken by way of custom to be its effect is not necessary*. A cause, and what is posited as its effect, both are distinct *events* that coexist without cementing connections between them.[3] This undermined the belief in natural laws, by picturing natural phenomena as being occasional aggregations of atoms.[4]

Al-Ghazālī recounts in his *Deliverance from Error* (*al-Munqidh mina'l-ḍalāl*) (1951) the difficulties in attempting to extricate truth from the confusion of contending sects. The different religious observances, communities, systems of thought, modes of leadership and governance, along with the multiplicity of sects, all appeared to him as an ocean in which the majority of people drown and only a minority might reach safety. Each separate group thinks that it alone is saved, and rejoices as such. On his view, this sense of perdition scatters the truth. Having scrutinized every creed, his aim was to distinguish truth, as a sound tradition, from falsity as heretic innovation. He sought a true comprehension of beings as they are, by drawing a distinction between opinions and the principles underpinning them. True knowledge necessitated that the phenomenon that is disclosed is knowable without doubts or illusions abiding besides it. He believed that the truth of Islam must be infallible and immune to scepticism. He thusly aimed at evaluating the truth-conditions of empirical sense-perception, necessity in mathematics, and reasoned logical demonstrations. He sought the truth of first principles, and wondered about what judges intellectual apprehension in a trustworthy manner beyond demonstration. This pointed him to the mystical experience as "an illumination that is cast into the heart" (Montgomery 1953, 25–26) by divine mercy, in a "withdrawal from the house of deception and a return to the mansion of eternity" (Montgomery 1953, 12), which is a cure from the disease of doubt.

Al-Ghazālī grasped theology (*kalām*) as an endeavour to defend orthodoxy against the heretics. In so doing, the theologians based their arguments on premises they affirmed with an unquestioning acceptance of scripture, and their dialectics were devoted to making explicit the contradictions of their opponents. Theology was not adequate for him in terms of arriving at results that were sufficient to universally dispel confusion. He then set study the philosophical legacies of materialists (*dahrīyyūn*), naturalists (*tabī'īyyūn*), and theists (*ilāhīyyūn*). Nonetheless, he believed that these and their followers amongst Muslim philosophers, such as Ibn Sīna (Avicenna; 980–1037 CE) and al-Fārābī (872–95 CE), carried the residues of unbelief in their onto-theologies.

3. Mysticism

Despite being the most accomplished amongst the Muslim theologians of his era, al-Ghazālī recognized the dangers of blindly following authoritative instructions. He eventually turned his quest towards the experiential ways of mysticism (Sufism); to be in constant recollection of reflections on the divine presence; striving for immediate ecstatic experiences of *dhawq* (taste) in states akin to drunkenness within an ascetic life that forsakes worldly affairs. He walked the mystic path away from vain desires, and busied himself with spiritual exercises in retreats to improve his character and cleanse his heart for the solitary meditations on God. The anxieties about his family during his retirement, and the quotidian needs of livelihood, altered his purpose and impaired the quality of his solitude. He reports that phenomena were unveiled to him that he judged as being unfathomable. He learnt via experiential situations what he felt was the best life, namely of the mystic way (*ṭarīqa*) in its pure character through the purification of the heart from what is other than God in adoration and complete absorption (*fanā'*). The mystics in their waking state relate that they have visions and unveilings, which they achieve in their nearness to God; some conceive this as inherence (*ḥulūl*), others as union (*ittiḥād*) or as a connection (*wuṣūl*) with divinity. Yet, as al-Ghazālī cautions, whomever has attained such mystic state need do no more than say: "Of the things I do not remember, what was, was; think it good; do not ask an account of it" (Montgomery 1953, 29). Al-Ghazālī eventually concedes that he came to truth partly through immediate experience, and in part via demonstrative knowledge, in addition to a leap of faith. The theological inclination towards argumentation contrasts as such with the situational lived experience of mysticism, gnosis, and theosophy. These pathways of spirituality rather accentuate the existential psychosomatic effects of the ecstatic disclosure of truth not by way of intellection but via *un-veiling the veiled* (*kashf al-maḥjūb*), as driven by love (*'ishq*), and morally aided by mentorship in spiritual exercises that assist in lifting the occultation (*satr*). This may point to epiphany (*tajallī*; ἐπῐφᾰνειᾰ) or παρουσία as a coming into presence of a disclosure of *the hidden truth* via apparitions. Such conception of the un-concealment of the veiled (*kashf al-maḥjūb*) as the happening of truth resonates with ἀλήθεια (*alḗtheia*) in lifting the forgetfulness of λήθη (*lḗthē*). The happening of ἀλήθεια is a retrieval of what has been forgotten, retreated, withdrawn, by way of bringing it back into presence via remembrance (*dhikr*). This points to a return to things themselves via an experiential clearing *qua* opening (*fatḥ*) that removes the occultation (*satr*).[5] An experiential situated un-veiling of truth in presence is implied herein rather than a logical demonstration. The happening of un-concealment in *gnosis* (γνῶσις) is a personal esoteric mode of knowing truth via divine revelation. The notion of un-veiling has a theological character in the manner God is described within the Islamic tradition as having seventy-thousand veils of light and darkness (Al-Majlisī 1983). The beatific vision (*mushāhada*) is sought through an intuitive lifting of the veil (*mukāshafa*). Truth is *kashf* as the un-veiling of what veils itself.[6] According to al-Ghazālī, the mystical truth cannot be experienced or taught to the majority of the people; rather a literal approach in following the directives of the legalistic rulings is more befitting to everydayness within a commonwealth.[7] Such gnostic attitudes are ultimately questioned by orthodoxy in Islam on the grounds that they purport knowing the unknowable.

4. Philosophy

Ibn Sīnā (Avicenna) held that true knowledge necessitates communion (*ittiṣāl*) with the cosmic Active Intellect (*al-ʿaql al-faʿʿāl*) that holds the Platonic forms (*ṣuwar*) and governs the sublunary realm. His ontological reflections on the question of being (*al-wujūd*) were mediated via the modalities of necessity (*al-wujūb*), contingency *qua* possibility (*al-imkān*), and impossibility (*al-imtināʿ*).[8] Based on this, the impossible being (*mumtaniʿ al-wujūd*) cannot exist, and the affirmation of its existence is a contradiction; as for the contingent *qua* possible being (*mumkin al-wujūd*), it is that whose existence or non-existence is neither impossible nor necessary; it is ontologically neutral in the sense that affirming its existence or negating it does not entail a contradiction. There is nothing inherent in the essence of the contingent *qua* possible that gives priority to its existence over its non-existence; it is essentially what exists or does not exist not due-to-itself, but due-to-what-is-other-than-itself (*bi-ghayrih*). The contingent is actualized due to what is other than itself, its existence is distinct from its essence; hence, its being happens to it from a source other than itself, which is its existential cause. As for necessary being (*wājib al-wujūd*) it exists *per se* in such a way that it essentially cannot but exist, and the affirmation of its non-existence entails a contradiction, since it is impossible for it not to be. However, necessary being (*wājib al-wujūd*) can be as such either *due-to-itself* (*bi-dhātih*) or *due-to-what-is-other-than-itself* (*bi-ghayrih*). The Necessary-Being-due-to-Itself (*wājib al-wujūd bi-dhātih*) is beyond the Aristotelian categories (*al-maqūlāt*); it is without definition or description, and its essence is none other than its existence. Necessary-Being-due-to-Itself is *one and only*, given that if there are two as such, then they are not each due-to-itself *per se*; since *differentia* (*faṣl*) is posited besides them to distinguish them from one another. Being as such, they are co-dependent, or require *differentia* as what is other than themselves beside themselves, while Necessary-Being-due-to-Itself need not other than itself for it to be.

Ibn Sīnā's conception of the necessary being *per se* also resonated with the earlier ontology of al-Fārābī in *al-Madīna al-fāḍila* (*Virtuous City*); namely in how the necessary being *per se* is posited as a first existent (*mawajūd awwal*) and a first cause (*sabab awwal*) of all existents. It is perfect (*kāmil*); eternal (*sarmadī*); without material, formal, teleological, or efficient cause (*mādiyya, ṣuwariyya, ghāʿiyya, fāʿiliyya*); having no equal (*shabīh*), counterpart, or opponent (*ḍid*); and is without definition (*ḥadd*). Such negative onto-theology admits with it the affirmation of the divine attributes of omniscience, sagacity, truth, and life (*'ālim, hakim, haqq, hayy*), as well as asserting the unicity of the divine intellect-intellection-intelligible (*'aql 'āqil ma'qūl*; νοῦς νόησις νοητόν) (Al-Fārābī 1968, §1–5).

Ibn Rushd (Averroes; 1126–1198) also reflected philosophically on the notion of truth but in a more direct engagement with the Islamic law. He held that Muslims are religiously urged to have demonstrative knowledge about God and all beings, and to consequently understand and apply demonstration in reasoning, which differs from dialectical, rhetorical, and fallacious approaches (Ibn Rushd 1961). He argued that demonstration leads to truth, and it thusly becomes an obligation in Islam to study philosophy and its instruments of syllogism, logic, and mathematics. Harm from philosophizing is as such accidental and not by essence. Demonstration and scripture do not conflict with one another in the quest to know the truth. If an apparent conflict arises, then allegorical interpretation comes close to bearing witness to the demonstration. If scriptural apparent literal meanings contradict each other in exegesis, then allegorical hermeneutics reconciles them. However, unanimity is never determined with certainty, but it can be achieved in practical selected issues within a delimited period. If scripture has both an apparent and a hidden meaning, then the latter should not be revealed to anyone who is incapable of understanding it. The situation is different in praxis, since the truth should be disclosed to all people, and consensus over it is easier in comportment than when dealing with doctrines.

5. Logic

The notion of truth necessitates (*awjab*) the use of a certified ascertained statement (*khabar yaqīn*) in the form of a meaningful proposition as supported by evidenced proven demonstration (*burhān*). This passes by way of a correspondence or correlation (*muwāfaqa* or *muṭābaqa*) between a factual state of affairs, as existing in the world, and what is said about it in affirming its existence. Such thesis is not simply upheld in epistemic and logical terms, but it also rests on a religious appeal to ethical correctness and honesty (*ṣiḥḥa wa ṣidq*), whereby the moral character of the speaker warrants also the trusting of the truthfulness of their sayings. This aspect was also affected by the care given by the logicians in judging the truth-value and structure of propositions and syllogism, given their disciplinary quarrels with the Arabic grammarians over the best rules that govern the sound use of language (Street 2004; Rescher 1964, 196–210). In view of these dimensions, the logicians scrutinized the sophists' attempts to entrap their adversaries in a debate through a false *qua* liar genre of paradoxes, (ἔτι ὁ σοφιστικὸς λόγος [ψευδόμενος] ἀπορία) wherein the syllogistic chain of reasoning ends in perplexity (Aristotle 1926, *Nicomachean Ethics*, Z.2, 1146a). If a sophist states that "all sophists are liars," then to assert the truth of this statement becomes paradoxical. If it were true then it is possible that it is false as well, since it is stated by a liar. The paradox emerges over the veracity of the declarative statement (*khabar*) (Alwishah and Sanson 2009; Al-Baghdādī 1981, 13, 217), wherein an un-satisfied truth-condition is embedded in the pseudo sentence as a genre of *insolubilia* (Crivelli 2004).

A declarative statement is true if it agrees with the subject about which it is declared, and is false in the opposite case; it cannot be true and false at the same time. A statement is false when made by an honest person who never lies who says "I am a liar"; and a statement is true when made by a dishonest person who always lies when saying "I am a liar." The same declarative sentence "I am a liar" can respectively be false and true. An exclusive bivalence entails that "there is no declarative sentence that is both true and false together, except one," namely the liar paradox (Alwishah and Sanson 2009, 100–104). It is ambiguous to have together two contradictories (*jam' al-naqīḍayn*), wherein no declarative sentence can be both true and false, whether affirmative or negative. However, the declaration: "all that I say at this moment is false," is a saying (*qawl*) that is either true or false. If it is true, then it must be true and false; and if it is un-true, then it is necessary that one of the sentences said at the moment is true, as long as the one saying it utters something. Albeit, saying nothing other than this sentence, would entail that what is said is true and false. If the liar's proposition declares itself to be false, then what it declares about itself, namely that it is false, is true, and thus the paradox due to a misuse of predication (Alwishah and Sanson 2009, 99, 107, 112, 120–125; Ibn Sīnā 1958, 98; Ibn Rushd 1972, 157).

The imagined representation (*taṣawwur*) has to have a certified verification of its truthfulness (*taṣdīq*). The veracity of knowing a given state of affairs proceeds in propositional terms through affirmation (*ithbāt*) and negation (*nafy*) in a logical syllogism (*qiyās*). The starting point rests on *a priori* first principles of the intellect (*al-awā'il al-'aqliyya*).

Logic is a canonical Ὄργανον (*āla qānūniyya*) that measures the veracity of thinking and how what is in the mind corresponds with what it denotes as being present there as an existent fact in the world (Ibn Sīnā 1958, 167–180). This resonates with Aristotle's *Metaphysics*; namely,

> to say that *what is* is not, or that *what is not* is, is false; but to say that "*what is*" is, and "*what is not*" is not, is true; and therefore also he who says that a thing *is*, or *is not*, will say either what is true or what is false. But neither "*what is*" nor "*what is not*" is said *not to be or to be*.
>
> *(Aristotle 1924, Metaphysics, Γ.7, 1011b; Θ.10, 1051b)*

Moreover, in the *Sophist* Platonic dialogue of Theaetetus with a *stranger* around the correspondence thesis of truth (Plato 1921, *Sophist* 262e-263d), the *stranger* brings Theaetetus to agree that every sentence must have some truth or falsity by which it is qualified. The true one states a fact, and the false states what is other than a fact, since the false sentence speaks of things that are not as if they were.

6. Science

Besides the deliberations in philosophy, the mathematical disciplines offered their own methods of demonstration and hypothetical-deductive reasoning via experimental controlled testing (*al-i'tibār; al-tajriba*) as structured through geometrical modelling. This was embodied in the procedures of Ibn al-Haytham (Alhazen; 965–1041) as set in his *Book of Optics* (*Kitāb al-manāẓir; De aspectibus* or *Perspectiva*).[9] His research belonged to the Archimedean-Apollonian mathematical "school of Baghdad" (9th–10th centuries CE) (El-Bizri 2007). His optics was based on mathematizing physics and offering a geometrical critique of Aristotelian natural philosophy. This necessitated the devising of geometrically modelled experimentations as controlled methods of proof, which transcended the mere reliance on logical forms of demonstration.

Ibn al-Haytham investigated the optimal conditions under which vision can be a reliable source of observational data in controlled experiments. He distinguished pure sensation (*mujarrad al-ḥiss*), which only senses physical light *qua* light, and colour *qua* colour, from the psychological workings in vision in terms of recognition (*ma'rifa*), judging discernment (*tamyīz*), comparative inferential measure (*qiyās*), as aided by imagination (*takhayyul*), memory (*dhikr*), and at times by acquired prior knowledge. He argued that sensation in connection with vision was ultimately effected by "the last sentient" (*al-ḥāss al-akhīr*) in the anterior part of the brain (*muqaddam al-dimāgh*). Vision is as such a physiological-neurological nested cluster of physical phenomena that pertain to the material effect of physical light on the anatomy of healthy eyes, and through them passes as sensations that are transmitted through properly functioning optical nerves to the last sentient at the front of the brain. Vision was not merely a phenomenon that resulted from the ocular functioning of the eyes as photoreceptors. Accordingly, the analysis of visual perception could not have rested only on geometric or physiological optics, but it would have required the cognitive psychological analysis of consciousness via embodied experiential situations.

In delineating the optimal conditions of vision, Ibn al-Haytham stressed that the viewed object must be bright enough and positioned at a moderate distance from the perceiver's eyes in a clear and homogeneous transparent medium (*shafīf*). Moreover, the visible object must be in a plane shared with the eyes, and its body should have a proper volume, not too small or too large, as well as allowing the trapping of some light rays, given that a highly transparent body is virtually invisible. In addition, the observer should have sufficient time to see the object with two healthy eyes that are able to perform effective concentrations in scrutiny and contemplation, and within experiments to isolate the variables. He argued that when sight perceives individuals of the same species repeatedly, a universal form (*sūra kulliya*) of that species takes shape in the imagination and is recollected by recognition, while consequently assisting in grasping the quiddity (*māhiyya*) of the corresponding visible object and its inspected seen properties; hence resulting in prior knowledge that aids imagination and memory. He also presented detailed phenomenological observations in experimental contexts regarding the role of embodied experiencing in vision, whereby the proper body of the observer contributes to estimating the visible distances and sizes of objects in the perceptual field when sharing a common spatial terrain with them. His experiential analysis showed how the manifestation of a thing in its plenitude through its visible aspects, as detected in a continuum of manifold appearances, occurs via contemplation and bodily spatial-temporal displacements. This is illustrated through perspective, whereby a thing is never seen in its entirety, since the appearing of some of its sides entails that its remaining surfaces are unseen. Hence, a partial un-concealment of an opaque object in vision is always associated with the concealment of some of its visual aspects. A stereoscopic distinction is posited here between authentic *qua* proper appearances (relating to a concrete act of seeing where the sides of the visible object are perceived immediately) and the inauthentic *qua* imagined appearances (designating the imaginary surplus accompanying the authentic appearances in the constitutive perception of the object of vision in its totality). The full silhouette of a thing is constituted via spatial-temporal bodily displacement in verifying the essential unity of its authentic and inauthentic appearances. It is with such exactitude in observations, controlled experimentation, and geometric modelling that the truth of visible attributes are verified, in the epistemic aim of overcoming the relativizing of knowledge.

Notes

1 The Arabic term for "truth" is "*al-ḥaqq*." Its etymology refers to *what prescribes* in the form of a law or ordinance. It is also one of God's ninety-nine beautiful names (*asmā' Allāh al-ḥusnā*) as noted in the *Qur'ān* (6:62, 22:6, 23:116, 24:25). See MacDonald and Calverley (2012) and the entries "*al-ḥaqq*" and "*al-ḥaqīqa*" in Ibn Manẓūr (1955–1956a, 1995b).

2 The *Shī'ī* traditions emulate the hereditary imamate in the lineage of Prophet Muḥammad's household (*ahl al-bayt*) (Kirmānī 2007).

3 This resonates with the maxim of Protagoras: ἄνθρωπον εἶναι, τῶν μὲν ὄντων ὡς ἔστι, τῶν δὲ μὴ ὄντων ὡς οὐκ ἔστιν (*man is the measure of all things, of the existence of the things that are and the non-existence of the things that are not*) (Diels and Kranz 1910, 80 B 1). See Plato (1921, *Theaetus* 152a); Sextus Empiricus (2005, M VII); Burnyeat (1997).

4 This rested on atomist physics (the atom conceived as "*the part that cannot be partitioned*" (*al-juz' al-ladhī lā yatajazza'*). See Alnoor Dhanani (1994). The *kalām* physical theory pictures the invisible atoms (akin to Epicurean "minimal-parts") as discrete "space-occupying" entities that have miniscule magnitudes, in a non-Aristotelian rejection of αἰτιολογία.

5 Resonating with Heraclitus' fragment: φύσις κρύπτεσθαι φιλεῖ (*nature loves to hide*) (Diels and Kranz 1910, B123).
6 Ibn Ṭufayl (1105–1185 CE) offered an auto-didactic unveiling of truth via storytelling (Ibn Ṭufayl 2003).
7 Al-Suhrawardī's (1154–1191 CE) *illuminationism* (*ishrāq*) advocated *true witnessing* (*mushāhada ḥaqqiyya*). Al-Suhrawardī (1952, 18, 85–87, 104).
8 Ibn Sīnā (1874, 262–263; 1960, 65; 1975, 35, 36–39, 43–47, 350–355; 1978, I.5–7, 110–122; 1985, 255, 261–265, 272–275, 283–285). I investigated Ibn Sīnā's ontology in some of my other publications (El-Bizri 2001, 2006, 2014b, 2016, 2018).
9 Ibn al-Haytham (1972, 1983, 1989, 2002); see especially Ibn al-Haytham (1983 §I.2 [1–26], I.8 [1–11]; §II.3 [67–126]). I also investigated related aspects (El-Bizri (2005, 189–218; 2014a, 25–47).

References

Al-Baghdādī, 'A. Q. (1981), *Kitāb uṣūl al-dīn*, Beirut: Dār al-āfāq al-jadīda.
Al-Fārābī (1968), *Ārāʾ ahl al-madīna al-fāḍila*, edited by A. N. Nader, Beirut: Dār al-mashriq.
Al-Ghazālī (1951), *Deliverance from Error*, translated by W. M. Watt, London: George Allen and Unwin.
———. (2000), *Tahāfut al-falāsifa* [*Incoherence of the Philosophers*], translated by M. E. Marmura, Provo, Utah: Brigham Young University.
Al-Majlisī, M. B. (1983), *Biḥār al-anwār al-jāmiʿa li-durar akhbār al-aʾimma al-aṭhār* [*Oceans of Lights: A Compendium of the Pearls of the Narrations of the Pure Imāms*], vol. 55, Beirut: Muʾassasat al-wafāʾ.
Al-Suhrawardī, S. (1952), *Opera metaphysica et mystica*, vol. II, edited by H. Corbin, Tehran: Institut Franco-Iranien.
Alwishah, A. and D. Sanson (2009), "The Early Arabic Liar: The Liar Paradox in the Islamic World from the Mid-Ninth to the Mid-Thirteenth Centuries CE," *Vivarium* 47: 97–127.
Aristotle (1924), *Metaphysics*, edited by W. D. Ross, Oxford: Clarendon Press.
———. (1926), *Nicomachean Ethics*, vol. XIX, translated by H. Rackham, Loeb Classical Library 73, Cambridge, MA: Harvard University Press.
Burnyeat, M. F. (1997), "The Sceptic in His Place and Time," in *The Original Sceptics: A Controversy*, edited by M. F. Burnyeat and M. Frede, Indianapolis: Hackett.
Crivelli, P. (2004), "Aristotle on the Liar," *Topoi* 23: 61–70.
Dhanani, A. (1994), *The Physical Theory of Kalām: Atoms, Space, and Void in Basrian Muʿtazilī Cosmology*, Leiden: Brill.
Diels, H. A. and W. Kranz (1910), *Die Fragmente der Vorsokratiker*, Berlin: Weidmannsche Buchhandlung.
El-Bizri, N. (2001), "Avicenna and Essentialism," *Review of Metaphysics* 54: 753–778.
———. (2005), "A Philosophical Perspective on Alhazen's *Optics*," *Arabic Sciences and Philosophy* 15: 189–218.
———. (2006), "Being and Necessity," in *Islamic Philosophy and Occidental Phenomenology on the Perennial Issue of Microcosm and Macrocosm*, edited by A-T. Tymieniecka, Dordrecht: Kluwer Academic Publishers, 243–261.
———. (2007), "In Defence of the Sovereignty of Philosophy: al-Baghdādī's Critique of Ibn al-Haytham's Geometrisation of Place," *Arabic Sciences and Philosophy* 17: 57–80.
———. (2008), "God: Essence and Attributes," in *The Cambridge Companion to Classical Islamic Theology*, edited by T. Winter, Cambridge: Cambridge University Press, 121–140.
———. (2014a), "Seeing Reality in Perspective: The 'Art of Optics' and the 'Science of Painting,'" in *The Art of Science: From Perspective Drawing to Quantum Randomness*, edited by R. Lupacchini and A. Angelini, Dordrecht: Springer, 25–47.
———. (2014b), *The Phenomenological Quest Between Avicenna and Heidegger*, Albany: SUNY Press.
———. (2016), "Avicenna and the Problem of Consciousness," in *Consciousness and the Great Philosophers*, edited by S. Leach and J. Tartaglia, London: Routledge, 45–53.
———. (2018), "Avicenna and the Meaning of Life," in *The Meaning of Life and the Great Philosophers*, edited by S. Leach and J. Tartaglia, London: Routledge, 95–103.
Ibn al-Haytham [Alhazen] (1972), *Opticae thesaurus Alhazeni*, edited by F. Risner, New York: Johnson Reprint Corporation.
———. (1983), *Kitāb al-manāẓir* [*Book of Optics*], vols. I–III, edited by A. I. Sabra, Kuwait: National Council for Culture, Arts and Letters.
———. (1989), *The Optics of Ibn al-Haytham, Books I-III, on Direct Vision*, translated by A. I. Sabra, London: Warburg Institute.

————. (2002), *Kitāb al-manāẓir* [*Book of Optics*], vols. IV–V, edited by A. I. Sabra, Kuwait: National Council for Culture, Arts and Letters.

Ibn Manẓūr (1955–1956a), "*al-ḥaqq*," in *Lisān al-ʿarab* [*The Tongue of the Arabs*], Beirut: Dār ṣādir.

————. (1955–1956b), "*al-ḥaqīqa*," in *Lisān al-ʿarab* [*The Tongue of the Arabs*], Beirut: Dār ṣādir.

Ibn Rushd [Averroes] (1961), *On the Harmony of Religion and Philosophy*, translated by G. Hourani, London: Luzac.

————. (1972), *Talkhīṣ al-safsaṭa*, edited by M. S. Salīm, Cairo: Dār al-kutub.

Ibn Sīnā [Avicenna] (1874), *Kitāb al-hidāya*, edited by M. ʿAbdū, Cairo: al-Hayʾa al-ʿāmma liʾl-maṭābiʿ al-amīriyya.

————. (1958), *al-Safsaṭa*, in *al-Shifāʾ*, *al-Manṭiq*, edited by F. Ahwānī, Cairo: al-Maṭbaʿa al-amīriyya.

————. (1960), *al-Ishārāt waʾl-tanbīhāt*, vol. III, edited by S. Dunya, Cairo: Dār al-maʿārif.

————. (1975), *Kitāb al-shifāʾ*, *Metaphysica II*, edited by G. C. Anawati, I. Madkour and S. Zayed, Cairo: al-Hayʾa al-miṣrīyya al-ʿāmma.

————. (1978/1985), *La métaphysique du Shifāʾ*, vols. I–II, translated by G. Anawati, Paris: Vrin.

————. (1985), *Kitāb al-najāt, Metaphysics I*, edited by M. Fakhry, Beirut: Dār al-afāq al-jadīda.

Ibn Ṭufayl (2003), *Ḥayy ibn Yaqẓān*, translated by L. E. Goodman, Chicago: University of Chicago Press.

Kirmānī, Ḥ. (2007), *Kitāb al-maṣābīḥ fī ithbāt al-imāma. Master of the Age: An Islamic Treatise on the Imamate*, edited and translated by P. Walker, London: I. B. Tauris.

MacDonald, D. B. and E. E. Calverley (2012), "*Ḥakk*," in *Encyclopaedia of Islam*, second edition, edited by P. Bearman, T. Bianquis, C. E. Bosworth, E. van Donzel and W. P. Heinrichs, Leiden: E. J. Brill.

Montgomery, W. (1953), *The Faith and Practice of al-Ghazālī*, London: G. Allen & Unwin.

Plato (1921), *Theaetetus, Sophist*, translated by H. N. Fowler, Cambridge, MA: Harvard University Press.

Rescher, N. (1964), *Studies in the History of Arabic Logic*, Pittsburgh: University of Pittsburgh Press.

Sextus Empiricus (2005), *Adversus mathematicos, M VII, Against the Logicians*, translated by R. Bett, Cambridge: Cambridge University Press.

Street, T. (2004), "Arabic Logic," in *Greek, Arabic and Indian Logic, Handbook of the History and Philosophy of Logic*, edited by J. Woods and D. Gabbay, Dordrecht: New-Holland, 523–596.

3

AFRICAN PHILOSOPHY

Anke Graness

1. Introduction

This chapter will not be able to describe all allegedly relativistic theories produced on the African continent. Instead, it focusses on relativism in the philosophy of Africa south of the Sahara in the twentieth century. The long-time of slavery and colonial subjugation, and the mythification of racist ideas of the superiority of the white race that went with them, has shaped all discourse in Africa in the twentieth century and up to the present, and is the background against which these theories are to be understood. The starting point of African philosophy in the twentieth century is the confrontation with the claim that Africans have no philosophy and even lack the capacity to think rationally, logically, and critically; and that African cultures are "traditional," "pre-reflective," "pre-scientific," or "emotive." Famous historical instances of such claims can be found in Kant (1775), Hegel (1837), Evans-Pritchard (1937), or Hollis and Lukes (1982). Numerous works of modern African philosophy attempt to correct such prejudices.

2. Ontological relativism

The debate about the nature of the African person, culture, and philosophy, and particularly the nature of rationality has shaped the discourse of African academic philosophy since the 1920s. Many scholars consider the debate between representatives who regard rationality as culturally universal and those who regard rationality as culturally relative as the core issue of modern African philosophy (Masolo 1994; Hallen 2009). A famous example of the latter approach is the literary-philosophical movement "Négritude," which was established in European metropolises in the 1920s and 30s by academically trained African philosophers who began to reassert their rationality and their right to describe and to represent their continent. They refused to be defined and represented by "Westerners" through an anthropological, colonial gaze. Of particular importance were such influential representatives of Négritude as Léopold Sédar Senghor (1906–2001) from Senegal, as well as the Caribbean writers Aimé Césaire (1913–2008), Léon-Gontran Damas (1912–1978), and Paulette Nardal (1896–1985). The movement was decisively influenced by, and closely intertwined with, the work of authors in the African diaspora, including such Haitian Renaissance authors as Jacques Roumain (1907–1944), and such African-American Harlem Renaissance authors as the philosophers Alain LeRoy Locke (1886–1954)

29

and W. E. B. Du Bois (1868–1963), the poets Langston Hughes (1902–1967) and Claude McKay (1890–1948), and the anthropologist and novelist Zora Neale Hurston (1891–1960). The close intellectual cooperation was primarily due to the prominence and wide-ranging influence of Pan-African ideals and shared ideas of Black pride, a consciousness of an African culture, and an affirmation of a distinct Black identity (Vaillant 1990, 93–94).

Léopold Sédar Senghor, philosopher, poet, and the first president of independent Senegal, is one of the most important African personalities of the twentieth century and one of the fathers of Négritude. He defines Négritude as "the sum total of the values of the civilization of the African world" (Senghor 1965, 97), a metaphysics of a Black identity and African personality. For him, Négritude was a weapon of defence, a reaction to the colonial and historical denigration of African culture and personality. Particularly in his poetry, Senghor praises everything Black and African; he praises the African woman, African metaphysics and African ways of apprehending reality. The primary task of Négritude was to revive Africans' consciousness of their personal and cultural worth, that is, the movement's central issue is identity. At the same time, Senghor elaborates Négritude as a philosophy of culture, implying that it embraces an ontology, an aesthetics, an epistemology, and a politics. Ontologically, it is founded on the notion of "vital force" or "élan vital" (and not of "being" as in European philosophy): The African universe is a hierarchy of forces organised according to their strengths; it starts from God and includes the ancestors, living humans, the not-yet-born, animals, plants, and minerals. According to Senghor, in contrast to European epistemology, African epistemology draws no line between the self and the object. Senghor writes,

> [the African] is moved, going centrifugally from subject to object on the waves of the Other. . . . Thus the Negro African . . . abandons his personality to become identified with the Other, dies to be reborn in the Other. He does not assimilate. He is assimilated. . . . Subject and object are face to face in the very act of knowledge.
>
> *(1964a, 72–73)*

Instead of Descartes' "I think, therefore I am," the Negro-African could say "I feel, I dance the Other, I am" (1964a, 73). Senghor asserts that the African does not analyse the object in the same manner as the European would do; rather, he touches it, feels it, smells it. Senghor famously declared: "Classical European reason is analytical and makes use of the object; African reason is intuitive and participates in the object" (1965, 33–34). In Senghor's view, the African conceptualises, interprets, and apprehends reality in a different way than the European; an African epistemology, or way of knowing, flows from African ontology. Senghor argues that modes of knowledge or forms of thought as constituted by each race are different since they are rooted in the psycho-physiological make-up of each race. For him each race is unique and has a unique contribution to make to the evolution of humanity. Accordingly, while seeking to preserve and proclaim the value of African culture, Senghor aims to highlight what this culture has to contribute to the assembly of humanity: "Every ethnic group possesses different aspects of reason and all the virtues of man, but each has stressed only one aspect of reason, only certain virtues" (1964a, 75). All aspects have to come together in what he termed the "Civilization of the Universal," a synthesis of the contributions of all the cultures of the earth (Senghor 1964b, 9). He conceives a future of humankind determined by a sort of *métissage* in which everyone is enriched by the mutuality of the positive contributions of all. But this will be achieved in such a way that no one culture loses its specificity, while giving to and accepting from other cultures.

Despite its claim of universalism, Négritude remains essentially an ahistorical, particularistic or relativistic cultural and philosophical conception, a point that was criticised early on by universalists such as Frantz Fanon. Négritude's return to a pre-colonial past for the development of a Black self-confidence cannot be used as a model for future social change, argues Fanon. Instead of showing the social reality and changing it for the better, Négritude naturalises differences and concentrates on looking into a mythical past and glorifying its own culture (Fanon 1952). The Nigerian poet and Nobel laureate Wole Soyinka offers an even more strident critique in a 1962 speech: "A tiger does not proclaim his tigritude, he pounces" (Jahn 1968, 265–266). In other words, when you pass where the tiger has walked before, and you see the skeleton of an antelope, you know that some "tigritude" once existed there.

An ontological approach to African culture and "being African" in this world is also characteristic of so-called ethnophilosophy, a strongly ethnographic approach to philosophy. Like Négritude, ethnophilosophers claim that African thought and African philosophy possess a nature utterly different than the thought and philosophy of other cultures. Moreover, they assume that all of Africa's cultures share certain core concepts, values, and beliefs. Ethnophilosophy encompasses all those theories that deal with the reconstruction of a so-called traditional African philosophy on the basis of proverbs, grammars, and social institutions and assume that the Africans (or the Bantu, the Akan, the Yoruba, the Wolof, etc.) have a collective, immutable philosophy. The main points of reference of such theories are "traditional" African societies and "traditional" African values – both of which are of course empirically hard to verify. Ethnophilosophy's sources are proverbs, studies of African oral literature, language analysis, and the beliefs and values enshrined in such African institutions as religions, political systems, and law. The ethnophilosophical approach presents an unchanging African philosophy free from historical and socio-political contexts, a genuinely once-for-all, given entity. Interestingly, one of the most influential examples of ethnophilosophy is a book by a white Belgian Franciscan priest and missionary, Placide Frans Tempels (1906–1977): *La philosophie bantou* (1945; English translation, 1959). In his work, he tries to reconstruct an "African or Bantu philosophy" based on the language, grammar and proverbs of the Baluba. He concludes essentially that there is a kind of collective, traditional "Bantu philosophy" inherent in the eternal immutable soul of the African (chapter 4 of the book "4. Bantu ontology" (see Tempels 1959)), an ontology that equates the concept of being with the concept of life-force. Tempels states that the idea of life-force is central to the life and world view of the Bantu, for whom reality is not static nor objective, but a dynamic network of "living forces." However, while Tempels, contrary to colonial prejudices, grants the Bantu a philosophy, he states on a most paternalistic vein that Africans are themselves incapable of articulating it – unless they first encounter European philosophy.

Tempels' propositions have been challenged by the Rwandan philosopher Alexis Kagame (1912–1981), who set out critically to verify and reformulate Tempels' claims concerning Bantu ontology. According to Kagame, Bantu philosophy differs from European philosophy in that it relies on a concept of "élan vital," taking vital force as the essence of being. Kagame outlines four fundamental categories of Bantu thought, namely *muntu* or human being (i.e. conscious being); *kintu* or animals, plants (i.e. unconscious being); *hantu* or space and time; and *kuntu* or modality. Kagame argues that those terms are indicative of implicit thought processes and vehicles of an explicitly philosophical discourse based upon terms in the Rwandan oral tradition. His initial consideration is that in an oral culture, such as the Bantu, philosophical concepts are reflected in the structure of words, in proverbs or other literary genres such as fairy tales, narratives, and poems, or in religion and social and cultural institutions. From his reflections, he then derives a Bantu cosmology, a Bantu psychology, a Bantu theology and a Bantu ethics. As with Tempels, for Kagame Bantu philosophy is philosophy without philosophers, the collective philosophy of an entire ethnic group (Kagame 1976, 7, 286).

Two further seminal works which offer an essentialist interpretation of African culture and philosophy are *The Mind of Africa* (1962) by Ghanaian philosopher W. E. Abraham (1934–) and *African Religions and Philosophy* (1969) by Kenyan philosopher John S. Mbiti (1931–). Mbiti argues that African philosophy consists of certain beliefs and values that all African people share. According to Mbiti, "Africans are notoriously religious" (Mbiti 1969, 1), and every important element of African culture is inextricably bound up with religion. Mbiti coined a well-known maxim to express the shared African value of relationality, which is the cardinal concept in the African view of what it means to be human: "I am because we are, and since we are, therefore I am" (1969, 108–109). Moreover, his concept of "African time" was the source of controversy, since he makes the remarkable claim that African languages generally have no term for the distant future. Basing his discussion on analyses of African languages, Mbiti tries to show that the African concept of "future" encompasses a period no longer than two years, a claim which was strongly rejected by numerous African philosophers (e.g. Masolo 1994, 111–119). For many critics, such theories unwittingly justify the alleged difference between Europeans and Africans stated by earlier European anthropologists. Nevertheless, such relativistic, ontological theories exert a strong influence in African philosophical debates even today (e.g. Kasozi 2011).

3. Language and alternative epistemologies

In contrast to an ontological and essentialising relativism, which proposes that there is a way of knowing that is uniquely African and maintains that rationality is culturally bound, authors like Paulin Hountondji (1942–), Marcien Towa (1931–2014), Peter Bodunrin (1936–1997), and Kwame Anthony Appiah (1954–) argue that rationality is universal across cultures. The most famous representative of this approach is the Ghanaian philosopher Kwasi Wiredu (1931–). Wiredu explicitly rejects relativism (Wiredu 1980, 176–177). A critic of the ethnophilosophical approach, he argues that rationality is an anthropological constant. He upholds the universalist claim that all human beings share the same cognitive capabilities and basic rational attributes. His argument is based on the fact that all humans possess language, that is, a system of rules whose adherence presupposes rationality, and that even simple actions presuppose a certain amount of rational thinking. The principles of induction and non-contradiction are exemplary of the logical universals inherent in the use of language (1996, 21–33, 85–87). However, in his essay collection *Cultural Universals and Particulars* (1996), Wiredu also points out linguistic contrasts between Akan and English which lead to, among other conclusions, the idea that the correspondence theory of truth cannot be expressed in the Akan language (Wiredu 1996, 107). He argues that there is no word in Akan equivalent to the English word "fact," and, moreover, that there is no one word in Akan for "truth." "To say that something is true the Akans say simply that it is so, and truth is rendered as what is so. . . . This concept they express by the phrase *nea ete saa . . .* 'a proposition which is so'" (1996, 107). Here the word *saa* means "so." Whereas in English one has the word "true" and the word "so," in Akan one has only *saa* (is so). And the English word "fact" is in Akan also simply expressed by the phrase *nea ete saa* (what is so). Consequently, the correspondence theory of truth leads to a tautology in Akan. Wiredu argues that this does not indicate any insufficiency in the Akan language, but rather points to the fact that in Akan the correspondence theory does not offer any enlightenment about the notion of "being so." He reasons therefore that the Akan language enables us to see "that a certain theory of truth is not of any real universal significance unless it offers some account of the notion of being so" (1996, 111). Moreover, Wiredu states, this trait of the Akan language shows that there are some philosophical issues which can be formulated in English but not in Akan. Wiredu concludes that the problem of the relation between truth and fact arises out of the nature of a language

(1996, 108) and, consequently, that some philosophical problems are not universal (1996, 109). What follows, he maintains, is that such intercultural or inter-linguistic comparisons should be made more frequently in philosophy. Another example to illustrate linguistic differences and their consequences for philosophical problems is Descartes' *cogito, ergo sum* (1996, 140–141). Wiredu argues that in Akan the concept of existence is intrinsically spatial: "to exist is to be there, at some place" (1996, 141). This is diametrically opposed to Descartes' concept, in which the ego exists as a spiritual, non-spatial, immaterial entity, and the mind is the entity responsible for thinking. Wiredu shows that *cogito, ergo sum* cannot be expressed properly in Akan (even though it can be paraphrased and also be understood), because Akan has an exclusively spatial-naturalistic understanding of "existent." He states that it is not possible in Akan to use the word "being" without any further determination. In addition, Akan has an exclusively processual or functional understanding of "mind" in the sense of the thought process itself. For this reason, the body-mind separation in the Akan world view makes no sense.

Wiredu concludes that there is a need for in-depth study of everyday as well as scientific and literary usages of African languages. African philosophers in particular should use their own African languages to think through concepts such as Being, God, Truth, etc. He explicitly demands:

> Try to think them through in your own African language and, on the basis of the results, review the intelligibility of the associated problems or the plausibility of the apparent solutions that have tempted you when you have pondered them in some metropolitan language.
>
> *(1996, 137)*

Another contribution to the literature on African epistemology is the classic book *Knowledge, Belief and Witchcraft* (1986/1997) by the American philosopher Barry Hallen (1941–) and the Nigerian philosopher Olubi Sodipo (1935–1999). In this work, they attempt to conceptually analyse three key words or concepts central to Yoruba thought, namely *aje*, *mò*, and *gbàgbó*, which putatively translate in English to "witchcraft," "knowledge," and "belief" respectively. On the basis of their conversations with Yoruba *onísègùn* (sages who are specialists in traditional medicine), they come to the conclusion that the Yoruba expressions *mò* and *gbàgbó*, which are usually translated "to know" and "to believe," occupy a significantly different semantic field in Yoruba than their English equivalents, knowledge and to believe (Hallen and Olubi Sodipo 1986/1997, 40–85). Hallen and Sodipo, following W.V. O. Quine's distinction between observation sentences and standing sentences, argue that the Yoruba make a distinction between knowledge and belief or opinion. According to them, whereas *mò* is derived from first-hand information, observation, and sense-experience which may be verified or falsified, *gbàgbó* is got through second-hand information, though *gbàgbó* could later become *mò* after rigorous empirical testing and verification. Hallen and Sodipo conclude that propositional attitudes are not universal, and that different languages imply alternative epistemological, metaphysical, and moral systems.

4. Truth as opinion

Of interest in respect of the concept of relativism is Wiredu's own theory of truth, as formulated in his book *Philosophy and an African Culture* (1980), which sometimes earned him the charge of relativism. In his book, Wiredu states that there is no difference between truth and opinion ("truth as opinion," 1980, 124). Wiredu argues that whatever is referred to as "truth" is more correctly interpreted as opinion or point of view, since history shows that what human beings consider to be "true" can be argued to be false from an alternative perspective. By asserting that "truth" is opinion, Wiredu rejects what is referred to in academic philosophical circles as the "objectivist theory" of truth, which he describes as holding that "once a proposition is true, it is true in itself and for ever. Truth, in other words, is timeless, eternal" (1980, 114). And yet, according to Wiredu (1980, 115), such an objectivist theory of truth implies that truth is "categorically distinct" from opinion. However, he argues, truth arises from human agency: from perception and rational inquiry, as opposed to deriving from some transcendent reality. Thus, whatever is called "the truth" is necessarily *someone's* truth. For an item of information to be considered "true," it must be discovered, defended, and known by human beings in a particular place, at a particular time. "We must recognize the cognitive element of point of view as intrinsic to the concept of truth" (1980, 115). However, the fact that truth arises from human endeavour and effort does not mean knowledge will reduce to the merely subjective or relative. "What I mean by opinion is a firm rather than an uncertain thought. I mean what is called a considered opinion" (1980, 115–116). This notion of "considered opinion" is of fundamental significance in Wiredu's overall understanding of truth.

That there can be no truth separate from considered opinion is certainly a problematic statement. By using the statement "to be true is to be opined" (1980, 114f.), Wiredu seeks to overcome the distinction between the object of knowledge and the knowing subject in the sense of Berkeley's *esse est percipi*. This puts him in dangerous proximity to subjective relativism, especially when he states: "There are as many truths as there are points of view" (1980, 115). However, Wiredu does not want to slip into subjective relativism and argues against it:

> It is the insistence on the need for belief to be in accordance with the canons of rational investigation which distinguishes my view from relativism. Truth is not relative to point of view. It is, in one sense, a point of view . . . born out of rational inquiry, and the canons of rational inquiry have a universal application.
>
> *(1980, 176–177)*

For Wiredu truth is dependent on the process of rational inquiry, and rational inquiry depends on circumstances. According to his theory, there are many different truths for different societies as well as for different generations of a society. In order to further distinguish his definition of truth from subjective relativism, Wiredu also calls opinion "rationally warranted belief" (1980, 216–232), that is, truth becomes rational opinion or functional feasibility. In this view, individual opinion is not arbitrary, but rationally grounded, and rationality is an anthropological constant common to all human beings. "Truth" – which he also defines as "what is the case" – refers to a human conception confirmed by a current perception in the real world. It says something about whether human ideas can explain the phenomena of the real world of perception or not. Therefore, it is always bound to the particular perspective from which the world is perceived.

Besides Wiredu, so also Nigerian philosopher Theophilus Okere (1935–) and Eritrean philosopher Tsenay Serequeberhan (1952–) refer to the "historicity and relativity of truth," which "always means truth as we can and do attain it" (Serequeberhan 1994, 118; see also Okere 1983). South African philosopher Mogobe B. Ramose offers a similar argument: truth may be defined as "the contemporaneous convergence of perception and action. Human beings are not made by the truth. They are the makers of truth" (Ramose 1999, 44). And he continues: "Seen from this perspective, truth is simultaneously participatory and interactive. It is active, continual, and discerning perception leading to action. As such, it is distinctly relative rather than absolute" (1999, 45).

5. African feminist theory and cultural relativism

Apart from epistemological considerations, one can also find relativism in a rather ontological or cultural sense in African feminist theory. Nigerian anthropologist Ifi Amadiume argues that Africa had a unique matriarchal system of social values that also had an effect of neutralising biological gendering. She characterises the gender system of the pre-colonial Nnobi culture (part of the Igbo in Nigeria) as flexible or neutered, that is, social roles were not rigidly masculinised or feminised (Amadiume 1987, 185). And Nigerian writer and feminist Catherine Achonolu conceptualises a specific African "Motherism" as an alternative to Western feminism (1995). The Nigerian philosopher Nkiru Nzegwu criticises such generalisations about Africa, which in her view do not do justice to the cultural diversity of the continent. She develops a non-matriarchal, non-gendered portrait of Igbo society, where seniority and lineage are more fundamental determinants of social status than gender (Nzegwu 2006). Nzegwu argues that pre-colonial Igbo society was a genderless society that was displaced by Western colonial powers who introduced gender roles and gender discrimination where previously none had existed. Prior to colonial rule, Igbo society was characterised by a dual-sex system based on separate male and female lines of government (2006, 15, 192ff.), in which political, economic, and social relations were distinct but interdependent and balanced in power (2006, 15, 192ff.). From this social system, Nzegwu derives a concept of complementary equality. In contrast to the Western feminist model of gender equality, which defines equality as equal rights for men and women (men's rights being the yardstick), the complementary model of equality is oriented towards the universal equality of all humans but at the same time takes biological differences between men and women into consideration. Out of respect for those differences, a society that embraces complementary equality must create conditions for men and women that meet the specific interests and needs of both sexes (2006, 199ff.).

6. Ubuntu

Recently, the broad academic discourse on the South African concept of *ubuntu* tends to drift towards (ontological) cultural relativism. *Ubuntu*, currently one of the best-known African indigenous concepts, has benefitted from increasing discussion and awareness, even outside the African continent, since the 1990s. In South Africa, it is a central concept not only in current academic discourse but in the public sphere as well. After the end of apartheid, *ubuntu* became a key abstract term used to frame the process of transition from an apartheid regime to a new "rainbow nation" and an African renaissance. There is no consensus on what *ubuntu* actually means, and the precise content of the concept is still contested. The translations range from "humanity" and "charity" to "common sense" and "generosity." Often *ubuntu* is seen as a concept enshrined in a traditional philosophy of life, although one needs to further differentiate between *ubuntu* as a moral quality of a person or as a way of living. The core meaning of the concept *ubuntu* is frequently expressed using the Zulu-Xhosa aphorism "*umuntu ngumuntu ngabantu*" – "A human being is a human being through other people." meaning that every human being needs other people in order to be human; every person is part of a whole, integrated into a comprehensive network of mutual dependencies. The aphorism expresses "the African idea of persons: persons exist only in relation to other persons. The human self . . . only exists in relationship to its surroundings; these relationships are what it is. And the most important of these are the relationships we have with other persons" (Shutte 2001, 23). Thus, the aphorism refers to the deep relational character of *ubuntu* and underlines at the same time that human beings (*umuntu*) are a "being becoming" (Ramose 1999, 36–37) in an already existing community. The human being is seen as an organic part of a community and the community as the necessary precondition for any human being. Two interrelated central aspects of *ubuntu*, which are widely accepted by many authors, are expressed here, namely the non-static and evolving nature of human beings, and the importance of the community. A widespread general view considers *ubuntu* to express a uniquely African world view or ethics. Here, *ubuntu* is usually described as an ethics or a philosophical concept rooted in the pre-colonial knowledge systems of Africa, a concept which belongs to the "essence" of a specific African mode of being, which, in contrast to the individualistic world view of "the West,"[1] is described as being more community oriented. The South African philosophers Augustine Shutte and Mogobe B. Ramose explicitly refer to an *ubuntu* ethics (Shutte 2001; Ramose 2003b, 324–330). Thaddeus Metz uses features of *ubuntu* ethics to work out an African moral theory worth taking seriously as a rival to dominant Western ethical conceptions (Metz 2007; Metz and Gaie 2010; for a recent attempt in this direction, see Chuwa 2014). Characteristic features of *ubuntu* ethics are compassion towards others, respect for the rights of minorities, conduct that aims at consensus and understanding, a spirit of mutual support and cooperation, hospitality, generosity, and selflessness. Ramose attributes these standards to the linguistic peculiarities and the epistemological structures of the Bantu language. In accordance with the approach of Placide Tempels (1945) and Alexis Kagame (1956, 1976), neither of whom explored the concept of *ubuntu* deeply in their analysis, Ramose describes the Bantu terms *muntu* (human), *kintu* (thing), *hantu* (space and time), and *kuntu* (modality) as the four basic categories of African philosophy. He adds a fifth: *ubuntu*. For him, *ubuntu* is a normative ethical category that defines the relationship between the other four categories. Moreover, it is the fundamental ontological and epistemological category in the "African thought of the Bantu-speaking people," which expresses the indivisible unity and totality of ontology and epistemology (Ramose 2003b, 324–325). For these reasons, he calls *ubuntu* "the root of African philosophy. The being of an African in the universe is inseparably anchored upon *ubuntu*" (2003a, 230).

In addition to its usage as an ontological concept, *ubuntu* recently has been conceptualised both as a postcolonial theory or "critical humanism" (Praeg 2014). Representatives of such an approach explicitly criticise an essentialising understanding of *ubuntu* and consider the concept to be a glocal phenomenon. Thus, the debate over relativistic and universalistic approaches in African philosophy continues.

Note

1 The opposition between Africa and "the West," which is a common feature of various discourses, is usually used to highlight a presupposed (cultural) difference. However, the concept of "the West" is very problematic because it denotes a compass direction, not a geographical entity. Usually the concept of "the West" is used to refer geographically to Europe and North America and to associate these geographical entities with a certain enlightened, secular, and idealised "scientific" mode of thought. Furthermore, it is presupposed that individual freedom is the central ethical idea of "the West," whereas the key ethical idea of Africa is communitarianism. The inner plurality of these geographical entities (Africa, Europe, and North America) in terms of culture, religion, world view, historical development, philosophical schools, etc., is totally neglected in such an approach.

References

Abraham, W. E. (1962), *The Mind of Africa*, Chicago: Chicago University Press.
Acholonu, C. O. (1995), *Motherism: The Afrocentric Alternative to Feminism*, Owerri, Nigeria: AFA Publications.
Amadiume, I. (1987), *Male Daughters, Female Husbands: Gender and Sex in an African Society*, Atlantic Highlands, NJ: Zed Books.

Chuwa, L. (2014), *African Indigenous Ethics in Global Bioethics: Interpreting Ubuntu*, Dordrecht: Springer.

Evans-Pritchard, E. E. (1937), *Witchcraft, Oracles, and Magic Among the Azande*, London: Oxford University Press.

Fanon, F. (1952), *Peau noire, masques blancs*, Paris: Seuil.

Hallen, B. (2009), *A Short History of African Philosophy*, second edition, Bloomington: Indiana University Press.

Hallen, B. and J. Olubi Sodipo (1986/1997), *Knowledge, Belief, and Witchcraft: Analytical Experiments in African Philosophy*, second edition, Stanford: Stanford University Press.

Hegel, G.W.F. (1837), *Werke*, Vol. 9: *Vorlesungen über die Philosophie der Geschichte*, Berlin: Duncker und Humblot.

Hollis, M. and S. Lukes (eds.) (1982), *Rationality and Relativism*, Cambridge: Cambridge University Press.

Jahn, J. (1968), *A History of Neo-African Literature*, London: Faber.

Kagame, A. (1956), *La philosophie bantu-rwandaise de l'être*, Brussels: ARSC.

———. (1976), *La Philosophie Bantu Comparée*, Paris: Présence africaine.

Kant, I. (1775), "Von den verschiedenen Racen der Menschheit", in *Gesammelte Schriften*, edited by I. Kant, Vol. II: *Vorkritische Schriften II*, Berlin: Preussische Akademie der Wissenschaften 1905, 1757–1777.

Kasozi, F. M. (2011), *Introduction to an African Philosophy: The Ntu'ology of the Baganda*, Freiburg: Alber Verlag.

Masolo, D. M. (1994), *African Philosophy in Search of Identity*, Bloomington: Indiana University Press.

Mbiti, J. S. (1969), *African Religions and Philosophy: African Writers Series*, London, Ibadan and Nairobi: Heinemann.

Metz, T. (2007), "Towards an African Moral Theory," *Journal of Political Philosophy* 15: 321–341.

Metz, T. and J. B. R. Gaie (2010), "The African Ethic of Ubuntu/Botho: Implications for Research on Morality," *Journal of Moral Education* 39(3): 273–290.

Nzegwu, N. (2006), *Family Matters: Feminist Concepts in African Philosophy of Culture*, Albany: State University of New York Press.

Okere, T. (1983), *African Philosophy: A Historico-Hermeneutical Investigation of the Conditions of Its Possibility*, Lanham: University Press of America.

Praeg, L. (2014), *A Report on Ubuntu*, Pietermaritzburg: University of KwaZulu-Natal Press.

Ramose, M. B. (1999), *African Philosophy Through Ubuntu*, Harare: Mond Books.

———. (2003a), "The Philosophy of *Ubuntu* and *Ubuntu* as a Philosophy," in *Philosophy from Africa*, second edition, edited by P. H. Coetzee and A. P. J. Roux, Cape Town: Oxford University Press Southern Africa, 230–238.

———. (2003b), "The Ethics of Ubuntu," in *Philosophy from Africa*, second edition, edited by P. H. Coetzee and A. P. J. Roux, Cape Town: Oxford University Press Southern Africa, 324–330.

Senghor, L. S. (1964a), *On African Socialism*, New York: Fredrick A. Praeger Publishers.

———. (1964b), *Liberté 1: Négritude et humanisme*, Paris: Seuil.

———. (1965), "Discours devant le Parlement du Ghana, February 1961," in *Prose and Poetry*, selected and translated by J. Reed and C. Wake, London: Heinemann.

Serequeberhan, T. (1994), *The Hermeneutics of African Philosophy: Horizon and Discourse*, New York and London: Routledge.

Shutte, A. (2001), *Ubuntu: An Ethic for a New South Africa*, Pietermaritzburg: Cluster Publications.

Tempels, P. (1945), *La philosophie bantoue*, Elisabethville, Congo: Lovania.

Tempels, P. (1959), *Bantu Philosophy*, Paris: Présence Africaine.

Towa, M. (2011), *Léopold Sédar Senghor: Négritude ou Servitude?* Yaounde: Editions CLE.

Vaillant, J. (1990), *Black, French and African: A Life of Léopold Sédar Senghor*, Cambridge, MA: Harvard University Press.

Wiredu, K. (1980), *Philosophy and an African Culture*, Cambridge: Cambridge University Press.

———. (1996), *Cultural Universals and Particulars: An African Perspective*, Bloomington: Indiana University Press.

PART 2

Relativism in Western philosophical traditions

4

RELATIVISM IN ANCIENT GREEK PHILOSOPHY

Tamer Nawar

1. Introduction

"Relativism" is said in many ways and there are several ancient views which might be regarded as being relativistic, such as that for certain Fs nothing is F simpliciter, but only relatively to something else; that apparently conflicting judgements may be correct simultaneously; that truth-bearers may have different truth-values at different times (sometimes called "temporal relativism"); and that there are no objective perspectives. I discuss these strands of thought, and some others, in ancient Greek and Roman philosophy.

2. Plato and Protagorean relativism(s)

Several early Greek philosophers seem to have been concerned with relativity. Thus, for instance, Heraclitus might be taken to suggest that apparently monadic predicates are better regarded as relational (so that "α is F" is elliptical for something like "α is F relative to β"; cf. Heraclitus DK B61). Moreover, some conventionalist ethical views that took normative properties to be relative to certain local conventions (cf. Bett 1989; Nawar 2018) come close to the kind of view described as "moral relativism" by Gilbert Harman (1975),[1] and several early puzzles concerning the impossibility of falsehood and contradiction (e.g. Euthydemus 286a1–287b1; cf. Denyer 1991) seem to indicate that apparently conflicting judgements may be correct simultaneously. However, the ancient figure most regularly described as a relativist is Protagoras of Abdera who famously claimed:

> (MEASURE DOCTRINE) Man is the measure of all things, of those that are that they are and of those that are not that they are not.
>
> *(Plato,* Theaetetus *152a2–4)*

This claim is often taken to articulate Protagoras' relativism. However, what Protagoras himself might have meant by these remarks and whether Protagoras should be viewed as a relativist is difficult to determine.[2] While some are optimistic that Plato's Protagoras gives us insight into the historical figure, it often seems that we are limited to making sense of Plato's Protagoras, Aristotle's Protagoras, or the (MEASURE DOCTRINE) as it appears in some particular work(s).

Plato's *Theaetetus* offers arguably the most detailed and influential extant ancient discussion of the (MEASURE DOCTRINE). There, the (MEASURE DOCTRINE) is initially elucidated by appealing to cases wherein the same wind seems cold to one person but not another, and yet it is plausible that the wind is neither cold nor not-cold in itself and that both parties are correct (cf. *Theaetetus* 152b1–c4). Discerning precisely which views Plato attributes to Protagoras is difficult, and modern readers of the *Theaetetus* have given significant attention to the following three questions:

(1) How, precisely, should the (MEASURE DOCTRINE) be understood in the *Theaetetus*?
(2) How should the argument(s) raised against the (MEASURE DOCTRINE) by Plato's Socrates at *Theaetetus* 170a–171d be understood?
(3) What is the relation between the (MEASURE DOCTRINE) and the so-called Secret Doctrine in the *Theaetetus*?[3]

Concerning (1), Plato offers a number of glosses on the (MEASURE DOCTRINE) that give some indication as to how it should be understood but which do not seem to be equivalent (e.g. *Theaetetus* 152a6–8, 152c2–3, 160c7–9, 161d3–e3, 170d4–6, e4–6, 171d9–e8; cf. *Cratylus* 385e4–386d2). Some possible ways of understanding the (MEASURE DOCTRINE) in the *Theaetetus* are as follows:

(MD$_1$) $\forall x \forall y \forall F$ (y appears F to x *iff* y is F for x)
(MD$_2$) $\forall x \forall y \forall F$ (x judges y to be F *iff* y is F for x)
(MD$_3$) $\forall x \forall p$ (x judges that p *iff* p is true for x)
(MD$_4$) $\forall x \forall p$ (x judges that p *iff* p is true)

MD$_1$ and MD$_2$ (which many readers take to be equivalent in the *Theaetetus*) do not explicitly invoke truth (but might nonetheless be regarded as instances of some kind of relativistic view(s); cf. Waterlow 1977; Ketchum 1992). MD$_3$ explicitly invokes *relative truth* while MD$_4$ explicitly invokes *non-relative* truth and the term "infallibilism" seems to be the most appropriate label for views like MD$_4$ (Fine 1994, 1996).[4] How one understands the (MEASURE DOCTRINE) has implications for how one answers (2) and (3) (and vice versa) and there are several further interpretative options beyond those just mentioned.[5] Moreover, each of MD$_1$–MD$_4$ can be understood more narrowly or more broadly. For instance, if one takes MD$_1$ or MD$_2$ to quantify only over so-called perceptual properties, then one will understand the (MEASURE DOCTRINE) more narrowly.

Concerning (2), much discussion has focused on one particular argument at *Theaetetus* 171a6–c9 (often called the "*peritropē*" or "self-refutation" argument). There, Plato's Socrates argues that because there are people who judge that the (MEASURE DOCTRINE) is not true, Protagoras himself must agree that the (MEASURE DOCTRINE) is not true. Some (e.g. Fine 1994, 1998a) have taken the argument to be directed against MD$_4$. If the (MEASURE DOCTRINE) is construed as per MD$_4$ (i.e. as a kind of infallibilism according to which every judgement is true) and one allows that there are judgements that MD$_4$ is not true, then accepting MD$_4$ seems to lead to a contradiction. When the argument at *Theaetetus* 171a6–c9 is construed in this way, it has often been thought that the argument is straightforwardly successful (cf. Fine 1994, 1998a).

Others have thought that the argument at *Theaetetus* 171a6–c9 should not be construed as an argument against MD$_4$, but as an argument against MD$_3$ (or something like MD$_3$). When construed that way there is less agreement as to how the argument proceeds and whether it is successful, but any discussion of these issues should engage with the highly influential work of

Myles Burnyeat on this topic (1976a, 1976b; cf. Nawar 2013b). Burnyeat thinks the (MEAS-URE DOCTRINE) articulates a claim about relative truth (as per MD$_3$) and glosses such talk of relative truth in terms of private worlds in such a way that "*p* is true for *x*" seemingly has the same meaning as "*p* is true in *x*'s (private) world" (precisely what this amounts to and whether it offers a substantive form of relativism requires discussion).

Precisely how Burnyeat's own construal of the self-refutation argument should be understood is not always entirely clear, but it seems to proceed roughly as follows (cf. Wedin 2005). Suppose that a proponent of MD$_3$ recognises that there are those who do not judge that MD$_3$ is true. Now, if there are those who do not judge that MD$_3$ is true, then (as per Burnyeat's "world's gloss") there are worlds in which it is not true that $\forall x \forall p$ (*x* judges that *p iff p* is true in *x*'s world). That being so, there are worlds in which or individuals *of whom* $\forall x \forall p$ (*x* judges that *p iff p* is true in *x*'s world) does not hold and who are thereby not Protagorean measures. Thus, from accepting MD$_3$, i.e. $\forall x \forall p$ (*x* judges that *p iff p* is true for *x*), and the fact that some do not judge MD$_3$ to be true, it *seems* to follow that $\exists x \exists p \neg$ (*x* judges that *p iff p* is true in *x*'s world), i.e. that $\neg \forall x \forall p$ (*x* judges that *p iff p* is true in *x*'s world), i.e. that $\neg \forall x \forall p$ (*x* judges that *p iff p* is true for *x*).

Many readers are inclined to think that Burnyeat's reconstruction of the argument takes some illicit step or makes certain dubious assumptions (Fine 1998b; Castagnoli 2004; Wedin 2005; Chappell 2006; cf. Erginel 2009). As Gary Matthews (perhaps the first to notice the problem) puts it in an unpublished paper: "isn't Burnyeat slipping from the idea of something's not being true *in* Socrates' world to the idea of something's not being true *of* it?" (Matthews cited in Fine 1998b, 152). While there is extensive literature on the topic, there is still room for significant work on these issues and several others in the *Theaetetus* (cf. McDowell 1973; Burnyeat 1990; Nawar 2013a) as well as in other dialogues that touch upon pertinent issues, such as the *Sophist* (cf. Notomi 1999; Crivelli 2012), the *Cratylus* (cf. Sedley 2003; Ademollo 2011), and the *Euthydemus* (cf. Nawar 2017).

3. Aristotle: Protagorean measures, virtue, and temporalism

Aristotle's thought concerning relativistic issues has various strands. In discussing Protagoras (or Plato's Protagoras as he appears in the *Theaetetus*; cf. McCready-Flora 2015), Aristotle typically takes the (MEASURE DOCTRINE) to articulate a kind of *infallibilism* as per MD$_{4*}$, i.e. as the claim that $\forall x \forall p$ (if *x* judges that *p*, then *p* is true) (e.g. *Metaphysics* 1009a6–1009a9, 1062b12–1062b15). Given that there are frequently contradictory appearances and judgements, Aristotle points out that Protagoras is thereby committed to embracing contradictions and criticises those attracted to claims like MD$_{4*}$ accordingly (*Metaphysics* Γ.4–6; cf. Gottlieb 1994; Wedin 2004a, 2004b; Priest 2006, 7–42).

Aristotle takes Protagoras to be wrong in claiming that *any person or appearance whatsoever* is a measure. However, while mere appearance is not factive (i.e. it is not the case that if it appears to *x* that *p*, then *p* is true), genuine perception is factive (i.e. if *x* perceives that *p*, then *p* is true) (*Metaphysics* 1010b1–3). Accordingly, if claims like MD$_{4*}$ are taken to include within their domain only those who have genuine *perception* or *scientific understanding* (making accurate judgements about those matters that they genuinely perceive and scientifically understand), then the claims turn out to be true and perhaps even trivially true. One might thereby be inclined to see *some grain* of truth in Protagoras' claims (*Metaphysics* 1053a35–b3).

In his ethical writings, Aristotle has sometimes been thought to hold that the ethically excellent person is a measure in some substantive sense (e.g. *Nicomachean Ethics* 1113a22–1113b2, 1166a11–13, 1176a10–1176a19). Precisely how this should be understood requires clarification (cf. Gottlieb 1991) and so too does the extent to which Aristotle takes virtue to be *relative*.[6] It is widely recognised that Aristotle is not optimistic about formulating informative and precise general ethical rules (e.g. *Nicomachean Ethics* 1094b14–1094b22, 1104a3–1104a10) because what kind of actions and feelings are appropriate depends upon many particulars of the situation confronted by the agent. As a result, Aristotle's notion of virtue is often said to be *situation*-relative. However, it is not entirely clear precisely how virtue is sensitive to various parameters. In *Nicomachean Ethics* 2.6, Aristotle famously claims that virtue of character aims at "a mean relative to us" (*Nicomachean Ethics* 1106a26ff.). It is sometimes thought that Aristotle thereby takes virtue of character to be not only situation-relative, but also *agent*-relative in such a way that the actions and feelings required by virtue depend not only upon features of the situation confronting the agent, but also upon features of the individual agent (e.g. Leighton 1992). If that is right, then Aristotle's notion of virtue is relativistic in some significant sense. Such interpretations have encountered resistance (e.g. Brown 1997), but the issue merits further discussion.

In his "theoretical" writings, Aristotle supposes that the same judgement (*doxa*) and the same statement (*logos*) possess a truth-value relative to times and can change truth value over time (e.g. *Categories* 4a23–b2; *Metaphysics* 1051b13–1051b17). It seems that Aristotle takes an utterance of "Socrates is sitting" made in the morning and an utterance of "Socrates is sitting" made in the evening to say the same thing and what is said is true at *t iff* Socrates is sitting at *t*. What is said or thought (i.e. the *content*) does not possess a truth-value absolutely or *simpliciter*, but instead possesses a truth-value *relative to a time* and it can go from being true to being false or vice versa. This sort of view, which takes truth to be relative to times, is nowadays sometimes called "temporal relativism" or "temporalism." It seems to have been widespread among ancient and medieval philosophers (cf. Prior 1957; Hintikka 1973; Barnes 2007) and has proved attractive to some (but not many) moderns (e.g. Prior 1967; Kaplan 1989; Recanati 2007). Whether such views should be regarded as a substantive form of relativism is not clear (cf. MacFarlane 2005, 2014; Recanati 2007), but here it suffices to note that temporal relativism has implications for at least two interesting sets of issues in Aristotle.

First, there are issues concerning how Aristotle conceives of truth, change, and the relations between truth-bearers and the world. Some of Aristotle's remarks (e.g. *Categories* 4a23–b2) suggest that Aristotle thinks that judgements and statements can change truth-value without undergoing *real change* (and instead undergoing mere Cambridge changes). In this respect truth and falsehood resemble relatives (*ta pros ti*) and this provides grounds for suspecting that Aristotle does not take truth to be a so-called real property. These issues have received lucid treatment by Paolo Crivelli (2004) but merit further attention.

Second, there are issues pertaining to Aristotle's famous discussion of future truths in *De Interpretatione* 9. Aristotle there aims to respond to those who take a kind of determinism (or, perhaps, fatalism) to follow from accepting the principle of bivalence. Interpretation is controversial, but according to one widespread family of interpretations (to which the terms "standard," "traditional," and "anti-realist" have been applied), Aristotle thinks that the future is open and rejects both determinism and an unrestricted application of the principle of bivalence. On this sort of reading, claims about future contingents, such as an utterance of "there will be a sea-battle on August 1st 2019" (by someone in 2018) are such that they are neither true nor false at the time of utterance (assume that some possible futures are such that the claim is true and some possible futures are such that the claim is false). Instead, what is said becomes true or false at some later time (perhaps at some context of assessment prior to the battle when the battle cannot occur or the battle cannot but occur).

Understanding Aristotle's views on these matters not only has important implications for our understanding of Aristotle's views about bivalence and the excluded middle, modality, and determinism, but is also important for understanding his views about truth, belief retention (cf. Richard 1981), and *content* (notably the content of utterances of future-tense declarative sentences of various kinds, e.g. "there will be a sea-battle," "there will be a sea-battle tomorrow," and "there will be a sea-battle on August 1st 2019"). While these issues have long been the subject of much discussion, progress in logic, semantics, and other areas of philosophy allows us to discuss these issues more fruitfully and so these issues continue to merit attention.[7]

4. Hellenistic and post-Hellenistic philosophy

Some of the issues concerning temporal relativism and content recur in the Master Argument of Diodorus Cronus and Hellenistic discussions of determinism, freedom, and future truth.[8] For instance, the Stoics were determinists and accepted an unrestricted version of the principle of bivalence while nonetheless arguing that various things nowadays discussed as falling under "free will" (such as things being up to us, moral responsibility, and so on) were compatible with determinism (cf. Bobzien 1998). Like most other ancients, the Stoics assumed that truth-bearers can possess truth relative to times and can change truth-value (e.g. Cicero *De Fato* 7.13–14, 11.26–12.28). However, unlike most other parties in the relevant debates, the Stoics offered a detailed semantic theory. The Stoics gave central importance to *lekta* ("sayables") and *axiōmata* ("assertibles") and maintained that an *axiōma* is what is signified by utterances of the right kind (e.g. a meaningful declarative sentence); "in itself complete" (*autotelēs*); and the primary bearer of truth or falsehood (e.g. Diogenes Laertius 7.65). Several details of Stoic views pertinent to temporal relativism (such as what, precisely, completeness [*autoteleia*] amounts to) require clarification.

Turning to the Epicureans, we find that Epicurus claimed that all perceptions (*aisthēseis*) are true, or that all appearances (*phantasiai*) are true (e.g. Plutarch *Adversus Colotem* 1109a8–1109b1; Sextus Empiricus *Adversus Mathematicos* 7.203–210; Lucretius *De Rerum Natura* 4.499). Precisely what the Epicureans might have meant by such claims (and similar claims, such as that all *perceptibles* (*aisthēta*) are true; cf. Sextus Empiricus *Adversus Mathematicos* 8.9), why they were motivated to hold such views, how they hoped to defend them, and how they conceived of truth and falsehood merit attention.

While "*alēthes*" (translated as "true" previously) may have more than one sense in these contexts (cf. Bown 2016), on one common reading the Epicureans should be understood as claiming that misrepresentation in perceptual-type experiences is impossible (and thereby make claims akin to MD$_{4*}$). There are two closely connected strands of thought as to how such claims might be defended. First, the Epicureans might think that what we are immediately aware of in perceptual-type appearances are not things in the world like tables or chairs but streams of atoms projected from them (Epicurus *Letter to Herodotus* 46–53; Sextus Empiricus *Adversus Mathematicos* 7.206–210; 8.63). Second, while we might form mistaken judgements about things in the world (such as tables or chairs), this is due to going beyond what is given to us in sense-perception (which concerns the-table-as-it-appears-to-me or the table-appearing-round-to-me) (Lucretius *De Rerum Natura* 4.360–468; Plutarch *Adversus Colotem* 1109b–1110e; Sextus Empiricus *Adversus Mathematicos* 8.63).[9] The precise details of Epicurean views on these issues, and how they conceived of perceptual content (and conceptual content) merit further attention.

It might seem that the Epicureans should simply allow that some perceptual-type appearances are false. However, the Epicureans' refusal to do so seems to have been motivated by the thought that all perceptual-type appearances have the same epistemic status and that preferring certain perceptual-type appearances to others was thereby unwarranted (cf. Lucretius *De Rerum Natura* 4.478–99). Accordingly, the Epicureans defend the truth of *all* perceptual-type appearances by appealing to epistemic hope. Roughly: given that all perceptual-type appearances have the same epistemic status, if any perceptual-type appearance is false, then there are no appropriate means of telling true perceptual-type appearances from false ones; and if there are no appropriate means of telling true perceptual-type appearances from false ones, then we are in a pickle (e.g. there will no knowledge). However, we are not in a pickle, therefore, etc. (Epicurus *Key Doctrines* 23; Cicero *De Finibus* 1.19.63–4; Diogenes Laertius 10.50–2).

The Academic sceptics were more than happy to make a *modus ponens* out of another school's *modus tollens* (cf. Cicero *Academica* (*Lucullus*) 2.25.79–28.90, 32.101). They drew upon claims of the kind just mentioned (e.g. if any perceptual-type appearance is false, then there are no appropriate means of telling true perceptual-type appearances from false ones) and appealed to the existence of false perceptual-type appearances to argue that knowledge is beyond our reach and that we should not assent to anything (cf. Nawar 2014).

The Pyrrhonian sceptics differed from the Academic sceptics in several important respects but also appealed to all perceptual-type appearances having the same epistemic status with an eye towards inducing suspension of judgement (*epochē*). According to Sextus Empiricus, suspension of judgement is a result of "the opposition of things" (*Outlines of Pyrrhonism* 1.31) and Sextus appeals to the so-called sceptical "modes" (*tropoi*) to illustrate what he has in mind. Relativistic issues and puzzles about relativity loom large in Sextus' work and are especially prominent in the Ten Modes of Aenesidemus, which are said to ultimately derive from a more generic mode of relativity (*Outlines of Pyrrhonism* 1.39).

Aenesidemus' Ten Modes appeal to conflicting appearances of various kinds which are relative to different parameters (*Outlines of Pyrrhonism* 1.36–163; Diogenes Laertius 9.7888). Thus, for instance, x might appear F when perceived by one agent or from a certain position, but appear $F\star$ (where "F" and "$F\star$" express incompatible properties) when perceived by another agent or from another position. One cannot, the thought goes, appropriately adjudicate between these conflicting appearances (e.g. *Outlines of Pyrrhonism* 1.112–16) and so one should suspend judgement over whether x is really F or $F\star$. Moreover, "since everything is relative, it is clear that we shall not be able to say what each existing object is like in its own nature and purely, but only what it appears to be like relative to something" (*Outlines of Pyrrhonism* 1.140). There is, then, at least one strand of Pyrrhonian argument which argues to the conclusion that there is no objective perspective from which things might be correctly assessed. However, there are important questions concerning the precise logical structure of the Ten Modes, the sceptic's doxastic stance (for the Pyrrhonian should presumably suspend judgement about the conclusion just mentioned), how suspension of judgement is meant to come about, what one is meant to suspend judgement about, and the relation between the Ten Modes and other "parts" of Pyrrhonian scepticism (several of which seem to have been later developments). Traditional ways of understanding the Ten Modes (e.g. Annas and Barnes 1985) have been challenged in recent years (e.g. Morison 2011) and several relativistic issues in Pyrrhonian scepticism merit further attention (cf. Machuca 2015; Vogt 2015).

Discussion of relativistic issues continued into later antiquity and one finds discussions of puzzles concerning relativity among later Platonists and the ancient commentators (cf. Harari 2009) which have some bearing on relativism. As with many issues in post-Hellenistic philosophy, these matters stand in need of substantial attention. Augustine, who played an important role in communicating ancient thought to the Latin middle ages, is sometimes thought to be an important figure in the history of thought about subjectivity (e.g. Burnyeat 1982). Whether this is correct or not, there is also a need for clarifying several issues relevant to relativistic concerns which arise from Augustine's discussion of scepticism, disagreement, and perception (cf. Nawar 2015b, forthcoming-b).

Notes

1 "Moral relativism is a soberly logical thesis, a thesis about logical form, if you like. Just as the judgment that something is large makes sense only in relation to one or another comparison class, so too, I will argue, the judgment that it is wrong of someone to do something makes sense only in relation to an agreement or understanding" (Harman 1975, 3). Such views would nowadays usually be regarded as *contextualist* (rather than as *relativist*), but "relativism" is a flexible term.
2 Cf. Zilioli (2007); Ophuijsen et al. (2013).
3 Concerning (3), here it suffices to say that the Secret Doctrine seems to support or be supported by the (MEASURE DOCTRINE), but its precise nature and relation to the (MEASURE DOCTRINE) is controversial. Cf. Fine (1994); Lee (2005); van Eck (2009).
4 The term "subjectivism" is sometimes applied to MD_4 (e.g. Burnyeat 1976b), but that term is also used for other positions.
5 For instance, some readers do not think that the (MEASURE DOCTRINE) articulates a biconditional claim, but simply a conditional claim. Accordingly, there are simple conditional variants:

(MD_{1*}) $\forall x \forall y \forall F$ (if x judges y to be F, then y is F for x)
(MD_{2*}) $\forall x \forall y \forall F$ (if y appears F to z, then y is F for x)
(MD_{3*}) $\forall x \forall p$ (if x judges that p, then p is true for x)
(MD_{4*}) $\forall x \forall p$ (if x judges that p, then p is true)

I return to these later.
6 In the *Categories*, Aristotle classes virtue as being among the *relative things* (*ta pros ti*, e.g. *Categories* 6b15). While this is orthogonal to my discussion here, ancient notions of *relatives* are important for understanding many aspects of Plato and Aristotle, from the partitioning of the soul to issues about infinitude (cf. Duncombe 2013, 2015, 2016, forthcoming; Nawar 2015a).
7 For discussion of some of these issues, cf. Waterlow (1982); Sorabji (1980); Gaskin (1995); Crivelli (2004); Wedin (2004a); Nawar (forthcoming-a).
8 As was long ago noticed by Benson Mates, Diodorus Cronus "uses 'truth' as though it were a temporal predicate" (Mates 1949, 237). How the Master Argument should be reconstructed is controversial and continues to merit attention (cf. Prior 1955, 1967; Denyer 1981).
9 Cf. Striker (1977); Long and Sedley (1987, 78–86); Everson (1990). Epicurean views may find antecedent in Cyrenaic theories (cf. Plutarch *Adversus Colotem* 1121a1–1121c6; Tsouna 1998; Fine 2003).

References

Ademollo, F. (2011), *The Cratylus of Plato: A Commentary*, Cambridge: Cambridge University Press.
Annas, J. and J. Barnes (1985), *The Modes of Scepticism: Ancient Texts and Modern Interpretations*, Cambridge: Cambridge University Press.
Barnes, J. (2007), *Truth, Etc.*, Oxford: Clarendon Press.
Bett, R. (1989), "The Sophists and Relativism," *Phronesis* 34: 139–169.
Bobzien, S. (1998), *Determinism and Freedom in Stoic Philosophy*, Oxford: Clarendon Press.
Bown, A. (2016), "Epicurus on Truth and Falsehood," *Phronesis* 61: 463–503.
Brown, L. (1997), "What Is 'the Mean Relative to Us' in Aristotle's Ethics?" *Phronesis* 42: 77–93.
Burnyeat, M. F. (1976a), "Protagoras and Self-Refutation in Later Greek Philosophy," *Philosophical Review* 85: 44–69.
———. (1976b), "Protagoras and Self-Refutation in Plato's *Theaetetus*," *Philosophical Review* 85: 172–195.
———. (1982), "Idealism and Greek Philosophy: What Descartes Saw and Berkeley Missed," *Philosophical Review* 91: 3–40.
———. (1990), *The Theaetetus of Plato*, Indianapolis: Hackett.
Castagnoli, L. (2004), "Protagoras Refuted: How Clever Is Socrates' 'Most Clever' Argument at *Theaetetus* 171a-c?" *Topoi* 23: 3–32.

Chappell, T. D. J. (2006), "Reading the περιτροπή: *Theaetetus* 170c–171c," *Phronesis* 51: 109–139.

Crivelli, P. (2004), *Aristotle on Truth*, Cambridge: Cambridge University Press.

———. (2012), *Plato's Account of Falsehood: A Study of the Sophist*, Cambridge: Cambridge University Press.

Denyer, N. (1981), "Time and Modality in Diodorus Cronus," *Theoria* 47: 31–53.

———. (1991), *Language, Thought and Falsehood in Ancient Greek Philosophy*, London: Routledge.

Duncombe, M. (2013), "The Greatest Difficulty at *Parmenides* 133c–134e and Plato's Relative Terms," *Oxford Studies in Ancient Philosophy* 45: 43–61.

———. (2015), "The Role of Relatives in Plato's Partition Argument: *Republic* 4, 436b9–439c9," *Oxford Studies in Ancient Philosophy* 48: 37–60.

———. (2016), "Aristotle's Two Accounts of Relatives in *Categories* 7," *Phronesis* 60: 436–461.

———. (forthcoming), *Ancient Relativity: Plato, Aristotle, Stoics, and Sceptics*.

Eck, J. van (2009), "Moving Like a Stream: Protagoras' Heracliteanism in Plato's *Theaetetus*," *Oxford Studies in Ancient Philosophy* 36: 199–248.

Erginel, M. (2009), "Relativism and Self-Refutation in the *Theaetetus*," *Oxford Studies in Ancient Philosophy* 37: 1–45.

Everson, S. (1990), "Epicurus on the Truth of the Senses," in *Companions to Ancient Thought 1: Epistemology*, edited by S. Everson, Cambridge: Cambridge University Press, 161–183.

Fine, G. (1994), "Protagorean Relativisms," *Boston Area Colloquium in Ancient Philosophy* 10: 211–243.

———. (1996), "Conflicting Appearances," in *Form and Argument in Late Plato*, C. Gill and M. M. McCabe, Oxford: Clarendon Press, 105–133.

———. (1998a), "Plato's Refutation of Protagoras in the *Theaetetus*," *Apeiron* 31: 201–234.

———. (1998b), "Relativism and Self-Refutation: Plato, Protagoras, and Burnyeat," in *Method in Ancient Philosophy*, edited by J. Gentzler, Oxford: Clarendon Press, 137–163.

———. (2003), "Subjectivity, Ancient and Modern: The Cyrenaics, Sextus, and Descartes," in *Hellenistic and Early Modern Philosophy*, edited by J. Miller and B. Inwood, Cambridge: Cambridge University Press, 192–231.

Gaskin, R. (1995), *The Sea Battle and the Master Argument: Aristotle and Diodorus Cronus on the Metaphysics of the Future*, Berlin: De Gruyter.

Gottlieb, P. (1991), "Aristotle and Protagoras: The Good Human Being as the Measure of Goods," *Apeiron* 24: 25–45.

———. (1994), "The Principle of Non-Contradiction and Protagoras: The Strategy of Aristotle's *Metaphysics* IV 4," *Proceedings of the Boston Area Colloquium in Ancient Philosophy* 8: 183–198.

Harari, O. (2009), "Simplicius on the Reality of Relations and Relational Change," *Oxford Studies in Ancient Philosophy* 37: 245–274.

Harman, G. (1975), "Moral Relativism Defended," *Philosophical Review* 84: 3–22.

Hintikka, J. (1973), *Time and Necessity: Studies in Aristotle's Theory of Modality*, Oxford: Clarendon Press.

Kaplan, D. (1989), "Demonstratives: An Essay on the Semantics, Logic, Metaphysics, and Epistemology of Demonstratives and Other Indexicals," in *Themes from Kaplan*, edited by J. Almog, J. Perry and H. Wettstein, Oxford: Oxford University Press, 481–563.

Ketchum, R. (1992), "Plato's 'Refutation' of Protagorean Relativism: *Theaetetus* 170–1," *Oxford Studies in Ancient Philosophy* 10: 73–105.

Lee, M. K. (2005), *Epistemology after Protagoras: Responses to Relativism in Plato, Aristotle, and Democritus*, Oxford: Clarendon Press.

Leighton, S. R. (1992), "Relativizing Moral Excellence in Aristotle," *Apeiron* 25: 49–66.

Long, A. A. and D. N. Sedley (1987), *The Hellenistic Philosophers*, Cambridge: Cambridge University Press.

MacFarlane, J. (2005), "Making Sense of Relative Truth," *Proceedings of the Aristotelian Society* 105: 321–339.

———. (2014), *Assessment Sensitivity: Relative Truth and its Applications*, Oxford: Clarendon Press.

Machuca, D. (2015), "Pyrrhonian Relativism," *Elenchos: Rivista di Studi Sul Pensiero Antico* 36: 89–114.

Mates, B. (1949), "Diodorean Implication," *Philosophical Review* 58: 234–242.

McCready-Flora, I. (2015), "Protagoras and Plato in Aristotle: Rereading the Measure Doctrine," *Oxford Studies in Ancient Philosophy* 49: 71–127.

McDowell, J. (1973), *Plato: Theaetetus*, Oxford: Clarendon Press.

Morison, B. (2011), "The Logical Structure of the Sceptic's Opposition," *Oxford Studies in Ancient Philosophy* 40: 265–295.

Nawar, T. (2013a), "Knowledge and True Belief at *Theaetetus* 201a–c," *British Journal for the History of Philosophy* 21: 1052–1070.

———. (2013b), "M. Burnyeat: Explorations in Ancient and Modern Philosophy," *Philosophy* 88: 621–626.

————. (2014), "The Stoic Account of Apprehension," *Philosophers' Imprint* 14: 1–21.

————. (2015a), "Aristotelian Finitism," *Synthese* 192: 2345–2360.

————. (2015b), "Augustine on the Varieties of Understanding and Why There Is No Learning from Words," *Oxford Studies in Medieval Philosophy* 3: 1–31.

————. (2017), "Platonic Know-How and Successful Action," *European Journal of Philosophy* 25: 944–962.

————. (2018), "Thrasymachus' Unerring Skill and the Arguments of *Republic* I," *Phronesis* 63: 359–391.

————. (forthcoming-a), "Dynamic Modalities and Teleological Agency: Plato and Aristotle on Skill and Ability," in *Productive Knowledge in Ancient Philosophy: The Concept of Techne*, edited by T. Johansen, Cambridge: Cambridge University Press.

————. (forthcoming-b), "Augustine's Defence of Knowledge Against the Sceptics," *Oxford Studies in Ancient Philosophy*.

Notomi, N. (1999), *The Unity of Plato's Sophist: Between the Sophist and the Philosopher*, Cambridge: Cambridge University Press.

Priest, G. (2006), *Doubt Truth to Be a Liar*, Oxford: Oxford University Press.

Prior, A. (1955), "Diodorean Modalities," *The Philosophical Quarterly* 5: 205–213.

————. (1957), *Time and Modality*, Oxford: Oxford University Press.

————. (1967), *Past, Present, and Future*, Oxford: Clarendon Press.

Recanati, F. (2007), *Perspectival Thought: A Plea for (Moderate) Relativism*, Oxford: Oxford University Press.

Richard, M. (1981), "Temporalism and Eternalism," *Philosophical Studies* 39: 1–13.

Sedley, D. (2003), *Plato's Cratylus*, Cambridge: Cambridge University Press.

Sorabji, R. (1980), *Necessity, Cause and Blame: Perspectives on Aristotle's Theory*, London: Duckworth.

Striker, G. (1977), "Epicurus on the Truth of Sense Impressions," *Archiv für Geschichte der Philosophie* 59: 125–142.

Tsouna, V. (1998), *The Epistemology of the Cyrenaic School*, Cambridge: Cambridge University Press.

van Ophuijsen, J. M., M. van Raalte and P. Stork (eds.) (2013), *Protagoras of Abdera: The Man, His Measure*, Leiden: Brill.

Vogt, K. M. (ed.) (2015), *Pyrrhonian Skepticism in Diogenes Laertius*, Tübingen: Mohr Siebeck.

Waterlow, S. (1977), "Protagoras and Inconsistency: *Theaetetus* 171a6-c7," *Archiv für Geschichte der Philosophie* 59: 19–36.

————. (1982), *Passage and Possibility: A Study of Aristotle's Modal Concepts*, Oxford: Clarendon Press.

Wedin, M. (2004a), "On the Use and Abuse of Non-Contradiction: Aristotle's Critique of Protagoras and Heraclitus in *Metaphysics* Gamma," *Oxford Studies in Ancient Philosophy* 26: 213–239.

————. (2004b), "Aristotle on the Firmness of the Principle of Non-Contradiction," *Phronesis* 49: 225–265.

————. (2005), "Animadversions on Burnyeat's *Theaetetus*: On the Logic of the Exquisite Argument," *Oxford Studies in Ancient Philosophy* 29: 171–191.

Zilioli, U. (2007), *Protagoras and the Challenge of Relativism: Plato's Subtlest Enemy*, Aldershot: Ashgate.

5

MEDIEVAL PHILOSOPHY

John Marenbon

1. Introduction

Histories of relativism usually jump from Ancient Greece to Montaigne.[1] Even when a few medieval figures are mentioned, the Middle Ages are still abruptly dismissed with the objection that, as Maria Baghramian puts it, "the dominant belief in a singular and absolute revealed truth within a Christian framework, on the whole, made the medieval period inhospitable to relativism" (Baghramian and Carter 2018, §3).[2] The common approach is wrong on two counts. First, Montaigne is not in any but a very limited and derivative way a relativist (see Fricker 2013, 795–798; Marenbon 2015b, 359–361). Second, the Middle Ages were, at least in the Latin West, a time when various forms of relativism flourished. There are three reasons why historians have missed something so obvious. First, rather than taking into account all sorts of relativist formulations, they have tended to concentrate on the very strong, sweeping ones usually debated today. Second, a finer-grained description of types of strong relativisms is needed for examining past theories than is usually used in considering contemporary ones. Third, in different periods, relativist arguments sometimes serve very different purposes.

Relativism about scientific knowledge, moral relativism and cultural relativism will be examined in turn, with the focus on the Christian Latin tradition First, however, a passage will be examined that poses sharply the question of what should be considered relativist in medieval texts. (It also has the distinction of being the one medieval text to be cited explicitly in Baghramian and Carter's (2018) survey.)

2. Boethius and the modes of cognition principle

In his *Consolation of Philosophy* (c. 524), Boethius is trying to show that the argument from divine prescience to there being no contingent events can be rejected. The difficulty is that "if those things which are uncertain in their outcome are foreseen as if they were certain, this is the darkness of opinion, not the truth of knowledge" (Boethius (2005) *De Consolatione Philosophiae*, V.4.23–25, 149, henceforth cited only by abbreviation and page number of the 2005 edition). This conclusion is justified by what seems to be a truism about knowledge: it is not knowledge "to judge something other than as it is" (V.4.23–25, 149). Since God does foreknow all events, it follows that none of them can be contingent. Boethius responds by rejecting the apparent truism. It is wrong, because of the Modes of Cognition Principle:

> Everything that is cognized is cognized, not according to its own power, but rather according to the capacity of those that cognize it.
>
> *(V.4.23–25, 149; cf. Marenbon 2013)*

That is to say, as the following paragraphs make clear, cognition is relativized to cognizers. ("Cognize" – rendering Boethius' "*cognosco*" – is equivalent to "know" in a wide sense, where the object need not, as for "know" in a narrow sense, be a proposition, but can be, for instance, the red colour I see or imagine, or the concept of human being I consider.)

Boethius' main example is of how we sense the same human being as an enmattered shape; we imagine its shape without matter; with our reason we go beyond the shape to grasp the universal form of human being; and by intelligence (God's mode of cognition) the "pure form" is contemplated (V.4.27–30, 149). There is, then, no unrelativized way that the human being is: rather, in relation to the senses it is an enmattered shape, in relation to the imagination a shape without matter, and so on. Boethius adds, however, that "the superior power embraces the lower one, but the lower one does not at all rise to the higher one" (V.4.31; 149–50). Does this mean that cognition is not relative after all, but just that cognitive powers are graded and the higher the power, the better it cognizes? No, because it remains true that, for example, the senses cognize this human as an enmattered shape, and the reason as a universal. By saying that "the superior power embraces the lower one," Boethius probably means no more than that the reason is able to know about the way the senses cognize (see V.4.34–361, 150; cf. Marenbon 2003, 134). Indeed, the Modes of Cognition Principle must be read as a genuine relativization, if it is to explain, as it needs to do, how the same future event can be both contingent, so far as our knowledge of it is concerned, and necessary as known by God.

This passage in Boethius is important, because the *Consolation* was the most widely read philosophical work in Western Europe from c. 800 to c. 1500 or later, both in Latin and in translation into almost every vernacular (see Marenbon 2003, 164–182). Moreover, Boethius' Modes of Cognition Principle was adapted and used in discussing divine prescience and future contingents by many thinkers, including Aquinas (Commentary on *Sentences* I, d. 38, q.1, a.5 ad4; *De veritate* q.2. a.12, ad17; *Summa Theologiae* I, q.14, a.13, ad2).

3. Relativism about scientific knowledge

It was another Boethius (of Dacia – that is, Denmark) who, about 750 years later, produced one of the most explicit medieval statements of a relativist position. As an Arts Master in the University of Paris in the 1260s and 1270s, Boethius' job was to teach a curriculum that had just become an Aristotelian one, based on almost the whole range of Aristotle's works. There were a few questions on which Aristotle's views went against Church teaching, such as the eternity of the world, a principle of Aristotelian science but contrary to Christian doctrine. What should an Arts Master, himself a Christian in a Christian university, teach? Boethius' treatise *On the Eternity of the World* aims explicitly to "bring into harmony the view of the Christian faith and the view of Aristotle and some other philosophers" about the eternity of the world (Boethius of Dacia (1976) *Topica; Opuscula*, 335:6–8).

Boethius considers the practitioners of three different philosophical disciplines: mathematics, metaphysics and natural science. None of them, working within his discipline, can demonstrate that the world, as Christian faith holds, had a beginning. But the position of natural scientists is particularly difficult, because it is not just impossible to show that the world has a beginning from the principles of natural science: it is contrary to these principles. The natural scientist must therefore deny that the world has a beginning and also other articles of the Christian faith, such as the resurrection of the dead. But, Boethius asks, since, the Christian faith is "most true," will not the natural scientist thereby be saying something false? No, he argues. Just as A and B can both be speaking the truth when A says, "Socrates is white" and B denies "Socrates is white with respect to something" (for instance, with respect to his hat), or when A says "The world has a beginning" and B says "The world has no beginning through natural causes and natural principles," so there is no contradiction between holding that the world has a beginning and that what natural scientists say when they deny that the world has a beginning is true. The reason is that

> The natural scientist denies that the world . . . has a beginning as a natural scientist, and that means that he denies that from natural principles it has a beginning. For whatever the natural scientist denies or accepts, he does or accepts it from natural causes and principles.
>
> (Boethius of Dacia Topica; Opuscula, 352:465–70)

Boethius, then, holds that every statement a natural scientist makes professionally carries the automatic qualification "according to the principles of natural science." Indeed, he believes in general that "no *artifex* (practitioner of one of the Arts Faculty disciplines) can effect, accept or deny anything except from the principles of his science."[3]

Boethius' view might seem untenable, since it seems to limit practitioners of the various disciplines, such as natural science, to making statements about these disciplines, rather than about the world. But his position need not be construed so narrowly. Within the framework of a discipline, say natural science, statements *are* about the world. The discipline as a whole must be seen as relative to its principles, but these principles are moulded by the world, not arbitrary, and the way the science is developed on their basis is also highly responsive to how the world is, although none of the Arts Faculty disciplines provides an unerring guide to reality, outside the framework set by its principles.

Christian doctrine is a different matter. As mentioned previously, Boethius describes it as "most true" and he explicitly describes the position that the world has a beginning as not only "the truth of the Christian faith," but also as "the truth without qualification" (Boethius of Dacia *Topica; Opuscula*, 351:421–22). Boethius thus limits his relativism – but not very much. He insists that the spheres of scientific enquiry and faith-based reasoning are kept strictly separate. Although true, as we can be sure by faith, the proposition that the world had a beginning must be rejected as false by natural scientists if they are to do their science well. This science will not be entirely accurate, because it can take account only of natural causes, and not the non-natural first cause, but it is the best science we can have (Boethius of Dacia *Topica; Opuscula*, 351:445; 352:455).

Boethius' treatise achieved some notoriety in the 1270s, but was then lost until the twentieth century. No other Arts Master put forward so elaborately reasoned a justification for relativism, but in the following centuries many of them followed in practice exactly what Boethius had set out in theory. They developed in full the Aristotelian sciences they taught, even when in doing so they ended by arguing for a position clearly contrary to Christian doctrine, and then they added the qualification that this conclusion is that reached by reason, but faith reaches a different conclusion. So long as they included a brief summary of the Christian teaching and accepted its truth, the Arts Masters were usually left in peace to work in their own sphere, just as Boethius had wished (see Marenbon 2015a, 153–155). But in late medieval Italy, where, more than elsewhere, Arts Masters took advantage of this freedom, the Church authorities finally decided to clamp down: in 1513, the Bull *Apostolici regiminis* required that Masters teaching philosophy try to provide rational refutations of any arguments with conclusions contrary to Christian doctrine "since," as it declared, "they all can be refuted" (Mansi 1772, 842DE). The Bull rejects the relativism on which the Arts Masters' freedom rested: truths accessible through faith should, it insists, be brought to bear on philosophical reasoning: when a philosophical argument leads to a conclusion contrary to any of them, it must be wrong on its own terms and so, in principle, an Arts Master should be able to refute it.

Pomponazzi wrote *On the Immortality of the Soul* (1516) in the aftermath of *Apostolici regiminis*. He says there that he accepts the Christian doctrine that the soul is immortal, and that Aquinas gives the best account of it; but he questions whether, as Aquinas held, it was Aristotle's view and whether it can be established by reason, without revelation. He goes on to put forward powerful criticisms of Aquinas' arguments, and to make and defend the view that the human soul is unqualifiedly mortal (though qualifiedly immortal just in that humans can grasp unchanging intellectual realities). Some historians consider that Pomponazzi really wished to reject the Christian doctrine of immortality, but had to pretend to accept it, to avoid censorship and persecution. It is more plausible, however, to see *On Immortality* as a challenge, not to Christianity, but to the Bull of 1513, and a defence of the relativism that protected the Arts Masters' autonomy.[4] Pomponazzi insists explicitly and strongly on the separation between the two spheres of argument. "Every art," he says, "ought to proceed through the things appropriate and fitting for the art" (Pomponazzi (1525) *Tutti i trattati peripatetici*, 15.5, 1100, henceforth cited only via abbreviation and page number). The immortality of the soul cannot be demonstrated in philosophy, and attempts to do so produce fallacious arguments. It can, however, be shown through the means proper for the doctrine of the faith: revelation and the Bible (15.5, 1100). In the *Apologia*, a defence of his position against critics, Pomponazzi underlines the separation of the spheres. From the premise that the human soul is immortal, a whole set of consequences, all of them truths of Christian doctrine, such as the existence of heaven and hell, and the resurrection of the dead, can, he claims, be deduced. But these are obviously incompatible with natural philosophy (Aristotelian science), and so the premise from which they follow, the immortality of the soul, must also be incompatible with it (Pomponazzi (1525) *Tutti i trattati peripatetici*, cited via abbreviation and page number).

4. Moral relativism

Moral realism, it may seem, had impregnable defences in the Middle Ages. First, it received strong metaphysical support from the acceptance that God is the supreme good. Second, it was generally held that, as well as the revealed laws of the Old and New Testaments, there is a natural law, the precepts of which all people at all times have been able to grasp. Third, theories of the virtues were common throughout the period and, in principle, they set out common ethical ideals for everyone. In fact, however, the virtues were not always conceived in such a universal way, and the division between natural and revealed law turned out to have awkward consequences with regard to reward and punishment. These two gaps in the fortifications of moral realism left room for relativistic ways of thinking about ethics.

Running through the Middle Ages, from Martin of Braga's late sixth-century *Rules for an Honest Life* to Engelbert of Admont's *Mirror of the Virtues* (1308–13) and beyond, there was a tradition of presenting the virtues in a deliberately this-worldly way, for the benefit of the laity, leaving theological considerations out of the picture. Without any intention to go against Christian teaching, and any considered theoretical stance, such works promoted a sort of uncontroversial, *de facto* relativizing of ethical ideals to social roles (see Von Moos 2012, 147–149).

The question of pagan virtues led to a more explicit split between two sorts of virtue. Few medieval thinkers were willing to follow Augustine's severe view that the virtues of the Greek and Roman heroes, heroines, poets and philosophers they so much admired were mere shams. Most preferred a middle course. They found useful a distinction made by Plotinus between "political" virtues and virtues of higher sorts.[5] These political virtues were considered to be in relation to human things alone, whereas the "catholic virtues" distinguished from them by some of the twelfth-century theologians concerned God, the duties laid down by him and the final enjoyment of Him in heaven. There developed, on the same lines, a distinction between two types of wisdom, courage, temperance and justice: the acquired cardinal virtues, open to every human being to gain, and the infused cardinal virtues, available to Christians only, through baptism. From the late thirteenth century onwards, theologians came to regard these infused cardinal virtues as an unnecessary complication, but they still made a distinction between virtuous Christians and pagans, by building into their accounts of virtue the relation between the virtuous acts and their ends. So, for example, in the early fourteenth century William of Ockham argued that a pagan philosopher might exercise the virtue of abstinence for the conservation of nature, whereas the Christian does so also to honour God. Ockham makes clear that the virtues practised by pagans are not just a deficient form of those Christians have, but a parallel system, by insisting that pagans can show even heroic virtue – in dying to defend justice, for instance – in the same way as Christian martyrs, who die for the sake of their faith (*De connexione virtutum*, q.7, q.2; William of Ockham 1984, 336–337).

In this case and the others, the parallel schemes of this-worldly virtues, open to all, and God-directed virtues open to Christians alone do not seem to be measurable against each other or by a common standard. It would make no sense, for instance, to ask whether a heroically virtuous pagan is more or less virtuous than virtuous, but not heroically virtuous Christian. There is, rather, an absolute gap between the schemes, since through their virtues Christians merit salvation, whereas all these theologians were agreed that pagan virtues, however great, did not straightforwardly help towards salvation. But the gap is not unbridgeable, since all these theologians wanted to avoid the conclusion that God would let an entirely virtuous pagan, unable to know anything of Christianity, be damned. They appealed either to a notion of "implicit faith" or to the possibility of enlightenment in the truths needed for salvation by a divinely sent messenger or by internal inspiration (see Marenbon 2015a, 168–176).

5. Cultural relativism

Just as the traditional view that there was no medieval relativism is ill-founded, so is the judgement, often made by historians of anthropology, that there were no medieval anthropologists or ethnologists.[6] In the last century, anthropology has been closely connected with cultural relativism, both because – though now the fashion is changing – most anthropologists took this standpoint, and because arguments for cultural relativism have frequently been made using the diversity of beliefs, practices and norms revealed by ethnographers. Was there a similar link in the Middle Ages?

The answer is not straightforward. The outstanding ethnographical texts of the twelfth and thirteenth centuries, Gerald of Wales' accounts of the Welsh and Irish (see Bartlett 1982), and the description of the Mongols by John of Piano Carpini and William of Rubruk (see Marenbon 2015a, 113–120), seem far from any relativism. Take John, for example. Whereas most of his contemporaries saw the Mongols as sub-human brutes, John takes great trouble to understand how their society is organized and what are their aims. This is a first step towards cultural relativism, but John does not go further. Although he records their behaviour accurately, he judges the Mongols according to the standards of his own culture, noting that they regard as "sins" and punish by death actions such as spitting out food, whereas they do not consider that acting against commandments, by fornicating or harming others, is a sin (John of Piano Carpini (1989) *Historia Mongolorum*, III.7–8, 239–40).

Nearer the end of the Long Middle Ages, when European observers came to know in detail about the natives of South America and their ways of life, a less weak relativism began to inform their approaches. Bartolomé de las Casas (1484–1566) was both a vigorous advocate for the native Americans and a careful observer of their beliefs and customs. He defends their paganism as a better paganism than that of the ancient Greeks and Romans, because, not in spite of, the human sacrifices it demands. He uses the following principle: the more valuable the sacrifice people are willing to make to God, the higher their conception of him. Their willingness to sacrifice what is most valuable of all, humans, shows that the Incas have the highest conception of God (Las Casas (1967) *Apologética historia sumaria* III.143, 183; II, 43, 242–44, cited by abbreviation and page number). Las Casas is accepting that, relative to the frame of pagan religious ethics, human sacrifice is highly commendable. Las Casas' relativism, though, is limited in two ways. First, Las Casas completely rejects this whole frame of pagan religious ethics as being based on idolatry, inspired by the devil; and he holds that pagans can, and have, avoided idolatry.[7] Second, the principle by which he elevates, relatively, the Incas' religion belongs Las Casas' own scheme of Christian and, he would claim, universal values, although he applies it within the framework of pagan ethics. This type of relativism, where by being related to its context, a practice initially abhorrent to the reader is seen to fit a universal ethical ideal, is also central to a work written a century and a half earlier, but which cannot count as ethnography, because it is based entirely on second-hand material: *The Book of Sir John Mandeville*. The author rewrites his mains source so that what the source reported as the weird and disgusting practices of peoples sometimes only partly human in appearance – drinking cow's urine, for instance, or killing and eating sick relatives – are seen, once their purpose is understood as rational and kindly, according to values he expects his readers to share (see Marenbon 2015a, 120–126).

By contrast, a cultural relativism nearer to that of twentieth-century anthropologists is found in the account of his life with the cannibalistic Tupí of Brazil, published by Jean de Léry in 1578. De Léry does not theorize, but in his account of the capture, entertainment, slaughter and ritual eating of an enemy, he enters so completely into the scheme of values of the Tupí that – although like Las Casas' he detects diabolic influence – it becomes difficult for readers not to accept the practice on the Tupís' terms. Montaigne used de Léry without acknowledgement in his essay on cannibals, and much of the impression of relativism it gives at first sight is due to the observations borrowed from him.

6. Areas to be explored

The central claim in this chapter is that relativism, so far from being unknown, was widespread in the Middle Ages. Given its enforced brevity, it has omitted more than it has included. Here are some notes about what is left out, and some suggestions about where more material may be found.

(1) The Western tradition of medieval philosophy has four branches: Latin, Greek, Arabic and Jewish. Only Latin philosophy is discussed here. There are few signs of relativism in medieval Greek philosophy from the current and very inadequate general surveys. Jewish philosophy is a more promising field for investigation, because of the strong adherence by many of its exponents, both writers in Arabic and Hebrew, to Aristotelian science. It may be,

however, that their inclination to develop relativistic views were limited by the comparative ease with which the Torah could be interpreted to fit philosophical views.

(2) Protagorean relativism – the most important ancient variety of relativism – is mentioned in passing by Aristotle in his *Metaphysics*, a text well known and commented on hundreds of times from the thirteenth century onwards. So far, study of medieval discussions of this passage has indicated that writers, influenced by Aristotle's presentation, do not focus on it as a treatment of relativism, but there is much more of this material to be explored.[8]

(3) The Latin tradition includes works written in the vernaculars, often as poetry, with philosophical content. Some of these may be important sources for medieval relativism – including two of the most widely read poems: the *Roman de la Rose*, as completed by Jean de Meun in the 1270s, and Dante's *Divina Commedia*. Arguably, the *Roman* implies a relativism about truth and morality by juxtaposing different frameworks of judgement and undercutting the unique reliability of any of them.[9] And it can be argued that the *Commedia* suggests there are two different frameworks for judgement of behaviour, one that of the virtues, the other that of divine decision. A pagan can be entirely virtuous, Dante holds, and yet damned – and we cannot even understand, through reason, how this can be so. Dante's is a stronger relativism than that of Ockham and the theologians, because he does not allow any way that the virtuous pagans could have saved themselves (see Marenbon 2015b, 356–359).

7. Conclusion

Medieval relativism had different uses from the contemporary variety. Most working scientists today regard relativism with suspicion. By contrast, from the thirteenth to the sixteenth centuries, for many Arts Masters relativism was a way of protecting their autonomy, so that they could develop Aristotelian sciences in their own terms, irrespective of Christian teaching. Moral relativists now tend to hold that there are many systems of moral values, and there is no absolute reason to prefer one of them over the others. Medieval moral thinkers were almost always convinced that there is an absolute scheme of values, by adherence or departure from which a person's eternal destiny would be decided. But some believed that this scheme was not easily open – or even not open at all – to all humans to follow, and they devised schemes of relative morality for those who were excluded. With at least one exception late in the period, ethnography did not lead medieval thinkers towards the cultural relativism common among twentieth-century anthropologists, but towards a weaker, universalizing relativism, in which common values are discerned beneath practices that superficially differ sharply from our own.

Do these differences, especially the fact that all these relativisms are, to some extent, limited, mean that "medieval relativism" is an oxymoron? According to Martin Kusch, philosophers now usually have three criteria for relativism: about frames – in a given domain, judgements or beliefs are "true or false, justified or unjustified, only relative to systems of standards"; about plurality – there is one more than one such system; and about neutrality – there is no neutral way of adjudicating between the systems.[10] Almost all the examples of medieval relativism discussed here meet the frame and plurality criteria. Some of them also meet the neutrality criterion. It is, for example, not neutrally, but only from within Christian ethics that pagan ethics can be judged deficient, or by revelation that Aristotelian natural science is shown to be deficient.

There is a further, egalitarian criterion, says Kusch, held by some relativists: that all the systems are equally valid. No variety of medieval relativism fulfils this criterion, but this does not mean there was no relativism at the time. Relativism comes in degrees: weak relativism fulfils the first two criteria, strong relativism the first three and very strong relativism all four. The absence of *very* strong relativism in the Middle Ages has misled historians into ignoring the importance of both weak and strong relativism in this long period.

Notes

1 See, for example, Baghramian (2010, 36), Irlenborn (2016, 30; where, in a paragraph, the possibility of medieval relativism is excluded), Wong (1661), Fricker (2013, 798–799), jumps from Protagoras to Hobbes. I query this approach in Marenbon (2015b), which the present chapter develops and overlaps in parts.
2 Baghramian had previously made the same point in *Relativism* (2004, 50) without any references to medieval authors.
3 Boethius of Dacia (*Topica; Opuscula*, 347:335–348:336; cf. 353:482–484). For Boethius' wide use of this principle across different disciplines, see Ebbesen (2009).
4 I give a defence of such a view, along with references to scholars against and in support of it, in Marenbon (forthcoming).
5 The theory was transmitted by the widely ready Commentary on the *Dream of Scipio* by the fifth-century Latin Neoplatonist, Macrobius: Macrobious (1970) I.8.5–8, 37–38 (cited by abbreviation and page number to the 2007 edition). For a general account, see Bejczy (2011), and cf. Marenbon (2015a, 160–167).
6 See especially J. Rubiés (2009) – a collection; Khanmohamadi (2014); Fazioli (2014, 336–355).
7 Aristotle, the Brahmins in India and the Chinese avoided idolatry: Las Casas *Apologética historia sumaria* III.71, 99; I.371, 631.
8 See Aristotle, *Metaphysics* (IV.5.1 1009a6–10) and Denery (2009).
9 For an excellent presentation, see Morton (2018), although he leans more to presenting Jean as a sceptic who holds that *no* framework is reliable.
10 Kusch, "ERC Advanced Grant Research Proposal 2013: The Emergence of Relativism," available at www.academia.edu/10181904/ERC_Proposal_on_Relativism_Long_Version_B2_

References

Baghramian, M. (2004), *Relativism*, London: Routledge.
———. (2010), "A Brief History of Relativism," in *Relativism, a Contemporary Anthology*, edited by M. Krausz, New York and Chichester: University of Columbia Press, 31–50.
Baghramian, M. and J. A. Carter (2018), "Relativism," in *The Stanford Encyclopedia of Philosophy*, Winter 2018 edition, edited by E. N. Zalta, https://plato.stanford.edu/archives/win2018/entries/relativism/.
Bartlett, R. (1982), *Gerald of Wales: 1146–1223*, Oxford: Oxford University Press.
Bejczy, I. (2011), *The Cardinal Virtues in the Middle Ages: A Study in Moral Thought from the Fourth to the Fourteenth Century*, Leiden: Brill.
Boethius, Anicius Manlius Severinus (2005), *De Consolatione Philosophiae. Opuscula Theologica*, edited by C. Moreschini, Munich and Leipzig: Saur.
Boethius of Dacia (1976), *Topica; Opuscula* (Opera 6,2; Corpus philosophorum danicorum medii aevi 6,2), edited by N-J. Green-Pedersen, Copenhagen: Gad.
Denery, D. (2009), "Protagoras and the Fourteenth-Century Invention of Epistemological Relativism," *Visual Resources* 25(1–2): 29–51.
Ebbesen, S. (2009), "Boethius of Dacia: Science Is a Serious Game," in *Topics in Latin Philosophy from the 12th-14th Centuries: Collected Essays of Sten Ebbesen*, II, Farnham and Burlington: Ashgate, 153–162.
Fazioli, K. P. (2014), "The Erasure of the Middle Ages from Anthropology's Intellectual Genealogy," *History and Anthropology* 25: 336–355.
Fricker, M. (2013), "Styles of Moral Relativism: A Critical Family Tree," in *The Oxford Handbook of the History of Ethics*, edited by R. Crisp, Oxford: Oxford University Press, 794–817.
Irlenborn, B. (2016), *Relativismus*, Berlin and Boston: De Gruyter.
John of Piano Carpini (1989), *Storia dei Mongoli*, edited by Ernesto Menestò et al., Spoleto: Centro italiani di studi sull'alto medioevo.
Khanmohamadi, S. (2014), *In Light of Another's Word: European Ethnography in the Middle Ages*, Philadelphia: University of Pennsylvania Press.
Kusch, M. "ERC Advanced Grant Research Proposal 2013: The Emergence of Relativism," www.academia.edu/10181904/ERC_Proposal_on_Relativism_Long_Version_B2_.

Las Casas, Bartolomé (1967), *Apologética historia sumaria*, edited by E. O'Gorman, Mexico City: Università Autónoma de México.

Macrobius (1970), *Commentarii in Somnium Scipionis*, edited by J. Willis, Leipzig: Teubner.

Mansi, J. D. (1772), *Sacrorum conciliorum nova et amplissima collectio* XXXII, Paris.

Marenbon, J. (2003), *Boethius*, New York: Oxford University Press.

———. (2013), "Divine Prescience and Contingency in Boethius' *Consolation of Philosophy*," *Rivista di storia della filosofia* 68(1): 9–21.

———. (2015a), *Pagans and Philosophers: The Problem of Paganism from Augustine to Leibniz*, Princeton, NJ and Oxford: Princeton University Press.

———. (2015b), "Relativism in the Long Middle Ages: Crossing the Ethical Border with Paganism," *Hau: Journal of Ethnographic Theory* 5(2): 345–365.

———. (forthcoming), "Italian Universities, Arts Masters and Interpreting Pomponazzi's *De immortalitate animae*," in *Scholasticism: Individuals and Institutions*, edited by A. Fitzpatrick and J. Sabapathy, London: Royal Historical Society.

Morton, J. (2018), *The Roman de la rose in Its Philosophical Context: Art, Nature, and Ethics*, Oxford: Oxford University Press.

———. (1525), *Tutti i trattati peripatetici*, edited by F. Raimondi and J. Valverde, Milan: Bompiani, 2013.

Rubiés, J-P. (2009), *Medieval Ethnographies: European Perceptions of the World Beyond*, Farnham and Burlington, VT: Ashgate.

Von Moos, P. (2012), "Du miroir des princes au Cortegiano. Engelbert d'Admont (1250–1331) sur les agréments de la convivialité et de la conversation," in *Formes dialoguées dans la literature exmeplaire du Moyen Âge*, edited by Marie-Anne Polo de Beaulieu, Paris: Champion, 103–162.

William of Ockham (1984), *Opera Theologica* VIII, edited by Girard Etzkorn, Francis Kelley and Jospeh Wey, St. Bonaventure: Franciscan Institute.

Wong, D. (1661), "Relativisme moral," in *Dictionnaire d'éthique et de philosophie morale* II, edited by Monique Canto-Sperber, Paris: Presses universitaires de France, 1996.

6

RELATIVISM IN EARLY MODERN PHILOSOPHY

Martin Lenz

1. Introduction

In this chapter, I aim to show that crucial elements of relativism emerge through an increasing dissociation of absolute and human standards in the early modern period. But what is relativism? Relativism is not a specific view but rather bundle of different doctrines. Accordingly, it is difficult to pin down its origins. Even though relativistic leanings were expressed throughout the history of philosophy, relativism is often assumed not to have emerged fully prior to the nineteenth century. In keeping with a working definition by Martin Kusch (2013), I would like to start from the following general characterisation of relativism:

> Relativism is the view that (1) a belief or an idea is called true, justified or apt in relation to a standard, and (2) that there is more than one standard. Moreover, (3) there is no neutral way of choosing between these standards. Some relativists hold (4) that all standards are equally valid.

Given that early modern authors do not address relativism explicitly, approaching their writings with this characterisation has to be genealogical. Although it is doubtful that there is an early modern author endorsing all these claims, many seem to hold variants of claims (1) and (2). An *absolute standard* that many accept can be found in the assumption of essences or divine ideas. My idea of cats might be said to conform to or divert from the essence of cats. Given that we don't (always) grasp essences, we might be said to have established provisional *non-absolute standards* amongst humans. Even if I don't grasp essences, my idea might be said to conform to or divert from a conventional standard of knowledge. Debates about the relation between these standards eventually open the door to relativistic assumptions in that they allow for seeing standards as cohering or competing.

   My approach benefits from a relatively small number of studies on the history of early modern forms of relativism, most notably the works of Maria Baghramian (2004) who focuses particularly on Montaigne (1533–1592) and the French Enlightenment, and John Marenbon (2015) who discusses the "long middle ages" (ca. 200–1700) and slightly deflates Montaigne's status as a radical relativist.[1] A persistent problem in capturing the early modern history of relativism is that relativist motivations are often conjoined with scepticism. Unlike relativism, scepticism does not relativise the validity of beliefs to standards but questions the possibility of knowledge in general. Complementing the existing literature, I shall highlight some non-sceptical approaches that have not been covered in this regard.

To introduce the conceptual setting in which much of the debates took place, I will begin by outlining the basics of early modern theories of ideas and show how standards are related in these theories. Distinguishing four types of relations, I will then sketch some individual positions from the least to the most relativistic.

2. Absolute standards and human ideas

Many early modern philosophers endorse some theory of ideas or concepts. If we look for relativist assumptions, it is helpful to start from discussions of ideas. Different communities or even subjects might be said to have ideas of things or actions that diverge from those of other subjects, communities or from the divine mind. At least in the wake of Descartes and Locke, many relativistic assumptions concern ideas. "Idea" is most commonly used as a technical term for a mental representation of a thing or of set of things. So if I perceive or know of cats or a cat, one might say that I have an idea of a cat. Terminologically speaking, ideas play the role that concepts played in the Aristotelian tradition. Ideas, then, are posited to explain our mental states, cognition, knowledge and categories as well as the meaning of linguistic signs.[2] Moreover, depending on how ideas are taken to be produced, structured and stabilised, they might be said to be relative to the subjects in which they originate or to the communities in which they are communicated.

A first key issue in this context is the *origin of ideas*. It might be helpful to distinguish between two opposed models of how ideas originate. The first model might be called a "rationalist" one; it dominated Neo-Platonist or Augustinian thought. According to this model, ideas originate in the divine intellect. For purposes of simplicity let's say they are the blueprints or essences for the creation of things. By contrast, ideas in our human minds are faint copies of divine ideas. If we want to cognise the true nature of things, we must turn inward and try to reach the divine through our ideas. Malebranche's doctrine of seeing things in God loses much of its initial oddity when considered in the light of this model. The second model might be called an "empiricist" one; it dominated Aristotelian thought. According to this model, our ideas originate mainly from sense experience. If we want to cognise the true nature of things, we must investigate things and try to grasp their essential properties. This model is fairly pervasive throughout early modern theories of ideas. Most authors don't go for one of these models but for a nuanced combination of the two. Although Descartes, for instance, has rationalist overtones, he equally allows for the empirical formation of ideas. And although Locke clearly endorses an empiricist understanding of ideas, he seems to employ a model of archetypes and copies, at least when it comes to certain kinds of ideas (Hight 2008, ch. 4 and ch. 7).

For our purposes, it's crucial to see these models at work when distinguishing between different standards or norms of truth, knowledge, cognition, morality or beauty. However, both models allow for cognitive deviation. Accordingly, an inadequate idea might be one *deviating from the divine idea* or one *deviating from the (essence of a) thing* or its properties. After all, it seems natural to assume that essences are following divine ideas and creation. If one or both of these standards apply, we might assume that we are dealing with an *absolute standard*.

There are cases, however, in which these standards do not apply or do not apply directly. Even if I take for granted that cats have an essence, I might know nothing about it. A common consequence is the distinction between divine and human knowledge and pertinent sets of ideas. Many theories of ideas employ a distinction between perfect divine and imperfect human knowledge. While God is seen as having a complete net of ideas comprising everything there is to be known, humans are often portrayed as having incomplete and thus inadequate sets of ideas. Accordingly, divine knowledge serves as an absolute standard of knowledge, in relation to which knowledge of "lesser" beings (angels, humans, other animals) counts as deficient. Given the remoteness of absolute standards, the intermediate status of human knowledge was sometimes taken as a standard in its own right, thus allowing for a domain of knowledge relative to human capacities and ends.

3. Human make-up and customs: stabilisers or standards?

As we have noted, (the essences of) things or ideas in God's mind can generally serve as absolute standards. However, human ideas are not only considered in relation to these standards. As such, ideas are considered fleeting entities. They come and go. And were they not coupled with emotions, most of them would pass without our taking notice. Therefore, we should pause to think what it is that *stabilises* ideas. Most authors adhere to the widespread assumption that ideas or their physical triggers are stored in our memory (Sutton 1998, ch. 1–2). But since ideas are not stable in themselves, they require something else for their stability. A common stabiliser of ideas provided through linguistic convention. If ideas of cats are commonly expressed through the word "cat," the tie to the word and the fact that other people use this word, too, give stability to the fleeting idea of a cat.

For our purposes, it's crucial that the medium of stabilisation can also fulfil the role of a standard. Although my idea of a cat might be said to be right in relation to cats or divine ideas according to which cats were created, my idea can still deviate from or conform to the ideas that other human minds produce in conjunction with the spoken sound "cat." The use of others might reinforce a certain kind of stereotype and make me forget about other ideas that I might have had. Thus, the common use of language serves as a standard that can be, at least in principle, considered independently from the essential or divine standard. I would like to consider three such stabilisers that are often considered in early modern discussions:

(1) Biological, physiological and cognitive make-up: The fact that humans have certain needs, certain bodily dispositions and cognitive mechanisms, makes our minds prone to have, or even favour, certain ideas over others. If we assume that humans have a specific make-up, different from that of other animals and angels, it is (i) plausible to assume that our categorisations are to some extent affected by this make-up and that (ii) they entail specifically different categorisations. In other words, beings with other needs carve up the world differently.

(2) Linguistic conventions: The fact that ideas can be expressed through language turns linguistic signs into triggers of ideas (for listeners) and thus stabilises them in common use. (i) Since ideas are often taken as complex bundles (lists of properties, as it were), words also give *structure* to ideas. (ii) The fact that words are associated with bundles or sets of ideas enables *variation in associations* in the same speaker and different speakers. (iii) A further question is to what extent different languages and variations in conventions give rise to different structures of ideas. If our general make-up might result in different preferences and ideas in comparison to other species, linguistic conventions can result in different ideas in comparison to other linguistic communities.

(3) Custom: Linguistic conventions can be seen as a subset of customs in thought, action and interaction. Thoughts and actions (along with their different conceptualisations) are equally stabilised by repetition or association. Having a certain idea repeatedly will establish and stabilise the physical trace that it is related to. Likewise, the repeated co-presence of ideas

can lead to mutual reinforcement. Thus, salt might make you think of pepper. Especially with increasing travel and the pertinent literature, the variations in customs became a topic of concern.[3]

Early modern philosophers clearly recognised these factors as responsible for the stabilisation of ideas. But the stabilisers could be said to be standards in their own right. Accordingly, an idea might be considered in relation to two sets of standards: the divine or essential and the stabilising standard.

One might of course ask how these stabilising factors relate to one another. Thus we might ask how our linguistic conventions relate to our biological or cognitive make-up. The crucial question for us, however, is to what extent these stabilisers constitute *standards* in their own right when related to the absolute standard. Once they are recognised as standards, they might be said to

(1) conform to the absolute standard,
(2) deviate from the absolute standard,
(3) be remote from the absolute standard,
(4) be independent from the absolute standard.

If relativism is taken as the view that there can be more than one standard in relation to which our ideas can be judged (as apt or correct), then the discussion of the relation between stabilising factors and traditionally absolute standards can be seen as a discussion of relativism.

4. The relation between absolute and human standards

The different possibilities of relating absolute and stabilising standards to one another generate different positions with regard to relativism. Whereas (1) and (2) treat stabilising standards as subordinate to the absolute standard, (3) and (4) allow for a decoupling and thus for the emergence of a relative standard. Nevertheless, the discussion of (1) and (2) is an important step in considering relativist possibilities. Therefore, I would like to begin with looking at (1) and (2) before discussing more decidedly relativist accounts.

Many early modern philosophers start out from the assumption that our stabilising standards deviate from the absolute standards. This is particularly true of Spinoza, who thinks that the inadequacy of human ideas is a function of their deviation from divine ideas.[4] But apart from this technical discussion, the deficiency of human ideas in opposition to the divine is a trope, handed down to us by such assumptions as the biblical Fall and the subsequent fallibility of human knowledge. However, this doesn't preclude that human is seen as conforming to essential standards. Let's consider instances of this case first.

4.1. Conformity

One of the most prominent discussions about (1), that is, the conformity of stabilising standards and absolute standards, is rooted in the quest for the language of the biblical Adam or some sort of original language. However, in the early modern period this language planning movement received the status of a scientific project, initiated by the Royal Society. The aim was to establish a language that matches things. Although the project was abandoned eventually, it opened up intriguing debates on the very possibility of such conformity (Lewis 2007, 6–22).

Yet, the *assumption* of conformity between these standards seems to be common presupposition among language users. Locke explicitly discusses the "double conformity" of our ideas to things and common use of language as a "tacit" or "secret" supposition. It constitutes, says Locke, "the rightness of our Knowledge, and the Propriety or Intelligibleness of our Speaking." But he then hastens to point out that this presupposition, while motivating our trust in speech, is often not fulfilled (see Locke 1689, II.32.8; Lenz 2010, 402). As will become clear later, Locke himself is agnostic about essences and endorses option (c).

4.2. Deviation

As has been noted, the assumption that our stabilising standards deviate from the absolute standard is rather widespread and forms the basis of most philosophical considerations. Most of the ideas (at least of natural things) that are taken to provide the basis of our cognitive life and linguistic practice count as inadequate. Perhaps one of the most obvious places is the discussion of so-called secondary qualities, such as sight, sound, touch, taste or smell. Since Galileo and Descartes, it is common to assume that we are systematically mistaken when attributing such properties to objects. When talking about the sweetness of the apple, we are in fact talking about experiences owing to the effects on our sense organs.[5] Thus, what we grasp through the senses seems to carve out a domain of knowledge relative to our sense organs.

Another common way of referring to our deviating standards is to speak of the "vulgar" view.[6] Francis Bacon's talk of the "idols of the marketplace" can be seen as an attempt to systematise the assumption that our linguistic practice embodies a certain (inadequate) standard and indeed shapes our thought. In the *Advancement of Learning*, Bacon writes:

> [A]lthough we think we govern our words, and prescribe it well, *Loquendum ut vulgus, sentiendum ut sapientes*, yet certain it is that words, as a Tartar's bow, do shoot back upon the understanding of the wisest, and mightily entangle and pervert the judgment. . . . [T]herefore, it must be confessed, that it is not possible to divorce ourselves from the fallacies and false appearances because they are inseparable from our nature and condition of life.
>
> *(Bacon 1859, 396–397)*

As we can see, Bacon argues against the Aristotelian assumption that we could dissociate our thought from language by speaking "like the vulgar" but thinking "like the wise." While we might assume to be in control of our language, language actually shapes thought. Whereas many Aristotelians assume that we can overcome such fallacious shaping, Bacon dismisses this optimism, claiming that the fallibility of language (and therefore of thought) is rooted in "our nature and condition of life." According to Bacon, then, the vulgar standard shaped by language is owing to a more fundamental standard, that is, our make-up or "our nature."

But what does it mean that our nature is inherently deficient? The post-lapsarian state of fallibility is still measured against an absolute standard. The language planners of the seventeenth century try to develop a proper language by devising an artificial language or reforming existing languages. But even if such projects were doomed to fail, the assumption of one uncontaminated standard seems to prevail in a number of early modern philosophers such as Spinoza and Leibniz.

Yet, the assumption that the vulgar view is owing to our nature seems to assign a fairly robust status to human mediocrity. This raises the question whether there is more than one standard. For even if one accepts an absolute standard of things' essences or divine ideas, the assumption that human practices follow a different standard puts pressure on the viability of the absolute standard. This pressure opens the door to relativism.

4.3. Remoteness

Until now we have considered the realm of the divine or things as providing an absolute standard to which human practices might relate. To say that there is more than the one absolute standard might sound counterintuitive, since everything would ultimately reduce to the absolute standard. Nevertheless, there is wiggle room. First, we should consider the possibility that the realm of the divine or things might not provide a standard *for us*. After all, divine ideas or things' essences might be of such a nature that they are not compatible with being cognised by us. A helpful illustration is a Heracliteian world in constant flux. As already Plato notes in the *Theaitetus*, the very idea of using linguistic signs aptly is inconceivable in such a world, since they would have to be as changeable as the things themselves (*Theaitetus*, 157b, 180a–183b). Secondly, we might say that human standards are at least quasi-standards in that they are sufficiently robust to endow us with rules of conceptual propriety. This might seem like saying "we are wrong according to the absolute standard, but it's fine since we're systematically wrong." Although this might sound silly at first, it thrives on the idea that conceptual stability might be more viable than truth. The upshot is that our own standards outweigh the absolute standard. In contrast to the standard we adhere to, the absolute standard is something *remote*.

A first, if tentative, illustration of such a view can be given by pointing to Descartes' "provisional moral code." Descartes claims that we lack clear and distinct ideas about morality and thus advances what he calls a "morale par provision."[7] According to the first of these rules, Descartes claims he has "to obey the laws and customs of my country" (Descartes 1988, 122–124). In doing so he clearly refers to what I called one of the stabilising standards: the laws and customs of a country. He never definitely clarifies whether or how this relates to an absolute standard. But the term "provisional" indicates that this and other rules might be subject to change. The qualification "national" allows for diversity of customs of different nations. So this could be a case where the absolute standard is assumed but is too remote to be induced, leaving us with one and potentially different standards of morality.

And yet, admitting to cultural or national standards in opposition to unavailable absolute standards or even cultural diversity does not automatically entail a sort of moral relativism. Many authors in Descartes' and later times were well aware of such diversity while holding on to the idea of a universal moral law.[8] Descartes however does not revert to such a universal moral law. Therefore, we can take his remarks as an acknowledgement of moral relativity. Yet, pace Baghramian, neither Descartes nor the Cartesians are committed to anything as strong as subjectivism *tout court*. A subjectivist reading ignores that Descartes' relativises morality not to subjects but to nations, and that the truth of many of our ideas are ultimately safeguarded by God (see Baghramian 2004, 41).

Another instance of a moral relativism can be seen in Spinoza. In contrast to Descartes, he argues that terms such as "good" and "bad" are merely owing to our ways of thinking and comparing things (Miller 2005). When we use such terms they don't relate to anything real independently of those using the terms. On the other hand, he admits that we cannot really abandon these terms, as we need them for practical guidance in many respects. At first sight, then, Spinoza seems to be an antirealist or instrumentalist about (moral) value. But this is not entirely right. Ultimately, Spinoza seems to hold an intriguing combination of subjectivist-objectivist relativism. On the one hand, values are relative to each individual insofar as certain things are only good for certain individuals (which can include groups). In this sense, they are subjective. On the other hand, values are objective in that these things are valuable for the essential self-preservation of these individuals. Their value does not merely depend on whether I take them to be valuable but on the nature of the individual. Thus, it is the objective make-up of a being that determines value. Still, it is not a given that this standard is available to me. I might in fact live my life in accordance with conventions and follow false beliefs about what is good for me.

Unlike Descartes and Spinoza, Locke seems to adhere to an absolutist natural law theory of morality.[9] Locke's relativism does not concern morality but first and foremost our ideas of the natural world. If our mind forms ideas of natural things such as stones, trees or rabbits, what makes them apt ideas is not that they conform to the real essences of things. Locke is agnostic about real essences: he doesn't deny them but thinks they are not cognitively available to us. Our ideas should be seen as conforming, not to the ineffable real essences but to the nominal essences of things. Nominal essences are abstract ideas associated with a name. Thus, my occurrent idea of cats is apt if it conforms to the nominal essence of cat, that is, a list of properties defining cats. In Locke, the name stabilises the complex idea in two ways. On the one hand, it functions as a mental abbreviation: "cat" is easier to memorise than a potentially indefinitely long list of properties. On the other hand, the name's currency in the linguistic community determines what I memorise or forget in the first place. This way, the establishment of names and thus categories depends on the community. What is more, Locke explicitly declares that different conventions of linguistic communities will affect our standards of categorisation:

> But to return to the *Species* of corporeal Substances. If I should ask any one, whether *Ice* and *Water* were two distinct *Species* of Things, I doubt not but I should be answered in the affirmative: And it cannot be denied, but he that says they are two distinct *Species*, is in the right. But if an *Englishman*, bred in *Jamaica*, who, perhaps, had never seen nor heard of *Ice*, coming into *England* in the Winter, find, the Water he put in his Bason at night, in a great part frozen in the morning; and not knowing any peculiar name it had, should call it harden'd Water; I ask, Whether this would be a new *Species* to him, different from Water? And, I think, it would be answered here, It would not to him be a new *Species*, no more than congealed Gelly, when it is cold, is a distinct *Species*, from the same Gelly fluid and warm; or than liquid Gold, in the Furnace, is a distinct *Species* from hard Gold in the Hands of a Workman. And if this be so, 'tis plain, that our *distinct Species, are nothing but distinct complex* Ideas, *with distinct Names annexed to them.*
>
> *(Locke 1689, III.6.13)*

Although real essences determine what there is, it is the linguistic community that determines how we categorise it and give shape to the ocean of being. Thus, we might end up relating the same phenomena to *competing* categories without grounds to choose between them beyond the customary practice. Given that Locke accepts real essences, it is no surprise that he describes our standards of categorising as mediocre. At the same time, he sees this mediocrity not as a state to be overcome for a better one; rather he assumes that this mediocrity is apt and indeed designed in relation to our biological and cognitive make-up and needs, while angels and non-human animals follow standards akin to their make-up (Lenz 2019).

4.4. Independence

As we have seen, some thinkers of the seventh century take divine or essential features as remote standards. They are not denied, but they don't function as an immediate corrective for our own standards. Once this is established, it becomes possible to assume that the divine or essential, if they are to be assumed at all, have no bearing on our standards. This seems to lead to conceiving of our human standard or standards as independent from any absolute corrective.

A fairly obvious view of a standard that is independent from the absolute divine standard is provided in Descartes' position on causal explanation in natural philosophy. It is often said that Descartes expelled final causes from nature. What he in fact claims is that we should not cite final causes in physical explanations because we cannot know divine purposes (Descartes 1988, 202). If this is correct, he distinguishes two standards, those of physics and theology, as relating to the same phenomena, and then gives reasons for choosing physics as the preferred explanatory framework. This dissociation was a widespread move already in medieval debates and can be considered what Marenbon calls "relativism of autonomy" (Marenbon 2015, 348–352).

Another view of independence pertains to human cognition or action in general. Perhaps the most prominent early modern author who might endorse some relativistic commitments in this sense is David Hume. Although there has been much discussion whether Hume endorses moral relativism, I believe that his basic views on the human mind commit him if not to relativism, certainly to the independence of human from any absolute standards.[10] While Locke explains our mediocre standards as being in line with divine design, Hume does not appeal to theological premises. He proclaims that "custom is the great guide." But how can custom function as a stabiliser? Arguably, the principle of sympathy, that is, the assumption that our mental states are affected by those of other people, functions as a social glue of cognitive and other activity. The often unreflected transmission of beliefs, emotions and other sentiments stabilises commonality. To be sure, already Bacon discusses such modes of intersubjective agreement. But in Hume there is no hint of a given absolute standard.

An interesting case is Hume's discussion of the vulgar view and the philosophical view. In the *Treatise*, Hume refers to the view that our perceptions are mental representations of objects as "philosophical," while the view that perceptions are not intermediaries but the objects counts as "vulgar." Now Hume claims that both of these "systems" are ultimately incoherent (see *Treatise* 1.4.2).[11] The way to go is a reflective view which gives credit to the vulgar inclinations without taking them to be justifiable. What we see here is the acknowledgement of diverse views and the denial of an absolute standard. Now Hume does not go as far as endorsing the view that we have two valid descriptions of the same phenomenon. What he does provide, however, is an argument for the establishment of customary standards irrespective of an absolute standard. Shared customs provide this and require us to relate all our actions to others.

5. Conclusion

Should human forms of action and cognition be seen in line with some absolute standard or do they establish standards in their own right? What we have seen is a development, going from a comparison between absolute and human standards to an increasing independence of the human standard. Whether we want to call these approaches relativistic depends on our definitions. What I find crucial, however, is the fact that the dissociation of standards opens the door to relativism by allowing for a detachment from the essential or divine as an absolute standard. The detachment from the absolute required that the human standard be established as an independent domain. If relativism is the view that there is more than one non-absolute standard, it is still questionable whether there are forms of relativism in the early modern period. There is of course the distinction between human and non-human standards, but despite the acknowledgement that some animals have more fine-grained senses, the perspective of non-human animals was not taken as a general competitor. We also find acknowledgement of diversity between human customs. But that does not entail that they are rooted in different categorical standards.

Locke, for instance, considers this possibility, but he does not conclude that there is an incommensurability between different standards. In principle, such diversity remains translatable for Locke. Yet what we do find is a growing acknowledgement that the establishment of human customs requires social interaction between individuals. If there are no absolute standards, whilst we still find ourselves in agreement, this agreement cannot simply be taken to originate from exposure to the same essences or the same make-up. But if the shape of human customs is taken as a result of interaction and history, the diversity of actions and histories gradually comes into view as a relativistic explanandum.

As has been noted, research on relativism emerging in early modern thought is still a relatively young project. In addition to further study on individual debates and authors, it would be particularly intriguing to look more closely at the relation between early modern and renaissance debates on human nature and forms of diversity as well as the reception of early modern philosophers in the nineteenth and twentieth centuries.

Notes

1 Blom (2011) also focuses on the French Enlightenment; Irlenborn (2016) also focuses on Montaigne and scepticism.
2 See, on general features in theories of ideas, Hight (2008).
3 See, for an extensive discussion of such factors, Lenz (2010, 68–73, 127–157). While the rise of linguistic relativism is commonly associated with Wilhelm von Humboldt or Benjamin Lee Whorf, linguistic stabilisation and influence on thought is explicitly considered in late scholastic and other seventeenth-century authors. See Meier-Oeser (2004).
4 This does not necessarily entail that divine ideas are actually separate from human ideas. One might consider the divine mind as an idealised set of ideas. See Renz (2018, ch. 1).
5 See Durt (2012) for a thorough discussion of secondary qualities.
6 The vulgar view is often understood as a kind of naïve realism or an unawareness of mental representations (ideas), resulting in a supposed conflation of appearance (via representation) and thing.
7 See Araujo (2012, 131) for an overview of various interpretations.
8 On the later French Enlightenment, see Baghramian (2004, 44–46), who shows that the natural law tradition still had a stronghold despite concessions of diversity.
9 A rare exception is Zinaich (2006), who argues that Locke turned to moral relativism in the *Essay*.
10 Here, I follow the interpretive approach of Craig (1987, 69–77), who claims that Hume generally abandons the doctrine that man was created in the image of God.
11 See Rocknak (2012) for a detailed interpretation of the vulgar-philosophical dialectic.

References

Araujo, M. de. (2012), *Scepticism, Freedom and Autonomy: A Study of the Moral Foundations of Descartes' Theory of Knowledge*, Berlin and New York: De Gruyter.
Bacon, F. (1859), *The Advancement of Learning II*, in *Works* 3, edited by J. Spedding et al., London: Longman.
Baghramian, M. (2004), *Relativism*, London: Routledge.
Blom, P. (2011), *Böse Philosophen: Ein Salon in Paris und das vergessene Erbe der Aufklärung*, Hanser: München.
Craig, E. (1987), *The Mind of God and the Works of Man*, Oxford: Oxford University Press.
Descartes, R. (1988), *The Philosophical Writings of Descartes*, 3 vols., translated by J. Cottingham, R. Stoothoff and D. Murdoch, vol. 3 including Anthony Kenny, Cambridge: Cambridge University Press.
Durt, C. (2012), "The Paradox of the Primary-Secondary Quality Distinction and Husserl's Genealogy of the Mathematization of Nature," PhD thesis, Santa Cruz: University of California.
Hight, M. (2008), *Idea and Ontology: An Essay in Early Modern Metaphysics of Ideas*, University Park, PA: Pennsylvania State University Press.
Hume, D. (1739), *A Treatise of Human Nature*, edited by D. F. Norton and M. J. Norton, Oxford: Oxford University Press, 2000.
Irlenborn, B. (2016), *Relativismus*, Berlin and New York: De Gruyter.

Kusch, M. "ERC Advanced Grant Research Proposal 2013: The Emergence of Relativism," www.academia.edu/10181904/ERC_Proposal_on_Relativism_Long_Version_B2_.

Lenz, M. (2010), *Lockes Sprachkonzeption*, Berlin and New York: De Gruyter.

———. (2019), "Locke's Life-World: The Teleological Role of Secondary Qualities," *History of Philosophy Quarterly* 38/1: 39–59.

Lewis, R. (2007), *Language, Mind and Nature: Artificial Languages in England from Bacon to Locke*, Cambridge: Cambridge University Press.

Locke, J. (1689), *An Essay Concerning Human Understanding*, edited by P. H. Nidditch, Oxford: Clarendon Press, 1975.

Marenbon, J. (2015), "Relativism in the Long Middle Ages Crossing the Ethical Border with Paganism," *HAU: Journal of Ethnographic Theory* 5(2): 345–365.

Meier-Oeser, S. (2004), "Sprache und Bilder im Geist: Skizzen zu einem philosophischen Langzeitprojekt," *Philosophisches Jahrbuch* 111(2): 312–342.

Miller, J. (2005), "Spinoza's Axiology," *Oxford Studies in Early Modern Philosophy* 2: 149–172.

Platon (1995), "Theaitetus," in *Opera I*, edited by E. A. Duke et al., Oxford: Clarendon Press.

Renz, U. (2018), *The Explainability of Experience: Realism and Subjectivity in Spinoza's Theory of the Human Mind*, Oxford: Oxford University Press.

Rocknak, S. (2012), *Imagined Causes: Hume's Conception of Objects*, Dordrecht: Springer.

Sutton, J. (1998), *Philosophy and Memory Traces: Descartes to Connectionism*, Cambridge: Cambridge University Press.

Zinaich, S. Jr. (2006), *John Locke's Moral Revolution: From Natural Law to Moral Relativism*, Lanham: University Press of America.

7

RELATIVISM IN GERMAN IDEALISM, HISTORICISM AND NEO-KANTIANISM

Katherina Kinzel

1. Introduction

The emergence and development of relativist themes and arguments in nineteenth century German philosophy has a complex trajectory. This trajectory was shaped by the encounter between different scholarly traditions, among them speculative German idealism, hermeneutics, professional historiography, sense-psychology, folk-psychology, neo-Kantianism, and phenomenology. The shared concern of the traditions that I will be focusing on in this chapter is history: how can we understand and conceptualize the essential historicity of human existence and how can we make sense of the variations between different historical epochs and cultures. The type of relativism at stake here is historical relativism, which might be considered a variant of cultural relativism. The historical relativist believes that historical epochs or cultures differ concerning the systems of beliefs and values that they adhere to, and that these different systems of beliefs and values cannot be ranked. For such a view to become plausible, different motives and ideas have to come together: the idea that beliefs and values are not static and eternal, but subject to historical change; the observation that the beliefs and values of past cultures were different from present ones, and sometimes radically so; the idea that beliefs and values are relative to culture or collective ways of life; the claim that there is no linear progress in history; the argument that later epochs are not warranted in judging earlier ones to be inferior; and the view that history is a contingent, undirected progress. Focusing on these and other relativist ideas about history and culture in nineteenth-century German philosophy I want to convey three general observations.

First, although many German philosophers and historians formulated theses and arguments that we today might view as ingredients to historical relativism, throughout most of the century, these thinkers did not worry too much about the problem. The main concern for speculative idealists and their opponents was not whether beliefs and values were relative to historical cultures in a way that forestalls a neutral adjudication between them. Instead, their problem was how the universal and the particular could be reconciled: how could one make sense of historical change and the resulting plurality of historical cultures without losing track of the unity of the historical process and the universal meaning that it embodied. Focusing on this problem meant that even when theses about the relativity of knowledge or moral values were formulated, these theses were usually joined to conceptions of theodicy or historical progress. And these conceptions held relativist implications in check.

Second, it was only towards the late nineteenth century that relativism emerged as an explicit philosophical concern. In particular, it was in the encounter between the hermeneutic tradition and neo-Kantian philosophy of values that the problem of historical relativism took a more definite shape. "Relativism" entered the debate as a pejorative term and was closely associated with psychologism and historicism. The relativist was someone who failed to observe the boundary between the factual and the normative and who reduced the absolute values underpinning moral and epistemic judgments to contingent facts about human psychology, or to historical facts about what was acceptable in this or that culture. It is interesting to note, however, that those who were criticized as relativists by the neo-Kantians rejected the label themselves. Those inspired by the hermeneutic tradition were confident that historical understanding based on universal human psychology could bridge the gaps between different cultures and systems of belief and would thus avoid the dangers of relativism.

Third, in the shift that occurred towards the late nineteenth century, the relation between historical ontology and historical methodology was reconceptualized. This reconceptualization may well have been one of the intellectual factors enabling the emergence of historical relativism as an explicit concern. Thinkers of the first phase saw ontology and methodology as closely interrelated. For these authors questions about historical method had to be answered on the basis of a general account of human history, its patterns and driving forces, while conversely, the correct methodological approach to the historical facts would reveal the unity and essence of history. For the philosophers of the second phase, this link was broken, or at least it became more problematic. The methods of the historical disciplines and the epistemic status of historical knowledge had to be clarified without reference to the ontology of the historical world. Historical ontology was considered bad metaphysics. It was in the context of a more purified methodological debate, in which history was no longer guaranteed to be understood as a unified and meaningful process, that historical relativism could take shape.

In what follows, I will substantiate these three observations, beginning with the conception of history that was formulated in speculative idealism, most importantly by Georg Wilhelm Friedrich Hegel (1770–1831).

2. German idealism

While Immanuel Kant (1724–1804) had treated history more or less as a footnote to his philosophical system, the speculative idealists that followed him – Johann Gottlieb Fichte (1762–1814), Friedrich Wilhelm Joseph von Schelling (1775–1854), and Hegel – gave development and process a more central place in their conceptions of reason. As a result, the philosophy of history began to take on a systematic significance. Despite the many differences between their idealist projects, the speculative idealists shared three central commitments.

The first commitment they took on from Biblical universal history: the unfolding of the history of mankind is a unified process, and this process does not reduce to the seemingly erratic and haphazard course of particular actions and events. Instead, history is directed and goal oriented.

Second, the moving agent and "subject" of historical development is reason itself. The development of reason follows an intrinsic logic or structure, and the unfolding of this structure determines the course of history. The philosopher of history differs from the empirical historian by taking reason as his starting point. This allows him to see connections and totalities rather than particulars, ideas and essences rather than brute facts, and the necessary and rational rather than the merely contingent.

Third, the central locus of the self-expression of reason is the relationship between the individual and the state, because it is here that the problem of human freedom receives a solution that is at the same time *geistig* (mental-spiritual) and concrete. Human freedom realizes itself in the historical institutions of the state, and the development of humankind towards freedom proceeds through necessary stages. By revealing the necessary patterns of the historical process, the philosopher of history not only makes sense of the past, but he is also able to situate and interpret the present.

Hegel's philosophy of history has become the most influential example of an idealist philosophy of history. Although Hegel is an absolutist, in the sense that he thinks of the historical process as the self-realization and simultaneous self-discovery of "absolute spirit," his philosophical reflections contains many relativist themes.

In *Elements of the Philosophy of Right* (1820) Hegel writes of "philosophy" as being "its own time comprehended in thoughts." Philosophy as the highest form of human reason thus has a historical dimension. This thought does not collapse into relativism though, since Hegel thinks of historical time as structured by the progressive realization of reason itself. For Hegel, ontology and methodology mutually reinforce one another: philosophy of history is empirical history as seen from the standpoint of reason. The philosophical perspective reveals reason to be operative in the historical world in a way that enables progress towards ever higher levels of self-consciousness.

According to Hegel, history has a final goal. On the one hand, Hegel describes this goal as the self-consciousness of absolute spirit. This is why in Hegel's philosophical system, the philosophy of history is situated at the transition from "objective spirit" to "absolute spirit": expressed in the historical reality of state and nation, objective spirit passes over into the forms of absolute spirit in religion, art, and ultimately philosophy. Second, Hegel describes the final goal of the world as the realization of freedom. The philosophy of history thus construes the stages of the development of spirit as stages in the development of human freedom, and explicates the systematic links between self-consciousness, rationality, and freedom.

Focusing on the second aspect, we can see that Hegel relativizes *Sittlichkeit* (ethical life) to particular historical national communities, while simultaneously making an argument for how the particular and relative are essentially related to and ultimately overcome by universal freedom. He argues that the idea of freedom becomes an active force in human history only once ethical life receives a concrete institutional form in the nation state and becomes conscious of itself in a concrete *Volksgeist* (national spirit). Accordingly, Hegel captures the dynamic development of world history through the lens of the relation between particular national spirits and the universal *Weltgeist* (world spirit): the former is a limited expression of the latter. Each historical community or nation is built on a specific principle that it expresses and develops in its ethical and cultural life. By rendering its principle explicit, the national spirit makes it available to the self-consciousness of the historical community. And yet, once the principle is fully expressed, it loses its vividness and immediacy, and the national spirit exhausts itself. The community enters a stage of decay, and ultimately gives way to the emergence of a new principle. Hegel interprets the resulting succession of different national spirits as the process in which the world-spirit overcomes its particular appearances. Historical change has thus both a negative and a positive dimension. In the negative dimension, world-history reveals the finitude and relativity of all forms of ethical life and practical freedom. The positive dimension is that the world spirit is engaged in a process of *Aufhebung* (sublation): by overcoming its particular expressions, the world-spirit preserves their essential principle in a universal concept of freedom.

According to Hegel, the process in which freedom is realized is predetermined by the structure of the *Begriff* (concept) of freedom itself. While Hegel does not deny the role of chance in history, he defines chance as that which is unrelated. Hence, chance is – by definition – excluded from the relational nexus of world history in which the concept of freedom unfolds itself. Hegel captures the structure of this process in terms of progressive stages. Each stage of the historical process expresses, to at least some degree, the final end of the world. There is some disagreement as to whether Hegel's talk of a "final end" should be understood in a temporal sense, that is, whether the end of history means an end of historical development. But even if we do not subscribe to the temporal interpretation, we can note that for Hegel history is unified and closed with respect to the final end.

Hegel's philosophy of history involves notable relativist elements. Most importantly, ethical life and consciousness are conceived as dynamic, and as relative to particular historical communities. But, at the same time, the conceptualization of history in terms of progressive stages allows for a hierarchy between different historical cultures. Moreover, the fact that history is a closed totality with a final end negates all relativism: the dynamic relation of expression and sublation that occurs between the particular and the universal ultimately realizes the absolute.

3. Professional history

When Hegel was delivering his *Lectures on the Philosophy of World History* in 1822, 1828, and 1830, the study of history had already undergone a process of increasing disciplinary professionalization. Wilhelm von Humboldt's (1767–1835) reform of the Prussian educational system had turned history from an *Hilfswissenschaft* (auxiliary science) in the service of law and theology to an autonomous academic discipline. The newly founded discipline implemented its own methodological standards, most importantly the principle of source criticism. Source criticism was supposed to make history empirical and objective. Professional historians rejected the postulate of reason that had shaped the historical understanding of the German idealists. This also called into question the idea of stages and progress. The historians agreed with the philosophers that the course of human history follows general developmental patterns, and that there is a universal meaning to the world-historical process encapsulated in the development of "ideas" and "spirit." Nevertheless, the historians insisted that the universal meaning of history could not be known by means of philosophical reflection. Instead it had to emerge as the result of a piecemeal investigation that started from historical particulars and was based on the critical study of the available sources. At the University of Berlin, an institutional-intellectual conflict erupted between Hegel and his followers on the one hand, and what would later be called the "historical school" of Friedrich Schleiermacher (1768–1834), Friedrich Carl von Savigny (1779–1861), and Leopold von Ranke (1795–1886), on the other.

The main fault-lines of this conflict are exemplified in the Hegelian historian Heinrich Leo's (1799–1878) critical review of Ranke's debut work *History of the Latin and Teutonic Nations 1494–1514* (1824). Leo attacks Ranke for sacrificing insight into what is essential at the altar of historical particularities (Leo 1828). In his response, Ranke argues that history is very much after the general and essential. And yet, the general is to be found and represented in the particular. In a different context, Ranke scolds Hegel for reducing human life to mere shadows or schemes. He warns that the linear conception of progress reduces the life and value of each historical epoch to that of being a mere predecessor of later stages.

Ranke himself believes that all historical periods are "equal to God." They have an intrinsic value that, although not always obvious from the finite human point of view, is acknowledged from God's eternal perspective. There is a divine plan in the world, even if it might not be fully accessible to the human observer. And only by emulating divine benevolence – by attending to diverse historical realities with an unbiased mind – can the historian approach an understanding of human history (Ranke 1854).

Ranke's appreciation of particulars and his idea that all epochs are equal to God go some way towards formulating a doctrine of historical relativism. But at the same time, these claims depend on the assumption of an absolute perspective: God's benevolent view of human history. Ranke radicalizes the relativist elements already present in Hegel by jettisoning the idea of progress. Still, he never abandons a theologically grounded absolutism that envisions history as a meaningful unity.

Not all historians were in agreement when it came to the question of progress in history. For example, Johann Gustav Droysen (1808–1884) and Heinrich von Sybel (1817–1895) believed in history as a progressive process. They conceived of this progress in somewhat restricted, nationalist terms, and tended to equate the progress towards human freedom with the process of German national unification (Sybel 1874). It is an open question to what extent their views on progress left the door open for a "nationalist" formulation of relativism (roughly, this would be the view that beliefs and values are relative to national histories). But however things may stand with respect to nationalist relativism, with Droysen's reformulation of the hermeneutic principle of "understanding," a new form of historical relativism entered the stage: relativism about historical knowledge.

Unlike Ranke, Droysen did not think that the study of history should aspire to objectivity. In his *Historik* lectures, which he held seventeen times from 1857 onwards, Droysen develops a subjectivist hermeneutics of historical knowledge. With an eye to Hegel, he conceptualizes history as the realm of the self-expression of *sittliche Mächte* (ethical powers) and ethical ideas. For Droysen, understanding the past means recapturing general ethical ideas in their concrete historical realisations. But the formative work of ethical ideas in their expressions can only be grasped if the historian taps into his own subjectivity. In order to organize the otherwise disparate empirical particulars in a consistent interpretation, the historian needs to maintain a firm standpoint. And this standpoint, in turn, is shaped by the historian's own subjectivity, as well as by his age, religion, and nation. Droysen concludes that historical truth is only a "relative truth," a truth relative to the historian's situated standpoint in the present. The radical implication is that history needs to be constantly rewritten in response to new developments (Droysen 1868, 1977).

Although nineteenth-century German historians continued to think of history as a unified process with a universal meaning, they were more sceptical than the German idealists about the prospect of this meaning being revealed from the finite human perspective. Nevertheless, we find that they did not worry too much about relativism either. When Ranke claims all epochs to be equal to God, or when Droysen argues that historical truth is standpoint-dependent, these are statements of confidence, not admissions of defeat. For Ranke, the meaning of history can only be grasped if the equal validity of all historical cultures is acknowledged. But this gives the historian a theological vocation: it is the task of the historian to emulate God's absolute and benevolent perspective. For Droysen, historical knowledge is perspectival and bound to the present. But this insight provides license for the historian to take on a political role in the service of the German nation-state. By writing history from the standpoint of the present, the historical can intervene in the historical course of events. In short, the professional historians of the period were fully convinced of the ethical calling of the study of history. This seems to have made them somewhat blasé about the relativist questions that had emerged in their own conceptions of history and historical method.

4. Hermeneutics and neo-Kantianism

The full extent of the challenge of historical relativism was only acknowledged towards the end of the nineteenth century, and in a different context. Wilhelm Dilthey (1833–1911) and the Baden neo-Kantians, Wilhelm Windelband (1848–1915) and Heinrich Rickert (1863–1936), engaged in a life-long controversy about the demarcation between the natural and the human-historical sciences. Many issues were at stake in this conflict: the question of historical method, the relation between psychology and history, the fact-value distinction, the historical nature of philosophical reasoning, and the possibility of understanding past cultures. In the context of this debate the problem of historical relativism took shape as well.

In his *Introduction to the Human Sciences* (1883) Dilthey argues that the *Geisteswissenschaften* (human sciences) require an epistemological foundation suitable to secure their autonomy from the natural sciences. His "critique of historical reason" sets out to provide such an epistemological foundation. But Dilthey departs from Kantian criticism in fundamental respects. Most importantly, he opposes what he conceives as the stifling dualisms of Kantian philosophy. He rejects the distinctions between sensibility and intellect, between the transcendental and the empirical, as well between theoretical and practical philosophy. In Dilthey's view, reason cannot be separated from the forces of life. Thinking, emotion, and volition form an interconnected and developing nexus. And it is only by reconsidering reason as dynamic, historical, and integrated with life that philosophy can do justice to the enormous successes of the various disciplines studying the human-historical world.

In *Einleitung* Dilthey bases the distinction between the human and the natural sciences on a distinction between "inner" and "outer experience." The facts of the natural sciences take the form of hypotheses that are checked against "outer experience." In contrast, the facts of the human sciences are "given originaliter from within." They are experienced not as isolated elements, but as a living continuum and nexus. Dilthey emphasizes that the various disciplines of the human sciences – psychology, law, economics, theology, literature, history, etc. – form a continuum that is based on the interconnected facts of lived experience. The different disciplines work together to explain the external organization of society in state and law, the historical development of cultural systems such as art and religion, and the individuation of human types along lines of gender, nationality, profession, and so on.

However, when it comes to the task of providing the human sciences with a firm epistemological basis, one discipline plays a special role for Dilthey: psychology. Dilthey proposes a novel "descriptive and analytic" psychology that does not explain psychological processes by reference to hypothetical "elements" and natural laws, but instead seeks to recapture the integrated contents of "lived experience." Dilthey reasons that since lived experience is inherently social, the faithful description and analysis of this form of experience can also provide the basis for objective knowledge in the human sciences (Dilthey 1894). But despite maintaining the possibility of objective knowledge in the human sciences, Dilthey imposes no firm distinction between facts and values. In line with his anti-dualist thinking, he emphasizes that values and ideals emerge from the very reality that is studied by the human sciences, and concludes that facts about this process can also form the basis of normative evaluation.

Finally, Dilthey rejects philosophy of history, as well as sociology as based on false generalizations. He grants that there are stable social structures and patterns on the level of individual nations, and in different cultural systems, and that these can be studied objectively. But he denies that these patterns form a unified, global process that could be captured in terms of universal laws, teleological progress, or other grand narratives. In this respect Dilthey is more radical than many other nineteenth-century historicists: he rejects universal history. Nevertheless, Dilthey remains committed to some form of universalism. He embraces universalism on the level of human psychology, arguing that the "psychic nexus" which connects intellect, emotions, and volitions builds out the same structures in all human beings, and hence also serves as the basis of mutual understanding across ages and cultures.

In his mature work, *The Formation of the Historical World in the Human Sciences* (1910), Dilthey identifies "understanding" as the basis of both socio-historical reality and of the human sciences studying this reality. Dilthey now argues that the possibility of understanding is not based on "inner experience," but rather on the fact that all human beings participate in the objective "manifestations of spirit," that is, in a world of shared social institutions and cultural meanings.

Dilthey does not see himself as a relativist, and it is only in his "philosophy of worldviews" that he explicitly engages with the problem. But many of Dilthey's contemporaries, most notably the Baden neo-Kantians, thought that the "critique of historical reason" harboured a relativist spirit. Windelband and Rickert opposed Dilthey on nearly all fronts. Where Dilthey saw a continuum between empirical and transcendental, sensibility and intellect, factual and normative, the neo-Kantians demanded firm boundaries between these domains. When Dilthey distinguished the natural and the human sciences by thinking about types of experience and the objects they involved, the neo-Kantians sought to demarcate different sciences in purely formal terms. And while Dilthey thought of psychology as central to the epistemology of the human sciences, the neo-Kantians classified psychology with the natural sciences and denied its relevance for the historical disciplines.

Perhaps the most important issue in the disagreement is the distinction between the factual and the normative. In "Kritische oder genetische Methode?" (1883), Windelband argues that failing to observe this distinction leads to relativism. When disciplines like history or psychology seek to answer questions about values and normativity by purely empirical means, the results are devastating: the attempt to derive what is valid thought from actual reasoning ends up treating all thoughts as equally valid – after all, both correct and incorrect reasoning follows psychological laws. Likewise, the attempt to derive moral evaluations from history ends up treating all behavior as equally morally acceptable, since both good and bad moral behaviors have at some point been accepted historically. Psychologism and historicism are destructive of normativity. They are forms of relativism.

Windelband allows, of course, room for empirical history as a legitimate enterprise carried out within the constraints set by its empirical method. He distinguishes the historical from the natural sciences not by reference to different types of experience, but by isolating different goals of the knowledge acquisition. According to Windelband, the natural sciences are "nomothetic," in that they seek knowledge of regularities and general laws. The historical disciplines, in contrast are "idiographics"; they are concerned with particular and unrepeatable individual realities. Although Windelband suggests that only the unique individual, but not the repeatable, has a human value, he leaves open what precisely is the relationship between the idiographic method and the philosophy of values (Windelband 1894).

In his *The Limits of Concept Formation in Natural Science* (1896), Rickert takes up Windelband's distinctions and develops them into a more systematic account of historical method. According to Rickert, the starting point for demarcating the natural from the historical sciences is the question of "concept-formation": how do different sciences develop a conceptual grasp of reality? According to Rickert, the natural sciences form their concepts by "generalization," while the historical sciences proceed by "individualization," meaning that their concepts express an individual, unique and unrepeatable – and in this sense historical – content. Rickert characterizes the procedure of individualization as "value-related": historical individuality can only be grasped as a meaningful unity if it is related to a value.

Like Windelband, Rickert thinks of his analysis as purely formal. He accuses Dilthey of committing a twofold mistake. First, Dilthey bases the epistemological account of the historical sciences on the nature of the historical object, and second, he misconstrues that object as psychological. On Rickert's account, the formal method shows the ultimate object of history to consist in the "unreal meaning configurations" which attach to the empirical realities studied by the historian. And he insists that such unreal meaning cannot be thought of as immanent to psychological life. For Rickert, the correct way to understand the object of the historical sciences is in terms of culture. Culture has both a material and an immaterial component, but it is the latter that bestows meaning upon the former. By classifying psychology as a "generalizing" and thus as a natural science, he removes psychology from the domain of meaning which features centrally in the historical sciences.

Although both history and philosophy deal with meanings and values, Rickert is keen to keep them separate too. He introduces a distinction between theoretical and practical "value-relation" and claims that only the former is relevant to historical method. The task of practical evaluation is then left to philosophy. A universal history that reveals the ethical meaning of history is only possible if it is grounded in a system of absolute values, which in turn, only philosophy can provide. Hence, Rickert makes philosophy's transcendental values relevant to the interpretation of the meaning of history. Empirical history, by itself, cannot reveal this meaning.

In this way, Rickert preserves the distinction between the factual and the normative that had also been at the heart of Windelband's philosophy of values. He insists that the theoretical value-relating method of history can only reconstruct particular individualities. It does not, by itself, provide us with a unified grasp of the totality of history. A universal history of progress is only possible on the basis of a philosophical system of transcendental values. And these values cannot be historical entities, or else the result would be a destructive historical relativism (Rickert 1924). Values and normativity are in the domain of philosophy, not history.

Dilthey reacts to the charge of relativism raised by his neo-Kantian contemporaries in his philosophy of worldviews. The philosophy of worldviews, which Dilthey also describes as a "philosophy of philosophy," is his attempt to make sense of the historicity of philosophical systems while also avoiding historical relativism (Dilthey 1960). According to Dilthey, worldviews emerge as attempts to solve the puzzles of life. Dilthey identifies three domains in which worldviews are formulated and developed – religion, poetry, and philosophy, with philosophy differing from the other domains in being concerned with rigorous justification and generalization. Dilthey identifies three basic configurations of philosophical thinking that reoccur throughout philosophy's history, three "types" of philosophical worldviews: naturalism, idealism of freedom, and objective idealism. By expressing one of these three basic types, each philosophical system participates in the universal structure of philosophical reasoning. Dilthey thinks that in uncovering the common basis of all philosophical worldviews, he has reaffirmed the sovereignty of reason above the conflicts between different worldviews. He claims to have avoided relativism: while the worldviews themselves are only partial and relative, the perspective that encompasses them all is not. Of course, many of Dilthey's critics remained unconvinced. For Edmund Husserl (1859–1938) and Ernst Troeltsch (1865–1923) the philosophy of worldviews continued to present a paradigmatic example of historical relativism (Husserl 1910, Troeltsch 1922).

5. Conclusions

As we have seen, the trajectory of relativist ideas in nineteenth-century German philosophy is complex and multilayered. It is shaped by the encounter between different intellectual traditions, among them German idealism, hermeneutics, professional history, and neo-Kantianism. In general, we can observe two shifts from earlier debates about history and historical knowledge surrounding German idealism and professional history to the later debates that arose between Dilthey and the Baden neo-Kantians.

First, while in the earlier contexts questions of methodology are always entangled with questions of historical ontology, these two layers come apart in the late nineteenth century. Second, and possibly as a result of the first shift, Dilthey, Windelband, and Rickert share a skeptical attitude towards universal history. They agree that the reconstruction of objective knowledge about particular histories is well within the reach of empirical history. But they cast doubt on whether history can be comprehended as a unified whole with a universal meaning.

These developments paved the way for a clearer acknowledgment of relativism as a philosophical problem. Note however that relativism did not become clearly defined as a position or doctrine. Relativism was closely associated with, if not identical to, psychologism and historicism. It was not clearly distinguished from skepticism, nihilism, and anarchism either. To some, relativism seemed to emerge as a violation of the fact-value distinction and could only be avoided on the basis of a philosophy that maintained absolute values. Others thought that relativism could be countered by a universalist perspective on history, or by hermeneutic understanding. In the early twentieth century, worries about relativism were frequently expressed together with political worries about the "anarchy of values." Under the label of "historicism," historical relativism came to stand in for an unmanageable plurality of conflicting worldviews that threatened political stability in the present.

Acknowledgments

Research leading up to this paper was made possible by the ERC ("The Emergence of Relativism", No 339382).

References

Dilthey, W. (1883), "Introduction to the Human Sciences," in his *Selected Works I*, edited by R. Makkreel and F. Rodi, translated by M. Neville, Princeton, NJ: Princeton University Press, 1989, 47–240.

———. (1894), "Ideas for a Descriptive and Analytic Psychology," in his *Selected Works II*, edited by R. Makkreel and F. Rodi, translated by R. Makreel and D. Moore, Princeton, NJ: Princeton University Press, 2010, 115–210.

———. (1910), "The Formation of the Historical World in the Human Sciences," in his *Selected Works III*, edited by R. Makkreel and F. Rodi, translated by R. Makkreel and J. Scanlon Princeton, NJ: Princeton University Press, 2002, 101–212.

———. (1960), "Die Typen der Weltanschauung und ihre Ausbildung in den metaphysischen Systemen," in his *Gesammelte Schriften VIII*, edited by G. Karlfried, Göttingen: Vandenhoek & Ruprecht, 73–118.

Droysen, J. G. (1868), *Outline of the Principles of History*, Boston: Ginn & Co., 1897 [partial translation of *Historik*].

———. (1977), *Historik: Historisch-kritische Ausgabe*, vol. 1, edited by P. Leyh, Stuttgart-Bad Cannstatt: Frommann-Holzboog.

Hegel, G. W. F. (1820), *Elements of the Philosophy of Right*, edited by A. W. Wood, translated by H. B. Nisbet, Cambridge: Cambridge University Press, 1991.

———. (1822–23), *Lectures on the Philosophy of World History. Volume 1: Manuscripts of the Introduction and the Lectures of 1822–3*, edited and translated by R. F. Brown and P. C. Hodgson with the assistance of W. G. Geuss, Oxford: Oxford University Press, 2011.

Husserl, E. (1910), "Philosophy as Rigorous Science," in *Phenomenology and the Crisis of Philosophy*, edited by Q. Lauer, New York: Harper, 1965, 71–147.

Leo, H. (1828), "Review of Ranke," *Geschichten der romanischen und germanischen Völker, Ergänzungsblätter zur Jenaischen Allgemeinen Literatur-Zeitung*, 1290–1336.

Ranke, L. von (1824), *History of the Latin and Teutonic Nations 1494–1514*, translated by P. A. Ashworth, London: George Bell and Sons, 1887.

———. (1854), "On Progress in History" (from "On the Epochs in Modern History, lectures delivered before King Maximilian II of Bavaria"), in *The Theory and Practice of History*, edited by G. Iggers, translated by W. A. Iggers, London: Routledge, 2011, 20–23.

Rickert, H. (1896), *The Limits of Concept Formation in Natural Science: A Logical Introduction to the Historical Sciences*, Cambridge: Cambridge University Press, 2009.

———. (1924), *Die Probleme der Geschichtsphilosophie: Eine Einführung*, Heidelberg: Winter.

Sybel, H. von (1874), *Vorträge und Aufsätze*, Berlin: Hofmann.

Troeltsch, E. (1922), *Der Historismus und seine Probleme. Erstes Buch: Das logische Problem der Geschichtsphilosophie*, Mohr: Tübingen.

Windelband, W. (1883), "Kritische oder genetische Methode?" in his *Präludien: Aufsätze und Reden zur Philosophie und ihrer Geschichte Bd. 1*, Tübingen: Mohr, 1924, 99–135.

Windelband, W. (1894), "Geschichte und Naturwissenschaft" in his *Präludien: Aufsätze und Reden zur Philosophie und ihrer Geschichte Bd. 1*, Tübingen: Mohr, 1924, 136–160.

8

NIETZSCHE AND RELATIVISM

Jessica N. Berry

1. Introduction

If there is a heavyweight title in the division "Most Infamous Relativists Who Never Were," it surely goes to Friedrich Nietzsche. However, both the casual and the professionally dedicated reader of Nietzsche's corpus (not to mention the sizeable body of literature it has generated) could be forgiven for making the association. The last decades of the twentieth century were effectively dominated by interpretations like those offered by Jacques Derrida (1978) on the Continent and Arthur Danto (1965/2005) in Anglophone circles, under the sway of which Nietzsche became best known for the bold assertion that there are no facts: "No; facts is precisely what there is not, only interpretations!" (*KSA* 12:7[60]).[1] The Nietzsche of these readings is a cantankerous radical who churlishly resists the idea of any stable, mind-independent reality that could anchor truths or be known by us objectively: "The world with which we are concerned is false, i.e., it is not a fact but a fable and an approximation [. . .]; it is 'in flux' as something in a state of becoming, as a falsehood always changing but never getting near the truth: for – there is no 'truth'" (*KSA* 12:2[108]).[2] Statements about the world of our experience are condemned inevitably to failures of reference, and therefore to falsehood, while we ourselves are locked in the prisons of our own idiosyncratic perspectives.

The scholarship has largely moved on, as we shall see. But Nietzsche's reputation as a philosophical rebel without a cause has been tough to shake, as the authors of the entry on "Relativism" in the widely consulted *Stanford Encyclopedia of Philosophy* confirm: "Nietzsche is possibly the single most influential voice in shaping relativistic sensibilities in twentieth-century Continental philosophy." Such titillating passages as those just quoted, "irrespective of how Nietzsche himself intended them," they continue,

> have been taken to express a core contention of relativism that no single account of truth or reality can occupy a privileged position, for such accounts are only one of many perspectives that prevail at a given time in history. We cannot appeal to any facts or standards of evaluation independently of their relation to the perspectives available to us; we can do little more than to insist on the legitimacy of our own perspective and try to impose it on other people through our "will to power."
>
> *(Baghramian and Carter 2018 §3)*[3]

Such was the received view among some two or more generations of scholars. But that Nietzsche's statements shaped these sensibilities "irrespective of how Nietzsche himself intended them" contains a qualification not to be ignored. It is or should by now be agreed that attributions of the most extreme or vulgar versions of relativism to Nietzsche are sustainable only on the basis of fragmentary textual evidence he chose not to publish, of published "aphorisms" stripped, sometimes violently, of the contexts in which they appear, or both.

2. Nietzsche and alethic relativism

Of all the fragmentary sources that have contributed to the radical interpretation, it is safe to say that Nietzsche's early, unpublished essay, "On Truth and Lies in a Nonmoral Sense" (*TL*) has been the most influential. It generated a good deal of secondary literature in the late twentieth century and fueled postmodern readings that have long been associated with the relativistic claim that there is no final court of appeals for interpretations of texts and of many other things besides. "What is truth?," Nietzsche asks in this essay. And in a much-quoted passage, he answers,

> A moveable host of metaphors, metonymies, and anthropomorphisms; in short, a sum of human relations which have been poetically and rhetorically intensified, transferred, and embellished [. . .]. Truths are illusions which we have forgotten are illusions; they are metaphors that have become worn out and lost their sensuous force.
>
> *(TL, p. 84)*

Unchecked by the fact that it was edited and polished but never submitted to a publisher in any final form,[4] scholarly interest in this partial essay has generated a wide variety of interpretations, many attributing to Nietzsche an out-and-out denial of the possibility of knowledge, even an abandonment of the principle of non-contradiction, and, in short, a sort of nihilism about truth.

A sea-change in the scholarly attention to this essay and, indeed, to the whole of Nietzsche's views on truth and knowledge, came with the appearance of Maudemarie Clark's *Nietzsche on Truth and Philosophy* (1990), which brought much-needed sobriety and philosophical clarity to the study of his texts and generally raised the bar for textual evidence necessary to support good interpretations of them.[5] Clark argued that Nietzsche's position developed and matured over time, and that he would ultimately abandon the half-baked views of his youth, but she nevertheless recognized both the "considerable influence" on prevailing interpretations of Nietzsche's thought exerted by the essay "On Truth and Lie," and also the importance of meeting its challenge head-on: "*TL* cannot be dismissed on the grounds that it belongs only to the *Nachlaß*," she conceded, and, moreover,

> I do not believe we can plausibly explain away the evidence of Nietzsche's denial of truth in *TL*. This early essay therefore constitutes an important piece of evidence (the most important, I believe) for the radical interpretation . . . of Nietzsche's position on truth.
>
> *(Clark 1990, 64)*

The position she attributed to the early Nietzsche has become widely known as the "Falsification Thesis (FT)"; it captures Nietzsche's denial of truth in the notion that "our merely 'human' knowledge necessarily falsifies what the world is really like in itself" (Leiter 1994, 335).[6]

On this reading, Nietzsche abandons neither the principle of non-contradiction nor the conception of truth as correspondence between our beliefs and things as they are in themselves.

In fact, Clark concludes, "On Truth and Lie" *presupposes* both commitments, which are jointly necessary for making any sense of the text at all (1990, 83). The declaration that "truths are illusions," then, is best understood as picking up on Nietzsche's mention of "anthropomorphisms" in the earlier-quoted passage, as a way of relativizing truth to historically contingent and ever-shifting linguistic conventions. "Truth" would require correspondence to things as they are in themselves (indeed, to a Kantian thing-in-itself, on Clark's reading), but our beliefs and the propositions that express them exhibit no such correspondence. The merely "human truths" to which we have access, therefore, are the deflated currency of an unstable, if not false, economy: "coins," as Nietzsche says, "which have lost their embossing and are now considered as metal and no longer as coins" (*TL*, p. 84).

Vigorous disagreement persists, unsurprisingly, about Nietzsche's degree of commitment to the falsification thesis; most commentators have embraced some version of the developmental hypothesis offered by Clark, though some see greater continuity across Nietzsche's corpus. Of those, some have expressed doubts that Nietzsche was ever committed to a falsification thesis and the metaphysical correspondence theory it requires (Berry 2006), while others have argued that he held the thesis but abandoned it earlier than Clark originally supposed (Nehamas 2015), and still others that Nietzsche was and remained committed to it late into his productive career (Anderson 1996). As the naturalist reading of Nietzsche has approached consensus in the Anglophone literature,[7] too, many scholars have, like Clark (1990, 1998) herself, insisted that Nietzsche's appreciation of the value and potential of science is incompatible with his alleged early denial of truth and therefore demanded his eventual abandonment of the falsification thesis (Leiter 2002/2015). Yet others have argued that, especially when we come to appreciate fully the historical context in which Nietzsche's understanding of the sciences is situated and the influence of contemporaries such as Ernst Mach and Afrikan Spir, we will see that he can accept both naturalism, even empiricism, and also falsification (Hussain 2004; cf. Green 2002, 2015; Riccardi 2013).

In addition to the fragmentary and partial support for relativist readings found in Nietzsche's *Nachlaß*, he has also appeared to be a "relativist by association" with figures both before and after him. Nietzsche was trained as a Classical philologist, and influences from Greek antiquity pervade his thought. He is a particularly vigorous assailant of the absolutism in Platonic philosophy, and his interest in figures like Protagoras of Abdera and the historian Thucydides (Mann and Lustila 2011), as well as the respect he accords the Greek Sophists (Mann 2003; cf. Brobjer 2005), all testify to this. But it is surely his esteem for the pre-Platonic philosopher Heraclitus of Ephesus, for whom he reserves some of the highest praise and regards almost as a kindred spirit (see, e.g., *TI* "Reason" 2; *EH* "Books"; *BT* 3), that has appeared to offer the greatest support for relativistic readings. The scholarship on Nietzsche and Heraclitus has tended to focus on the latter's racy reputation as a promoter of an ontology of radical flux, and on that basis to attribute to Nietzsche not only a critique of Platonic "being," but an ontological view that would eliminate altogether anything that could serve as a stable referent of any linguistic expression. The appearance of support for a radically relativistic reading of Nietzsche, however, is only superficial. These interpretations tend to take for granted, and therefore to oversell, the Heraclitean commitment to a doctrine of flux (*Flusslehre*), and to neglect, and therefore to undersell, Nietzsche's estimation of other important features of Heraclitus' thought and legacy. In the end, there are reasons to doubt the attribution of a flux doctrine even to Heraclitus; but even if he is the radical ontologist he is so frequently taken to be, it does not obviously follow that Nietzsche, for all his praise of Heraclitus, must parrot his views.[8] One can successfully resist the rigidity of Platonic absolutism without positing a reality so chaotic as to render "a world we cannot *think*" (Meyer 2014, 8).

In addition to ascriptions of relativism in Nietzsche's intellectual forebears, there are a number of relativistically inclined twentieth-century inheritors of his project, among whom Richard Rorty has probably done the most to earn Nietzsche a reputation for epistemic relativism and a pragmatist view of truth. One consequence of Rorty's "anti-representationalist" view "is the recognition that no description of how things are from a God's-eye point of view, no skyhook provided by some contemporary or yet-to-be-developed science, is going to free us from the contingency of having been acculturated as we were" (Rorty 1991, 13). This commonplace does indeed have some analog in Nietzsche's thought; in particular, as we shall see, with respect to his "perspectivism." However, Rorty's anti-representationalism, his constructionism (i.e., his view that the facts are dependent on our descriptions) and his pragmatism (i.e., the semantic view that would make "truth" a success term), positions for which he is so well known, are also deeply and perhaps inextricably connected to his stance as a "liberal ironist." If the most powerful motivations for accepting relativism are normative ones, that is, that it is the best way of promoting the liberal virtues of toleration, equality, democracy, and understanding, as Rorty seems to think, then they will also be among the best reasons for dissociating Nietzsche from the relativist tradition represented by Rorty. It risks comic understatement to say Nietzsche does not shrink from asserting the superiority of his views over others, even to the point of chauvinism. Nietzsche is no friend of liberal democracy; he regards the universal equality of human beings as a prejudice of the modern age (*GM* I 4), one that has so corrupted the good judgment of otherwise clear thinkers that they have become incapable of making distinctions, of affirming or negating (*BGE* 207). Inability to recognize "orders of rank" is, for Nietzsche, symptomatic of nihilism.[9]

3. Nietzsche's perspectivism

Thus, there is no reason any longer to associate Nietzsche, certainly not the mature Nietzsche, with alethic relativism, taken as the view that there is no truth except relative to a framework, so that one and the same proposition may be true for one person and false for another.[10] Not only is the notion philosophically problematic, insofar as it is open to charges of self-refutation, but in Nietzsche's case there is ample evidence that he retains familiar notions of truth and of a world that can indeed serve as a check, at least in principle, on what it is possible, even reasonable, for human beings to believe. That reality consists only of whatever has the eternal and unchanging character of Platonic essences, for instance, is a mischaracterization of things, a false description, and one, Nietzsche constantly maintains, that mendaciously undermines the value of our worldly, human, all-too-human existence. It would be impossible to make good sense of Nietzsche's critique of conventional morality, and in particular Christianity, which includes regular accusations of priestly "mendacity" and "dishonesty," if we dispense with the idea that at least some worldviews are not only pernicious but false. When Nietzsche claims, therefore, "not [to] consider the falsity of a judgment as itself an objection to a judgment" (*BGE* 4), we must note that his declaration presupposes the distinction, in perfectly ordinary terms, between truth and falsity. We are being encouraged, here as in many other of Nietzsche's texts, to question our attachment to the *value* of truth, and challenged "to acknowledge untruth as a condition of life" (*BGE* 4), but not to deny its very *existence*.

Epistemic relativism, too, construed as the thesis that the cognitive norms governing knowledge, evidence, or justification co-vary with linguistic, historical, or cultural frameworks, especially if it entails a commitment to the equal validity of all points of view,[11] is no part of Nietzsche's philosophical outlook. In fact, Nietzsche's mature philosophy contains a powerful critique of the tradition. As a part of this critique, he frequently levels charges of irrationality at his predecessors, whose arguments he reveals to be *post hoc* rationalizations, infected by bias and driven by "ulterior motivation" (*BGE* 22), and who make unfounded claims to objectivity, neutrality, and cool logical reasoning. Much of the first chapter of *Beyond Good and Evil* is devoted to just such accusations (see, e.g., *BGE* 5, 6, 8, 12), which lose their force altogether if it turns out that Nietzsche thinks one man's justification is as good as the next.

What Nietzsche *does* maintain is that human cognitive activity is inescapably conditioned by our perspectives, understood as points of view consisting of our prior beliefs and knowledge, our cognitive abilities, as well as our values, instincts, affects, and drives. In a famous passage of the *Genealogy*, Nietzsche exhorts us to be on guard

> against the dangerous old conceptual fabrication that posited a "pure, will-less, painless, timeless subject of knowledge"; let us guard ourselves against the tentacles of such contradictory concepts as "pure reason," "absolute spirituality," "knowledge in itself": here is always demanded that we think an eye that cannot possibly be thought, an eye that must not have any direction [. . .]; thus, what is demanded here is always an absurdity and a non-concept of an eye. There is *only* a perspectival seeing, *only* a perspectival "knowing."
>
> *(GM III 12; see also BGE 16)*

This passage (which, again, is predicated on Nietzsche's understanding of the unacceptability of contradiction) contains Nietzsche's blunt rejection of "the view from nowhere," or the epistemic value dominant in philosophy (and elsewhere) that an ideal or true representation of reality would eliminate all traces of the knowers whose representation it is. Nietzsche's assertion of perspectivism here serves, among other things, to probe our commitment to eradicating subjectivity. Objectivity, as it is understood in the philosophical tradition Nietzsche challenges, betrays a desire to erase ourselves from our own picture of reality, and so constitutes further evidence of a pervasive moral ideal, the ascetic ideal that is the target of his campaign against morality, lurking where we least expect to find it.

That Nietzsche attacks objectivity as an unconditional epistemic ideal, however, does not mean that he dispenses altogether with some notion of it, or that he does away with ways of ranking or privileging perspectives. In the famous passage just quoted, he continues: "*the more* affects we allow to speak about a matter, *the more* eyes, different eyes, we know how to bring to bear on one and the same matter, that much more complete will our 'concept' of this matter, our 'objectivity' be" (*GM* III 12). What he rejects as both absurd and undesirable is the idea that we could "disconnect the affects one and all" from our cognitive apparatus. To want to anesthetize ourselves in such a way, "to *castrate* the intellect" (*GM* III 12), would be tantamount to willing the end of our lives – at least, the end of ourselves as the kinds of animals we are, as opposed to some sort of "pure spirits" or disembodied knowers. Thus, he invokes the idea of perspective again explicitly in the Preface to *Beyond Good and Evil* when he says, "talking about spirit and the Good [in itself] like Plato did meant standing truth on its head and disowning even *perspectivism*, which is the fundamental condition of all life."

Getting to the bottom of his "perspectivism" has been a major driver of scholarly interest in Nietzsche's work at least since Danto. Once again, we find in the literature no shortage of disagreement about the details of Nietzsche's "doctrine of perspectivism," but as the scholarship has itself matured, progress is on the horizon. For one thing, recent consensus distances perspectivism from relativism. Even Hales and Welshon, who have offered one of the most elaborate interpretations of it as a multi-faceted view informing Nietzsche's understanding not only of truth and knowledge but also of logic, ontology, causality, and selfhood, distinguish Nietzsche's "alethic perspectivism" (which indexes truth itself to perspectives) from the relativism familiar in earlier accounts: "We think it obvious that Nietzsche is not a 'come as you are' relativist" (2000, 190). He can, in other words, admit trans-perspectival truths and therefore resist the conclusion that one view is as good as another. And interpreters who now accept that Nietzsche's "perspectivism" need not commit him to a metaphysical or semantic theory of truth are right to do so (Berry 2005). The force of perspectivism, as an important weapon in Nietzsche's critical arsenal, is primarily epistemic: perspectives are ineradicable, contingent, and limiting; they do not eliminate knowledge, but only the dream of knowing from an Archimedean vantage point (*HH* I 9).

Clearer understandings and reconstructions of Nietzsche's philosophy of mind and moral psychology have recently shed a good deal of light on his concept of "drives" and their role in shaping (or determining) human behavior (see, e.g., Katsafanas 2013, 2016); on the relative roles of conscious and unconscious thought in his hypotheses about our deliberative processes and actions (see, e.g., Fowles 2018; Katsafanas 2005, 2015, 2016; Anderson 2002; Riccardi 2015); and on his views on human freedom and agency (see, e.g., Acampora 2013; Gemes and May 2009; Jenkins 2003; Katsafanas 2018; Miyasaki 2016). It has recently been argued that no further advance in our understanding of Nietzsche's remarks about "perspective" will really be made until these recent innovations in our understanding of Nietzsche's moral psychology and philosophy of mind are applied to his perspectivism, and the psychological mechanisms behind it worked out to the fullest extent the texts will allow (Anderson 2018).

4. Moral relativism

Nietzsche is also skeptical about the objectivity of morality, as commentators now widely agree, even as they disagree about the nature and scope of that skepticism. As he announces it in the *Genealogy*, Nietzsche's task is to undertake a critique of *all* morality (*GM* P 6). Now, although he should no more be thought of as a moral or cultural relativist than as a relativist of any other sort, this critical enterprise does leave readers with similar challenges: "If Nietzsche denies the kinds of objectivity upon which morality's claims to authority rest," for instance, "he may thereby deprive his own positive ideal of a legitimate claim to objectivity and authority" (Janaway and Robertson 2012, 10).[12] A sizeable literature on Nietzsche's metaethics has arisen in response to this challenge, and it remains a growth industry (see, e.g., Janaway and Robertson 2012).

If there is any emergent consensus here, it is that Nietzsche's critique of morality is not bottomless, in the way a relativistic assessment might be. Richard Schacht characterized Nietzsche's approach nicely early on:

> It does not involve the denial that there are any objective considerations transcending cultural formations or subjective determinations by reference to which particular moralities may be justified and assessed, even though it does deny unconditional validity to any of them. It links them to the contingently obtaining and varying – but nonetheless definite – psycho-physiologically grounded "conditions of life" of human beings. . . . They are indeed "relative"; but what they are relative to are circumstances pertaining to the actual constitutions of human beings of different sorts. Nietzsche's "naturalization of morality" thus involves the incorporation of moral theory into philosophical anthropology, in the context of which it loses its autonomy but acquires legitimacy.
>
> *(Schacht 1983, 463–464)*

In short, Nietzsche is best read as maintaining a conception of human flourishing, or health, which is quite likely to be construed in terms of maximizing the feeling of growth, life, and power (*GM* III 7), or perhaps cheerfulness, joy, or free-spiritedness, and to which any morality that survives his critique would have to answer. Put succinctly, the goal of the critique is "to translate humanity back into nature" (*BGE* 230), and he makes clear that not all moral systems will survive the translation.

5. Conclusion

As the scholarship on Nietzsche has matured, so has the literature on relativism, resulting in a number of nuanced distinctions now available to us for specifying what we mean by the term well above and beyond the idea that "anything goes." It makes less sense now than ever before to explore the connections between Nietzsche and relativism without a lot of further qualification. But as we think through these subtleties, we should bear in mind that many of the terms we can now use to make progress in, say, the semantics of truth-predication, and so many of the problems we take to be central in contemporary philosophy, were unavailable or invisible to Nietzsche, and not trivially so. So in assessing Nietzsche's relationship to relativism or any other contemporary domain of philosophical research, the importance of avoiding anachronism cannot be overstated. Not only does Nietzsche show no clear anticipation of or interest in such problems as the semantics of our linguistic operators for truth, we have good reasons for thinking that in some cases he would straightforwardly deny their "deep" problematic nature. Ken Gemes has argued compellingly "that Nietzsche is ultimately not interested in (theories of) truth" (1992, 41), and we would be wise to extend this observation to (theories of) epistemology, metaphysics, and metaethics. Indeed, if the most pressing problem facing modernity is the loss of our bearings with respect to whatever it is that makes life genuinely valuable, he might well think that some of our preoccupation with such puzzles contributes more to the deepening of the problem than to its solution.

Notes

1 Translated as *WP* §481. References to Nietzsche's texts refer to chapters or major parts (where applicable) and section numbers (not page numbers, unless otherwise indicated) of the following translations, using these conventional abbreviations:

KSA = *Kritische Studienausgabe* (1967)
WP = *The Will to Power* (1967)
TL = "On Truth and Lies in a Nonmoral Sense" (1999)
A, EH, and *TI* = *The Antichrist, Ecce Homo, Twilight of the Idols* and Other Writing (2005)
BGE = *Beyond Good and Evil* (2002)
GM = *The Genealogy of Morality* (1998)
HH = *Human, All Too Human* (1998)

2 Translated as *WP* §616.
3 For a fuller account of the Nietzsche–relativism association, as it still appears to many outside of Nietzsche studies, see Baghramian (2004, 75–81).
4 In his review of the expanded edition of Danto (1965/2005), Jonathan Cohen characterizes a newly added essay on *TL* as "a waste of Danto's time . . . as if he had pulled something out of Nietzsche's wastebasket" (2010, 81).
5 That Clark has not exempted herself from those standards is evident in the refinement of her own position over the years. See, e.g., the recent exchange on falsification and perspectivism between Clark (2018) and Alexander Nehamas (2017, 2018), whose influential book *Nietzsche: Life as Literature* (1985) also helped to motivate the serious treatment of Nietzsche's view of truth and knowledge and to shape the subsequent discussion.
6 See Clark (1990, ch. 3) for a full discussion of how to make sense of this "denial." Though Clark refers several times to "the falsification thesis," Leiter's compact formulation here represents its entry into the Nietzsche literature as a term of art.

7 Leiter (2002/2015) has become the touchstone of the naturalist reading; see also Schacht (1983).
8 See Berry (2013, 91–98) for discussion.
9 On the importance of the concept "order of rank" in Nietzsche, see Guay (2013).
10 I set aside here the question whether all varieties of relativism are ultimately reducible to alethic relativism. On this issue, see Baghramian (2004, ch. 4, esp. 121).
11 As, e.g., Paul Boghossian (2006) has argued it does.
12 In the same volume, Railton (2012, 21) helpfully teases apart *several* of these challenges, including a version that links the problems engendered by Nietzsche's critique of morality directly to his perspectivism: If Nietzsche's critique is true only "from a certain perspective," what claim to authority would it have for those who do not share his standpoint?

References

Acampora, C. D. (2013), "Nietzsche, Agency, and Responsibility: 'Das Thun ist Alles,'" *Journal of Nietzsche Studies* 44(2): 141–157.
Anderson, R. L. (1996), "Overcoming Charity: The Case of Maudemarie Clark's Nietzsche on Truth and Philosophy," *Nietzsche-Studien* 25: 307–341.
———. (2002), "Sensualism and Unconscious Representations in Nietzsche's Account of Knowledge," *International Studies in Philosophy* 34(3): 95–117.
———. (2018), "The Psychology of Perspectivism: A Question for Nietzsche Studies Now," *The Journal of Nietzsche Studies* 49(2): 221–228.
Baghramian, M. (2004), *Relativism*, New York: Routledge.
Baghramian, M. and J. A. Carter (2018), "Relativism," *The Stanford Encyclopedia of Philosophy*, Winter 2018 edition, edited by E. N. Zalta, https://plato.stanford.edu/archives/win2018/entries/relativism/.
Berry, J. N. (2005), "Perspectivism as *Ephexis* in Interpretation," *Philosophical Topics* 34(1): 19–34.
———. (2006), "Skepticism in Nietzsche's Earliest Work: Another Look at 'On Truth and Lie in an Extra-Moral Sense,'" *International Studies in Philosophy* 38(3): 33–48.
———. (2013), "Nietzsche and the Greeks," in *The Oxford Handbook of Nietzsche*, edited by J. Richardson and K. Gemes, Oxford: Oxford University Press, 83–107.
Boghossian, P. (2006), *Fear of Knowledge*, Oxford: Oxford University Press.
Brobjer, T. (2005), "Nietzsche's Relation to the Greek Sophists," *Nietzsche-Studien* 34: 256–277.
Clark, M. (1990), *Nietzsche on Truth and Philosophy*, Cambridge: Cambridge University Press.
———. (1998), "On Knowledge, Truth, and Value: Nietzsche's Debt to Schopenhauer and the Development of His Empricism," in *Willing and Nothingness: Schopenhauer as Nietzsche's Educator*, edited by C. Janaway, Oxford: Oxford University Press, 37–78.
———. (2018), "Perspectivism and Falsification Revisited: Nietzsche, Nehamas, and Me," *The Journal of Nietzsche Studies* 49(1): 3–30.
Cohen, J. R. (2010), "*Nietzsche as Philosopher* (Review)," *The Journal of Nietzsche Studies* 40: 81–82.
Danto, A. (1965/2005), *Nietzsche as Philosopher*, expanded edition, New York: Columbia University Press.
Derrida, J. (1978), *Spurs: Nietzsche's Styles*, translated by B. Harlow, Chicago: University of Chicago Press.
Fowles, C. (2018), "Nietzsche on Conscious and Unconscious Thought," *Inquiry* 62(1): 1–22.
Gemes, K. (1992), "Nietzsche's Critique of Truth," *Philosophy & Phenomenological Research* 52: 47–65.
Gemes, K. and S. May (eds.) (2009), *Nietzsche on Freedom and Autonomy*, Oxford: Oxford University Press.
Green, M. S. (2002), *Nietzsche and the Transcendental Tradition*, Urbana, IL: University of Illinois Press.
———. (2015), "Was Afrikan Spir a Phenomenalist? And What Difference Does It Make for Understanding Nietzsche?" *Journal of Nietzsche Studies* 46(2): 152–176.
Guay, R. (2013), "Order of Rank," in *The Oxford Handbook of Nietzsche*, edited by J. Richardson and K. Gemes, Oxford: Oxford University Press, 485–508.
Hales, S. and R. Welshon (2000), *Nietzsche's Perspectivism*, Urbana, IL: University of Illinois Press.
Hussain, N. J. Z. (2004), "Nietzsche's Positivism," *European Journal of Philosophy* 12(3): 326–368.
Janaway, C. and S. Robertson (2012), *Nietzsche, Naturalism, and Normativity*, Oxford: Oxford University Press.
Jenkins, S. (2003), "Morality, Agency, and Freedom in Nietzsche's *Genealogy of Morals*," *History of Philosophy Quarterly* 20(1): 61–80.
Katsafanas, P. (2005), "Nietzsche's Theory of Mind: Consciousness and Conceptualization," *European Journal of Philosophy* 13(1): 1–31.

———. (2013), "Nietzsche's Philosophical Psychology," in *The Oxford Handbook of Nietzsche*, edited by J. Richardson and K. Gemes, Oxford: Oxford University Press, 727–755.

———. (2015), "Nietzsche on the Nature of the Unconscious," *Inquiry* 58(3): 327–352.

———. (2016), *The Nietzschean Self: Moral Psychology, Agency, and the Unconscious*, Oxford: Oxford University Press.

———. (2018), "Nietzsche's Account of Self-Conscious Agency," *Philosophical Explorations* 21(1): 122–137.

Leiter, B. (1994), "Perspectivism in Nietzsche's *Genealogy of Morals*," in *Nietzsche, Genealogy, Morality*, edited R. Schacht, Berkeley: University of California Press, 334–357.

———. (2002/2015), *Nietzsche on Morality, Routledge Philosophy Guidebooks*, second edition, London: Routledge.

Mann, J. E. (2003), "Nietzsche's Interest in and Enthusiasm for the Greek Sophists," *Nietzsche-Studien* 32: 406–428.

Mann, J. E. and G. L. Lustila (2011), "A Model Sophist: Nietzsche on Protagoras and Thucydides," *The Journal of Nietzsche Studies* 42(2): 51–72.

Meyer, M. (2014), *Reading Nietzsche Through the Ancients: An Analysis of Becoming, Perspectivism, and the Principle of Non-Contradiction*, Berlin: De Gruyter.

Miyasaki, D. (2016), "Feeling, Not Freedom: Nietzsche Against Agency," *Journal of Nietzsche Studies* 47(2): 256–274.

Nietzsche, F. (1967a), *Kritische Studienausgabe*, edited by G. Colli and M. Montinari, Berlin: Walter de Gruyter, cited as *KSA*.

———. (1967b), *The Will to Power*, translated by W. Kaufmann and R. J. Hollingdale, New York: Vintage, cited as *WP*.

———. (1998a), *The Genealogy of Morality*, translated by M. Clark and A. J. Swensen Indianapolis, IN: Hackett, cited as *GM*.

———. (1998b), *Human, All Too Human*, translated by R. J. Hollingdale, Cambridge: Cambridge University Press, cited as *HH*.

———. (1999), "On Truth and Lies in a Nonmoral Sense," in *Philosophy and Truth: Selections from Nietzsche's Notebooks of the Early 1870's*, edited and translated by D. Breazeale, Amherst and New York: Humanity Books, 79–97, cited as *TL*.

———. (2002), *Beyond Good and Evil*, translated by J. Norman, Cambridge: Cambridge University Press, cited as *BGE*.

———. (2005), *The Antichrist, Ecce Homo, Twilight of the Idols and Other Writings*, translated by J. Norman, Cambridge: Cambridge University Press, cited as *A*, *EH*, and *TI*.

Nehamas, A. (1985), *Nietzsche: Life as Literature*, Cambridge, MA: Harvard University Press.

———. (2015), "Did Nietzsche Hold a 'Falsification Thesis'?" *Philosophical Inquiry* 39(1): 222–236.

———. (2017), "Nietzsche on Truth and the Value of Falsehood," *The Journal of Nietzsche Studies* 48(3): 319–346.

———. (2018), "Perspectivism and Falsification: A Reply to Maudemarie Clark," *The Journal of Nietzsche Studies* 49(2): 14–20.

Railton, P. (2012), "Nietzsche's Normative Theory? The Art and Skill of Living Well," in C. Janaway and S. Robertson (2012), 20–51.

Riccardi, M. (2013), "Nietzsche's Sensualism," *European Journal of Philosophy* 21(2): 219–257.

———. (2015), "Inner Opacity: Nietzsche on Introspection and Agency," *Inquiry* 58(3): 221–243.

Rorty, R. (1991), "Solidarity or Objectivity," in *Objectivity, Relativism, and Truth: Philosophical Papers*, vol. 1, Cambridge: Cambridge University Press.

Schacht, R. (1983), *Nietzsche, the Arguments of the Philosophers Series*, New York: Routledge.

9

MARX AND MARXISM

Lawrence Dallman and Brian Leiter

1. History and knowledge

Some passages in Marx seem to suggest that the era in which observers live constrains what they can know. Consider this passage from *Capital*, Vol. 1:

> Greek society was founded upon slavery, and had, therefore, as its natural basis, the inequality of men and of their labour powers. The secret of the expression of value, namely, that all kinds of labour are equal and equivalent, because, and so far as they are human labour in general, cannot be deciphered, until the notion of human equality has already acquired the fixity of a popular prejudice. This, however, is possible only in a society in which the great mass of the produce of labour takes the form of commodities, in which, consequently, the dominant relation between man and man, is that of owners of commodities. The brilliancy of Aristotle's genius is shown by this alone, that he discovered, in the expression of the value of commodities, a relation of equality. The peculiar conditions of the society in which he lived, alone prevented him from discovering what, "in truth," was at the bottom of this equality.
>
> *(Marx 1996, 70)*

What Marx refers to as "the secret of the expression of value" concerns how goods serving different purposes (e.g. shoes and houses) can be exchanged, with "value" as the name for whatever it is that is equal between the two quantities exchanged. Marx explains such transactions with his labor theory of value, according to which the value of a commodity is determined by the quantity of human labor time a society must invest in order to produce it. But slave labor is an invisible investment, so Aristotle cannot even grasp the labor theory of value due to the relations of production characteristic of his society.

Marx seems to endorse a kind of *historicism* about knowledge, according to which the set of knowable truths varies relative to historical parameters. There are at least two ways to take this suggestion. According to the *Strong Historicist*, the set of knowable facts varies across epochs because truths themselves are historically relative. An approach of this sort has been taken by historicist Marxists, with clear relativistic implications (cf. Kołakowski 1978, 215–239). According to the *Weak Historicist*, it is not the truths that vary with historical epoch, but our ability to acquire knowledge of those truths. Aristotle cannot acquire knowledge of the labor theory of value because the relations of production in slave society obscure it. But even in Aristotle's time, the labor theory of value correctly describes commodity values under capitalism. In Marx, this Weak Historicism depends on his theory of history (which identifies the mechanisms responsible for historical change) and his theory of ideas (and "ideology"), which identifies the mechanisms by which historical circumstances determine the shape of human thought.

2. The theory of history

In the 1859 preface to *A Contribution to the Critique of Political Economy*, in *The Communist Manifesto*, and in fragments of *The German Ideology*, Marx lays out two central theses about history. The first thesis is that, at any historical juncture, the material circumstances of life play the *primary* role in explaining the non-material circumstances of life. "Material circumstances," here, are the means ("forces of production") and methods of organization ("relations of production") by which people produce the necessary elements of their existence – food, clothing, shelter – as well as the kinds of food, clothing, and shelter they produce (cf. Marx and Engels 1975b, 82). "Non-material circumstances" include facts about the politics, law, religion, culture, and philosophy of an era (cf. Marx 1986b, 263). Though this formulation of the first thesis leaves some questions unanswered – in particular what it is for something to play a *primary* explanatory role (as opposed to some other kind of explanatory role) – it specifies the claim in sufficient detail for present purposes.

Forces of production include human labor power, industrial machinery and technology, and raw materials (i.e. elements whose productive power is tapped in production), all of which operate under a set of *relations of production* (i.e. the ways the process of production is organized and facilitated, especially with respect to the assignment of property rights). The dominant system of relations is the *mode of production* of an epoch. Under the capitalist mode of production, for instance, capitalists own raw materials and machines or technology, purchase the labor power of workers for wages, organize production as best suits their needs, and sell the products ("commodities") for a profit. When Marx writes, in the preceding cited passage, that the contemporary era is one in which men relate to one another as "owners of commodities," he means that their economic relationships are structured by the capitalist mode of production. Marx understands differences between modes of production to have a strong bearing on the character of human life: "what individuals are depends on the material conditions of their production" (Marx and Engels 1975b, 32). He takes these differences to cut so deep that some (although not all) aspects of human nature vary between epochs (cf. 1975b, 3).

Marx's second thesis is that transitions from one system of relations to another occur only when growth in the forces of production outstrips the limits of prevailing relations of production. According to this model, historical transformations occur as a result of a disequilibrium between forces and relations of production. Growth among the forces upsets the equilibrium, and transformation of the productive relations reestablishes it. That is, technological growth, such as the invention of the spinning jenny, makes it such that existing relations of production, which once functioned to facilitate and enhance production, come instead to stymy it, and new relations are adopted to maximize productivity. Further, because non-material circumstances are determined by material circumstances, transitions of this kind have broad consequences for the shape of society. Marx recognizes the descriptive limits of this model (cf. Marx 1986a, 46), but deploys it throughout his career.

3. The theory of ideas

Many important aspects of human thought count among the non-material circumstances Marx takes to be determined by material circumstances: "The totality of these relations of production constitutes the economic structure of society, the real foundation, on which arises a legal and political superstructure and to which correspond definite forms of social consciousness" (Marx 1986b, 263). Marx inherits the idea of "forms of consciousness" from Hegel. A form of consciousness is not simply a particular set of beliefs, but something closer to a *Weltanschauung* or conceptual scheme: a general framework for describing and making sense of the world. For Hegel, of course, history is a process of growth and redemption, in which successive epochs are governed by progressively more rational basic principles. These principles serve as schemata for ordering society, but they also serve as the bases for forms of consciousness. It is the job of philosophers, Hegel thinks, to carve out the hidden structures of these forms of consciousness. Marx rejects Hegel's theodicy, and the primary explanatory role it affords to the rational progress of ideas, but accepts the claim that history produces a sequence of forms of consciousness. Marx, however, understands this sequence as explained by, rather than as having the power to explain, the course of economic history.

Marx understands consciousness as realized, in the first instance, in language (cf. Marx and Engels 1975b, 43–44, 150) and he understands consciousness as essentially social, such that semantic facts have the publicly accessible, intersubjective character we usually take them to have (cf. 1975b, 36; Marx 1986a, 414). Unsurprisingly, he also takes the human capacity for language to spring from the need to communicate about practically salient features of a shared environment, and he takes this set of practically salient features to be partly determined in its extension by existing relations of production (cf. Marx and Engels 1975b, 44 and fn. 43). Finally, he understands at least some concepts to represent such features of the environment veridically. (We will turn later to how veridical representation is possible, and to Marx's reflections on non-veridical representation.)

According to Marx, prevailing relations of production play a major role in shaping our thinking. We can always distinguish between private labor (e.g. building a new fence around one's own home, or pursuing gardening as a hobby), and economically significant labor. Under capitalism, however, all economically significant labor comes to be understood as commodity-producing labor (i.e. producing goods for profit through wage-labor). This stands out, in particular, when multiple systems of productive relations compete for dominance within a single epoch. Only one system of relations can be dominant at a time. Competing systems are ultimately subsumed under the dominant system: "That we now not only describe the plantation-owners in America as capitalists, but that they *are* capitalists, is due to the fact that they exist as anomalies within a world market based upon free labour" (Marx 1986a, 436). It is the form of consciousness given rise to by this dominant system, the mode of production, that takes on the character of "popular prejudice." Its presuppositions, and not those of its rivals, confront us as the commonsense realities of our time.

4. Marx's weak historicism

Recall that Marx's claim about Aristotle is that he is unable to grasp the (ostensible) truth of the labor theory of value because he lives in a slave society. Recall also that the puzzle about value concerns how two goods serving different purposes can be exchanged at a fair rate (e.g. many shoes for one house). Aristotle notes that without some equalizing element, no exchange will be possible between otherwise incommensurable goods. Though he appears to recognize the possibility of non-monetary exchange, Aristotle identifies this equalizing element with *monetary value*. He takes the latter, however, to be a mere "convention" adopted out of "need" (Aristotle 1999, 1133a). Thus, Aristotle claims, "though things so different cannot become commensurate in reality, they can become commensurate enough in relation to our needs" (1999, 1133b). The genius of Aristotle, Marx maintains, just consists in his recognizing the presence of some such equalizing element in exchange, even if he is ultimately unable to give proper account of it.

In Aristotle's Greece, the prevailing mode of production is one in which different kinds of labor (e.g. the labor of a slave, and the labor of a citizen) play very different social roles. Only under capitalism, where all goods are treated as commodities (i.e. as the profit-making products of wage labor), does all labor come to be interchangeable: in general, the value of a commodity does not depend on *whose* labor-power produces it. Thus, it never occurs to Aristotle that what is equal between two otherwise incommensurable goods might be the quantity of interchangeable human labor time invested in each. This never occurs to him because the notion of "labor" operative in the popular prejudice of his day is that proper to the slave system: a conception on which *not* all labor is interchangeable. Suppose that some Greek merchants *did* hire workers on the basis of wages to produce goods for profit. Still, the Greek notion of value reflects the realities of the slave system, and the products of this work would be afforded value *as though* they were produced by slave labor. Likewise, under capitalism, products are afforded value *as though* they were produced by wage labor, even when they are in fact produced by slaves (cf. Marx 1997, 119, 385).

We can treat the labor theory of value as an account of value *in general* (i.e. under any mode of production), or we can treat it as accounting only for commodity value *under capitalism*. If we treat it as a general theory of value, then the principal reason Aristotle cannot grasp "the secret of the expression of value" is that, in Aristotle's Greece, that secret is simply false. Even if it is true of commodity values, the labor theory of value is *false* where the value of goods produced in slave societies is concerned. If we instead restrict the labor theory of value to the domain of commodity values, then Aristotle's failure can be explained in terms of inadequate evidence: Aristotle cannot know the nature of commodity value much as we cannot know how many threads were in Muhammad's cloak when he conquered Mecca. In short, Marx is committed only to the claim that both available evidence and popular prejudice themselves change in time, imposing constraints on what can, in practice, be known. Marx's historicism, therefore, is Weak Historicism.

5. The paradox of ideology

If all thinking, for Marx, is shaped by non-cognitive determinants, why suppose any subset of thinking to be closer to the truth than any other? After all, Marx claims that "[j]ust as one does not judge an individual by what he thinks about himself, so one cannot judge a period of transformation by its consciousness," suggesting instead that "this consciousness must be explained" (Marx 1986b, 263). But if every form of consciousness is subject to debunking explanation, then the form of consciousness from which such explanations are articulated must itself be subject to debunking explanation. This gives rise to a self-refuting global conceptual relativism: the ostensibly universal claim that all conceptual schemes are explained by non-cognitive determinants reduces, on the assumption of its own truth, to a scheme-relative truth. We will call this *possible* implication of Marx's theories the *paradox of ideology*.

Some readers of Marx bite the bullet on global conceptual relativism (cf. Resnick and Wolff 1987), but such interpretations are unfaithful to Marx's writings. Marx does not hold that all thinking which admits of explanation by non-cognitive determinants is by dint of that fact limited to merely scheme-relative truth. Rather, he distinguishes different processes by which forms of consciousness are given shape as different ways of appropriating the world (cf. Marx 1986a, 38). Some of these produce representations of *reality*, while others produce representations of mere *appearances* (i.e. popular prejudice, "common sense"). Marx is committed, metaphysically, to a version of the natural kinds theory, according to which it is possible in principle for a conceptual scheme to carve out the *correct* sortal concepts and categories, thereby representing the world as it is. When Marx claims to treat the "real living individuals themselves" (Marx and Engels 1975b, 37) as the basis of his social analysis, he means that the theoretical vocabulary he takes on board for describing the foundations of society is one that carves the relevant natural kinds. But how can Marx's own science of society escape the charge of ideology?

6. The theory of science

Science (*Wissenschaft*), according to Marx, aims to move past how things appear in ordinary life, to "uncover the essence which lies hidden behind commonplace appearances, and which mostly contradicts the form of commonplace appearances (as for example in the case of the movement of the sun about the earth)" (Marx 1994, 86). Theories that simply systematize the ordinary appearances (e.g. that the sun rises in the West and sets in the East) are not genuine science. Real explanatory science identifies the unseen mechanisms responsible for generating the appearances (e.g. that it is the earth's spinning on its axis as it revolves around the sun that explains why it should *appear* that the sun revolves around the earth). Likewise, such theorizing often introduces new classes of entities into our ontology (e.g. when we explain macroscopic features of gasses in terms of the postulated particles of the molecular theory). It is only through this conception of science that Marx avoids the paradox of ideology.

Real science is not perniciously ideological for Marx. Various structural features of scientific discourse help to ensure this. Science requires a community of inquirers, whose contributions inform and correct one another (cf. Marx 1986a, 527). In his own scientific work, Marx makes clear that he understands scientific progress to consist in the explanation, from a new theoretical framework, of the successes and failures of previous efforts (cf. Marx 1988, esp. 348–376). Moreover, he takes it that, in the fullness of time, such processes of criticism will converge upon a single unifying theory for each subject matter (Marx and Engels 1975a, 296). Many readers of Marx, including both Engels and Lenin, have denied this, supposing that theoretical convergence functions only as a regulative ideal in science (cf. Engels 1987, 80 and Lenin 1972, 154). Marx, however, derives from Feuerbach the idea that, though individual inquirers may be limited in their capabilities, the scientific community (and species) as a whole are capable in principle of arriving at *total* knowledge of empirical matters of fact (cf. Feuerbach 2012, 189).

Most importantly, Marx takes good scientific explanations to be distinguished by their practical efficacy: they should, in principle, allow those who accept them to exercise some measure of control over their historical circumstances (cf. Marx 1975, 337; Marx and Engels 1975b, 8). This *practical* model of theory success understands the adequacy of a theory in terms of the role adopting that theory plays in producing desired outcomes. To punch through the appearances, on this view, is to describe the real mechanisms responsible for how things are, where *real* mechanisms are those whose manipulation reliably produces expected outcomes. For reasons having to do with his debts to the classical German tradition, however, Marx is unable to fully explicate this criterion (cf. Dallman forthcoming), and Engels's flatfooted efforts to elaborate upon Marx's treatment do little to ameliorate the problem (cf. Engels 1990, 367–368). Nevertheless, for Marx, scientific inquiry is that mode of appropriating the world that aims at carving the natural kinds (i.e. the real mechanisms). It is through reliance on the results of science, even in the face of how things *seem*, therefore, that we escape popular prejudice and the paradox of ideology.

Some interpreters of Marx have supposed that members of the ruling classes are *essentially* taken in by appearances, while members of the working classes are essentially alert to the historical contingency of prevailing relations of production, and thus uniquely positioned to see past appearances and transform those conditions (cf. Lukács 1968). Marx does take the working class under capitalism to possess a special revolutionary potential, but he understands workers to be, like everyone else, dependent upon science for their access of reality. The power of appearances to distract and compel extends to the members of every class. Denial of this fact leads, in the most extreme case, to the Lysenkoist distinction between bourgeois and proletarian science, according to which the results of scientific inquiry are to be evaluated not in terms of their practical efficacy or their ability to carve the natural kinds, but in terms of their stemming from (ostensibly) proletarian institutions, and their cohering with (ostensibly) proletarian principles.

7. Ideology and morality

An ideology is an inferentially related set of beliefs, the dominance of which can be explained both in terms of its mirroring the commonsense appearances, and in terms of this serving the interests of the ruling class. It serves the ruling class by legitimizing their control of the productive forces, either by suggesting that this control is *right*, or by suggesting that it is *necessary*. Accordingly, ideological forms of consciousness usually concern moral, legal, and political subject-matter, though some purportedly scientific theories (especially in the social sciences) are ideological as well. Ideologies involve *false* beliefs (e.g. false beliefs about whose interests are really being served, and/or about how the beliefs in question really arose). Genuine science is not ideological, for the reasons just discussed. It is true that, under capitalism, "the *scientific factor* is for the first time developed, applied, and called into existence on a scale which earlier epochs could not have imagined" (Marx 1994, 34). This, however, is because those on the winning side of capitalist productive relations have a strong vested interest in knowing what is true.

Under capitalism, the *bourgeois* class is that which owns the forces of production or, in the case of human labor power, rents it and thus owns its output for some period of time. Members of the *proletarian* class, by contrast, own only their labor power, which they must sell to the bourgeoisie in order to survive. In any epoch, the class that owns the major forces of production (or is able to purchase them) is the *ruling* class. Naturally, it is in the interest of the ruling class to maintain whatever system of relations affords them opportunity to rule: this world, for the bourgeoisie, is "the best of all possible worlds" (Marx 1996, fn. 92). By contrast, it is in the interest of all non-ruling classes, notwithstanding the private circumstances of individuals, to upset and replace any system that would condemn them to subservience. Ideology functions to confuse these interests, to represent what is only in the interest of the ruling class as in everyone's interest. The bourgeoisie, as the ruling class under capitalism, expounds a system of beliefs that "voices their particular interests as universal interests" (Marx and Engels 1975b, 180).

Can there be non-ideological moralities or moral judgments? Some scholars have argued that moral judgments are always to be understood relative to prevailing relations of production, such that there is no sense in which capitalism, for Marx, is unjust (cf. Wood 2004, ch. 9). In the *Critique of the Gotha Programme*, Marx claims that "present-day distribution" is "the only 'fair' distribution on the basis of the present-day mode of production" (Marx and Engels 1989, 84). We should not hear this as suggesting that present-day distribution is in fact just. Marx's use of scare quotes communicates that the measure of fairness at issue is not genuine. Rather, with this remark, he suggests that, to the extent that there *is* a general interest in class-based society – though it is mere appearance – it does not vary between classes: it springs directly from popular prejudice, and thus reflects the interests of the ruling class. Marx suggests that those unhappy with present conditions study the mechanisms responsible for maintaining them, and act intelligently to undo them. Reflection upon the meaning of "fair" or "just" does nothing to serve these ends (cf. Leiter 2015).

Marx's general views on morality are sparsely developed, principally for the reason that he cared very little about moral questions. He makes clear in *The German Ideology* that "communists do not preach morality at all" (Marx and Engels 1975b, 247). It seems clear, however, that Marx himself does have some "moral" opinions about capitalism: he attacks those who work to keep capitalism intact, in part because capitalism frustrates the interests of the vast majority, rendering them miserable. Why should that attitude not be seen as an ideological moral attitude?

Crucially, Marx's own "moral" attitudes acknowledge their class specificity, and do not take advantage of others' ignorance of that class specificity. In *The Manifesto of the Communist Party*, Marx and Engels describe "the proletarian movement" as simply being "in the interest of the immense majority" (Marx and Engels 1976, 495). They deride the German "True" Socialists (though they may just as well have been thinking of Jürgen Habermas) for thinking that social-ism reflects "the requirements of Truth; not the interests of the proletariat, but the interests of Human Nature, of Man in general, who belongs to no class, has no reality, who exists only in the misty realm of philosophical fantasy." They deride Critical-Utopian Socialists for "consider[ing] themselves far superior to all class antagonisms," and claiming to take up the cause of "every member of society, even that of the most favoured" (Marx and Engels 1976, 511, 515). They write:

> The theoretical conclusions of the Communists are in no way based on ideas or principles that have been invented, or discovered, by this or that would-be universal reformer. They merely express in general terms actual relations springing from the existing class struggle, from a historical movement under our very ideas.
>
> *(Marx and Engels 1976, 498)*

Marx's "moral" objection to capitalism is class-interest-specific, a judgment in the interests of the vast majority as opposed to the ruling class. But this "moral" attitude is *not* ideological because its acceptability does not depend on its *not* being class-interest-specific. Indeed, there is no mistake about its genesis either: "The proletarian movement is the self-conscious, independent movement of the immense majority, in the interest of the immense majority" (Marx and Engels 1976, 495). That leaves completely open what moral views would arise in a society not marked by class conflict.

8. Materialism and idealism

In *The German Ideology*, Marx distinguishes two fundamentally opposed viewpoints: the mate-rialist and idealist worldviews. Idealist explanatory strategies, Marx claims, tend to reinforce the appearances, while materialist explanatory strategies tend to subvert them. Some readers have understood this opposition to describe a basic antagonism between bourgeois and proletarian perspectives in philosophy (cf. Althusser 2001). On its most sophisticated formulation, this sug-gestion makes sense of the materialism/idealism dispute in terms of opposed epistemic stances (cf. Suchting 1986). Those adopting the materialist worldview, it might follow, can say nothing to categorically demonstrate the illegitimacy of taking up the opposed, idealist worldview; the dispute bottoms out in disagreement, and therefore in a sort of *stance relativism*. Proponents of this interpretation might understand the dispute between idealists and materialists as a *political* dispute (i.e. one to be resolved through political militancy within the academy, rather than through rational persuasion). There is little evidence, however, to suggest that Marx himself conceived of the materialism/idealism dispute in these terms. The suggestion more probably traces back only so far as Lenin (cf. Lenin 1966). Accordingly, whatever independent interest the idea may have, it does nothing to show that Marx himself was a relativist.

9. Conclusion: a future for Marx

Marx is not, at least on our account, a relativist. He differs from most non-relativists, however, in his close attention to the problems that give rise to relativism. It is almost certainly because Marx affords relativism such honest reflection – though never by that name – that he is so frequently misread as a relativist. This, it would seem, is the cost of doing honest philosophy: Kant is just as often misread as a skeptic, Nietzsche as a nihilist, and Wilfrid Sellars as an anti-naturalist. Where Marx in particular is concerned, too much has been written based only on isolated texts, and with too little attention to the background of theoretical commitments Marx develops over the course of his career. Concerns bearing on the topic of relativism are central to this theoretical background. Accordingly, those for whom the "new relativism" has spurred renewed interest in the "old relativism" would do well to spend more time with Marx, not only as a political thinker, but as a powerful exponent of a historically unique form of philosophical naturalism.

References

Althusser, L. (2001), "Lenin and Philosophy," in *Lenin and Philosophy and Other Essays*, translated by B. Brewster, New York: Monthly Review Press, 11–43.
Aristotle (1999), *Nicomachean Ethics*, second edition, translated by T. Irwin, Indianapolis: Hackett.
Dallman, L. (forthcoming), "Marx's Naturalism: A Reconstruction and Critique," PhD dissertation, Chicago: University of Chicago.
Engels, F. (1987), *Collected Works*, vol. 25, London: Lawrence and Wishart.
———. (1990), *Collected Works*, vol. 26, London: Lawrence and Wishart.
Feuerbach, L. (2012), "Principles of the Philosophy of the Future," in *The Fiery Brook: Selected Writings*, translated by Z. Hanfi, London: Verso, 175–246.
Kołakowski, L. (1978), *Main Currents of Marxism: Its Rise, Growth, and Dissolution*, translated by P. S. Falla, vol. 2, Oxford: Clarendon Press.
Leiter, B. (2015), "Why Marxism Still Does Not Need Normative Theory," *Analyse & Kritik* 37(1–2): 23–50.
Lenin, V. I. (1966), "On the Significance of Militant Materialism," in *Collected Works* 33, Moscow: Progress Publishers, 227–236.
———. (1972), *Materialism and Empirio-Criticism*, Peking: Foreign Languages Press.
Lukács, G. (1968), "Reification and the Consciousness of the Proletariat," in *History and Class Consciousness: Studies in Marxist Dialectics*, translated by R. Livingstone, Cambridge, MA: MIT Press, 83–222.
Marx, K. (1975), *Collected Works*, vol. 1, London: Lawrence and Wishart.
———. (1986a), *Collected Works*, vol. 28, London: Lawrence and Wishart.
———. (1986b), *Collected Works*, vol. 29, London: Lawrence and Wishart.
———. (1988), *Collected Works*, vol. 30, London: Lawrence and Wishart.
———. (1994), *Collected Works*, vol. 34, London: Lawrence and Wishart.
———. (1996), *Collected Works*, vol. 35, London: Lawrence and Wishart.
———. (1997), *Collected Works*, vol. 36, London: Lawrence and Wishart.
Marx, K. and F. Engels (1975a), *Collected Works*, vol. 3, London: Lawrence and Wishart.
———. (1975b), *Collected Works*, vol. 5, London: Lawrence and Wishart.
———. (1976), *Collected Works*, vol. 6, London: Lawrence and Wishart.
———. (1989), *Collected Works*, vol. 24, London: Lawrence and Wishart.

Resnick, S. A. and R. D. Wolff (1987), *Knowledge and Class: A Marxian Critique of Political Economy*, Chicago: University of Chicago Press.

Suchting, W. A. (1986), "Marx, Lenin and the Concept of Materialism," in *Marx and Philosophy: Three Studies*, London: Palgrave Macmillan, 53–80.

Wood, A. W. (2004), *Karl Marx,* second edition, New York: Routledge.

10

THE MANY FACES OF ANTI-RELATIVISM IN PHENOMENOLOGY

Sonja Rinofner-Kreidl

1. Preliminaries

Relativism is motivated by observations concerning subjective, or otherwise framework-related, conditions of gaining knowledge, accessing objects, or using concepts and languages. Relativists' aim is to challenge a taken-for-granted notion of objectivity and an allegedly absolute stance concerning for instance epistemological, metaphysical, or metaethical claims. Phenomenologists philosophize in an experience-based manner; this includes a methodically reflected first-person perspective. It thus should not come as a surprise that arguments about relativism loom large in phenomenological circles. At the same time, it is important to emphasize that phenomenologists disagree with each other concerning – amongst other things – the issue whether epistemological questions are primary with respect to other areas; or whether phenomenology should be agnostic towards metaphysics. This has implications for how phenomenologists respond to relativism: thus one finds conflicting views on which forms of relativism are most important; or how best to tackle relativist views. Still, the overall attitude of "classical phenomenologists" towards various types of relativism is unanimously critical and negative (cf. Husserl 2002a; Scheler 1973, 2007; Spiegelberg 1935). In the present context, "classical phenomenology" refers to the work of first-generation phenomenologists, that is, writings that were published up until the 1940s.

Relativist positions conflict with the phenomenological commitments to acknowledging first-hand evidences and to striving for objectivity (where such objectivity might take different forms, depending on the subject matter). Of course, phenomenologists and relativists both reject *certain* forms of strong or absolute notions of objectivity, say, forms based on mistaken ideas of scientism. Still, phenomenologists and relativists have different goals. Relativists ultimately aim to defend forms of subjectivism and intersubjectivism, or conventionalism or social constructivism. In other words, relativists deny that attacks on scientific forms of objectivity leave room for the idea of objectivity *independent of given (individual or collective) epistemic subjects, languages, customs, morals, cultural frameworks*, and so on.[1]

Phenomenologists warn against the dangers of (inter-)subjectivism as much as they rail against scientistic objectivism. Instead, they argue in favor of diversity by intending to establish a system of different types of objectivity according to different kinds of objects or matters of affairs. Doing so requires distinguishing between *subject-relatedness* and *subject-dependence*.

Phenomenologists emphasize the former when analyzing how different types of objects and states of affairs are "intended" by the subject. And yet, phenomenologists deny that subject-relatedness coincides with subject-dependence. That is, they rather emphasize the differences between, for instance, "(merely) subjective," "subject-related," or "subject-dependent" aspects. In this vein, phenomenologists resist fuelling and continuing the "merely frivolous inquiry of a dialectic that argues back and forth between skeptical negativism or relativism and logical absolutism" (Husserl 1929, 200 [§80]). As long as we fail to consider the reasons and partial legitimacy of different ways of positioning us towards aspects of experience, we remain stuck between "a wrong skeptical relativism and a no less wrong logical absolutism, mutual bugbears that knock each other down and come to life again like the figures in a Punch and Judy show" (Husserl 1929, 278 [§105]).[2] In other words, phenomenologists refuse the dialectic between subjectivism and objectivism. They deny that rejecting objectivism inevitably ends up in (a problematic brand of) subjectivism.

This is especially clear in Husserl's case. His guiding idea is that different claims to validity (be they, e.g., epistemic or axiological) concerning objects must be investigated as so many modes of "intentional reference to" and "constitution of" different types of objects. For Husserl it is crucial to note that reference to different types of objects involves both object-related and subject-related aspects. The former as different logical forms of meaning, the latter, for instance, as sensations, modes of attentiveness or habitually practiced mental activities.

From the phenomenological point of view, the problem with relativism is then the following: relativism pays attention only to the subject-related aspects and ignores the object-related ones. And this destroys all claims to validity.

2. Refuting, undermining, and bypassing relativism

Within classical phenomenology, one can distinguish between three strategies for warding off relativism:

(REFUTING RELATIVISM) (RR) One shows that relativism is committed to ideas that make it self-refuting;
(UNDERMINING RELATIVISM) (UR) One argues that relativism conflates acceptable with unacceptable forms of "subjectivity" and "relativity";
(BYPASSING RELATIVISM) (BR) One demonstrates that the overall phenomenological project simply leaves no space for raising relativistic worries.

Relativism can of course take many forms and target, or seek to undermine, many conceptual distinctions. Such distinctions are for instance, *form vs. content, genesis vs. validity, eidos vs. individual object* or, more generally, *ideas vs. matters of fact, quid iuris vs. quid facti, a priori vs. a posteriori*. Consider the opposition between genesis and validity. In undermining it, relativists often use *debunking arguments*. Relativists seek to show that a subject's contingent circumstances – her upbringing or social standing, her psychological constitution or biologically predetermined dispositions – fully account for her beliefs. Consequently, *internal demands of rationality* are held to be unnecessary. Opponents of relativism argue that the debunkers conflate reasons and causes.[3] The example brings out a further important facet of Husserl's perspective on relativism: relativism is inseparable from a *naturalistic stance towards the world*. This claim is of course important to Husserl's attack on psychologism in logic, epistemology, or value theory (cf. Rinofner-Kreidl 2004).

3. Husserl's critique of psychologism in his
Logical Investigations

Husserl's LI starts off from a detailed attack on logical psychologism.[4] Husserl first introduces a (RR)-style argument focused on the conditions of the possibility of theories in general. Husserl contends that these conditions centrally include the existence of ideal ("logical") objects, and that in this sense "('logical') objectivism" – or "logical idealism" – are forced upon us as soon as we theorize at all.[5] Defending logical idealism is the basic anti-skeptical and anti-relativist move in the LI.

Psychologism in nineteenth-century logics, Husserl's target in the *Prolegomena*, is a form of relativism. Some of its advocates assume that the principles and laws of logic (e.g., the principle of excluded middle) can be explained *without any substantial change of meaning* in terms of psychological laws of thinking. Some psychologistic thinkers take these principles and laws to be rules we should follow. That is, for them logic essentially is normative. To be sure, Husserl does not deny that normative (or applied) logic is relevant in various practical contexts. He nonetheless insists that logic has purely theoretical foundations ("pure logic"). The latter necessarily lie beneath all actual and possible applications and forms of theories. Disregarding the priority of pure logic with respect to normative logic hence is the first presupposition of logical psychologism Husserl considers untenable. Moreover, psychologistic thinkers conflate the purely formal relations holding between logical objects with the causal connections between acts (or experiences) of thinking. Proponents of psychologism thus maintain that logical laws and principles refer to particular psychological experiences of subjects that use these laws and principles. "Logicists" (like Husserl) deny that logical laws and principles say anything about someone's experiences. (See Husserl's explanation concerning the principle of non-contradiction in §36 of the *Prolegomena*.) A third psychologistic presupposition denies the distinction between sciences of real entities and sciences of ideal objects (e.g., concepts, judgements, the laws and principles of pure logic). This amounts to an illegitimate shift between different domains of objects (*metabasis eiis allo genos*) when talking about the ultimate foundations of all possible theories (cf. Husserl 1900a, §§40, 45–46).

Apart from specifying the untenable presuppositions of psychologism, Husserl also holds that psychologism has highly problematic, yet far-reaching consequences.

(1) If psychologism were true, logical principles and laws were merely inductively based or probably true. Yet they are valid in an absolute and universal manner.
(2) If psychologism were true, validity claims would have to be relative to individual subjects (individual relativism) or humankind on the whole (species relativism or anthropologism) (cf. Husserl 1900a, §§34–36).

According to Husserl, as individual or species-relativism, psychologism actually is *a form of skepticism*. This is owing to the fact that psychologistic relativism *if formulated in a sufficiently strong manner* unavoidably results in a self-refutation (cf. Schmitt 2018). (The phrase "if formulated in a sufficiently strong manner" indicates that Husserl's refutation of relativism is based on non-relativistic presuppositions which proponents of relativism would straightforwardly deny. Cf. Mohanty 1997, 181.) In the *Prolegomena* Husserl argues that psychologism, as a self-refuting skepticism, contravenes the ideal conditions of any possible theory whatever. It violates (i) the "subjective-noetic" conditions (e. g. that evident judgements are discernible from non-evident judgements) as well as (ii) the "objective-logical conditions" (e.g., that every science is

99

constituted by a unitary context of justification and thereby presupposes valid laws of deduction) (cf. Husserl 1900a, §32). A specific concern of Husserl is to argue that redefining the notion of truth in terms of a relativistic notion ("true for x") breaks up this unitary context of justification. Doing so denies, in self-defeating way, the realm of objective being as correlate of possible true statements (cf. Husserl 1900a, §§36, 39).

Though I cannot enter into a detailed discussion here, it has to be noted that the presuppositions and consequences of psychologism, as enumerated and criticized by Husserl, have met fierce criticism, both in contemporary and recent debates. (For an overview of the main counterarguments that have been marshalled against Husserl's critique of psychologism, see Kusch 2015.)

Let us take stock. In the first volume of his LI Husserl takes for granted that the positing of ideal (logical) objects is warranted. He is, however, eager to provide arguments that speak in favor of acknowledging ideal objects. He does so, among other things, by expounding the detrimental consequences that result from reducing logical objects to mental entities or occurrences. Following this strategy, Husserl talks about logical objects and defends the claim that they are irreducible to actual occurrences in someone's mind. He does so by means of an indirect proof (*reductio ad absurdum*), showing that the very opposite view is absurd. Part of this (RR)-argument involves showing that logical psychologism as a brand of relativism undermines itself.

Husserl's arguments in the *Prolegomena* are modified and enriched in the second volume of LI (*Investigations Concerning a Phenomenology and Theory of Knowledge*). The (RR)-strategy is presupposed in the second volume, but it is now supplemented and strengthened by an (UR)-strategy. For example, in the second investigation, entitled "The Ideal Unity of the Species and Recent Theories of Abstraction," Husserl does not only talk about logical objects, explaining why stipulating their irreducibility is a necessary condition for any knowledge claim. He also provides a detailed study on the question which kinds of problems require what types of logical objects.

One central problem in this context is abstraction, that is, the question how the formation and logical function of concepts are to be analyzed. Husserl's aim is to show the inadequacy of empirical theories of abstraction which undertake to explain the identity of concepts by recourse to the similarity of those objects that fall within the extension of these concepts (cf. Husserl 1900a, §§1–12, §§24–28).[6] For meeting this critical goal, Husserl walks the reader through a synopsis of Locke's, Berkeley's, Hume's, J. St. Mill's, and others' proposals.

The second logical investigation shows how Husserl is engaged in an effort to reach original, that is, intuitive evidence or insight into how species are given as ideal unities. Husserl carries out a fine-grained descriptive analysis within his overall intentionalist framework. Its guiding question is: How is the subjective (performance of an) experience related to the objective intentional content ("meaning") and the intended object? (Here the experience at issue is assumed to refer to the intended objects by means of the intentional content.)

In this context, the (UR)-strategy is meant to show that as soon as we can offer a correct and sufficiently distinct descriptively based explanation of how these aspects of intentionality (instantiated in any consciousness whatever) interact with one another, there is nothing left that stands in need of further clarification. There is no leeway left for raising relativistic doubts and no evidence-based motivation for doing so. In this vein, the (RR)-line of reasoning takes a peculiar *phenomenological* twist as soon as Husserl incorporates it into the comprehensive investigations of the second volume of LI (cf. Metzger 1966, 121–162). In doing so, (UR) enters the stage. Carrying out phenomenological-descriptive work goes beyond *arguing* in favor of or against certain positions. Its goal rather is to *show* or to instruct someone how to *immediately* (*intuitively*) *grasp* the evidential force of certain ideas that may then function as key references of arguments (cf. Husserl 2002b, 320–322; Scheler 1957, 391f.). Of course, any such "training" concerning how to gain evidences is of limited range. It must not overstep the bounds of reason. Nobody can be forced to see the rational structure of all experiences, objects, or states of affairs. In this situation, Husserl insists that the burden of proof lies on those who repudiate given evidences. However, since the opponents are free to claim, and actually do claim, that the burden of proof has to be distributed differently, it is obvious that debates on relativism easily get locked in a stalemate. Both parties mutually charge each other with a *petitio principii*. Presenting evidence as the ultimate source of any justification, as Husserl, Scheler and other phenomenologists do, does not rule out the concession that conflicts of evidence, as specific forms of disagreement (Spiegelberg 1935, 85; Scheler 1957, 393f.), also stand in need of thorough investigation.

In the investigations I, II, V, and VI of the second volume of LI Husserl establishes a genuinely phenomenological take on the issues involved in the debate on logical psychologism. Throughout these investigations, he defends the objectivity of logical objects by focusing on the ways in which subjective acts of thinking are related to them. This amounts to endorsing a qualified notion of objectivity: it is recovered by descriptive analysis rather than being stipulated. This sense of objectivity becomes intelligible by understanding how operations involving logical objects or meaning contents make such objectivity possible. Husserl elucidates the notion of objectivity by considering how logical objects are involved in the intentional structure of consciousness. That is, he explains objectivity by entering an investigation that others (e.g., G. Frege) would consider as merely subjectively orientated.

4. Husserl's critique of social consensus theories and subjectivism/ hedonism in his lectures on ethics 1920–1924

In the first volume of his LI Husserl attacks psychologism in terms of the (RR)-strategy. One important aspect of this attack is the rejection of attempts to relativize the notion of truth (cf. Husserl 1900a, § 36). Correspondingly, Husserl rejects the idea that the justificatory force of reasons could be considered relative to certain persons or groups of persons. "A radical relativist holds that an action is right *for someone* if and only if it is justified by the reasons available to that person (or, even more radically, if and only if that person thinks it is right)" (Levy 2002, 3).

Although phenomenologists deny that the justificatory force of reasons could be *dependent on* certain individuals, groups or species, they are nonetheless ready to acknowledge the fact that, from the perspective of individual persons or groups, reasons may be *available* only within a limited range. Limitation is not at all an exceptional case. We usually expect limitations of this sort, for instance, in case of complicated mathematical proofs and a variety of scientific theories, yet also in case of art works and art criticism and on many other occasions as well. However, the *availability* of reasons, which depends on a variety of contingent (historical, sociological, biological, cultural) conditions, does not touch upon (i.e., does not modify) their *validity*. At this juncture, the meta-level issues mentioned previously again impose themselves. Relativists deny that availability conditions and validity claims could be neatly separated. They argue that differences with regard to the availability of reasons do have an impact on the validity claims conveyed by means of these very reasons. Whether it is reasonable to stick to the distinction between (subject-dependent) availability of reasons and (subject-independent) validity of reasons is a fundamental matter: ultimately it concerns the question whether we can defend the *quid facti / quid iuris* distinction. Again, we find ourselves pushed back towards two crucial issues: (i) Relativists and non-relativists seem to be engaged in deep disagreements. (ii) It is therefore hard to see how their dispute could be decided in a sensible manner without committing a *petitio principii*.

Husserl focuses on the distinction between the availability of reasons and their justificatory force ("validity") for instance in his reflections on ethics in the 1920s. These reflections also show that, more than two decades after the publication of LI, Husserl still sticks to his earlier view that relativism and skepticism are tightly connected. The former easily leads into the latter. The reasons for this become obvious as soon as we ask for the consequences of abandoning the distinction between the availability of reasons and their validity. If this distinction gets blurred or abandoned, we have to concede that the validity of any statement must be considered indeterminate and indeterminable. This is because we then cannot know (and can *never* know) in advance whether and to what extent reasons speaking in favor of a given statement will be available to varying populations of epistemic subjects (individual persons, groups, larger communities). And if we accept this as true, we then have to accept the skeptical consequence that the validity conditions of no one statement can ever be correctly, and once and for all, specified.

In his lectures on ethics (1920–24) Husserl introduces a conceptual distinction that clearly marks his opposition to the preceding skeptical relativism: *general validity qua factual consensus vs. general validity as grounded in (moral) insight* ("Allgemeingeltung als faktischer Konsens" vs. "Allgemeingültigkeit als in ethischer Vernunfteinsicht gründend"). He elucidates the importance of this distinction when discussing hedonism.

> Undoubtedly the idea of a right – as something that in some way ought to be – implies general validity. From the general prevalence of hedonistic aspirations one might infer that human beings generally regard pleasure as a good and proper goal. But if we infer from the general prevalence of this aspiration that it is generally valid, are we then not falling prey to the equivocation of the words 'prevalence' (Geltung) and 'validity' (Gültigkeit) . . . ? The questio iuris can be raised concerning aspirations, but in a strictly analogous sense it can also be raised with respect to judgements, assertions, and beliefs. Isn't it obvious that the general prevalence of belief amongst human beings – never mind whether it is strictly prevalent or prevalent only roughly – does not tell us anything about the justification of the belief, its truth, or its theoretical validity in and of itself?
>
> *(Husserl 2004, 42, translation by the author)*[7]

This statement is in full agreement with Husserl's earlier line of reasoning in the LI. For example, in § 39 of his *Prolegomena* Husserl argues as follows.

> Universal likeness of content, and constant functional laws of nature which regulate the production of such content, do not constitute a genuine universal validity, which rather rests upon ideality. If all creatures of a genus are constitutionally compelled to judge alike, they are in empirical agreement, but, in the ideal sense demanded by a supra-empirical logic, there might as well have been disagreement as agreement. To define truth in terms of a community of nature is to abandon its notion.
>
> *(Husserl 1900a, 87)*

The same goes for the concept of justification. Relating to the distinction between a (complete or incomplete) actual empirical agreement of judgements, on the one hand, and ideal validity, on the other, it does not make a difference whether we assume this empirical agreement to be reached by means of a common brain-physiological functioning, or whether we consider the agreement as following from specific patterns of social interaction and communication.

For Husserl (and other phenomenologists) the ultimate significance of the preceding distinction between large-scale actual agreement (*Allgemeinheit*) and general validity (*Allgemeingültigkeit*) is to ward off value subjectivism. Hedonism is a peculiar manifestation of value subjectivism.

Actual agreement among those who are members of a group or society (e.g., with regard to average appearance of pleasure and pain) does not suffice for giving warrant to the statement that something is true or good. What is required rather is to gain reasonable insight, an insight that possibly goes beyond or contradicts actual agreement. Acknowledging reason's force and gradual independence of the varying circumstances in which it actually is, or can become, enacted, means that its normativity is not grounded in and, consequently, is not restricted by the willingness to agree with others. Reason's normativity rather is owing to a reasonable subject's ability to commit herself to the idea of truth. A subject's motivation to enter this commitment does not depend on her taking account of the fact that many others actually do so as well. (Arguing along these lines, does of course not deny that being committed to the idea of truth might have benefits in terms of human sociability. Yet it refrains from considering the former as instrumental with regard to the latter. Reason's significance goes beyond enhancing social harmony.) This basic situation, according to Husserl, is the very same with regard to epistemic, ethical-practical and axiological issues. Actual feelings of pleasure do not render the object referred to positively valuable. Rather, it is the object's own value which renders the subject's positive responses directed towards it appropriate or inappropriate.

Though the topic of hedonism appears to be quite special, Husserl's treatment of it shows how it fits into his overall criticism of relativism and naturalism. Hedonism, as conceived here, denies the existence of an objective value that manifests itself in human actions, achievements and characters (and non-human natural appearances as well). Instead, for the hedonist, value depends on whether and to what extent something is actually appreciated by the evaluating subject. The subject's appreciation depends on a variety of factors that must be considered external to the valuable object assessed. These factors vary from person to person, from group to group, and from culture to culture.

Husserl refutes hedonism by pointing out that it would make all meaningful talk of values impossible. Any object or action could be said to manifest whatever value due to the psychological and social dispositions of those responding to them. And thus, our ability to grasp the evaluative aspects of actions and objects would be mysterious and useless. Accordingly, the value-related aspect of reality would remain inaccessible and inscrutable. Any discussion about the manifestation of different values in different objects and actions would have to be considered futile. In other words, hedonism leads to a global skepticism concerning the evaluative aspects of reality.

To make matters worse, let us leave aside the argument from pragmatic self-refutation and suppose that human beings indeed were able to live in conformity with such a hedonistic attitude. It would still remain mysterious why we should adopt this attitude or how we ever could feel motivated to do so. If it turned out that global value skepticism, strictly speaking, was irrefutable, it may still be the case that nothing speaks in its favor. That said, again it is true that showing the inadequacies of value skepticism is not tantamount to proving value absolutism (cf. Spiegelberg 1935, 15, 58, 90f.). Ultimately, it may therefore turn out that the theoretical interventions of both relativists and non-relativists share a moment of arbitrariness due to the lack of any conclusive proof on both sides (cf. Mohanty 1997). Still, on this viewpoint it is not true that the lack of any conclusive proof of absolute values *in terms of mind-independently or subject-independently existing value entities* (cf. Mackie 1977, ch. 1) forces us to consider the relativist outlook as just as sensible as the non-relativist. The crucial step for phenomenologists is to safeguard objectivity. As explained previously, doing so does not follow the idea of a rigidly unified, homogeneous notion of objectivity. Neither does it hinge upon the defense of an axiological absolutism if this meant to assert a separate ontological positing of values. A phenomenological framework allows for non-relativist position-takings that do not usher in strong programs of anti-relativism in terms of stipulating ontologically independent values (Spiegelberg 1935, 90–100; Rinofner-Kreidl 2016).

What is crucial from a phenomenological point of view is that evaluative, that is, axiological and practical experiences (or acts), do not represent black boxes or merely natural events we passively undergo like shudders of fever, feelings of hunger, pain or bodily joy. Evaluative experiences can be analyzed and need to be analyzed. It is necessary to enter descriptive analysis. (Note that this mirrors the theoretical move we stated previously with regard to Husserl's transition from the first volume of his LI to the second.)

5. Conclusion: the (RR)/(UR)/(BR) – big picture: phenomenology's overcoming of relativism

The (UR)-strategy ultimately depends on acknowledging given evidences. At the same time, phenomenologists recognize that evidences are not infallible. In many cases there can be rational disagreement on the range of a given evidence and the precise determination of its justificatory force under given circumstances. This is because specific types of evidence pertain to specific types of objects. Hence the (UR)-strategy leaves room for dispute. Still, phenomenologists do not conflate the leeway of rational disagreement in particular cases with a general suspicion towards evidence. At this point it becomes palpable that relativists and non-relativists do not just disagree about given evidences in single cases or a multitude thereof. They also differ on the conditions and consequences of disagreement and the requirements of a possible agreement. In other words, they disagree about the significance of their disagreement. While relativists stress that no instance of "evidential giveness" could ever deliver *objective* warrant for any relations of ideas or judgements, phenomenological non-relativists vehemently deny this to be the case. They insist on the justificatory force of evidence. What is more, phenomenologists take it that here the relativist position is back in the line of fire of the (RR). Rejecting the principle of evidence amounts to abandoning a necessary condition of any possible theory whatever (i.e., that evident and non-evident judgements can be clearly distinguished). Positively stated, the logical connection between (UR) and (RR) is obvious: the principle of evidence functions as a necessary presupposition of any possible phenomenological-descriptive analysis by means of which an (UR) strategy can be carried out.

Finally, (BR) indicates that the descriptive practice, which comes to the fore when the (UR)-strategy prevails, turns into an object of methodological concern. This means, for instance, that one (i) acknowledges the need for a sophisticated theory of evidence and intuition within an overall phenomenological understanding of these and other components of a phenomenological theory. Equally important with regard to the overall aim of offering a viable alternative to relativistic theories is to (ii) get interested in meta-level issues like the distribution of burdens of proof, especially when the opponents are stuck in deep disagreements. Part of this exercise should be to reflect upon one's own presuppositions or otherwise limited grasp of ideas and problems (e. g. by figuring out the possible pitfalls of accurate descriptions). Obviously, (i) and (ii) are meant to strengthen the immanent coherence and plausibility of a phenomenological theory.

Having said so much, it should be clear why (RR), (UR), and (BR) are logically inseparable and practically intertwined strategies. From a phenomenological point of view, (UR) should take the lead although pondering its theoretical foundations brings us back to (RR) and the Husserlian framework of pure logic. The reason for the predominance of (UR) becomes obvious as soon as we realize that the (RR)-strategy, if isolated from (UR) and (BR), could also be adopted by logical objectivists like Bernard Bolzano or Gottlob Frege who do not have any interest in establishing a phenomenological investigation of logical *acts* (or any other types of acts as well). The logical idealism of Husserl's LI becomes an integral part of his later transcendental idealism. Yet neither in LI nor later on does Husserl act as a logical absolutist if this indicates a lack of genuine interest in the possible multitude of realizations of ideas and the pertaining types of experiences in which these ideas are grasped, judged, compared, and so on. As argued previously, the common ground of a variety of phenomenological responses to relativism is to show how unity and multitude or unity and difference function as inseparably linked up aspects in any possible (e.g., logical or evaluative) experience.

Acknowledgements

I am grateful to Aaron Ben Ze'ev and Martin Kusch who read a more extensive draft and made very accurate suggestions on how to reorganize the material. I would also like to thank Martin for his linguistic corrections.

Notes

1 Cf. Hollis and Lukes (1982); Gowans (2004); Lukes (2008); Krausz (2010); Hales (2011); Baghramian and Carter (2018). Baghramian (2004, 277 f.) warns against an automatic equating of relativism with subjectivism.
2 See a similar reminder in recent debates, favouring value pluralism as an additional option: "that relativism and reactionary absolutism are not the only ethical options available to us" (Baghramian 2004, 277).
3 For a characterization of historical relativism in terms of debunking, see Mandelbaum (1967, 19, 31–37). A thorough discussion of debunking arguments is presented in Nagel (1997). The manner in which Nagel rebuts debunking arguments largely coincides with Husserl's line of reasoning. See e.g. Rinofner-Kreidl (2011).
4 Note that the English translation does not stick to the arrangement of materials in the German original. Husserl (1900a) does not only include the *Prolegomena*, but also the first two investigations of the second volume (for the others cf. Husserl 1900b). Any reference to the "first/second volume of LI" in the text, however, follows the original ordering (vol. 1: Prolegomena; vol. 2: six investigations).
5 As is well known, Husserl often mentions Bernard Bolzano's theory of science (*Wissenschaftslogik*, 1837) and Rudolf Hermann Lotze's *Logic* (1843/1874) as important forerunners of this idea.
6 Second investigation.
7 Cf. Scheler (2007, 102–106); Soffer (1991, 129, 132–136, 187f.). It is illuminating that even those phenomenologists who emphasize the indispensable role of moral consensus ultimately introduce normative aspects that cannot be reduced to consensus. "On the one hand, I submit that the individual agent's moral insight and 'conscience' (including its wider meaning of 'sense of obligation') are not self-subsistent primary data but subject to a wide-ranging social consensus and indeed representative of it. On the other hand, I maintain that all collective states of mind, moods and pressures do not by any means deserve the name of moral consensus; that, on the contrary, moral consensus needs to be sifted out and ascertained by the critical tool of an independent and non-consensual once-for-all *distinction* between *moral and non-moral principles*, concepts or types of experience" (Kolnai 1970, 99).

References

Baghramian, M. (2004), *Relativism*, London and New York: Routledge.
Baghramian, M. and J. A. Carter (2018), "Relativism," in *The Stanford Encyclopedia of Philosophy*, Winter 2018 edition, edited by E. N. Zalta, https://plato.stanford.edu/archives/win2018/entries/relativism/.
Bolzano, B. (1837), *Theory of Science*, Oxford: Oxford University Press, 2014.
Gowans, C. W. (2004), "Moral Relativism," in *The Stanford Encyclopedia of Philosophy*, Summer 2019 edition, edited by E. N. Zalta, https://plato.stanford.edu/archives/win2018/entries/relativism/.
Hales, S. D. (ed.) (2011), *A Companion to Relativism*, Chichester, West Sussex: Wiley-Blackwell.
Hollis, M. and S. Lukes (eds.) (1982), *Rationality and Relativism*, Cambridge, MA: MIT Press.

Husserl, E. (1900a), *Logical Investigations*, vol. I, translated by J. N. Findlay from the Second German edition of *Logische Untersuchungen*, London and New York: Routledge, 1970.

———. (1900b), *Logical Investigations*, vol. II, translated by J. N. Findlay from the Second German edition of *Logische Untersuchungen*, London and New York: Routledge, 1970.

———. (1929), *Formal and Transcendental Logic*, translated by D. Cairns, The Hague: Martinus Nijhoff, 1969.

———. (2002a), "Philosophy as Rigorous Science," translated by M. Brainard, In *The New Yearbook for Phenomenology and Phenomenological Philosophy*, edited by B. Hopkins and S. Crowell, London and New York: Routledge, 249–295.

———. (2002b), *Logische Untersuchungen. Ergänzungsband. Erster Teil: Entwürfe zur Umarbeitung der VI. Untersuchung und zur Vorrede für die Neuauflage der Logischen Untersuchungen (Sommer 1913)*, edited by U. Melle, Dordrecht, Boston and London: Kluwer Academic Publishers (Husserliana XX/I).

———. (2004), *Einleitung in die Ethik. Vorlesungen Sommersemester 1920 und 1924*, edited by H. Peucker, Dordrecht, Boston and London: Kluwer Academic Publishers (Husserliana XXXVII).

Kolnai, A. (1970), "Moral Consensus," in his *Ethics, Value, and Reality: Selected Papers of Aurel Kolnai*, New York: Routledge, 2008, 144–164.

Krausz, M. (ed.) (2010), *Relativism: A Contemporary Anthology*, New York: Columbia University Press.

Kusch, M. (2015), "Psychologism," in *The Stanford Encyclopedia of Philosophy*, Winter 2015 edition, edited by E. N. Zalta, https://plato.stanford.edu/archives/win2018/entries/relativism/.

Levy, N. (2002), *Moral Relativism: A Short Introduction*, Oxford: Oneworld Publications.

Lotze, R. H. (1843/1874), *Logic*, Clarendon Press, 1884.

Lukes, S. (2008), *Moral Relativism*, New York: Picador.

Mackie, J. L. (1977), *Ethics. Inventing Right and Wrong*, Harmondsworth: Penguin Books.

Mandelbaum, M. (1967), *The Problem of Historical Knowledge: An Answer to Relativism*, New York, Evanston and London: Harper Torchbooks.

Metzger, A. (1966), *Phänomenologie und Metaphysik. Das Problem des Relativismus und seiner Überwindung*, Pfullingen: Verlag Günther Neske.

Mohanty, J. N. (1997), "Husserl on Relativism in the Late Manuscripts," in *Husserl in Contemporary Context: Prospects and Projects for Phenomenology*, edited by B. C. Hopkins, Dordrecht, Boston and London: Kluwer Academic Publishers, 181–188.

Nagel, T. (1997), *The Last Word*, New York and Oxford: Oxford University Press.

Rinofner-Kreidl, S. (2004), "What is Wrong with Naturalizing Epistemology? A Phenomenologist's Reply," in *Husserl and the Sciences: Selected Perspectives*, edited by Richard Feist, Ottawa: Ottawa University Press, 41–68.

———. (2011), "Motive, Gründe und Entscheidungen in Husserls intentionaler Handlungstheorie," in *Die Aktualität Husserls*, edited by V. Mayer, C. Erhard and M. Scherini, Freiburg und München: Karl Alber, 232–277.

———. (2016), "Disenchanting the Fact/Value Dichotomy: A Critique of Felix Kaufmann's Views on Value and Social Reality," in *The Phenomenological Approach to Social Reality: History, Concepts, Problems*, edited by A. Salice and H. B. Schmid, Switzerland: Springer, 317–348.

Scheler, M. (2007), *Ressentiment*, new edition, translated by L. B. Coser and W. W. Holdheim, Introduction by M. S. Frings, Milwaukee, WI: Marquette University Press.

———. (1973), *Formalism in Ethics and Non-formal Ethics of Values: A New Attempt Toward the Foundation of an Ethical Personalism*, translated by M. S. Frings and R. L. Funk, Evanston, IL: Northwestern University Press.

———. (1957), "Phänomenologie und Erkenntnistheorie," in *Schriften aus dem Nachlass, Band I: Zur Ethik und Erkenntnislehre*, Bern: Francke Verlag, 379–430 (GW Band 10).

Schmitt, D. (2018), *Das Selbstaufhebungsargument. Der Relativismus in der gegenwärtigen philosophischen Debatte*, Berlin and Boston: Walter de Gruyter.

Soffer, Gail (1991), *Husserl and the Question of Relativism*, Dordrecht, Boston and London: Kluwer Academic Publishers.

Spiegelberg, H. (1935), *Antirelativismus*, Zürich und Leipzig: Max Niehans Verlag.

11

HISTORY, DIALOGUE, AND FEELING

Perspectives on hermeneutic relativism

Kristin Gjesdal

1. Introduction

In *Truth and Method* (1960), Hans-Georg Gadamer formulates a response to nineteenth-century hermeneuticians such as Friedrich Schleiermacher and Wilhelm Dilthey. In his discussions of Schleiermacher and Dilthey, Gadamer agrees with their fundamental postulate that history involves a plurality of perspectives. He also agrees that this creates a challenge for hermeneutic work. He disagrees, though, with their response to this challenge. For Schleiermacher and Dilthey, the plurality of perspectives creates a need for a hermeneutic methodology. This methodology, however, is unlike that of the natural sciences in that it involves an element of empathy or feeling. In Gadamer's view, this is a serious mistake: no hermeneutic methodology, be it empathetic or not, can get us beyond the threat of relativism posed by the various historical perspectives. Instead, we need to realize that the different horizons all contribute to, and emerge as part of, the constitutive totality of a shared tradition that establishes the ultimate condition of possibility for intelligibility and meaning. Gadamer supports these arguments – on the variety of perspectives and the unity of tradition – by reference to a notion of dialogical understanding. By taking on a charitable, conversational attitude and seeing the works of the past as a potential source of truth, understanding, and *Bildung*, the individual interpreter relates to the past by keeping alive its meaningful material and, at the same time, applying it in ever new contexts of interpretation. This continuing process of application is, for Gadamer, what tradition ultimately is. Understood in this way, hermeneutic engagement with tradition facilitates self-understanding as well as understanding of a historical-cultural other.

In the wake of *Truth and Method*, critics of Gadamer's work have pointed out that the tradition does not necessarily facilitate meaningful conversations of this kind, but can also shelter structures that systematically distort our understanding of ourselves and of others.[1] This chapter is motivated by a related, yet different worry. I address Gadamer's reconstruction of phenomenology, especially the work of Martin Heidegger, as an effort to overcome traditional epistemology. In his orientation towards Heidegger, Gadamer overlooks a range of positions that are of relevance to hermeneutic thought. In particular, he overlooks a generation of early phenomenologists, whose contributions, gesturing back to Schleiermacher and Dilthey, plead for a model of intersubjectivity as rooted in self-feeling, human embodiment, and an empathic orientation towards the hermeneutic other. In this context, the work of Edith Stein is particularly relevant.

For Stein, the relativity of individual perspectives is not overcome by reference to a shared context of historical meaning and culture. Instead, her models offer an analysis of the phenomenological conditions of the self's relation to itself and others that centers on empathy.

2. Conversation and understanding

In articulating his philosophical hermeneutics, Gadamer leans on Heidegger – Heidegger's early as well as later work. This also applies to Gadamer's criticism of the understanding of nineteenth-century philosophy and what he takes to be its epistemological approach to hermeneutics. In *Being and Time*, Heidegger sketches his understanding of the modern period as haunted by a narrow epistemic focus. As Heidegger argues, philosophers, from René Descartes, via Wilhelm Dilthey, to Edmund Husserl, have sought to cast philosophy in light of the (quasi-scientific) ideal of a clear-cut methodology that serves as a reference point for knowledge and objectivity.[2]

In the period leading up to and including the publication of *Being and Time*, Heidegger draws a contrast between a phenomenology that is still indebted to this tradition and his own post-Husserlian hermeneutic philosophy.[3] He seeks to extend the scope of hermeneutics from being concerned with "correct" or "objective" interpretation of texts and meaningful material to being an existential-phenomenological analysis of the human being as a self-interpreting animal.[4] It is the turn to a phenomenological analysis of the human being-in-the-world that can help us to get beyond what Heidegger views as overly narrow epistemological approaches. Once we see the human being as an understanding and interpreting being-in-the-world, we realize that the world itself is disclosed as meaningful and this meaning, in turn, is historically mediated. As Heidegger argues, there is no point of reference that escapes the structures of historical practices and meaning-ascription.

Gadamer supports the young Heidegger's hermeneutic ambitions.[5] Hermeneutics, he argues, is an attempt to articulate a sound conception of human understanding – one that takes as its starting point the idea that the world is disclosed to us as understanding, interpretative, and historical beings.[6] For Gadamer, though, the full realization of this point requires a step beyond Heidegger. In particular, it requires a study of the constitution of the human self, as a nexus of meaning, with reference to a *thou*. This is the gist of Gadamer's early work on Socratic-Platonic dialogue (1991, 11, 51–65).

Right from his work in the 1930s (i.e., three decades before the publication of *Truth and Method*), Gadamer views human reason and understanding as dialogically constituted.[7] He sees the dialogical structure of reason as an antidote to relativism. It is through a dialogical engagement, secured by a shared orientation towards truth, that we can overcome the limitations of our own (relative) perspectives through a fusion, thus also an expansion, of horizons. For Gadamer, tradition, too, works through such fusions of horizons. Indeed, this is ultimately what tradition is: It is an ongoing, ever-renewing process through which the interpreter engages with the perspectives disclosed by the great works of the canon, thus also realizing their historical meaning-potential in ever new contexts of interpretation. By actively (dialogically) furthering a tradition, an interpreter learns about her culture, the world, and about herself, as a member of the tradition of which the work is part.

Like Heidegger, Gadamer seeks to illuminate a fundamental dimension of meaning that is prior to the individual interpreter and his or her self-understanding and, in a certain sense, also prior to the relativity of a given historical perspective (since, in his view, a particular historical perspective emerges against the background of a unifying tradition). Unlike Heidegger, however, Gadamer does not refer to a metaphysics of Being (*metaphysica generalis*, as Heidegger has it in *Being and Time*), but seeks, instead, to spell out our dialogical application of, relation to, and understanding of ourselves in light of the larger, ever-evolving context of tradition. The tradition provides an experiential horizon of meaning. Hermeneutic labor makes explicit the implicit background of this horizon. The meaning of traditional texts does not rest with an original context of production (this view is, for Gadamer, a case of bad historicism). In applying the resources of the past within a contemporary context, we do not relate to a meaning context that is foreign to us (i.e., that is constitutively other and, as such, calling for a method of understanding). Instead, we actively and productively own up to and take over that which is our own.

Nineteenth-century philosophers of interpretation had worried that the historical mediation of thought generates a relativist challenge. They had responded to this challenge by articulating methodological guidelines through which an interpreter can maximize her interpretative skills and sensitivity. Gadamer, by contrast, turns the argument around: he insists that historicity is not only the source of, but also the solution to, the threat of relativism. The tradition provides a continuum against which any differences in perspectives can be (dialogically) negotiated and overcome.

Can, then, this appeal to a continuum of tradition be defended? Do we primarily understand ourselves and each other in light of our participation in the continuous mediation of the tradition? And is tradition genuinely dialogical and conducive to self-understanding for everyone? Or does it give priority to some individuals while potentially hurting or stunting the self-realization of others? I would like to ask these questions by way of a turn to a pre-Heideggerian phenomenological positions that Gadamer does not consider. The position I have in mind is that of Edith Stein. Stein does not take tradition to provide a unity that is prior to the I-thou relation, i.e., as serving to bolster against hermeneutic relativism. In her view, tradition can impede our relation to ourselves and others. Stein insist that individual plurality and differences in perspectives cannot be led back to a shared meaning-content of a "we." As a result, an interpreter must seek to do justice to the irreducibly individual aspects of ourselves and others. The hermeneutic challenge, in other words, is not necessarily to overcome the relativity of individual perspectives by reference to a shared "we." Instead, she draws on the resources found in the early hermeneutic movement that Gadamer sought to overcome: the notion of an immediate human feeling or sympathy, i.e., an appeal to an (inter)subjective faculty that is culturally neutral and more fundamental than the meaning-structures handed down to us through the tradition.

3. Empathy revisited

Edith Stein is an early phenomenologist whose work can offer a perspective on Gadamer's response to hermeneutic relativism (the plurality of perspectives in history and in between individuals). Stein has not received the attention she deserves, especially not in the context of hermeneutics. Of particular importance is the return to the idea of *Einfühlung*. Stein develops this notion in her 1917 *On the Problem of Empathy*.[8] The idea of feeling-with, or putting oneself in the place of others, is key to the hermeneutic tradition (Herder, Schleiermacher, and Dilthey) that we find prior to Heidegger's ontological turn.

Stein distinguishes between empathy and sympathy, or fellow-feeling (Stein 1917, 15). Sympathy is abstract and relates to the humanity of the other (as a generally shared category). For her purpose, what matters is our directedness towards the other as a living, individual body (*Leib*). The living body (*Leib*) is different from that of non-human physical objects (*Körper*). It is the ability to direct ourselves towards the other as a living body that is at the heart of Stein's notion of *Einfühlung*.

Just as the self is characterized by its intentionality, so the other, too, is a nexus from which the world is experienced. This other is concrete: she senses, thinks, feels, and wills (1917, 5). She is part of the world. Yet unlike non-human parts of the world, she is a part of the world to whom, just as in my own case, the world is disclosed. My I, too, is disclosed to her as part of her world. She is intentionally oriented towards the world, and her intentionality is that of an embodied being. The living body is, for Stein, an I. As she puts it, soul and living body co-constitute the psycho-physical individual (1917, 50).

Stein centers on how a living, embodied I can adequately relate to and understand other living and embodied I's. Her 1917 study seeks to describe empathy phenomenologically as that which, at a fundamental level, opens us up to others (1917, 22). Stein does not, in other words, deny that we relate to others in an array of different ways. Her point, rather, is that empathy discloses the other to the I as, precisely, a living, embodied, and experiencing self, i.e., as something we can relate to or, rather, as somebody with whom we can empathize. This relation establishes a basic bond or commitment to the other, which, in turn, can give rise to an infinite spectrum of different I-thou relations.

With respect to hermeneutics and relativism, it is of importance that Stein claims that empathy discloses a sense of the other that retains her individuality, and that individuality, in turn, is co-constituted with embodiment. Unlike Theodor Lipps, who had argued that empathy represents the experience of joining the other in a higher unity (thus offering a model that, albeit psychologically grounded, structurally anticipates Gadamer's appeal to the tradition), Stein maintains that empathy retains the concrete, physical distinctness of self and other. Indeed, it is the very fact that the other is embodied, and thus discloses herself as an individual, that enables empathy (1917, 11). If I, say, empathize with the sorrow of a friend, this is not a question of me simply (re)experiencing her sorrow (thus being one with her), but rather of me being sad because I empathize with how she, as an (other) individual, is sorrowful. Likewise with joy or pleasure: in empathically relating to the other's joyfulness, I do not make it mine, but I take joy in her joy (1917, 16–17). For Stein, this leads to a particular enrichment of one's own experience (1917, 18). Not only do we learn about the world, but empathy also helps us understand ourselves in that we get to experience the boundaries of our selves and, in Stein's language, what we are and what we are not.

For Stein, empathy is a way of opening oneself to the other without, for that reason, sublating individuality into something shared. This makes empathy different from re-experience, but also from the way I relate to others through imitation, association, or analogy (1917, 24ff). In empathy, the other does not stand forth as an individual because I relate to her. Rather, I relate to her because she presents herself as an embodied individual (1917, 38). The other's physical appearance is therefore not a symbol of who she is (1917, 61, see also 76–79). Instead, the other is given in and through her physical presence.

For Stein, empathy is rooted in what she calls the "con-primordiality of the other's body." As such, empathy does not require a shared context or continuum of meaning. As a matter of fact, empathy, for Stein, extends beyond the realm of human meaning to animals and plants, i.e., parts of nature that are not self-conscious. Further, we can empathize with the other's bodily states even though they are not conscious states. I can empathize with an other feeling tired, sluggish, energetic, and so on. These are not states that are exclusive to self-conscious beings, nor are they mediated by tradition. They are, rather, physically modified expressions of life. Like other phenomenologists of her generation – including Else Voigtländer, whose early work on self-feeling is overshadowed by her later political activity and affiliations[9] – Stein allows the insights of vitalism to resonate in her view, yet unlike Voigtländer, she does base her position in Nietzschean psychology. Further, even though the very notion of *Einfühlung* draws on the full capacities of the human being and relates to it as one, Stein maintains a dualist position on the mind-body relationship (1917, 91).

Stein concludes her study of empathy with a discussion of Dilthey's hermeneutics. In particular, she turns to his argument about the inseparability of the domains of nature and normativity in the cultural-historical sciences. Unlike Dilthey, Stein insists on the difference between causality (rational lawfulness, as she puts it) and value (1917, 113–114). She suggests that Dilthey is right in seeing empathy as foundational to our understanding of the human sciences. Yet, and related to the preceding point, she argues that Dilthey misunderstands the relationship between types and tokens (in this case: individuals) in the human sciences. For Stein, in seeking to understand others, we do not see them as realizing a general type (e.g., that of the nineteenth-century intellectual or twentieth-century phenomenologist). Rather, we empathize with others as individuals.

The extent to which I understand an other – how much of his or her experiential structure I can bring to my fulfilling experience (1917, 115) – depends on who I am. While I can understand things I have not myself experienced (i.e., the joy of winning the lottery), I cannot understand that which conflicts with my own experiential structure (1917, 115). This, however, does not imply that we should take our own experience as a standard. If so, we end up, as Stein puts it, being locked into "the prison of our individuality." In such a state, others "become riddles for us, or still worse, we remodel them into our image and so falsify historical truth" (1917, 116). For Stein, we encounter others – be they present or past – as concrete individuals whose perspective is shaped in interaction with an actual, historical world. Yet, it is not as representatives of a time or a culture, but as concrete individuals that others are given to me in and through my empathic directedness. For Stein, the capacity for empathy enables understanding and interpretation, not the other way around.

As Stein argues, we can understand and misunderstand others. Our ability to understand an other might depend on the participation in a shared meaning or tradition. However, what makes such understanding possible – empathy – cannot, as such, be a product of this kind of meaning (such an argument would be circular). Hence, in our primary – empathetic – relation to others, a shared tradition does not play in. It is simply by presenting me with a living body that the other calls on my empathy. Thus empathy bypasses factors such as shared education, tradition, culture, ethnicity, and language. I simply empathize with the other *qua* another individual.

This dimension of Stein's thought, while centering on the relationship between an I and another, shows how the appeal to a shared tradition masks the challenge of relativism – solving it, however, it does not. There is always the risk that we, by placing the other in a shared tradition or context, we overlook their individual point of view and fail to empathize accordingly.

4. Concluding thoughts

Phenomenology develops in dialogue with the larger, hermeneutic tradition. The early phenomenologists had referred back to the resources of nineteenth-century theories of interpretation, especially their appeal to feeling and empathy. Later on, with Gadamer, hermeneutic feeds off Heidegger's effort to overcome Dilthey and larger nineteenth-century lineage in hermeneutic thought. From the very beginning, however, the phenomenological tradition entails a number of strategies and attempts to grapple with the issues of self, other, and intersubjectivity – *and* the related problem of hermeneutic relativism. It includes a series of competing positions on how an I relates to itself and the minds and utterances of others. While Gadamer, throughout his work, seeks to dissolve the problem of relativism through an orientation toward the shared backdrop of tradition, early phenomenologists like Stein, by furthering the phenomenology of embodied and affective life that had been initiated by Herder, Schleiermacher, and Dilthey, offer models that allow us to celebrate individuality as a key concept in hermeneutic thought.

Notes

1 These arguments were developed, in particular, by Jürgen Habermas and Karl Otto Apel. For a discussion of this point, see Warnke (1987). See also Warnke (2019).
2 Heidegger (1927, 94–96). For the notion of a phenomenological "destruction," as it goes back to Heidegger's work in 1919 and, ultimately, to Husserl's phenomenology, see Kisiel (1993, 493–494).
3 See for example the lectures from the Summer Semester of 1923 (Heidegger 1999).
4 Hermeneutics is, as he puts it in 1923, "the unified manner of the engaging, approaching, accessing, interrogating, and explicating of facticity" (1999, 6). This, further, is contrasted with "the modern meaning" of hermeneutics in that "in no sense does it have the meaning of such a broadly conceived doctrine *about* interpretation" (1999, 11). In the appendix to these lectures, hermeneutics is directly linked to the activity of "historical destruction," which is pitched as "a more radical possibility" (1999, 83; see also 59–60, 81).
5 Gadamer discusses Heidegger's influence in a number of places. For the importance of the early lectures, including the 1923 Aristotle lectures, see Gadamer (1997, 7–9). As he presents it in *Truth and Method*, Heidegger "burst asunder the whole subjectivism of modern philosophy" (Gadamer 1994, 257). Gadamer, further, speaks of Heidegger's philosophy as a "completely fresh beginning" (1994, 258).
6 In a 1963 essay, Gadamer puts it as follows: "Today the style [of Heidegger's *Being and Time*] of those 'years of decision' has lost its magic, but the task has remained the same, namely, to preserve within an increasingly technical age and its anti-historical ideal the great heritage of Western thought that phenomenology and existential philosophy had appropriated with a new passion" (1977, 141).
7 See for example Gadamer (1991, 44).
8 Stein discusses her approach to the human sciences in particular in Stein (1922).
9 Else Voigtländer (1882–1946) was the first woman phenomenologist to earn a PhD. She did so in 1909, and in 1910 she published *Vom Selbstgefühl* (see Voigtländer 1910). Voigtländer had worked with Theodor Lipps and Alexander Pfänder in Munich. She initially situated herself within the psychological branch of the early phenomenological movement and developed an interest in psychoanalysis, though her engagement with Freud's work was short-lived and mostly critical (see Voigtländer 1911). Since the Prussian law of 1908 barring women from Habilitation (the virtually indispensable condition for a professorship) was changed only in 1920, Voigtländer was excluded from the start from pursuing a traditional academic carreer. She went to work as a research assistant for Adalbert Gregor (1878–1971), already an established psychologist, and co-authored several works with him. Their works from this period include disturbing discussions of eugenics and race. As a civil servant during the Weimar Republic, Voigtländer worked, starting in 1926, as the director of the women's wing of Waldheim Zuchthaus (Saxony). She retained this position during the Third Reich and was a member of the NSDAP from 1937 onwards. Voigtländer died in 1946. For an overview of (women) contributors to early phenomenology and the larger emphasis, within this movement, on feeling and emotion, see Vendrell Ferran (2008). For a pioneering analysis of Voigtländer's early philosophy, see Vendrell Ferran (2006). Relatively little is known about Voigtländer's biography and much more work needs to be done on it. For a brief but intriguing biographical sketch, see Hackl and Sack (2017). I thank George Heffernan for drawing my attention to Voigtländer's biography, her work in psychology, and her directorship at Waldheim. I thank Gabriele Hackl for her help in shedding a first light on the complicated story of Voigtländer's life and work after she left philosophy, and George Heffernan and Ingrid Vendrell Ferran for an important and on-going exchange about her life and work.

References

Gadamer, H.-G. (1960), *Truth and Method,* translated by J. Weinsheimer and D. G. Marshall, New York: Continuum, 2003.
———. (1977), "The Phenomenological Movement," in *Philosophical Hermeneutics*, edited and translated by D. E. Linge, Berkeley: University of California Press.
———. (1991), *Plato's Dialectical Ethics: Phenomenological Interpretations Relating to the Philebus*, translated by R. M. Wallace, New Haven: Yale University Press.
———. (1994), *Truth and Method*, translated by J. Weinsheimer and D. G. Marshall, New York: Continuum.
———. (1997), "Reflections on My Philosophical Journey," in *The Philosophy of Hans-Georg Gadamer*, edited by L. E. Hahn, Library of Living Philosophers, vol. XXIV, Chicago: Open Court, 3–63.

Hackl, G. and B. Sack (2017), *Das Frauenzuchthaus Waldheim (1933–1945)*, Leipzig: Leipziger Universitätsverlag.

Heidegger, M. (1927), *Being and Time*, translated by J. Stambaugh, Albany: SUNY Press, 1996.

———. (1999), *Ontology: The Hermeneutics of Facticity*, translated by J. van Buren, Bloomington: Indiana University Press.

Kisiel, T. (1993), *The Genesis of Heidegger's Being and Time*, Berkeley: University of California Press.

Stein, E. (1917), *On the Problem of Empathy*, translated by W. Stein, Washington, DC: ICS Publications, 1989.

———. (1922), *Philosophy of Psychology and the Humanities*, translated by M. C. Baseheart and M. Sawicki, Washington, DC: ICS Publications, 2000.

Vendrell Ferran, I. (2006), *Die Emotionen. Gefühle in der realistischen Phänomenologie*, Berlin: Akademie Verlag.

———. (2008), "Emotionen und Sozialität in der frühen Phänomenologie. Über die Möglichkeiten von Frauen in der ersten Phase wissenschaftlicher Schulenbildung," *Feministische Studien* 26: 48–64.

Voigtländer, E. (1910), *Vom Selbstgefühl. Ein Beitrag zur Förderung psychologischen Denkens*, Leipzig: R. Voigtländer Verlag.

———. (1911), "Über die Bedeutung Freuds für die Psychologie," in *Münchener Philosophische Abhandlungen. Theodor Lipps zu seinem sechzigsten Geburtstag gewidmet von Früheren Schülern*, Leipzig: Barth, 294–316.

Warnke, G. (1987), *Gadamer: Hermeneutics, Tradition and Reason*, Stanford: Stanford University Press.

———. (2019), "Hermeneutics and Critical Theory," in *The Cambridge Companion to Hermeneutics*, edited by M. N. Forster and K. Gjesdal, Cambridge: Cambridge University Press, 237–259.

12

RELATIVISM IN THE CONTEXT OF NATIONAL SOCIALISM

Johannes Steizinger

1. The received view: NS as relativism

Relativism and NS are often associated with each other in the philosophical literature. This association is used as a critical argument against both. The weakest form of this argument runs as follows:

> Relativism involves the idea that there are many radically different, yet equally valid, epistemic or moral systems. The "equal validity" idea commits the relativist to a strong form of tolerance even of morally and epistemically highly problematic systems of thought. Hence, relativism does not provide us with the normative resources to criticize irrational views such as Nazi racism. We need a normative universalism to confront racist ideologies.
>
> *(Böhler 1988; Tugendhat 2009; Kellerwessel 2014)*

The different versions of the aforementioned anti-relativist argument share a common core: Their advocates embed their identification of NS with relativism in a broader claim about the nature of philosophy. They argue that philosophy has to be based on reason and that this in turn commits one to absolute truth. Moreover, these authors defend the possibility of objective knowledge about reality and believe in a universal foundation of morality.

The systematic criticism of relativism in the context of NS is often supported by the historical argument that the relativism of post-Hegelian philosophy paralyzed the moral consciousness of German intellectuals during the Weimar Republic. Their inability to mobilize universal moral principles is regarded as a reason for the rise of NS (Apel 1988). This historical argument can even take a stronger form. Some authors argue that the relativism of post-Hegelian philosophy is a prerequisite of Nazi ideology. Here Nazi ideology is classified as a radical kind of relativism that emerges from the general path of German philosophy after Hegel (Böhler 1988; Wolin 2004; Kellerwessel 2014).

The identification of NS with relativism has a long history and is still popular today. The most influential account stems from Georg Lukács (1885–1971) in his polemical treatise *Die Zerstörung der Vernunft* (*The Destruction of Reason*, 1954). The Neo-scholastic Josef de Vries (1898–1989) confronted Nazi philosophers with the charge of relativism already in the 1930s (de

Vries 1935a, 1935b). Many of the critics, past and present, accuse especially historicism and *Lebensphilosophie* (philosophy of life) of having advocated a "dangerous" relativization of truth, knowledge, and values. On this view, the "relativistic nineteenth century" created a philosophical framework that enabled the flourishing of irrational beliefs, arbitrary maxims, and nihilistic attitudes. Ideologies such as NS are regarded as the ultimate step in the "destruction of reason" (Lukács 1954; see also de Vries 1935a, 1935b; Lieber 1966; Apel 1988; Böhler 1988; Wolin 2004).

Recent accounts highlight the destruction of moral rationality by the alleged relativism of Nazi racism. They read Nazi ideology as a biological determinism that attributes mutually exclusive sets of values to the assumed races. The particular values of a race are chosen arbitrarily and are understood only instrumentally: they serve the survival and flourishing of the respective race. The alleged racial hierarchy has no normative foundation and is thus completely arbitrary too. This "extreme relativism" of NS is defined as the opposite to moral rationality and is considered as an attack on philosophy itself (Böhler 1988; Tugendhat 2009; Kellerwessel 2014). Earlier accounts emphasize the opposition of NS to rationality in general. Lukács characterizes Nazi ideology as a modern myth that is nothing more than demagogic and nihilistic propaganda designed to deceive the population. Here NS is portrayed as the consequence of the decay of philosophy that was caused by relativism. Following Lukács, Hans-Joachim Lieber (1923–2012) explicitly defines NS as the "end of philosophy" (Lieber 1966, 93).[1]

There are several reasons why the argument equating NS with relativism is problematic. *First*, it rests on strong background assumptions about the nature of philosophy and morality. Most presentations of the argument simply take absolute standards for granted.

Second, "relativism" is used exclusively in a pejorative sense. And this pejorative use of the concept of relativism makes the historical argument problematic. The historical accounts are uncharitable and not based on the detailed study of existing sources. Historicism and *Lebensphilosophie* are used as mere whipping boys.[2]

Third, the argument equating NS with relativism rests on a poor understanding of Nazi ideology. Recent historical research shows that Nazi ideology can neither be reduced to deceitful propaganda nor to simple biologism. The mere fact that many professional philosophers contributed to Nazi ideology should already make us doubt the equation of National Socialism with "the end of philosophy."

Fourth, and most importantly, most critical anti-relativists do not consider the actual debate about relativism in the context of NS. Since Nazi philosophers were accused of being relativists by their contemporaries such as de Vries (1935a, 1935b), they engaged seriously with the problem of relativism. The actual contributions of Nazi philosophers to this debate reveal their self-understanding and are therefore an important source for defining the relation of NS to relativism. The critical analysis of this engagement also shows the meaning and use of the concept of relativism in the historical context. In the next section, I examine this historical context.

2. The historical debate: NS versus relativism

Nazi intellectuals were confronted with philosophical problems such as relativism because of the comprehensive character of their political claims. NS considered itself as a political revolution that realizes a new image of the human. Recent historical research confirms the self-understanding and contemporary perception of NS as a *weltanschauliche Bewegung* (*ideological movement*; see Kroll 1998; Raphael 2014).

The Nazis adopted the concept *Weltanschauung* (*worldview*) in order to highlight the comprehensive character of their movement. Because of its thoroughgoing political character, the Nazi *Weltanschauung* can be characterized as an ideology. Nazi ideology has to be seen as a set of basic beliefs and convictions which left much scope for interpretation. Although key concepts like race had to be accepted as guidelines of thinking and acting, different interpretations of such ideological core elements coexisted and competed even in the inner circle of Nazi leadership. In short, there was no unified and mandatory ideological system of NS. Nevertheless, Nazi ideology was not a chimera. The "combination of fluidity and flexibility with a set of convictions and core arguments" (Raphael 2014, 74) shows, instead, that a political ideology works best as controlled plurality. Take the example of the concept of race: once you had accepted its key role for understanding whatever phenomenon interested you, you could still engage in heated debates on its meaning and significance. The range of views developed in the writings of NS leaders reached from bluntly biological conceptions (e.g. Darré [1902–1946]) to metaphysical interpretations of race (e.g. Rosenberg [1893–1946]). Such obvious tensions were never removed and created the impression that NS was always in need of further explication.

Philosophers in particular took up the task of elaborating, justifying, and explaining what NS truly is. Many German philosophers put their philosophy into political service. The gesture of general agreement with the political change and the willingness to work in the direction of the leader (*dem Führer entgegenarbeiten*) were even more widespread.[3]

NS and philosophy were linked by the concept of *Weltanschauung*. This vague concept generally connected philosophy and politics in the early twentieth century. The concept referred to comprehensive theories about the world that were meant to guide human actions. There was no sharp distinction between philosophy and *Weltanschauung*.

Weltanschauungsphilosophie (philosophy of worldviews) became the dominant force in German-speaking philosophy during the Weimar Republic. This is demonstrated by the influence of Oswald Spengler's (1880–1936) *Untergang des Abendlandes* (*Decline of the West*, 1918/1922) on public discourse, on political debates, and in academic philosophy (see e.g. Kusch 1995, 227f.). Spengler connected a general theory of culture with a critical diagnosis of its current state. He claimed that Western culture had reached the final stage of cultural life. This state of decline could only be managed by the individual authority of a dictator. Spengler embraced the political idea of Caesarism and regarded Mussolini's fascism as its modern realization (see Felken 2006, 1256f.). This political conservatism was accompanied by a philosophical commitment to relativism.[4] Spengler believed that all forms of human expression were determined by their historical context (Spengler 2006, VII, 31f.). Human history consisted of the life of cultures. Spengler defined cultures as distinct organisms with individual, incommensurable characteristics. He thus combined historicism with cultural relativism.

Other authors referred to the legacy of historicism by taking up Wilhelm Dilthey's theory of worldviews within academic philosophy.[5] Erich Rothacker published a new *Einleitung in die Geisteswissenschaften* (*Introduction to the Human Sciences*, 1920) that was followed by his *Logik und Systematik der Geisteswissenschaften* (*Logic and System of the Human Sciences*, 1926). In this period, Rothacker claimed that the methodological debates about scientific principles reflected fundamental "struggles of life" (1926, 112). He argued that these debates could be reduced to the fundamental differences of metaphysical *Weltanschauungen*. Rothacker regarded *Weltanschauungen* as rooted in decisions about life. They expressed practical demands and ethical postulates that could not be reconciled (1926, 137, 139). Rothacker highlighted the troubling consequences of his understanding of science. On his view, there was no science without presuppositions (*voraussetzungslose Wissenschaft*). Such positions were often accused of being relativistic.

In his treatise from 1926, Rothacker explicitly addressed said problem of relativism. He distinguished between a positive and a negative kind of relativism. Rothacker suggested that "negative relativism" was a kind of skepticism that had to be rejected because of its "decomposing" effects (1926, 147f.). "Positive relativism," on the other hand, simply acknowledged that humans had to choose among different possibilities of life. The choice between a plurality of values was guided by practical interests. Rothacker embraced this pragmatic principle and defined himself as a positive relativist. He regarded this position as a legacy of historicism. He affirmed the historicist's emphasis on historical particularity also in the realm of politics. Rothacker praised cultural particularism and concluded in the last section of his systematic treatise: "Only on this ground, we philosophers are entitled to love our country" (1926, 171). Again, philosophical relativism was connected with political conservativism.[6]

Nazi ideology thus emerged in the context of a political discourse that associated some of its key convictions with relativism. This might lead one to expect that Nazi intellectuals referred to relativism in a positive way. This expectation is, however, not supported by historical evidence. The intellectual discourses of the Nazi era reveal a commitment to anti-relativism: For instance, the popular *Meyers Dictionary* of 1942 defined relativism as an attitude that represents the modern liberal-individualistic perplexity and lack of principles. "Jews" allegedly belonged here because of their disposition and intention to "decompose" society (Meyers Lexikon 1942, 290). Nazi scientists also rejected Einstein's theory of relativity because of its alleged "decomposing" impact on society (see Herbert 2001, 12–14, 213; Danneberg 2013, 74ff.). Nazi accounts of contemporary philosophy defined relativism as a dangerous consequence of the modern spirit for which intellectual tendencies such as individualism, liberalism, historicism, and pragmatism were responsible.[7] They often referred to German-Jewish philosophers such as Georg Simmel as prime examples of relativistic attitudes.[8] In 1936, the historical-philosophical class of the Prussian Academy of Sciences announced a prize question on the topic: "the inner reasons of philosophical relativism and the possibility of its overcoming." The prize question presented relativism as philosophical problem that had to be overcome.[9]

This overview shows that the philosophical, scientific, and public debates during NS presented relativism as a problem. Moreover, anti-relativism featured prominently in antisemitic propaganda. It is thus not surprising that Nazi intellectuals presented their worldview as the overcoming of relativism. The Nazi ideologue Alfred Rosenberg characterized the idea of the "relativity of the universe" as an "illness of our time" that was overcome by the "organic truth of NS" (1938, 694). Nazi philosophers such as Alfred Baeumler or Ernst Krieck were convinced that their position overcame the opposition between absolutism and relativism. They developed argumentative strategies to present NS as a third way in philosophy.[10]

These anti-relativistic convictions of Nazi intellectuals expose the weakness of the received view of NS as relativism. Nazi ideology did not follow the relativistic strand of political conservativism, but adopted an anti-relativistic stance that was widespread in German-speaking philosophy (see, e.g., Kinzel, this volume).

This understanding of relativism shaped an important political debate during Weimar Republic. Anti-relativism was a key motivation for Carl Schmitt's (1888–1985) political conservativism. Schmitt claimed that liberal democracy was rooted in the relativism of modern thinking. He regarded the liberal politics of tolerance as a relativistic principle and identified this constitution with the decline of political authority. For Schmitt, Hans Kelsen (1881–1973) represented the relativistic mind-set of political liberalism (Schmitt 1922, 47–55; see also Schmitt 1931; 1916, 59, 65, 67–70).[11] Schmitt's anti-relativism was explicitly connected with an antisemitic sentiment. Schmitt saw his controversy with Kelsen as a "struggle against the Jewish spirit," his "true enemy" (see Gross 2016).

In what follows, I want to take a closer look at two philosophical approaches to relativism during NS. Bruno Bauch and Rothacker connected their philosophies with NS and engaged with the problem of relativism. Moreover, their philosophical endeavors represent the two main tendencies of German-speaking philosophy before and during NS. Hans Sluga (1993) has convincingly shown that the diverse landscape of early twentieth-century philosophy was divided into two general camps: The "philosophical conservatives," on the one hand, saw their philosophizing as the recovery of a great past. The neo-Kantian Bauch was one of the most influential proponents of a distinctively "German" philosophy that was grounded in the idealist tradition of Kant, Fichte, and Hegel. The "philosophical radicals," on the other hand, demanded a radical new beginning and developed varying proposals for this task. Rothacker reinvented philosophy multiple times during his philosophical career and became a leading proponent of the "philosophical radicals" in the 1930s.

3. Bruno Bauch's objectivist nationalism

The neo-Kantian Bauch was an influential figure in early twentieth-century philosophy. His infamous lecture *Vom Begriff der Nation* (*On the Concept of the Nation*) – given in 1916 and published as an article in 1917 – caused a scandal that divided the philosophical community. The publication of Bauch's nationalist and antisemitic views in *Kant-Studien* (*Kant Studies*), the journal of the *Kant*-Gesellschaft (*Kant-Society*), prompted protests of German-Jewish philosophers in particular. In response, Bauch founded the *Deutsche Philosophische Gesellschaft* (DGP; *German Philosophical Society*) as an alternative to the Kant-Society. In one of its first programmatic statements, the society announced that it aimed to cultivate, deepen, and preserve the "German character" in philosophy.[12] This program attracted all kinds of conservative, nationalist, and racist philosophers.

Bauch believed that humanity consisted of distinct nations. He claimed that fostering nationalist attitudes was a prerequisite for the flourishing of humanity (1917, 160–162; see also 1935, 228). Bauch thus emphasized the general character of his nationalist views. In his article from 1917, he developed a *völkisch* particularism. He construed nations as distinct and mutually exclusive entities. The individual character of a nation could not be understood by members of another nation. Bauch explicitly denied "Jews" the ability to understand the cultural achievements of Germans.

Bauch claimed that his concept of the nation followed the German tradition, especially Fichte (1917, 139f., 149, 150–153). He held that nations were the synthesis of a natural fact (*natürliche Gegebenheit*) and a cultural fact (*kultürliche Gegebenheit*). The nature of a nation consisted of biological, spiritual, and linguistic components. Bauch believed that human communities were rooted in biological groups. This biological link shaped the physical appearance of nations. He insisted that there were physical types that could be easily distinguished by their skin color, their face shape, their body type, and their skull shape (1917, 141, 144). He also claimed that the members of a nation shared a way of thinking, feeling, willing, and acting. This spiritual link constituted the distinct soul of nation (1917, 141–144). Finally, the specific character of a nation was expressed in their common language (1917, 157f.).

The cultural unity of a nation was realized in history. Bauch claimed that nations developed self-consciousness through their history. They became aware of their historical destiny and expressed their national consciousness in cultural goods. From the perspective of culture, nationhood was thus a task that had to be fulfilled. Construed as an end, nationhood participated in the transcendent realm of values. Hence, nations possessed an objective meaning through their cultural existence (1917, 144, 148f., 156–158).

The particularist foundation of Bauch's *völkisch* nationalism clearly had a relativistic tendency. Bauch distinguished a plurality of distinct nations and postulated an incommensurability between them. This view could prepare the ground for cultural relativism. Bauch, however, believed that his objectivist understanding of the cultural aspect of nations blocked the paths towards relativism. This conviction becomes apparent in a lecture from October 1933 about *Wert und Zweck* (*Value and Purpose*) that was published in 1934/35.[13] Bauch gave the lecture at the twelfth annual conference of the DGP. The conference was an occasion for the DPG to pledge its allegiance to the Nazi regime – Hitler indeed sent his greetings.[14]

In his article from 1934/35, Bauch fiercely attacked critics of a transcendent realm of values. He characterized the deniers of the independence of objective values as subjectivists. On Bauch's view, all relationist concepts of values fell prey to the charge of subjectivism. He identified subjectivism with relativism. Bauch regarded pragmatism, utilitarianism, and materialism as modern kinds of philosophical relativism. He also referred to Plato's devastating critique of Protagoras and agreed with its dehumanizing images: Bauch claimed that the denial of the objectivity of values revealed the personal inferiority of the relativists. Their subjectivism degraded them to the rank of "baboons," "tadpoles," or "pigs" with whom a discussion was superfluous (Bauch 1934/35, 43f., 49f.).

Bauch connected his fierce polemic against relativism with a critique of his time. He praised NS as the political overcoming of the "pragmatist-materialist demon" (*pragmatisch-materialistische Ungeist*, 1934/35, 52) that haunted modernity. Intellectual developments such as individualism, atheism, or liberalism were characterized as "decomposing" the value foundation of society. The antisemitic character of Bauch's critique of relativism was obvious in his discussion of science. Bauch held that pursuing truth was the sole task of all scientific endeavors. At the same time, he accused "Einstein, Freud, and co." of trying to turn science into a tool for personal interests (1934/35, 53). He contrasted the alleged egoism of "Jewish" scientists with the German attitude to science: Germans were capable of doing science for its own sake. They acknowledged the unconditional value of truth (1934/35, 52).[15]

4. Ernst Rothacker's realist perspectivism

Rothacker believed that one had to understand humanity through its cultural achievements. His thesis from 1912 takes Lamprecht's historiography and Wundt's *Völkerpsychologie* (ethnic psychology) as starting points for a biologistic concept of culture. After WWI, Rothacker's philosophical interests changed significantly. He turned to Dilthey's theory of worldviews and developed a historicist understanding of humanity. In the 1920s – the middle period of his work – Rothacker committed to "positive relativism" and explicitly took sides in politics. From 1919 to 1928, he was member of the *Deutsche Volkspartei* (German People's Party) which combined a skeptical attitude towards Weimar Republic with a nationalist and business-friendly agenda. Around 1930, Rothacker began to develop his cultural anthropology and published two major treatises that set out his new views: *Geschichtsphilosophie* (*Philosophy of History*) in 1934 and *Probleme der Kulturanthropologie* (*Issues of Cultural Anthropology*) in 1942 (republished in 1948). The second fundamental revision of his philosophical thought was accompanied by a political turn to NS.

From the beginning, Rothacker associated the revision of his philosophical position with the rise of NS. His *Geschichtsphilosophie* from 1934 started with the claim that the upcoming "folkish consciousness" (*volkstümliche Bewußtsein*) constituted the beginning of a new epoch. Rothacker stated that his philosophy revealed the anthropological foundation of this historical transformation. Moreover, he concluded his *Geschichtsphilosophie* with an explicit political statement. The chapter "*Im dritten Reich*" argued that his philosophy belonged to the new world order by showing its conformity with key features of Nazi ideology.[16]

Völkisch particularism was also at the core of Rothacker's cultural anthropology. Human history was characterized by the struggle of particular communities for the realization of their life and their world (Rothacker 1934, 38). Philosophy had thus to provide insight into the "life laws of *Völker*" (*Lebensgesetze der Völker*, 1934, 5). Rothacker defined *Völker* as the "bearers and creators" of all "moral, cultural, and spiritual life" (1934, 38). *Völker* were characterized by both biological and cultural aspects. Rothacker held that *Völker* rested on biological groups: families, clans, and tribes were their biological sources, which ensured the future of a community. The members of a community also shared a way of life and possessed common spiritual goods. A community established a specific attitude by its struggle with the natural and social environment (1934, 38).

Rothacker held that cultures were rooted in shared ways of life. Cultures emerged, when the specific attitude of a community was completed in a "characteristic form" (*durchgeprägte Form*) that shaped all areas of life. Rothacker defined the "characteristic form" of *Kulturvölker* (cultural people) as particular, collective, public, homogeneous, comprehensive, and productive (1934, 73–79, 1948, 68f.).

Rothacker's particularist understanding of cultures obviously had a relativistic tendency. It was based on the conviction that there was no neutral response to the challenges of life. Rothacker referred to the pragmatic principle of active selection that grounded the "positive relativism" of his middle period. Immersed in a concrete situation, humans could only develop perspectival knowledge and partial solutions. Their knowledge and their activities were guided by their practical interests and specific desires. The partiality of knowledge and actions impacted the character of the emerging attitudes: these basic orientations in life possessed a *weltanschaulich* character. Rothacker assumed that our beliefs are dependent on the contexts we lived in. Moreover, the different ideological orientations of human life give rise to plurality of "won worlds" (*erkämpfte Welten*, 1934, 99f.). Rothacker held that each culture created a particular world.

Contrary to his middle work, the mature Rothacker explicitly rejected the label relativism for his perspectivist view. He illustrated the non-relativistic feature of his mature perspectivism by a simple example: What a farmer spontaneously interprets as "wood," the forester takes as a forest, the hunter as a hunting ground, and the fugitive as a hiding place (Rothacker 1934, 85f.; see also 1948, 161, 170f.). Rothacker argued that these divergent perspectives reveal different aspects of the same reality and do not contradict each other (Rothacker 1934, 86, 1948, 173, 177).

Rothacker used the realist core of his mature perspectivism to rank communities. He assumed that some human groups developed better responses to the practical challenges of life. Cultures were thus an achievement of communities. Human history was shaped by the cultural existence of certain *Völker* who constituted the peak of human excellence. Only these communities possessed historical significance. All other human groups were forgotten – and justifiably so (see, e.g., Rothacker 1934, 53). This view shows us the main reason for the rejection of relativism by Nazi philosophers: They were convinced that there was a hierarchy of *Völker* and believed in an objective justification of their ranking.

5. Conclusion

This chapter started with a critical examination of the widespread reproach that relativism and NS are connected historically as well as systematically. My investigation of the historical context revealed a rather different picture: Nazi intellectuals presented NS as the overcoming of the problem of relativism. The prevalence of anti-relativist motifs in Nazi debates is especially significant because a similar kind of fascism committed to relativism at that time.

The anti-relativistic tendency of Nazi thought was confirmed by the views of two major philosophers who associated their philosophy with NS. Bauch welcomed NS with a lecture that included a fierce polemic against relativism. His polemics also demonstrated that modern anti-relativism could rely on popular motifs from the philosophical tradition. Anti-relativist sentiments were thus a strong motivating factor for the philosophical collaboration with NS.

The development of Rothacker's philosophy was characterized by significant changes. Rothacker committed to the "positive relativism" of historicism during the 1920s. The revision of his philosophical views was, however, accompanied by a political turn to NS and a withdrawal from relativism in the 1930s.

Both cases showed that Nazi philosophers had to reconcile the relativistic tendency of their *völkisch* particularism with the anti-relativistic assumptions of their views. They believed in the superiority of certain *Völker* and races. Bauch argued that nationalism belonged to the transcendent realm of objective values. Rothacker insisted on the realist foundation of his perspectivism.

My chapter also demonstrates the pejorative meaning of term relativism in the context of NS. Nazi intellectuals and Nazi critics shared a specific understanding of relativism: They all considered relativism as a fundamental problem that had to be overcome. Relativism often covered alleged intellectual "ills" such as subjectivism, historicism, skepticism, materialism, or nihilism. Moreover, "relativism" was mostly used in a polemical sense: relativists were always the others, the philosophical and/or political enemies. To label someone a relativist was almost tantamount to making him an enemy.

Acknowledgements

This work was supported by the European Research Council (ERC) under Grant 339382.

For critical comments and helpful suggestions, I am indebted to audiences in Vienna as well as to the editor.

Notes

1 All translations are my translations.
2 For careful examinations of this historical context, see Steizinger (2017a), Kusch (2019), Kinzel (this volume).
3 For examples, see Sluga (1993), Wolters (1999), Sandkühler (2009), Sieg (2013). The research on philosophy during National Socialism is focused on Martin Heidegger. His case is, however, not exceptional from a historical point of view.
4 Note that Benito Mussolini (1883–1945) characterized fascism as a relativistic movement because of its anti-scientism and voluntarism. With his belief in the ultimate authority of pure power, the fascist despised all fixed and stable categories (Mussolini 1921).
5 On Dilthey's theory of worldviews, see Steizinger (2016, 2017a, 2017b), Kinzel (2019).
6 For a detailed account of Rothacker's life, including his political engagement, see Stöwer (2012).
7 See, e.g., Krieck (1936, 3, 7f., 1938, 11), Rosenberg (1938, 694f.), Del Negro (1942, 10–13), Baeumler (1943, 27f., 67ff.).
8 See, e.g., Del Negro (1942, 11f., 42f.). For careful accounts of Simmel's relativism, see Steizinger (2015), Kusch (2019).
9 For a detailed account of the prize question of 1936, see Steizinger (2019).
10 For detailed accounts of this strand of Nazi ideology, see Steizinger (2016, 2018, 2019).
11 Kelsen indeed considered the relativistic denial of absolute truth and values as the prerequisite of liberal democracy. He believed that philosophical absolutism goes hand in hand with political absolutism and is thus the philosophical henchman of despots, dictators, and autocrats (Kelsen 1920, 1948).
12 See Deutsche Philosophische Gesellschaft (1918/19). This statement can be found on the back cover of the first issue of the *Beiträge zur Philosophie des deutschen Idealismus* (Contributions to the Philosophy of German Idealism), the journal of the DPG (latter renamed to *Blätter für Deutsche Philosophie* (Pages for German Philosophy)). For detailed accounts of the DPG, see Sluga (1993), Sieg (2013).
13 Bauch included his lecture as a chapter in his *Grundzüge der Ethik* (Fundamentals of Ethics). See Bauch (1935, 287–310).

14 For more details about the DPG conference, see Sluga (1993, 154–164).
15 The contrast between "Jewish" egoism and German selflessness was a key motif of Nazi antisemitism (see Steizinger 2018, 147f.).
16 The literature on Rothacker's association with NS is problematic because most authors identify Nazi ideology with biological racism. Böhnigk (2002) believes that he can convict Rothacker's cultural anthropology of a commitment to biological racism. Yet Rothacker was no biological racist, although he discussed the concept of race and clarified its role within his cultural anthropology (Rothacker 1934, 132–145). Rothacker's dismissal of biologist concepts of humanity is not tantamount to distancing himself from NS. His intellectual connection with NS should thus be not disregarded, like Fischer (2008) does. Stöwer's (2012) biography gives a detailed account of Rothacker's political association with NS. Yet Stöwer (2012, 350f.) explains this engagement by the social prejudices and cultural stereotypes Rothacker held. He does not see that Rothacker's cultural anthropology developed a *völkisch* particularism that corresponded with an important strand of Nazi ideology.

References

Apel, K.-O. (1988), "Zurück zur Normalität? Oder könnten wir aus der nationalen Katastrophe etwas Besonderes gelernt haben? Das Problem des (welt-)geschichtlichen Übergangs zur postkonventionellen in spezifisch deutscher Sicht," in Kuhlmann (1988), 91–142.

Baeumler, A. (1943), *Alfred Rosenberg und der Mythos des 20. Jahrhunderts*, München: Hoheneichen.

Bauch, B. (1917), "Vom Begriff der Nation. Ein Kapitel zur Geschichtsphilosophie," *Kant-Studien* 21: 139–162.

———. (1934/35), "Wert und Zweck," *Blätter für Deutsche Philosophie* VIII: 39–59.

———. (1935), *Grundzüge der Ethik*, Stuttgart: Kohlhammer.

Böhler, D. (1988), "Die deutsche Zerstörung des politisch-ethischen Universalismus. Über die Gefahr des – heute (post-) modernen – Relativismus und Dezisionismus," in Kuhlmann (1988), 166–216.

Böhnigk, V. (2002), *Kulturanthropologie als Rassenlehre: nationalsozialistische Kulturphilosophie aus der Sicht des Philosophen Erich Rothacker*, Würzburg: Königshausen und Neumann.

Danneberg, L. (2013), *Wissenschaftsbegriff und epistemischer Relativismus im Nationalsozialismus*, http://fheh.org/wp-content/uploads/2016/07/relativismusld.pdf (Accessed March 16, 2018).

De Vries, J. (1935a), "Wissenschaft, Weltanschauung, Wahrheit," *Stimmen der Zeit* 129: 93–105.

———. (1935b), "Rationale oder irrationale Weltanschauung," *Stimmen der Zeit* 129: 380–392.

Del Negro, W. (1942), *Die Philosophie der Gegenwart in Deutschland*, Leipzig: Felix Meiner.

Felken, D. (2006), "Nachwort," in Spengler (2006), 1250–1270.

Fischer, J. (2008), *Philosophische Anthropologie. Eine Denkrichtung des 20. Jahrhunderts*, Freiburg: Karl Alber.

Gross, R. (2016), "The 'True Enemy:' Antisemitism in Carl Schmitt's Life and Work," in *The Oxford Handbook of Carl Schmitt*, edited by Jens Meierhenrich and Oliver Simons, Oxford: Oxford University Press, 96–116.

Herbert, C. (2001), *Victorian Relativity: Radical Thought and Scientific Discovery*, Chicago: University of Chicago Press.

Kellerwessel, W. (2014), "Universalism and Moral Relativism: On Some Aspects of the Modern Debate on Ethics and Nazism," in *Nazi Ideology and Ethics*, edited by W. Bialas and L. Fritze, Newcastle: Cambridge Scholars, 367–387.

Kelsen, H. (1920), *Vom Wesen und Wert der Demokratie*, Tübingen: Mohr.

———. (1948), "Absolutism and Relativism in Philosophy and Politics," *The American Political Science Review* 42: 906–914.

Kinzel, K. (2019), "The History of Philosophy and the Puzzles of Life: Windelband and Dilthey on the Ahistorical Core of Philosophical Thinking," in Kusch et al (2019), 26–42.

Krieck, E. (1932), *Nationalpolitische Erziehung*, 20th edition, Leipzig: Armanen-Verlag, 1936.

———. (1938), *Völkisch-politische Anthropologie*, vol. 3, Leipzig: Armanen-Verlag.

Kroll, F-L. (1998), *Utopie als Ideologie. Geschichtsdenken und politisches Handeln im Dritten Reich*, Paderborn: Schöningh.

Kuhlmann, W. (ed.) (1988), *Zerstörung des moralischen Selbstbewußtseins: Chance oder Gefährdung?* Frankfurt am Main: Suhrkamp.

Kusch, M. (1995), *Psychologism*, London and New York: Routledge.

———. (2019), "Simmel and Mannheim on the Sociology of Philosophy of Philosophy, Historicism and Relativism," in Kusch et al (2019), 165–180.

Kusch, M., Kinzel, K., Steizinger, J. and Wildschut, N. (2019), *The Emergence of Relativism: German Thought from the Enlightenment to National Socialism*, London and New York: Routledge.

Lieber, H.-J. (1966), "Die deutsche Lebensphilosophie und ihre Folgen," in *Nationalsozialismus und die deutsche Universität*, edited by Freie Universität Berlin, Berlin: De Gruyter, 92–108.

Lukàcs, G. (1954), *Die Zerstörung der Vernunft. Der Weg des Irrationalismus von Schelling zu Hitler*, Berlin: Aufbau.

Meyers Lexikon (1942), Vol. 9, eighth edition, Leipzig: Bibliographisches Institut.

Mussolini, B. (1921), "Nel solco delle grandi filosofie: relativismo e fascism," *Il Popolo d'Italia*, November 22.

Raphael, L. (2014), "Pluralities of National Socialist Ideology: New Perspectives on the Production and Diffusion of National Socialist *Weltanschauung*," in *Visions of Community in Nazi Germany. Social Engineering and Private Lives*, edited by M. Steber and B. Gotto, Oxford: Oxford University Press, 73–86.

Rosenberg, A. (1930), *Der Mythus des 20. Jahrhunderts*, 125th–128th edition, München: Hoheneichen, 1938.

Rothacker, E. (1912), *Über die Möglichkeit und den Ertrag einer genetischen Geschichtsschreibung im Sinne Karl Lamprechts*, Leipzig: R. Voigtländer Verlag.

———. (1926), *Logik und Systematik der Geisteswissenschaften*, in *Handbuch der Philosophie*, Section II, Contribution C, edited by A. Baeumler and M. Schröter, München und Berlin: Oldenbourg.

———. (1934), "Geschichtsphilosophie," in *Handbuch der Philosophie*, Section IV, edited by A. Bäumler and M. Schröter, München und Berlin: Oldenbourg.

———. (1948), *Probleme der Kulturanthropologie*, Bonn: H. Bouvier u. Co.

Sandkühler, H. J. (2009), *Philosophie im Nationalsozialismus*, Hamburg: Felix Meiner.

Schmitt, C. (1916), *Theodor Däublers "Nordlicht": Drei Studien über die Elemente, den Geist und die Aktualität des Werkes*, München: G. Müller.

———. (1922), *Politische Theologie. Vier Kapitel zur Lehre von der Souveränität*, Berlin: Duncker & Humblot, 2009.

———. (1931), "Übersicht über die verschiedenen Bedeutungen und Funktionen des Begriffs der innerpolitischen Neutralität des Staates," in *Der Begriff des Politischen*, Berlin: Duncker & Humblot, 2002, 97–101.

Sieg, U. (2013), *Geist und Gewalt. Deutsche Philosophen zwischen Kaiserreich und Nationalsozialismus*, München: Carl Hanser.

Sluga, H. (1993), *Heidegger's Crisis: Philosophy and Politics in Nazi Germany*, Cambridge, MA: Harvard University Press.

Spengler, O. (1922), *Der Untergang des Abendlandes*, München: Deutscher Taschenbuch Verlag, 2006.

Steizinger, J. (2015), "In Defense of Epistemic Relativism: The Concept of Truth in Georg Simmel's Philosophy of Money," *Proceedings of the 38th International Wittgenstein-Symposium* XXIII, 300–302.

———. (2016), "Politik versus Moral. Alfred Baeumlers Versuch einer philosophischen Interpretation des Nationalsozialismus," *Jahrbuch zur Geschichte und Wirkung des Holocaust* 20: 29–48.

———. (2017a), "Reorientations of Philosophy in the Age of History: Nietzsche's Gesture of Radical Break and Dilthey's Traditionalism," *Studia philosophica* 76: 223–244.

———. (2017b), "Vorbild, Beispiel und Ideal. Zur Bedeutung Goethes in Wilhelm Diltheys Philosophie des Lebens," in *Goethe um 1900*, edited by C. Haas, J. Steizinger, and D. Weidner, Berlin: Kulturverlag Kadmos, 27–49.

———. (2018), "The Significance of Dehumanization: Nazi Ideology and Its Psychological Consequences," *Politics, Religion & Ideology* 19: 139–157.

———. (2019), "National Socialism and the Problem of Relativism," in Kusch et al (2019), 233–251.

Stöwer, R. (2012), *Erich Rothacker: Sein Leben und seine Wissenschaft vom Menschen*, Göttingen: V&R University Press.

Tugendhat, E. (2009), "Der moralische Universalismus in Konfrontation mit der Nazi-Ideologie," *Jahrbuch zur Geschichte und Wirkung des Holocaust* 13: 61–75.

Wolin, R. (2004), *The Seduction of Unreason: The Intellectual Romance with Fascism from Nietzsche to Postmodernism*, Princeton, NJ: Princeton University Press.

Wolters, G. (1999), "Der 'Führer' und seine Denker. Zur Philosophie des 'Dritten Reichs,'" *Deutsche Zeitschrift für Philosophie* 47: 223–251.

13

RELATIVISM AND PRAGMATISM

Anna Boncompagni

1. Introduction

Although a connection between pragmatism and relativism is often taken for granted in the philosophical debate on relativism, classical pragmatists – Charles S. Peirce, William James, and John Dewey – were not particularly concerned with the issue. Rather, they generally aimed at avoiding "isms." Trying to place them in either side of the dispute is therefore inevitably a bit of a stretch. The task of clarifying the relationship between pragmatism and relativism is further complicated by the fact that neither has a generally agreed-upon definition.

Given the many existent proposals in the literature for identifying the key features of relativism (see Baghramian and Coliva 2019 for the most recent one), in what follows I will not put forward a new proposal but concentrate instead on what is generally considered its most basic and challenging version: alethic relativism. I take its central claim to be non-absolutism about truth, or the idea that the truth of a proposition is not absolute because it crucially depends on parameters or frameworks of reference. Additionally, for a relativist there are no overarching standards for evaluating these frameworks: a proposition can be true in one framework and false in another one and both verdicts can be equally valid. In other words, there may be competing truths, and no truth is "truer" than any other in absolute terms. I will focus on alethic relativism as it allows for a fruitful discussion of the pragmatist conception of truth, arguably the most controversial tenet of pragmatism. I will however include other forms of relativism in the examination whenever the authors considered deal (or are taken to deal) with them.

As for the definition of pragmatism, opinions diverge on whether a single and comprehensive definition is possible. Classical pragmatism is commonly associated with the names of Peirce, James, and Dewey, and I will therefore deal with them in the first place, but also leave some room for more recent exponents. In particular, I will include in the discussion two thinkers who, besides addressing relativism in some of their crucial writings, explicitly connected this discussion with their interpretation of pragmatism itself: Richard Rorty and Hilary Putnam. These choices inevitably leave aside other representatives who also have some links with relativism. One is Ferdinand Canning Scott Schiller, a German-British early pragmatist whose views are probably the closest to Protagorean alethic relativism (Schiller 1908). I will not include Schiller because in spite of the role he had in spreading American pragmatism on the Old Continent, he is not considered nowadays among the most significant representatives of this tradition (Haack 2004, 23–24). Other thinkers that come to mind are Willard V. O. Quine, Donald Davidson, and Nelson Goodman; but their discussion of truth and relativism is not directly linked to their (alleged) pragmatism, and given limits of space I prefer not to include them in this chapter (see Baghramian and Coliva 2019 for an overview).

Perplexities in regard to the nature of pragmatism as a movement are not only to be found in retrospect: the dialogue between Peirce and James is itself revealing of the tensions that have accompanied pragmatism since the very first appearance of the term. It was James who introduced it into the philosophical arena in 1898 (James 1907, 255–270), but he credited Peirce with the original idea, citing his 1878 paper "How to Make Our Ideas Clear" (CP 5, 388–410). Peirce was disappointed by James' appropriation of the term and later coined the new expression "pragmaticism" to characterize his own view (CP 5, 414). However, what both philosophers highlighted is, roughly, that pragmatism is a *method* for the clarification of the *meaning of concepts* by reference to their (practical or potential) *consequences*. This is essentially the so-called pragmatic or pragmatist maxim:

> Consider what effects, which might conceivably have practical bearings, we conceive the object of our conception to have. Then, our conception of those effects is the whole of our conception of the object.
>
> *(CP 5, 402)*

At first sight, connecting meaning to practical consequences is a theoretical move that *might* have relativistic implications. If this is the core of pragmatism, then, it *might* be the case that there is a close association between pragmatism and relativism. The following pages will explore this "might" in more detail.

2. Peirce

The maxim cited earlier appears in Peirce's reasoning concerning how to attain clearness about concepts. He devoted much time to the elucidation of what the "practical bearings" of concepts amounted to. He wanted to distance himself from the idea that they were simply the sensations and behavior produced by the concept, and insisted instead on their *potential* nature, on the "would-bes" that they involved (CP 5, 18; 5, 438). The emphasis on potentiality and generality is also reflected in Peirce's notion of reality, which is intertwined with his notions of truth and the community. In one of his first essays, he observed that we develop the concept of the real precisely when we realize that we, as individuals, might be mistaken and need the corrective input of others:

> And what do we mean by the real? It is a conception which we must first have had when we discovered that there was an unreal, an illusion; that is, when we first corrected ourselves. . . . The real, then, is that which, sooner or later, information and reasoning would finally result in, and which is therefore independent of the vagaries of me and you. Thus, the very origin of the conception of reality shows that this conception essentially involves the notion of a COMMUNITY, without definite limits, and capable of a definite increase of knowledge.
>
> *(CP 5, 311)*

Peirce underlines that reality is independent of the thought of the individual, but at the same time, it is *not* independent of the thought of the community. A crucial role is played here by what we might call the "in the long run" clause.

> The opinion which is fated to be ultimately agreed to by all who investigate, is what we mean by the truth, and the object represented in this opinion is the real.
>
> *(CP 5, 407)*

Yet, one might object: how do we know that the opinion toward which the largest community would converge in the long run is effectively true? It is in the very nature of inquiry, Peirce says, that it leads to truth in the long run. Indeed, once inquiry is conducted by the members of a scientific community who make their results public and publicly correctable, the inductive process naturally leads them towards a true opinion (CP 2, 769). To be sure, Peirce acknowledged that we cannot be *absolutely certain* that the community will effectively come to "an unalterable conclusion upon any given question"; yet, "we are entitled to assume . . . in the form of a *hope* that such conclusion may be substantially reached concerning the particular questions with which our inquiries are busied" (CP 6, 610).

The concept of hope allows Peirce to combine fallibilism with a confidence in the workings of the community of inquirers. It also signals, at least for some commentators, that Peirce's notions of reality and of truth should not be taken as substantive, but rather as *regulative* (Hookway 2012; Misak 2004): the hope in the convergent process of scientific inquiry is a regulative principle that we adopt in our inquiries, and by doing so, we effectively contribute to the success of the collective enterprise. Moreover, Hookway (2012) argues, Peirce recognized that what we are engaged in is not a general inquiry into the nature of reality, but rather always a particular investigation guided by some particular questions and interests. In Hookway's interpretation, this "saves" Peirce from the commitment to an absolute conception of reality, a conception Bernard Williams famously attributed to the anti-relativism of science (Williams 1978). If this is correct, then it would be misleading to interpret Peirce as a precursor of Williams' idea,[1] or in other words, an advocate of anti-relativism, at least in the scientific domain.

Yet, it would be equally misleading, I think, to interpret Peirce as a precursor of relativism (not that this is Hookway's point). In fact, the same reasoning that leads Peirce to root logic in the "social principle" (CP 2, 654) also leads him to the idea of an indefinite growth of the community:

> [L]ogicality inexorably requires that our interests shall not be limited. They must not stop at our own fate, but must embrace the whole community. This community, again, must not be limited, but must extend to all races of beings with whom we can come into immediate or mediate intellectual relation. It must reach, however vaguely, beyond this geological epoch, beyond all bounds. He who would not sacrifice his own soul to save the whole world, is, as it seems to me, illogical in all his inferences, collectively.
>
> *(CP2, 654)*

For Peirce then, the existence of particular interests is not only compatible with, but in fact requires, a progressive widening of one's perspective, and ultimately the widest possible horizon.

3. James

Personal interests and needs play a much more important role in James' perspective. James did not shy away in front of the possibility of there being more than one way of conceptualizing the world, and did not feel the urge to think that this multiplicity ought to finally converge towards one single view:

> There is nothing improbable in the supposition that analysis of the world may yield a number of formulae, all consistent with the facts. . . . Why may there not be different points of view for surveying it, within each of which all data harmonize, and which the observer may therefore either choose between, or simply cumulate one upon another?
>
> *(1897, 66)*

In particular, it is in choosing between different perspectives that personal inclinations matter. This is the central claim of James' "The Will to Believe" (James 1897), the publication of which was met with considerable criticism, including criticism coming from Peirce (CP5, 3). James' controversial point was that, given that no evidence is available for either the existence or the non-existence of God, we may decide to believe in God, if this proves useful and helps us live a better and more satisfactory life.

Criticism only increased when James published *Pragmatism* (1907) and *The Meaning of Truth* (1909), where the conception of truth as usefulness or "what works" took center stage:

> [A]ny idea upon which we can ride, so to speak; any idea that will carry us prosperously from any one part of our experience to any other part, lining things satisfactorily, working securely, simplifying, saving labor; is true for just so much, true in so far forth, true *instrumentally*.
>
> *(1907, 34)*

Changeability, pluralism, and usefulness are some of the points on which he insists:

> The truth of an idea is not a stagnant property inherent in it. Truth *happens* to an idea. It *becomes* true, is *made* true by events. Its verity *is* in fact an event, a process: the process namely of its verifying itself, its veri-*fication*. . . . This function of agreeable leading is what we mean by an idea's verification.
>
> *(1907, 97)*

> Our account of truth is an account of truths in the plural, of processes of leading, realized *in rebus*, and having only this quality in common, that they *pay*.
>
> *(1907, 104)*

James' reference to the "in the long run" notwithstanding (1907, 106), critics were very harsh. Bertrand Russell for instance affirmed that for James "a truth is anything which it pays to believe" (Russell 1910, 118), and that "if . . . we agree to accept the pragmatic definition of the word 'truth,' we find that the belief that A exists may be 'true' even when A does not exist" (1910, 148).

James replied that pragmatism acknowledges the notion of the "Truth absolute," with a capital "T" ("An ideal set of formulations towards which all opinions may be expected in the long run of experience to converge" (1909, 143)) and considers that of agreement with reality as a "matter of course" (1909, 3, 104; see also 1907, 42, 111–112). What pragmatism wants to add to this definition, James continues, is a clarification, in the spirit of the pragmatic maxim, of what this means, that is, what practical difference it makes *to us*, in our everyday life. "Satisfaction," moreover, must not be intended in the abstract but in the concrete, i.e. in what concrete human beings find satisfactory; and among the things they find satisfactory are believing in other human beings' minds, in independent physical realities, in past events and in eternal logical relations (1909, 104).

James' view, therefore, is more developed and complex than is usually thought (Putnam 1997), and is probably best described as pluralist rather than relativist. Yet, he also directly engaged with the issue of relativism, though not extensively. In fact, in "Abstractionism and 'Relativismus'" (James 1909), he responded to the charge of reproducing the Protagorean doctrine. After conceding (or at least not denying) that "we pragmatists are typically relativists," he rejected the criticism that relativists can neither believe their own principles nor admit any ideal opinion on which everyone would agree; to the contrary, he argued that "[t]o admit, as we pragmatist do, that we are liable to correction . . . *involves* the use on our part of an ideal standard" (1909, 142), and explained:

> No relativist who ever actually walked the earth has denied the regulative character in his own thinking of the notion of absolute truth. What is challenged by relativists is the pretense on anyone's part to have found for certain at any given moment what the shape of that truth is.
>
> *(1909, 143)*

James therefore on the one hand lays claim to a form of relativism, and on the other hand denies that the upshot of relativism is that there is no regulative standard of truth at all. A way to make sense of this position, while keeping the definition of relativism centered on non-absolutism about truth, is by emphasizing James' insistence on the *personal* aspect of the regulative principle ("in his own thinking," he says) and on its changeability. It is, in other words, James' individualism (together, of course, with other features of his thought, such as holism) that allows reading his position as relativist, accepting his self-attribution as such, notwithstanding his acknowledgment of a regulative ideal of truth.

4. Dewey

John Dewey was working on the theme of truth in the same years that saw the publication of James' *Pragmatism* and *The Meaning of Truth*. Dewey's starting point was the concrete situations experienced by human beings in everyday life; in this case, the specific conditions, the use, the relevance, and the purpose of thinking (MW 4, 60–61).

In general terms, Dewey claims that ideas are correct insofar as they help solve a problematic situation, removing what is undesirable and inconsistent (MW 4, 63). In this practical framework, "to make an idea true is to modify and transform it until it reaches this successful outcome" (MW 4, 66). In accordance with James' point that the truth of an idea is the process of its verification, Dewey claims that "verification and truth are two names for the same thing. We call it 'verification' when we regard it as a process; . . . We call it 'truth' when we take it as a product" (MW 4, 67). The criterion for the truth of an idea, in this fashion, is its "capacity of operation" (MW 4, 73).

In 1911, Dewey examines the issue in more detail, taking its cue from the common sense intuition that "truth" has to do with something important for human beings and that it is closer to a social virtue than an epistemological relation (MW 6, 14–20). He starts from the nature of propositions:

> [E]very proposition . . . is a hypothesis concerning some state of affairs; . . . it is a means of setting on foot activities of inquiry which will test the worth of its claim. Truth,

then, can exist only in the testing of the claim, in making good through the subsequent acts it prescribes....Truth is a matter of its career, of its history: ... it becomes or is *made* true (or false) in process of fulfilling or frustrating in use its own proposal.

(MW 6, 38–39)

The resemblances with the Jamesian perspective are clear, but whereas James emphasized the role of usefulness and personal interests, Dewey emphasizes the similarity between the ordinary sense of "truth" and the scientific method based on hypotheses and verification. Like James, Dewey underlined that in his account of truth the idea of correspondence is not denied, but specified: it has the sense of a mutual correction or mutual aid of the two parts involved, so that an anticipation meets its fulfilment, or a tool fits its end, leading to the successful resolution of a problem (MW 6, 45; see also LW 14, 179–180).

Are there relativistic elements or themes in this view? Rather than align Dewey with relativism, I would suggest his view lends itself to contextualism and instrumentalism. But in order to get a fuller picture, it is worth taking into consideration other aspects of his philosophy.

In *Logic: The theory of inquiry* (1938, see LW, 12; see also LW 14, 168–189), Dewey introduced the notion of "warranted assertibility" as a new and preferable way to deal with epistemic concepts such as belief and knowledge, which by then he considered too ambiguous (LW 12, 16). By "warranted assertibility" Dewey meant the distinctive property of "progressively stable belief" (LW 12, 17) resulting from successful inquiry. It is inquiry, in its ongoing and self-correcting character, that guarantees the assertibility of ideas, or "knowledge in its strictest and more honorific sense" (LW 12, 145). Therefore, if true knowledge is relative to anything, it is relative to inquiry itself, which, in Dewey's account, is characterized by the method and techniques of science – a view broadly indebted to Peirce for a number of key concepts (cf. LW 12, 3, 19, 46, 158, 464–465).

In this light, I find it problematic to attribute to Dewey a form of alethic relativism. Yet, the question remains whether he advanced relativist ideas in social, cultural, and moral matters. It is Dewey himself that speaks of relativism in his *Ethics* (second edition, 1932). After a historical detour into the various ways in which the doctrine of individualism manifests itself, he affirms that the purpose of his survey was "to suggest the *relativism* of social formulae in their ethical aspect" (LW 7, 336), explaining:

> No single formula signifies the same thing, in its consequences, or in practical meaning under different social conditions. That which was on the side of moral progress in the eighteenth and early nineteenth centuries may be a morally reactionary doctrine in the twentieth century; that which is serviceable now may prove injurious at a later time.
>
> *(LW 7, 336)*

Again, perhaps he is making a contextualist point rather than a strong relativistic claim, and in fact his approach is usually described as moral contextualism or moral particularism; yet, he names it "relativism." Similarly, in a later work, he observes that pragmatism "is charged with promotion of relativism," and explains that the "relativity" of pragmatism is the same that one can find in scientific inquiry, concluding:

> Not "relativity" but absolutism isolates and confines. The reason, at bottom, that absolutism levels its guns against relativity in a caricature is that search for the connection of events is the sure way of destroying the privileged position of exemption from inquiry which every form of absolutism secures wherever it obtains.
>
> *(LW 15, 162–163)*

Moreover, in the paper "Context and Thought" (1931) – a reflection inspired by Malinowsky – Dewey invites philosophers not to "freeze the quotidian truths relevant to the problems that emerge in [their] own background of culture into eternal truths inherent in the very nature of things" (LW 6, 13) and observes that "a standpoint which is nowhere in particular and from which things are not seen at a special angle is an absurdity" (LW 6, 15).

Therefore, following Larry Hickman, I would not refrain from attributing a "moderate version of cultural relativism" to Dewey, provided that this version refuses a stronger claim: that the truth values of judgments are relative to standpoints *and* that no standpoint is in any sense privileged; the pragmatic method in fact does provide a way for evaluating and testing different points of view (LW 6, 14, Hickman 2007, 30, 42–43).

5. Richard Rorty and Hilary Putnam

In *Philosophy and the Mirror of Nature* (1979, 328–333), Rorty described the dispute between Galileo Galilei and Cardinal Bellarmine in a way that has been taken as paradigmatic of (epistemic) relativism, especially after Paul Boghossian's (2006) criticism of it. Is it really possible, asked Rorty, to say that the arguments put forth by Bellarmine against Galileo – namely, the scriptural descriptions of the heavens – were illogical or unscientific? Rorty answers in the negative, claiming that our very categories of scientific rationality are as historical and contingent as any other. In commenting on Rorty's claim, Boghossian (2006, 62–63) attributes to it the central tenets of relativism, including the idea that there are many radically different but equally valid ways of knowing the world.

In spite of the intuitiveness with which Rorty's position is associated with relativism, he has notoriously denied being a relativist, insisting instead that, *as a pragmatist*, he was committed to a form of ethnocentrism but *not* of relativism. In fact, he thought that relativism, or the view that "every belief on a certain topic is as good as any other," is obviously easy to refute (1980, 727–728). More precisely:

> "Relativism" is the traditional epithet applied to pragmatists by realists. Three current views are commonly referred to by this name. The first is the view that every belief is as good as every other. The second is the view that "true" is an equivocal term, having as many meanings as there are procedures of justification. The third is the view that there is nothing to be said about either truth or rationality apart from descriptions of the familiar procedures of justification which a given society – ours – uses in one or another area of inquiry. The pragmatist holds the ethnocentric third view. But he does not hold the self-refuting first view, nor the eccentric second view.
>
> *(Rorty 1991, 23)*

It is tempting to summarize Rorty's position as a refusal of alethic and epistemic relativism and a commitment to cultural relativism. But he would dismiss this characterization for two reasons. The first is that he thought that relativism concerning "real theories," as opposed to philosophical theories (for instance, proposals for political change, as opposed to Kantianism or Platonism) *is* disturbing and worth resisting (1980, 728–729). The second reason is more radical: pragmatism, for Rorty, is not interested in theories and does not hold *any* epistemology, let alone a relativistic one (1991, 24). Rorty's ethnocentrism, then, amounts to a fully aware acceptance of our dependence on the categories defined by our belonging to a historical, social, and cultural community, where such an acknowledgment comes – and must come – without theory and without epistemology.

Rorty's pragmatism, in this sense, goes further than classical pragmatism in accepting some consequences of our belonging to a cultural tradition: it goes further in that it interprets the aspiration to philosophical theories itself, relativism included, as a manifestation of that belonging.

Rorty's position developed through a constant dialogue with Hilary Putnam, who addressed the theme of relativism, usually in connection with realism, on many occasions in his multifaceted work. During his "internal realism" phase in particular, Putnam proposed the notion of "conceptual relativity": he claimed, in a Kantian spirit, that "we can reject a naïve 'copy' conception of truth without having to hold that it's all a matter of the *Zeitgeist*" (Putnam 1981, x). We can do so by conceiving of truth as a sort of "(idealized) rational acceptability, some sort of ideal coherence of our beliefs with each other and with our experiences *as those experiences are themselves represented in our belief system*" (Putnam 1981, 49–50).

Conceptual relativity, Putnam explains, allows for different ways of defining our most basic categories; for instance, the way in which we define "objecthood" varies depending on whether we are using common sense or the theoretical apparatus of quantum physics. But this does not imply that if we use common sense we live in one world, and if we use quantum physics we live in another: the two descriptions are still descriptions of one single world. Neither does it imply that reality is hidden or noumenal; rather, "it simply means that you can't describe the world without describing it" (Putnam 1992, 122–123).

The point where conceptual relativity comes rather close to relativism, is in what Putnam calls a "quite controversial" corollary, namely, "[t]he doctrine that two statements which are incompatible at face value can sometimes both be true" (1990, x). Yet, this is not relativism, Putnam affirms (1981, 119); in fact, conceptual relativity is accompanied by the recognition that there *is* a transcultural notion of justification: rationality and justification are not themselves defined by any single paradigm, but rather are presupposed by the very activity of criticizing and inventing paradigms (1981, 125).

In clarifying his view, Putnam also criticizes Dewey's "objective relativism"[2] according to which, in Putnam's words, "certain things are right – *objectively right* – in certain circumstances and wrong – *objectively wrong* – in others, and the culture and the environment constitute relevant circumstances" (1981, 162). This perspective, he argues, faces a serious problem: it cannot handle cases such as that of the Nazi (1981, 168), because their culture and environment would make their practices acceptable. We need, instead, to be able to call the Nazi's goals *irrational*. And we can do this, he claims, only if, besides appealing to the values of our own culture and tradition (as Rorty would do), we also possess a *Grenzbegriff*, a limit-concept of the ideal of truth (1981, 215–216).

6. The contemporary debate

In the light of this examination, we can see that the relationship between pragmatism and relativism is rather complex and nuanced. While it is surely incorrect to simply equate pragmatism with a form of relativism, it is equally incorrect to negate any possible link between the two.

In the contemporary debate, there is a tendency to shape the history of pragmatism as divided into two main strands: a Peircean, broadly anti-relativist strand, and a Jamesian-Deweyan-Rortyan broadly relativist strand, where relativism is considered the enemy of good philosophy (see for instance Misak 2013). As pragmatism itself teaches, however, dichotomies usually impoverish debate. In this case, in particular, the most interesting insights that pragmatism has to offer to the discussion of relativism are actually in aspects that the different threads of this tradition *share*. The pragmatic method of connecting meaning to consequences, which is commonly considered to be the core of pragmatism, has bearings both on the concept of truth and the evaluation of different frameworks and forms of life. Regarding the former, the pragmatist insight is that truth is something that we *need* and *use* in our everyday exchanges with the world, helping us to orientate ourselves not only in the here and now, but in the long run. Regarding the latter, the pragmatist insight is that we can (and usually do) choose among different ways of conceptualizing the world, as well as between different ways of living, based on the practical difference they make, in a broad sense. Depending on how one defines relativism, these insights can be read in a relativist light or not. In any case, if a form of relativism is to be read in pragmatism, *this* relativism sounds neither paradoxical nor particularly dangerous.

Notes

1 The target of Hookway's criticism is Hilary Putnam (see Putnam 1992, 84). However, Williams himself cited Peirce; see Williams (1978, 229, 231–232, 237).
2 The expression is not Dewey's own, but was proposed by Murphy (1927).

References

Baghramian, M. and Coliva, A. (2019), *Relativism*. London: Routledge.

Boghossian, P. (2006), *Fear of Knowledge. Against Relativism and Constructivism*, Oxford: Oxford University Press.

Dewey, J. (1978), *The Middle Works of John Dewey, 1899–1924,* 15 vols., Carbondale and Edwardsville: Southern Illinois University Press. [cited as MW volume number, page number]

———. (1985), *The Later Works of John Dewey, 1925–1953,* 17 vols., Carbondale and Edwardsville: Southern Illinois University Press. [cited as LW volume number, page number]

Haack, S. (2004), "Pragmatism, Old and New," *Contemporary Pragmatism* 1(1): 3–41.

Hickman, L. (2007), *Pragmatism as Post-Postmodernism: Lessons from John Dewey*, New York: Fordham University Press.

Hookway, C. (2012), "Truth, Reality and Convergence," in *The Pragmatic Maxim*, Oxford: Oxford University Press, 127–149.

James, W. (1897), *The Will to Believe and Other Essays in Popular Philosophy*, Cambridge, MA: Harvard University Press, 1979.

———. (1907), *Pragmatism*, Cambridge, MA: Harvard University Press, 1975.

———. (1909), *The Meaning of Truth*, Cambridge, MA: Harvard University, 1975.

Misak, C. (2004), *Truth and the End of Inquiry*, Oxford: Oxford University Press.

———. (2013), *The American Pragmatists*, Oxford: Oxford University Press.

Murphy, A. E. (1927), "Objective Relativism in Dewey and Whitehead," *Philosophical Review* 36(2): 121–144.

Peirce, C. S. (1935–1958), *The Collected Papers of Charles Sanders Peirce*, vols. I–VI, edited by P. Weiss, C. Hartshorne, Cambridge, MA: Harvard University Press; vols. VII–VIII, edited by A. W. Burks, Cambridge, MA: Harvard University Press [cited as CP volume number. paragraph number].

Putnam, H. (1981), *Reason, Truth, and History*, Cambridge: Cambridge University Press.

———. (1990), "A Defense of Internal Realism," in *Realism with a Human Face*, edited by J. Conant, Cambridge, MA: Harvard University Press, 30–43.

———. (1992), *Renewing Philosophy*, Cambridge, MA: Harvard University Press.

———. (1997), "James' Theory of Truth," in *The Cambridge Companion to William James*, edited by H. Putnam and R. A. Putnam, Cambridge: Cambridge University Press, 166–185.

Rorty, R. (1979), *Philosophy and the Mirror of Nature*, Princeton, NJ: Princeton University Press.

———. (1980), "Pragmatism, Relativism, and Irrationalism," *Proceedings and Addresses of the American Philosophical Association* 53(6): 719–738.

———. (1991), "Solidarity or Objectivity?" in *Objectivity, Relativism, and Truth*, edited by R. Rorty, Cambridge: Cambridge University Press, 21–46.

Russell, B. (1910), *Philosophical Essays*, London: Longmans, Green & Co.

Schiller, F. C. S. (1908), *Plato or Protagoras?* Oxford: Blackwell.

Williams, B. (1978), *Descartes: The Project of Pure Inquiry*, London: Routledge, 2015.

14

RELATIVISM AND POSTSTRUCTURALISM

Gerald Posselt and Sergej Seitz

1. Introduction: relativism as a classificatory scheme

Relativism as a label for the categorization of different philosophical and theoretical positions did not emerge until the nineteenth century. Since then, however, it has become a powerful theoretical device for the retrospective classification, reinterpretation, and reconstruction of a wide variety of philosophical positions since early Greek philosophy: from skepticism and nihilism across perspectivism, historicism, and psychologism up to pluralism and postmodernism. Thus, the introduction of the label "relativism" involved the claim that huge parts of the history of philosophy could be rewritten through the lens of relativism (see Freudenberger 2010). This also includes poststructuralism as an array of influential theoretical and methodological approaches within the last third of the twentieth century. The term "poststructuralism" was first coined in North America in the 1980s to label various branches of contemporary francophone philosophy. Thus, "poststructuralism" does not refer to a coherent unity of thinkers and theoretical approaches, but functions as an umbrella term that subsumes authors as different as Gilles Deleuze, Jacques Derrida, Michel Foucault, or Jean-François Lyotard who are loosely related to each other by their critical engagement with structuralism (see Hoy 2010; Angermüller 2015).[1]

These thinkers have been repeatedly confronted with charges of relativism that are usually accompanied by the misleading equation of poststructuralism with postmodernism as well as with the claim that poststructuralism advocates a radical form of constructivism (see Hacking 1999; Vasterling 1999; Babka and Posselt 2016). Paul Boghossian states that poststructuralism "attempts to evade commitment to any absolute epistemic truths of any kind" by holding a "doctrine of equal validity" of conflicting truth claims, conceptual schemes, moral and ethical standards, cultural habits, etc. (Boghossian 2006, 94).[2] Similarly, Jürgen Habermas argues that Foucault reduces "validity claims . . . to the effects of power," which leads to a self-refuting relativism insofar as "truth claims [are] confined to the discourses within which they arise" (Habermas 1987, 276, 279).[3] Moreover, poststructuralism is said to imply a form of "ethical indifference" (Honneth 2007, 99) and to confine itself to exhibiting the constructed and perspectival character of certain positions without providing normative criteria for an effective critique of ethical and political problematics (see Jaeggi 2009, 281).[4]

Against this background, the juxtaposition of relativism and poststructuralism poses (at least) three interrelated questions: (1) the question of whether poststructuralist theories advocate a form of relativism, and if so, in what way;[5] (2) the question of whether (the contemporary discourse on) relativism provides a suitable conceptual framework for the comprehension of poststructuralism; and (3) the question of whether poststructuralist accounts, in turn, could be used for the critical analysis of the implicit epistemic and normative presuppositions that are at stake in the discourse on relativism. Although it seems possible to label poststructuralism as a form of relativism, we argue that such a classification ultimately misses the crucial theoretical and methodological insights to be learnt from poststructuralism. In contrast, we demonstrate that poststructuralism, consistently thought out, provides a productive provocation for the discourse on relativism to critically reflect on its presuppositions and implications.

2. From structuralism to poststructuralism

Structuralism as a theoretical approach and methodology within the humanities can be retraced to the work of the Genevan linguist Ferdinand de Saussure. Saussure conceives of language as a system of signs in which every element is defined by its differences from other elements. Consequently, "in language there are only differences. Even more important: difference usually presupposes the existence of positive terms between which the difference is set up; but in language there are only differences *without positive terms*" (Saussure 2011, 120). By prioritizing difference over identity, Saussure at the same time breaks with the main tenets of Western philosophy. Against the traditional assumption that differences arise between previously given entities, structuralism claims that identity does not precede difference, but is an effect of the processes of differentiation. In this sense, structuralism takes an anti-essentialist stance: linguistic elements are exclusively determined by the structural relations within the system. The sign does not possess a *meaning in itself* but only a *value* in contrast to all the other signs as well as in relation to its specific systemic position. This approach paved the way for the generalization of structuralism as a universal method. Insofar as structuralism does not analyze positively given entities but differential relations, it is not restricted to a particular object or set of phenomena but can be applied to all kinds of symbolic systems such as kinship relations (Claude Lévi-Strauss), literature and popular culture (Roland Barthes), psychoanalysis (Jacques Lacan), or critique of ideology (Louis Althusser).

With regard to the question of whether structuralism can be seen as a form of relativism, the answer is twofold: Insofar as it takes the value of every element to be dependent upon its differential position in the system, structuralism certainly constitutes a form of relativism. However, although the value of the sign appears to be "relative," this does not apply to the structure itself. In fact, structuralism aims at detecting underlying universal structures and principles that are common to different symbolic systems (from language across culture and society to the human psyche).

Poststructuralism can, in a first approach, be understood both as a critique and as a radicalization of structuralism's basic assumptions. The focus of poststructuralism's critique of structuralism is the latter's thinking in clear-cut dichotomies (such as langue/parole, synchrony/diachrony, signifier/signified, etc.) that disregards the hierarchies implicit in these oppositions as well as the associated primacy of synchrony over diachrony, and of language as a system over the concrete individual speech act. These hierarchies lead not only to the omission of the material condition of speech but also to the neglect of the temporality and historicity of symbolic structures. Furthermore, poststructuralism questions structuralism's assumption of a closed structure with a locatable center in order to investigate language as an object for scientific analysis. The reasoning here is as follows. If the meaning/value of each element within a structure is solely determined by its differing from other elements, one must subscribe to the notion of a structure with defined boundaries and a definite center. Otherwise, the meaning/value of an element would be infinitely deferred or suspended. However, precisely by presupposing an organizing principle – a "center" – that guarantees the unity of the system and the possibility of sense and meaning, structuralism introduces – against its own anti-essentialist aspirations – a "new essentialism" of a different kind (Laclau and Mouffe 2001, 112f.).[6]

3. Deconstruction and genealogy

In contrast to structuralism, poststructuralist accounts underline the constitutive absence of a center that could function as a transcendental basis or ultimate foundation for the conclusive determination of meaning. Thus poststructuralism reaffirms and radicalizes structuralism's difference-oriented approach and its anti-essentialist stance by performing a *decentering* of structure (Derrida 1978). Furthermore, poststructuralism takes structuralism's implicit "centrism" as a symptom of a general tendency in Western philosophy. Consequently, the main target of Derrida, Foucault, and others is less the linguistic theory of Saussure and his followers than the set of unacknowledged premises and presuppositions in the history of philosophy.[7]

In Derrida's philosophy of deconstruction, language is no longer conceived of as a centered and self-enclosed system based on binary oppositions, but as a continuous process of differentiation and reiteration. Derrida closely investigates Western philosophy's basic dichotomies such as presence/absence, speaking/writing, mind/body, reason/madness, culture/nature, man/woman, universal/particular, or absolute/relative (see Derrida 1997). In his critical engagement with the philosophical canon Derrida demonstrates that such basic distinctions are never neutral; rather, they are immanently hierarchized and normatively charged, such that one part of the opposition appears as the derivative and deficient part of the other (absence as lost presence; writing as ossified, dead voice; madness as the loss of reason; woman as inferior man, etc.). Derrida then shows that the dominating term (e.g. speech) is necessarily dependent upon its purportedly derivative, secondary counterpart (e.g. writing). But deconstruction does not stop here. Its aim is neither to do away with these concepts and oppositions nor to simply neutralize or reverse the hierarchies. Instead it seeks to transform the conceptual system in which such hierarchies are unreflectingly taken for granted as naturally and universally given, as well as to demonstrate that such oppositions are often *undecidable* or *aporetic*. Consequently, deconstruction performs a "double gesture" that puts "into practice a *reversal* of the classical opposition *and* a general *displacement* of the system" (Derrida 1988, 21).

Foucault's historical-philosophical project of genealogy performs both a decentering of *power* and a decentering of the *subject*. Power, according to Foucault, cannot be simply understood as sovereign domination. Rather, power is essentially diffuse, subtle, and relational. It does not act upon social relations from outside or above, but is immanent to them and constantly pervades them. As such, power is not only repressive, but also productive, insofar it is intertwined with practices of subjectivation and the production of knowledge (Foucault 1978). Relations of power are part of the processes through which subjects and forms of knowledge are constituted; in turn, power relations change their form and functioning with the transformation of epistemic discourses and modes of subjectivation. In the course of extensive historical analyses, Foucault analyzes the different ways in which phenomena that are crucial for our present self-understanding (such as sexuality, punishment, diseases, delinquency, psychiatric deviance, etc.) are constitutively intertwined with economic forms of production, discursive forms of knowledge, practices of subjectivation, as well as social and political power relations. In *Discipline and Punish* (1995), Foucault exemplarily shows how the introduction of disciplinary techniques of power and social control – as it manifests itself in the replacement of the traditional system of physical punishment with the modern prison system in the nineteenth century – goes along with the emergence of new sciences such as psychology, psychiatry, or criminology as well as with new modes of subjectivation and technologies of the self.

4. Poststructuralism as a form of relativism?

In view of poststructuralism's decentering of structure and its deconstruction of last principles or ultimate foundations, it seems to represent a form of relativism. Whereas classical structuralism maintains the notion of universal and ahistorical structures, thereby apparently evading the charge of radical or global relativism, the poststructuralist critique of structuralism's implicit essentialism appears to lead to the self-contradicting fallacies of global relativism.

The concept of *global relativism* refers to the thesis that any truth and validity claim is relative to some frame of reference, such as society, culture, history, etc. Such a position is contradictory and self-refuting; for its truth would imply its own falsehood, as already Plato held against Protagoras's *homo-mensura*-doctrine.[8] In other words: If global relativism were true, it would itself only be "relatively" true, that is, true with respect to some specific set of conditions – and therefore invalid as a universal epistemological position. In contrast, *local relativism* seems to avoid this deadlock, insofar as it specifies the domain (e.g., morals, aesthetics, etc.) or the scope to which the diagnosis of relativity applies, thereby evading the trap of "includ[ing] its own statement in the scope of what is to be relativized" (Baghramian and Carter 2018). While global relativism formulates a general thesis about the relativity of all bases of knowledge and norms, local relativism attempts to designate domains where absolute, non-relativizable foundations for truth are possible.

If we accept this schematic account, the charge of global relativism might be leveled against poststructuralism. Insofar as key poststructuralist thinkers hold that there are no pre-given entities or ultimate foundations, but rather infinite processes of differentiation (Derrida) or different regimes of truth (Foucault), there seems to be no absolute truth but at best relative "truths" in the plural, in the sense of context-dependent perspectives and interpretations. Although such criticism has proved to be productive for the clarification of poststructuralism's epistemic and normative implications, poststructuralist approaches do not advocate such relativism, nor do they put into question the possibility of truth and knowledge or the insistence of ethical demands or political obligations. This can clearly be shown in the cases of Derrida and Foucault.[9]

As already pointed out, Foucault's genealogy can be understood as a historical-philosophical form of inquiry into the complex genesis of present notions, institutions, and concepts. In doing so, genealogy by no means "relativizes" its objects of analysis; it asks how these "objects" (including ourselves) have *become* what they are, as well as which institutional frameworks, economic and material conditions, power relations, and practices of subjectivation are involved and mobilized in their respective genesis. It would therefore be a gross misunderstanding to infer from Foucault's inquiry into the complex relations of power, knowledge, and subjectivity that knowledge, truth or subjectivity are mere effects of power: "when I talk about power relations and games of truth, I am absolutely not saying that games of truth are just concealed power relations – that would be a horrible caricature" (Foucault 1997a, 296; altered translation). Foucault rather focuses on the specific historical entanglements and dependencies between power, truth and subjectivity, "without ever reducing each of them to the others" (Foucault 2011, 9).

Equally, neither the validity of scientific claims nor the possibility of scientific truth is impugned by the claim that forms of knowledge are interconnected with historically specific social, economic and political relations, institutions, disciplines and power structures. In fact, such criticism presupposes the false alternative between *validity* and *genesis*, between *truth* and *history* – an alternative that assumes a transcendent foundation for truth and validity, thereby locating truth and validity outside of the epistemic relations of justification and reasoning.[10] In contrast, genealogical critique – as the exhibition of the manifold interrelated epistemic, ethical, economic, social, and political "origins" of objects and phenomena – subscribes to logical reasoning and the possibility of drawing coherent conclusions from the envisaged historical and empirical data. In this respect, it claims to be a historically founded and justified analysis of the object's genesis, given the available material.

Moreover, even though genealogy refrains from an external moral or epistemic evaluation of the envisaged historical constellations, it is by no means a purely descriptive endeavor. As a "critical attitude," genealogy attempts to analyze the ways in which our relation to the world, to others, and to ourselves is, from the start, conditioned by historically sedimented social, material, and discursive practices, conceptual schemes, and modes of subjectivation (Foucault 1997b). Genealogy problematizes the value of norms, practices, and institutions by making visible the historicity and the transformability of what would otherwise appear as naturally and invariably given. It is precisely by this gesture that genealogy opens naturalized phenomena and relations to critique in the first place (Foucault 1995, 31; see also Saar 2002).

In this sense, Foucault's analyses of the genesis and development of the modern punishment system, of psychiatry, or of medicine do not purport to show that the different historical shapes of the punishment system, of psychiatry, etc. are "equally valid," but rather approach the topic from a different angle altogether. Instead of being concerned with the 'validity' of different developmental stages of a given system, genealogy poses the genuinely critical question of whether we want "to be governed like that, by that, in the name of those principles, with such and such an objective in mind and by means of such procedures" (Foucault 1997b, 44). In doing so, genealogy pursues the emancipatory aim of contesting naturalized and reified power structures implicit in these practices, schemes, and discourses, by making explicit the ways in which our present self-understanding is bound up with relations of power, forms of knowledge, and modes of subjectivation.

Just as Foucault's genealogy amounts neither to relativism about science nor to relativism about cultures or social institutions, Derrida's deconstruction, as an endeavor of disclosing and making explicit the systematic exclusions, implicit hierarchizations and unacknowledged aporias at work in Western philosophy, leads neither to an arbitrariness of interpretations nor to an indeterminacy of meaning.[11] *Undecidability* or *aporia* implies neither a suspension of logic or judgment nor an overturning of the relation of logic and rhetoric.[12] For Derrida, logic and rhetoric are complementary, like rhetoric and dialectics are according to Aristotle (1991, 354a). Logical reasoning must rely on rhetoric for convincingly articulating its truth claims,[13] just as rhetoric necessarily relies on logical forms of reasoning and argumentation. Accordingly, and in stark contrast to skepticism, deconstruction's focus on undecidability is not concerned with detecting aporias, in the sense of an equilibrium of arguments (Greek *isosthenia*), in order to then suspend judgment. Derrida rather argues that every decision, every judgment necessarily presupposes a moment of undecidability. Contrary to skepticism, he therefore regards aporia not as a hindrance for ethical and political responsibility, but rather as a precondition. Derrida's point is that without facing undecidability, there would be no *decision* and thus no responsibility at all, insofar as every decision would be reducible either to a mere application of rules or to a purely arbitrary act.

Derrida illustrates this aporetic moment with the example of a judge. On the one hand, justice (as a universal claim) demands that the judgment is formed according to a historically conveyed heritage of legal rules and norms. The judge cannot decide arbitrarily but has to form her judgment by applying the general rule of law. On the other hand, if the judgment were reduced to a mere application of rules, it could again not be called just. For justice also demands that the judge decides responsibly in regard of the singularity of every case (see Derrida 1992). In other words, (the) universality (of the law) does not rule out subjective responsibility but requires it necessarily. In this sense, the undecidable, for Derrida, is not – as for skepticism – the faltering between pre-given options that would lead to abstention of action and the suspension of responsibility, but rather the ethico-political urgency of acting and judging in a situation that cannot be reduced to mere calculation. At the same time, decision-making is all but random, since every decision faces a demand to universality while being at the same time performed against the background of a historical heritage of norms, values, and rules. Thus, although Derrida criticizes classical conceptions of universality, he does not reject the notion of universality as such. However, in contrast to the Kantian notion of universality as an abstract law, Derrida emphasizes that claims to universality are inevitably articulated from a particular position and within a specific context.

In fact, contrary to the view that poststructuralism doubts not only the possibility of universal justification but already the meaningfulness of aiming at it (see Wimmer 2001, 203), universality plays a pivotal role in many poststructuralist theories. While thinkers such as Jacques Derrida, Judith Butler, Ernesto Laclau, or Slavoj Žižek criticize substantial and procedural notions of universality as, for example, in discourse ethics, they by no means question the necessity and inevitability of articulating universal claims. Instead they point to an aporia inherent in universality as both necessary and impossible. It is *necessary* insofar as "each particular position, in order to articulate itself, involves the (implicit or explicit) assertion of its own mode of universality" (Žižek, in Butler et al. 2000, 315); it is *impossible* insofar as any claim to universality is necessarily articulated from a particular position and thus already contaminated with forms of particularity. Consequently, there is no pure universality detached from particularity. However, this is not to say that all universality is "false," by "secretly privileging some particular content, while repressing or excluding another" (Žižek, in Butler et al. 2000, 101). The constitutive interdependency of universality and particularity rather leads to a notion of universality as an open process that reflects the contingency of its foundations and the related inclusions and exclusions.[14] What poststructuralist accounts criticize is thus not universality itself, but rather the totalizing gesture of universality that negates the fact that any universal claim necessarily assumes a particular form, while any particular claim is performed within the horizon of universality.

Importantly, this leads us to an immanent criterion to criticize and evaluate different or conflicting truth claims, conceptual schemes, moral standards, cultural habits, social structures, and political institutions. An epistemic, social, ethical, or political scheme or system is "unjust" if it does not take into account the aporetic entanglement of universality and particularity, but rather strives to totalize its claim to universality. As a result, structuralism's "centrism," as outlined previously, as well as the various metaphysical "centrisms" in the philosophical tradition, are not simply descriptively or epistemically incorrect; they are also "wrong" by enforcing totalization and reification in areas as different as epistemology, ethics, or politics. In epistemology, centrism involves a totalization of conceptual schemes and theoretical systems by ignoring the constitutive exclusions inherent to them. In ethics, centrism amounts to a reification of intersubjective relations by negating or assimilating otherness. In politics, centrism leads to a notion of society as an all-encompassing, self-identical totality by negating dissent and difference. This becomes manifest in various versions of nationalism, xenophobia, racism, or sexism: they all constitute violent forms of centering or re-centering social relations around a notion of pure identity by obliterating difference within identity as well as particularity within universality.

5. Relativism and poststructuralism revisited

It is decisive to note that poststructuralist thinkers reject not only the charge of relativism but question the very theoretical and conceptual framework on which the discourse of relativism is based, especially the guiding difference of absolutism/relativism. According to Derrida (1999, 78), "relativism is, in classical philosophy, a way of referring to the absolute and denying it; it states that there are only cultures and that there is no pure science or truth," a position Derrida himself explicitly rejects.[15] Thus, as Derrida emphasizes, relativism as well as its conceptual counterpart absolutism (or fundamentalism) are tied up together by the general assumption that knowledge must be grounded in an absolute, indubitable foundation; they only differ in the question of whether or to what extent such grounding is possible. Consequently, absolutism as well as relativism remove the foundation of truth from the field of epistemic justification. In other words: both relativism and absolutism adhere to the view that truth – if such a thing exists – would have to be grounded in a "self-evident" principle or foundation exempt from reasoning and justification.[16] This observation also applies to forms of local relativism, since local relativists only modify the scope of relativism, so that "absolute" truths may be found in confined, particular areas, such as scientific discourse. As a consequence, both local and global relativism are inherently tied up with absolutism precisely by opposing and rejecting it. While global relativism undermines its own claims by falling prey to a performative self-contradiction, absolutism as well as local relativism, albeit to different degrees, postulate a justification-independent foundation for truth in empirical evidence, brute facts, the *cogito*, consciousness, etc. (see Puhl 2000, 170). In doing so, absolutism as well as (global and local) relativism bracket not only the constitutive justificatory nature of knowledge, but also undermine the possibility of critique by confining it to the field of epistemology.

Poststructuralism, in turn, underlines the conflictual and the ethico-political character of all truth and validity claims, thereby emphasizing the need of reasoning and argumentation in the name of responsibility and justice, precisely because there is no ultimate grounding. However, to say that there are no absolute foundations does not mean that there are no foundations at all. Rather, we are confronted with "contingent foundations" that have to be reaffirmed and grounded ever anew. The crucial question is then "what the theoretical move that establishes foundations *authorizes*, and what precisely it excludes or forecloses" (Butler 1995, 39). Poststructuralism thus contests the possibility of an ultimate ground, not the necessity of partial and always provisional grounding attempts (Marchart 2007, 14). In this sense, poststructuralism, far from advocating a doctrine of "anything goes" or "nothing matters," or a suspension of all grounds and reasons, makes a decisive plea for argumentation, questioning, and reasoning that takes account of the ethico-political dimension implicit in any kind of judgment.

In turn, absolutism as well as relativism can be conceived as attempts to do away with justification: either by positing some ultimate ground or Archimedean point withdrawn from all questioning (this is the gesture of absolutism) or by pointing to the contingency and, thus, equal validity of conflicting norms, values, standards, cultures, etc. (as relativism does). As a side effect, the discourse on relativism misleadingly "epistemologizes" poststructuralist accounts by pinning them down to the Gretchen question "What is your opinion on truth?," thereby disregarding poststructuralism's critical implications as well as its intrinsic ethico-political stakes.

Against this background, one of the crucial insights of poststructuralism consists in emphasizing the necessity of a critical attitude and argumentative practice that cannot retreat to a safe shelter of ultimate foundations without denying the necessity (and urgency) of contingent foundations and of articulating universal claims. In stark contrast to the "relativist" claim that all grounds are "contingent" and therefore "equally valid," poststructuralism sees "contingency" itself as a normative criterion to evaluate conflicting truth claims, social structures, moral norms, and political institutions. It is precisely in this sense that "the only democratic society is one which permanently shows the contingency of its own foundations" (Laclau 2000, 86), which is also the criterion to critically evaluate existing democracies, societies, moral systems, etc. At the same time, poststructuralist approaches engage in an ethical and political practice of reasoning and justification that recognizes the need to work through aporias and to responsively face up to specific historical constellations of epistemic standards, ethical norms, as well as political and juridical institutions by which we have become what we are without being determined by them. Thus, poststructuralism does not only evade the classificatory scheme of relativism but also provides a productive provocation to the discourse on relativism, by questioning the ways in which it remains bound up with the oppositions between absolute and relative as well as universal and particular without taking into account the complex interrelations and normative hierarchizations at work in these terms.

Acknowledgements

This chapter has been supported by the Austrian Science Fund (FWF) Research Project P26579-G22 "Language and Violence." Our special thanks go to Melissa Augustin, Arno Böhler, Anne-Kathrin Koch, and Martin Kusch for their valuable comments and remarks. We are also grateful for the opportunity to present a draft of this chapter at the ERC project seminar "The Emergence of Relativism" at the University of Vienna (November 9, 2018). We thank all participants for the productive discussion and many helpful suggestions.

Notes

1 As Slavoj Žižek emphasizes, "the very term 'poststructuralism,' although designating a strain of French Theory, is an Anglo-Saxon and German invention. The term refers to the way the Anglo-Saxon world perceived and located the theories of Derrida, Foucault, Deleuze, etc. – in France itself, nobody uses the term 'poststructuralism'" (Žižek 1991, 142). Therefore, when we speak of poststructuralism, invisible quotation marks are always added. Although the term allows one to highlight certain heuristic tendencies and basic threads, any thorough analysis has to reconstruct the tensions and differences between these authors. We focus on Derrida and Foucault as exemplary figures. Let us therefore just note that although Lyotard is regularly accused of advocating a postmodernist form of relativism, he is probably the least 'relativistic' of the aforementioned thinkers. As Honneth (2007) argues, Lyotard's conception of *differend* (1988) implicitly relies on normative criteria that are compatible with Habermas' discourse ethics.
2 This account of poststructuralism also plays a prominent role in the current debates on "post-truth" and "alternative facts." (See Latour 2004; Lepore 2016.)
3 See also Taylor (1984), who accuses Foucault of a "monolithic relativism."
4 On the accusation that poststructuralist accounts undermine the possibility of political agency, see the discussions in Benhabib et al. (1995).
5 On various classifications of the different forms of relativism, see Baghramian and Carter (2018), Freudenberger (2010), Krausz (2010), Kusch (2016).
6 For a more detailed account of the history of structuralism, see Dosse (1997); for an in-depth discussion of Saussure and Derrida, see Posselt and Flatscher (2018).
7 Accordingly, poststructuralist theories diagnose and criticize various forms of centrism in Western thought: *phonocentrism* (privileging spoken language over writing), *logocentrism* (privileging sense over sensibility, meaning over materiality), *phallocentrism* (privileging symbolic masculinity over femininity), *androcentrism* (privileging male over female perspectives), and *eurocentrism* (privileging the European culture over non-European traditions, Occident over Orient).

8 This is the general claim that man is the measure of all things, which leads to the view that truth and norms are only relatively valid (in relation to a certain perspective, culture, traditional value system, etc.). A convincing argument that "epistemological relativism is not the self-defeating, impotent and arbitrary view it is often made out to be" is provided by Kusch (1991, 200–210). On global relativism and self-refutation, see Kölbel (2011).

9 In his defense of "Foucault's relativism," Martin Kusch (1991, 210–226) demonstrates that Foucault's "genealogical criticism of power and rationalities" neither undermines the possibility of critique nor leads to normative indifference – quite the contrary. Although we largely agree with Kusch's reconstruction of Foucault's position, we argue that the absolute-relative dichotomy is already an object of genealogical and deconstructive critique.

10 As Pierre Bourdieu concisely puts it, "the ubiquitous opposition between relativism (or historicism) and absolutism, that is, the opposition between truth and history, is a fiction. . . . It has to be accepted that . . . law, science, and art . . . can be only historically and socially founded, without thereby rejecting their claims to universality" (Bourdieu 1991, 95; our translation).

11 On the difference between "indeterminacy" and "undecidablity," see Derrida (1999, 79): "When I say that there is nothing outside the text, I mean there is nothing outside the context, everything is determined." See also Hoy (2010, 530).

12 Notorious in this respect is Habermas's claim that "Derrida does not belong to those philosophers who like to argue" (Habermas 1987, 193), accompanied by the allegation that "Derrida is particularly interested in standing the primacy of logic over rhetoric . . . on its head" (1987, 187) – an accusation that Derrida vigorously rejects (Derrida 1988, 157). See also Posselt (2019).

13 This thesis has in fact already been put forward by Quintilianus (1996, XII 2, 5): "Whenever philosophers would defend the significance of their philosophy they should consider that they thereby make use of weapons of rhetoric and not of philosophy."

14 While thinkers such as Butler, Derrida, Laclau, or Lyotard agree on conceiving of universality as a process and not as a totality, they differ considerably in the description and conceptualization of this process. See also Zerilli (2002).

15 This also applies to Foucault. He is critical of "scientism," "that is, a dogmatic belief in the value of scientific knowledge," but also of "a skeptical or relativistic refusal of all verified truth": "What is questioned is the way in which knowledge circulates and functions, its relations to power. In short, the 'régime du savoir'" (Foucault 1983, 212).

16 Even though a card-carrying relativist such as David Bloor would probably reject this account, he perfectly illustrates that absolutism and relativism are mutually dependent when he claims that "if you are a relativist you cannot be an absolutist, and if you are *not* a relativist you *must* be an absolutist" (Bloor 2007, 252). In contrast, we argue that positing the absolutist-relativist dichotomy as absolute and unquestionable is itself part of the problem. For it undermines the possibility of reflecting on the implicit premises and conceptual hierarchies, as well as the mechanisms of exclusion and defamation that go along with this opposition.

References

Angermüller, J. (2015), *Why There is No Poststructuralism in France: The Making of an Intellectual Generation*, London: Bloomsbury.

Aristotle (1991), *The "Art" of Rhetoric*, translated by J. H. Freese, Cambridge, MA: Harvard University Press.

Babka, A. and G. Posselt (2016), *Gender und Dekonstruktion. Begriffe und kommentierte Grundlagentexte der Gender- und Queer-Theorie*, unter Mitarbeit von Sergej Seitz und Matthias Schmidt, Wien: facultas/UTB.

Baghramian, M. and J. A. Carter (2018), "Relativism," *The Stanford Encyclopedia of Philosophy*, Winter 2018 edition, edited by E. N. Zalta, https://plato.stanford.edu/archives/win2018/entries/relativism/.

Benhabib, S., J. Butler, D. Cornell and N. Fraser (1995), *Feminist Contentions: A Philosophical Exchange*, London and New York: Routledge.

Bloor, D. (2007), "Epistemic Grace. Antirelativism as Theology in Disguise," *Common Knowledge* 13(2–3): 250–280.

Boghossian, P. A. (2006), *Fear of Knowledge: Against Relativism and Constructivism*, Oxford: Clarendon Press.

Bourdieu, P. (1991), "Les juristes. Gardiens de l'hypocrisie collective," in *Normes juridiques et regulation sociale*, edited by F. Chazel and J. Commaille, Paris: Librairie générale de Droit et de Jurisprudence, 95–99.

Butler, J. (1995), "Contingent Foundations: Feminism and the Question of 'Postmodernism,'" in S. Benhabib, J. Butler, D. Cornell and N. Fraser (eds.) (1995), 35–58.

Butler, J., E. Laclau, and S. Žižek (2000), *Contingency, Hegemony, Universality: Contemporary Dialogues on the Left*, London: Verso.

Derrida, J. (1978), "Structure, Sign, and Play in the Discourse of the Human Sciences," in *Writing and Difference*, translated by Alan Bass, Chicago: University of Chicago Press, 351–370.

———. (1988), *Limited Inc*, edited by G. Graff, translated by S. Weber and J. Mehlmann, Evanston, IL: Northwestern University Press.

———. (1992), "Force of Law: 'The Mystical Foundation of Authority,'" in *Deconstruction and the Possibility of Justice*, edited by D. Cornell, M. Rosenfeld and D. G. Carlson, London and New York: Routledge, 3–67.

———. (1997), *Of Grammatology*, corrected edition, translated and with an Introduction by G. C. Spivak, Baltimore, MD and London: Johns Hopkins University Press.

———. (1999), "Hospitality, Justice and Responsibility: A Dialogue with Jacques Derrida," in *Questioning Ethics: Contemporary Debates in Philosophy*, edited by R. Kearney and M. Dooley, London and New York: Routledge, 65–83.

Dosse, F. (1997), *History of Structuralism*, translated by D. Glassman, Minneapolis, MN: University of Minnesota Press.

Foucault, M. (1978), *History of Sexuality, Volume I: An Introduction*, translated by R. Hurley, New York: Pantheon Books.

———. (1983), "The Subject and Power," in *Beyond Structuralism and Hermeneutics*, edited by Michel Foucault, Chicago: University of Chicago Press, 208–228.

———. (1995), *Discipline and Punish: The Birth of the Prison*, New York: Vintage Books.

———. (1997a), "The Ethics of the Concern for Self as a Practice of Freedom," in *Ethics: Subjectivity and Truth*, edited by P. Rabinow, translated by R. Hurley et al., New York: The New Press, 281–302.

———. (1997b), "What is Critique?" in *The Politics of Truth*, edited by S. Lotringer and L. Hochroth, translated by L. Hochroth and C. Porter, New York: Semiotext(e), 41–81.

———. (2011), *The Courage of the Truth: The Government of Self and Others II. Lectures at the Collège de France 1983–1984*, edited by F. Gros, translated by G. Burchell, Basingstoke: Palgrave Macmillan.

Freudenberger, S. (2010), "Relativismus," in *Enzyklopädie Philosophie*, edited by H. Jörg Sandkühler, Hamburg: Meiner.

Habermas, J. (1987), *The Philosophical Discourse of Modernity: Twelve Lectures*, translated by F. Lawrence, Cambridge, MA: MIT Press.

Hacking, I. (1999), *The Social Construction of What?* Cambridge, MA: Harvard University Press.

Honneth, A. (2007), "The Other of Justice: Habermas and the Ethical Challenge of Postmodernism," in *Disrespect: The Normative Foundations of Critical Theory*, Cambridge: Polity Press, 99–128.

Hoy, D. C. (2010), "One What? Relativism and Poststructuralism," in *Relativism: A Contemporary Anthology*, edited by Michael Krausz, New York: Columbia University Press, 524–535.

Jaeggi, R. (2009), "Was ist Ideologiekritik?" in *Was ist Kritik?* edited by R. Jaeggi and T. Wesche, Frankfurt am Main: Suhrkamp, 266–295.

Kölbel, M. (2011), "Global Relativism and Self-Refutation," in *A Companion to Relativism*, edited by S. D. Hales, Malden, MA: Wiley-Blackwell, 11–30.

Krausz, M. (ed.) (2010), *Relativism: A Contemporary Anthology*, New York: Columbia University Press.

Kusch, M. (1991), *Foucault's Strata and Fields: An Investigation into Archaeological and Genealogical Science Studies*, Dordrecht and Boston: Kluwer.

———. (2016), "Wittgenstein's On Certainty and Relativism," in *Analytic and Continental Philosophy: Methods and Perspectives. Proceedings of the 37th International Wittgenstein Symposium*, edited by H. A. Wiltsche and S. Rinofner-Kreidl, Berlin and New York: De Gruyter, 29–46.

Laclau, E. (2000), "Identity and Hegemony: The Role of Universality in the Constitution of Political Logics," in Butler et al (2000), 44–89.

Laclau, E., and Mouffe, Ch. (2001), *Hegemony and Socialist Strategy: Towards a Radical Democratic Politics*, second edition, London: Verso.

Latour, B. (2004), "Why Has Critique Run Out of Steam? From Matters of Fact to Matters of Concern," *Critical Inquiry* 30(2): 225–248.

Lepore, J. (2016), "After the Fact: In the History of Truth, a New Chapter Begins," *The New Yorker*, March 21, www.newyorker.com/magazine/2016/03/21/the-internet-of-us-and-the-end-of-facts.

Lyotard, J.-F. (1988), *The Differend: Phrases in Dispute*, translated by G. Van Den Abbeele, Minneapolis, MN: University of Minnesota Press.

Marchart, O. (2007), *Post-Foundational Political Thought. Political Difference in Nancy, Lefort, Badiou and Laclau*, Edinburgh: Edinburgh University Press.

Posselt, G. (2019), "Rhetorizing Philosophy: Toward a 'Double Reading' of Philosophical Texts," *Philosophy & Rhetoric* 52(1): 24–46.

Posselt, G. and M. Flatscher (2018), *Sprachphilosophie: Eine Einführung*, unter Mitarbeit von Sergej Seitz, second edition, Wien: facultas/UTB.

Puhl, K. (2000), "'Eine rationale Kritik der Rationalität'. Wahrheit und Relativismus bei Michel Foucault," in *Das Ende der Eindeutigkeit. Zur Frage des Pluralismus in Moderne und Postmoderne*, edited by B. Boisits and P. Stachel, Wien: Passagen, 167–181.

Quintilianus, M. F. (1996), *Institutio Oratoria*, translated by H. E. Butler, vol. 4, Cambridge, MA: Harvard University Press.

Saar, M. (2002), "Genealogy and Subjectivity," *European Journal of Philosophy* 10(2): 231–245.

Saussure, F. de (2011), *Course in General Linguistics*, edited by C. Bally and A. Sechehaye, translated by W. Baskin, New York: Columbia University Press.

Taylor, C. (1984), "Foucault on Freedom and Truth," *Political Theory* 12(2): 152–183.

Vasterling, V. (1999), "Butler's Sophisticated Constructivism: A Critical Assessment," *Hypatia* 14(3): 17–38.

Wimmer, R. (2001), "Universalisierung," in *Historisches Wörterbuch der Philosophie*, edited by J. Ritter and K. Gründer, Basel: Schwabe, 200–204.

Zerilli, L. (2002), "Review: Contingency, Hegemony, Universality: Contemporary Dialogues on the Left by Judith Butler, Ernesto Laclau, Slavoj Žižek," *Political Theory* 30(1): 167–170.

Žižek, S. (1991), *Looking Awry: An Introduction to Jacques Lacan Through Popular Culture*, Cambridge, MA: MIT Press.

PART 3

Relativism in ethics

15

MORAL AMBIVALENCE

David B. Wong

1. Introduction

One of the major tasks of contemporary moral philosophy is to deal with the implications of disagreement over values for issues concerning moral objectivity, universality, pluralism, and relativism. Of special relevance is the kind of disagreement that resists explanation as disagreement over factual matters that can be stated without use of normative terms. Reflecting on this kind of disagreement can lead to the phenomenon of moral ambivalence. If we are party to a disagreement and take the trouble to try to understand the other side's reasons, we may come to conclude that reasonable and knowledgeable people could have made different judgments than we have made, and any prior conviction about the superiority of our own judgments gets shaken (Wong 2006, 21). The importance of moral ambivalence is denied by those who take the universalist position that there is a single true morality. Moral universalism is most often associated with the meta-ethical position of robust realism – the claim that there are moral facts existing independently of how people have conceived of morality. This chapter discusses why moral ambivalence cannot be defused in the way universalists and robust realists suggest, and how lends support to a kind of moral pluralism or qualified relativism.

2. Arguments against the significance of moral ambivalence

Some who question the significance of moral ambivalence claim that relatively few people have this experience (e.g., Gowans 2007). Yet this claim seems to be made purely on the basis of anecdotal observation together with the assumption that laypeople possess a uniformly objectivist attitude towards morality. This attitude supposedly leads laypeople to treat all moral disagreements as conflicts in which at least one side must have a false or incorrect position. In fact, empirical study of attitudes toward moral disagreement indicates moral ambivalence to be a fairly pervasive feature of laypeople's response to moral disagreements, especially disagreements that are persistently difficult to resolve.

When they queried participants in a study of attitudes toward the truth and/or correctness of various ethical statements, psychologists Geoffrey Goodwin and John Darley (2008) found that participants varied considerably in assigning truth according to the content of the statement.

Most strikingly, although they generally agreed with the permissibility of abortion, assisted death, and stem cell research, only very small percentages assigned truth-values to statements expressing their positions (2008, 1346). In a second experiment, Goodwin and Darley asked participants not whether an ethical statement was true but whether there was one correct answer as to whether it was true or not. The participants were far more willing to say there was a correct answer as to the wrongness of robbing a bank than they were to say the same about the goodness of anonymous giving and highly unlikely to claim a correct answer as to the permissibility of assisted death (2008, 1352). Goodwin and Darley use the tendency to regard an ethical statement as true or false and there being one correct answer or not as measures of the extent of objectivity assigned to ethical statements by participants in the experiments. The participants' reluctance to attribute truth to an ethical statement may stem from their recognizing the extreme difficulty of resolving disagreements such as abortion together with the assumption that conflicting positions cannot both be true. Moreover, when given the opportunity to say that there *is* more than one correct answer as to whether an ethical statement is true or not, significant numbers of participants took that opportunity.

A study by Sarkissian et al. (2011) noted that most cases of moral disagreement portrayed in experimental studies depict implicitly or explicitly the disagreeing parties as peers from the same cultural group. They presented to participants cases of this type but also cases in which the disagreement was between someone described as a peer to the participant and a person of a foreign culture with different value orientations. A majority of participants thought there could only be a single correct answer in the former kind of cases, but not when cases were of the latter kind. The study's conclusion is that laypeople might tacitly believe moral judgments are relative to a cultural framework. Another possible conclusion, the one adopted here, is that many laypeople's theories of morality may be labile and are influenced by the kind of moral disagreements they presented with. Such possibilities need to be explored to fully evaluate familiar arguments from the philosopher's armchair. Of course, the fact that laypeople hold a certain view about morality doesn't mean that it's right. But a very common move moral philosophers employ is to argue that relativist positions cannot explain the lay assumption that there is always or nearly always a correct answer to moral disagreements. This argument is undermined if the objectivist attitude attributed to laypeople is more of a reflection of the majority of moral philosophers.

To get further in probing the significance of moral ambivalence, one has to look into *how* people arrive at this experience. One could arrive at ambivalence simply because one has already adopted a subjectivist or conventionalist attitude toward morality – that it is simply a matter of the individual's pro or con attitudes or the norms for behavior embedded in one's culture. On the other hand, one might arrive at moral ambivalence by actually investigating the basis of people's conflicting views in disagreements that seem difficult to resolve, and most importantly, by seeing what the best reasons are that are and could be given for each view, given that in any serious disagreement that concerns many people over time, many sorts of reasons, good, bad, and indifferent, will be given. Moral ambivalence gains relevance in proportion to the care and attention paid to the best reasons that could be given for each side in a disagreement. It gains relevance to the extent that a serious attempt to determine who is right results in an indefinite conclusion.

Consider another very common "defusing" strategy of defending against the idea that serious and difficult-to-resolve disagreements undercut the objectivity of morality. David Enoch points to a number of factors that could explain disagreement that are compatible with the truth of a robust realism that holds in independently existing moral facts and properties. Parties to moral disagreements do not reason carefully enough; they find it hard, or do not want, to imagine what it is like to occupy someone else's position; their interests wrongly influence their beliefs.

Perhaps, says Enoch, there are cases of moral disagreement in which there really is no fact of the matter as to who is right, i.e., what is right on the issue in dispute is simply indeterminate. This last concessionary move must be asserted to apply to only a small number of moral disagreements. Such proffered explanations have become the stock argumentative moves of realists. In his explanations of how people come to have false moral beliefs, Enoch pays special attention to the distorting effects of self-interest that might account, e.g., for why most people reject stringent duties to contribute to famine relief of the sort for which Peter Singer (1972) has argued. The rejection is to protect one's moral innocence, to fend off the deeply discomforting conclusion that one might be a morally monstrous person (2009, 25–26). The urge to protect one's fairly comfortable life is among the largely unconscious, unacknowledged forces that influence what moral beliefs we adopt and hold onto.

While this is a possible and prima facie plausible explanation for why most people reject stringent duties to help others, Enoch states no possible moral counter-reasons for acknowledging a space for the individual to achieve goals that do not fit into a strictly impersonal and consequentialist ethic. He does not consider the argument that we owe more to those with whom we are in some personal relationship, parents who nurtured us or children we brought into existence. He simply declares that he knows of no plausible counterargument to the impartialist position. This is not to deny the force of arguments that the reality of strangers and their desperate situations ought to weigh heavily on one's moral conscience. But such arguments need to be weighed against moral considerations pertaining to the moral dimensions of our relationships with particular people. While Enoch is right to get into the specifics of how people arrive at their positions on stringent duties to strangers, he does not go far enough. The richness and complexity of moral life is bleached out in the course of making a move in a debate that remains distressingly abstract.

When normative positions are asserted, it usually is not long before brute moral intuitions are deployed to support these positions, without much consideration, if any, of intuitions that might come into conflict with them. By a "brute moral intuition," I mean thoughts as to what is right that are taken to have no need of further justification but whose truth should be evident upon reflection. One of the benefits of the movement toward "experimental philosophy" is that it brings out into the open the question of how representative philosophers can claim their intuitions to be. Philosophers have long deployed their intuitions without sufficient critical self-reflection on the way they reflect their social, economic, political, and culturally like-minded circles. Thus, instead of the usual short a priori arguments in metaethics that presume there is a common shared conception of what morality is about, what if the brute moral intuition asserted by one side is challenged with interpretations of opposing positions that are less easy to dismiss as wrong or false but rather connect with values the importance of which is commonly appreciated? What if a measure of sympathetic imagination is employed to try to understand why others might believe something that one is inclined to abruptly dismiss?

Here these questions will be applied to disagreements stemming from differences in value priorities accorded to autonomy and relationship. The moralities that give highest priority to relationship will be defended against those who would assert the priority of autonomy as the only correct position to have. This choice is made partly because sociologically, the more popular position taken by the subculture of people who read this essay is the autonomy side. It is the relationship side that needs the most defending. A very different argument defending the viability of the autonomy side might have to be made to audiences from most of the rest of the world.

A type of morality that is centered on the good of relationship and community stands in contrast to a type centered on autonomy and rights (Wong 2006). This does not mean that one type of morality needs to exclude the values associated with the other type. The difference is typically a difference in the priority assigned to one or the other set of values in case of conflict. Many Americans hold a morality centered on autonomy and rights, but they also value relationships. And among those who place central moral value on relationships and community, there is concern for the individual when her interests conflict with those of others in her group. There is also concern for something that can legitimately be called autonomy, although this kind of autonomy might not be all that is valued by moral traditions in which this value takes center stage. Relationship-centered moralities are often stereotyped by advocates of rights-oriented moralists. It will reveal something of the deeper normative structure of these moralities to show how this stereotype is profoundly mistaken.

3. Do relationship-centered moralities subordinate the individual to the group?

One feature of the stereotype of relationship-centered moralities is that they subordinate the individual to the group. Indeed, when social scientists and cultural psychologists began studying what is here called "relationship-centered moralities" as a distinct configuration, they often called them "collectivist," followed by a definition that mentions subordination of the individual to the group (e.g., Triandis 1989, 1995). The stereotype also includes the presence of strong and rigid social hierarchies: father or the head of the family, the patriarch of the clan, tribal chief, and king. The stereotype may hold that "collectivist" moralities may have once served a good social function, perhaps under conditions that necessitated close cooperation between members of a group, but that development and new forms of technology have made such intensive forms of interdependence and hierarchy unnecessary. According to the stereotype, relationship-centered moralities became discredited when the intrinsic worth of the individual became more and more apparent, where this worth might be interpreted, for example, as the capacity for self-direction. Those who remain wedded to relationship-centered ways of life may not have had experience of other more liberating and self-directed ways. Or they may benefit from being at or near the top of hierarchies that are endorsed under these moralities, and self-interest biased their moral beliefs toward these moralities. Now, however, the enlightened are in a position to clearly recognize the inherent dignity and worth of the individual and that such worth grounds rights that require for the individual opportunities and liberties to be protected against the demands of others.

An alternative to this stereotype of relationship-centered moralities have stressed not that the individual is subordinated to the group but that the person is conceived differently. Studies in cultural psychology indicate a difference in the degree to which persons are conceived in terms of their relationships with each other (e.g., Markus and Kitayama 1991; Shweder and Bourne 1982). People's behavior and attitudes may be understood more or less as responses to the particular others with whom they are interacting, or the type of situation they are. In other words, cultures vary in the extent to which the kind of person one is seen as depending on the others one is with or on the type of situation. Such tendencies in description suggest a "relational" conception of a person's identity that is constituted in good part by traits that must be specified as patterns of action and reaction toward particular people or particular situations. I am a certain way with my family and another way with my colleagues and still another way with strangers I encounter on the street, and these different ways go into the person I am. At the other end of spectrum stands an individual with an identity that is detachable from the particular people one happens to be interacting with or from the present situation. The traits going into the identity of such a self are specified to a much greater degree independently of the particular people one happens to be with or the kind of situation one is in.

What difference does this relational conception make when it replaces the stereotype of a "collectivist" self that is subordinated to the group? It implies that a person cannot be conceived as fully separate from the group who is then subordinated to it. Others in that group may already be a part of me. Furthermore, this conception of the person as overlapping in identity with others has normative implications for what constitutes the good of the individual and how that good relates to the good of others. One's relationship with others can form a part of one's good *as an individual*. One has a compelling interest in their welfare and in one's relationship with them. Sustainable and morally viable relationships depend the individuals' gaining satisfaction and fulfillment from being in those relationships. Rather than the individual's good being subordinated to others, the individual's good overlaps with the good of others. For example, the good of each member of a family may include or overlap with the good of other members of the family. When one family member flourishes, so do the others.

But what, it may be asked, is to be done in cases where an individual's interests come into conflict with those of other members? Not all our interests are interests in relationships, and of course we have multiple interests in different relationships. There are times when we cannot satisfy all these interests. One picture of what should be done comes from the texts of Confucianism – a long and internally diverse moral tradition of thought centered around the value of relationships. The picture is a continual process of balancing and negotiation – harmonization – between the interests of individuals and those of the others to whom the individual is related. Negotiation between interests is carried on in the light of the interdependence of individuals and the various communities to which they belong, and also the interdependence of the goods towards which they aim. An individual's interests may sometimes have to yield to those of others. A partial compensation for this is that a central part of the individual's good lies in being, for instance, a member of the family. On the other hand, the good of the family cannot be achieved without consideration of an individual's important interests. If those interests are urgent and weighty, they must become important interests of the family and can sometimes have priority in case of conflict. Sometimes differences have to be split in compromise. Sometimes yielding to others must be balanced against having priority at other times. In thinking what we must done, we are guided by the thought that our own good involves the good of others and our relationships to them. We mutually adjust conflicts in light of the interdependence of our goods.

A story from 5A2 of the *Mencius* illustrates these points. It is about Shun the sage-king, who is presented as an exemplar to those aspiring to live the Confucian *dao* or way. He was exemplary for his ability to get people to work together and for his extraordinary filial piety. His father and stepmother treated him cruelly and accordingly to legend even tried to kill him three times. But he remained filial and according to legend succeeded in winning their love. This background us to understand the story of the time when he wanted to get married. It was at a time in his relationship to his parents when he knew that if he were to ask permission to marry, he would be denied. He decided to marry anyway without telling them. This is ordinarily an extremely unfilial act. But Mencius defends what Shun did, saying that if Shun had let his parents deny him the most important of human relationships, it would have embittered him toward his parents. Mencius' reasoning is that Shun's good as an individual depends on both his desired marriage relationship and his relationship to his parents. For him to conform to his parents' wishes is not only to deny him the first relationship but also to adversely affect the second. For the sake of both relationships he must assert his own good, which in the end is not separate from the good of his parents.

4. Are relationship-centered moralities rigidly hierarchical?

Consider the stereotype of "rigid" hierarchy. This is the picture of well-defined social roles with duties that define what it means for someone in a particular social role to be subordinate to someone else in a corresponding social role. The continuing tradition of Confucian thought up the present day has led many who are basically sympathetic to Confucianism to partial or wholesale rejection of many of the hierarchies accepted by the early Confucians. Sinyee Chan's argument (2000) that the core ethical content in the *Analects* and the *Mencius* can be separated from acceptance of gendered hierarchies of social and political roles is an excellent case in point. There is nothing in the ideal of the moral excellent person that goes against the idea of women being fully able and eligible to satisfy that ideal. There is no pretense to have found some natural qualities of women per se that make them less suitable for pursuit of that ideal.

The stereotype of rigid hierarchy, for another example, has children unconditionally submit to the wishes of their parents. Whether one should always obey parents' wishes has always been a fraught question in the Confucian tradition. The *Analects*, partly because of Confucius' frequently indirect manner of addressing questions posed to him, and partly because of textual ambiguities, gives no straightforward answer. Other texts in the tradition provide a wide range of different answers, but it would suffice here to highlight Xunzi's discussion of this question. In his chapter *Zidao* or "The Way to Be a Son," Xunzi lays down three conditions for when one should disobey parents: first, when obeying them would be to harm them and disobeying would be to benefit them; second, when obeying would be to disgrace them and disobeying would be to bring honor to them; and third, when obeying them would be doing something beastly and when disobeying would be doing something requiring cultivation and decorum. Xunzi concludes his discussion by saying, "Follow the Way, not your father." Xunzi goes on to argue for ministers who are contentious with their rulers for the sake of the state's welfare, and for individuals who are contentious with their friends for the sake of what is right. He concludes, "To be careful about the cases in which one obeys another – this is called filial piety, this is called fidelity" (ch. 29.55, translation from Hutton 2014).

5. Do relationship-centered moralities devalue autonomy?

Another stereotype of relationship-centered moralities is that they do not value individual autonomy. Seeing beyond the stereotype requires recognition of the multiple dimensions of the concept of autonomy. The type of autonomy most valued in Confucianism is moral autonomy. It is speaking truth to power and not submitting to whatever power commands, as is exemplified by Xunzi's views about the moral necessity of frankness. It is also independence from the approval and disapproval of popular opinion as reflected by Confucius' act (in *Analects* 5.1) of giving his daughter in marriage to a man convicted of a crime because he believed the man to be innocent. The social stigma of having a criminal history was enormous at that time, and Confucius' act symbolized his emphatic rejection of the corrupt manner in which the law was administered. Moral autonomy also involves authenticity. In *Analects* 17.13, Confucius speaks of the "village worthy" who is the "thief of virtue." As explained by Mencius (7B37) the village worthy simulates the appearance of a genuinely good person in order to gain its social advantages. The implied contrast is one who focuses on trying to be truly worthy, whether or not this is recognized. In sum, Confucian moral autonomy is judging and acting according what is humane and right and proper in the face of the countervailing pressures of power and social favor and disfavor. The more one wholeheartedly identifies with the project of realizing these values, the more one is able to reliably judge and act upon them, the more one realizes the kind of moral independence Confucians prize.

Confucian autonomy in its classical form does not include what one might call "personal autonomy": having a range of choice about how to live one's life that is protected against the influence or what some might call the interference of others. But Confucianism, like any other viable moral tradition, can evolve. Recently, contemporary forms of Confucianism, most notably Joseph Chan's (2014), have proposed making room for a pluralism of conceptions of the good life, and thus making room for "personal autonomy," defined as enabling an individual to develop a personal identity, fashion his or her character, and chart a unique path in life. Such a concept of personal autonomy functions not as a moral right but as a valuable aspect of a good life. It is a matter of degree – one can be more or less autonomous in the personal sense, and this value need not be absolute. "Personal autonomy competes with, and at times can be outweighed by, other values such as well-being and other ethical ideals" (Chan 2014, 198).

This has merely a partial sample of what one should do to take seriously the phenomenon of moral ambivalence. One must descend from abstract arguments in meta-ethics to investigate what other great traditions of moral thought have found to be lives worthy of aspiration and of compelling excellence. One must take the trouble to reconsider stereotypes of moral perspectives that make it too easy to dismiss them and to escape moral ambivalence. Often, there is not just the brute assertion of an intuition but also absence of consideration of the alternatives to the intuition. Serious comparative study of another tradition can unseat such tendencies to rest on brute intuition or simple neglect. The fair and sympathetic interpretation of another moral tradition not only takes study but also the willingness to weigh its claims against one's own experience of life and that of others. Of special relevance are the satisfactions of relationship that a morality such as Confucianism accords the highest priority. One should also have experience of the satisfactions of a way of life that affords the individual, more than Confucianism does, substantial protections against the expectations and requirements of others and even their well-meaning interventions.

6. How a pluralism of moralities might be part of the best explanation of moral ambivalence

Perhaps morality is not about moral properties that are *sui generis* or an irreducible part of the fabric of the world, but a cultural invention that has interpersonal and intrapersonal coordination functions in human life. The interpersonal function is to promote and regulate social cooperation. The intrapersonal function is to foster a degree of ordering among potentially conflicting motivational propensities, including self- and other-regarding motivations, and this ordering serves to encourage people to become constructive participants in the cooperative life and to live worthwhile lives (the latter function, to be clear, does not reduce to the former). The content of moral norms is constrained not only by the functions they must fulfill but also by the nature of the beings they govern. This nature includes a diverse array of self- and other-regarding motivational propensities plausibly considered to have a biological, innate basis in most human beings. Morality's functions, plus the nature of the beings it governs, constrain the content of its norms. Different moralities must share some general features if they are to perform their functions of coordinating beings having particular kinds of motivations.

For example, a strong case can be made that all moralities adequately serving the function of fostering social cooperation must contain a norm of reciprocity – a norm of returning good for good received. Such a norm is a necessity because it helps relieve the psychological burden of contributing to social cooperation when it comes into conflict with self-interest. There is also a constraint one could call "justifiability to the governed," which implies that justifications for subordinating people's interests must not rely on falsehoods such as the natural inferiority of racial or ethnic groups or of women. This follows from the way that morality came to be conceptualized as a distinctive way of fostering social cooperation on a largely voluntary basis that can be accepted without coercion or deception. A third example of a universal constraint stems from the ubiquity of disagreement itself: one might call "accommodation" the willingness to maintain constructive relationship with others with whom one is in serious and even intractable disagreement. The particular sources of disagreement discussed here – differences over how to interpret values that might even be shared among people, and differences as to how to prioritize values in case of conflict – make serious and potentially intractable disagreement ubiquitous, even in the most harmonious of societies. Social cooperation would come under impossible pressure if it always depended on strict agreement over moral norms (see chapters 2 and 9, Wong 2006, for a discussion of these constraints).

The shared general features help account for the commonalities experienced in moral ambivalence: the recognition that the moral paths others have taken in response to conflicts of values is not only intelligible but can be understood to have normative force. The diverging paths experienced in moral ambivalence arise from the fact that the constraints only *place boundaries on* the range of moralities that adequately fulfill the coordinating functions, but they do not select only one specific morality. The resulting theory is something that can be called "pluralistic relativism" (Wong 2006). There is more than one true or adequate kind of morality, but the range is a limited plurality. Not all moralities are true.

References

Chan, J. (2014), *Confucian Perfectionism*, Princeton, NJ: Princeton University Press.

Chan, S.Y. (2000), "Gender and Relationship Roles in the *Analects* and the *Mencius*," *Asian Philosophy* 10(2): 115–232.

Enoch, D. (2009), "How Is Moral Disagreement a Problem for Realism?" *Journal of Ethics* 13: 15–50.

Goodwin, G. P. and J. M. Darley (2008), "The Psychology of Meta-Ethics: Exploring Objectivism," *Cognition* 106: 1339–1366.

Gowans, C. (2007), Review of *Natural Moralities: A Defense of Pluralistic Relativism*, *Notre Dame Philosophical Review* 4/14/2007, https://ndpr.nd.edu/news/25269-natural-moralities-a-defense-of-pluralistic-relativism/ (Accessed December 6, 2014).

Hutton, E. L. (2014), *Xunzi: The Complete Text*, Princeton, NJ: Princeton University Press.

Markus, H. R. and S. Kitayama (1991), "Culture and the Self: Implications for Cognition, Emotion, and Motivation," *Psychological Review* 98(2): 224–253. http://content.apa.org/journals/rev/98/2/224.html.

Sarkissian, H., J. J. Park, D. Tien, David, J. Wright and J. Knobe (2011), "Folk Moral Relativism," *Mind and Language* 26(4): 482–505.

Shweder, R. A. and E. A. Bourne (1982), "Does the Concept of the Person Vary?" in *Cultural Conceptions of Mental Health and Therapy*, edited by A. J. Marsella and G. M. White, Dordrecht, NL: D. Reidel, 97–137.

Singer, P. (1972), "Famine, Affluence, and Morality," *Philosophy and Public Affairs* 1(3): 229–243.

Triandis, H. C. (1989), "The Self and Social Behavior in Differing Cultural Contexts," *Psychological Review* 96(3): 506–520.

———. (1995), *Individualism and Collectivism*, Boulder, CO: Westview Press.

Wong, D. B. (2006), *Natural Moralities: A Defense of Pluralistic Relativism*, New York: Oxford University Press.

16

MORAL RELATIVISM AND MORAL DISAGREEMENT

Alexandra Plakias

1. Introduction

This chapter examines some of the routes from moral disagreement to relativism. The first part of the chapter looks at the main strands of argument from moral disagreement to relativism;[1] the second half assesses the evidence concerning the nature and extent of moral disagreement, and how this bears on each type of argument. As we'll see, some arguments depend heavily on empirical claims about the nature and extent of moral disagreement, while others claim that the mere possibility of moral disagreement is enough to motivate moral relativism. However, it's not always easy to say whether moral disagreement on an issue *is* possible, and our best evidence for claims about possible disagreements often turns out to be empirical evidence of actual disagreement. The chapter concludes with suggestions for further areas of research.

2. Disagreement, diversity, and convergence

Before we start, it will be useful to distinguish a few related topics. The title of this chapter refers to *disagreement*, but the term is ambiguous. Many philosophers maintain that disagreement only occurs when parties take up conflicting attitudes towards some proposition, belief, or other content. In this sense, the phenomenon is relatively narrow: it's not clear that it makes sense to talk about disagreements between cultures or groups, or across historical eras. Those are better described as instances of moral *diversity* – cases where various cultures and groups hold a wide variety of moral views, and engage in diverse moral practices. "Diversity" is a more encompassing term, and allows us to remain neutral with respect to questions about the content of moral beliefs, utterances, values, and practices (which is useful because these questions are controversial, and it's hard to take a view on them without begging the question for or against relativism!). Where an argument doesn't rely on this narrower sense, I'll use "diversity"; otherwise, I'll use "disagreement." We'll say more about these concepts, and their roles in various arguments, in Section 3.4.

Not all moral disagreements matter equally. People get things wrong in all sorts of ways, and mistakes about matters of empirical fact, or errors in reasoning, lead to moral disputes that lack any philosophical significance. These may be easy to resolve, or difficult – Descartes' belief that animals are automata has been solidly refuted, but the debate over fish pain rages on – but where a disagreement is caused by an epistemic fault on the part of one (or both) parties, it tells us more about the disputants and their beliefs than it does about morality itself. After all, no one denies that morality is difficult, so it's not surprising that we often get things wrong.

This is why philosophers often focus on *faultless* disagreements, in which neither party is mistaken. Whether these are even possible, much less actual, is also controversial (as we'll see), because here again we lack a theory-neutral way of assessing its possibility – in saying that faultless moral disagreement is possible, we seem to be affirming that realism is false. In saying it's impossible, we seem to beg the question against realism. Faultless disagreement is often cited as the diagnostic criteria for anti-realism:[2] if two parties can disagree about some matter (whether moral, aesthetic, epistemic, or scientific) and neither party is mistaken, then there is no objective fact about the matter at hand. So faultless disagreement is a loaded term, and it will be useful to have a more neutral descriptor in hand. I'll use the term *fundamental disagreement* to refer to moral disagreements in which neither party is guilty of irrationality, ignorance of non-moral facts, or any other identifiable epistemic shortcoming[3] relevant to, *but not including*, the question under dispute. (In practice, it can be difficult to distinguish these – where does the subject of a moral disagreement begin, and the relevant/related facts or questions end? Is the personhood of a fetus the issue at hand in debates over abortion, or is it just a related, but distinct, issue? – but we'll put this to one side for the time being.)

With these clarifications and concepts in hand, let's look at some of the arguments from disagreement to relativism. We'll examine four types of argument – some empirical, some a priori – before turning, in Section 4 to the evidence concerning moral disagreement.

3. The arguments

3.1. *From moral diversity to relativism via IBE*

One of the most influential formulations of the argument from disagreement comes from Mackie, who refers to it as "the argument from relativity." Mackie takes as his starting point, "the well-known variation in moral codes from one society to another and from one period to another, and also the differences in moral beliefs between different groups and classes within a complex community" (1977, 36). He goes on to note that, while this is itself a descriptive fact, "it may indirectly support second-order subjectivism" (1977, 36), because the differences we observe in moral judgments

> make it difficult to treat those judgments as apprehensions of objective truths . . . the actual variations in the moral codes are more readily explained by the hypothesis that they reflect ways of life than . . . that they express perceptions, most of them seriously inadequate and badly distorted, of objective values.
>
> *(1977, 36–37)*

Mackie offers two competing hypotheses: on the realist explanation, there are mind-independent moral facts, but diversity reveals that most of our beliefs about these facts are mistaken – many individuals and cultures are getting (and have gotten) it wrong. We'll look at the epistemic implications of this argument in more detail in Section 3.2. But the argument needn't be interpreted solely as a point about moral error; we can also read Mackie as claiming that moral diversity shows that mind-independent moral facts have no *explanatory role* to play in our moral judgments.[4] That's because, in explaining how we've ended up with such diverse moral codes, the realist offers a story about how we acquire moral beliefs which doesn't require exist-ence of mind-independent moral facts.

On the relativist explanation, we are making roughly correct judgments about the moral facts, and the apparent conflict is explained by variations in the facts themselves: the best expla-nation is *not* that one culture happened to land on the correct moral code while others did not, but that each culture is (roughly) correct about the moral code, because what morality requires varies according to culture. This in turn would best be explained by the idea that morality is a social or cultural construction – "people approve of monogamy because they participate in a monogamous way of life," (1977, 37) rather than the other way around. Morality is a codifica-tion of our existing practices and "way of life," rather than an apprehension of facts determined independently of our (individual or collective) attitudes.[5]

Responses to Mackie's argument have tended to point out explanations of diversity he over-looks – for instance, the many ways error and irrationality creep into our moral judgments. Fol-lowing Doris and Plakias (2008), call these "defusing explanations." Whether, and how widely, these explanations apply to moral diversity and disagreement is an empirical question, and we'll assess this in Section 3.3. It's also unclear how to interpret the scope of Mackie's claim. It's unlikely that the same explanation applies to all moral diversity and disagreement. How many cases can the realist cede to the relativist? If some instances are best explained by factual error, or irrationality or self-interest, while others are best explained as a deep divergence in values, should we count this as a draw? We won't arrive at an answer to this question here, but note that to the extent that a realist is successful at explaining disagreements in terms of an epistemic shortcoming, they lend further credence to skeptical worries about our ability to successfully arrive at moral knowledge. In other words, the more disagreements are due to failures of ration-ality, the worse we seem to be at moral reasoning. This brings us to the epistemic argument from moral disagreement.

3.2. *From disagreement to relativism, via skepticism*

If moral realism is true, moral questions have a uniquely correct answer. But disagreement persists about many moral questions, even between experts who have studied the issues for years. Given this disagreement, how can individuals remain justified in their moral beliefs? The skeptical argument from disagreement says that they cannot.[6] Faced with disagreement between equally well-informed, thoughtful individuals, it seems that we should suspend judgment in our moral beliefs. In sum, the conjunction of moral realism plus fundamental disagreement equals skepticism.

This version of the argument doesn't require that the disagreement in question be faultless, only that we can't tell where the fault lies. The argument is most naturally presented in a first-person form: if I am faced with an epistemic peer who disagrees with me about a moral issue, it seems that I am required to suspend judgment. But we can generalize it to disagreements in which we're not personally or presently engaged: many of us find ourselves surrounded by peo-ple who, by and large, share our moral views. We may even go out of our way to avoid interact-ing with people who hold conflicting moral views. Surely this shouldn't immunize us against the epistemically deleterious effects of disagreement. If that's right, then the argument is equally effective from a first- and third-person point of view.

The epistemic argument is directed against realism; it becomes an argument for relativism if we defuse its force by denying absolutism and adopting some form of relativism instead. Since moral skepticism is especially unappealing, moral relativism becomes a relatively palatable option.[7] Because relativists can allow for both parties to such a disagreement to be correct, it doesn't put pressure on either to alter their judgment another person's apparently faultless, yet divergent, judgment has no epistemic bearing on one's own justification.

This is a specific application of the more general debate over the correct epistemic response to disagreement: conciliationists argue that we should reduce our confidence in the face of disagreement; steadfast views argue that we can maintain confidence, unmoved by disagreement. But there are features to the moral case that put extra pressure on the steadfast view: first, moral judgments may be opaque, because they're based on affective processes to which we lack introspective access. Haidt (2001) argues that most moral reasoning is a sort of post-hoc confabulation; moral judgment is driven by quick, automatic processes. This doesn't mean that we *can't* produce reasons justifying our moral judgments, but it does mean that it may be *more difficult* to rule out epistemically disreputable influences on moral judgments than on, say, judgments about mathematical calculations. A second difference is that in the case of moral judgments, the practical stakes are often very high. Whether practical stakes influence the conditions for knowledge is itself controversial,[8] but it does seem to recommend caution and epistemic humility. It also gives us additional reasons to suspect our judgments, since we ourselves sometimes have a stake in the outcome. Interestingly, this latter point is sometimes made by realists by way of explaining why we shouldn't be too troubled by metaphysical arguments from apparently faultless disagreement; the realist notes that there are lots of ways to get a moral judgment wrong, many of which are subtle and hard to detect, so it's likely there's a fault somewhere, even if we can't see where.[9] If that's right, though, it tells in favor of a conciliationist approach to moral disagreement, and lends an extra bite to the skeptical argument against realism and in favor of relativism.

If morality is to play its practical role, it must be epistemically accessible to us as we are. Moral facts should be action-guiding, and not just in principle. That's not to say that moral knowledge must be *easy*, but it obviously can't be impossible. Exactly how to strike the balance between the two – how difficult moral knowledge can be – is a tricky question, and I don't think we should expect an answer here. Suffice to say, responding to instances of apparently faultless moral disagreement by saying that the parties are ignorant of the moral facts, is a potentially costly move for the realist – especially since some of the most entrenched and intractable moral disagreements concern central, paradigmatically moral issues such as capital punishment, abortion, and the ethics of eating animals.

3.3. From disagreement to semantic relativism – and back!

3.3.1. From faultless disagreement to relativism

This argument is similar to the preceding, but it doesn't rely on inference to the best explanation. Rather, the argument asks us to consider the possibility of two individuals who disagree over some moral issue and are making no mistake: a faultless moral disagreement. If it is possible for two individuals to give different, apparently conflicting answers to some question, and for neither to be mistaken, the issue in question is relative.

The biggest difference between this version of the argument from disagreement and the first version we discussed is that this version is a priori. We needn't evaluate competing explanations of a phenomena; rather, we need only establish that faultless disagreement is possible. In one sense, this is easier, since we don't have to gather actual instances of disagreement and see how they're best explained. But in another sense, this is harder: how are we to establish the (im)possibility of faultless moral disagreement?

3.3.2. From faultless disagreement against relativism

The relativist thinks that participants in a faultless disagreement must not really be contradicting each other, else the disagreement wouldn't be faultless. The realist thinks one of them must be at fault, else they would not be engaged in disagreement.

The challenge for the realist is to explain where the fault might be; we saw some strategies for this in Section 3.1. The challenge for the relativist is to explain how something faultless can nonetheless be a disagreement. If relativism is motivated by the observation that some disagreements seem faultless, it is threatened by the observation that where there is no fault, there doesn't seem to be disagreement in the first place. In other words, the very notion of faultless disagreement seems incoherent.

In response, relativists have offered various approaches to moral semantics that explain how two individuals might simultaneously offer incompatible statements that are nonetheless both true. These approaches, sometimes known as "new relativism," or "semantic relativism," attempt to reconcile the intuition that disagreement requires shared content, in the form of a single proposition, about which subjects can disagree, with the intuition that both parties to a dispute can be getting it right. We will not discuss the details of such views here except to note that their motivation relies heavily on intuitions about what "genuine" disagreement requires and where it occurs. And when it comes to moral disagreement, our intuitions may be muddied by various factors: the stakes of such disagreements can be high, and we have a significant practical interest in getting others to agree with us. For this reason, our intuitions about which disagreements are "real" and worth pursuing may not be a good guide to the semantics of moral claims – they may instead reflect the pragmatic circumstances in which the disagreement takes place. This assumes we *have* such intuitions to begin with, which is an empirical claim; we'll return to it in Section 4.2, when we look at some preliminary evidence regarding intuitions about the relationship between disagreement and content.

3.4. From disagreement to relativism, via incommensurability

Philosophers who resist the idea that there can be genuinely faultless disagreements – who want to insist on the truth of relativism, and are willing to concede the ability to capture "genuine" disagreement (see the discussion in Section 3.3.2) point to moral *diversity* as the relevant empirical phenomenon. Whereas disagreement relies on similarity – two parties must be arguing about a single proposition, or belief, or norm – diversity invokes difference. Moral diversity is therefore a broader phenomenon; it is found in cases where the conceptual, cultural, and historic differences are too vast to compare moral beliefs and practices. For example, one it might seem strange to say that we have a *disagreement* with the ancient Greeks over the permissibility of slavery; the institutions that each culture is judging, and the set of social practices in which its meaning is embedded, are too far removed from one another to merit the label. Or consider a case from Rovane (2013, 41–43), who describes an American woman visiting India and meeting Anjali, a woman with an arranged marriage, who devotes herself to domestic duties and to taking care of her family. The American has worked in investment banking and made lots of money. She sees her family occasionally but they are not close, and she doesn't offer them financial support, nor do they expect her to. While these women can communicate about their lives, and while each can see that the other is doing as she ought according to her culture's values, neither can genuinely view the other's life as an option for herself – the American can appreciate that when Anjali speaks of obligations, she means what is obligatory in her culture. But neither woman can accept the other's point of view. The normative frameworks in which they exist are incommensurable, and individuals cannot move between them. Rovane is inspired by Williams' "relativism of distance," according to which we can have only "notional confrontations" with distant cultures, because the moral concepts and values in which their beliefs are embedded, and against which their beliefs have meaning, are not open to us. But Williams confines his relativism to our encounters with temporally distant societies, arguing that "today all confrontations between cultures must be real confrontations" (1985, 163).

Is historical incommensurability any less metaethically significant than contemporary incommensurability? Suppose it turns out that we lack a shared conceptual framework with historical societies, but have one with contemporary society – does this undermine realism? To the extent that it shows that moral facts vary across time and societies, it shows that moral facts are not strongly mind-independent; rather, they depend on socially and culturally local practices, beliefs, and attitudes. Indeed, we might go further and argue that even historical relativism shows us something about the metaphysics of moral facts: namely, that there are various moral "worlds" within the actual world we inhabit.

4. The evidence

Having surveyed a few of the routes from moral disagreement to moral relativism, it's time to turn our attention to the evidence for moral disagreement: how much and what kind exists, and how should we go about investigating it? In this section, we'll look at empirical evidence bearing on the existence of moral disagreement. We'll start by looking at some traditional approaches to the issue, consisting of cross cultural and ethnographic investigation. Then, we'll look at a second, newer approach: experimentally investigating moral intuitions and their origins, as well as intuitions about disagreement itself. We'll conclude by discussing directions for future research.

4.1. Traditional approaches to moral disagreement

Observations of cross-cultural diversity have played a role in arguments from disagreement since the ancient Greeks; in the contemporary era, some philosophers have gone so far as to take up the role of ethnographer themselves. For example, the mid-twentieth-century philosophers John Ladd and Richard Brandt spent time with the Navajo and Hopi, respectively, cataloging their ethical beliefs and practices. Brandt, in particular, was explicit about the need for information bearing on whether a disagreement could truly be described as fundamental: "We have . . . a question affecting the truth of ethical relativism which, conceivably, anthropology can help us settle. Does ethnological evidence support the view that 'qualified persons' can disagree in ethical attitude?" (1954, 238). Brandt continues, "Some kinds of anthropological material will not help us – in particular, bare information about intercultural differences in ethical opinion or attitude"[10] (1954, 241).

While methodologically instructive, these early efforts yielded relatively scant results. Brandt himself pointed to Hopi attitudes towards animal suffering as a disagreement with other Americans, a claim that may surprise contemporary readers impressed by the suffering of factory-farmed livestock.

Critics have pointed out the dangers of this kind of fieldwork, and about the dangers of generalizing about a culture's moral views. Moody-Adams (1997) is particularly pessimistic about the prospects for an empirically grounded argument from relativism to disagreement: "The problem is that it is profoundly difficult to construct a reliable description of the moral practices of an entire *culture* – a description of the sort that could license judgments contrasting one culture's basic moral beliefs with those of other cultures," in large part because of "facts about the opacity of human conduct, the complexity of cultures, and the resultant difficulties of determining who has authority to represent the defining moral principles of a culture" (1997, 41–43). She also points to disagreement among ethnographers, whose accounts of cultures themselves conflict, as reason to be skeptical of claims to characterize "the" moral values or practice of a culture (1997, 47–49).

Moody-Adams' critiques are important reminders of the hazards of ethnography (and history). She draws our attention to the fact that cultures are subject to critique from within, and the moral practices of a culture may be the subject of disagreement among/between its own citizens. While this is a useful caution, it depends in part on the existence of *intra*cultural moral disagreement – and this may be a more fruitful source of data for the relativist.

4.2. *Disagreement intuitions and semantic arguments*

A much more recent source of evidence about disagreement and relativism comes from experimental work on people's intuitions about moral disagreement. Psychologists and experimental philosophers have investigated people's willingness to grant the possibility of faultless moral disagreement, taking this willingness as evidence for a kind of folk moral relativism. For example, Goodwin and Darley (2008) gave subjects a range of statements in ethical, factual, and aesthetic domains and asked them to rate their agreement with each statement, as well as to indicate whether they thought it was a true statement, a false statement, or an opinion or attitude. Subjects were also asked whether, if another person disagreed with them about the claim, one of the two must be mistaken. What they found was that subjects treated ethical claims as intermediate between factual and aesthetic ones: whereas in the factual case, they agreed someone must be mistaken, and in the aesthetic case, they tended to deny this, the moral statements fell somewhere in between, suggesting that moral statements were treated as less-than-fully objective. Interestingly, there was significant variation *within* the moral domain; when a claim concerned a highly controversial matter such as euthanasia or stem cell research, subjects treated it as less objective than when it concerned a relatively settled issue like the wrongness of bank robbery. This suggests that the folk conception of moral facts and objectivity is linked with convergence: the more convergence there is about a question, the more likely people are to view it as objective. Exactly what the nature of this link is remains to be determined – is convergence a symptom of objectivity, or is it constitutive?

One limitation of these data is that they don't necessarily indicate relativism so much as antirealism, since they're compatible with a kind of folk noncognitivism. Indeed, one might wonder whether any empirical data *could* distinguish relativism from noncognitivism, especially when the difference between the views turns on subtle distinctions about semantics and the mental content associated with moral judgment which may not be accessible to subjects, nor discernable to experimenters. For example, some writers (see, e.g., Gill 2009; Sinnott-Armstrong 2009) have cited evidence from neuroimaging showing that emotion/affect is implicated in moral judgment as support for noncognitivism. But since relativists don't deny – and indeed often welcome – a role for emotion, this evidence is inconclusive at best.

A second complication comes from nascent research into people's intuitions about the relationship between disagreement and content. As we saw in Section 3.3, philosophers tend to assume that for two individuals to genuinely disagree, at least one party's claim must be false – they can't both be true. Therefore, we can infer, from our intuition that a dispute involves a genuine disagreement, that there must be some "exclusionary content" at stake. Khoo and Knobe (2016, 111) refer to this as the "exclusionary inference":

> One first observes that speakers in moral conflict cases disagree, and one needs an explanation for this fact. It seems that the best explanation is that these speakers are making claims with exclusionary content. Therefore, one should infer that these speakers are indeed making claims with exclusionary content.

However, in a series of experiments, the authors failed to find empirical support for such an inference. Instead, they found that subjects' intuitions about moral disagreement and their intuitions about exclusionary content diverged[11] – they were willing to treat a case as a disagreement, while denying that one of the parties' claims must be false. This suggests that people are willing to countenance disagreement even in the absence of exclusion. More work is needed here, but if these results hold up, we may need to retreat from the exclusion inference. Alternatively, we might question the reliability of intuition as a guide to "real" or "genuine" disagreement. In either case, this study suggests one avenue for future investigations into the limitations of intuitions about disagreement as a guide to semantic content.

Experimental research is one way to investigate people's intuitions about the meaning of "moral disagreement," but we can also look to interviews and ethnography for further information. An instructive example here is Hallen and Sodipo's (1986; see also Hallen 2000) work in Yoruba epistemology. Interested in philosophical debates over the indeterminacy of translation, the authors set out to investigate Yoruba concepts of knowledge and belief. They did so by interviewing elder, respected members of the Yoruba community (the Onisegun), asking them about various terms used in folk epistemic discourse. They describe their methods: "Data was collected in the context of guided, cross-cultural, discussions rather than in question and answer sessions" (1986, 10). During these discussions, "the *onisegun* introduce numerous concrete examples. . . . What is important, however, is that they . . . provide action-oriented illustrations of . . . the criteria of these two concepts [of knowledge]" (1986, 75).

It is perhaps impractical to demand that contemporary experimental philosophers engage in fieldwork of their own. But Hallen and Sodipo's work demonstrates that there is still much to be gained from qualitative investigation. How do people use the term "disagreement"? Do we have multiple concepts associated with the term, and do these require shared content? When do we think a conflict is "real," or "genuine," or "significant"? Because philosophical arguments for and against relativism turn on these questions, empirical investigations into the way the concept(s) of disagreement functions in everyday language matters, and surveys about intuitions are just one of several methods that may be useful here.

4.3. The psychology of moral disagreement

Another source of evidence about the extent and persistence of moral disagreement comes from studying the origins of, and influences on, moral intuitions. In Section 3.2, we briefly mentioned Haidt's (2001) work showing that moral judgment relies on affect-laden intuitions, and that reasoning's role is often post-hoc, rather than causal. Other work has discovered correlations between individual sensitivity to disgust and moral judgments about homosexuality as well as political orientation – subjects who are more disgust-sensitive are more likely to disapprove of homosexuality and endorse conservative political views.[12] This illustrates one way in which hypotheses about the psychology underlying moral judgment bears on the prospects for resolving moral disagreement: if the causal processes responsible for moral judgments are inaccessible to conscious reflection and reasoning, then we might be pessimistic that continued moral argument will bring us closer to consensus. Changing people's underlying attitudes, rather than reflective beliefs, might be a more effective route to moral convergence. This in turn reminds us that, just as not all kinds of moral disagreement are philosophically significant, nor are all kinds of moral convergence.

5. Conclusion

The arguments surveyed here represent just a few of the many arguments from disagreement that figure in the metaethical debates between realism and relativism. There are also a number of related issues, such as disagreements in politics and psychology, that bear on moral disagreement. For example, what is the correct theory of emotion? If emotions – particularly moral emotions such as disgust and anger – involve cognitive appraisals of their elicitors, this suggests that conflicts of emotion are themselves a form of (rather than just a corollary of) moral disagreement. Likewise, the relationship between political and moral disagreement is similarly unclear; if relativism is true, we confront the question of how to make political accommodations for divergent moral outlooks and conflicting practical goals.

Indeed, as moral and political polarization becomes increasingly urgent from a practical perspective, one area for future research is not moral disagreement, but moral agreement. Doris and Plakias (2008) refer to "superficial" moral disagreements as those caused by factual error or ignorance, with the implication being that the underlying disagreement is not really moral at all. In evaluating moral convergence (and the various routes to it) we will confront questions about whether the moral convergence we see is merely superficial, merely verbal, or a genuine agreement on values and beliefs.

Notes

1 The format of this first section owes a debt to David Enoch's (2009) paper, "How is Disagreement a Problem for Realism?" Enoch's paper is worth consulting for those interested in detailed realist responses to some of the arguments discussed here; since his topic is arguments *against* realism rather than *for* relativism, he discusses a different and more extensive list of arguments from those presented here.

2 See, for example, Wright (1992); Smith (1994).

3 This clause precludes us from insisting that there must be some fault if a disputant is unable to, for example, simply *see* the wrongness of the case.

4 See Harman (1977, ch. 1) for a version of this argument that doesn't rely on disagreement.

5 Mackie explicitly does *not* intend this to be an argument for relativism; he is arguing against moral realism in favor of error theory, a position we won't discuss here. But this is because of his premise that any adequate, vindicatory analysis of moral discourse will be committed to the existence of mind-independent objective universal moral facts, a premise the relativist naturally rejects!

6 See McGrath 2008.

7 I have put aside expressivist/noncognitivist views here, due to space constraints. But note that if one begins as a realist, relativism is a closer neighbor than expressivism, so it is less of a "retreat." There is of course much more to be said on this issue.

8 Stanley (2005), Hawthorne and Stanley (2008).

9 See, e.g., Shafer-Landau (2003, ch. 1).

10 Brandt's use of "qualified persons" corresponds roughly with our conception of an idealized agent: one who is informed, rational, has had time to deliberate, and so on. The phrase functions to exclude disagreements that are based on epistemic defects.

11 Interestingly, the finding applied in moral cases, but not in cases involving non-moral, factual disputes.

12 See Inbar et al. (2009); Inbar, Pizarro and Bloom (2009).

References

Brandt, R. (1954), *Hopi Ethics*, Chicago: University of Chicago Press.

Doris, J. and A. Plakias (2008), "How to Argue about Disagreement: Evaluative Diversity and Moral Realism," in *Moral Psychology*, vol. 2, edited by W. Sinnott-Armstrong, Cambridge, MA: MIT Press, 303–331.

Enoch, D. (2009), "How Is Moral Disagreement a Problem for Realism?" *Journal of Ethics* 13(1): 15–50.

Gill, M. B. (2009), "Indeterminacy and Variability in Meta-Ethics," *Philosophical Studies* 145(2): 215–234.

Goodwin, G. P. and J. M. Darley (2008), "The Psychology of Meta-Ethics: Exploring Objectivism," *Cognition* 106(3): 1339–1366.

Haidt, J. (2001), "The Emotional Dog and Its Rational Tail," *Psychological Review* 108(4): 814–834.

Hallen, B. (2000). *The Good, the Bad, and the Beautiful: Discourse About Values in Yoruba Culture*, Bloomington, IN: Indiana University Press.

Hallen, B. and J. O. Sodipo (1986), *Knowledge, Belief, and Witchcraft: Analytic Experiments in African Philosophy*, Stanford: Stanford University Press.

Harman, G. (1977), *The Nature of Morality: An Introduction to Ethics*, Oxford: Oxford University Press.

Hawthorne, J. and J. Stanley (2008), "Knowledge and Action," *Journal of Philosophy* 105(10): 571–590.

Inbar, Y., D. A. Pizarro and P. Bloom (2009), "Conservatives Are More Easily Disgusted Than Liberals," *Cognition and Emotion* 23(4): 714–725.

Inbar, Y., D. A. Pizarro, J. Knobe and P. Bloom (2009), "Disgust Sensitivity Predicts Intuitive Disapproval of Gays," *Emotion* 9(3): 435–443.

Khoo, J. and J. Knobe (2016), "Moral Disagreement and Moral Semantics," *Noûs* 52: 109–143.

Mackie, J. L. (1977), *Ethics: Inventing Right and Wrong*, London: Penguin Books.

McGrath, S. (2008), "Moral Disagreement and Moral Expertise," in *Oxford Studies in Metaethics: Volume 4*, edited by R. Shafer-Landau, Oxford: Oxford University Press, 87–108.

Moody-Adams, M. M. (1997), *Fieldwork in Familiar Places: Morality, Culture, and Philosophy*, Cambridge, MA: Harvard University Press.

Rovane, C. (2013), *The Metaphysics and Ethics of Relativism*, Cambridge, MA: Harvard University Press.

Shafer-Landau, R. (2003), *Moral Realism: A Defence*, Oxford: Oxford University Press.

Sinnott-Armstrong, W. (2009), "Mixed-Up Meta-Ethics," *Philosophical Issues* 19(1): 235–256.

Smith, M. (1994), *The Moral Problem*, Oxford: Blackwell.

Stanley, J. (2005), *Knowledge and Practical Interests*, Oxford: Oxford University Press.

Williams, B. (1985), *Ethics and the Limits of Philosophy*, Cambridge, MA: Harvard University Press.

Wright, C. (1992), *Truth and Objectivity*, Cambridge, MA: Harvard University Press.

17

MORAL RELATIVISM

Peter Seipel

1. Introduction

Moral relativism comes in a variety of forms.[1] Understood in one sense, moral relativism is an empirical or anthropological claim, roughly as follows:

> (DESCRIPTIVE MORAL RELATIVISM) (DMR) Many moral disagreements between cultures appear to be fundamental.

Saying that a disagreement is fundamental means it involves an underlying conflict in moral values (Brandt 2001; Levy 2003). DMR is a claim about the existence and scope of fundamental disagreement. A second version of moral relativism is normative. Here is a rough characterization:

> (NORMATIVE MORAL RELATIVISM) (NMR) If a fundamental disagreement between cultures is not or cannot be rationally resolved, members of the divergent cultures morally ought to refrain from judging and interfering with the actions of each other.

NMR is a claim about what we should do. Proponents of NMR maintain that we morally ought to tolerate members of other cultures in cases of intractable disagreement. A third form of moral relativism is neither empirical nor normative but rather metaethical in nature:

> (METAETHICAL MORAL RELATIVISM) (MMR) Moral judgments are not true or false in an absolute or universal sense but rather are true or false relative to cultures.

MMR says that whether something is right or wrong, good or bad, virtuous or vicious, and so on depends on cultures. No moral judgment is true or false absolutely speaking. A judgment may be true in relation to one culture but false in relation to another.

My aim in this chapter is to spell out a disagreement argument for MMR (hereafter the disagreement argument). I will not say anything further about NMR. I will discuss DMR, however, as the disagreement argument is based on DMR and the claim that the best explanation of widespread fundamental disagreement is that moral judgments are true or false only relative to cultures. Not all relativist arguments are based on disagreement. Whereas some arguments for MMR focus on incommensurability (MacIntyre 1988; Rovane 2013), others consider the nature of moral agency (Velleman 2013). Still others appeal to disagreement but emphasize an analysis of moral judgments (Harman 1975; Prinz 2007) or an account of some feature of ordinary moral experience (Wong 2006). Arguably, however, the disagreement argument is the most widely discussed and perhaps the strongest version of the argument for MMR.[2] For this reason, I will focus on it here.

2. MMR

Like moral relativism, MMR comes in different forms. Some relativists maintain that the truth-value of moral judgments is relative to individuals (Dreier 1990; Prinz 2007). Others, however, reject individualistic forms of relativism in favor of the view that a moral judgment is true or false in relation to groups or cultures (Harman 1996; Rachels 2009). Sometimes known as "cultural relativists," these latter relativists maintain that moral judgments have at least some intersubjective authority. I consider their view in the present context.

MMR further subdivides into appraiser or agent relativism and content or truth relativism. Appraiser relativists say that whether a moral judgment is true or false depends on the values of the appraiser – the person who makes or evaluates the judgment. For agent relativists, the truth-value of the judgment depends on the values of the agent – the person being evaluated (Lyons 1976). Most proponents of MMR are appraiser relativists. Some also defend a form of content relativism, the view that the content or meaning of a moral judgment varies between contexts. In this chapter, however, I consider truth relativism. Truth relativists maintain that the content of a moral judgment is always the same but its truth-value varies from one context to another.

MMR is incompatible with moral objectivism, the view that moral judgments are true or false in an absolute or universal sense. For moral objectivists, the truth-value of a moral judgment does not depend on one culture or another but rather is the same across all cultures. Relativists agree with objectivists that moral judgments are true or false but maintain that the truth-value of these judgments is relative rather than absolute. MMR is not the only alternative to moral objectivism. Whereas moral skeptics deny that we are justified in accepting or rejecting some or all moral judgments, non-cognitivists argue that moral judgments are neither true nor false. I will return to the debate between objectivists and relativists later.

How prevalent is MMR? Although the view has a growing number of supporters, most ethicists reject it.[3] Many philosophy professors, however, claim that college students tend to espouse some version of moral relativism (Bjornsson 2012, 373; Harman 1998, 210; Loeb 2003, 34; Stich and Weinberg 2001, 641). Recent psychological experiments provide some confirmation of this idea. Some ordinary people espouse MMR with respect to some moral issues (Goodwin and Darley 2008; Wright et al. 2013), especially in cases of apparent intercultural disagreement (Sarkissian et al. 2011). Nonetheless, the empirical literature on (MMR) remains extremely limited.[4]

3. DMR

We are now in a position to consider the disagreement argument. Earlier we noticed that proponents of this argument defend MMR as the best explanation of DMR, the claim that there appears to be widespread fundamental disagreement between cultures. Is DMR correct? Although it may seem obviously true, DMR is actually highly contentious. Some critics argue that there is more cross-cultural moral agreement than relativists appear to recognize in their discussions of DMR (Rachels 2009). Others attempt to refute DMR on *a priori* grounds, arguing that widespread moral disagreement is impossible because genuine disagreement requires considerable agreement as a precondition (Davidson 2004; Myers 2004). Both argumentative strategies are controversial.[5]

A third criticism of DMR takes its starting point from the concept of culture. Critics argue that relativists tend to view cultures as homogenous and self-contained entities. In reality, however, cultures are diverse and contain a variety of conflicting views. Because they often lack sharply defined boundaries and are also subject to nearly constant change, distinguishing between them in the way that DMR requires is at best difficult or at worst impossible (Moody-Adams 1997).

In reply, relativists tend to grant that cultures have indistinct boundaries and are not monolithic but deny that this undermines DMR. Many appeal to an analogy between language and culture.[6] Like cultures, languages lack sharply defined boundaries. Words in one language may belong to another. Examples in Italian and English include paparazzi, cappuccino, bruschetta, and graffiti, to mention just a few. Like cultures, moreover, languages change over time as new words are developed or borrowed from other languages, such as *il* computer or *il* iPhone in Italian. Finally, like cultures, languages are not homogenous. Many contain regional differences in grammar and pronunciation. Still, it is plausible to speak of particular languages and to assess statements as correct or incorrect relative to a language. Given that we can make these linguistic assessments, Neil Levy argues that "nothing prevents us from making moral judgments relativized to a culture or sub-culture" (2003, 172).

But even if can make such judgments, opponents of DMR argue that the deeper issue is that of determining the fundamental values of a culture. Who is to count as representing these values? Why should we consider one person a representative member of the culture and not another? Michelle Moody-Adams calls this "a difficulty peculiar to the study of cultures: that of deciding who – if anyone – has the 'authority' to represent the defining principles, especially the basic principles, of a given culture" (1997, 43). Here the difficulty is not only that cultures contain conflicting views but also, and more importantly, that a particular practice may not reflect the moral views of a majority of the members of a culture. Some practices may be merely ideological, in the sense that they exist only to serve the interests of a powerful group. Such practices presumably do not support DMR, as it is generally agreed that there is a difference between morality and ideology.[7]

Consider the much-discussed example of a medieval Japanese practice known as *tsujigiri*. Samurai warriors believed that a sword should be able to cut a person in half with a single stroke. A defective sword was thought not only to bring dishonor on a Samurai but also to disappoint his emperor and offend his ancestors. So, rather than risk using a new sword in battle, warriors tried out their swords on random passersby. Does this example support DMR? Moody-Adams does not think so, as she doubts that "endangered passersby would have generally *consented* to be sacrificed to the ritual." Given the "familiar human distaste for unnecessary suffering," she writes, they must have found it objectionable (1997, 82; cf. Midgley 2003, 84–87). But if that is so, then the practice presumably lacks a genuine moral justification – it existed only to serve the interests of an oppressive elite.

Discussing this case, Levy agrees that a practice must be generally accepted to support DMR and also grants that Japanese peasants did not consent to being cut in half. But he denies that the members of a culture must consent to being harmed by a practice to accept it. Some World War I soldiers appear to have accepted the risk of dying in the trenches but they arguably did not consent to being killed. In a similar way, says Levy, even if Japanese peasants did not consent to being split in half, they may have accepted the risk of premature death at the hands of a Samurai.

Just as World War I soldiers risked their lives for their countries, the peasants may simply have believed that "it is better for the society as a whole that *tsujigiri* be practiced" (2003, 176).

Suggesting that Japanese peasants may have consented to *tsujigiri* is one thing, however. Showing that they actually accepted it is something else entirely. Recognizing this point, Levy writes that "whether the Japanese peasants actually regarded *tsujigiri* as permissible or not is an empirical question which can only [be] settled by way of detailed historical and anthropological investigation" (2003, 177). Even granting that such investigation is necessary to support DMR, however, it may not be sufficient. In a recent article, Sonia Sikka (2012) argues that consent by itself does not enable us to determine whether a practice reflects morality or ideology. Sometimes consent may be due to inequalities in the distribution of power, as less powerful groups are silenced or manipulated by elites. Establishing DMR, therefore, requires us to determine not only whether most members of a culture consent to a given practice but also how the consent was in fact generated. We need, says Sikka, "to consider the actual conditions of cultural production" (2012, 57; cf. Fricker 2013, 794–795).

If this is correct, then showing that a practice is widespread and even generally accepted is not enough to support DMR. Relativists must also show that the practice has a moral rather than a merely ideological basis. But that is surely no easy task, not least since marginalized groups like medieval Japanese peasants tend to be underrepresented in the historical record (Sikka 2012, 57). Further research is needed to determine whether this task can be completed.

4. Inference to the best explanation

Even if DMR is true, philosophers generally agree that widespread disagreement does not, by itself, establish that morality is relative. Relativists must also show that MMR best accounts for DMR. One popular strategy here is to maintain that morality is a social construction. For example, Gilbert Harman (1975) famously argues that human beings construct moralities through implicit agreements. Why do fundamental disagreements arise? The reason is simple: the terms of these agreements are often vague or inconsistent. People may disagree not only about the implications of their agreements for particular situations but also about what changes should be made to address defects in the agreements.[8]

Supporters of the disagreement argument follow Harman and other relativists in defending MMR by appeal to inference to the best explanation. Granting that explanatory success is a reliable indicator of truth, moral objectivists have three possible counter-strategies:

(1) Show that DMR is false;
(2) Deny that DMR calls for explanation; or
(3) Provide an equally plausible or better explanation of DMR.

We considered the first strategy in the previous section. What about the second? Commentators on both sides of the relativism/objectivism debate have wondered whether every moral disagreement requires explanation.[9] It remains unclear, however, why we should think that some moral disagreements are inexplicable. Worse, it is troubling if the best objectivist response to the disagreement argument is simply to deny that DMR is explanatorily significant.

To provide a stronger response to the disagreement argument, objectivists must consider the third counter-strategy – they need to provide "defusing explanations" of DMR, explanations that promise to defuse the threat that disagreement poses to their view (Doris and Plakias 2008, 311). Multiple defusing explanations have been proposed in the literature. A familiar explanation blames many apparent moral disagreements on conflicting non-moral beliefs.[10] Another attributes these disagreements to the distorting influence of self-interest.[11] Other defusing explanations attempt to account for DMR in objectivist terms by appealing to differences in material conditions between cultures,[12] the youthfulness of secular ethics,[13] and the difficulty of resolving some moral problems.[14]

Clearly objectivists have a number of options for responding to the claim that MMR provides the best explanation of DMR. One commentator even remarks that "the striking fact" about the disagreement argument is that "alternative [i.e., defusing] explanations are so easy to come by" (Enoch 2009, 25). But how plausible are these explanations? Because this is in part an empirical question, answering it requires us to engage in empirical investigations. We must consider actual cases of serious moral disagreement in light of the best available evidence from empirical disciplines such as psychology and history (Doris and Stich 2005, 132; Loeb 1998, 284). To date, however, most objectivists have been content to describe defusing explanations in general terms.[15] Few have engaged in the difficult work of using empirical research to make sense of actual disputes.[16] Some simply offer broad speculations, as in Richard Boyd's often-cited remark that "agreement on non-moral issues would eliminate almost all disagreement about the sorts of moral issues which arise in ordinary moral practice" (1995, 339).

Recognizing the need to consider the empirical record, some relativists and other critics of objectivism have begun examining empirical research on real-world disagreements. For example, Doris and Stich (2005) and Doris and Plakias (2008) discuss differences in attitudes towards violence in the American North and South. Drawing on the work of Nisbett and Cohen (1996), Doris et al. argue that a "culture of honor" persists in the South. Unlike Northerners, Southerners tend to respond to insults with violence and value male reputations for strength. According to Doris et al., standard defusing explanations are unable to account for this apparent disagreement over the permissibility of violence. Both Northerners and Southerners agree on relevant non-moral facts, such as what constitutes an insult, and neither side seems to have non-question-begging reasons for thinking the other is irrational. Self-interest also does not appear to be responsible for the conflict, as there is little reason to suspect that "Southerners' economic interests are served by being quick on the draw" or that Northerners' interests are "served by turning the other cheek" (Doris and Plakias 2008, 320). Thus, according to Doris et al., common defusing explanations do not apply to this case.[17] If relativists can provide a better explanation, then we will have some reason to accept (MMR).

In another study, Abarbanell and Hauser (2010) examined the moral beliefs of a group of rural Mayan subjects in Mexico. Subjects considered several dilemmas. In one dilemma, a man can save five people by choosing not to warn a pedestrian about to walk into the path of an oncoming truck. In another dilemma, a man can save five people by pushing the person next to him in front of the truck. Most Western subjects agree that the second case is worse than the first. Westerners tend to view harms caused by action as worse than harms caused by omission (Cushman et al. 2008). Even young children exhibit this distinction in their moral judgments (Cushman and Young 2011). Strikingly, however, Mayans do not seem to recognize it: they think harms caused by omission are just as impermissible as harms caused by action.

Westerners and Mayans appear to disagree, then, in their assessments of the permissibility of some harms. Do defusing explanations account for this disagreement? Fraser and Hauser (2010) do not think so. Westerners and Mayans have similar views about the causal responsibility of agents who act as opposed to those who omit. Moreover, although Westerners and Mayans tend to differ in their supernatural beliefs, Mayans do not think that the decision to refrain from calling out to the pedestrian can cause harm through supernatural forces. Mayan subjects also apparently understand both dilemmas. Even when researchers ask them to consider extremely simplified versions of the cases, they still deny the act/omission distinction, though Western children who consider the cases do not. For these reasons, Fraser and Hauser conclude that the disagreement between Mayans and Westerners supports a non-objectivist account of morality such as MMR.

Relativists may thus be able to draw on empirical research to support the disagreement argument and refute defusing explanations of DMR. Several points must, however, be kept in mind. First, the research we have considered has limitations. For one thing, work on the North-South and Mayan-Western disagreements is underdeveloped and may fail to replicate. For another, neither Doris and Plakias nor Fraser and Hauser have considered all possible defusing explanations. Doris and Plakias grant that other defusing explanations "might plausibly be offered" (2008, 321) and Fraser and Hauser concede that the disagreement between Mayans and Westerners may be explained in terms of a difference in material circumstances (2010, 556).[18] These are significant concessions. No doubt inferences to the best explanation are always incomplete in the sense that new explanations may come to light in the future (Wong 1984, 171). But more work is needed to establish that MMR provides a better explanation of the disagreements under consideration than objectivism.

Second, even if MMR provides the best account of these disagreements, the number of cases in the existing literature is still relatively small.[19] Further research is needed to determine whether relativists can make sense of a wide range of moral disputes using evidence from empirical disciplines. Moreover, since experts tend to have more training and relevant evidence than ordinary people, relativists should also consider expert disagreement about cross-cultural moral issues (Sneddon 2009).

Third, it is important to note that widespread moral disagreement is not the only phenomenon that calls for explanation in the debate over MMR. Relativists must also account for other features of ordinary moral experience, such as the supposed appearance that morality is objective (Sayre-McCord 1991, 167).[20] If they fail to explain some feature, then we may have reason to reject MMR even if we grant that MMR provides the best account of DMR. None of this is to suggest that relativists are unable to make sense of ordinary moral experience. But the point remains that moral disagreement is only part of the relevant evidence in assessing the plausibility of MMR. Relativists may need to provide a stronger explanatory case for their view.

5. Conclusion

We have considered a disagreement argument for MMR. This argument moves from a claim about the appearance of many cross-cultural fundamental disagreements to the conclusion that moral judgments are true or false only in relation to cultures. Proponents of the argument claim that the best explanation of widespread moral disagreement is that the truths in ethics are culture-bound. Much of the discussion of the argument has focused on the plausibility of this explanatory claim and (DMR). To provide stronger support for their view, relativists must continue to engage with empirical research on actual disputes. In the debate over (MMR), as in some other philosophical debates, armchair speculation is not sufficient.

Notes

1 Ethicists often distinguish between the three versions of moral relativism discussed here. See Brandt (2001, 25), Gowans (2015), and Wong (1991, 442).
2 For a similar suggestion, see Levy (2003, 165) and Sayre-McCord (1991, 163). Wong (2011, 411) writes that "the most frequently traveled path to [MMR] starts from the existence of moral disagreement and runs through an argument to the best explanation." For discussion of the disagreement argument, see Gowans (2000, 13–33), Harman (1991, 1996, 8–31), Olinder (2016, 531–532), and Wong (1984).

3 But see Harman (2000, 82).

4 For a helpful overview of the literature, see Gowans (2015) and Miller (2011, 348–351).

5 For discussion of the first strategy, see especially Harman (1996, 8–9) and Sinnott-Armstrong (2006, 200). For the second, see Gowans (2004, 144–146) and Prinz (2007, 195–199).

6 For example, see Levy (2003, 170–173), Tannsjo (2007, 129), and Wong (2011, 423–425).

7 As Miranda Fricker remarks, "a moral culture is more than an incumbent ideology, more than what is got away with by the powerful" (2013, 807); cf. Levy (2003, 174–175) and Wong (2006, 59).

8 One virtue of this explanation is that it may be able to account for the widespread belief that morality contains a stricter duty not to harm than to help. For discussion, see Harman (1996, 24–25). For an alternative relativist explanation, see Wong (1984, 2011).

9 See Williams (1985, 132–133) and Enoch (2009, 23).

10 See for example Appiah (2010, 236–237), Boyd (1995, 339), Brink (2000, 161), Gowans (2000, 23), Levy (2003, 165–167), Rachels (2009, 41), Shafer-Landau (2003, 219), and Vavova (2014, 323). For discussion, see Prinz (2007, 192) and Sinnott-Armstrong (2006, 199).

11 See Boyd (1995, 338), Brink (2000, 163), Enoch (2009, 25–26), Loeb (1998, 283), Nagel (1986, 148), Shafer-Landau (2003, 219), and Tersman (2006, 27).

12 See especially Rachels (2009, 41–42), Levy (2003, 167–168), and Loeb (1998, 283). Wong (1984, 118–119) discusses this explanation but argues that it is mistaken.

13 See Brink (2000, 205–206), Nagel (1986, 145–149), Parfit (1984, 453–454), Singer (1995, 14–16), and Smith (1994, 188).

14 "In some cases," writes Judith Jarvis Thomson, "the source of the apparent intractability is the fact that the issues in dispute are just plain hard, hard in that deciding what to think about them has implications for a wide variety of kinds of actions" (1996, 205; cf. Baghramian 2004, 287).

15 Brink (2000, 204–208) and Shafer-Landau (2003, 218–219) are representative examples.

16 Objectivists are not the only philosophers who prefer to work from the comfort of the armchair. See Ballantyne's (2018, 373–374) call for "field work" in epistemology, for example.

17 For discussion, see Bloomfield (2008) and Fraser and Hauser (2010, 546–551).

18 For discussion of some defusing explanations not considered by Doris and Plakias, see Fitzpatrick (2014) and Tersman (2006, 32–33).

19 Additional examples include differences in attitudes about fairness (Machery et al. 2005); differences between Chinese and American undergraduates in responses to utilitarian thought experiments (Doris and Plakias 2008); and the debate over whether, and to what extent, individuals in rich countries like the United States must donate to foreign aid agencies (Seipel forthcoming).

20 For discussion of the objective-seeming character of ordinary morality, see for example Bjornsson (2012), Smith (1994, 6), and Shafer-Landau (2003, 4).

References

Abarbanell, L. and M. Hauser (2010), "Mayan Morality: An Exploration of Permissible Harms," *Cognition* 115: 207–224.

Appiah, K. A. (2010), "More Experiments in Ethics," *Neuroethics* 3: 233–242.

Baghramian, M. (2004), *Relativism*, London: Routledge.

Ballantyne, N. (2018), "Is Epistemic Permissivism Intuitive?" *American Philosophical Quarterly* 55: 365–378.

Bjornsson, G. (2012), "Do 'Objectivist' Features of Moral Discourse and Thinking Support Moral Objectivism?" *Journal of Ethics* 16: 367–393.

Bloomfield, P. (2008), "Disagreement about Disagreement," *Moral Psychology*, vol. 2, edited by E. Sinnott-Armstrong, Cambridge, MA: MIT Press, 333–338.

Boyd, R. (1995), "How to be a Moral Realist," in *Contemporary Materialism*, edited by P. Moser and J. D. Trout, London and New York: Routledge, 181–228.

Brandt, R. B. (2001), "Ethical Relativism," in *Moral Relativism: A Reader*, edited by P. Moser and T. Carson, Oxford: Oxford University Press, 25–31.

Brink, D. (2000), "Moral Disagreement," in *Moral Disagreements*, edited by C. Gowans, London: Routledge, 157–167.

Cushman, F., J. Knobe and W. Sinnott-Armstrong (2008), "Moral Appraisals Affect Doing/Allowing Judgments," *Cognition* 308: 281–289.

Cushman, F. and L. Young (2011), "Patterns of Moral Judgment Derive from Nonmoral Psychological Representations," *Cognitive Science* 35: 1052–1075.

Davidson, D. (2004), *Problems of Rationality*, Oxford: Clarendon Press.

Doris, J. and A. Plakias (2008), "How to Argue about Disagreement: Evaluative Diversity and Moral Realism," in *Moral Psychology*, edited by W. Sinnott-Armstrong, vol. 2, Cambridge, MA: MIT Press, 303–331.

Doris, J. and S. Stich (2005), "As a Matter of Fact: Empirical Perspectives on Ethics," *The Oxford Handbook of Contemporary Philosophy*, edited by F. Jackson and M. Smith, Oxford: Oxford University Press, 114–152.

Dreier, J. (1990), "Internalism and Speaker Relativism," *Ethics* 101: 6–26.

Enoch, D. (2009), "How Is Moral Disagreement a Problem for Realism?" *Journal of Ethics* 13: 15–50.

Fitzpatrick, S. (2014), "Moral Realism, Moral Disagreement, and Moral Psychology," *Philosophical Papers* 43: 161–190.

Fraser, B. and M. Hauser (2010), "The Argument from Disagreement and the Role of Cross-Cultural Empirical Data," *Mind and Language* 25: 541–560.

Fricker, M. (2013), "Styles of Moral Relativism – A Critical Family Tree," *Oxford Handbook of the History of Ethics*, edited by R. Crisp, Oxford: Oxford University Press, 793–817.

Goodwin, G. and J. Darley (2008), "The Psychology of Meta-Ethics: Exploring Objectivism," *Cognition* 106: 1339–1366.

Gowans, C. (2000), "Introduction: Debates about Moral Disagreements," in *Moral Disagreements*, edited by C. Gowans, London: Routledge, 1–44.

———. (2004), "*A Priori* Refutations of Disagreement Arguments Against Moral Objectivity: Why Experience Matters," *Journal of Value Inquiry* 38: 141–157.

———. (2015), "Moral Relativism," in *The Stanford Encyclopedia of Philosophy*, Fall 2015 edition, edited by E. N. Zalta, http://plato.stanford.edu/entries/moral-relativism/.

Harman, G. (1975), "Moral Relativism Defended," *Philosophical Review* 84: 3–22.

———. (1991), "Moral Diversity as an Argument for Moral Relativism," *Perspectives on Moral Relativism*, edited by D. Odegard and C. Stewart, Milliken, ON: Agathon Books, 13–31.

———. (1996), *Moral Relativism and Moral Objectivity*, Cambridge: Blackwell.

———. (1998), "Responses to Critics," *Philosophy and Phenomenological Research* 58: 207–213.

———. (2000), "Is There a Single True Morality?" in *Explaining Value*, edited by G. Harman, Oxford: Oxford University Press, 77–99.

Harman, G. and J. J. Thomson (1996), *Moral Relativism and Moral Objectivity*, Oxford: Blackwell.

Levy, N. (2003), "Descriptive Relativism: Assessing the Evidence," *The Journal of Value Inquiry* 37: 165–177.

Loeb, D. (1998), "Moral Realism and the Argument from Disagreement," *Philosophical Studies* 90: 281–303.

———. (2003), "Gastronomic Realism – A Cautionary Tale," *Journal of Theoretical and Philosophical Psychology* 23: 30–49.

Lyons, D. (1976), "Ethical Relativism and the Problem of Incoherence," *Ethics* 86: 107–121.

Machery, E., D. Kelly and S. Stich (2005), "Moral Realism and Cross-Cultural Normative Diversity," *Behavioral and Brain Sciences* 28: 830–831.

MacIntyre, A. (1988), *Whose Justice? Which Rationality?* Notre Dame: University of Notre Dame Press.

Midgley, M. (2003), *Heart and Mind: The Varieties of Moral Experience*, third edition, London: Routledge.

Miller, C. (2011), "Moral Relativism and Moral Psychology," in *The Blackwell Companion to Relativism*, edited by S. D. Hales, Oxford: Blackwell Publishing, 346–367.

Moody-Adams, M. (1997), *Fieldwork in Familiar Places: Morality, Culture, and Philosophy*, Cambridge, MA: Harvard University Press.

Myers, R. (2004), "Finding Value in Davidson," *Canadian Journal of Philosophy* 34: 107–136.

Nagel, T. (1986), *The View from Nowhere*, New York and Oxford: Oxford University Press.

Nisbett, R. and D. Cohen (1996), *Culture of Honor: The Psychology of Violence in the South*, Boulder, CO: Westview Press.

Olinder, R. (2016), "Some Varieties of Metaethical Relativism," *Philosophy Compass* 11: 529–540.

Parfit, D. (1984), *Reasons and Persons*, Oxford: Clarendon Press.

Prinz, J. (2007), *The Emotional Construction of Morals*, Oxford: Oxford University Press.

Rachels, J. (2009), "The Challenge of Cultural Relativism," in *Exploring Ethics*, edited by S. M. Cahn, Oxford: Oxford University Press, 245–255.

Rovane, C. (2013), *The Metaphysics and Ethics of Relativism*, Cambridge, MA: Harvard University Press.

Sarkissian, H., J. Park, D. Tien, J. C. Wright and J. Knobe (2011), "Folk Moral Relativism," *Mind and Language* 26: 482–505.

Sayre-McCord, G. (1991), "Being a Realist About Relativism (In Ethics)," *Philosophical Studies* 61: 155–176.

Seipel, P. (forthcoming), "Why do We Disagree About Our Obligations to the Poor?" *Ethical Theory and Moral Practice*.

Shafer-Landau, R. (2003), *Moral Realism: A Defense*, Oxford: Oxford University Press.

Sikka, S. (2012), "Moral Relativism and the Concept of Culture," *Theoria* 59: 50–69.

Singer, P. (1995), *How Are We to Live?* Amherst: Prometheus Books.

Sinnott-Armstrong, W. (2006), *Moral Skepticisms*, Oxford: Oxford University Press.

Smith, M. (1994), *The Moral Problem*, Cambridge: Basil Blackwell.

Sneddon, A. (2009), "Normative Ethics and the Prospects of an Empirical Contribution to Assessment of Moral Disagreement and Moral Realism," *Journal of Value Inquiry* 43: 447–455.

Stich, S. and J. Weinberg (2001), "Jackson's Empirical Assumptions," *Philosophy and Phenomenological Research* 62: 637–643.

Tannsjo, T. (2007), "Moral Relativism," *Philosophical Studies* 135: 123–143.

Tersman, F. (2006), *Moral Disagreement*, Cambridge: Cambridge University Press.

Vavova, K. (2014), "Moral Disagreement and Moral Skepticism," *Philosophical Perspectives* 28: 302–333.

Velleman, J. (2013), *Foundations for Moral Relativism*, Cambridge: Open Book Publishers.

Williams, B. (1985), *Ethics and the Limits of Philosophy*, Cambridge, MA: Harvard University Press.

Wong, D. (1984), *Moral Relativity*, Berkeley: University of California Press.

———. (1991), "Relativism," *A Companion to Ethics*, edited by P. Singer, Oxford: Blackwell, 442–450.

———. (2006), *Natural Moralities*, Oxford: Oxford University Press.

———. (2011), "Relativist Explanations of Interpersonal and Group Disagreement," in *The Blackwell Companion to Relativism*, edited by S. D. Hales, Oxford: Blackwell Publishing, 411–430.

Wright, J., P. Grandjean and C. McWhite (2013), "The Meta-Ethical Grounding of Our Moral Beliefs: Evidence for Meta-Ethical Pluralism," *Philosophical Psychology* 26: 336–361.

18

RELATIVISM AND MORALISM

John Christian Laursen and Victor Morales

1. Introduction

Some kinds of relativism might encourage moralism, and some kinds might undermine it. Some kinds of moralism might rest on and imply relativism, and some kinds might reject it. This chapter explores the connections between the two philosophical concepts, bringing out their similarities and differences, and their mutual implications.

There are of course many forms and interpretations of relativism, and we will return to some of them later. The most general distinction between types of relativism is between epistemological or cognitive relativism, which is to say relativism about knowledge claims, and relativism about morals. The latter may include knowledge claims about morals, so perhaps moral relativism is best understood as a subcategory of knowledge claims in general. But not all claims about morals are knowledge claims: one can justify a moral claim on the basis of faith or emotion or gut feeling. Relativism usually contrasts with absolutism, or the conviction that one does indeed know things with certainty, either about the world or about morals. Other terms for positions opposed to relativism include "objectivism," "monism," "universalism," and "realism." We will use the term "absolutism" here in a broad sense, including these other terms and the arguments in favor of them. One can be absolutist about one of the dimensions mentioned previously and not about the other, such as certain about scientific claims but not about morals, or vice versa.

Perhaps not everyone will be familiar with the definition of or the philosophical literature on moralism. One collection of essays on moralism defined it as "a style of speaking, writing, and thinking that is too confident about its judgments and thus too positive in its orientation to others" (Bennett and Shapiro 2002, 4). Another author defines moralism as "the vice of overdoing morality" (Coady 2006, 1). People who engage in moralism are characterized as "moralists" or "moralistic." Where to draw the line will never be easy, but one can always ask oneself, "Am I overdoing the morality here, weighing the moral dimension too much?" It will be easier to judge that other people are doing so. A rich vocabulary of common words for characterizing moralistic people includes "judgmental," "busybodies," "sanctimonious," "self-righteous," "priggish," and so forth. One recent author defines moralism as "the pettifogging, mindless, legalistic, excessively judgmental, or punitive policing of our own or others' compliance with moral duties" and "a defensive, ungenerous, and unforgiving stance toward others" (Satkunanandan

2015, 1, 11). In some literatures, moralism is contrasted with realism (e.g. Williams 2005; see Taylor 2012).

Moralism may be found at one end of a continuum with immoralism, amoralism, or psycho-pathology at the other end and good healthy morality that is not taken too far in the middle. Good healthy morality would then not be excessive, not overly judgmental, not self-righteous, and not sanctimonious. Moralists might say that they are not excessive because it is not pos-sible to overdo morality, and that one should be judgmental. Anti-moralism is the critique of moralism, from any of several possible positions. There will be differences of opinion about who is and who is not a moralist. One person's healthy morality will be another person's moralism. Judgments about moralism are thus examples of relativism: they are not objective or absolute nor accepted by everyone.

It might be assumed that absolutist theories of truth or morality are more likely to justify moralism than relativist theories. After all, if one is absolutely sure of one's truths and morals, one has a strong case for pushing them to the limit. But a theory of relativism might back up moralists by assuring them that they are entitled to take their knowledge claims or moral truths as far as they want on the ground that no one else's knowledge claims or moral truths are enti-tled to any more respect. That is, within a relativist framework, one person may be as entitled as another to take his or her own claims seriously. It might be assumed that relativists must be tolerant because they recognize the relativity of their knowledge. But that is not true: they can be racists and xenophobes, too (Lukes 2008, 39; see also Long 2011). And on the other hand, diversity and value pluralism do not require relativism: they can be valued for their own sakes (Lukes 2008, 153). Thus, relativism may support or undermine moralism, and moralism may support or undermine relativism.

Since we are studying the interfaces of two sets of contrasting concepts, at the most abstract level, there are four possibilities. One can be a relativist and an anti-moralist, a relativist and a moralist, an absolutist and anti-moralist, or an absolutist and moralist. The philosophy of relativ-ism and of moralism intersect fruitfully with the theology of relativism and moralism, so we are going to make some references to the debates in philosophical theology as well.

2. Relativism and anti-moralism

Perhaps the most consistent and skeptical pluralism is to oppose both absolutism and moralism. Relativists might oppose moralism if they believe that contingency and relativity mean that one has no right to impose one's morality, even on oneself; and anti-moralists might believe that overly strong claims to truth might have an inevitable tendency to justify moralism. We shall explore several such thinkers here.

George Santayana was an eclectic philosopher who carried out a life-long campaign against moralism, even if he fell into moralism himself from time to time and concerning some issues. He was a relativist in philosophy and morals in the sense that he stood above all philosophi-cal and political positions, adopted some of them temporarily when convenient, and declined to take a stand on most of them (Laursen and Román 2015). His autobiography and his great novel, *The Last Puritan* (1935), can be read as thorough-going critiques of moralism (Román and Laursen 2016). In the novel, most of the characters are moralists in one way or another, most often as a matter of class bias. The main character, Oliver, suffers from self-destructive moralism. Santayana's novel can stand in as the type of many nineteenth and twentieth century novels that were critiques of moralism, understood as taking some social, family, or sexual morality too far, to the point of cruelty.

In his autobiography, Santayana refuses to endorse a political position with any dogmatic or prescriptive passion. He worries that those who fancy their positions "to be exclusively and universally right" are subject to an "illusion pregnant with injustice, oppression, and war" (Santayana 1963, 227). "I dislike all the quarrels and panaceas of the political moralists [and] turn my back on the disaffected and on the fanatics of every sect" (1963, 502). Santayana characterized the very moralistic Boston of his time, the "Transcendentalists," and his colleagues at Harvard such as William James and Josiah Royce as the "genteel tradition," and rejected it across the board (Santayana 2009).

Michael Oakeshott drew on Hegelian philosophy for a relativistic philosophy that claimed that different "modes of experience" have their own rules and standards. History, science, and practice (which includes morals) were each modes of experience with their respective independent logic and standards (Oakeshott 1933). His critique of moralism was couched as a critique of moral ideals: "every moral ideal is potentially an obsession; the pursuit of moral ideals is an idolatry" (Oakeshott 1991, 476). "Too often the excessive pursuit of one ideal leads to the exclusion of others, perhaps all others; in our eagerness to realize justice we come to forget charity, and a passion for righteousness has made many a man hard and merciless" (1991, 476). This "is a form of the moral life which is dangerous in an individual and disastrous in a society" (1991, 476). We may conclude that "the pursuit of moral ideals has proved itself (as might be expected) an untrustworthy form of morality" (1991, 486). This is Oakeshott's judgment about one kind of moral experience, based on his own moral experiences. The fact that his relativistic philosophy has identified other modes of experience does not stop him from judging this one. And his judgment might be too harsh: others might judge his anti-moralism to be a moralism (Laursen 2013).

Wendy Brown is a self-described postmodernist who also criticizes moralism, arguing instead for political pragmatism. In her view, there is a "contemporary tendency to personify oppression in the figure of individuals," and "theoretical as well as political impotence and rage ... is often expressed as a reproachful political moralism" (2001, 21). The problem is that often "one finds moralizers standing against much but for very little" (2001, 28). Her conclusion is that moralism "marks both analytical impotence and political aimlessness" (Brown 2001, 29).

Dennis Nineham, an influential Oxford theologian, cited Nietzsche, R. G. Collingwood, and T. E. Hulme in support of his critique of moralism and his argument for cultural relativism in theology. A Nietzschean and Collingwoodian sense of history meant a kind of relativism:

> What I have in mind is more like what the Germans often have in mind when they use the word *Historismus*. It is like what I think Nietzsche meant when he said that in the nineteenth century mankind developed, or recognized, a sixth sense, the historical sense ... or what R.G. Collingwood meant when he wrote that "the really new element in the thought of today as compared with that of three centuries ago is the rise of history."
>
> *(1975, 3)*

In challenging the notion of everlasting ideas he wrote that it is wrong to take these ideas as "as much part of the givenness of things as the proximity of the nearby mountains or the periodic flooding of the local river" (1975). For Nineham, theologians and churchgoers misinterpret Christian doctrine as the infallible, never changing truth of God that was revealed once and is applicable to all at all times. In doing so they have moralized certain truths and principles without taking into consideration how a given culture – its language and values – within a specific period helped shaped those views (1975, 8). Nineham concluded that (1) the historical truth of God cannot be known; (2) what has really occurred is the reading of a story from one culture through "the spectacles of another"; and (3) "that so far as the real Jesus can be discerned, he

[Jesus], like the New Testament witnesses to him, belong essentially to the culture of his time and place; there is no real reason to think that the outlook of historical Jesus will have been such as to be any more immediately acceptable today than that of, let us say, the historical Paul" (Nineham 1975, 14). Thus, most moralistic Christian positions are unjustified.

3. Relativism and moralism

Some relativists support moralism thinking that authenticity or consistency require a whole-hearted commitment to one's own moral position, even if one recognizes its contingent and relative status. Conversely, moralists might be comfortable enough with relativism to believe that the relative status of one's truths or morals does not undermine one's commitment to them. In all cases, judgment as to whether someone is a moralist might be accepted or denied by the subject. It might be an observer's judgment that the subject takes morality too far.

Friedrich Nietzsche is often given credit for spurring on the development of philosophical relativism (e.g. Lukes 2008, 2–3). His "perspectivism" is a sort of relativism, claiming that everything we know is known from a given perspective, and that there is nothing outside of perspectives. Nietzsche is an anti-moralist where moralism is identified with Christianity, and yet his critique of, and alternative to, Christian morality can plausibly be described as moralistic: while he is unwilling to acknowledge anything positive about Christianity, he seems to admire an anti-modern aristocratic morality to an extent that might be labeled "moralistic."

Even Santayana, already mentioned under the heading of "relativist anti-moralist," occasionally veered towards moralism himself: examples include his anti-liberalism and anti-Semitism (Román and Laursen 2016, 8–9). But there was no philosophical justification of these positions, neither absolutist or relativist.

Ludwig Wittgenstein can stand for the type of a philosophical relativist for whom everything we say is a part of a non-universal language game. He wrote that our languages depend on our way of life to such an extent that "if a lion could talk, we could not understand him." The lion's life is too different from ours (1953, 223). And yet, Wittgenstein also held high moral standards that can be understood as moralistic. In these he was a kind of relativist, too, because he declined to apply his moralistic standards to other people. When Fania Pascal, asked him "You want to be perfect?," "he pulled himself up proudly, saying '*Of course* I want to be perfect'" (Pascal 1984, 37). His failure to live up to his own perfectionist moral standards made him miserable, and that is a pragmatic argument against such moralistic standards.

Jesse Prinz's work on the emotional construction of morality implies both the relativist principle that our morals come from our different emotional makeups, and that our emotions can drive us to moralism whether we want it or not (Prinz 2007). If morals come from emotions and emotions admit of degrees, it seems likely that strong emotions could drive us to moralism.

Michael Wreen has recently proposed a definition of relativism according to which it "is the view that there are two or more equally valid or true but conflicting moral codes" (2018, 338). It is not a universalism because it is not the case that there is only one such code, but holders of each of the two or more codes could be absolutists, thinking that they are exclusively right. On Wreen's account, in a relativist situation you could have moralists insisting too much on each of several conflicting moral codes. Maybe that is what we do find in much of history.

4. Absolutism and moralism

As mentioned previously, absolutism is the opposite of relativism, but we must consider it here for the light that opposites and critiques throw on the concept of relativism. Proponents of relativism often use Aristotle and Kant as examples of forms of absolutism that should be rejected. Without going into too much detail concerning either philosopher, we must draw attention to some crucial features.

There certainly are aspects of Kant's philosophy meriting the label "absolutist." Kant wrote that his own philosophy was the last and final philosophy that anyone would need, that it "would leave no task to our successors, save that of adapting it in a didactic manner" (KrV Axx)[1] it was "a capital to which no addition can be made" (KrV Bxxiv). He also thought that he had found the foundations of morality valid for all times and places. And there are aspects of his morality that others might deem moralistic. He judged that the Latin saying "Fiat iustitia, pereat mundus" ("let justice be done although the world be lost") (VIII:378)[2] was true. He also thought that people have a duty to endure "even the most intolerable abuse of supreme authority" (VI:320) and that punishment was necessary to the point that in order to avoid "bloodguilt" the last murderer in prison must be executed before a civil society is dissolved (VI:333). From the point of view of a more humane morality, this is a cruel and unnecessary moralism. Relativist David Velleman shows that in fact Kant's assumptions about human agency and moral purposes are not shared around the world (Velleman 2015, 32ff.).

A contemporary sometime-absolutist who also occasionally falls into moralistic extremes is Martha Nussbaum. From Aristotle she takes a philosophical moral absolutism which says that there is nothing relative about the good of life and the bad of death (Nussbaum 1987). (Nussbaum moved toward a mixed position in some of her writings, as do many virtue ethics theorists, recognizing relativism, sometimes only tacitly (Gowans 2011).) She skated rather near to moralism when she added that if Socrates taught a skeptical relativism, he was indeed guilty of corrupting the youth (Nussbaum 2000, 192). In defense of Socrates, whether or not he was a skeptic or relativist, absolutists have done a lot of corrupting of youth, too. Nussbaum mentions Heidegger as an absolutist, and he can be accused of corrupting the youth because of his collaboration with the Nazis. In any case, Velleman shows that Aristotelian assumptions about moral agency are not shared everywhere (Velleman 2015, 41ff.).

John Lennox, a philosopher of science and Christian apologist, takes the absolutist position by arguing that biblical truths are alive and well. He attacks the false assumption peddled by atheists that science has disproven the existence of God and the universal principles he espouses. For Lennox, science is not – and never was – at odds with religion. The notion that it is at odds is a fabrication of thinkers with the hidden agenda to impose their atheistic worldview. Lennox reminds his readers that the fathers of science – from Galileo and Kepler to Newton and Maxwell – were all theists, and most were Christian (Lennox 2007, 110). The scientific method as originally proposed, he claims, never sought to undermine the existence of God. Early proponents of science left alone questions that it could not answer, like: "why are we here?" Lennox contends that science was weaponized against religion by atheists who sought to strip the transcendent from reality. But the "Enlightenment ideal of the observer – one completely independent . . . doing investigations and coming to dispassionate, unbiased, rational conclusions . . . is nowadays regarded as an idealized myth," he writes (Lennox 2007, 112). Scientists bring prejudices to their investigations and use them to advance their particular worldview. Lennox chastises philosophers like Bertrand Russell for promoting a reductionist approach to science which claims that all knowledge is scientific. This view, Lennox asserts, is at odds with how philosophers from Aristotle and Descartes to Spinoza and Berkeley understood reality, leaving room for transcendence.

Like Lennox, theologian and professor of philosophy and ethics John Hughes is an absolutist who moralizes about the prevailing approach to apologetics. Hughes explains that apologetics has always sought to "prove" the things of faith to people by employing "simple rational arguments" (2011, 4). He believes that philosophical theology has relied too heavily on this "rationalist foundationalist" framework and argues that theology must abandon this framework if it wants to succeed in persuading people. He argues that the logical approach lost its currency when Hume and Kant showed the problem in rooting the "proof" of God's existence in cosmological and teleological arguments. The rationalists' system was thrown into greater disarray when postmodern thinkers such as Nietzsche, Heidegger, and Derrida questioned the existence of universal values and reason (Hughes 2011, 5, 7). Nevertheless, philosophical theology has continued on the same course, sometimes making problematic declarations such as the claim by philosopher Richard Swinburne to calculate the statistical probability of the existence of God and resurrection of Christ. For Hughes, claims like this are absurd because they reduce Christ to "any other ordinary 'thing' in the world" (2011, 4). If the ultimate questions are beyond the empirical world, and if postmodernism has shown that "proofs" of God are no longer possible, apologetics must either take a new methodological course or become irrelevant. Hughes draws on the philosopher Alasdair Macintyre – who recognized that the critiques of rationalist foundationalism by the postmodernists may be valid but insisted that reason not be abandoned altogether – to argue for a return to the study and use of rhetoric (2011, 4). "If [apologists] are authentic then their rhetoric will persuade by virtue of their *inherent* beauty and goodness, rather than because of some added spin or window-dressing" (2011, 9). For Hughes, reason still has a place in apologetics; it just takes a back seat to faith.

Similarly, theologian Garry Graves, a moralist and absolutist concerning biblical truths, chastises Christians that have fallen prey to the postmodern "verbal con game." For Graves, the threat to absolute truth and values comes from within the religious ranks. There is an irreconcilable conflict between two diametrically opposed groups: postmodern Christians and traditional Christians (Graves 2018, 53). Graves fumes over the influence of Nietzsche, Wittgenstein, Foucault, and Derrida on postmodern Christians. He points to philosophical concepts such as the *Übermensch*, language games, and the deconstruction of authority working in unison to hoodwink postmodern Christians and "diminish the significance of biblical authority" (2018, 50). The church has willingly injected the opioid of worldly pursuits such as relativism, entertainment, fashion, etc., into its bloodstream instead of Bible-based theology. The concern for Graves is that "[w]hen Christians cease to view the Bible as authoritative and sufficient, any number of ideas may replace established biblical principles" (2018, 56). Notwithstanding, "traditional Christians recognize the deception by remembering that Satan may appear as an angel of light," Graves – in a moralistic tone – proclaims (2018, 53).

Are most religious people absolutists and moralists? If you know you are right about your religious truths you might think you should not be ambivalent about your moral truths. You might think you should not be a relativist because it undermines morality. Relativism is inadequate because there is a moral imperative to believe in the absoluteness of one's morality. And you should not worry about being a moralist because it is not possible to overdo the right morals. Could these be what we call fanatics? But some religious people have insisted that they relied only on faith, not truth or knowledge, and others have claimed only that their faith was theirs, and need not be accepted by others.

5. Absolutism and anti-moralism

It might be assumed that absolutists have to be moralists, but there are ways they can be anti-moralists. We have already mentioned that Aristotle is often interpreted as a philosophical absolutist. And he certainly had high confidence in his moral judgments. But there is a way in which he can be understood as an anti-moralist. One of the pillars of his thought was the doctrine of the mean, arguing that extremes should be avoided. Thus there is a philosophical reason for avoiding extremism. Naturally, one person's mean is another person's extreme, so there will always be questions about how to characterize any position. But at least the doctrine of the mean can be taken as an effort by a philosophical absolutist to avoid moralism. Satkunanandan has drawn attention to other elements of Aristotle's virtue ethics that militate against moralistic calculations in ethics (2015, 94). She also found that even the absolutists Plato and Kant recognized an extraordinary morality as a critique of moralism, although in doing so she may be undermining claims that they are absolutists as well (2015, 93ff., 125ff.). One lesson here is that complex texts can often be interpreted in both absolutist and relativist terms, and as both moralistic and anti-moralistic.

Bernard Williams was an absolutist and anti-moralist of a specific kind (Lukes 2008, 22). In what he called a "relativism of distance" he argued that although we can judge contemporary cultures by our moral standard, it makes no sense to apply it to past cultures (2005, 62–74). There are absolute human rights and "universal paradigms of injustice and unreason" that are "self-evident" (2005, 23, 26), but we cannot judge past violations of them. Distance in time is what makes the difference: Williams does not explain why distance in space does not imply a relativism as well. Williams was also a major critic of moralism in politics, rejecting both what he called "moralistic liberalism" and "communitarian relativism" and preferring what he called "realism" (2005, 6, 24). As he put it, "my view is in part a reaction to the intense moralism of much American political and indeed legal theory" (2005, 12).

Wojciech P. Grygiel is an absolutist philosopher of religion who is also an anti-moralist. For Grygiel, the question of the existence of the Divine has been settled but what remains open is the framework that shapes the way in which the Divine is understood. "Theology should always be open towards explanative enrichment as the exploration of the *loci theologici* in hand yield new theological data," he wrote (2016, 31). His anti-moralism comes out in his insistence that theology take a page from the sciences, which have – like theology – relied on abstract concepts but rejected conceptual dogmatism. In other words, as new discoveries in the field arose, obstacles – old frameworks – were removed to make way for the new data, which altered our understanding of reality. As such, "no theological doctrine should be regarded as final and the true obstacle to theology's quest for truth is the enthronization of any conceptual system as absolute and ultimately indispensable," he insists (Grygiel 2016, 17).

Like Grygiel, Rob Bell is an absolutist concerning the elements of the Christian faith, especially when it comes to the concepts of salvation, heaven, and hell; but he is an anti-moralist in his reading of the Christian mission. Bell believes that people's understanding of salvation, heaven, and hell have been hijacked by a reading that cannot be found in the biblical text. "Think of the cultural images that are associated with heaven: harps and clouds and streets of gold, everybody dressed in white robes . . . and St. Peter is there, like a bouncer at a club, deciding who does and doesn't get to enter," he declares (2011, 39). Such readings of theological concepts have distorted religion for believers. Scrutinizing the original use of the concepts in Greek, Bell concludes that heaven is here on earth and Christians should work to construct that heaven here on earth. This sort of anti-moralism strips Christians of their monopoly over salvation and heaven, and similar philosophical arguments can be found in many contemporary religions. Not everyone who believes in a religious truth believes they must impose it on others.

Since very little philosophical work has been done on the relationship between relativism and moralism, there is plenty of room for future work. Pinning down exactly what moralism is and how to recognize it from different perspectives, which may or may not require relativism, is one place to start.

Notes

1 Reference to Kant's *Critique of Pure Reason (Kritik der reinen Vernunft)* via the standard abbreviation "KrV" and page references to the first edition ("A") and second edition ("B"), e.g. "KrV A3/B6."
2 Reference to Kant's *Perceptual Peace (Zum Ewigen Frieden)* and *Metaphysics of Morals (Metaphysik der Sitten),* or, more specifically, its first part *The Metaphysical Elements of Justice (Metaphysiche Anfangsgründe der Rechtslehre)* via the standard edition *Gesammelte Schriften* (also known as *Akademieausgabe*), cited by volume and page number, e.g. "VIII:378."

References

Bell, R. (2011), *Love Wins*, New York: Harper One.

Bennett, J. and M. J. Shapiro (eds.) (2002), *The Politics of Moralizing*, New York: Routledge.

Brown, W. (2001), *Politics Out of History*, Princeton, NJ: Princeton University Press.

Coady, C. A. J. (ed.) (2006), *What's Wrong with Moralism?* Special Issue of *Journal of Applied Philosophy* 22(2005): 101–210; reprinted as *What's Wrong with Moralism?* Malden: Blackwell.

Gowans, C. (2011), "Virtue Ethics and Moral Relativism," in *A Companion to Relativism*, edited by S. D. Hales, Malden: Blackwell, 391–410.

Graves, G. (2018), "Postmodern Challenges for the Family: An Examination of Detrimental Influences and Biblical Responses," *The Journal of Mid-America Baptist Theological Seminary* 5: 49–61.

Grygiel, W. P. (2016), "Theology Under Siege: Reflections of a Troubled Philosopher and a Believer," *Theological Research* 4: 7–33.

Hughes, J. (2011), "Proofs and Arguments," in *Imaginative Apologetics: Theology, Philosophy and the Catholic Tradition*, edited by A. Davison, London: SCM Press, 3–11.

Laursen, J. C. (2013), "Michael Oakeshott, Wendy Brown, and the Paradoxes of Anti-Moralism," *Agora: Papeles de filosofía* 13: 67–80.

Laursen, J. C. and R. Román (2015), "George Santayana and Emotional Distance in Philosophy and Politics," *Revista de Filosofía* 40: 7–28.

Lennox, J. (2007), "Challenges from Science," in *Beyond Opinion*, edited by R. Zacharias, Nashville: Thomas Nelson, 106–133.

Long, G. M. (2011), "Relativism in Contemporary Liberal Political Philosophy," in *A Companion to Relativism*, edited by S. D. Hales, Malden: Blackwell, 309–325.

Lukes, S. (2008), *Moral Relativism*, New York: Picador.

Nineham, D. (1975), *New Testament Interpretation in an Historical Age. The Ethel M. Wood Lecture*, London: Athlone Press.

Nussbaum, M. (1987), "Non-Relative Virtues: An Aristotelian Approach," *Midwest Studies in Philosophy* 13: 32–53.

———. (2000), "Equilibrium: Scepticism and Immersion in Political Deliberations," in *Ancient Scepticism and the Sceptical Tradition*, edited by J. Sihvola, Helsinki: Societas Philosophica Fennica, 171–198.

Oakeshott, M. (1933), *Experience and Its Modes*, Oxford: Oxford University Press.

———. (1991), *Rationalism in Politics and Other Essays*, Indianapolis: Liberty Fund.

Pascal, F. (1984), "A Personal Memoir," in *Recollections of Wittgenstein*, edited by R. Rhees, Oxford: Oxford University Press, 12–49.

Prinz, J. (2007), *The Emotional Construction of Morals*, Oxford: Oxford University Press.

Román, R. and J. C. Laursen (2016), "Santayana's Critique of Moralism," *Limbo: Boletín internacional de estudios sobre Santayana* 36: 5–39.

Santayana, G. (1935), *The Last Puritan. A Memoir in the Form of a Novel*, Cambridge, MA: MIT Press, 1994.

———. (1963), *Persons and Places*, Cambridge, MA: MIT Press, 1987.

———. (2009), *The Genteel Tradition in American Philosophy* and *Character and Opinion in the United States*, New Haven: Yale University Press.

Satkunanandan, S. (2015), *Extraordinary Responsibility: Politics Beyond the Moral Calculus*, Cambridge: Cambridge University Press.

Taylor, C. (2012), *Moralism: A Study of a Vice*, Montreal: McGill Queen's University Press.

Velleman, J. D. (2015), *Foundations for Moral Relativism*, second edition, Cambridge: Open Book.

Williams, B. (2005), *In the Beginning Was the Deed: Realism and Moralism in Political Argument*, Princeton, NJ: Princeton University Press.

Wittgenstein, L. (1953), *Philosophical Investigations*, translated by G. E. M. Anscombe, New York: Macmillan, 1958.

Wreen, M. (2018), "What Is Moral Relativism?" *Philosophy* 93: 337–354.

19

INCOMMENSURABILITY, CULTURAL RELATIVISM, AND THE FUNDAMENTAL PRESUPPOSITIONS OF MORALITY

Mark R. Reiff

1. Introduction

Morality, in its most general sense, is a set of principles that provide guidance on how we should behave. Some of these principles are very specific. For example, many people believe morality requires that we pay men and women equally for comparable work. But the more specific the principle, the more likely it is that it has been derived from some more general principle: in this case, people who are similarly situated should be treated similarly or, even more generally, all persons have equal moral worth. The more general the principle, the more it needs to be "cashed out" or given further content before it can be applied. Given that general principles can often be cashed out in many ways, this process can be controversial. But even when there is widespread disagreement on how a general principle should be given greater content, there is often widespread agreement on the general principle with which we start. There is agreement, in other words, on the fundamental principles on which morality is based.

But where do these fundamental principles themselves come from? On what are they based? Why should we agree to them? "We hold these truths to be self-evident," the American *Declaration of Independence* says, suggesting that our most fundamental principles are based on certain objective moral facts. It may be debatable how we come to know these facts, but the thought is that they are derived from certain non-moral facts about human nature, physicality, and circumstances, such as limited altruism, physical vulnerability, and limited resources (Hart 1994, 193–199). As G. A. Cohen pointed out, however, the significance of any facts which ground a principle must itself be determined by some other principle that explains why that fact is significant (Cohen 2003). Accordingly, facts – moral or otherwise – cannot be the ultimate starting point for morality. Everything, Cohen claimed, has to begin with a principle.

Cohen said very little about where these first principles are supposed to come from. Nevertheless, the same reasoning Cohen employed to show that facts cannot be our starting point tells us something important about his claim that principles are prior to facts. There must be some principles that are not themselves "justified" by some other principle, for otherwise morality would simply be a black hole of reasoning, with its starting point totally obscured. H. L. A. Hart made the same point with regard to the nature of law many years ago: the most fundamental principle of law is the rule of recognition, which tells us what counts as "law" and what does not (Hart 1994, 94–110). Hart's point, however, was not that everything can be derived from a single principle – the rule of recognition can itself be a collection of principles rather than a single principle. Rather, his point was that this principle or set of principles is not itself based on any other principle, it must simply be accepted or rejected. It is true that some of Hart's critics have argued that this acceptance or rejection of the rule of recognition, given that it is merely a principle of law, must itself be based on a moral principle, but Hart's more general point is rarely questioned. Everything has to begin somewhere, and morality is no different in this than anything else. Once we have certain fundamental principles to work from, morality can proceed from there, but the choice of these principles is made before what we know as morality can exist. Whatever may be the case for law, morality must begin with principles that are neither moral principles themselves nor derived from facts or other moral principles.

On what basis, then, might we decide to accept or reject these principles? Well, we could do so on the basis of our religious beliefs, received or otherwise, or simply on the basis of habit or convention, which if long-standing enough we might call "tradition" or even "culture." Or we could do so on the basis of our personal preferences. These preferences might in some sense even be hard-wired into us, and therefore the fundamental principles of morality could simply be those that we (or at least most of us) naturally embrace as a result of evolutionary biology. Or they could simply be an expression of self-interest, enlightened or otherwise, determined by what we think would give us or some group of persons the greatest advantage. Or they could be selected randomly, I suppose, although this seems unlikely, for in this case morality would not be as similar as it is from culture to culture.

2. Relativism and personal morality

For morality is indeed very similar from culture to culture. Of course, people often claim this is not true – they claim that morality is culturally relative, by which they mean it varies significantly from culture to culture (Rachels and Rachels 2012, ch. 2). But morality does not really vary very much. Intentionally or even negligently killing or injuring people, other than in self-defense and (in some circumstances) in war, is seen as morally wrong everywhere, with very few exceptions (honor killing might be one). So is taking the property of another, although there may be disagreement of what property is and by whom it can be owned. Wrongdoers are subject to liability for punishment and/or have a duty to pay compensation in all cultures. And almost everyone everywhere believes that promises should be kept unless it is unreasonable to require the promisor to do so. There may not be universal agreement on the exact scope of these moral principles, and as I said, people may differ on how these principles should be cashed out, especially in "hard cases." But the claim that this makes morality culturally relative to a meaningful extent, and therefore at least partially ungrounded, is overblown. It *is* grounded, in agreement. The reasons for this agreement may vary from individual to individual, and morality at the applied level may differ in its details from culture to culture, but it does not differ significantly at the more general, normative level. Even at the applied level, differences come into play only around the edges, not in the center, even though this may be of little comfort when those differences in application are relevant to something that we or someone we care about have personally done or had done to them. Moreover, these variations are as likely to occur within cultures as between them given the need to give further content to the applicable fundamental principle before it can be applied. In either case, the "central case" of morality, as Hart might say, where most people agree on not only what principles apply but how they should be interpreted, is actually very large (Hart 1994, 81, 123).

Note that this does not mean that there is necessarily a single, ultimate moral value, or that if there are a plurality of moral values, they are all commensurable or at least comparable according to some common scale. On the contrary, I have argued elsewhere that there must be widespread and significant incommensurabilities between moral values for morality as we know to exist (Reiff 2014). But this does not mean that morality is not largely the same at the fundamental level. It merely means that there are some number of fundamental principles rather than a single supervalue, and the value that each principle expresses cannot easily be translated into the value expressed by another. Even in combination with the fact that differences often arise with regard to how these fundamental principles should be cashed out, this does not change the fact that these fundamental principles themselves are more or less the same.

3. The more complex case of political morality

At least, this is the case with regard to personal morality, the code which determines how we each behave in our interactions with others. But what about political morality? These are the rules that govern the design and operation of the institutions of society, the scope and nature of the rights, responsibilities, and opportunities these institutions generate, and the distribution of the burdens and benefits of social cooperation within this institutional structure. These rules regulate different aspects of our lives than the rules of personal morality, and are at least arguably not entailed by them. Do people agree on these to the same extent that they agree on principles of personal morality? Is it possible that the fundamental principles in operation here are radically different, far more different than the rules that govern our personal conduct? If it were, then the claim that political morality is relative could be very significant. There could be widespread disagreement on how society should be organized because different groups of people start with very different fundamental principles. And if these different principles really are fundamental – that is, they are not themselves derived from some even more fundamental principle that everyone embraces, then reasoned argument about political morality could be impossible. Such arguments would simply degenerate into situations where people are talking past one another, unable to even conceive of what the other is going on about.

This, I am afraid, seems an apt description of what is going on in much of the world today. Almost everywhere, ideas associated with the illiberal right are experiencing a resurgence in popular support that seems utterly inconsistent with what had been thought to be the long-standing acceptance by many different societies and cultures of the principles of political liberalism. And by "liberalism," I am not referring to that collection of views usually associated with the moderate left. Rather, I am using the term more generally, to refer to that broad family of political theories which are based on ideas that came out of the Enlightenment, ideas which express themselves most strongly today in liberal capitalist democracies. These ideas, or "presuppositions" as I shall refer to them, establish certain ends and priorities for political life, but they are concepts rather than conceptions.[1] Like other fundamental principles of morality, they must be further specified before they can generate advice on what to do. There can accordingly be disagreement on what liberalism requires even though there is agreement on the fundamental presuppositions with which we start. As a result, liberals can be for or against abortion, for or against greater redistribution of income, for or against greater government regulation of the market, for or against free trade with other nation-states, for or against stricter limits on immigration, and on either side of any number of other hotly contested social, domestic, and foreign policy issues of the day. People who consider themselves progressives, social democrats, Christian democrats, libertarians, and even traditional conservatives can accordingly all be liberals in the sense that I am using the term.

4. How liberalism and perfectionism differ

But the fundamental presuppositions that liberals embrace are not the only presuppositions on offer. Some people embrace a set of illiberal presuppositions that form the basis of a competing family of political theories. Political philosophers generally refer to this family of political theories as "perfectionism," although many other names are also used to describe certain members of this family. Fascism, communism, racism, religious fundamentalism (no matter what the religion), ethnic nationalism, and a variety of other isms that are each very different from and often extremely hostile to one another are nevertheless all perfectionist because they each embrace a very rigid, comprehensive set of substantive beliefs and see the role of government as ensuring that everyone in their community embraces each and every one of these views as well. In the newspapers, these groups are often referred to as "populist" and their attitude toward government described as "authoritarian" or "totalitarian" depending on whether they are on the right or the left. But focusing on whether these groups are on the right or the left obscures the more important point to be noted here. Perfectionist groups all embrace the same set of fundamental presuppositions for the organization and conduct of political life, and these presuppositions are very different than those embraced by liberal groups. This does not mean that liberals and perfectionists will necessarily arrive at different answers to all the questions of political morality, for this depends on how each fundamental presupposition is cashed out. But it does mean they ask different questions – that there is disagreement at the most fundamental level about what political morality requires, what our political values are, what priorities we should assign between them, and what political morality even means.

For example,

- Liberals believe in toleration, while perfectionists reject this. Perfectionists believe that toleration is the first step toward self-destruction, for it allows society to be irreparably undermined by ideas to which it fundamentally objects and which it could have suppressed but instead allowed to take root and spread.

- Liberals believe that government should be neutral between reasonable but controversial conceptions of the good, while perfectionists believe that there is one true path to leading a moral life, and that government's primary role is to enforce this conception of the good and ensure that all members of society support and abide by it.

- Liberals favor liberty over authority, distrust concentrations of private power, and believe that the role of government should be strictly limited, although they disagree dramatically on how these ideas should be cashed out. Perfectionists, in contrast, believe that government must have sufficient authority to ensure conformity with their particular detailed conception of how people should act and what they should believe, even when this is highly controversial, something that is anathema to liberals because of their embrace of the principle of neutrality.

- Liberals believe that no one is above the law, and that the law may not be violated even in extreme conditions. Perfectionists, in contrast, attach higher priority to security than the rule of law because a perfectionist society always sees itself as both surrounded and infiltrated by enemies, and the rule of law merely gets in the way of addressing these existential threats.

- Liberals believe that the individual is the fundamental social unit, whereas perfectionists believe that the fundamental social unit is the community. Indeed, in a perfectionist society, the very idea of an "individual" makes no sense, for all members of the community are defined by their possession of a certain set of characteristics and their adherence to a specific set of views, and therefore all members of a perfectionist community are identical in all relevant respects. Those who differ in any way are by definition outside the community and must be classified as enemies.

- Liberals believe in the separation of religious and political authority, for the appeal to religious authority would otherwise end the possibility of political argument. Perfectionists, on the other hand, see nothing wrong with conflating these as long as religious authority can be used to further instantiate the particular conception of the good that the perfectionist society chooses to embrace.

- Liberals believe that all members of a political community should have an opportunity to participate in its political decision-making under conditions of full information, and that the purpose of public discourse and debate is to persuade others of the rightness of one's position by resorting to arguments that one's opponents could not reasonably reject. Perfectionist, in contrast, reject the idea of negotiation and compromise with their enemies, and with regard to their fellow citizens, they embrace the tactic of the noble lie – Plato's idea that the common people are incapable of seeing the truth even when it is laid out before them and therefore may be manipulated and led by falsehoods whenever it is necessary to do so.

- Liberals elevate facts over faith in their pure and practical reasoning, for the realm of reason is often the only domain in which different faiths can meet and find a common basis for discussion and agreement. Perfectionists, in contrast, see the scientific method and its corresponding exaltation of reason as a threat to authority, and therefore to stability and order. Truth is to be discovered through faith, whether it be religious faith or faith in "the market" or in a leader who can "Make America Great Again" or in the superiority of one race or gender or in some mythical conception of the American or English or whatever way of life. But no matter what the source of the belief, one simply believes and then follows this belief until it is fulfilled – lack of success is not a sign of error but of a lack of competence or commitment.

- Liberals believe that all people have equal moral worth. Perfectionists deny this. They believe that members of their own community have greater moral worth. Perfectionists accordingly do not give the lives of outsiders the same weight in their moral reasoning, although they may give them some, like wild animals or pets.

It is possible, of course, that some of the presuppositions I have included here could be derived from other presuppositions on the list, and therefore may not, strictly speaking, be fundamental (most theorists try to reduce the differences between liberalism and perfectionism to just one presupposition or perhaps two, usually tolerance and neutrality). It is also possible that I have left something fundamental out.[2] But coming to a view as to whether any refinements should be made here is not necessary in order to explore the issue in which we are interested. Which is this: given these differences between liberalism and perfectionism, is it possible to argue for one set of presuppositions – one version of the framework for political morality – over another? For if it is not, then political morality may indeed be more problematic than personal morality seems to be.

5. Arguing about political morality at the fundamental level

There are three possible argumentative strategies we could employ. First, we could make an argument from consequences, attempting to show how perfectionist beliefs time and time again produce bad consequences, far worse than those produced by liberalism. If this were true, there would arguably be a teleological moral argument for liberalism over perfectionism. Second, we could make an argument from coherence, arguing that the particular ends sought by perfectionists are inherently unattainable and/or the set of beliefs advocated by a particular perfectionist sect are internally inconsistent and therefore incapable of being instantiated as a set. In this case, embracing perfectionism – or at least certain forms of perfectionism – would simply be irrational and logic would demand we embrace something else. And third, we could argue from some higher order fundamental presupposition, one which both liberals and perfectionists embrace, and show that while the liberal set of ordinary fundamental presuppositions can be derived from it, the perfectionist set cannot, a deontological moral argument. Unfortunately, each of these argumentative strategies has some serious problems.

The problems with the argument from consequences should be obvious. It is not always clear what consequences a particular approach will have, and because perfectionists do not see empirical facts as providing conclusive reasons for action or belief, at least with regard to the organization and management of society (this is one of fundamental presuppositions set forth previously), disputes about the effects of these presuppositions are going to be impossible to resolve. For perfectionists, scientific "facts" do not exist. Truth is an expression of ideology, not science.[3] And since these truths are not based on science, using the scientific method to challenge them simply misses the point. Empirical arguments have no purchase here because the claimed "truth" these arguments are challenging are not based on scientific reasoning in the first place.

More importantly, however, remember that all consequentialist theories put the good before the right. That is, they determine what is right by asking what will maximize the good. To make an argument from consequences, one must accordingly begin with a theory of the good. But where is this conception of the good supposed to come from? Liberals and perfectionists embrace different sets of fundamental presuppositions precisely because they have different conceptions of the good. And if they do not agree on their overall conception of the good, it is going to be impossible to argue that by embracing one set of presuppositions rather than the other one groups is committing some kind of wrong. At least it is impossible to do so using arguments that the other cannot reasonably reject, the appropriate test for all of those who embrace liberalism (Scanlon 1982).

Even if we try to abstract out from a more comprehensive theory of the good and rely on what John Rawls called "a thin theory of the good" – a theory of the good that is instrumental to achievement of a wide variety of comprehensive theories of the good (Rawls 1999, 347–350, 380–386), there are still problems. Take, for example, the idea of social stability. Initially, it appears that social stability is something that everyone, perfectionist and liberals alike, should want. And we might be able to show that liberal societies tend to be more stable than perfectionist ones. But some perfectionist organizations are very stable (consider the Catholic Church), and just because some perfectionist societies are less stable than liberal ones is not a persuasive reason for rejecting a form of perfectionism that has not been fully instantiated before. Because the forms of perfectionism are constantly mutating, there will always be enough differences between whatever form is currently under examination and what has gone before to leave open the possibility that this time is different. Moreover, even if they each value stability, liberals and perfectionists do not agree on the priority that stability should have. So even if we were to convince a perfectionist that a particular liberal society was likely to be more stable than whatever perfectionist society they had in mind, they could still respond that a stable liberal society was worse than an unstable perfectionist one.

The same problems arise if we argue that perfectionism has a marked tendency toward violence, and use the avoidance of violence as our thin theory of the good. First, we still have the problem of using empirical evidence to show that this is true, given that perfectionists do not think that empirical evidence provides a sufficient reason for action or belief. Second, it is not clear that perfectionists think violence is bad. Under their conception of the good, violence is often necessary, since they see themselves surrounded by enemies (Schmitt 1996, 26). In any case, it is always heroic. "Whatever value human life has does not come from reason; it emerges from a state of war between those inspired by great mythical images to join battle," says the Nazi legal theorist Carl Schmitt (1988, 71). "War," says the Italian anti-liberal Julius Evola, "offers man the opportunity to awaken the hero that sleeps within him," and "the moment the individual succeeds in living as a hero, even if it is the final moment of his earthly life, weighs infinitely more on the scale of values than a protracted existence consuming monotonously among the trivialities of cities" Evola (1935, 21).[4] By satisfying men's natural desire to be destructive, the French anti-Enlightenment reactionary Joseph de Maistre claims, war leaves them feeling "exalted and fulfilled"[5] To most perfectionists, living in a perfectionist society under threat of violence is uncontroversially preferable to living in a liberal society in peace. Thus, it seems that no matter what its form, the argument from consequences is going to be of no use to us here.

Alternatively, we could try the argument from coherence. But coherence is simply a requirement of reason, not morality, and therefore has correspondingly less argumentative power. Hence the common saying, "a foolish consistency is the hobgoblin of little minds" (Emerson 1993, 24). Moreover, unless we can argue that only one set of moral principles is coherent, there are many and perhaps even an infinite number of possible sets of moral principles that could fall within either liberalism or perfectionism. Some of these sets will be more coherent, some less so. More than likely, every set of applied principles we can come up with based on these fundamental presuppositions is going to be inconsistent to some extent. So if coherence is going to be our super-fundamental principle, then no set of fundamental presuppositions of political morality may be correct. If the argument works, it works just as well against liberalism as it does against perfectionism. So it is difficult to see how coherence alone can be used to defend one set of fundamental presuppositions against another.

That leaves only one hope for finding a way to argue for liberalism over perfectionism. This is to find some overridingly fundamental presupposition, some super-fundamental presupposition, from which both set of ordinary fundamental principles are arguably derived but which can be shown to really support only one set. Despite years of searching, however, I have come to believe that such a principle does not exist. Equality cannot provide such a principle, for perfectionists expressly reject the idea that all persons have equal moral worth, just as they reject the principles of toleration and neutrality. Self-ownership cannot provide it either, because self-ownership has no meaning is a society which sees the community and not the individual as the fundamental social unit, as perfectionist societies all do. For the same reason, even the separateness of persons cannot provide such a super-fundamental principle, for given the perfectionist conception of the person, persons are not separate in a perfectionist community in the relevant sense. And these are the best candidates; the remaining candidates are even less promising.

What does this mean? Well, it does not mean that the fundamental principles of political morality are culturally relative, for culture actually seems to have little to do with it – perfectionist views can arise in any culture. But what it does mean is actually more troubling than this. Liberalism and perfectionist start with such different fundamental principles, they are not merely relative – that is, different from one another around the edges – they are incommensurable. In other words, they are neither derived from, nor may be compared against, some overriding super-principle or covering value (Reiff 2014). Whether we believe in liberalism or its antithesis, no moral argument can be made for one view or the other. Moral arguments can only be made once the choice between one framework or the other is made, and then only with regard as to how one's particular theory of liberalism or perfectionism is going to be cashed out.

Many people fail to recognize this. They are so embedded in the liberal way of thinking they cannot imagine that anyone might embrace a set of contrary presuppositions. They think that what is going on now is merely a family disagreement. If we can just get our facts straight, what seem like intractable disagreements will disappear. And it is true that many people are not purely liberal or perfectionist – that is, they embrace some presuppositions from one family of political theories and some from the other and simply ignore any contradictions this may create. Given this overlap, political argument may be possible for these people to some extent. But the number of people who are fully committed to some set of perfectionist presuppositions in what were widely considered liberal societies is growing. For more and more people, the fundamental presuppositions that inform their conception of political morality are accordingly incommensurable with the presuppositions formally embraced by the societies in which they live. And unfortunately, whenever this happens – whenever people feel that they can no longer describe their most closely held values in a way that those who disagree with them can appreciate and understand, things do not usually end well.

Notes

1 For further discussion of the difference between a concept and a conception, see Dworkin (1986, 70–72).
2 For further discussion of the differences between liberalism and perfectionism, see Reiff (2007).
3 See, e.g., Gregor (2005, 203–204) (discussing the views of Julius Evola) (footnote omitted); Orwell (1943, 199).
4 See also (Evola 2017, 50), "belonging to the Right means upholding the values of Tradition as spiritual, aristocratic, and warrior values."
5 Berlin (2001, 21), describing Maistre's views.

References

Berlin, I. (2001), "The Counter-Enlightenment," in *Against the Current: Essays in the History of Ideas*, Princeton, NJ: Princeton University Press, 1–24.
Cohen, G. A. (2003), "Facts and Principles," *Philosophy and Public Affairs* 31: 211–245.
Dworkin, R. (1986), *Law's Empire*, Cambridge, MA: Harvard University Press.
Emerson, R. W. (1993), *Self-Reliance and Other Essays*, New York: Dover.
Evola, J. (1935), "The Forms of Warlike Heroism," in *Metaphysics of War: Battle, Victory and Death in the World of Tradition*, London: Arktos, 2011, 21–27.
———. (2017), "What It Means to Belong to the Right," in *A Handbook for Right-Wing Youth*, London: Arktos, 49–53.
Gregor, J. (2005), *Mussolini's Intellectuals*, Princeton, NJ: Princeton University Press.
Hart, H. L. A. (1994), *The Concept of Law*, second edition, Oxford: Oxford University Press.
Orwell, G. (1981), "Looking Back on the Spanish War (1943)," in *A Collection of Essays*, New York: Houghton Mifflin, 188–209.
Rachels, J. and S. Rachels (2012), *The Elements of Moral Philosophy*, seventh edition, New York: McGraw Hill.
Rawls, J. (1971), *A Theory of Justice*, revised edition, Cambridge, MA: Harvard University Press, 1999.
Reiff, M. R. (2007), "The Attack on Liberalism," in *Law and Philosophy*, edited by M. Freeman and R. Harrison, Oxford: Oxford University Press, 173–210.
———. (2014), "Incommensurability and Moral Value," *Politics, Philosophy, and Economics* 13: 237–268.
Scanlon, T. M. (1982), "Contractualism and Utilitarianism," in *Utilitarianism and Beyond*, edited by A. Sen and B. Williams, Cambridge: Cambridge University Press, 103–128.
Schmitt, C. (1988), *The Crisis of Parliamentary Democracy*, Cambridge, MA: MIT Press.
———. (1996), *The Concept of the Political*, translated by G. Schwab, Chicago: University of Chicago Press.

20

EVOLUTIONARY DEBUNKING AND MORAL RELATIVISM

Daniel Z. Korman and Dustin Locke

1. A neglected response to the moral debunking argument

The moral debunking argument aims to show that a proper appreciation of the source of our moral beliefs threatens to undermine those beliefs. In broad brush strokes, the argument runs as follows:

(D1) Our moral beliefs are not explained by the moral facts and we realize this.
(D2) If so, then we're rationally committed to withholding our moral beliefs.
(D3) So, we're rationally committed to withholding our moral beliefs.

(D1) can be motivated in a variety of ways. One common strategy is to appeal to evolutionary considerations and, in particular, the selective pressures that shaped our moral beliefs. The belief that it's good to feed one's children, for instance, can be accounted for in terms of the adaptive value of attitudes that motivate one to care for one's children; and the (putative) moral fact that it truly is good to feed one's children has no role to play in this evolutionary explanation. In what follows, we'll be restricting our attention to this way of developing the debunking argument, though much of what we say won't hang on how (D1) is motivated. The idea behind (D2) is then that, in light of the explanatory disconnect between the moral facts and the factors influencing our moral beliefs, it would be an extraordinary coincidence if those beliefs turned out to be true, the realization of which should convince us to withhold belief.[1]

Numerous responses to the debunking argument can be found in the literature, including (but not limited to) the following three. According to the *naturalist* response, the moral facts do explain our moral beliefs, insofar as they are identical to certain of the mundane natural facts that explain our moral beliefs. According to the *constructivist* response, the explanation runs in the opposite direction: the moral facts are as they are because we have the moral beliefs that we do, which is why it's no coincidence that we have accurate moral beliefs. According to the *third-factor* response, while the moral facts neither explain nor are explained by our moral beliefs, what secures the noncoincidental accuracy of our moral beliefs is some further fact (a "third factor") that is responsible both for the moral beliefs and for the moral facts.[2]

Each response has its drawbacks. Naturalist reductions tend to have profoundly counterintuitive implications regarding the first-order moral truths; constructivists seem committed to saying that we can change the facts about what's right and wrong simply by changing our minds about what's right and wrong; and third-factor theorists seem to unabashedly beg the question against the debunker. (More on each of these later.)

Our aim here is to explore the prospects of a *relativist* response to the debunking argument. We begin by clarifying the relativist thesis under consideration (Section 2), and we explain why relativists seem well-positioned to resist the debunking argument in a way that avoids the drawbacks of existing responses (Section 3). We then show that appearances are deceiving. At bottom, the relativist response is no less question-begging than the third-factor response (Section 4), and – when we turn our attention to the strongest formulation of the debunking argument – the virtues of relativism turn out to be vices.

2. Relativism

Relativism, as we will understand it here, is the thesis that (i) no moral propositions are true simpliciter and (ii) for some moral proposition p and moral frameworks F and F', p is true relative to F and false relative to F'. A *moral framework* may be understood simply as a set of moral propositions, and a moral proposition is true relative to a given framework just in case that proposition is a member of that framework.

Since moral propositions are true only relative to a framework, beliefs can be accurate only relative to a framework. Unless otherwise specified, when we say that S's belief is accurate, this should be understood to mean that the content of S's belief is true relative to S's framework. What is it then for a given framework to *be* S's framework? In other words, what is it to *have* a given framework? Following the lead of Gilbert Harman (1996, ch. 1.3), we won't assume that S's framework is invariably going to be the set containing all and only the propositions S in fact believes, since that would make it impossible to have inaccurate moral beliefs. While there are a variety of ways one might fill in the details, for the sake of concreteness we'll assume that S's framework is the set of propositions that S *would* believe upon reaching reflective equilibrium.[3]

One might naturally assume that relativists are committed to saying that right actions are right and wrong actions are wrong because we have the moral frameworks we do – for instance, that one ought to feed one's children because we have a framework relative to which it's true that one ought to feed one's children. In fact, and perhaps surprisingly, they are committed to no such thing.[4] Propositions of the form *p because q* are themselves moral propositions, and as such can be assessed for truth only relative to a framework. What's true relative to our ordinary moral framework(s), the relativist will say, is *that one ought to feed one's children because they will die without our assistance.* The proposition *that one ought to feed one's children because we have the framework we do*, by contrast, comes out false relative to the framework of any normal person (relativists included).[5] Relativists can therefore happily deny that it is anyone's *having* a given belief or framework that makes moral propositions true. As we shall see, herein lies one of the distinctive advantages of relativism.

Relativism is often taken to be motivated by its explanation of fundamental moral disagreement.[6] So long as the disputants have different moral frameworks, the relativist can maintain that neither party is mistaken (simpliciter); each is saying something true relative to his or her own framework. At the same time, the relativist can affirm that the disagreement is genuine: the one party is denying the very thing that the other is affirming.[7] Relativism also makes it easy to see how moral knowledge is possible. Having accurate moral beliefs is just a matter of believing propositions that are in one's own framework. This isn't automatic – since one's framework consists of the propositions one *would* believe under certain idealized conditions – but it does put accurate beliefs within reach, without requiring anything like responsiveness to some transcendent realm of moral facts.

3. The relativist response to the debunking argument

Relativists seem to have a straightforward way out of the debunking argument, by denying the second premise:

(D2) If we realize that our moral beliefs are not explained by the moral facts, then we're rationally committed to withholding our moral beliefs.

The usual motivation for this premise, again, turns on the idea that, if indeed the moral facts aren't explaining our moral beliefs, then it would be a massive coincidence if our moral beliefs turned out to be accurate. Relativists will deny that any coincidence is required. For as we just saw, by relativist lights one's moral beliefs are guaranteed to be accurate so long as they're aligned with one's own framework. And while there is room for moral error, it certainly isn't a *coincidence* if our beliefs are largely aligned with our respective frameworks. Accordingly, it's no coincidence if our moral beliefs are largely accurate, and that is so regardless of the source of our moral beliefs, even if it's "blind" evolutionary forces aimed only at keeping us and ours alive.

Now let us see how the relativist response avoids the drawbacks of the other responses. The naturalist resists (D1) by identifying moral facts with the very natural facts that explain our moral beliefs. But such reductions are plagued by apparent counterexamples.[8] (You don't need our help to see the nasty consequences of, for instance, identifying the fact that something is the right thing to do with the fact that such actions tend to promote human flourishing.) Since relativists are able to put moral truths within our epistemic reach without requiring moral facts to explain our beliefs, there is no need for them to identify moral facts with any such (causally efficacious) natural fact.

Constructivists maintain that the moral facts are as they are because we have the moral beliefs that we do. For instance, a certain sort of Humean constructivist might say that, for any given person S, what makes it true that S is required to Φ is that S believes that she herself is required to Φ. Since explanations support counterfactuals, the constructivist is thereby committed to such "repugnant counterfactuals" as *if I were to stop believing that I'm required to feed my children, I wouldn't be required to feed my children.*[9] The relativist, as we saw previously, need say no such thing. For, according to the relativist, the fact that a given person *believes* the proposition that she is required to feed her children, or *has* a framework that includes that proposition, is no part of the explanation of why it's true (relative to her framework) that she is required to do so. Since the relativist can reject the constructivist's (counterfactual-supporting) explanations, she isn't saddled with the repugnant counterfactuals.[10]

Finally, there are the third-factor responses according to which some natural fact explains both the moral beliefs and the associated moral facts. The fact that feeding one's children promotes their survival, for instance, may be cited as the third factor that figures both in the evolutionary explanation of why we believe it's good to feed them and the explanation of why it *is* good to feed them. Or the fact that altruistic acts promote social cohesion may be cited as the third factor that explains both why we believe that we are required to act altruistically and why we *are* required to act altruistically. These lines of reasoning straightforwardly beg the question against the debunker, by relying on the moral belief that promoting survival or social cohesion is a good-making feature, which is precisely the sort of belief that the debunker is calling into question. The relativist, by contrast, doesn't rely on any specific first-order moral beliefs in her response to the debunker. What's doing the work for the relativist is just the *metaethical* assumption that having accurate moral beliefs (whatever they may be) is simply a matter of having beliefs that align with one's own moral framework.[11]

If indeed the relativist response to the debunking argument avoids all of the shortcomings of existing responses, this could potentially serve as an important secondary motivation for relativism, alongside its explanation of moral disagreement. On closer inspection, however, matters are not so simple, and relativism turns out to be ill-equipped to handle a fortified version of the debunking argument.

4. Is the relativist response question-begging and does it matter?

The relativist response to the debunking argument, as we just observed, does not directly invoke any of the first-order moral beliefs that the debunker means to be calling into question. But the relativist may still be accused of begging the question indirectly. In particular, one might object that the *rationale* for embracing relativism relies on first-order moral beliefs.

To see this, consider the motivations for relativism canvassed previously. One was that relativism yields the best explanation of fundamental moral disagreement, for instance when one person thinks it's wrong to kill someone for dishonoring one's family, whereas another thinks it's the right thing to do. The relativist is able to explain how her belief that honor killings are wrong can be accurate without thereby privileging her own belief. After all, the proponent of honor killings also has an accurate belief (true relative to his framework).

Here, though, the relativist is assuming that among the things to be explained is the *accuracy* of the belief that honor killings are wrong. This is, she is assuming that honor killings are wrong. In assuming this, she is indirectly relying on her first-order moral beliefs. Without that assumption, the relativist's explanation of moral disagreement is no improvement on the error theorist's. The error theorist can give precisely the same explanation of *why* the disputants disagree, in terms of their having substantially different moral frameworks, but will say that *both* are mistaken, since nothing is right or wrong. We have reason to prefer the relativist's account of moral disagreement to the error theorist's only if we take for granted that our moral beliefs are accurate and that this is something a theory of moral disagreement ought to account for.

Likewise for the contention that relativism yields the best response to the debunking argument, insofar as it is able to explain (quite easily) the striking correlation between adaptive moral beliefs and accurate moral beliefs. In assuming that there is such a correlation to be explained, the relativist is taking for granted that adaptive moral beliefs – for instance that it is good to feed one's children – are indeed accurate. So the rationale again ultimately takes for granted the accuracy of her first-order moral beliefs.

The relativist does help herself to her first-order moral beliefs (albeit indirectly), and is therefore no less open to the charge of question-begging than the third-factor theorists considered previously. That said, we don't think that there is anything so bad about relying on beliefs that one's interlocutor has called into question. The reason, in short, is that the mere fact that someone has presented you with an argument against some belief of yours is not by itself enough to remove whatever entitlement you have to rely on that belief, and in particular doesn't prevent you from relying on that belief when assessing whether you ought to accept the premises of that argument. Otherwise, anyone could be driven to global skepticism by the following inane argument: "Everything in the Harry Potter stories is true, and every proposition you believe is contradicted by something in the Harry Potter stories."[12]

5. The explanatory constraint

There is, however, a potentially more serious problem for relativists who help themselves to their moral beliefs in the face of the debunking argument. The problem isn't that they beg the question, but rather that they concede that moral facts have no role to play in explaining our moral beliefs.

To see why such explanatory concessions are problematic, imagine that, after consulting your magic 8-ball toy, you form the belief that your crush likes you back. Now you learn, much to your disappointment, that facts about who has a crush on who play no role in determining the 8-ball's outputs, and hence no role in determining what you believe on the basis of those outputs. Clearly, you should then withhold belief about whether your crush likes you back.

Extrapolating from this, the realization that you have some beliefs for reasons having nothing to do with the range of facts those beliefs purport to be about rationally compels you to abandon those beliefs. Getting the explanatory constraint exactly right is a delicate matter, but the following will do for our purposes here:

Explanatory constraint

If p is about domain D, and S believes that her belief that p isn't explained by D-facts, then S is thereby rationally committed to withholding belief that p.[13]

The relativist response runs afoul of the explanatory constraint. After all, the relativist concedes that the moral facts have no role to play in explaining why she has the moral beliefs that she does, but persists in relying on those very beliefs in reasoning her way to relativism.

The explanatory constraint also has ramifications for our understanding of the debunking argument. Thus far, following the literature, we have motivated (D2) in terms of coincidence: the realization that the moral facts don't explain moral beliefs undermines those beliefs *by way of* revealing that moral accuracy would at best be coincidental. The explanatory constraint eliminates the middleman by directly entailing (D2). To our mind the strongest defense of the debunking argument is one that makes no reference to coincidence and that reasons directly from an explanatory constraint to (D2).

Relativists who wish to resist this fortified version of the argument must therefore reject the explanatory constraint. But they must do so in a way that does justice to the insight that explanatory concessions have the power to undermine one's beliefs (as in the 8-ball case). We will explore two strategies that the relativist might pursue.

6. Loosening the explanatory constraint

The first strategy that relativists might pursue is to amend the explanatory constraint, permitting the explanatory arrows to run in either direction:

Bidirectional explanatory constraint

If p is about domain D, and S believes that her belief that p isn't explained by *and doesn't explain* D-facts, then S is thereby rationally committed to withholding belief that p.

There is some reason to believe that the constraint needs to be loosened in this way. Imagine a dictatorship in which, if the dictator so much as *believes* that something is illegal, it is thereby illegal. Suppose the dictator one day forms the belief that it's illegal to wear orange on Sundays. The fact that it's illegal doesn't explain his belief that it is. But he's certainly rational in believing that it's illegal. Plausibly, that's because there's an explanatory connection running in the other direction: his believing it's illegal makes it illegal.

We're not entirely convinced that the explanatory constraint *needs* loosening in order to handle this case. The dictator presumably believes it's now illegal to wear orange partly on the basis of his belief that the laws are determined by his beliefs. So long as this latter belief is explained by the fact that his beliefs determine the laws (which is itself a fact about the laws), the dictator's law beliefs satisfy the original explanatory constraint.

But let's just grant that the explanatory constraint does need to be loosened in this way. Still, shifting to the bidirectional constraint helps the relativist *only if* she is willing to affirm that our beliefs explain the moral facts. But that would then undermine one of the main attractions of the relativist response to the debunking argument, namely that it permits one to avoid the drawbacks of constructivism. For as soon as the relativist affirms that moral truths are made true by our beliefs – that is, as soon as she embraces constructivism – she will be saddled with the constructivist's repugnant counterfactuals (see Section 3).

Additionally, relativism and constructivism make for strange bedfellows. For suppose that the relativist does go in for constructivism. It will then be true relative to her framework that (e.g.) *we are required to feed our children because we believe we're required to feed our children*. But, relative to the framework of any other normal person, it's false that that's why we are required to feed our children (see Section 2). This leaves us with the uncomfortable result that the truth of constructivist relativism would not be enough to save the masses from debunking. Rather, those exposed to the debunking argument remain rationally required to withhold moral belief until they *accept* constructivism. None of the other responses to the argument has this odd result. For instance, on a nonrelativist constructivist view – on which moral propositions are made true (simpliciter) by our adoption of frameworks – one doesn't have to accept constructivism in order for one's beliefs to be noncoincidentally accurate; it's enough just that constructivism is *correct*.

7. The modal constraint

Relativists might instead insist that the explanatory constraint misdiagnoses what's going on in the 8-ball case. It's not the explanatory concession *per se* that undermines the beliefs, they might say. Rather, it's that explanatory concessions often commit one to conceding that one could easily have ended up with inaccurate beliefs, and it's this modal concession that does the defeating. Applied to the 8-ball case, the idea is that, when you realize that your crush beliefs aren't explained by the crush facts, you realize that, in forming beliefs on the basis of the 8-ball's outputs, you could easily have ended up with inaccurate crush beliefs. And it is this modal concession – that this could easily have happened – that rationally commits you to withholding belief about whether your crush likes you back.

This diagnosis of the 8-ball case opens up the possibility that explanatory concessions don't *always* defeat. Rather, they defeat when, only when, and because they rationally commit one to believing that one could easily have been mistaken (or, in other words, that the belief is not "safe"). More precisely:

Modal constraint

If p is about domain D, S's rational commitment to believing that [her belief that p is not explained by some D-facts] defeats her belief that p when, only when, and because she is thereby rationally committed to believing that she could easily have had an inaccurate belief about whether p.

If true, the modal constraint creates a toe-hold for relativists, for morality is arguably one of the domains where the explanatory concession doesn't rationally commit one to the modal concession. After all, the relativist will reason, our moral beliefs are more or less bound to be accurate, regardless of what's explaining them. For even had we had different beliefs, those beliefs would still have been true relative to the frameworks we would then have had. In this case, the explanatory concession does not force a modal concession.

Unfortunately for the relativist, the modal constraint is false. To see this, consider the following case (adapted from Locke 2014). Suppose Jack believes P:

(P) Protons cause streaks of type S in cloud chambers.

Jack believes P, but not because he has received the training of an ordinary physics student. Rather, it's because he asked some Martians (who had previously convinced him of their superior intellect) what causes those streaks, and they replied: protons. Later, however, Jack learns that the Martians told him that it's protons that cause those streaks, not because they themselves had done any physics, but simply because they liked the sound of the word "proton." You may even suppose, if you like, that there is some deep law of Martian psychology that makes them like the sound of "proton," and so it could not easily have happened that the Martians told Jack that such streaks were caused by anything else.

Upon learning all of this, it is clearly irrational for Jack to stand by his belief that P. But suppose that, rather than abandoning the belief, Jack attempts to assure himself that it is still in good standing by reasoning as follows:

> Yes, my belief that it's protons that cause streaks like that is not explained by the facts about protons. Still, given what I have just learned about Martian psychology, I could not easily have formed a different belief about whether protons cause streaks like that. Moreover, *protons do cause streaks like that in cloud chambers*, and – since this interaction is surely underwritten by natural laws – it could not easily have failed to be the case that protons cause such streaks. Putting the pieces together: I could not easily have been mistaken about whether such streaks are caused by protons.

Clearly the reasoning is illicit. And the obvious diagnosis of why it is illicit is that the reasoning relies on the belief that P (see the italics), which has already been undermined by his explanatory concession that he believes that those streaks are caused by protons for reasons having nothing to do with whether they in fact are caused by protons – just as our explanatory constraint would have it. The modal constraint, by contrast, leaves us with no explanation of why Jack's reliance on P is illicit. After all, just as in other cases of testimony from an otherwise reliable source, Jack is initially entitled to the testimonial belief that P.[14] And if the modal constraint is correct, then Jack's entitlement to P is undermined *only once* he is driven to the conclusion that his belief could easily have been false. Thus, by the lights of the modal constraint, there should be nothing illicit about relying on his not-yet-impugned belief in P when assessing whether P satisfies this modal condition.

In short: The Jack case shows that when explanatory concessions defeat, it isn't always by way of committing one to a modal concession. So the modal constraint is mistaken, and the relativist is again without any viable replacement for the explanatory constraint.

8. Relativism naturalized?

We have been assuming that the relativist will grant that moral beliefs are not explained by moral facts. There is, however, a different route open to the relativist, which is to follow the naturalist in *identifying* moral facts with some of the natural facts that explain our moral beliefs. Exploring this route requires taking a closer look at what "moral facts" would amount to in a relativist setting and how the reference to moral facts in the explanatory constraint is to be understood.

Relativists will of course regard moral facts as entities that obtain only relative to a framework – lest it end up being an absolute fact that it's good to feed one's children or that honor killings are wrong. For concreteness, let's assume they say that a moral fact that p obtains relative to S if and only if p is a member of S's framework. Relativists may then insist upon understanding the explanatory constraint as follows:

Relativistic explanatory constraint

If p is about domain D, and S believes that her belief that p isn't explained by D-facts that obtain relative to S's framework, then S is thereby rationally committed to withholding belief that p.

To show that this explanatory constraint is satisfied, the relativist will need to find natural facts that explain our moral beliefs with which to identify the moral facts that obtain relative to our frameworks. For instance, she may identify the moral fact that it's good to feed one's children with certain psychological facts about us, for instance that we are strongly disposed to have positive attitudes towards feeding one's children. Or she may identify it with the fact that feeding one's children tends to promote human flourishing. Since such natural facts are indeed part of what explains why we believe that p, the relativist seems able to satisfy the relativistic explanatory constraint.

There are two potential problems with this strategy. First, the relativist response will now inherit the drawbacks of the naturalist response alluded to previously, for instance that candidate reductions of moral facts are plagued by apparent counterexamples. Second, it would seem that the relativist *cannot* identify moral facts with natural facts – even natural facts about people's evaluative dispositions. Suppose, for the sake of illustration, that the moral fact that honor killings are wrong which obtains relative to S is identical to the natural fact that S is strongly disposed to have a negative attitude towards honor killings. The trouble is that, by relativist lights, the former fact obtains relative to some frameworks but not others. So, by Leibniz's Law, the latter fact also must obtain relative to some frameworks but not others. Plainly though, the latter fact, if it obtains at all, obtains simpliciter (or relative to all frameworks). So, on the envisaged relativist conception of moral facts, the fact that honor killings are wrong cannot be identified with this or any other natural, framework-invariant fact.

9. Conclusion

We have been exploring a relativist response to the debunking argument. At first glance, the relativist response seemed to enjoy a number of advantages over existing responses. It seems able to secure the noncoincidental accuracy of our moral beliefs without commitment to repugnant counterfactuals, without directly begging the question against the debunker, and without controversial reductions of the moral to the natural.

However, when the debunking argument is properly understood – as underwritten directly by an explanatory constraint – the advantages of the relativist response vanish. Like third-factor theorists, relativists rely on their antecedent moral beliefs (albeit indirectly) in the face of the debunking argument, which in turn are jeopardized by their explanatory concessions. They are able to accommodate a loosened version of the explanatory constraint only by collapsing into constructivism and inheriting all of its problems. And relativism, by its very nature, prevents them from accommodating the explanatory constraint by embracing a naturalistic reduction of the moral to the natural.

Acknowledgments

We would like to thank Charity Anderson, Sinan Dogramaci, Kenny Easwaran, Justin Fisher, Amy Flowerree, William Roche, Josh Thurow, and Joel Velasco for valuable discussion.

Notes

1 Arguments of this sort can be found in Ruse (1986), Joyce (2006, ch. 6), and Street (2006). The arguments are often wielded with a restricted scope, targeting only realists. We focus on the unqualified formulation, so as to shed light on why certain sorts of antirealists – in particular, moral relativists – are supposed to be able to resist the argument.
2 For representative defenses, see Copp (2008), Street (2006, §10), and Enoch (2010) respectively.
3 There also are other ways for relativists to fill in the details of what truth, frameworks, and accuracy amount to, as well as other fine points about the semantics of moral attributions (see, e.g., Brogaard 2008). We leave it to others to explore how different ways of developing a relativist thesis might affect what we say later and, in particular, the relativist treatment of the debunking argument.
4 See Woods (2018, §8); cf. Sosa (1999, 134) and Einheuser (2006) on existential relativism.
5 Likewise, while relativists may say that what makes a moral proposition p *true relative to F* is that p is in F, that is not to say that p is made true by anyone's *having* that framework.
6 See, e.g., Wong (1984) and Harman (1996).
7 Cf. MacFarlane (2014) and Lasersohn (2017) on faultless disagreement.
8 Not to mention the open question argument (Moore 1903) and the moral twin earth argument (Horgan and Timmons 1991).
9 More nuanced versions of constructivism are committed to more nuanced, but no less repugnant, repugnant counterfactuals.
10 At most, the relativist is committed to saying: *if I didn't believe p, p wouldn't be true relative to the framework I would then have had.* This, however, is not an affirmation of the repugnant counterfactual; rather, it just reports the triviality that p isn't true relative to frameworks it isn't in.
11 Cf. Street (2006, 163–164) on whether constructivism is question-begging.
12 See Korman and Locke (forthcoming, §4) for more on charges of begging the question.
13 This formulation of the constraint has the flexibility to accommodate beliefs about the future. Your belief that the sun will rise tomorrow isn't explained by the fact that it will, but it *is* explained by facts about sunrises (viz. past sunrises), thereby satisfying the constraint. Still, we are not entirely satisfied with this formulation. For some alternative formulations, see Setiya's K (2012, 96), Locke's CD (2014, 232), McCain's EF (2014, §§4.4 and 6.4), Lutz's EAD (2018, §2), and Korman's EC5 (2019, §8).
14 Those who doubt that we have a default entitlement to such testimonial beliefs may instead suppose that Jack was brainwashed to have an intuition or disposition to believe that protons cause those streaks, or to have experiences that (richly) present those streaks as caused by protons.

References

Brogaard, B. (2008), "Moral Contextualism and Moral Relativism," *Philosophical Quarterly* 58: 385–409.
Copp, D. (2008), "Darwinian Skepticism About Moral Realism," *Philosophical Issues* 18: 186–206.
Einheuser, I. (2006), "Counterconventional Conditionals," *Philosophical Studies* 127: 459–482.

Enoch, D. (2010), "The Epistemological Challenge to Metanormative Realism," *Philosophical Studies* 148: 413–438.

Harman, G. (1996), "Moral Relativism," in *Moral Relativism and Moral Objectivity*, edited by G. Harman and J. Jarvis Thomson, Oxford: Oxford University Press, 3–64.

Horgan, T. and M. Timmons (1991), "New Wave Moral Realism Meets Moral Twin Earth," *Journal of Philosophical Research* 16: 447–465.

Joyce, R. (2006), *The Evolution of Morality*, Cambridge, MA: MIT Press.

Korman, D. Z. (2019), "Debunking Arguments in Metaphysics and Metaethics," in *Cognitive Science and Metaphysics*, edited by B. McLaughlin and A. I. Goldman, Oxford: Oxford University Press, 337–363.

Korman, D. Z. and D. Locke (forthcoming), "Against Minimalist Responses to Moral Debunking Arguments," in *Oxford Studies in Metaethics*, Oxford: Oxford University Press.

Lasersohn, P. (2017), *Subjectivity and Perspective in Truth-Theoretic Semantics*, Oxford: Oxford University Press.

Locke, D. (2014), "Darwinian Normative Skepticism," in *Challenges to Moral and Religious Belief*, edited by M. Bergmann and P. Kain, Oxford: Oxford University Press, 220–236.

Lutz, M. (2018), "What Makes Evolution a Defeater?" *Erkenntnis* 83: 1105–1126.

MacFarlane, J. (2014), *Assessment Sensitivity*, Oxford: Oxford University Press.

McCain, K. (2014), *Evidentialism and Epistemic Justification*, New York: Routledge.

Moore, G. E. (1903), *Principia Ethica*, New York: Cambridge University Press.

Ruse, M. (1986), *Taking Darwin Seriously*, Oxford: Basil Blackwell.

Setiya, K. (2012), *Knowing Right from Wrong*, Oxford: Oxford University Press.

Sosa, E. (1999), "Existential Relativity," *Midwest Studies in Philosophy* 23: 132–143.

Street, S. (2006), "A Darwinian Dilemma for Realist Theories of Value," *Philosophical Studies* 127: 109–166.

Wong, D. B. (1984), *Moral Relativity*, Berkeley: University of California Press.

Woods, J. (2018), "Footing the Cost (of Normative Subjectivism)," *Methodology and Moral Philosophy*, edited by J. Suikkanen and A. Kauppinen, New York: Routledge.

21
MOTIVATING REASONS

Christoph Hanisch

1. Introduction

College Professor Camilla cancels her lecture that was supposed to take place this upcoming Monday. She had just looked up the online academic calendar and noticed that the annual departmental picnic is taking place that day, a fact that she had been unaware of. Camilla sends an email to her students, apologizing for notifying at short notice, and cancels the lecture. Camilla's action, cancelling the lecture, was not an instance of mere behavior or something that happened to Camilla. Rather, it is an action done for a reason. Why did Camilla cancel her lecture? When we explain an intentional action, we seek answers to such distinctive questions, namely those that ask for Camilla's reasons; the reasons that made her cancel the lecture, understood as the considerations in the light of which she deliberated, decided, and acted.

This chapter clarifies a category within the broader phenomenon of "reasons for action," namely motivating reasons. I present the main action-theoretical and ontological accounts, beginning with psychologism and factualism, followed by two recent contenders (representationalism and truthy-psychologism). The debate about the "identity thesis" provides the segue into the concluding reflections on the apparently relativistic character of motivating (as distinct from normative) reasons.[1]

2. Psychologism

With regard to Camilla's action of cancelling her lecture, it is uncontroversial that the reasons in play must be divided into normative reasons (NRs) and motivating reasons (MRs). NRs are considerations that justify Camilla's action in an objective, non-relativistic manner. More controversial, but still widely endorsed, is the claim that NRs are facts or states of affairs, broadly conceived of as good-making features of an action. In Camilla's case, the NR is constituted by the fact that the picnic collides with her lecture. This fact speaks, at least *prima facie*, in favor of the cancellation. Moreover, it does so independently of what Camilla's deliberative stance and "state of mind" look like from within her own, first-personal, perspective. The NR doesn't go away simply because Camilla is ignorant of its presence, ignores it, or fails acting on it for some other reason.

Things look different when we want to know what Camilla's reasons are for cancelling. We are now asking for the considerations that she took as favoring her action. Since NRs, even if they apply to an agent in the preceding sense, do not necessarily translate into motivating considerations, the distinct explanatory device of MRs must be introduced in order to rationalize or render intelligible Camilla's action. Since we are focusing on the distinctive phenomenon of intentional action, the explanation in question must incorporate as an indispensable feature Camilla's own deliberative standpoint.

Beginning with Donald Davidson (1963), the dominant theory of MRs had been psychologism. According to psychologism, MRs *are* mental states. Davidson calls an agent's "primary reason" a combination of two such states, namely a pro-attitude and a belief. The Humean picture of action underlying psychologism suggests that Camilla's MR is the combination of a certain desire-like state, e.g., the preference to avoid scheduling conflicts, and an instrumentalist belief, namely that cancelling the lecture is a way to satisfy this preference. This pair of mental states explains Camilla's action by providing an answer to our inquiry of why she cancelled the lecture: by means of reconstructing her primary reason, we are able to render intelligible Camilla's action. We do so insofar as we take into account her motivational states and allow these states to play a distinctive role that is unique to intentional action (as opposed to different kinds of "events," such as Camilla passing out in front of her computer due to being overworked). For Davidson this mode of rationalization is the prime desideratum of an account of "reasons-explanation" of actions.

Davidson is aware that pointing to a particular pair of pro-attitude and belief does not always suffice for an action explanation. Camilla might have had, simultaneously, a number of belief-desire pairs, each of which with the potential to explain her action. Davidson adds a causalist feature in order to address this problem. In explaining Camilla's action we need to know which of these mental states were causally efficacious with regard to Camilla's eventual action. Only that one particular primary reason is, according to classical psychologism, Camilla's MR for cancelling the lecture.[2]

3. Factualism

Davidson presents causalism/psychologism as an empiricist account that asks the question of an agent's reasons from the third-personal perspective of the observer. However, even Davidson's Humean approach acknowledges that the explanation of intentional actions in terms of MRs has to accommodate the role of the first-personal view point and of a subjective and perspectival normativity regarding the considerations in the light of which the agent deliberates and acts. At a crucial juncture Davidson says that, "[t]here is a certain irreducible – though somewhat anaemic – sense in which every rationalization [in terms of primary reasons] justifies: from the agent's point of view there was, when he acted, something to be said for the action" (Davidson 1963, 9).

Strengthening the role of the agent's perspective in reasons-explanations is the driving commitment of factualist alternatives to psychologism. In the 1990s, Arthur Collins (1997) and, more influentially, Jonathan Dancy (2000) started to undermine the hegemony of the Davidsonian framework. Factualists are united in pushing the anti-psychologistic counter that when we ask for an explanation of Camilla's action, we hardly find it enlightening to learn anything about Camilla's mental states. Developing the claim (somewhat acknowledged even by Davidson) that the agent's appreciation of her reasons must be part of any action explanation, Collins analyzes Camilla's own take on her action. This analysis shows that it is only in rare cases that an agent refs to her beliefs and desires in order to answer why she did what she did (for example, when someone sees her psychiatrist *because* she realizes that she has hallucinations).

Instead of mental states, Camilla refers to a worldly fact when she deliberates and when she reports why she cancelled. "Because the picnic is scheduled that day," is the answer that our request for a reason will trigger. It was in the light of that favoring consideration that Camilla acted, a consideration that becomes her MR for action because she, first-personally, endorses the "objective circumstance" consisting in the picnic being scheduled that day. The final clause is crucial for Collins' factualist account: When she reports her MR, Camilla is not pointing to her state of believing (as psychologism would have it); rather; she takes up a certain commitment towards, what she takes to be, a worldly fact. Put a bit more technically, Camilla always assigns the truth value "true" to the proposition that she presents to herself and others as *her* reason in the course of acting. Importantly, for the final section, the "objective" in Collins' "objective circumstances" precisely does not mean, "non-relative to Camilla's agential point of view." This will become clearer in the discussion of error cases.

Collins continues that under pressure from Davidsonians Camilla might indeed admit: "Yes, in a sense you are right. Strictly speaking I cancelled because I *believed* that the picnic was scheduled." Is Camilla's restatement of her MR in terms of her belief a conclusive indicator that psychologists are correct, after all? Factualists deny this. They highlight that Camilla is not withdrawing the endorsement of her fact directed statement. What Camilla does is to acknowledge her fallibility as an epistemic agent. Certainly, she might be incorrect about what she believes (because, trivially, nobody is either omniscient or gets it right all the time). However, neither does this switch towards "believing" constitute a move towards psychologism (Camilla is not suddenly pointing to any mental item qua MR), nor is Camilla retracting the endorsement of what she believes. Collins argues that any epistemic neutrality on Camilla's part regarding the objective circumstance of the picnic would actually undermine a constitutive condition of her *believing* what she reports in her reason-giving statement: "If [Camilla] refuses to represent herself, when she acted, as a person endorsing the objective claim [about the picnic], she does not restate her explanation, she repudiates it" (Collins 1997, 123).

Dancy's factualism, despite its anti-psychologism (that he shares with Collins), allows that mental states play an even necessary role in intentional action, namely as "enabling conditions." Had Camilla not *believed* that the picnic is scheduled, she wouldn't have cancelled her lecture. However, the correctness of such counterfactuals should not mislead us into suggesting that Camilla's mental state of believing plays a constitutive part in spelling out her motivating *reason* and in formulating the corresponding explanation of her action. Only perceived facts, as they present themselves as the contents of Camilla's beliefs, can accomplish that feat.

Even if this enabling-condition-move were promising, factualism is confronted with another notorious challenge, namely error cases. Imagine Camilla was mistaken about the picnic. She consulted last year's departmental calendar, got the picnic date wrong, and cancels her lecture in light of that consideration. What explains her action? We can now hardly appeal to the facts (there are none) that feature in Camilla's belief content in order to formulate an explanation. In error cases, it seems, psychologism presents the only plausible explanans, namely Camilla's state of believing (something incorrectly). Factualists realize the comprehensive threat posed by error cases. Were they to allow mental states to play, a now constitutive as opposed to merely enabling, role regarding MRs in error cases, this would render redundant the appeal to non-pyschologistic facts in the veridical cases too.[3]

Dancy develops two arguments in support of the bold claim that in error cases too it remains the fact that "the picnic is scheduled," that explains misguided Camilla's action. First, the "appositional account" suggests that a full statement of misguided Camilla's MR will acknowledge the incorrectness of her belief by "paratactically adding" a clause such as "as she supposed" to the consideration that Camilla took to favor the action. When Camilla gets it wrong, what explains her cancelling is that the "picnic is scheduled, as she supposed." Importantly, the added "as she supposed," is not a part of the MR. Rather, it "is a comment on that reason, one that is required by the nature of the explanation that we are giving. That explanation specifies the features *in the light of which* the agent acted" (Dancy 2000, 129).

The psychologistic rejoinder insists that all explanations must be factive, in the sense of referring to explanatory components that *are* indeed the case. Dancy is aware that a particular instance of an explanation like "the reason why it is the case that p is that q," appears to always necessitate the inference of "both that p and that q" (Dancy 2000, 131). But do explanations have to satisfy this unconditional factive criterion? In his second argument, Dancy attempts to clear space for a non-factive conception of explanation in terms of a number of complex reflections that cannot be reconstructed here for lack of space. They share the same punchline in that they claim that in the face of Camilla's incorrect belief, and in the grip of the pressure towards psychologizing the explanation of her action in terms of her mental states,

> we can avoid any apparent commitment on our part to things being as [Camilla] supposed by use of one of a number of special constructions such as "as [she] supposed." . . . [A] thing believed that is not the case can still explain an action.
>
> *(Dancy 2000, 134)*

Hence, Dancy hopes to fend off the onslaught of psychologistic intuitions in error cases by highlighting how his alternative explanatory paradigm focuses on the acting subject's *fact*-directed stance when she formulates her MRs. At the same time, Dancy must add the controversial twist that action explanation must be non-factive in *all* instances because successfully appealing to the fact that constitutes the MR never implies that the fact exists.

4. Representationalism

Focusing on MRs, I have so far avoided their complex relationship with the justifying counterparts, i.e., NRs. Already Dancy's theory of practical reasons rests on the observation that an understanding of MRs has to situate them within a comprehensive picture, clarifying their connection with normativity. Psychologism has a difficult time honoring this connection because it fails to satisfy what Dancy labels the "normative constraint" on MRs. In short, psychologism makes it *impossible* to act for (as distinct from merely "in accordance" with) a good (normative) reason (Dancy 2000, 103).

Why does this impossibility plague psychologists? As mentioned previously, NRs are considered to be facts or states of affairs, considerations that support a proposed action. NRs put objective constraints on an agent's deliberative processes; agents are criticizable for ignoring the demands that are implicated in NRs. Psychologists agree, Dancy reminds us, that NRs are "metaphysically different beasts from psychological states of the agent" (Dancy 2000, 106). Psychologists have underestimated that the resulting ontological bifurcation of MRs (mental states) and NRs (facts) comes at a high cost, namely the violation of the aforementioned intuitive constraint that "requires that a motivating reason, that in the light of which one acts, must be the sort of thing that is capable of being among the reasons in favor of so acting; it must . . . in this sense, be possible to act for a good reason" (Dancy 2000, 103).

Much of the contemporary debate about MRs is concerned with clarifying the ontological relationship between MRs and NRs. In particular, the question of the ontological nature of motivating and normative reasons has informed much of the original work in the last fifteen years. I conclude with two takes on the "identity thesis," the claim that when an agent acts for a NR, then there must be a motivating reason that is identical with the former (Heuer 2004).

The first view is representationalism, a version of what is now often called the "standard story," i.e., the ontological view implied by traditional psychologism (MRs are mental states, NRs are facts). Representationalists reject the identity thesis as too strict and replace it with an alternative and still intimate relationship. Camilla's belief that the picnic is on Monday, since it is a mental state, must be connected with a corresponding justifying fact in terms of a proper "basing relation." R. Jay Wallace (2006, 63–70), in an early formulation of representationalism, recognizes that by rejecting Dancy's strict rendering of the normative constraint, the challenge consists in providing an account of how precisely agents and the NRs for which they act are connected "in the right way" with their MRs.

Susanne Mantel (2018) has developed the standard view into a book long investigation of how agents can act for a good reason, even if the represented and the representing considerations are entities that are located in two distinct ontological realms. Mantel's representationalism consists of two tasks. First, the "normative competence account" is a dispositional model of acting for a NR. It comprises several "tracking dispositions," which an agent has to deploy in order to perform an action that is brought about in the right way and not accidentally or merely "in accordance" with the NR supporting it. The dispositions are epistemic, volitional, and executional competencies that agents actualize upon recognizing a (factualistically conceived) normative consideration. Furthermore, there is an overarching "normative competence," i.e., the unifying relation between the aforementioned dispositional aspects of agency. Acting for a good reason, therefore, is an event in which psychological elements play a crucial explanatory (not merely an enabling) role. However, *pace* the preceding discussed causal-psychological tradition, the switch to normatively charged dispositions promises to incorporate the favorable light in which the agent acts in a non-trivial manner.

Mantel's second concern is to reject the claim that, when an agent acts for a good reason, there must be a strictly *identical* MR at work in the course of the agent actualizing her normative competence. Mantel attacks the identity thesis on the basis of a complex ontological distinction. NRs are "individuated" in what she calls a "coarse-grained" way whereas MRs are of the opposite nature, namely "especially fine-grained" (Mantel 2018, 111–112). The NR that Camilla conceptualizes and responds to in her deliberation is of a very general sort, a consideration that speaks in favor of a large number of individual actions that comply with this one NR. On the other hand, Camilla's MR must "depend on the exact conceptualization that an agent employs" in her deliberation and action. Mantel concludes that MRs "should be construed as [Fregean] propositions [and] premises of practical thought" and NRs "as objective states of affairs or events," which rules out the strict ontological identity between the two (Mantel 2018, 5).

Mantel's representationalist alternative consists in the claim that NRs ("in the world") and MRs ("in the mind") are, their ontological distinctness notwithstanding, closely connected. It is in terms of a "relation of correspondence" that these constitutive items, involved in acting for a good reason, are tied together. Putting aside for our purposes Mantel's distinction between a deliberative and an explanatory version of the correspondence thesis, the unifying thread is the following, complex, representational relation: "The consideration that *p* is the content of a belief that *represents* the [NR] that *p*. Acting for the NR that *p* entails having the belief that *p*, the 'descriptive belief,' and the content of this belief motivates, i.e., it counts as a motivating consideration" (Mantel 2018, 125). Camilla, who acts for the NR that the picnic is next Monday, is motivated not by the fact that it is so scheduled, but by the content of her competently representing belief that it is.

5. Truthy psychologism

Veli Mitova (2017) has developed a defense of the thesis that MRs and NRs are of the same ontological kind by exploring an option that had not been considered hitherto.[4] "Extreme psychologism" (the first of the two cornerstones of Mitova's "truthy psychologism") accepts – with the factualist and in opposition to the standard story – the ontological homogeneity of reasons for action that she also labels the "Beast of Two Burdens Thesis." At the same time, Mitova exploits factualism's Achilles' heel, i.e., the problems that come with the commitment that MRs are always (supposed) facts; recall Dancy's struggle with error cases. Ergo, MRs and NRs must both be mental states.

Extreme psychologism preserves the plausible features that come with the Davidsonian tradition of action explanation that factualists had to renounce. Neither does it conflict with the view that explanations are factive, nor does it flout the well-entrenched intuition that MRs are causing our actions. In short, in both success as well as error cases, Camilla's action of cancelling the lecture is explained by non-deviant causal chains that have as their elements Camilla's mental states, her beliefs, about the picnic's date. This belief state is not merely an enabling condition for explaining her action; rather, it is the entity that effectively moved Camilla in her deliberation and resulted in the action that we try to render intelligible with the help of exactly that psychological explanans.

Mitova defends at length the controversial part of her proposal, which is the claim that NRs are mental states. She argues, "[t]he trick is to make our extreme psychologism 'truthy': when the beliefs . . . are *true*, they are *good* reasons for my actions" (Mitova 2016, 300). The resultant "truthy psychologism" is Mitova's complete metaethical and epistemological theory that adds the following thesis ("TRUTHY") to extreme psychologism: NRs "are factive; they are veridical psychological states" (Mitova 2017, 86). In an error case, such as confused Camilla, we again explain the action as caused by her belief, but there is no NR in support of her action. Crucially, what accounts for the absence of a NR is the *falsity* of Camilla's belief. Hence, truthy psychologism attempts to preserve the objective (non-relativistic) features of NRs by means of defending an account of how acting for a good reason is connecting our agency "to the world" and thereby "in the right way." The determinants that render Camilla's beliefs into NRs (or not, in error cases) are not "up to her," even if the beliefs themselves, of course, remain full-fledged mental states of Camilla's and, hence, have the potential to explain *all* of Camilla's actions in a unified, causalist and factive, framework.

By means of assigning conceptual priority to beliefs, truthy psychologism meets Dancy's normative constraint in an innovative manner (Mitova 2017, 228–229). Ontological sameness is preserved because the beliefs that motivate Camilla are the kind of entity that *can* in principle justify her action, regardless of whether or not these beliefs actually are veridical (they could have been). Some of these beliefs justify successfully, namely when the beliefs are correct ones. Understood as an alternative to representationalism, truthy psychologism presents an account of good action that rests on a more direct basing relation between the reasons-components involved. Since they all reside in the agent's belief system (without intermediate representational steps that then require further accounts of basing), Camilla acts for a good reason, in so far as her state of believing something true – as distinct from the propositional content of her belief – is the state of being motivated.

6. Motivating reasons' omnipresent and benign relativism?

Until recently, the debate about MRs had not been much concerned with relativism. One explanation for this silence in the literature is that MRs are (and have to be) obviously relative to an agent's deliberative point of view. Recall the formulations such as "the light in which she acted," "what the agent took to favor her action," "subjective considerations," etc. The discourse on MRs, its internal controversies notwithstanding, appears to agree at least on that much, given that the point of rendering actions intelligible is precisely that the acting subject's perspective (even if it incorporates incorrect beliefs) must *always* be part of this distinctive form of rationalization. From Smith's (1994) and Davidson's versions of psychologism to Collins' and Dancy's early factualism that seems to have been common ground. The resulting relativistic (or probably better, "perspectival"[5]) feature of MRs is therefore not merely omnipresent. It is benign and essential for a particular consideration to play its role in explaining intentional action, on the one hand, and for fulfilling its function in an agent's deliberative procedures, on the other.

The innocuous status of relativism regarding MRs can be illustrated in a more historical vein. Compare Davidson's with other varieties of psychologism such as the one calling into question the objectivity of the norms of logic and epistemology that constituted the early 20th century "psychologism dispute" (Kusch 1995). The conceptions of psychologism in play in these two distinct philosophical debates differ with respect to their relativistic implications: The relativism that comes with psychologism about MRs has not troubled many observers because there is the additional layer of NRs, which stands ready to *complete* the sphere of practical reasons in a non-relativistic manner. With respect to NRs, of course, relativistic worries are substantial. Whether or not NRs are conditioned by subjective perspectives, socio-historical contingencies, and cultural paradigms is a lasting debate.[6]

Now, it is this feature of the practical-reasons-debate that differentiates it from the psychologism dispute, where a relativistic grounding of logical laws and epistemic norms is considered a potent threat to their objective and universal truth and validity (in the sense of agent- and species-independence).[7] *This* psychologism, attacking the (supposedly) objective normative foundations of theoretical reasoning and scientific inquiry, is indeed on a par with the relativism challenge regarding NRs (which, again, is not our topic). Psychologistic theories about MRs, on the other hand, are not competing with objectivist contenders in this way. Davidson's "anemic" normativity is a reminder that psychologists about MRs (and most of their competitors) allow a relativistic notion of normative force into their account. Camilla must regard her consideration as a justifying ("favoring") one for the consideration to constitute a MR in this minimal sense of a perspectival correctness standard and constraint (also recall Collins' MR-constituting role of lining up one's practical perspective along what one takes to be "objective circumstances"). At the same time, this minimal, subjective, normativity of MRs is certainly not threatening the non-relativistic prescriptivity that is commonly associated with NRs, moral reasons in particular.

Is this thesis about the universal and benign relativism of MRs really that obvious? The sections on Mantel and Mitova illustrated, what one might call, an "epistemic turn" in recent debates about MRs. Understanding the relationship between MRs and NRs has become much more nuanced and, in the wake of it, concerns about relativism about belief and (even) knowledge spill over into the territory of MRs. One manifestation of this epistemic turn should be discussed in order to conclude that conceptions of "MR" are undergoing a transition towards more demanding "existence criteria" that no longer tolerate agnostic attitudes towards relativistic implications.

The unifying thread here is to render MRs more exclusive. Maria Alvarez (2010), for example, agrees with Dancy that both MRs and NRs are facts. However, in error cases such as misguided Camilla, the agent, while obviously motivated by "something," does not act for a motivating *reason*. This is the case regardless of the incorrect picnic date having appeared as a justifying consideration from within Camilla's perspective: "Sometimes people are motivated to act by something that is only an apparent fact, and hence only an apparent reason" (Alvarez 2010, 51).

A motivating consideration that is incorrect in terms of the worldly, non-relativistic, facts that it purports to incorporate, does not qualify as any type of reason to begin with. A similar wedge is driven by Derek Parfit (2011, 37) when he defends the view that MRs are of two (ontological) kinds. In error cases, Camilla's MR is "provided" by her false belief and since a false belief fails to track the facts, the reason again turns out to be a merely apparent one, i.e., not a reason at all.

These new proposals have generated important replies. For example, they seem to conflict with the intuition that misguided Camilla *did* act for a reason, namely for a *bad* MR – as distinct from merely some other motivating entity as Alvarez has it (Mitova 2017, 69). Regarding the topic of relativism, the rationale for concluding with these (epistemically) demanding notions of "MR" is that they highlight the epistemic turn: Given the dispute on the identity thesis and the recent factualist moves to restrict more stringently what kind of action-explanantia should be awarded the precious title of "MR," it appears much more controversial than it did in the earlier phases of the debate that a consideration that presents itself from within the acting agent's perspective counts as a *reason* merely in virtue of being taken by her to favor the action. If the preceding trend continues, then reporting, reconstructing, and incorporating the "light in which an agent acted," while still an important component of reasons-explanations, might lose its status as a sufficient device for doing so. Non-relativistic and supposedly objective criteria would then, at least, co-constitute the existence conditions of MRs (not merely of NRs), understood as a distinctive category of practical *reasons* for action.

Acknowledgements

I am much indebted to Martin Kusch and Veli Mitova for very helpful comments.

Notes

1 There are three other topics on practical reasons that are often intermingled with the concern of this chapter. First, Bernard Williams' (1981) theory of internal vs. external reasons statements that focuses on the role of a person's motivational states in determining under what conditions a statement about normative reasons meaningfully applies to her. Second, the contrast between internalism and externalism about moral motivation with its central question of whether or not moral judgments necessarily motivate the agent who makes that judgment (Smith 1994, 60–91). Third, the Humean theory of motivation with its central tenet that beliefs alone can never motivate in the absence of desires (Smith 1994, 92–129). I do not claim that these three metaethical issues (especially the first one) are utterly unrelated to the topic of this chapter, i.e., the technical concept of "motivating reasons." Still, the discussion in the text attempts to put them aside.

2 For the literature on deviant causal chains, one of the most prominent objections to causalism, see Stout (2010).

3 A further contender is disjunctivism (Haddock and Macpherson 2008; Stout 1996), modeled upon the influential approach in the philosophy of perception. See Wiland (2018) for an updated discussion of disjunctivism's prospects.

4 The remaining fourth option (NRs are psychological; MRs are facts) has not been seriously entertained. Mitova (2016, 300) summarizes that there "can be no theoretical motivation for getting things quite so backwards" as this option would have it.

5 I thank Veli Mitova for highlighting that the distinction between relativism and perspectivism is more complex than my exposition suggests. Perspectivism has recently gained prominence in the work on NRs (and their relationship to instrumental rationality and rational requirements). Expanding this program into the realm of MRs is a task yet to be undertaken. See Littlejohn (forthcoming); Hanisch (2017).

6 On moral relativism and reasons, see Harman (2000), Streiffer (2003), and Wong (2006).

7 The debate between Benno Erdmann and Edmund Husserl is a good example of how epistemic and logical psychologism constitutes a comprehensive relativistic challenge. See Kusch (1995, 49–52).

References

Alvarez, M. (2010), *Kinds of Reasons*, Oxford: Oxford University Press.

Collins, A. W. (1997), "The Psychological Reality of Reasons," *Ratio* 10: 108–123.

Dancy, J. (2000), *Practical Reality*, Oxford: Oxford University Press.

Davidson, D. (1963), "Actions, Reasons, and Causes," in *Essays on Actions and Events*, edited by D. Davidson, Oxford: Oxford University Press, 2001, 3–19.

Haddock, A. and Fiona Macpherson (eds.) (2008), *Disjunctivism: Perception, Action, and Knowledge*, Oxford: Oxford University Press.

Hanisch, C. (2017), "Two Conceptions of Practical Reasons," *The Ethics Forum* 11: 108–132.

Harman, G. (2000), *Explaining Value: And Other Essays in Moral Philosophy*, Oxford: Clarendon Press.

Heuer, U. (2004), "Reasons for Actions and Desires," *Philosophical Studies* 121: 43–63.

Kusch, M. (1995), *Psychologism: A Case Study in the Sociology of Philosophical Knowledge*, London: Routledge.

Littlejohn, C. (forthcoming), "Being More Realistic About Reasons: On Rationality and Reasons Perspectivism," *Philosophy and Phenomenological Research*.

Mantel, S. (2018), *Determined by Reason: A Competence Account of Acting For a Normative Reason*, London: Routledge.

Mitova, V. (2016), "Clearing Space for Extreme Psychologism About Reasons," *South African Journal of Philosophy* 35: 293–301.

———. (2017), *Believable Evidence*, New York: Cambridge University Press.

Parfit, D. (2011), *On What Matters, Volume One*, Oxford: Oxford University Press.

Smith, M. (1994), *The Moral Problem*, London: Basil Blackwell.

Stout, R. (1996), *Things That Happen Because They Should*, Oxford: Clarendon Press.

———. (2010), "Deviant Causal Chains," in *A Companion to the Philosophy of Action*, edited by T. O'Connor and C. Sandis, Malden, MA: Blackwell, 159–165.

Streiffer, R. (2003), *Moral Relativism and Reasons for Action*, London: Routledge.

Wallace, R. J. (2006), *Normativity and the Will: Selected Papers on Moral Psychology and Practical Reason*, Oxford: Oxford University Press.

Wiland, E. (2018), "Psychologism and Anti-psychologism About Motivating Reasons," in *The Oxford Handbook on Reasons and Normativity*, edited by D. Star, Oxford: Oxford University Press, 197–213.

Williams, B. (1981), "Internal and External Reasons," in *Moral Luck*, edited by B. Williams, New York: Cambridge University Press, 101–113.

Wong, D. (2006), *Natural Moralities: A Defense of Pluralistic Relativism*, Oxford: Oxford University Press.

PART 4

Relativism in political and legal philosophy

.

22

RELATIVISM AND LIBERALISM

Matthew J. Moore

1. Introduction

This chapter examines what role, if any, relativism plays in the political theory of liberalism. As the other chapters of this handbook attest, relativism is a large and complex area of thought. I will speak of relativism in fairly general terms, and will focus on the idea that moral truth may be relative to some frame or context. Liberalism is similarly internally complex. For our purposes, I will define liberalism as a theory of government that arises from a special concern for individual freedom.

I argue that there are two places where liberalism so understood approaches relativism: (i) in a disposition toward open-mindedness and seeing things from others' perspectives that liberal societies valorize as a political and intellectual virtue; (ii) when liberalism asserts that it can be a neutral framework within which citizens with very different value systems can live peacefully. In my estimate, liberalism does not ever tip over fully into relativism at either of these points, though it comes awfully close.[1]

2. Prelude: the liar's paradox objection

Inevitably, when someone suggests that some version of relativism (particularly with regard to truth or knowledge) might be the case, the immediate response is that this claim is an example of the *Epimenides* or *Liar's Paradox*: if you say that truth is relative to context, your statement contradicts itself, since there is at least one statement (yours) that is not relative to context. There is a large literature on this question,[2] including several chapters in the present collection. I do not pretend to be an expert on the Liar's Paradox, but it does seem to me that there are some practical observations that make it less threatening than it appears.

First, there is no paradox in relativism *being true*, but rather only with *saying that it is true* or saying that one can *know that it is true*. In other words, the paradox exists only at the level of speech or epistemology, and not at the level of ontology. This is as opposed to something like the Principle of Identity – the same object cannot be both X and not-X at the same time. That's an ontological claim, not an epistemological or speech-related claim, like the *Epimenides Paradox*.

Further, it isn't clear that the *Epimenides Paradox* poses a genuine barrier to our *knowing* that knowledge, truth, or morality are relative to some context. If some condition C is true of the world (and is susceptible of investigation, the gathering of evidence, and so on), then it should in principle be possible for us to know that C is true. For example, if all of our evidence points toward C being the case, and we have no (or minimal) evidence against C being the case, the epistemologically responsible thing to do would be to conclude that C is the case and that we know that fact, with the usual caveats about the limits of knowledge, probability of error, subsequent revision, etc. Thus, if there is no logical barrier to relativism being true, then it seems that there should not be a barrier to us being able to know that it is true. This doesn't resolve the paradox, but it does reveal how bizarre it is to claim (as its defenders must) that we cannot know or say something that may be true of our world and is the kind of thing that is (at least in principle) knowable.

Another approach is to note that the objection is in effect a claim that relativism makes a category mistake. To claim that relativism is true would require us to perch at an Archimedean point from which we can see precisely how truth maps onto the real, but of course such a point is unattainable because we ourselves are parts of the world of the real, and cannot get the necessary distance to make the judgment (assuming for the sake of making the point that truth is a correspondence, that there is a real to map onto, etc.). But it seems obvious that there's a mistake in our thinking here. Our knowledge *never* gets that kind of distance, and to demand that a claim do so to qualify as knowledge is to set the bar unreasonably (indeed, impossibly) high. Rather, our knowledge claims always occur within the world, and express our best attempts to describe what we experience from that perspective. To say that truth is relative is to make the claim that one would or could come to that conclusion from any and every perspective in the world, not that one can transcend perspective altogether.

Given these points, it seems to me that the *Epimenides Paradox* can be avoided in most discussions of relativism related to politics through a strategy that in a different context I have called "*faute de mieux* contextualism," and here we might adapt as *faute de mieux* relativism (Moore 2010). This is the claim that, based on the available evidence, and in contrast with the other plausible explanations, relativism appears to be the best explanation of our experience, that we are therefore justified in proceeding *as if* relativism were the case, and indeed that any other way of proceeding would be less justifiable on epistemic grounds.

Finally, none of this is meant to prove or even assert that relativism is the case, but rather merely to say that such a claim should not be dismissed out of hand as incoherent or logically impossible, and therefore not deserving of further discussion.

3. A version of liberalism

As mentioned previously, I define liberalism as a theory of government that arises from a special concern for individual freedom. Although it isn't logically entailed in that concern for freedom, liberalism is as a matter of fact also committed to the *equal* freedom of persons, though often more faithfully in theory than in practice.[3] Thus, each person's rightful exercise of freedom is limited by everyone else's equally rightful exercise, at least within a single society.

From this it follows that any interactions among persons must generally be voluntary to be rightful, and that the only legitimate bases for the exercise of authority by one person over another are consent, prior/greater right, and self-defense.[4] This leads to the idea that any use of power against a person must be acceptable to that person – or a similarly situated, reasonable person, in the case of the incompetent or profoundly unreasonable – and thus that an exercise of power/authority is only permissible if it can be justified to everyone affected by it.

Because exchange and social cooperation are so obviously beneficial, liberal theorists argue that they will always emerge (or be recreated or endorsed). Exchange is typically modeled on a "free market," where persons who have skills or goods may choose to exchange them for the skills or goods of others (see Locke 1690; Nozick 1974). Cooperation takes the form of government, which acts to protect the freedom of its constituents (and possibly others) from internal and external threats.

Given its focus on individual freedom, liberalism is committed to a policy of maximum tolerance: citizens may believe, say and do whatever they like, including agitating against liberalism itself, so long as their actions do not infringe on the rights of others.[5] This isn't relativism, and is perhaps better described as agnosticism: liberalism simply tries to avoid the question of whose views are correct. This leads to the typical self-understanding of liberalism: seeing itself as a neutral framework within which citizens with varying normative beliefs can co-exist peacefully.

This position has been so thoroughly criticized that it may be difficult to fully appreciate its powerful intuitive appeal: if people have varying and mutually conflicting normative values, it is not at all obvious that they will be able to cooperate peacefully. Non-liberal political theories rest on a willingness to exert power or authority without consent, perhaps resting on the assumption of the truth of some set of normative values that allow believers to disregard the dissent of non-believers. Any such system is anathema to liberals, since it limits the freedom of (dissenting) individuals without providing a justification that those individuals could accept. From this perspective, liberalism seeks to impose the fewest restrictions necessary to enable peaceful cooperation among persons embodying the largest possible range of normative beliefs. Even if a liberal theorist is willing to concede that liberalism rests on substantive claims, he or she would typically argue that those values are both reasonable and the only values that could support such normative heterogeneity.

4. The liberal disposition

One effect of liberalism's normative agnosticism is that it cultivates in liberal citizens a dispositional leaning toward relativism, though it rarely, if ever, tips all the way into relativism proper. This is encouraged both by liberalism's foundational normative commitment to the equal freedom of persons and by the lived experience of being a liberal citizen. For example, in practice liberal polities are democratic,[6] and in democratic societies citizens ideally experience both wins and losses on various policy battles, over time. This gives them the opportunity to learn to be both good losers and magnanimous winners, and to appreciate institutions that limit what the majority can do. Thus, for example, they learn to accept a loss on substance that preserves an institution that they value. Similarly, liberal citizens are used to hearing views different from their own, and are used to being required to treat the other citizens who hold those views with respect and toleration (as a condition of securing their ongoing, peaceful cooperation). Further, following Mill (*On Liberty*), the ability to understand and even argue sympathetically for a view with which you disagree is widely seen as a virtue both of liberal intellectual life and also of liberal citizenship. Together, these experiences (and, no doubt, many more) lead to a situation where the liberal citizen is inclined to say: "I don't agree with your policy preferences or your underlying values, but I can see how the preferences arise from the values, and if I had those values I would probably have those preferences." This is a partial, perhaps trivial, relativism: some less important beliefs (policy preferences) are relative to more important beliefs (normative values).

That basic sympathy with another citizen's ideas naturally widens, and the liberal citizen can imagine that she would have different normative beliefs if she lived in a different era, or in a different country, or if she occupied a different subject position or identity in her own time and place, and so on. Each step on that path leads her closer to relativism – that is, to the idea that some parts of truth and/or morality are relative to some kind of context. It seems to me that this does not typically lead to full-blown relativism, because even those who lean the furthest in this direction nonetheless tend to believe that some kinds of truth/knowledge are universal (the internal angles of a Euclidean-plane triangle sum to 180 degrees at all times and all places), and that there are some moral absolutes (torturing infants just to hear them scream is wrong, and no acceptable moral system could argue otherwise). Further, as Fricker argues, the inference from descriptive plurality to metaethical relativism, though tempting, is not obviously valid – that people disagree is not proof that there is no fact of the matter.[7]

5. Liberal neutrality

The view of liberalism as a neutral arbiter among conflicting normative views has clear limits. First, eventually liberalism will encounter non-liberals who wish to act on their non-liberal beliefs in ways threatening to liberalism, and liberalism will have to decide how to respond. Second, eventually critics (who may themselves be liberal) will point out that liberal government is not a metaphysically pure referee, but instead rests on particular substantive beliefs (individual freedom and equality) that are not universally shared, and that liberalism may not be able to impose those beliefs on unwilling others without self-contradiction.

These problems at least appear to lead liberalism to paradoxes. Take the case of a non-liberal citizen of a liberal polity who wishes to create a sub-culture based on a gender hierarchy in which women are subordinated to men. Liberals are inclined to allow that if it is mutually consensual, though that then raises the difficult question of false consciousness – whether someone's willingness to be subordinated is evidence that they are not fully free to choose otherwise (due to threats of violence, coercion, manipulation, and so on) – and possibly the necessity of an intrusive regime of ensuring that the subordinated women's participation is (and remains) truly voluntary. This is the general problem that faces versions of liberalism that try to preserve liberalism under conditions of plurality by permitting non-liberal "islands" within a liberal regime: if every citizen is to enjoy the protections of a liberal regime, then the non-liberal subcultures must ultimately be optional, and their perpetual optionality must be enforced by the liberal state (see Kukathas 2003; Deneen 2018).

Now take the case of a female child born into this sub-culture in which women are subordinated. It seems likely that a child socialized to believe that women should be subordinated will internalize non-liberal values that will not only lead to her becoming a non-liberal adult, but also to making it difficult or impossible for her to decide to leave her sub-culture when she reaches adulthood, if she were to want to do so. Does a liberal state have a duty to ensure that this child is educated in liberal values or in a liberal setting, against the wishes (that is, without the consent) of her parents? Does the state have the authority to act in the interests of preserving the autonomy of the child, at least until she reaches adulthood?[8] No matter which answer we choose, we seem to be sacrificing someone's freedom to preserve someone else's, and we seem to be allowing the development of non-liberal policies.

Liberal states face roughly the same problem when confronted with non-liberal regimes abroad. May or must a liberal state intervene in the internal affairs of a non-liberal state to protect the freedom and equality of (some of) the citizens of that other state against the actions of others of its citizens?[9] Again, no matter which policy we pursue, we seem to be sacrificing someone's freedom to preserve someone else's.

There are three basic strategies for responding to these problems: (i) admitting that liberalism is rooted in a contestable normative preference for equal individual liberty (perhaps understood as the greatest possible liberty for the greatest number of people, consistent with all people having equal liberty), and thus treating people committed to engaging in non-liberal actions as the functional equivalent of criminals who wrongfully restrict the freedom of others (even if we might recognize that their unacceptable behavior arises from good-faith, if sadly mistaken, beliefs); (ii) trying to make liberalism's foundational commitment to individual liberty less like a contestable normative view and more like a truly neutral framework, for example by showing that it is logically entailed in interpersonal communication (Habermas 2007), that it is what anyone involved in the creation of a society would insist on as a rational basis for maximizing self-interest (early Rawls 1971), or that it is not a normative truth but that acting as if it were true is a condition of possibility for cooperation despite difference (late Rawls 1985, 1987, 1999, Rorty 1998); (iii) arguing that it is possible to show that certain values (which turn out to be consistent with a liberal regime) are logically prior to or more fundamental than others (which conflict with such a regime) (value pluralist liberalism).[10]

The first strategy, pursued in earnest by thinkers like William Galston (2002) and Stephen Macedo (1995), and with ironic self-awareness by thinkers like Richard Rorty (1989), may work as a defense of liberalism (though I have doubts, since it seems to threaten endless debates over where to draw the line between pro-liberal defense of liberalism and anti-liberal oppression of non-liberals), but it does not tend toward relativism, and I will not pursue it further here.

The second strategy, especially in the form given to it by Rawls in *Political Liberalism* and "Justice as Fairness: Political, Not Metaphysical," leads to an interesting question (the "public reason" debate[11]) about how citizens may argue with one another in public. Rawls argues that while each citizen is presumably motivated by some set of normative commitments (their "comprehensive doctrine"), what makes cooperation possible is our agreement to a more limited set of principles that govern our collective activity. These principles are perhaps best understood as hypothetical imperatives – citizens do not have to believe that they are true, but if they want peaceful social cooperation, citizens have to act as if they were true. When we reason with each other in public, we can't reasonably expect our fellow citizens to be moved by appeals to our comprehensive doctrines, since those are not the basis of our mutual cooperation. Rather, we can only expect arguments to succeed if they are rooted in the hypothetical principles that we all share, which in effect means liberal principles. Again, we aren't assuming that liberal principles are metaphysically true, only that acting as if they were true is the condition of possibility for the social cooperation that we presumably all value.

This argument is subject to a number of criticisms – that it prevents citizens from appealing to their most important sources of motivation and normative guidance (Sandel 1982); that it creates a society that is incapable of instilling its motivating values into succeeding generations and thus cannot be stable (Deneen 2018); that it cannot work because it is always rational for someone to prefer to pursue their comprehensive doctrine at the expense of social cooperation. I want to focus on the point that this is, I think, the place where liberalism gets the closest to relativism, at least in terms of its explicit arguments. Rawls acknowledges – indeed, embraces – the fact that citizens will be motivated by their comprehensive doctrines. Thus, individual citizens are not relativists. But when citizens act politically, they must treat all comprehensive doctrines as equally irrelevant, and they must treat the values that guide their cooperation as merely hypothetical rather than categorical guides to decision making: they tell us how to choose, but not what is worth choosing. Thus, in a sense, the polity is relativistic – the values expressed in its policies are grounded ultimately in the method of decision-making we have adopted, and not in normative truths. Further, the fact that some policy may be "right" or "wrong" from the perspective of some comprehensive doctrine is strictly irrelevant to whether it should be pursued. In that sense, individual citizens are required to take a relativistic stance toward their own values, both when creating laws and when obeying them.

215

Finally, the third strategy for coping with the paradoxes of liberalism – that of the value pluralists – requires a little more explanation, since it at least appears to move towards relativism. Value pluralism is the belief that normative values can and do conflict, and that there is no non-controversial way of resolving such conflicts. Sartre famously writes of a young man torn between the duty to care for his aged mother and the duty to join the French Resistance, which illustrates the possibility of irresolvable conflict within an individual person's value system (Sartre 1973). There are also conflicts between individuals or value systems, such as between the religiously faithful and atheists, or between those who believe that animals are rights-bearers and those who do not. Strategies for resolving such conflicts would be: to identify an intrinsic, non-controversial ranking among normative values such that all apparent conflicts could be resolved through subordination of some values to others; to identify a value whose expression is in some way a prerequisite for the existence and/or expression of other values (which creates a partial hierarchy); or to argue that the existence of value plurality itself necessarily requires that we embrace a master value, such as toleration.

None of these approaches can work, for a set of related reasons. First, the fundamental premise of value pluralism is that no non-controversial, comprehensive hierarchy of values can be created. Of course there are partial hierarchies, both within value systems and between them, but by hypothesis those hierarchies can't solve all the important conflicts, such as that between liberty and equality. Of course, value pluralists may be wrong about this, but then the burden falls on the critic to show how apparent conflicts can be resolved. Second, it is always potentially rational for an individual to prefer to pursue the expression of his or her normative values over other goals, such as maintaining peaceful cooperation.

Finally, the idea that pluralism itself requires us to adopt toleration or a similar "procedural" master value is a confusion of ethics and metaethics. Value pluralism is a recognition that value conflicts cannot be resolved in a way that will persuade everyone. That may be true as a statement of metaethics, but it is effectively irrelevant to ethical deliberation in practical cases. Person X thinks murdering the innocent is fine, while Person Y thinks it is abhorrent. Both can recognize that their values conflict, and that there isn't a way to solve the conflict that will be persuasive to both of them. But that tell us nothing about what Y should do when X threatens to kill Y, or the otherwise-innocent Z. Y might respond in a variety of ways, but those responses will be determined by Y's ethical belief (what she believes her duty to be) not her metaethical beliefs (what she believes about why she and X disagree).

Thus, the value-pluralist route to resolving the internal paradoxes of liberalism appears not to work, either. It is not ultimately a relativistic approach, since it seeks to solve liberalism's paradox by identifying or creating a universal hierarchy among values, but it deserved our attention because its acceptance of the possibility of irresolvable value conflict at least appeared to bring liberalism closer to relativism.

Where does this leave us? We've looked at three strategies for reconciling liberalism's commitment to equal individual liberty with the fact that some citizens don't share that commitment, and indeed some actively oppose it. None of those strategies seems able to resolve the conflict: the first, that of admitting that liberalism is a substantive value system and defending it on its normative merits, seems to invite endless debates over where to draw the line between defending liberalism and oppressing non-liberal citizens; the second, that of reducing liberalism's normative claims to the minimum necessary to hold a society together, is subject to a variety of damaging criticisms reviewed previously; the third strategy, that of value pluralism, fails to solve the problem because any such resolution would require a ranking of values that value pluralism denies is possible and liberalism denies is desirable. None of these strategies for resolving the paradox at the heart of liberalism embraces relativism, and none of them succeeds in its aim. That suggests that the live choices in front of us are to embrace relativism, but at the cost of not being able to claim the special status for equal liberty that has been the hallmark of liberalism since the seventeenth century,[12] or to acknowledge that liberalism is internally incoherent because it cannot justify forcing its citizens to be free, even in terms of its own internal values.[13]

6. Conclusion

Because it is committed to equal individual freedom, and thus seeks to create a society that can embrace the largest number of normative belief systems with the minimum restrictions, liberalism has an inevitable tendency towards relativism. This is seen most clearly in the lived experience of being a liberal citizen, and in the attempt to justify the claim that liberalism could provide a neutral framework for cooperation under conditions of normative heterogeneity. Nonetheless, liberalism does not ultimately become relativism, for the simple reason that liberalism is a political theory premised on the equal protection of individuals' freedom – that is, on the belief that some values (freedom and equality) are of absolute importance. But by avoiding the potential logical paradox of becoming a value system without values, liberalism inevitably teeters into internal incoherence because it asserts that everyone must be treated as equal and free regardless of whether they would freely choose that condition.

Notes

1 Other recent work in this area includes Graham M. Long (2011).
2 An essay I found helpful was Kölbel (2011).
3 There is nothing contradictory about being interested in one's own freedom, or the freedom of members of a social group, while not being concerned about the freedom of others, as the existence of slavery in nominally liberal societies like the pre-Civil War United States demonstrates.
4 This is roughly the argument we see in John Locke's *Second Treatise of Government* (1690).
5 The classical statement of this view is given in John Stuart Mill's, *On Liberty* (1859).
6 Majoritarian democracy maximizes the protection of equal individual freedom by requiring that the largest possible share of the citizens support the actions of the group, without empowering a minority veto, as supermajority and consensus systems both do.
7 Her essay (Fricker 2013) is also an excellent, recent typology of relativisms.
8 Of course, these aren't merely hypothetical concerns: see *Yoder v. Wisconsin* regarding the state's duty to ensure that Amish children receive an education adequate to permit them to leave the Amish community, and *Mozert v. Hawkins* on whether parents may opt out of pro-toleration lessons in public schools.
9 On this question, see the work of Michael Ignatieff.
10 For a good, recent overview of the literature, see Mason (2018).
11 See especially Gaus (2003).
12 For consideration of an argument that liberalism might be able to attain coherence through embracing relativism, see Long (2011).
13 Although I don't have room to develop this argument in any depth here, it seems to me that liberalism also comes close to relativism in the theories of Edmund Burke (1790) and Michael Oakeshott (1991), who both argue that it is inappropriate to judge political institutions against abstract principles. Their arguments are subtle and complex, but my basic objection is that if there is no criterion for judging social/political systems other than their continued existence, then it would be possible to rightly conclude that diametrically opposed systems, or systems that involve practices abhorrent to most people (i.e., human sacrifice), are all equally morally acceptable.

References

Burke, E. (1790), *Reflections on the Revolution in France*, Indianapolis: Hackett Publishing Company, 1987.
Deneen, P. J. (2018), *Why Liberalism Failed*, New Haven: Yale University Press.

Fricker, M. (2013), "Styles of Moral Relativism: A Critical Family Tree," in *The Oxford Handbook of the History of Ethics*, edited by R. Crisp, Oxford and New York: Oxford University Press, 793–817.

Galston, W. A. (2002), *Liberal Pluralism: The Implications of Value Pluralism for Political Theory and Practice*, Cambridge: Cambridge University Press.

Gaus, G. F. (2003), *Contemporary Theories of Liberalism: Public Reason as a Post-Enlightenment Project*, London and Thousand Oaks, CA: Sage.

Habermas, J. (2007), *The Theory of Communicative Action*, vol. 1, Cambridge: Polity Press.

Kölbel, M. (2011), "Global Relativism and Self-Refutation," in *A Companion to Relativism*, edited by Steven D. Hales, Malden, MA and Oxford: Blackwell, 11–30.

Kukathas, C. (2003), *The Liberal Archipelago: A Theory of Diversity and Freedom*, Oxford and New York: Oxford University Press.

Locke, J. (1690), *Second Treatise of Government*, edited by C. B. Macpherson, Indianapolis: Hackett Publishing Company, 1980.

Long, G. M. (2011), "Relativism in Contemporary Liberal Political Philosophy," in *A Companion to Relativism*, edited by S. D. Hales, Malden, MA and Oxford: Blackwell, 309–325.

Macedo, S. (1995), "Liberal Civic Education and Religious Fundamentalism: The Case of God V. John Rawls?" *Ethics* 105(3): 468–496.

Mason, E. (2018), "Value Pluralism," in *The Stanford Encyclopedia of Philosophy*, Spring 2018 edition, edited by E. N. Zalta, https://plato.stanford.edu/archives/spr2018/entries/value-pluralism/.

Mill, J. S. (1859), *On Liberty*, Indianapolis: Hackett, 2011.

Moore, M. J. (2010), "Wittgenstein, Value Pluralism, and Politics," *Philosophy & Social Criticism* 36(9): 1113–1136.

Mozert v Hawkins County Public Schools, 647 F. Supp. 1194 (1986).

Nozick, R. (1974), *Anarchy, State, and Utopia*, New York: Basic Books, 2013.

Oakeshott, M. J. (1991), *Rationalism in Politics and Other Essays*, Indianapolis: Liberty Press.

Rawls, J. (1971), *A Theory of Justice*, revised edition, Cambridge, MA: Harvard University Press, 1999.

———. (1985), "Justice as Fairness: Political Not Metaphysicalm," *Philosophy and Public Affairs* 14: 223–251.

———. (1987), "The Idea of an Overlapping Consensus," *Oxford Journal of Legal Studies* 7(1): 1–25.

———. (1993), *Political Liberalism*, New York: Columbia University Press.

———. (1999), *The Law of Peoples*, Cambridge, MA: Harvard University Press.

Rorty, R. (1989), *Contingency, Irony and Solidarity*, Cambridge and New York: Cambridge University Press.

———. (1998), *Achieving Our Country: Leftist Thought in Twentieth-Century America*, Cambridge, MA and London: Harvard University Press.

Sandel, M. J. (1982), *Liberalism and the Limits of Justice*, Cambridge: Cambridge University Press, 2010.

Sartre, J. P. (1948/1973), *Existentialism and Humanism*, translated by Philip Mairet, London: Eyre Methuen Ltd.

Wisconsin v Yoder, 406 U.S. 205 (1972).

23

RELATIVISM AND RADICAL CONSERVATISM

Timo Pankakoski and Jussi Backman

1. Introduction

The relationship between conservative political thought and relativism is complex and tension-ridden. Many contemporary conservatives see cultural relativism as undermining communal life and, in the footsteps of Leo Strauss (1953, 9–34), point their finger at historicism (see Bloom 1987, 25, 34, 38–39), the tendency to relativize human thought and morality to their historical contexts. However, as Strauss (1953, 13–16) points out, modern conservatism and historicism have common roots in the Counter-Enlightenment's opposition to ideas underlying the French Revolution, such as natural law and universal progress. In its criticism of the unhistorical outlook of Enlightenment progressive rationalism, nineteenth-century German historicism "inevitably became a powerful ally of Conservatism" (Epstein 1966, 74). We argue that the key link between conservatism and relativism can be found in historicism.

Historicism was a predominantly German intellectual trend. Arguably, many key features of modern conservatism culminated in the German version of conservatism of the Weimar period often known as the "conservative revolution," but more comprehensively and accurately characterized as "radical conservatism" (Muller 1997; Dahl 1999). German thinkers took the intellectual developments behind classical conservatism to their "logical conclusion" (Mannheim 1925, 47), producing a radicalized, ardently antiliberal conservatism in which, we argue, conservatism's problematic relationship with relativism comes to a head.

We focus here on German radical conservatism and its shifting relationship with relativism. After mapping the common genesis of the German conservative and historicist traditions, we study relativistic aspects of twentieth-century radical conservatism and conservative aspects of Heideggerian philosophical hermeneutics, arguably the most important philosophical offspring of German historicism. We conclude with a brief overview of the legacy of these trends in contemporary conservative thought.

2. Historicism and conservatism

The roots of historicism go back to the eighteenth century. Already Giambattista Vico, often considered the father of modern philosophy of history, questioned timeless and abstract principles and emphasized the historicity of reason. Not all peoples advanced concurrently in Vico's developmental scheme of aristocracy, democracy, and monarchy; rather, governments must conform to the nature of the people governed (Vico 1744, 99, 440). Montesquieu (1748), in turn, emphasized the effects of the climate on citizens' mentality, social conventions, laws, and the form of government suitable for a given nation.

This type of political and cultural relativism was pivotal to classical conservatism's critique of the ideas of rational progress and the ideal constitution underlying the French Revolution. Chateaubriand (1797, 255, 297–298) maintained that any form of government could only be evaluated in terms of the needs of the nation and its natural constitution. Edmund Burke (1790, 27, 129) considered it "fanaticism" to regard, for example, monarchy or democracy as the only correct form of government and refused to reprobate any political form "merely upon abstract principle." Jerry Z. Muller (1997, 7, 11–12) links classical conservatism with "historical utilitarianism" according to which the merits of any societal institution are deduced from its historical survival, and with "historicism and particularism" according to which institutions are a product of human development rather than natural law or unchanging human qualities. It follows that expedient institutions vary widely across societies, eras, and traditions, and conservatism carries primarily a "procedural" emphasis on the need for institutions in general to regulate human action and keep egoistic aspirations at check, rather than a "substantive" commendation of particular institutions.

A more full-fledged form of historicism emerged within the German Enlightenment and *Sturm und Drang* proto-Romanticism. Friedrich Meinecke (1936, 235, 295) identifies a "new sense of history" in the late-eighteenth-century German movement represented particularly by Justus Möser, Johann Gottfried von Herder, and Johann Wolfgang von Goethe. Herder, who emphasized the historical and linguistic particularity of ideas, ranks as one of the founders of modern cultural relativism and philosophical hermeneutics, but was, despite his profound sense of national uniqueness, a liberal-minded believer in human progress. Möser, by contrast, combined a notion of historical particularity with a conservative theory of the reason-of-state (Meinecke 1936, 281–291) that effectively jettisoned ideas of natural law and the ideal state, positing that the state as a historical formation had an inalienable individual character and that its primary duties were to safeguard and nurture the particularity of its existence in rivalry with other states. The idea of historical individuality entailed by these approaches provided relativism with a novel theoretical basis (Meinecke 1924, 18–19, 377–380; 1936, 488–489).

Isaiah Berlin (1980, 71–72, 79–80, 87) argues that neither Enlightenment nor proto-Romantic historicists advocated strict cultural or moral relativism, but rather historical pluralism. What was common to all human beings manifested itself in dissimilar, potentially conflicting, and even incommensurable ways in different historical situations, but the irreducible plurality of values did not entail a lack of foundation or standards. This form of historicism is fundamentally akin to that of Hegel, for whom different historical moral, social, and political systems were merely stages in the self-elaboration of universal objective spirit in the world – an argument that, in the conservative Right-Hegelian reading, implied that contemporary Prussia could be seen as the most developed form of *Rechtsstaat*. According to Berlin (1980, 77–78, 87–88), strong relativism only emerged in the nineteenth century with the antirationalism of German Romanticism, Arthur Schopenhauer, and Friedrich Nietzsche.

In the most prominent German Romantics, such as Friedrich Schlegel and Novalis, we find, in addition to a new poetic subjectivism and relativism, historicism in the form of an "organic" view of the historical growth of unique cultural formations and nostalgia for premodern, tradition-bound communities. Mannheim (1925, 47, 127) reads Romantic organic historicism as "a product of the German conservative spirit" that initially emerged "as a political argument against the revolutionary breach with the past." Ernst Troeltsch (1922, 285), however, observes that when detached from its original context, the theory of cultural growth could amalgamate with either conservative or liberal politics. The politically conservative use of organic historicism is exemplified by the German historical school of jurisprudence of Friedrich Carl von Savigny and Karl Friedrich Eichhorn, which insisted, against Enlightenment ideas of natural right, that law emanated from a historically and culturally particular "national spirit" (Harstick 1974). Another example is Leopold von Ranke (1833/1836), the father of empiricist historicism in historiography, who opposed Hegelian teleology and the dissemination of the French revolutionary legacy through Napoleon's attempt at "universal monarchy," emphasizing the importance of organic, tradition-bound national identities.

According to Friedrich Jaeger and Jörn Rüsen (1992, 101–104), classical historicism was eventually dissolved as German idealism gave way to *Lebensphilosophie*: abstract life, conceived in vitalistic and quasi-biological terms, now replaced creative cultural spirit as the driving force of history. Rather than ethical, cultural, and intellectual progress, history increasingly appeared as a natural, biologically evolving life-process that underneath remained qualitatively unchanged. One can identify several steps from historicism toward the eventual affirmation of constant change, novelty, and mobilization without an ideal of ultimate teleological progress, characteristic of twentieth-century radical conservatism. These include Schopenhauer's metaphysics of the fundamentally aimless and purposeless will as the basic reality; Nietzsche's notion of truths as temporary instrumental perspectives of the eternally recurring will to power; Wilhelm Dilthey's grounding of all thought upon the continuous historical flow of lived experience; and Oswald Spengler's organic and cyclic model of history. This turn from spirit to life had obvious political implications: while classical historicism, guided by an idea of an ultimately unified humanity, still implied binding standards for states' actions, the apotheosis of life per se liberated radical nationalist politics from such bounds. Pure factual power now became a sufficient basis for political existence.

3. Radical conservatism and relativism

German radical conservatism was a loose intellectual movement of the Weimar era, inspired by the experiences of the world war. Rather than suggesting reactionary measures, it called for the creation of novel institutions by radical means in order to overcome the alleged egalitarian decay of modern liberal, democratic, and capitalist societies as well as the Marxist threat. It abandoned the traditional conservative postulate of "historical utility": the fact that certain institutions had survived no longer spoke in their favor (Muller 1997, 29). Radical conservatives upheld conservative ideals such as wholeness, community, or authority, but rather than seeking to conserve the old, they underlined the need to "create values worth conserving" (Moeller van den Bruck 1923, 182) – by way of a "conservative revolution," if needed. The key thinkers of German radical conservatism in the wide sense include Carl Schmitt, Hans Freyer, Ernst Jünger, Oswald Spengler, Martin Heidegger, Arthur Moeller van den Bruck, Edgar Julius Jung, Ludwig Klages, and Othmar Spann.

In accordance with the paradox inherent in conserving by revolutionizing, radical conservatism's relationship with relativism is complex. While rejecting liberal democracy on the grounds of its alleged link with relativism and nihilism, these authors often utilized the disruptive power inherent in historicism and relativism against any remnants of Enlightenment universalism, elevating irrational decision, life, or power per se, or seeking to ground the novel idea of community in the inherent plurality of world history. Many called for cultural and national unity and resented liberal value pluralism as "relativism." The sociologist Spann (1921, 63, 187–191) blamed the relativism inherent in the crisis of contemporary culture on individualistic social theories that allegedly led to nihilism and atomization, proposing in their stead a radically anti-democratic "universalistic" agenda relying on organic analogies. The political essayist Moeller van den Bruck (1923, 166–167) contrasted the conservative approach with the merely "relative" standpoints of liberal subjects, apparently always willing to change their views according to shallow opportunism.

Such accusations carried a grain of truth, as many contemporary liberal democrats in fact endorsed a doctrine of relativism. Hans Kelsen (1929, 101–105), for example, maintained that political stances could not be verified or refuted rationally and that relativism, combined with an open market of opinions and an electoral system based on majority rule, was therefore the optimal democratic framework. Spann (1921, 116–117) criticized Kelsen's political relativism for causing dissolution and anarchical factional struggles: liberal value relativism was a threat to political stability.

Carl Schmitt similarly rejected pluralism for shattering political unity into latent civil war and accused Kelsen's legal positivism of opening the door to relativism and nihilism (see Scheuerman 1999, 64–65). For Schmitt (1922, 10, 39–41), the pluralist chaos could not be tamed by abstract norms or inherited customs; only a sovereign decision *ex nihilo*, without rational justification, could secure the foundation of both law and political order. This idea is iterated in Schmitt's (1932, 26–27) concept of the political: any criterion can give rise to the political opposition between friend and enemy, and when a decision concerning the political enemy is in effect, it is instantaneous and intuitive, since the enemy is existentially alien and other. Schmitt's theory of decision and existential enmity has been criticized for implying the kind of nihilism he sought to avoid (Scheuerman 1999, 79), and the aforementioned aspects suggest reading him as a theorist of post-foundationalism, difference-based identities, and subjectivism – a list easy to supplement with relativism.

However, despite his radicalism, Schmitt operated within the tradition of historicist conservatism, emphasizing historical determination and particularity. In the early 1930s, he argued that key legal categories were determined by their particular national community and that all legal thought took place within "a historical, concrete, total order" (Schmitt 1934, 73). In his postwar historical work, Schmitt underscored the uniqueness of all historical truths and epochs (Lievens 2011); however, this does not imply a lack of foundations, but rather entails that singular historical events and situations are comprehensible in the context of a concrete order even when decisions lack normative foundations. Schmitt's (1932, 53) idea that the political world is a "pluriverse" rather than a universe is extendable into an overall theory of the global order and world history: there is always a plurality of such historically constituted units of decisions, and norms and orders are only comprehensible within their respective cultural and political spheres. Schmitt thus endorsed historical particularism and global pluralism, designed to manage the consequences of his antiuniversalistic notion of politics. This served his project of undermining liberal universalism in law and politics in favor of a "realistic" doctrine of international relations based on the equal right of states to wage wars and geopolitical segregation into several "great spaces" (Schmitt 1939–41).

Hans Freyer shared Schmitt's concept of the political, albeit with a post-Hegelian idealistic twist. Freyer saw politics as the historical implementation of cultural forms and ideas, emphasizing that the state represented a certain "unity of values" and that political deeds were necessitated by the historical moment (Freyer 1930, 105). These starting points suggest a monistic conception of the political community: for Freyer (1930, 108, 113), the objectively good in political life was that which served the development of the people and its historical mission. In the 1930s, for Freyer, the most topical political task of the mythical and singular *Volk* was to destroy the structures of "the industrial society" and liberate the state from particular "interests" – that is, to replace liberal democracy with authoritarianism (1931, 44–45). In his call for a mythical revolution from the right, Freyer represented radical, even revolutionary conservatism, justified by situational political ethics. However, a plurality of such historically conditioned ideas and particularistic political units always prevailed, and pluralism was the essence of world history (1925, 23, 194–195, 211–212). Freyer's modified Hegelianism allowed no overarching telos, but merely the tumult of numerous consecutive empires that rose and sunk based on their factual prowess, political will, and resolution. All power was contingent, historically conditioned, and valid only once. In keeping with German historicism and reason-of-state theory, Freyer found the basis for political order in historical particularity and cultural uniqueness.

Amongst the radical conservatives, Spengler (1918, 23–24, 46, 345, 364) took the relativistic impulse in historicism the furthest, denying eternal truths and universal morality and arguing that all standards were valid only within their respective cultural spheres. However, in his belief that the current "Faustian man" was best disposed to understand history in general, Spengler claimed universal validity for his own theory of history as cyclic and organic growth (see Falken 1988, 66–67). Spengler's epistemic and moral relativism built on the postulate that human beings existed only in terms of certain epochs and regions with their particular historical horizons (Spengler 1931, 30–31). For Spengler, "humankind" was a fiction incapable of possessing shared goals or living a truly universal history (Merlio 2009, 134–135). This Counter-Enlightenment stance implies a twofold political relativism: first, there were no universal ideals, truths, or justice in international relations, only factual power and effect; second, there was no single optimal form of government, but each state was unique and constantly changing (Spengler 1922, 368–370, 401). Authoritarianism or monarchism were thus not optimal forms per se, but only the most suitable ones for Germany (Spengler 1920, 71). Spengler's historicism was thus motivated by his conservative politics and extreme nationalism (Merlio 2009, 137).

4. Heideggerian hermeneutics and conservatism

Martin Heidegger is sometimes listed among the radical conservatives (Bourdieu 1988, 55–69; Dahl 1999, 134–135) and recently Aleksandr Dugin (2010, 23–26, 171–173) has designated Heidegger as *the* philosopher of the conservative revolution. In the historical narrative of the later Heidegger, the Western metaphysical tradition was currently culminating in a total technological domination and homogenization of reality, which, however, opened the possibility of a post-metaphysical, postmodern "other beginning" of Western thinking, involving a profound rethinking of the Greek beginning of philosophy. This idea of a cyclic movement through which nihilistic modernity was overcome by letting modernity radicalize and culminate itself, which allows us to turn back to the roots of our tradition in a novel sense, is indeed analogous to the basic model of a "conservative revolution."

Heidegger is in many ways an heir of nineteenth-century German historicism (see Barash 2003). A key objective of the hermeneutic phenomenology of Heidegger's *Being and Time* (1927) was to elaborate an account of the human *Dasein*'s dynamic, context-bound, and historically singular understanding of being (*Sein*) in order to articulate a radically temporal and historical "fundamental ontology." This approach, which would make concepts, truths, and, ultimately, meaningfulness in general, historically situated, had obvious relativistic implications, which Heidegger accepted, noting that "fear of relativism is fear of *Dasein*" (Heidegger 1924, 20). In the 1930s, Heidegger's focus increasingly turned from the individual *Dasein* to the *Volk* as the basic unit of politics, and here we encounter the familiar national particularism of the conservative tradition: for an authentic community of nations to emerge, each nation must assume "responsibility for itself" by discovering its particular historical "determination" (Heidegger 1933).

In 1935, Heidegger (1935, 40–41) depicted the Germans as a "historical" people caught in "pincers" between the United States and the Soviet Union – two ahistorical, multinational, and technological world powers with universalistic ideologies. However, Heidegger (1938–39, 318–319; 1941–42, 80) was soon disillusioned with Nazism, coming to see its biological racism and total warfare as just another, albeit extreme, avatar of modern technicity alongside liberalism and Bolshevism, thus revealing himself as a radical conservative rather than a committed National Socialist.

Remarkably, few of Heidegger's most prominent students were outright conservatives; some were quite the opposite, such as the Marxist and 1960s counterculture icon Herbert Marcuse. In France, the introduction of Heideggerianism through existentialism and poststructuralism coincided with a marked left-wing wave. Leo Strauss's work is often considered a key source of intellectual inspiration for American neoconservatives – but this status ultimately hinges on Strauss's relentless battle *against* relativism and Heideggerian "radical" historicism (Strauss 1953, 9–34; 1961, 251; Gottfried 2012, 43–57). Strauss endorsed a democracy that would *not* be founded on individualistic relativism, but rather on values conceived as universal and on a Platonic form of natural right. Precisely this made Strauss appealing to a section of the American right, but also fundamentally opposed him to the German conservative and historicist heritage. Hans-Georg Gadamer was famously accused by Jürgen Habermas (1967, 168–170) of ending up in Burkean-style conservatism associated with German historicism that rendered a Marxist-type critique of ideology unfeasible by insisting on the irreducible role of tradition in all understanding. In his reply, Gadamer (1967, 285), while defending the emancipatory potential of philosophical hermeneutics, credited conservatism with an important insight into the untenability of the Enlightenment's "abstract antithesis" between reason and the authority of the tradition.

Hannah Arendt, the most influential political theorist among Heidegger's heirs, certainly does not fit the narrow category of "conservatism." Nonetheless, her vivid interest in modern revolutions was balanced with a Roman-inspired emphasis on the political importance of authority, tradition, and preservation. She is perhaps most aptly characterized as a "reluctant modernist" (Benhabib 2003): for Arendt (1951, 305–364; 1963), modernity was at once a possibility for political refoundation and an atomizing and leveling force which dissolved local communities into a homogeneous mass society that, due to the loss of communal forms of "common sense," provided a breeding ground for totalitarian ideologies. Heidegger's influence is traceable in the latter, bleaker evaluation of modernization as a nihilistic obliteration of local differences and traditions. His philosophy has thus channeled at least some Counter-Enlightenment implications of German historicism into contemporary political thought.

5. Contemporary repercussions: multipolarity and ethnoparticularism

The postwar denazification of Germany weakened, but did not eradicate radical conservatism: Freyer, Schmitt, and Heidegger all published further and remained influential. In recent years, many radical conservative arguments have resurfaced in rightwing political theorizing. In particular, Francis Fukuyama's (1992) post-Cold War vision of liberal democracy as a Hegelian universal "end of history" provoked a new wave of historicist and culturally particularistic counterreactions. Among the most influential was Samuel Huntington's (1996) theory of the clash of culturally and regionally distinct "civilizations" as the dominant matrix for new geopolitics – a vision partly prefigured in Schmitt's (1939–41) model of a geopolitical ordering of "great spaces" based on different, particular political identities but with a degree of internal political homogeneity. Although rejecting Western universalism as "imperialism," Huntington

(1996, 310–311, 318) did not surrender to moral and cultural relativism, either – rather, he advocated the search for shared minimal moral standards amidst global cultural diversity and the recognition of the uniqueness of the West amongst other cultures.

Schmitt, Heidegger, and Huntington have inspired a new generation of conservative theorists, including, most prominently, Aleksandr Dugin and Alain de Benoist. The Russian Dugin, who has recently gained a questionable reputation as a chief ideologue of "Putinism," characterizes his recent political theory (Dugin 2009) as a "fourth" ideological alternative to the great ideologies of the twentieth century (liberalism, communism, and fascism). For Dugin, the global ideological hegemony of liberalism stemming from the "Atlantic" (Anglophone) civilization threatens the particular identities of traditionally nonliberal civilizations, such as the "Eurasian" cultural sphere dominated by Russia. Dugin's "fourth political theory" is a Schmittian and Huntingtonian multipolar geopolitical model that allows for vast differences between the major civilizational areas, each equipped with a particularistic political idea in tune with their particular traditions and ethnogenesis, yet presupposes a degree of internal cultural and political unity. While opposed to the late modern world of globalized and individualistic liberal capitalism, the fourth ideology openly exploits the relativist and historicist aspects of "postmodern" or poststructuralist thought against liberal unipolarity. Dugin (2009, 83–100) explicitly proclaims himself an heir of the German conservative revolutionary movement and Heideggerian philosophy of history.

The key figure of the French New Right, de Benoist (2004, 21, 63) criticizes the Western human rights doctrine, first, for being excessively universalistic in its attempt to govern the particular by abstract principles; second, for being subjectivistic in that it defines rights in terms of individuals and thus leads to relativism; and third, for absolutizing historically particular ideas into timeless truths and for imperialistically imposing the values of one culture upon others. However, he denies that this rejection of universal human rights would itself entail relativism, rather endorsing the "pluralist position" that humanity "presents incompatible value systems" (Benoist 1985, 99; 2004, 78). Every human being is equally a member of humankind, but this membership is "always mediated by a particular cultural belonging" (Benoist and Champetier 1999, 123). The pragmatic political corollary of this view is the doctrine of "differentialist ethnopluralism" or "ethnoregionalism": universalist "imperialism" must be rejected, each culture should stay confined to its geographical area, and Western Europe, too, has the right and duty to protect its cultural heritage (Spektorowski 2003). While traditional conservatism used pluralism to underpin traditional state-driven nationalism, here ethnopluralism is raised against liberal multiculturalism, in order to justify exclusionary political practices.

Despite its complexity, the alliance between relativism and political conservatism has proved surprisingly enduring. The Counter-Enlightenment counternarrative of irreducible differences between cultures and epochs and the concomitant idea of history as nonteleological change has coexisted with the Enlightenment narrative of the universal progress of humanity in one form or other for more than two centuries. Contrary to a common conception, neither narrative shows genuine signs of abating – on the contrary, both have recently been reaffirmed, with Steven Pinker (2018) as the most prominent recent herald of the latter, and may well continue to configure the ideological parameters of the future for an indefinite time.

References

Arendt, H. (1951), *The Origins of Totalitarianism*, second edition, San Diego: Harcourt Brace & Co, 1979.
———. (1963), *On Revolution*, London: Penguin Books, 1990.
Barash, J. A. (2003), *Martin Heidegger and the Problem of Historical Meaning*, second edition, New York: Fordham University Press.

Benhabib, S. (2003), *The Reluctant Modernism of Hannah Arendt*, second edition, Lanham, MD: Rowman & Littlefield.

Benoist, A. de (1985), *The Problem of Democracy*, translated by S. Knipe, London: Arktos, 2011.

———. (2004), *Beyond Human Rights: Defending Freedoms*, translated by A. Jacob, London: Arktos, 2011.

Benoist, A. de and C. Champetier (1999), "The French New Right in the Year 2000," *Telos* no. 115: 117–144.

Berlin, I. (1980), "Alleged Relativism in Eighteenth-Century European Thought," in *The Crooked Timber of Humanity: Chapters in the History of Ideas*, edited by H. Hardy, London: Murray, 1990, 70–90.

Bloom, A. (1987), *The Closing of the American Mind: How Higher Education Has Failed Democracy and Impoverished the Souls of Today's Students*, New York: Simon & Schuster.

Bourdieu, P. (1988), *The Political Ontology of Martin Heidegger*, translated by P. Collier, Stanford, CA: Stanford University Press, 1991.

Burke, E. (1790), "Reflections on the Revolution in France," in *Revolutionary Writings*, edited by I. Hampsher-Monk, Cambridge: Cambridge University Press, 2014, 1–250.

Chateaubriand, F.-R. de (1797), *An Historical, Political, and Moral Essay on Revolutions, Ancient and Modern*, London: Colburn, 1815.

Dahl, G. (1999), *Radical Conservatism and the Future of Politics*, London: Sage Publications.

Dugin, A. (2009), *The Fourth Political Theory*, translated by M. Sleboda and M. Millerman, London: Arktos, 2012.

———. (2010), *Martin Heidegger: The Philosophy of Another Beginning*, translated by N. Kouprianova, Arlington, VA: Radix, 2014.

Epstein, K. (1966), *The Genesis of German Conservatism*, Princeton, NJ: Princeton University Press.

Falken, D. (1988), *Oswald Spengler: Konservativer Denker zwischen Kaiserreich und Diktatur*, Munich: Beck.

Freyer, H. (1925), *Der Staat*, Leipzig: Rechfelden.

———. (1930), "Ethische Normen und Politik," *Kant-Studien* 35: 99–114.

———. (1931), *Revolution von Rechts*, Jena: Diederichs.

Fukuyama, F. (1992), *The End of History and the Last Man*, New York: Free Press.

Gadamer, H.-G. (1967), "Rhetoric, Hermeneutics, and the Critique of Ideology: Metacritical Comments on *Truth and Method*," translated by J. Dibble, in *The Hermeneutics Reader*, edited by K. Mueller-Volmer, New York: Continuum, 1985, 274–292.

Gottfried, P. E. (2012), *Leo Strauss and the Conservative Movement in America: A Critical Appraisal*, Cambridge: Cambridge University Press.

Habermas, J. (1967), *On the Logic of the Social Sciences*, translated by S. Weber Nicholsen and J. A. Stark, Cambridge, MA: MIT Press, 1988.

Harstick, H.-P. (1974), "Historische Schule," in *Historisches Wörterbuch der Philosophie*, vol. 3, edited by J. Ritter, Basel: Schwabe, 1137–1141.

Heidegger, M. (1924), *The Concept of Time*, translated by W. McNeill, Oxford: Wiley-Blackwell, 1992.

———. (1927), *Being and Time*, translated by J. Stambaugh, translation revised by D. Schmidt, Albany, NY: State University of New York Press, 2010.

———. (1933), "German Men and Women!" translated by W. S. Lewis, in *The Heidegger Controversy: A Critical Reader*, edited by R. Wolin, Cambridge, MA: MIT Press, 1993, 47–49.

———. (1935), *Introduction to Metaphysics*, translated by G. Fried and R. Polt, New Haven, CT: Yale University Press, 2000.

———. (1938–39), *Ponderings VII–XI: Black Notebooks 1938–1939*, translated by R. Rojcewicz, Bloomington, IN: Indiana University Press, 2017.

———. (1941–42), *The Event*, translated by R. Rojcewicz, Bloomington, IN: Indiana University Press, 2013.

Huntington, S. P. (1996), *The Clash of Civilizations and the Remaking of World Order*, New York: Simon & Schuster.

Jaeger, F. and J. Rüsen (1992), *Geschichte des Historismus: Eine Einführung*, Munich: Beck.

Kelsen, H. (1929), *The Essence and Value of Democracy*, edited by N. Urbinati and C. Invernizzi Accetti, translated by B. Graf, Lanham, MD: Rowman & Littlefield, 2013.

Lievens, M. (2011), "Singularity and Repetition in Carl Schmitt's Vision of History," *Journal of the Philosophy of History* 5: 105–129.

Mannheim, K. (1925), *Conservatism: A Contribution to the Sociology of Knowledge*, edited by D. Kettler, V. Meja, and N. Stehr, translated by D. Kettler and V. Meja, Abingdon: Routledge, 2007.

Meinecke, F. (1924), *Machiavellism: The Doctrine of Raison d'État and Its Place in Modern History*, translated by D. Scott, New Haven, CT: Yale University Press, 1962.

———. (1936), *Historism: The Rise of a New Historical Outlook*, translated by J. E. Anderson, London: Routledge & Kegan Paul, 1972.

Merlio, G. (2009), "Spenglers Geschichtsmorphologie im Kontext des Historismus und seiner Krisen," in *Spengler: Ein Denker der Zeitenwende*, edited by M. Gangl, G. Merlio, and M. Ophälders, Frankfurt am Main: Lang, 129–143.

Moeller van den Bruck, A. (1923), *Germany's Third Empire*, translated by E. O. Lorimer, London: Arktos, 2012.

Montesquieu, C.-L. de Secondat de (1748), *The Spirit of the Laws*, translated by A. M. Cohler, B. C. Miller and H. S. Stone, Cambridge: Cambridge University Press, 2008.

Muller, J. Z. (1997), "Introduction: What Is Conservative Social and Political Thought?" in *Conservatism: An Anthology of Social and Political Thought from David Hume to the Present*, edited by J. Z. Muller, Princeton, NJ: Princeton University Press, 3–31.

Pinker, S. (2018), *Enlightenment Now: The Case for Reason, Science, Humanism, and Progress*, New York: Penguin.

Ranke, L. von (1833/1836), "A Dialogue on Politics; The Great Powers," translated by T. Hunt von Laue and H. Hunt von Laue, in *Leopold Ranke: The Formative Years*, edited by T. Hunt von Laue, Princeton, NJ: Princeton University Press, 1950, 152–218.

Scheuerman, W. E. (1999), *Carl Schmitt: The End of Law*, Lanham, MD: Rowman & Littlefield.

Schmitt, C. (1922), *Political Theology: Four Chapters on the Concept of Sovereignty*, translated by G. Schwab, Chicago: University of Chicago Press, 2005.

———. (1932), *The Concept of the Political*, second edition, translated by G. Schwab, Chicago: University of Chicago Press, 2007.

——— (1934), *On the Three Types of Juristic Thought*, translated by J. W. Bendersky, Westport, CT: Praeger, 2004.

———. (1939–41), "The *Großraum* Order of International Law with a Ban on Intervention for Spatially Foreign Powers: A Contribution to the Concept of *Reich* in International Law," in *Writings on War*, translated and edited by T. Nunan, Cambridge: Polity Press, 2011, 75–124.

Spann, O. (1921), *Der wahre Staat: Vorlesungen über Abbruch und Neubau der Gesellschaft*, Leipzig: Quelle & Meyer.

Spektorowski, A. (2003), "The New Right: Ethno-Regionalism, Ethno-Pluralism and the Emergence of a Neo-Fascist 'Third Way,'" *Journal of Political Ideologies* 8(1): 111–130.

Spengler, O. (1918), *The Decline of the West, Volume 1: Form and Actuality*, translated by C. F. Atkinson, New York: Knopf, 1927.

———. (1920), "Prussianism and Socialism," in *Selected Essays*, edited and translated by D. O. White, Chicago: Regnery, 1967, 1–131.

———. (1922), *The Decline of the West, Volume 2: Perspectives of World-History*, translated by C. F. Atkinson, New York: Knopf, 1928.

———. (1931), *Man and Technics: A Contribution to a Philosophy of Life*, translated by C. F. Atkinson and M. Putman, London: Arktos, 2015.

Strauss, L. (1953), *Natural Right and History*, Chicago: University of Chicago Press, 1965.

———. (1961), *On Tyranny*, second edition, edited by V. Gourevitch and M. S. Roth, Chicago: University of Chicago Press, 1991.

Troeltsch, E. (1922), *Gesammelte Werke, Volume 3: Der Historismus und seine Probleme, Volume 1: Das logische Problem der Geschichtsphilosophie*, Tübingen: Mohr.

Vico, G. (1744), *New Science: Principles of the New Science Concerning the Common Nature of Nations*, third edition, translated by D. Marsh, London: Penguin, 2001.

24

COMMUNITARIANISM

Henry Tam

1. Introduction

Communitarian ideas have evolved through a series of attempts to steer a justifiable path away from authoritarian impositions of beliefs and rules on the one hand, and any form of "anything goes" outlook on the other.[1] In the West, the process can be traced back to Aristotle who sought to meet Socrates' contention that nobody seemed to have a sound reason for their beliefs, and at the same time reject Plato's vision of absolute truths known only to a philosophical elite.[2] In the East, the debates began when Mozi, who founded the Mohist School in China in the 5th century BC, aimed to displace top-down Confucian ideas and anarchic Daoist individualism with a philosophy that linked the acceptability of beliefs and practices to the experiences of past, present, and future communities (Mei 1929).

By the mid-19th century, the ideas associated with the defence and development of communities capable of finding answers through sustained cooperation had come to be known as "communitarian" in connection with the works of Robert Owen and others who shared his commitment to improve society by empowering people to share in the tasks and rewards of problem-solving (Bestor 1950; Harrison 1969). The Owenite approach built on an important strand of the Enlightenment that is often overlooked by later commentators, namely, the pursuit of understanding through a community of mutually respectful enquirers free from intimidation and manipulation. Distinct from the Kantian reliance on a rational self, unencumbered by any social attachment, to decide what was right or wrong, the communitarian quest was directed instead at enabling people to interact with each other as fellow members of common endeavours, with shared interpersonal bonds and interweaved interests in discovering better answers to how we should live. Pure reason could not reveal to us any absolute truth about the natural or moral world. But through practical exchanges and critical explorations carried out in suitably structured communities, we could learn and revise what ought to merit our assent.

When the Kantian version of "Reason"-centric Enlightenment began to lose credibility in the 19th century, it became commonplace to suggest that the entire Enlightenment project had failed. Some argued that submission to traditional hierarchies and values was the only way to avoid slipping into a chaos of permanent discord. Some urged people to follow nationalistic regimes that would offer them unquestionable ideals and commands. Others maintained that it was far better to leave individuals to think and act as they saw fit – they could live up to Nietzsche's vision of the *Übermensch*, making unilateral decisions without a care about what it might mean for anyone else; or fulfil Herbert Spencer's dream of *laissez faire*, wherein individuals choose for themselves without any meddlesome authority setting limits and obligations on all.

By the late 19th and early 20th century, a number of liberal (i.e., anti-authoritarian) thinkers responded to the spread of relativism by articulating communitarian ideas for social development. In the arguments put forward by three representative figures – Emile Durkheim from France, L. T. Hobhouse from Britain, and John Dewey from the US – we detect four common themes: the need to recognise community as well as personal interests as valid ends to pursue; the ordering and, where necessary, reconciliation of those ends and related means can only be adequately achieved through the community's members participating in the process; the required participation should be guided by civic education and a democratic culture that enable people to deliberate in an informed and un-coerced manner; and a publicly accountable government ought to play a critical role in ensuring that the conditions for open and equal participation are obtained.[3]

Durkheim argued for the development of organic solidarity (as opposed to mechanical solidarity) whereby people could interact freely within an agreed framework for joint endeavours (Durkheim 1893). He emphasised civic symbols, reinforcement of civil respect and discourse, and the cultivation of professional ethics (Cladis 1992). Hobhouse explained why liberal values could only be realised when people's freedom to live without coercion were not rendered meaningless by the freedom of the powerful to intimidate and exploit others (Hobhouse 1994). He highlighted the need for government to take action in a range of policy areas so that citizens could cooperate in attaining harmony in the pursuit of social and personal goals without many being left vulnerable because of their weak socio-economic positions (Freeden 1986; Simhony and Weinstein 2001). Dewey called for the cultivation of the "Great Community" wherein people could prioritise and revise their concerns through shared deliberations, and devise the means to address them (Dewey 1927). He focussed on educational methods, community journalism, and decentralised participatory opportunities to effect social cohesion that would embrace, not shun, critical individuality (Ryan 1995).

These communitarian views were highly influential in the first half of the 20th century. The traumas of two world wars, fresh memories of the devastation caused by socio-economic and nationalistic divisions, the desperate need to articulate and protect the common good, these factors brought about a broad consensus that sustained major reforms to build a more inclusive society. However, in the 1960s, post-war prosperity and political complacency were beginning to erode interests in civic unity and collective action. Cultural and market individualism advanced at the expense of communitarian solidarity. In academia, philosophical trends that favoured formal logical and linguistic analysis, emotivist reduction of ethics to expressions of feelings, or deconstruction of texts and theories, resulted in the marginalisation of the ideas of thinkers like Durkheim, Hobhouse, and Dewey as outmoded, vague, and irrelevant.

How could respect and support for the kind of caring, thoughtful, cooperative society promoted by communitarian advocates be revived? Would not relativism of one form or another always be poised to challenge it by exposing its lack of an indisputable foundation? In 1971, John Rawls' *A Theory of Justice* appeared and it purported to have formulated rationally undeniable principles that would in practice underpin many of the inclusive policies sought by communitarians. But in adopting a quasi-Kantian approach that would ground principles in rational minds untainted by any empirical factor, Rawls' attempt to keep relativism at bay was ultimately found wanting by communitarian-minded thinkers.

2. Communitarian critiques of the rationalist defence against relativism

Although the criticisms directed at Rawls' rationalist conception of justice by Alasdair Mac-Intyre (1981), Michael Sandel (1982), Michael Walzer (1983), and Charles Taylor (1985), have often been characterised as part of a liberal-communitarian debate, it would be more accurate to view them as communitarian objections to Kantian-Rawlsian attempts to overcome moral and political relativism in general. While these critics disagreed with the Rawlsian approach to defending liberalism, they were all sympathetic to liberal political goals such as socio-economic equality and redistributive justice.[4]

What they found unacceptable was the supposition that one can locate through pure reason a fulcrum, which exists independently of all actual communities, for determining how we should live. Rawls, in essence, conceived of rational beings as analytical minds abstracted from all desires, aspirations and knowledge of their personal circumstances. Behind this "veil of ignorance," all they will consider is what principles for governing society would make the most sense. Since by definition they have no preference for any particular state of the world to be realised, and they cannot foretell if they will be favoured or disadvantaged by the conditions in which they will find themselves, Rawls deduced that they would all sign up to two indubitable principles: first, "each person is to have an equal right to the most extensive total system of equal basic liberties compatible with a similar system of liberty for all," and second, "social and economic inequalities are to be arranged so that they are both (a) to the greatest benefit of the least advantaged, and (b) attached to offices and positions open to all under conditions of fair equality of opportunity" (Rawls 1971, 302).

For the communitarian critics, such a philosophical strategy would not provide a sound foundation for a just society.[5] Far from blocking off relativist challenges, it could all too easily be exposed as an arbitrary platform for ethical decisions. MacIntyre held that values were only possible in the context of a moral tradition that regards human beings as having a purpose or ideal to fulfil. In the absence of such a tradition, a postulate of the "rational person" devoid of any moral quality cannot offer any meaningful guidance. One can just as easily invoke a different construct of such a "rational" individual and arrive at anti-egalitarian principles.[6] Sandel targeted his objection at the incoherence of a notion of personal identity stripped of all its constitutive elements. Since how a person may think or make choices in life is inextricably connected with the beliefs, dispositions, sense of value that person has developed as a member of a community, it makes no sense to make a priori assumptions about what an unencumbered self, without any of these vital characteristics, would choose. Walzer argued that social assessments about what level of freedom should be extended to different activities, or how diverse goods should be distributed, are related to the social meanings people with a shared culture attached to those activities and goods. We cannot, therefore, brush aside all those meanings and pretend "rational" choices can be predicted without reference to them or any lived culture that can invest them with varying degrees of importance. For Taylor, we are members of linguistic communities through which we interpret beliefs. Rival claims are resolved by exchanges within the relevant communities, and not by anyone stepping back into some form of solipsist isolation. The individual mind locked in solitary "reasoning" is not a guarantor of indisputable truth, but is more a prisoner of arbitrary thoughts cut off from corrections by other interlocutors.

A common concern with the preceding communitarian criticisms is that they would open the door to all kinds of ideas and practices that can no longer be ruled out by a rationally grounded doctrine. One particular complaint, given the critical references made to the role of communities and cultures, focuses on the apparent inability to counter forms of community life such as those filled with traditional prejudices against women and minorities, and habituated into neglecting or exploiting the disadvantaged.

This is something anticipated by the communitarian critics, and each of them has a ready rebuttal. MacIntyre (1988, 1990) argued that moral traditions have always had to contest with one another, and those that persist with problems stemming from, for example, discrimination and oppression, will encounter "epistemological crises" that can only be dealt with through appropriate adjustment. Similarly, Taylor pointed to the clash of conflicting cultures as a fact of life, and believes that those that help people better understand and thus cope with their life experiences would offer greater "epistemic gains,"[7] would commend themselves over their rivals (1989). Walzer drew from historical examples of social critics to illustrate that there is no guarantee that prevailing cultures will go on irrespective of how they interact with the people living under them. Flaws such as hypocrisy, inconsistency, or callous ill treatment of particular individuals and groups, betraying a lack of commitment to respect for others, can be seized upon by members of the community in question to challenge its legitimacy (1987). From Sandel's perspective, it would be a mistake to think of the problem as though it exists outside time and space. What should or should not change is rooted in a particular community, and if the case of the US, for example, is to be considered, then he maintains that we should look at how the civic republican ethos that was at the heart of the founding of the country is being advanced or marginalised, and engage citizens in reflecting on and devising reform options (1996).

3. Communitarian responses to relativist divergence

While the communitarian critics considered in the last section remain convinced that their rejection of rationalist arguments would not deprive us of legitimate means to challenge arbitrary and unacceptable norms, they have not gone on to explain how such challenges can be effectively carried out in practice. By contrast, another group of communitarian thinkers have tackled this problem in much greater depth. In the works of David Miller (1989), Jonathan Boswell (1990), Philip Selznick (1992), Amitai Etzioni (1997), and Henry Tam (1998), close attention is given to the conditions under which divergent views within and between communities may come to be succeeded by a common understanding.[8]

A recurring theme in relativist arguments is that since there is no absolute Platonic idea of justice, indubitable Cartesian foundation, or unconditional Kantian imperative, people will have to be left to believe whatever they want to believe. Communitarians, however, regard such a viewpoint as having little bearing on the real world. To paraphrase David Hume, sceptical philosophers may talk a good talk about there being nothing to choose between one belief and another, but from the moment they wake up in the morning, they would not last long through the day if they were to suspend all beliefs or just randomly assume the validity of various propositions. The question that needs answering is how differences are actually resolved and satisfactory solutions are agreed upon.

Leaving aside differences which have no need to be reconciled – for example, subjective preferences in relation to support for sports teams, choice of cuisine, style of clothes[9] – there are many rival claims that call for universal assent. Some may say that if others refuse to follow "divine commands" as they have discerned, everyone will suffer. Some may argue that if their technological specifications were not incorporated into legal requirements, lives could be put at risk. Society has to decide when people ought to be left to act on their own beliefs, and when all must accept the adoption of particular beliefs as universally binding.

Miller argued that in order to explore what should apply to the whole society, we must go beyond simplistically aggregating individual opinions and interests (1989). He set out a nation-state framework which, with democratic safeguards and deliberative support, can enable citizens to engage in "politics as dialogue." Structures and guidance need to be put in place so that members are aware of the bond of citizenship they share, and that it is the coherence and evidential strengths of arguments that should sway them.[10] Diversity is to be encouraged, as it helps to ensure contrasting perspectives and concerns to come into the open; but those participating must be given a sense of efficacy and accountability, so that they take responsibility for contributing to the discussions and the decisions.

A key challenge facing any nation-state is the running of the economy. Boswell warned against leaving individuals without sufficient information or enough bargaining power to shape outcomes through a laissez faire market system (1990). He was also opposed to important policies being determined by powerful lobby groups. Instead, he proposed the development of cooperative partnerships at all levels of society, so that inter-connected citizen groups and intermediary institutions can keep each other informed and facilitate assessments made of policy options across the different levels. To do this, stable, transparent, and accountable organisations are required to build trust and shared deliberations. They must ensure socio-economic divisions are bridged to prevent the views of the marginalised from being overlooked. Technocratic expertise must itself be scrutinised by bodies that will engage and explain to their members the critical activities they carry out on their behalf.

The role of institutions in general in cultivating a shared moral outlook was extensively explored by Selznick (1992). He believed that people's sense of what to believe and what to pursue, was influenced by their interactions with others most frequently through the organisations they engage with. People can be become more polarised into factions or motivated to find common ground, depending on how the organisations they are involved with behave. It is when those organisations promote inclusiveness and accountability that people will be more likely to develop cooperative and mutually respectful relations. Experience has shown that without sustained support for civility, reasoned discourse, and integrity in honouring commitments, the claim that disagreement will proliferate becomes a self-fulfilling prophecy.

Of course, group consensus does not by itself give complete validation to a set of ideas or values. Etzioni developed a four-step process to explore the need for further revisions (1997). First, members of the wider community can through democratic voting and social consensus-building find out what may command general support. Secondly, any particular option obtaining current backing should be tested against more widely established societal values; for example, a majority in a neighbourhood demanding to inflict pain without any due legal process on anyone suspected of having committed a serious (but nonetheless alleged) crime, would not be acceptable.[11] Thirdly, even long-standing values may have ceased to be held by everyone, and cross-societal moral dialogues would be required to test to what extent attitudes may be ready to shift. Fourthly, there is room for global cross-cultural judgements, which contrary to abstract assumptions, can be grounded on the reality of common values found in different cultures around the world.

Tam's general theory of communitarianism explains how critical understanding and informed agreement for societal problem-solving can be better attained through the application of three related approaches to the activities of the state, business, and third sectors (1998). These approaches cover cooperative enquiry that will enable proposed claims to be judged and revised with reference to the extent to which informed participants deliberating under conditions of thoughtful and un-coerced exchanges would concur; mutual responsibility that will encourage members of any community to assist one another in pursuing those values that stand up to the test of reciprocity; and citizen participation that will ensure all those affected by any given power structure to participate as equal citizens in determining how the power in question is to be exercised.

4. Relativist challenges to communitarian consensus-building

Communitarian ideas for resolving conflicts over how we should live may nonetheless be rejected by relativists on two counts. There are many examples of significant differences remaining despite the best efforts being made to help people find common ground. Moreover, even where agreement is reached, it can be challenged that there is no basis for assuming that the consensus view is the correct one. Relativists are likely to insist that just because people agree about something being true/right/justified, it is still logically possible for it not to be. While rationalists may pick up the challenge by trying to find a way to establish truth/rightness/justification that cannot be doubted logically, communitarians will argue that the onus on proving that what has been agreed is not acceptable (because it is in fact false/wrong/flawed) falls on those who want to substantiate that claim. If there is no absolute reason to establish a claim, there is no absolute reason to reject it either.

What we have is the reality of communities attempting to find ways to settle disputes and develop shared solutions where possible. There is no utopian scenario of agreement being reached about everything, or an infallible authority emerging to close off all possible future enquiry. But it does not follow that contested claims will never get resolved, or seemingly intractable disputes cannot reach general settlement. Unless we want to end up like the sceptic who gets nowhere with a total suspension of belief, we have to try to apply the approaches communitarians propose, and dedicate real efforts in enabling people to deliberate without coercion or manipulation, explore with the aid of accumulated evidence and proven expertise, and ascertain what common ground can be established and what arrangements can be adopted to manage any residual divergence. Or as Wittgenstein might say, we must show the fly out of the solipsist bottle, before it can appreciate that some means of resolving disagreement work better than others in the world out there.[12]

Ultimately, the relativist challenges to communitarian ideas will be met, not by some remodelled rationalist solution which is impossible on relativism's own terms,[13] but by problem-solving techniques without which relativists, like everyone else, cannot get by in real life. A variety of conflict resolution and reconciliation techniques have been found to be highly effective in dealing with situations from neighbourhood disputes to long-standing violent conflicts (Long and Brecke 2003; Bar-Siman-Tov 2004). Restorative justice approaches have helped to change minds and behaviours in schools and the criminal justice system (Johnstone 2011). A wide range of processes in support of deliberative democracy have been developed to enable citizens to go beyond incoherent interests-aggregation to attain reflective consensus (Gutmann and Thompson 2004; Fishkin 2009). Participatory budgeting has helped communities with disparate concerns and conflicting priorities engage in facilitated discussions before signing up to new plans of action for all (Röcke 2014). Tam applied his communitarian ideas to policy development when he was the UK Government's Head of Civil Renewal, and took forward an extensive community empowerment programme which supported diverse communities in exploring and devising public policy solutions which might not otherwise have been formulated, let alone chosen (Tam 2011, 2019)

Some relativists will dismiss the preceding references to the many different types of consensus-building as *philosophically* irrelevant. For them, it proves nothing against their views that people can insist on believing whatever they want to believe. And of course, there are people who may want to hang on to their beliefs regardless of what evidence, concerns, or arguments others present to them. It poses a practical challenge as to how particular individuals can be persuaded to change their mind, and calls for policy responses to find ways to facilitate shared deliberations and consensus building.

Relativism may serve as an antidote to rationalism. However, in practice, if it is taken seriously as a basis for dismissing all forms of consensus-building as devoid of justification, and insisting that none of the outcomes agreed can ever in any sense be reasonably preferred by those who support them compared with the views of others who simply reject them, then it could in effect undermine everything from the setting of standards for medical treatment to trial by jury. Of course, relativists will declare that such consequentialist concerns have no greater justificatory legitimacy than anything else, in which case, they need to be reminded that their declaration – on their own relativist terms – has no legitimacy that anyone else needs to take account of either. What is left before us is the lived world where many disagreements cannot be ignored, and joint decisions often have to be made. And it is further research on communitarian approaches at the strategic level, and consensus-building processes at the interpersonal level, that holds the key to dealing with divergence over how we should live.

Notes

1 For a general introduction to the historical development of communitarian ideas, see Tam (2020).
2 Aristotle's group approach to the study of natural phenomena and political constitutions exemplified his belief in cooperative communities of enquiry, and underpinned his rejection of Plato's philosophy of absolute ideas accessible only to a few exclusive minds.
3 Durkheim, Hobhouse, and Dewey all rejected the false dichotomy of tightly bound authoritarian *Gemeinschaft* and impersonal/individualistic *Gesellschaft* put forward by Ferdinand Tönnies in his 1887 book.
4 Like Durkheim, Hobhouse, and Dewey before them, they were on the side of greater inclusion, and stood against the agenda of "market might is right" – whether that was promoted by plutocratic conservatives or laissez faire libertarians.
5 Rawls' strategy in *A Theory of Justice* was a variation of Kantian reasoning, which in turn echoed Cartesian and Spinozist rationalism in being designed to secure indubitable status beyond all possible empirical falsification. The tendency to equate this rationalist outlook with the Enlightenment is flawed when it ignores the pragmatic strand that ran through Francis Bacon to David Hume and Denis Diderot, and which was highly influential amongst the *philosophes*. We will not go into the extent Rawls revised his arguments in the light of communitarian criticisms, but see Mulhall and Swift (1992) for an extensive discussion.
6 Which is what Robert Nozick did (1974), and his arguments were also criticised by MacIntyre (1992).
7 Taylor has argued that even though two different outlooks may be strictly incommensurable in the sense that neither subscribes to any common criterion that can be applied to evaluate their comparative merit, it is still possible for one to be found to be superior to the other because of the respective impact they have on people's lives (1982).
8 Unlike MacIntyre, Sandel, Taylor, or Walzer whose writings have been referred to by commentators as "communitarian" although they prefer not to be classified as "communitarians" themselves; Miller, Boswell, Selznick, Etzioni, and Tam unequivocally describe the theories they put forward as communitarian.
9 At times incompatible views arise even with subjective preferences. One British politician in the 1980s infamously objected to British citizens supporting any non-British sports team. And there have been attempts to demand cultural conformity in the name of social integration. In such cases, the dispute would require resolution.
10 The role of the state in relation to citizens is here analogous to the role of the judge in relation to jurors. The former oversees the process for impartial deliberation, and the latter are required to reflect on what have been presented to them and deliver their verdict. No one claims it is infallible, but that does not render it arbitrary either.
11 Etzioni explains that the US Constitution embodies the deep societal values he is referring to, but stresses that in other countries, it may not be a formal constitution that captures their long-standing values.
12 Peter Strawson, drawing on Hume and Wittgenstein to respond to scepticism, wrote: "The correct way with the professional skeptical doubt is not to attempt to rebut it with argument, but to point out that it is idle, unreal, a pretense" (1985, 19).

13 Whatever rationalist attempt is made, a relativist retort will counter that it must ground itself in a circular argument, has no independent foundation, or relies on some prior premise which itself is based on another prior premise and so on *ad infinitum*.

References

Bar-Siman-Tov, Y. (ed.) (2004), *From Conflict Resolution to Reconciliation*, Oxford: Oxford University Press.

Bestor, A. E. (1950), *Backwoods Utopias: The Sectarian and Owenite Phases of Communitarian Socialism in America, 1663–1829*, Philadelphia: Pennsylvania Press.

Boswell, J. (1990), *Community and the Economy: The Theory of Public Cooperation*, London: Routledge.

Cladis, M. S. (1992), *A Communitarian Defense of Liberalism: Emile Durkheim and Contemporary Social Theory*, Stanford: Stanford University Press.

Dewey, J. (1927), *The Public and Its Problems*, New York: H. Holt & Co.

Durkheim, E. (1893), *The Division of Labour in Society*, Basingstoke: Macmillan, 1984.

Etzioni, A. (1997), *The New Golden Rule: Community and Morality in a Democratic Society*, London: Profile Books.

Fishkin, J. S. (2009), *When the People Speak: Deliberative Democracy and Public Consultation*, Oxford: Oxford University Press.

Freeden, M. (1986), *The New Liberalism: An Ideology of Social Reform*, Oxford: Clarendon Press.

Gutmann, A. and D. Thompson (2004), *Why Deliberative Democracy*, Princeton, NJ: Princeton University Press.

Harrison, J. F. C. (1969), *Robert Owen and the Owenites in Britain and America: The Quest for the New Moral World*, London: Routledge & Kegan Paul.

Hobhouse, L. T. (1994), *Liberalism and Other Writings*, edited by J. Meadowcroft, Cambridge: Cambridge University Press.

Johnstone, G. (2011), *Restorative Justice: Ideas, Values, Debates*, London: Routledge.

Long, W. and P. Brecke (2003), *War and Reconciliation: Reason and Emotion in Conflict Resolution*, Cambridge, MA: MIT Press.

MacIntyre, A. (1981), *After Virtue*, London: Duckworth.

———. (1988), *Whose Justice? Which Rationality?* London: Duckworth.

———. (1990), *Three Rival Versions of Moral Enquiry*, London: Duckworth.

———. (1992), "Justice as a Virtue: Changing Conceptions," in *Communitarianism and Individualism*, edited by S. Avineri and A. de-Shalit, Oxford: Oxford University Press, 51–64.

Mei, Y. P. (1929), *The Ethical and Political Works of Motse*, London: Arthur Probsthain.

Miller, D. (1989), *Market, State and Community*, Oxford: Clarendon Press.

Mulhall, S. and A. Swift (1992), *Liberals and Communitarians*, Oxford: Blackwell.

Nozick, R. (1974), *Anarchy, State & Utopia*, Oxford: Oxford University Press.

Rawls, J. (1971), *A Theory of Justice*, Cambridge, MA: Harvard University Press.

Röcke, A. (2014), *Framing Citizen Participation: Participatory Budgeting in France, Germany and the United Kingdom*, Basingstoke: Palgrave Macmillan.

Ryan, A. (1995), *John Dewey and the High Tide of American Liberalism*, New York: W. W. Norton & Co.

Sandel, M. (1982), *Liberalism and the Limits of Justice*, Cambridge: Cambridge University Press.

Selznick, P. (1992), *The Moral Commonwealth: Social Theory and the Promise of Community*, Berkeley: University of California Press.

———. (1996), *Democracy's Discontent*, Cambridge, MA: Harvard University Press.

Simhony, A. and D. Weinstein (2001), *The New Liberalism: Reconciling Liberty and Community*, Cambridge: Cambridge University Press.

Strawson, P. F. (1985), *Skepticism and Naturalism*, London: Methuen.

Tam, H. (1998), *Communitarianism: A New Agenda for Politics and Citizenship*, Basingstoke: Macmillan.

———. (2011), "Rejuvenating Democracy: Lessons from a Communitarian Experiment," *Forum* 53(3): 407–420.

———. (ed.) (2019), *Whose Government Is It? The Renewal of State-Citizen Cooperation*, Bristol: Bristol University Press.

———. (2020), *The Evolution of Communitarian Ideas: History, Theory & Practice*, Basingstoke: Palgrave Macmillan.

Taylor, C. (1982), "Rationality," in *Rationality and Relativism*, edited by M. Hollis and S. Lukes, Oxford: Blackwell, 87–105.

———. (1985), *Philosophical Papers*, Cambridge: Cambridge University Press.

———. (1989), *Sources of the Self: The Making of Modern Identity*, Cambridge: Cambridge University Press.

Tönnies, F. (1887), *Community and Society*, Mineola: Dover Publications, 2003.

Walzer, M. (1983), *Spheres of Justice*, New York: Basic Books.

———. (1987), *Interpretation and Social Criticism*, Cambridge, MA: Harvard University Press.

25

MULTICULTURALISM

George Crowder and Geoffrey Brahm Levey

1. Introduction

Multiculturalism is in general the idea that the coexistence of multiple cultures within societies is not only a fact about the modern world but also something to be valued.[1] Further, multiculturalism usually requires that the desirability of cultural diversity within a single society should be given public recognition in the form of minority cultural rights or state accommodation of cultural minorities.

It may seem that such a view connects readily with forms of relativism, in particular cultural relativism, according to which no single culture is ethically superior to any other. This is indeed how multiculturalism is understood in some contexts, not least by many of its conservative critics.[2] But such a picture is too simple. Many national multiculturalism policy regimes, such as those of the first multiculturalist democracies, Canada and Australia, are expressly wedded to liberal-democratic values. Much multicultural political theory is similarly committed to liberal values. In these cases, multiculturalism should be seen not as a form of relativism but as a fundamentally universalist doctrine, although one that welcomes a certain range of legitimate cultural diversity. There are also some influential justifications of multiculturalism, liberal and non-liberal, that may appear to embrace cultural relativism but which, in fact, do not.

Our discussion is structured as follows. We begin by looking more closely at the meaning of multiculturalism, noting weaker and stronger versions, and contrasts with other ideas, opposed or related. Next, we consider possible justifications of multiculturalism, notably cultural relativism and the "polyglot" argument. This leads us to the foremost liberal justification, offered by Will Kymlicka, and to some of the criticisms of his view. Thereafter, we discuss some alternative liberal defences of multiculturalism and some prominent non-liberal arguments for multiculturalism, all of which give greater credence to cultural identity, and all of which are sometimes construed, wrongly, as endorsing relativism. We conclude with some general observations.

2. What is multiculturalism?

Many societies have been "multicultural" in the sense that they have in fact contained multiple cultures or ways of life. The Roman and Ottoman empires, for example, showed considerable toleration for religious and ethnic minorities in their midst. Several modern states, such as

Belgium and Switzerland, have been structured around the accommodation of two or more cultural communities for most of their history. "Multiculturalism," however, is the relatively recent idea that cultural diversity involving *all* cultural identity groups within a single political society is not only a tolerable reality but one to be welcomed and affirmed.

Three distinctions are worth noting immediately between multiculturalism and opposing or cognate ideas. The obvious contrast is with assimilation, where all members of a society are encouraged or forced to adopt a single cultural identity. But multiculturalism should also be distinguished from both toleration and common rights of citizenship. Toleration entails non-interference with those with whom we disagree. For centuries the notion of toleration provided the model for the most progressive attitudes towards the treatment of minorities. In the liberal tradition, the beginnings of this tendency can be traced to the seventeenth-century movement for religious toleration, championed by thinkers such as John Locke in the wake of the European wars of religion (Locke 1689). In the twentieth century the rights of minorities were strengthened in liberal democracies by legislation designed to protect individuals from invidious discrimination based on background group characteristics (such as race, ethnicity, nationality, religion, and sexual orientation) in areas such as education and employment. However, multiculturalism, while building on these achievements, goes beyond toleration and common citizenship rights to advance a more positive, proactive and accommodating view of minorities. On the multiculturalist view, minorities are not merely to be left alone or entitled to the same suite of rights or provisions as all other citizens. Precisely because a society's norms and institutions were typically developed by the dominant majority to accord with its own cultural patterns, insisting on uniform rights is likely to disadvantage cultural minorities. Thus, the multiculturalist view is that minorities should be respected and diversity celebrated to the point of reforming the established institutions, laws, and practices. In this sense, multiculturalism is open to forms of differentiated citizenship.

The historical roots of multiculturalism lie in a combination of global and local developments over the past century or so. First, at the global level, moral and political dogmatism and evangelism have been undermined by an increased moral scepticism and toleration, due in part to a sober recognition of the costs associated with the modern experiences of slavery, colonialism, genocide and world war. Second, the spread of liberal-democratic ideas has encouraged minority social groups to reject inferior social and political status and to demand rights of equal treatment and participation in their societies' central institutions, where equal treatment is understood as taking into account people's background circumstances. Third, accelerating processes of economic and technological globalization have made it increasingly difficult to maintain cultures as separate and independent entities. Fourth, the end of the Cold War unleashed a new wave of "identity politics" based on ethnic or nationalist affiliations that had been suppressed during the global contest between capitalism and communism. As a result of these various factors, liberal-democratic states came under increasing pressure to accommodate their minorities and became increasingly open to the moral justification of that accommodation.

At the same time, circumstances in particular countries pressed in the same direction. In the United States, continued frustration of the civil equality of African Americans in the 1960s, produced, in reaction, a public pride in racial and ethnic identity. Canada sought a new politics that could keep a restive Québec within the federation. Britain was forced to grapple with the arrival of immigrants from South Asia and the Caribbean in the wake of its collapsed empire. Australia contemplated multiculturalism as it abandoned the "White Australia" policy and ramped up its immigration intake.[3] And so on.

Contemporary multiculturalism takes several forms. At its weakest, multiculturalism may be simply rhetorical, involving public statements to the effect that a society values diversity and respects its cultural minorities. A stronger form of multiculturalism would concede its minorities "special" or "group" rights, claimable only by members of those particular groups rather than by all citizens. Such special rights may extend certain legal exemptions to the group: for example, permitting Sikh motorcyclists to wear turbans rather than helmets. Special rights might also grant educational or other welfare opportunities to members of a group, as in the case of newly arrived immigrants who seek help with integrating into their host society. Alternatively, the goal may not be integration but the maintenance of a distinctive culture. In this connection many indigenous peoples claim special entitlements to use land and other natural resources for traditional purposes. In their strongest form, again typically invoked in the case of indigenous groups, multicultural policies may promote a group's self-determination and self-governance, perhaps even involving the institution of a parallel legal system.

3. Justifications

Multicultural policies are often highly controversial. By according some citizens entitlements that others do not have, multiculturalists may seem to depart from standard liberal-democratic notions of equality. Moreover, some minority groups subscribe to values and practices that violate widely held notions of human rights: for example, restrictions on religious and other freedoms, female genital mutilation and even "honour killing." Can such conduct be accommodated by even the most open-minded modern societies?

One way of justifying multiculturalist policies might be to appeal to cultural relativism. According to Paul Scheffer, "multicultural thinking represents a continuation of cultural relativism by other means" (2011, 197). Cultural relativism is the view that there are no universal moral standards, only the particular moral codes of particular cultures.[4] On this view, every moral judgment is necessarily made from within some such code, and there is consequently no neutral ground from which to judge that one code is superior to another. Consequently, all cultures are morally equal and all deserve equal respect. In the multiculturalism literature, writers emanating from anthropology and cultural studies tend to come closest to this line of argument (e.g. Hage 1998; Shweder 2013).

Although cultural relativism may sound humane and enlightened, it is in fact a dubious basis for either toleration or multiculturalism. If moral justification depends entirely on cultural perspective, then those cultures that reject toleration or respect for others cannot be criticized. At a deeper level, cultural relativism assumes that all cultures are valuable holistically and intrinsically. But no culture is without its faults, and many of those faults – historically, many cultures have been characterized by sexism, racism, imperialism, arbitrary class divisions – militate against the respect for diversity that is at the heart of multiculturalism.

More compelling defenses of multiculturalism have been mounted by liberal thinkers committed to universal notions of human rights and personal liberty. Indeed, while multiculturalism is now practiced in some non-liberal regimes (such as Singapore), multiculturalism as a public philosophy and policy arose in liberal democracies and from a basis in liberal-democratic principles.

Robert Goodin usefully distinguishes between two liberal accounts of multiculturalism, "polyglot" and "protective." Both see multiculturalism as justified on the basis of individual liberty, but they adopt different attitudes to the relation between liberty and culture. Polyglot multiculturalism aims to expand liberty by giving individuals more cultural options. On this account, as Goodin puts it, "the really great virtue of multiculturalism is that it provides a broad smorgasbord of mix-and-match options from which to choose" (2006, 295). In this version of multiculturalism, people "'borrow from' without fully 'living in' the other cultures around them"

(2006, 290). The result is that they live in an overarching culture which values diversity. Goodin presents polyglot multiculturalism as mainly benefiting members of the majority culture, although those from cultural minorities can also access and enjoy the wider range of cultural choices on offer.

The polyglot vision connects with the strong sense of multiculturalism in that it affirms rather than merely acknowledges diversity. In so doing, it links with theories of liberalism and multiculturalism that place particular emphasis on diversity and cultural and value pluralism (e.g. Spinner-Halev 2000; Galston 2002; Crowder 2002). However, it is also true that Goodin's polyglot multiculturalism, with its picking and choosing from a variety of cultures without necessarily being immersed in any of them, closely resembles cosmopolitanism (Waldron 1995).

The alternative liberal account identified by Goodin is "protective." In this case the goal is not to expand choice but to protect existing choices by preserving the cultural contexts within which they are currently made. Emphasizing the rights to survival of minority cultures, it amounts to "a defensive manoeuver on the part of beleaguered groups" rather than the promotion of diversity as such (2006, 295). The best-known theorist of multiculturalism, Will Kymlicka, presents a protective argument in this sense, grounded in the value of individual autonomy.

4. Autonomy and cultural rights

Kymlicka's theory of minority cultural rights has a similar shape to the egalitarian-liberal arguments of John Rawls and Ronald Dworkin in favour of economic redistribution (Rawls 1971; Dworkin 2000). Just as Rawls and Dworkin argue that people suffering from unfair economic disadvantage are entitled to redress, so Kymlicka argues that people suffering from unfair cultural disadvantage are entitled to forms of redress appropriate to their situation (Kymlicka 1989, 1995a, 2001, 2002, 2007).

For Kymlicka, membership of a flourishing culture possesses a fundamental value akin to that of income and wealth in the theories of Rawls and Dworkin. It provides the necessary context within which people can make sense of their life choices and realize their individual autonomy. The culture in which people are brought up is especially important in this way, since cultures are not like jobs that can be changed at will. Consequently, we all have good reason to preserve the culture in which we are raised.

When a cultural group finds itself in a minority, its members immediately find themselves at a disadvantage in relation to the majority – for example, in matters of language, education, professional qualifications, religious recognition and so on. In the absence of state action, that disadvantage is likely to continue. It follows that liberal states have a duty to compensate the members of minority cultures in appropriate ways for their disadvantage.

What kind of compensation will be appropriate depends crucially on what kind of minority group we are talking about. In Kymlicka's view, immigrant or "ethnic" minorities have chosen to move to a new society and to leave their institutions behind. They are entitled to "polyethnic" rights that enable them to "integrate" into the new society – that is, fit into it in their own unique way. "National" minorities, however, are those – notably including indigenous minorities in settler societies – which have been absorbed into a new society against their will and without ever agreeing to surrender the institutionalization of their culture. For national minorities the proper remedy is some level of self-determination.

However, Kymlicka argues that there is a limit to the extent to which cultural minorities, whether ethnic or national, should be protected, since there can be no protection for those practices that undermine individual liberty. This is the value on which the whole argument is based. Indeed, Kymlicka sees his argument as justifying the liberalization of non-liberal cultures, although he argues that this need not be coercive or intrusive because there is room for prudence and restraint in the means by which such a policy is pursued.

Kymlicka's view has been extremely influential, but it has also drawn a good deal of criticism. Some critics argue that his view does not do enough to accommodate non-liberal minorities. According to the libertarian account of Chandran Kukathas, Kymlicka's cultural rights (and their limits) require undesirably intrusive judgements by a large state (Kukathas 2003). Such judgements are futile because cultures are fluid and cannot be preserved in their current form, and unnecessary because adequate protection for legitimate cultural identifications is already provided in liberal democracies by the right of free association. As long as individuals are free to enter and leave cultural groups, that is all the protection that either the group or its individual members need. Kukathas's ideal is of a "liberal archipelago" of freely adopted cultural affiliations in which the state itself is just one institution among others.

On the other hand, some commentators argue that Kymlicka concedes too much to cultural traditions. Brian Barry, for example, sees much of the literature of multiculturalism as a damaging distraction from more important dimensions of social justice (Barry 2001). For Barry, special rights are a betrayal of the Enlightenment ideal of equal treatment. They fail to address the real causes of social disadvantage, which are usually economic, not cultural. The goal of progressive politics should be a strengthening of standard egalitarian-liberal rights that apply to all citizens equally, regardless of cultural identification.

Another criticism comes from the feminist theorist, Susan Okin (1999). Okin points out that Kymlicka's special rights are often invoked to protect traditional cultures that are strongly patriarchal in character. Such cultures place traditional restrictions on the roles that may be undertaken by women compared with those of men, amounting in effect to systems for controlling the lives of women. Okin acknowledges Kymlicka's limitation of cultural rights in the name of individual freedom of choice. But she sees this as applying more readily to the public realms of citizenship and work rather than to the private realm of the home and family. For Okin, the family is not only a site of much patriarchal oppression within its bounds but also a school that imparts oppressive attitudes towards women in wider society (Okin 1989). Her answer to the question "Is multiculturalism bad for women?" is a resounding "Yes!"

5. Authenticity, identity and recognition

As noted, some multiculturalists find Kymlicka's argument for multiculturalism overly restrictive. Feminist concerns notwithstanding, they seek greater accommodation of cultural identity and difference and thus can appear as cultural relativists even though they are not. Perhaps the most prominent liberal argument in this vein is the "recognition" theory of Charles Taylor (1994). Taylor contrasts a "politics of universalism" and a "politics of difference." The former emphasizes the equal dignity of citizens, is historically underscored by the modern value of individual autonomy, and leads to the "difference-blind" liberal state that respects equal rights based on common citizenship. In contrast, the "politics of difference," while growing out of the politics of universal dignity, is motivated by the modern value of authenticity and the uniqueness of individuals' and groups' identities. Here, recognition of one's identity is sought simply because it is one's own. In this context, Taylor argues, not recognising or misrecognising people's identity can do them real social and psychological harm. Hence, the shift to the multiculturalist state.

Thus far, Taylor's argument provides a powerful case for politically recognizing individuals' and groups' cultural identity based on the avoidance of a harm. Cultural relativism, as such, does not enter the picture. However, Taylor goes on to make two further arguments which sometimes have been taken to suggest cultural relativism.

First, whereas liberal multiculturalists generally assign instrumental value to cultures (witness Kymlicka's argument that cultures provide a context for individual choice), Taylor argues that cultures are intrinsically valuable. As they are the locus of the goods valued by individuals, they are not reducible to the value that individuals may singly place on the culture. Taylor's principal example is the survival of Francophone culture in his native Québec, whose importance, he argues, exceeds being available simply for the sake of those who might choose it (1994, 58). Second, Taylor considers the idea that "we owe equal respect to all cultures" and should "place all cultures more or less on the same footing," and the "presumption" that humanity's cultures are of "equal worth" (1994, 66). This consideration occurs in relation to demands for broadening the "canon" taught in humanities departments. Taylor notes that such demands spring less from an interest in broadening the culture for everyone (as polyglot multiculturalism might have it) than from the idea that nonrecognition of particular groups and cultures is to slight and devalue them.

Although these aspects of Taylor's analysis may seem to broach cultural relativism, they do not. His argument for cultural survival and the intrinsic value of cultures is not that cultures should be preserved even in the absence of a critical mass of members or adherents; rather, it is to appreciate the importance of a community wishing to maintain its culture. Moreover, while Taylor considers the presumption of the equal worth of cultures, he finds this idea to be problematic. He argues that if the idea contains any validity at all, it is the quite different (and much weaker) presumption that every longstanding societal culture has *something* of value to teach humanity. Even here, however, he insists that this presumption is provisional and subject to evidence and experience. Taylor also places national-cultural and liberal limits on the recognition of cultures. For example, cultural minorities in Québec must accept the precedence of the established Francophone culture at the same time that Francophone Quebecers must respect the fundamental liberal rights of all Quebecers, minorities included. Such strictures are incompatible with cultural relativism.

6. An individual right to cultural identity

Another liberal argument for multiculturalism which may appear to resemble cultural relativism is that individuals have a fundamental right to their own culture grounded in identity or a "right to one's way of life" (Margalit and Halbertal 1994). Flatly stated, such a right is powerful and presents the challenge of how it might be reconciled with other liberal rights, such as the freedom of association and dissociation. The Israeli philosopher Moshe Halbertal attempts to remove some of the tension by contrasting autonomy-based and identity-based approaches to education (1996, 111–112).

On the autonomy approach, the purpose of education is to familiarize children with alternative ways of life and to equip them with the capacity to assess these. On this model, attempts by traditional groups such as the Old Order Amish to prevent their children from receiving a liberal education lest they be lost to their community are unacceptable (*Wisconsin v. Yoder* 1972; Arneson and Shapiro 1990). On the identity approach, however, education is primarily concerned with transmitting a particular tradition and inculcating adherence to it. Halbertal cites the Kibbutz movement and the ultra-Orthodox community in Israel as operating "closed" education systems of this sort. He argues – not unlike Kukathas with his "liberal archipelgo" – that as long as they do not force or penalize individuals who do opt for an alternative way of life, such an educational practice should also be acceptable.

The difficulty arises insofar as "force" and "penalize" refer only to blatant coercion and physical threats against members who *express* a wish to leave. Such intimidation certainly happens; some Ultra-Orthodox communities in Israel dispatch "'morality squads' that follow, report on, threaten, and sometimes act violently against members who may deviate from the community's strict codes" (Margalit and Halbertal 1994, 493). However, defending education simply in terms of the transmission of a tradition and adherence to it overlooks how the educational content itself may be built on and aim at the entrenchment of extreme asymmetrical power relations, especially between the sexes, which then deny some members meaningful choice and options. Ultimately, Halbertal and Avishai Margalit argue that a "cultural minority cannot be granted *control* over its members' exit" (1994, 508, emphasis added). This stipulation would seem to include subtler forms of pressure, even, perhaps, the kinds of norms taught in a school curriculum. At this point it is arguable that a genuine capacity to exit really requires individual autonomy, in which case the identity approach collapses into the autonomy approach after all (Levey 2006; Crowder 2007).

7. Non-liberal defences of multiculturalism

Unbound by liberal precepts, non-liberal defences of multiculturalism may appear especially open to the charge of cultural relativism. Again, the appearance is often misleading. Bhikhu Parekh, for example, begins from the proposition that cultural diversity is a defining feature of humanity (Parekh 2006). He objects to liberalism as a form of monism that is closed in advance to alternative ways of life. Cultural groups should be allowed to learn from and mutually enrich each other. In the case of controversial minority practices, such as female circumcision and polygamy, what is required, Parekh argues, is a sympathetic intercultural dialogue governed by a society's "operative public values" (2006, 267). Such values are found in a society's constitution, laws and civic relations and "represent the shared moral structure of a society's public life" (2006, 270). Parekh acknowledges that the operative public values may include liberal values. In Britain, for example, they include "individual liberty, equality of respect and rights, tolerance, mutual respect, a sense of fair play and the spirit of moderation" (2009, 38). However, he maintains that these values have been interpreted and applied by the British in their own way. He also maintains that these values should be subject to intercultural dialogue and not be wielded like a blunt instrument.

In this dialogue, minorities are enjoined to explain the cultural significance of the practice in question, why it is important for sustaining their culture, and why it has beneficial consequences. In turn, the wider society may appeal to the significance of its own historical and religious traditions, demonstrate the centrality of the values it seeks to uphold for its way of life, and otherwise defend these values in a way that is intelligible to the minority concerned (Parekh 1996, 2006, ch. 9). The process is valuable because, in the manner of deliberative democracy, the encounter provides an opportunity for the parties to arrive at a better understanding of each other and thus to modify their original claims. Nevertheless, Parekh is clear that in the absence of such movement, the majority remains entitled to insist on its account of the operative public values. Critics also point out that the dialogic procedure is pre-set and not itself "multiculturalized" (Ulbricht 2015, 34–37). Some see in it decided commitments to Kantian ethics (Preiss 2011).

Tariq Modood provides another non-liberal argument for multiculturalism, and arguably goes further than Parekh in his acceptance of difference (Modood 2013; Levey 2019). Like Parekh, Modood acknowledges that multiculturalism developed in liberal societies but argues it should not be defined by or understood in terms of liberal principles. For him, multiculturalism is the political expression of minorities' "identity" and "difference" and their quest for being included in their own way within the wider society and the national story. Modood entertains a "variable geometry" of relationships in how minorities, and even minorities within minorities, might legitimately interact with the state (2013, 77). He is less ready than Parekh to concede precedence to the established culture, at least not without corresponding benefits for minorities. However, Modood also does not accept that "anything goes" in the name of culture. He rules out practices that violate the "fundamental rights of individuals" or which "cause harm to others" (2013, 62).

8. Conclusion

Semantically, multiculturalism lends itself to being construed as a species of relativism. Historically, however, multiculturalism developed as a public philosophy and policy in liberal democracies and from a quest to better honour liberal-democratic values and citizenship. Much multicultural political theory continues to follow that path. As such, it stands opposed to cultural relativism. Although non-liberal defenses of multiculturalism have subsequently been advanced, and strive to be more open to cultural diversity and difference, the most prominent fall short of endorsing cultural relativism.

Multiculturalism is most often associated with cultural relativism by its conservative critics. While cultural diversity produces genuine controversies and clashes between values, the identification of multiculturalism with cultural relativism is polemical and hyperbolic. This often reflects the fact that multiculturalism asks established majorities to relinquish some of their power and privilege and the established culture to make room for others, albeit within legal, moral and national-cultural limits.

Nevertheless, the past two decades have witnessed an international backlash against multiculturalism (Vertovec and Wessendorf 2010). This has been more pronounced in Europe but has also had echoes in North America and Australia. For the most part, the backlash has been occasioned by the rise of militant Islamism, terrorist attacks and concerns over the social and cultural integration of new waves of immigrants. Although multiculturalism – the word and the philosophy – has been accused of promoting each of these developments, there is little hard evidence of it (Levey 2011; Kymlicka 2012). In Europe, in particular, claims have been made that multiculturalism locks people into their ethnic boxes, is ill-equipped to accommodate the dynamic interaction between identity groups, and, therefore, a new model of "interculturalism" is required (Council of Europe Committee of Ministers 2008; Cantle 2012). These claims have been vigorously contested and debated (Meer et al. 2012; Levrau and Loobuyck 2018).

Notes

1 Concise treatments of multiculturalism in general can be found in Rattansi (2011) and Modood (2013). The political theory literature is examined by Kymlicka (2002, ch. 8), Festenstein (2005), Murphy (2012), Crowder (2013). Significant collections of essays include Horton (1993), Kymlicka (1995b), Joppke and Lukes (1999), Laden and Owen (2007), Balint and Guérard de Latour (2013).
2 See, for example, Bloom (1988) and Murray (2006). For a critical review of charges of relativism in the Canadian context, see Ryan (2010).
3 For comparative discussions of multiculturalism in different countries, see Kivisto (2002) and Fleras (2009).
4 On cultural relativism in ethics see Ladd (1973), Hollis and Lukes (1982), Krausz (1989) and Lukes (2008).

References

Arneson, R. and I. Shapiro (1990), "Religious Liberty and Democratic Autonomy: A Critique of *Wisconsin v. Yoder*," in *Nomos* 38: *Political Order*, edited by I. Schapiro and W. Kymlicka, New York: New York University Press, 365–411.

Balint, P. and S. de Latour (eds.) (2013), *Liberal Multiculturalism and the Fair Terms of Integration*, Basingstoke: Palgrave Macmillan.

Barry, B. (2001), *Culture and Equality*, Cambridge: Polity Press.

Bloom, A. (1988), *The Closing of the American Mind*, New York: Simon and Schuster.

Cantle, T. (2012), *Interculturalism: The New Era of Cohesion and Diversity*, Houndmills, Basingstoke: Palgrave Macmillan.

Council of Europe Committee of Ministers (2008), *Living Together as Equals in Dignity: White Paper on Intercultural Dialogue*, Strasbourg: CECM.

Crowder, G. (2002), *Liberalism and Value Pluralism*, London and New York: Continuum.

———. (2007), "Two Concepts of Liberal Pluralism," *Political Theory* 35: 121–146.

———. (2013), *Theories of Multiculturalism: An Introduction*, Cambridge: Polity Press.

Dworkin, R. (2000), *Sovereign Virtue: The Theory and Practice of Equality*, Cambridge, MA: Harvard University Press.

Festenstein, M. (2005), *Negotiating Diversity: Culture, Deliberation, Trust*, Cambridge: Polity Press.

Fleras, A. (2009), *The Politics of Multiculturalism: Multicultural Governance in Comparative Perspective*, New York: Palgrave Macmillan.

Galston, W. A. (2002), *Liberal Pluralism: The Implications of Value Pluralism for Political Theory and Practice*, Cambridge: Cambridge University Press.

Goodin, R. (2006), "Liberal Multiculturalism: Protective and Polyglot," *Political Theory* 34: 289–303.

Hage, G. (1998), *White Nation: Fantasies of White Supremacy in a Multicultural Society*, Sydney: Pluto Press.

Halbertal, M. (1996), "Autonomy, Toleration, and Group Rights: A Response to Will Kymlicka," in *Toleration: An Elusive Virtue*, edited by D. Heyd, Princeton, NJ: Princeton University Press, 106–113.

Hollis, M. and S. Lukes (eds.) (1982), *Rationality and Relativism*, Cambridge, MA: MIT Press.

Horton, J. (ed.) (1993), *Liberalism, Multiculturalism and Toleration*, New York: St. Martin's Press.

Joppke, C. and S. Lukes (eds.) (1999), *Multicultural Questions*, Oxford: Oxford University Press.

Kivisto, P. (2002), *Multiculturalism in a Global Society*, Oxford: Backwell.

Krausz, M. (ed.) (1989), *Relativism: Interpretation and Confrontation*, Notre Dame: University of Note Dame Press.

Kukathas, C. (2003), *The Liberal Archipelago: A Theory of Diversity and Freedom*, Oxford: Oxford University Press.

Kymlicka, W. (1989), *Liberalism, Community, and Culture*, Oxford: Oxford University Press.

———. (1995a), *Multicultural Citizenship: A Liberal Theory of Minority Rights*, Oxford: Oxford University Press.

———. (ed.) (1995b), *The Rights of Minority Cultures*, Oxford: Oxford University Press.

———. (2001), *Politics in the Vernacular*, Oxford: Oxford University Press.

———. (2002), *Contemporary Political Philosophy: An Introduction*, second edition, Oxford: Oxford University Press.

———. (2007), *Multicultural Odysseys*, Oxford: Oxford University Press.

———. (2012), *Multiculturalism: Success, Failure, and the Future*, Brussels: Migration Policy Institute.

Ladd, J. (1973), *Ethical Relativism*, Belmont: Wadsworth.

Laden, A. S. and D. Owen (eds.) (2007), *Multiculturalism and Political Theory*, Cambridge: Cambridge University Press.

Levey, G. B. (2006), "Identity and Rational Revisability," in *Identity, Self-Determination and Secession*, edited by I. Primoratz and A. Pavkovic, Aldershot: Ashgate, 43–58.

———. (2011), "Multiculturalism and Terror," in *Essays on Muslims and Multiculturalism*, edited by R. Gaita, Melbourne: Text Publishing, 19–45.

———. (2019), "The Bristol School of Multiculturalism," *Ethnicities* 19: 200–226.

Levrau, F. and P. Loobuyck (2018), "Multiculturalism – Interculturalism Symposium," *Comparative Migration Studies* 6: 1–13.

Locke, J. (1689), *A Letter Concerning Toleration*, edited by J. Horton and S. Mendus, London: Routledge, 1991.

Lukes, S. (2008), *Moral Relativism*, New York: Picador.

Margalit, A. and M. Halbertal (1994), "Liberalism and the Right to Culture," *Social Research* 61: 491–510.

Meer, N., T. Modood and R. Zappata-Barrero (eds.) (2012), *Multiculturalism and Interculturalism: Debating the Dividing Lines*, Edinburgh: Edinburgh University Press.

Modood, T. (2013), *Multiculturalism: A Civic Idea*, second edition, Cambridge: Polity Press.

Murphy, M. (2012), *Multiculturalism: A Critical Introduction*, London: Routledge.

Murray, D. (2006), *NeoConservatism: Why We Need It*, New York: Encounter Books.

Okin, S. M. (1989), *Justice, Gender, and the Family*, New York: Basic Books.

———. (1999), *Is Multiculturalism Bad for Women?* Edited by J. Cohen, M. Howard and M. C. Nussbaum, Princeton, NJ: Princeton University Press.

Parekh, B. (1996), "Minority Practices and Principles of Toleration," *International Migration Review* 30: 251–284.

———. (2006), *Rethinking Multiculturalism: Cultural Diversity and Political Theory*, second edition, London: Palgrave Macmillan.

———. (2009), "Being British," *The Political Quarterly* 78: 32–40.

Preiss, J. B. (2011), "Multiculturalism and Equal Human Dignity: An Essay on Bhikhu Parekh," *Res Publica* 17: 141–156.

Rattansi, A. (2011), *Multiculturalism: A Very Short Introduction*, Oxford: Oxford University Press.

Rawls, J. (1971), *A Theory of Justice*, Oxford: Oxford University Press.

Ryan, P. (2010), *Multicultiphobia*, Toronto: University of Toronto Press.

Scheffer, P. (2011), *Immigrant Nations*, translated by L. Waters, Cambridge: Polity Press.

Shweder, R. A. (2013), "The Goose and the Gander: The Genital Wars," *Global Discourse* 3: 348–366.

Spinner-Halev, J. (2000), *Surviving Diversity: Religion and Democratic Citizenship*, Baltimore: Johns Hopkins University Press.

Taylor, C. (1994), "The Politics of Recognition," in *Multiculturalism*, edited by A. Gutmann, Princeton, NJ: Princeton University Press, 25–73.

Ulbricht, A. (2015), *Multicultural Immunisation: Liberalism and Esposito*, Edinburgh: Edinburgh University Press.

Vertovec, S. and S. Wessendorf (eds.) (2010), *The Multiculturalism Backlash: European Discourses, Policies and Practice*, London: Routledge.

Waldron, J. (1995), "Minority Cultures and the Cosmopolitan Alternative," in *The Rights of Minority Cultures*, edited by W. Kymlicka, Oxford: Oxford University Press, 93–119.

Wisconsin v. Yoder, 406 U.S. 205 (1972).

26

CRITICAL THEORY AND THE CHALLENGE OF RELATIVISM

Espen Hammer

1. Introduction

Both institutionally and in terms of its theoretical commitments and research, the Frankfurt School has been influential far beyond the academic confines of twentieth-century social and political thought. Providing ample opportunity for collaboration between philosophy and the social sciences, since its founding in the Weimar Republic its mission has been that of disclosing social pathologies and revealing potentials for emancipatory action. From a largely Marxist framework, influenced by the social conflicts and intellectual challenges posed by the turbulent post-World War I years, the school – through its two or perhaps three generations, depending on how one counts – has gradually turned more liberal as theorists such as Jürgen Habermas started to replace such luminaries as Max Horkheimer, Theodor W. Adorno, Herbert Marcuse, Erich Fromm, and Leo Löwenthal. While the theoretical and practical commitments associated with the term "critical theory" have remained relevant to a number of researchers up until and beyond the turn of the millennium, considered institutionally the Frankfurt School may well be said to have come to an end around 1970.

The aim of this chapter is to identify and discuss how the Frankfurt School responded to the question of relativism. While relativism was never the most central topic of the first generation, it features quite regularly in the works of Horkheimer and Adorno. Of course, Marxism itself, forming, as it were, the horizon of the early Frankfurt school's research, has notoriously been haunted by the question of relativism. On the one hand, Marxists tend look at knowledge in terms of its ideological and functional role in society; hence knowledge is supposed to be relative to class and class-interest. On the other hand, they reserve for their own theory (Marxism) an epistemological privilege as existing beyond, and being unaffected by, ideology. Having defined the problem of relativism as it emerges within the context of the Frankfurt School's theoretical commitments, I first briefly discuss this particular problem of Marxist thinking. My suggestion will be that, by introducing "immanent critique" as their preferred method, the early Frankfurt School remained somewhat skeptical of Marxism's universality-claim, viewing it largely as a theory in need of continuous revision as new social constellations emerge. I then discuss Horkheimer's early essay on Montaigne. As Horkheimer argues, an inclination towards relativism has historically gone hand in hand with a class-related tendency to withdrawal and passivity in the face of social uncertainty. Advocating, like Horkheimer, an anti-relativist position consistent with socially and politically informed thinking and action, Adorno puts forward an account of objectivity informed, in particular, by his view of art. In Habermas, finally, the question of relativism attains a centrality not seen among members of the first generation critical theorists. For Habermas, the alternative to universal validity is skeptical defeatism, giving rise to a number of philosophically, culturally, and politically regressive orientations. Defending universalism and objectivism – and rejecting relativism – becomes indistinguishable from that of supporting modernity's enlightenment essence.

In general, the account one finds across much of the Frankfurt School centers on "immanent transcendence" – the idea that claims to universal validity can be made, but always from within particular social and historical positions, and from the vantage-point of particular vocabularies. While no doubt a difficult, if not necessarily self-contradictory, commitment to sustain, both Horkheimer, Adorno, and Habermas offer plausible reasons for thinking that the appeal to immanent transcendence can defeat relativism and indeed provide critical theory with its desired rational potential. However, I ultimately suggest that Habermas's emphasis on rational reconstruction in the Kantian sense represents a challenge to the first-generation critical theorists' desire to conduct an effective critique of social pathology. At the end of the day, critical theory in Habermas's account no longer seems focused on conducting criticism of specific social ills and injustices.

2. Relativism and immanent critique

In its first two decades, from the early 1920s inception until well into the 1940s, the Frankfurt School, analyzing social movements and progressivist setbacks in the turbulent and ideologically charged era of fascism and communism, was largely a Marxist undertaking. While expanding and refining the classical notion of ideology, and also criticizing determinist visions of revolutionary action in the name of rational self-assertion, the school's central members – Horkheimer, in particular – never questioned the essential framework of Marxist thought.

However, the question of relativism haunts Marxism from the beginning, posing problems for the Frankfurt School. Relativism, of course, comes in many varieties and has proved notoriously hard to formulate consistently. At the very least it tends to involve an objective or universal truth-claim about the relativity of all knowledge – and as such it may seem either logically or performatively self-contradictory. One possible formulation would be the following (see Bernstein 1983, 8–15): Whatever one is able to count as *truth* is relative to the epistemic framework in which it has been put forward. Likewise, relativism about *justification* would claim that whatever one is able to count as a justification for a given proposition is relative to the epistemic framework in which the justification has been made. Typical versions of such frameworks include language, vocabularies, practices, traditions, and various facts about the subjective nature of experiential uptake. According to the standard relativist view, there can neither be a non-relative point of view from which to make claims, nor do human agents have recourse to a position from which to adjudicate objectively between claims arising in incommensurable or incompatible frameworks. As relativists deny that we can know the world as it is independently of our frameworks and perspectives (there is no "absolute" point of view), they are often skeptics about knowledge. Claims to knowledge that cannot purport to be universally true are at best well justified within their relevant frameworks. Yet being well justified in this sense cannot – and should not – satisfy realists about knowledge. Moreover, it doesn't entail objective truth.

According to a well-known statement by Marx and Engels in the 1845 *The German Ideology*,

> we will not proceed on the basis of what men say and imagine about themselves,
> nor on the basis of imagined and conceptualized men; we will begin with real, active

men, and from their real life process we will expose the development of the ideological reflections and echoes of this life process. (. . .) Consequently, morality, religion, metaphysics, and the other ideological constructs and forms of consciousness that correspond to them no longer retain the appearance of independence.

(Marx 1994, 125)

Whatever one might make of such vague terms as "life process" and "life," this is evidently an expression of relativism, and as such it sits uneasily with Marxism's theoretical universality-claims. According to Marx and Engels, ideological knowledge of the kind that serves to legitimize ruling social interest – and quite possibly all knowledge – must be considered as being relative to what they elsewhere call a society's "mode of production," the way in which it organizes, through labor and the division of labor, its perpetual transformation of raw nature into items of human value.

If agents are indeed incapable of making claims that are able to transcend the socio-economically determined context of ideology and attain a position from which to freely formulate universal claims, then, obviously, the very Marxian commitment to critique – and to formulating a universal theory of ideology itself – may seem to be in jeopardy. While Marx may not have worried much about this problem, it emerges as quite central in much subsequent Marxist thinking. Among the researchers of the Frankfurt School, for whom the rise of fascism came to suggest the extraordinary power of ideology, it is elevated to the center stage of theoretical concern.

In first-generation Frankfurt School Critical Theory, the most widespread approach to the problem of relativism (and ideology in particular) takes the form of conducting *immanent criticism*. For the early critical theorists, our standards of reflective acceptability and the social and cultural ideals in terms of which one criticizes societies are, as Raymond Geuss (1981, 63) puts it, "just part of our tradition and have no absolute foundation or transcendental warrant." In conducting social critique, the theorist addresses, Geuss (*ibid*.) continues, "the members of *this* particular group in the sense that it describes *their* epistemic principles and *their* ideal of the 'good life' and demonstrates that some belief they hold is reflectively unacceptable for agents who hold their epistemic principles and a source of frustration for agents who are trying to realize this particular kind of 'good life.'" In other words, given their ideal of a rational, satisfying existence (an ideal that itself may call for theoretically informed acts of self-reflection), what kinds of beliefs would these agents need to endorse in order to stand a reasonable chance of attaining it? Ultimately, immanent criticism is undertaken in the service of *self-reflection*: it is supposed to help agents clarify the beliefs it is rational for them, given their goals, to espouse.

An example of such immanent critique, pertinent to the problem of ideology, might focus on the concept of freedom. An agent may have a conception of freedom involving some sort of rationally motivated self-actualization, free of domination. However, when scrutinizing that agent's actual beliefs about what it is that would count as freedom within the framework of his own actual social condition, one often finds tensions and contradictions. The agent may think that being fully free according to the stipulations of what counts as freedom is to act rationally on the job-market: it is to successfully manage the constraints of the flexible capitalist economy. However, as Adorno points out, such an individual ought to realize that such behavior requires a great deal of conformity to externally imposed standards and expectations. Ultimately, the person who appears to be most free – the "entrepreneur," etc. – may in fact be rather unfree. By following this set of reflections to their conclusion, we encounter, Adorno (1966, 152) claims, "a contradiction . . . between the definition which an individual knows as his own and his 'role,' the definition forced upon him by society when he would make his living."

Immanent critique may lead to greater awareness of one's actual commitments and practices, and in some fortunate cases motivate their successful reform. However, it may also bring agents to reflect upon their own fundamental standards: What is it that *really* constitutes freedom? Perhaps one's current understanding of what counts as freedom needs rethinking. Perhaps it needs a more socially inflected definition. While following a trajectory set by the actual position and commitments of specific agents, questions arising from immanent critique are capable of revealing new vistas for critical self-understanding, disclosing unexpected paths of liberation and emancipation.

Immanent critique does not purport to provide absolute, universal truth, independently of any framework. It is not a "science" in the way in which Marxists (see for example Althusser 1965, 63) occasionally claim for Marxism the status of being a science. However, it may rationally claim to furnish agents – individually or collectively – with a procedure whereby their everyday entanglement in ideology can be overcome and, ideally, replaced by insight of a more objective nature. Typically, contemporary proponents of immanent critique, such as Rahel Jaeggi (2018), see rational activity as a historically evolving process, incrementally advancing from one position to the next.

In several essays from the 1930s, Horkheimer addresses the question of relativism and its relation to immanent critique. One essay (Horkheimer 1968, 201–259), in particular, stands out as representative of this trend: the extensive 1938 "Montaigne and the Function of Skepticism," published in the seventh volume of the Institute's *Journal of Social Research*. Locating Michel de Montaigne in the historical context of the turbulent sixteenth century, Horkheimer argues that his well-known skepticism and relativism must be viewed as more than simply an epistemological stance. The aim, rather, is practical. In effect, Montaigne sets aside his positive political views in order to avoid feeling forced to engage with his contemporaries and their social problems. Surrounded by his books, Montaigne entertains the idea that values and standards of truth and justification are always internal to given cultural settings and practices. While the social critic may report his or other people's beliefs and evaluations to his fellow men, he cannot claim any universal or objective validity. Thus, the critic becomes more akin to an observer or anthropologist, someone who is able to articulate commitments while still being fully aware of their limited acceptability. In his famous essay on cannibalism, Montaigne (1958, 152) writes that "each man calls barbarism whatever is not his own practice; for indeed it seems we have no other test of truth and reason than the example and pattern of the opinions and customs of the country we live in." Since this is evidently not sufficient to escape skepticism – and since Montaigne (1958, 782) also espouses a fair degree of skepticism with regard to both the senses and reason as sources of objective knowledge – the conclusion follows that one ought to lead a "gliding, obscure, and quiet life," avoiding any stance that would violate status quo. One should simply blend in as much possible and "go with the flow."

According to Horkheimer, whereas the ancient skeptics (Pyrrho and to some extent Epicurus) appealed to skepticism in order to create conditions for ἀταραξία, the balance of the unaffected, unperturbed soul, the individualistic property owner of early bourgeois and capitalist society, viewing the world solely from the vantage point of his material interest and dismissing any context-transcending validity claims, desires only the modest pleasures of private life. As such, skeptical relativism is "reactionary." It resigns itself to the circumstances. More damaging, however, in the twentieth century, as huge swaths of the German population replaced political reasoning with nationalist mythology and fidelity to the *Führer*, a related sort of withdrawal made possible the cynicism with which the rise and consolidation of fascism became possible. Hitler's supporters didn't ask whether his policies were morally right or wrong. The only relevant question was whether they expressed the will of the nation and the "Aryan race."

The only intellectually satisfying way to avoid this self-defeating stance, Horkheimer avers, consists in exploring, as the first-generation Frankfurt School critical theorists would do, the possibilities of conducting immanent criticism. While fascism must be met with political resistance, it can be criticized from within, appealing to internal incoherency.

Horkheimer's associate, Adorno, with whom he co-wrote the influential *Dialectic of Enlightenment*, also viewed relativism as form of resignation. The bourgeois subject, Adorno claims in his 1966 *Negative Dialectics*, rightfully points out that the intellect, by being conditioned, is limited. Yet it wrongly concludes that all that is left for the individual in terms of rationally responding to the world is to adopt a reactive attitude: "The attitude behind that thesis (of relativism – E. H.) is one of disdaining the mind and respecting the predominance of material conditions, considered the only thing that counts" (Adorno 1966, 36). Adorno further claims that asserting the relativity of all cognition presupposes a point of view outside the contextual circumstances in which claims to knowledge are actually being advanced. When comparing, as one would then would try to do, those claims to an absolute or "abstract" standard, they would fall short. However, since such a transcendent point of view is not available to human beings, it follows that relativism becomes impossible to formulate coherently. Moreover and just as importantly, when thought engages seriously with some definite issue, its veneer of "subjective accidentality will dissolve" (Adorno 1966, 36). While conditioned, human reflection can be true or false – transcending, as it reflects intellectually and philosophically, at least the immediate the context from it emerges and attaining genuine objectivity. Ultimately, the view of social life as being composed of "divergent perspectives," allowing no unifying, binding vision to be articulated, is typical, Adorno maintains, of agents under capitalism where ownership of the means of production is exclusively in private hands: to each his own. To conduct social criticism is to be self-reflective, aiming to break with complacency and appeals to the merely given.

3. Relativism and the question of objectivity

The program of immanent critique, in which standards or norms already accepted by members of the social system under scrutiny are used to examine the coherence of their position, has inspired much of the most powerful Marxist social critique, including that of the early Frankfurt School. Such an endeavor, however, finds itself faced with a serious challenge. What if the standards or norms themselves are rationally unacceptable? Indeed, what if those standards or norms form a system of commitments that systematically and without any recourse to epistemic transparency serve to justify bad or evil practices?

Starting from the admittedly controversial assumption of society as being indeed a totally integrated system of bad or evil practice, in several of his writings Adorno (1967, 29) calls for a form of "transcendent critique" with which to complement immanent critique. For Adorno, such transcendent critique would appeal to standards and values *that exceed* the socially given horizon of belief and self-understanding. Certain modernist art-practices, for example, would for Adorno (1970) instantiate modes of sense-making that are foreign to our everyday modalities of relating to agents and objects. Ultimately, such art-practices are viewed as place-holders for a radically different mode of rationality, in particular as it relates to the subject-object relation.

While Adorno's call to supply critical theory with an account of transcendent criticism may seem justified, it rests on at least one problematic assumption. It is far from obvious that art itself, and especially the arcane modernist art-practices that Adorno highlights, can inform the procedures one would associate with the pursuit of theoretical and practical validity. As Adorno admits, art presents itself as illusion (*Schein*); rather than informing us about facts, its central task is to provide the senses with an organized appearance the apprehension of which gives rise to aesthetic experience. Moreover, even if the engagement with appearance in this sense may, as Adorno suggests, provide opportunities for more cognitively binding encounters, the sense-making that Adorno has in mind when referring to art – "mimetic," predominantly non-conceptual awareness of the "non-identical" – may seem to border on the ineffable or even downright mystical.

This, at least, is the claim that Adorno's assistant and later central figure of second-generation critical theory, Jürgen Habermas (1984, 382–383) makes in his 1981 *Theory of Communicative Action*, a work that for the first time made the threat of skepticism and relativism a central concern for the Frankfurt School.

Habermas, however, had already paved the way for an alternative grounding of social critique in the 1972 "Postscript" to his earlier work *Knowledge and Human Interests*. Here, he (Habermas 1987, 377) advocates for a distinction – not properly attended to earlier in his career when he was still seeing himself as continuing the tradition of immanent critique – between immanent critique, or what he calls "self-reflection," and rational reconstruction (along broadly Kantian lines) of purportedly universal conditions of validity.

In subsequent writings, the project of reconstruction, rather than extending the Hegelian tradition of conducting immanent critique, focuses on abilities that make possible the formation of rational powers capable of making claims to universal validity. While Adorno had responded to the threat of relativism by reconstructing an alternative notion of reason altogether (displayed primarily in art), Habermas takes inspiration from the Kantian search for transcendental conditions of knowledge. If human reason in both its everyday and specialized use can be shown to be capable of transcending its immanent conditions, then not only would Habermas be able to present his own version of the older Frankfurt notion of "transcendence from within," but he would also provide an account of universality that potentially could withstand the skeptical and relativist concerns. For Habermas, such concerns are not only epistemologically and meta-theoretically relevant to the project of providing a critical theory of society, but are also of significance for society at large in its attempt to reject particularist doctrines and defend the value of reason.

In reconstructing such conditions, Habermas turns to language in its everyday use. Following John Austin and John Searle, he claims that the use of language is best theorized as a social achievement in which speakers mutually seek to understand each other on the basis of criticizable claims to universal validity implicit in speech. Habermas divides speech acts into three fundamental kinds. In so-called *constatives*, agents lay claim to truth; in so-called *regulatives*, they lay claim to normative rightness; and in so-called *expressives*, they lay claim to sincerity. While agents always find themselves within particular contexts, each with its own standard of rationality, the reasons they relate to in dialogue, generating a normatively binding claim to universal validity, contain the ability to challenge those standards. On this view (see Habermas 1996, 20–21), for a claim to count as universally valid, it must purport to withstand all relevant objections – in short, it must be ideally justifiable to all.

In his polemic against Richard Rorty (1991, 24), who as a radical contextualist views "truth" as no more than an expression of commendation, made within a community of interpreters who share standards of justification, Habermas (1993, 136) insists on the normativity of reason. To say that a statement is sufficiently justified, and therefore valid, is to say that it ought to be accepted as valid by an ideal community of rational investigators. Of course, statements may in fact be rebutted. The point is that for anyone to claim that they are valid, they must presuppose the concept of justifiability to all, and therefore make reference to a universal forum of investigators capable of transcending particular horizons of reasoning. Making validity-claims must therefore be distinguished from the mere reporting of "what we do and say." By reducing truth to the solidarity achieved by articulating such "we-intuitions," Rorty fails to do justice to what we mean by truth. It cannot coherently be maintained that truth is relative to a particular community's notion of what would count as justified according to its standards. Truth, thanks to its basis in rational speech, transcends context.

Habermas's strongest argument (stated in Habermas 1993, 137; see also Putnam 1987, 222–244; Price 2011, 451–470) in favor of distinguishing between warranted assertibility and truth (or validity) proper (transcendent, universal truth) may be that without the latter, the notion of improving current commitments through rational critique would make no sense: we would have no conceptual space for conceiving of such improvement. Without the normativity of truth, differences of opinion, even when locally justified (justified "for myself or my community"), would just slide past one another without the kind of consequence that mutual commitment to truth proper has, namely that disagreement calls for further argument and an eventual resolution in which one or both parties agree to revise their view.

Like Adorno, Habermas also argues that relativism is forced to appeal, often implicitly, to an absolute notion of truth. Its confinement-imagery – that knowledge claims cannot transcend contextual conditions – can only make sense when contrasted with the world as it is "in itself." However, since the relativist denies himself any recourse to the "God's eye point of view" from which such a contrast can be drawn, his position is unstable and incoherent.

Ultimately, Habermas's view offers a third way between contextualism and objectivism. He agrees with the contextualist that human agents cannot coherently be said to be able to occupy a point of view from which the world can be known as it is "in itself." Human knowledge and reasoning are conditioned by social context, natural facts, and vocabulary: there is no God's eye point of view. However, he also agrees with the objectivist that the very act of reasoning refers agents to a context-transcending notion of validity. The third way focuses on "transcendence from within" – that reasoning always moves from a real to an idealized and approximated stand-point.

4. Challenges to Habermas's view

While in many ways powerful, Habermas's account of rational context-transcendence makes extensive reference to idealizing presuppositions made by rational speakers. As they engage in dialogue, they must have a sense of what rational agreement – the touchstone of truth or universal validity – requires. According to Habermas, the requirement encompasses a stipulation regarding the freedom of each participant to make his her or reasons known as well as a stipulation regarding the equal standing of every discourse partner. Only discourses aspiring to be free of asymmetries such as power-relations and mechanisms of exclusion are able to count as genuinely truth-tracking. The presuppositions underwrite the mutual attribution of rationality among speakers if and only if Habermas can show that they are universal. However, despite the inspiration from Kant's transcendental argumentation, he has notoriously struggled to demonstrate their universality. Part of why this is the case has been his (Habermas 2001) insistence that the presuppositions of rational speech, in addition to having an a priori component, must be vulnerable to empirical considerations and evidence. It must be possible to support the demonstration of these purportedly unavoidable presuppositions with empirical propositions. However, if the presuppositions are dependent on empirical fact, then they stand in danger of losing their universal status.

Another complicating factor arises from the fact that in his most recent work, Habermas (2005, 252) has started to doubt that the epistemic approach to theoretical truth, familiar from his earlier writings, can do justice to the way in which the world itself, or facts, constrains what may count as truth. While normative rightness (since it refers to a world of norms themselves created through discourse) can be accounted for in terms of rational acceptability, claims about the world are faced with a "harder" and more non-negotiable sense of constraint. Unfortunately, the realistic turn in Habermas calls for a number of clarifications not yet offered. While Habermas would resist the notion of an absolute conception of reality, he has not yet explained how "correspondence" may constitute truth. For his new theory to escape familiar problems of skepticism and relativism, he would need to engage with wide-ranging debates over truth-bearers, facts, the notion of correspondence, and so on.

A more general challenge to Habermas's position may be that, in replacing the self-reflective project of first-generation Critical Theory with a quasi-transcendental, formal approach to validity, he seems to be missing out on the explicitly critical dimension that once used to be central to the Frankfurt School. For Habermas, social critique takes place in discourses aimed at universal validity. Typically, such discourses have as their goal to provide principles of a very general nature. While such undertakings play an important role in institution-building, they can hardly be said to be characteristic of efforts to criticize more contextually defined commitments. Moreover, in order for social critique to have a purchase it must be able to engage with identities and self-interpretations. While Habermas assigns to rational discourse the task of tackling claims to theoretical and practical validity (since they permit such objectivity), he sees so-called ethical discourse as being involved with questions of the good life, either for a given individual or for a particular group or polity. Ethical discourse, however, since it deals with issues that concern only the self-understanding of particular individuals or groups, does not lend itself to universality. The (mainly evaluative) claims made in it are, as it were, locked inside the framework of a particular cultural horizon. Habermas thus restricts the conception of immanent transcendence to theoretical and practical discourse, assigning to ethical discourse a cognitively inferior status as it cannot escape particularity.

Summing up, it can be said that, while the question of relativism has played a large role in the history of the Frankfurt School, it ultimately motivated Habermas to provide major revisions to the foundation of critical theory. Those revisions have made the theory more responsive to contextualist objections. It is not, however, evident that the critical impetus that informed the research activities of the first generation (Horkheimer, Adorno, and others) has continued to be as important in its successor. The abiding question of what it is that makes *Kritische Theorie* critical is today more difficult to answer than ever.

References

Adorno, T. W. (1966), *Negative Dialectics*, translated by E. B. Ashton, New York: Continuum, 1973.

———. (1967), *Prisms*, translated by S. and S. Weber, Cambridge, MA: MIT Press, 1997.

———. (1970), *Aesthetic Theory*, translated by R. Hullot-Kentor, Minneapolis, MN: University of Minnesota Press, 1997.

Althusser, L. (1965), *For Marx*, translated by B. Brewster, London and New York: Verso, 2005.

Bernstein, R. (1983), *Beyond Objectivism and Relativism: Science, Hermeneutics, and Praxis*, Philadelphia: The University of Pennsylvania Press.

Geuss, R. (1981), *The Idea of a Critical Theory: Habermas and the Frankfurt School*, Cambridge: Cambridge University Press.

Habermas, J. (1984), *The Theory of Communicative Action*, vol. 1, translated by T. McCarthy, Boston: Beacon Press.

———. (1987), *Knowledge and Human Interests*, translated by J. Shapiro, Cambridge: Polity Press.

———. (1993), *Postmetaphysical Thinking: Philosophical Essays*, translated by W. M. Hohengarten, Cambridge, MA: MIT Press.

————. (1996), *Between Facts and Norms: Contributions to a Discourse Theory of Law and Democracy*, translated by W. Rehg, Cambridge, MA: MIT Press.

————. (2001), *Moral Consciousness and Communicative Action*, translated by C. Lenhardt and S. Weber Nicholsen, Cambridge, MA: MIT Press.

————. (2005), *Truth and Justification*, translated by B. Fultner, Cambridge, MA: MIT Press.

Horkheimer, M. (1968), *Kritische Theorie: Eine Dokumentation*, Frankfurt am Main: Fischer Verlag.

Jaeggi, R. (2018), *Critique of Forms of Life*, translated by C. Cronin, Cambridge, MA: The Belknap Press of Harvard University Press.

Marx, K. (1994), *Early Political Writings*, translated by J. O'Malley, Cambridge: Cambridge University Press.

Montaigne, M. de (1958), *The Complete Essays of Montaigne*, translated by D. M. Frame, Stanford: Stanford University Press.

Price, H. (2011), "Truth as a Convenient Friction," in *The Pragmatism Reader: From Peirce through the Present*, edited by R. E. Talisse and S. F. Aikin, Princeton, NJ and Oxford: Princeton University Press, 451–470.

Putnam, H. (1987), "Why Reasons Can't Be Naturalized," *After Philosophy – End or Transformation?* edited by K. Baynes, J. Bohman and T. McCarthy, Cambridge, MA: MIT Press, 222–244.

Rorty, R. (1991), *Objectivity, Relativism, and Truth: Philosophical Papers*, vol. 1, Cambridge and New York: Cambridge University Press.

27

RELATIVISM IN FEMINIST POLITICAL THEORY

Charlotte Knowles

1. Introduction

As has often been observed, there is not a single, unified feminist theory. There are many waves and within and between those waves, many branches and many disagreements (Bailey 1997). But a broadly unifying concern, is a desire to increase women's options in order to give them more – and more valuable – choices.[1] Choice is often thought to be a key indicator of feminist liberation, both theoretically and politically, where it has perhaps been most effectively mobilised in the fight for reproductive rights and the rhetoric of "women's right to choose." But what happens when feminist liberation and choice come into conflict? That is, when women appear to choose things which do not further feminist liberation.[2]

The conflict between respecting women's choices and remaining committed to feminist liberation, brings the issue of relativism into focus in feminist political theory. How should we adjudicate between, for example, Andrea Dworkin's claim that pornography is "the graphic depiction of vile whores . . . central to the male sexual system," that it "does not have any other meaning" and so should be condemned (Dworkin 1981, 200); and the perspective of women who work in the sex trade and "refuse the label of 'victim' . . . offer[ing] an alternative view of feminism that emphasize[s] their right to pursue their own desires"? (Snyder-Hall 2010, 257). Feminists have often puzzled over whether there is something inherently problematic about women choosing to adopt roles or engage in practices that have traditionally been critiqued on the grounds that they reinforce female subordination; or whether in condemning such choices and ways of life we are denying women's agency,[3] blaming the victim[4] or participating in unwarranted paternalism.[5]

In an applied global context, these commitments divide between global radical feminists who "assert the universality of 'patriarchal' violence against women . . . [and] postcolonial feminism, which asserts the diversity of forms of women's oppression" (Jaggar 2005, 56). Those in the former camp are often committed to universal values of gender justice that they believe can, and should, be applied cross-culturally; whilst those in the latter tend to take a more relativist position, and argue that judgments about women's choices must be made relative to their individual, social, cultural and historical context. This chapter explores the issue of relativism as it manifests in feminist debates around choice and gendered cultural practices. I begin by offering an overview of the tension between cultural relativism and feminism, before moving on to explore the issue of relativism *within* feminist political theory, focussing on feminist critiques and defences of gendered cultural practices such as FGM (female genital mutilation) and "veiling."[6] I then move on to explore broader issues these debates raise in feminist theory around questions of agency and adaptive preference, before offering some suggestions of productive avenues for future research.

2. Is multiculturalism bad for women?

In 1999 Susan Moller Okin asked "is multiculturalism bad for women?" At the core of Okin's worry was that appeals to multiculturalism are often used to conceal, defend or overlook gender injustice. She observes that expectations for immigrants and indigenous peoples to assimilate into majority cultures are now "often considered oppressive," and notes the Left's support for "multiculturalist demands for flexibility and respect for diversity, accusing opponents of racism or cultural imperialism" (1999, 9). However, she goes on to argue that we "have been too quick to assume that feminism and multiculturalism are both good things which are easily reconciled" (1999, 10). Okin writes persuasively about the harms to women of permitting practices such as polygamy on cultural grounds. She discusses US legal cases in which perpetrators have received reduced sentences for domestic abuse and even received downgraded charges (for example from murder to manslaughter) by invoking cultural defences, which she argues are underpinned by the idea that "women . . . are ancillary to men and should bear the blame and the shame for any departure from monogamy" (1999, 19). For Okin, it is unacceptable that we excuse and overlook gender injustice on cultural grounds. Instead we must recognise that the demands of (cultural) relativism will often conflict with feminist principles, and thus to be a feminist will often mean calling into question, and even rejecting, relativist claims.

This commitment is shared by Martha Nussbaum, who argues for an international feminism that can "make normative recommendations that cross boundaries of culture, nation, religion, race and class" (Nussbaum 2000, 34). For Okin and Nussbaum it appears clear that – as feminists committed to gender justice – we have a duty to intervene and help to put an end to harmful gender practices, even when they belong to cultures which are not our own. In order to be able to do this, Nussbaum argues, we need to make central to our political practice "universal norms of human capability" that can "provide the underpinning of a set of constitutional guarantees in all nations" (2000, 35). Nussbaum argues that relativism is clearly false, and in a multicultural society has no bite (2000, 49). Far from seeing her position as paternalistic or in conflict with respecting women's agency and the choices they make, she sees respecting choice and endorsing universal values as necessarily interlinked, since she equates respecting choice with "endors[ing] explicitly at least one universal value, the value of having the opportunity to think and choose for oneself" (2000, 51). However, this raises the question of what it is to think and choose for oneself, something which proves particularly difficult to establish when the women who are subject to practices that Western feminists might deem problematic or harmful, are also the women who defend or endorse these practices.

To bring this tension into focus, let us consider the issue of clitoridectomy, or as it is often called FGM (female genital mutilation – notably an already loaded turn of phrase in this debate). Okin argues that the explicit function of this practice is to control women and their sexuality (1999, 14). However, we might wonder if this claim is complicated when we realise that "female genital cutting practices are typically controlled and organized by women" (Wade 2012), and thus it appears – at least in some sense – women's choice to engage in and continue this practice. How then should we evaluate FGM when it appears to be something harmful to women, but equally it appears to be something women choose?

Nussbaum holds that when we are considering practices or customs that appear harmful, such as FGM, we should not necessarily take women's own evaluation of their situation as the final word on the matter: "[e]ven when women appear to be satisfied with such customs, we should probe more deeply" (2000, 42). Nussbaum's argument relies on the idea that women's defence of FGM may be the result of an adaptive preference. The idea that:

> individuals in deprived circumstances are forced to develop preferences that reflect their restricted options. A woman's perception of herself and her world may be so skewed by her circumstances and cultural upbringing that she may say and believe that she genuinely prefers certain things that she would not prefer if she were aware of other possibilities.
>
> *(Narayan 2005, 34)*

The notion of adaptive preference allows us to explain why women may not only unwillingly perform or engage in practices such as FGM, but why they may defend and speak in favour of such practices, even in the absence of any obvious or identifiable oppressing agent coercing them to do so.

To say that the defence of FGM reflects an adaptive preference is to claim that the preference to support or participate in FGM is not a genuine preference of the agent in question, and so on these grounds it does not need to be respected. But this introduces a new question: how do we distinguish between those choices and preferences which are genuinely an agent's own and those that are adaptive?

The dominant way to identify an adaptive preference is to argue that it in some way signals a deficiency in autonomy. For example, one might argue that women are often socialised into understanding themselves as inferior to men, and so come to lack a sense of their own self-worth, which would be necessary in order for their decision to participate in certain forms of life to count as autonomous.[7] Similar arguments have been made on the grounds that adaptive preferences demonstrate a lack of procedural autonomy: the way in which the preference was formed was not fully autonomous, and so the preference should not be understood to be genuinely the agent's own.[8]

The approach from adaptive preference can be seen as a way to dismiss relativist claims within feminist political theory by denying that choices, practices or ways of life that conflict with the goals of feminist liberation are things that are genuinely chosen by women. But this approach has some serious consequences for how we view the agency of women in oppressive circumstances, as Uma Narayan has argued.

3. Adaptive preferences, women's agency and the dupes of patriarchy

In her oft cited 2002 essay "Minds of Their Own: Choices, Autonomy, Cultural Practices, and Other Women," Narayan argues that to label women's compliance in practices such as FGM, but also veiling and the practice of "purdah" (female seclusion), "adaptive preferences" is to deny women's agency and present women in the Global South – as this is where adaptive preferences are most commonly "diagnosed"[9] – as the "dupes of patriarchy."

Narayan distinguishes between what she takes to be the two dominant accounts of women in the Global South implicit in Western feminist analyses. Firstly, "the prisoner of patriarchy" who "has various forms of patriarchal oppression imposed on her entirely against her will and consent – similar to how a prisoner is subject to constraints on liberty" (Narayan 2002, 418). And secondly, "the dupe of patriarchy":

> Although patriarchal violence is coercively imposed on the prisoner of patriarchy, it is virtually *self-imposed* by the dupe of patriarchy because she is imagined to completely subscribe to the patriarchal norms and practices of her culture. Her attitudes are envisioned as completely shaped by the dominant patriarchal values of her cultural context.
>
> *(2002, 418–419)*

Whilst the former imagines women from the Global South to be identical with "Western" women in their values, their understanding of, and response to, particular practices – an approach which reflects a "West is best" mentality[10] – the latter conception of Southern women as the dupes of patriarchy "presents a totalising from of 'difference' on one's others" (2002, 419). Both approaches deny the agency of the women in question and "share the problem of imagining one's others as *monolithic in their responses*, failing to recognize that one's others have a *variety* of responses to the practices that shape their lives" (2002, 419).

As Serene Khader observes, "the Western imaginary characterises women with APs [adaptive preferences], not only as poor, but also as belonging to 'backward cultures'" (Khader 2012, 316 n. 8).[11] If these prejudices are already in play when one approaches relativist debates around issues such as FGM, veiling or purdah, which appear to conflict with universal values of gender justice, the adaptive preference theorist may end up doing a further injustice to women from the Global South: Denying their agency in circumstances where their agency may already be under threat. In a similar vein, we may ask why veiling is commonly viewed as an adaptive preference, while wearing make-up is not. Or why participating in FGM is seen as an action resulting from an oppressive cultural context, whilst breast enhancement is so often defended as a choice.[12]

To return to the issue of FGM, one approach to guard against the over-eager diagnosis of adaptive preferences is to re-evaluate the way in which Western feminists characterise certain "foreign" practices, beginning with the terminology used to describe them. Lisa Wade has suggested that to brand all genital cutting "mutilation" is misleading, arguing that it is an unhelpful characterisation that marks the practice as barbaric before any discourse around it can occur (Wade 2012).

Similarly, Lila Abu-Lughod comments on discovering the idea of the burqa as "portable seclusion," a (re)characterisation which allowed her to understand such "enveloping robes as 'mobile homes' . . . a liberating invention enable[ing] women to move out of segregated living spaces whilst still observing the basic moral requirements of separating and protecting women from unrelated men" (Abu-Lughod 2002, 785). To view a practice or a way of life only through "Western eyes" may overlook the specificities or nuance of that practice, as well as the significance that it has in a particular culture.[13] Wade emphasises that in cultures where female genital cutting is practiced, a cut body is seen to be more aesthetically pleasing. Women who are cut are seen to be more marriageable, whereas those who are uncut may suffer being social outcasts. Attention to the status of a gendered practice within a culture helps us to understand why women may consent to, or comply with, certain practices, even when that practice appears to harm or disadvantage them. It presents an argument against simply branding the compliance with such practices "adaptive preferences" and riding rough shod over the agency of the women in question.

Narayan argues that rather than presenting women as the dupes of patriarchy we should understand women who, in other circumstances might be diagnosed as having adaptive preferences, as "bargaining with patriarchy." This means that "the decisions many women make with respect to 'cultural practices' ought . . . to be understood as a choice of a 'bundle of elements,' some of which they want and some of which they do not, and where they lack the power to 'undo the bundle' so as to choose only those elements that they want" (Narayan 2002, 422). Narayan's analysis lets us appreciate that despite women's restricted options or their oppressive social context, they can still possess agency and weigh up the costs and benefits of engaging (or not engaging) with a particular practice, even under conditions of oppression (2002, 427).

Narayan is against coercive state intervention, drawing attention to the fact that in many cases it does not serve women's interests, or necessarily have their interests at heart – even if this is the cited reason for a particular course of action (2002, 426). Narayan's analysis emphasises that when anti-relativist appeals to feminist liberation are made, we must consider what other motivations may be in play.[14] She also stresses that heavy handed, anti-relativist policies – even if well intentioned – can often have the opposite effect from the one they intend. As Narayan observes, state outlaw of the veil in Iran and Turkey in the 1930s actually reduced the freedom of many women, who felt they could not go out without the veil, and became "dependent on male relatives for the public tasks they previously carried out themselves" (2002, 426).

From Narayan's more relativist perspective, if we are truly concerned with feminist liberation and combatting gender injustice in all its forms, analyses that rely on a notion of adaptive preference and question whether choices to comply with oppression are really choices at all, will not be the way to go. Nevertheless, from a feminist perspective, it is important to be able to critique practices which appear to reinforce gender hierarchy, since it allows us to make the case that "it does not follow from the fact that women perpetuate their oppression that their oppression is justified" (Khader 2012, 302f.). But if the cost of doing so is denying women's agency, for many this will be too high a price to pay.

4. Diagnosing adaptive preferences: occupational hazards and the context of the agent

What, then, does this mean for the issue of relativism within feminist political theory? Have we reached an impasse between contextually sensitive relativist analyses and those theorists committed to universal values of gender justice? Or is there a way of questioning oppressive gender practices and women's compliance with them, without denying women's agency?

One clear starting point is that judgements about what is or is not compatible with feminist liberation will need to be made with a deep understanding of the cultural context in which a (potentially problematic) practice is embedded. The importance of understanding the social and cultural context of "other women" is emphasised in the three "occupational hazards" Ingrid Robeyns identifies in diagnosing adaptive preferences.

Firstly, we must resist the temptation to psychologise structural constraints (Robeyns 2017, 140). Rather than simply suggesting that a particular woman or group of women are the dupes of patriarchy, we should aim for a deeper understanding of their social context and the limitations and constraints therein that may help to explain their behaviour. However, as Khader warns, and Alison Jaggar also observes, this should not mean essentialising non-Western cultures and attributing gender injustice solely to the non-Western nature of a particular culture (Khader 2017; Jaggar 2005). It is a mistake to imagine that we can draw such strict lines between "other cultures" and our own. To imagine ourselves as in some way "separate" misrepresents our history,[15] as well as the globalised world in which we live.

Robeyns' second point is that we must be careful not to misidentify possible trade-offs between various dimensions of well-being (2017, 140), a point similar to Narayan's conception of "bargaining with patriarchy." Khader emphasises a related point in her own analysis, arguing that women may comply with a practice not because they lack agency, but because it achieves some other end. For example, the woman who "depriv[es] herself [of food] only because of conditions where keeping her male relatives happy is the best way to ensure access to income, safety and so on" (Khader 2012, 310). Within this context, we should also be aware of the transitional value of particular practices. For example, in a context where female seclusion is the norm, the burqa is not a mark of oppression, but something which affords women greater freedom in the form of "portable seclusion," and as such can be read as a move toward gender justice.[16]

Thirdly, Robeyns warns, we must be cautious in diagnosing adaptive preferences when we may have just failed to recognise a form of flourishing different from our own (2017, 140). Saba Mahmood makes this point, focusing on piety as a form of self-realisation. Read through a Western lens, piety may be interpreted only in terms of submission to male dominance and thus as something incompatible with feminist liberation. Mahmood challenges this assessment on the grounds that in the context of the women's piety movement in Egyptian mosques, piety can be seen as an expression of agency and self-creation in and of itself (Mahmood 2001, 211). Mahmood's point is not simply that we should recharacterise certain practices so that they can be seen as compatible with Western narratives of resistance, liberation and agency. Rather, she brings into question the binary of passive submission and active resistance, and the idea that resistance signals agency, while submission is code for passive conformity (2001, 212).

5. An additional consideration: complicity and the active nature of submission

Although such contextual considerations are key when diagnosing adaptive preferences – and in the best analyses are heeded[17] – they do not ultimately help to overcome the potential impasse between those sympathetic to relativist critique in feminist political contexts, and those committed to anti-relativist universal values of gender justice. These considerations alone do not provide grounds on which to resolve debates over whether something counts, or does not count, as a form of flourishing or as an expression of agency. Indeed, there seems no easy way to resolve these issues so as to please all parties. Accordingly, rather than focussing on agency and "genuine choice" as the decisive factor in the assessment of a particular practice or way of life, I suggest debates over relativism and cultural practices in feminist political theory may be better served by acknowledging that even if something is genuinely chosen, this does not insulate that choice from critique.

Although this may be an idea alien to the liberal tradition, in the fields of phenomenology and existentialism it is a familiar thought. Martin Heidegger argues that the human way of being is characterised in part by a tendency to flee from our own freedom (1927, 229). Jean-Paul Sartre offers an analysis of bad-faith: the way in which we commit ourselves to practices and ways of life as if it were not our free choice to do so (1943, 83). Simone de Beauvoir observes that women often comply with their own oppression in such a way that it will seem to them to be the expression of their freedom (1949, 684). Taking these insights on board enables us to develop a notion of what I shall call "complicity": the idea that people often actively make choices that alienate them from their own freedom, reduce their options, or maintain a commitment to an existing way of life simply because it is familiar.[18] Moreover, this concept can open up new ways of thinking about the tension between choice and liberation in feminist contexts.

The adaptive preference theorist explains adaptive preferences in terms of a lack of autonomy and genuine choice. The relativist defends the same choice by arguing that it *is* an expression of genuine agency. An impasse is reached. Introducing the notion of complicity shifts the debate, establishing a middle ground between passive submission and active resistance. In so doing it disrupts the binary logic currently dominating the debate, but without leading to the conclusion that active submission is simply a different form of flourishing.[19] The notion of complicity incorporates Narayan's insight that women can exercise agency even under conditions of oppression, but without having to concede that women's participation in their own oppression justifies that oppression. Thinking about complicity involves acknowledging that we often actively limit our own possibilities in a problematic way, and so encourages critique without denying agency.[20] Moreover, such an approach highlights that we should widen our purview with regard to the practices and ways of life that might be worthy of critique, and so does not simply mean endorsing "Western" conceptions of flourishing at the expense of those in other cultures.

Taking seriously the notion of complicity as a common human tendency, means that we should not only concentrate on those ways of life which are thought to be passively submitted to. As we have seen, in practice this has meant focussing on women in the Global South: Those who are stereotyped as "ignorant, poor, uneducated, tradition-bound, domestic, family-oriented, victimized"; in contrast to Western women who are presented as "educated, modern, as having control over their own bodies and sexualities, and the freedom to make their own decisions" (Mohanty 2003, 22). Thinking about choice and liberation through the lens of complicity, will mean recognising the way in which we may all often alienate ourselves from our own freedom, or limit our options, rather than projecting such tendencies only on to women in the Global South or those in extremely oppressive situations. Moreover, even if such a passive-Third-World/agential-Western binary did hold, it would not exclude the possibility that Western women can, and do, participate in their own oppression, or that we should leave such participation unchallenged out of respect for "genuine choice."[21]

Analysing complicity will mean understanding the ways in which we create and limit ourselves in and through our practices and the ways of life we adopt,[22] and not only focussing on an external analysis of the practices we engage in or the context in which we exist. Analysing complicity will not involve speaking for others, but nor will it mean positing an "authentic Other" who can speak more truly about their own oppression and subordination, and who is imagined to judge more transparently about themselves and their situation than others can.[23] Analysing complicity will mean widening the field of critique and bringing the tensions between choice and liberation, between universal principles and relativist analysis, closer to home and usefully complicating the way in which we think about issues which may seem far simpler when we view them as primarily affecting women "over there."

Notes

1 The importance of choice is emphasised in seminal second wave texts (Beauvoir 1949) and is central to third-wave feminism (Snyder-Hall 2010).
2 I mean feminist liberation in a broad sense, as signalling a commitment to "bringing about an end to gender hierarchy" (Levey 2005, 127).
3 See Narayan (2002).
4 See Superson (1993).
5 See Snyder-Hall (2010).
6 Since the late 1990s questions around women's "problematic choices" have primarily been conducted with reference to women in the Global South. On this point see Khader (2012) and Jaggar (2005). However, prior to this there was a great deal of work on women's choices in secular and Euro-American contexts. See for example Dworkin (1983); the debate between Sommers (1990) and Freidman (1990); Wendell (1990); Superson (1993); Cudd (1994). I focus here on "foreign" cultural practices because these dominate the contemporary debate.
7 On this point see Okin (1998) and Superson (2005).

8 See Khader (2009) for an overview of various positions within this camp. For Khader's own account of adaptive preferences, which falls into neither camp, see Khader (2011).

9 Jaggar (2005) and Khader (2012) also make this point.

10 For more on this point, see Jaggar (2005).

11 On this point, see also Mohanty (2003).

12 Narayn makes a similar point (2002, 426); see also Clare Chambers (2008).

13 Or indeed the way in which it functions as a transitional practice, as I discuss further in the next section.

14 On this point, see also Jaggar (2005, 62); Okin (1999) and Abu-Lughod (2002).

15 Martha Nussbaum highlights the ways in which British colonial rule in India often served to decrease women's power and promote gender injustice (2000, 47).

16 Although this does not mean equating secularism with liberation. For more on this point, see Khader (2017).

17 Here one must applaud Okin's engagement with empirical data and Nussbaum's work with women in India, as Khader also notes (2012).

18 The explanation for these actions in the phenomenological-existential tradition is that it allows us to flee from the self-responsibility we must assume if we face up to our own freedom. This tradition does not assume that agents naturally gravitate towards freedom, as does the liberal tradition.

19 As Mahmood seems to suggest (2001).

20 It does, however, foreground other questions about blame and responsibility which will need to be addressed. On this point see Superson (1993); Cudd (1994, 2006); Hay (2005).

21 Such analysis is not designed to blame those in oppressive circumstances for their behaviour, but to draw attention to those in less oppressive circumstance who nevertheless act to reduce their options or uphold oppressive practices, despite having the means to resist. An analysis of the complicity of women in privileged, Euro-American contexts seems particularly urgent in the current political climate with the rise of movements like "The Honey Badgers" (the women's wing of the men's rights movement) and the vocal backlash from women defending their own objectification and downplaying the harms of sexual harassment. On this point, see Knowles (2019) and Hay (2005).

22 See Beauvoir (1949) and Knowles (2017, 2019).

23 Spivak also warns against positing an authentic Other (1993, 90).

References

Abu-Lughod, L. (2002), "Do Muslim Women Really Need Saving? Anthropological Reflections on Cultural Relativism and Its Others," *American Anthropologist* 104(3): 783–790.

Bailey, C. (1997), "Making Waves and Drawing Lines: The Politics of Defining the Vicissitudes of Feminism," *Hypatia* 12(3): 17–28.

Beauvoir, S. D. (1949), *The Second Sex*, translated by C. Borde and S. Malovany-Chavallier, New York: Vintage, 2011.

Chambers, C. (2008), *Sex, Culture and Justice: The Limits of Choice*. University Park, Pennsylvania: Pennsylvania University Press.

Cudd, A. (1994), "Oppression by Choice," *Journal of Social Philosophy* 25(1): 22–44.

———. (2006), *Analyzing Oppression*, Oxford: Oxford University Press.

Dworkin, A. (1979), *Pornography: Men Possession Women*, London: The Women's Press, 1981.

———. (1983), *Right Wing Women*, New York: Pedigree.

Freidman, M. (1990), "Does Sommers Like Women? More on Liberalism, Gender Hierarchy, and Scarlett O'Hara," *Journal of Social Philosophy* 21(2–3): 75–90.

Hay, C. (2005), "Whether to Ignore Them and Spin: Moral Obligation to Resist Sexual Harassment," *Hypatia* 20(4): 94–108.

Heidegger, M. (1927), *Being and Time*, translated by J. Macquarrie and E. Robinson, Southampton: Basil Blackwell, 1962.

Jaggar, A. (2005), "'Saving Amina': Global Justice for Women and Intercultural Dialogue," *Ethics and International Affairs* 19(3): 55–119.

Khader, S. (2009), "Adaptive Preferences and Procedural Autonomy," *Journal of Human Development and Capabilities* 10(2): 169–187.

———. (2011), *Adaptive Preferences and Women's Empowerment*, Oxford: Oxford University Press.

———. (2012), "Must Theorizing About Adaptive Preferences Deny Women's Agency?" *Journal of Applied Philosophy* 29(4): 302–317.

———. (2017), "Transnational Feminisms, Nonideal Theory, and 'Other' Women's Power," *Feminist Philosophy Quarterly* 3(1): 1–23.

Knowles, C. (2017), "*Das Man* and Everydayness: A New Interpretation," in *From Conventionalism to Social Authenticity: Heidegger's Anyone and Contemporary Social Theory*, edited by H. B. Schmid and G. Thonhauser, London: Springer.

———. (2019), "Beauvoir on Women's Complicity in Their Own Unfreedom," *Hypatia* 34 (2): 242–265.

Levey, A. (2005), "Liberalism, Adaptive Preferences, and Gender Equality," *Hypatia* 20(4): 127–143.

Mahmood, S. (2001), "Feminist Theory, Embodiment, and the Docile Agent: Some Reflections on the Egyptian Islamic Revival," *Cultural Anthropology* 16(2): 202–236.

Mohanty, C. T. (2003), *Feminism Without Borders: Decolonizing Theory, Practicing Solidarity*, Durham: Duke University Press.

Narayan, D. (2005), "Conceptual Framework and Methodological Challenges," in *Measuring Empowerment: Cross Disciplinary Perspectives*, edited by D. Narayan, Washington: The World Bank, 3–38.

Narayan, U. (2002), "Minds of Their Own: Choices, Autonomy, Cultural Practices and Other Women," in *A Mind of One's Own: Feminist Essays on Reason and Objectivity*, second edition, edited by L. Antony and C. Witt, Boulder, CO: Westview, 418–432.

Nussbaum, M. (2000), *Women and Human Development: The Capabilities Approach*, Cambridge: Cambridge University Press.

Okin, S. M. (1998), "Feminism and Multiculturalism: Some Tensions," *Ethics* 108: 661–684.

———. (1999), *Is Multiculturalism Bad for Women?* Princeton, NJ: Princeton University Press.

Robeyns, I. (2017), *Wellbeing, Freedom and Social Justice: The Capability Approach Re-examined*. Cambridge: Open Book Publishers.

Sartre, J. P. (1943), *Being and Nothingness*, translated by Hazel Barnes, London: Routledge, 2003.

Snyder-Hall, R. C. (2010), "Third-Wave Feminism and the Defense of 'Choice,'" *Perspectives on Politics* 8(1): 255–261.

Sommers, C. (1990), "Do These Feminists Like Women?" *Journal of Social Philosophy* 21(2–3): 66–74.

Spivak, G. C. (1993), "Can the Subaltern Speak?" in *Colonial Discourse and Post-Colonial Theory: A Reader*, edited by P. Williams and L. Chrisman, New York: Columbia University Press.

Superson, A. (1993), "Right-Wing Women: Causes, Choices and Blaming the Victim," *Journal of Social Philosophy* 24(3): 40–61.

———. (2005), "Deformed Desires and Informed Desire Tests," *Hypatia* 20(4): 109–126.

Wade, L. (2012), "A Balanced Look at Female Genital 'Mutilation,'" *The Society Pages*, https://thesocietypages.org/socimages/2012/12/10/a-balanced-look-at-fgm (Accessed June 12, 2018).

Wendell, S. (1990), "Oppression and Victimization: Choice and Responsibility," *Hypatia* 5(3): 15–46.

28

RELATIVISM AND RACE

E. Díaz-León

1. Introduction: race does not travel

The idea that someone's race is relative to their context, or somehow depends on features of their context, is a familiar one in the literature. In particular, many scholars of race have claimed that race does not "travel," that is, that someone's race can vary from place to place, or from time to time, even if features of that person such as visible traits and ancestry remain the same. Here are two illustrations of this idea:

> Race does not travel. Some men who are black in New Orleans now would have been octoroons there some years ago or would be white in Brazil today. Socrates had no race in ancient Athens, though he would be a white man in Minnesota.
>
> *(Root 2000, 631–632)*

> This is part of the significance of the "critical" in contemporary critical race theory: to make plausible a social ontology neither essentialist, innate, nor transhistorical, but real enough for all that.
>
> *(Mills 1998, xiv)*

Michael Root (2000) seems to suggest that some people who count as black in the context of contemporary New Orleans would count as something other than black at different places and times, such as contemporary Brazil, or New Orleans several decades ago. Likewise, Charles Mills (1998) aims to develop a social ontology of race according to which race is not trans-historical, that is, it can vary across time.

In this chapter, my aim is to explore the idea that race varies from context to context. In particular, I want to examine different formulations of this claim, and assess their plausibility.

Ron Mallon (2004) argues that it is a desideratum of a characterization of social construction about race that it allows that race does *not* travel. He puts forward a characterization of the idea that race does not travel, and then he examines several formulations of social construction about race in order to assess whether they are compatible with the idea that race does not travel. That is to say, it is a desideratum for a formulation of social constructionism about race that it entails (or at least is compatible with) the idea that someone's race varies from context to context. Mallon argues that some standard versions of social constructionism, but not all, have the consequence that properties that are socially constructed in the corresponding way do not travel, that is to say, someone can have the property at one context but lose the property if they "travel" to another context. And therefore, this is a criterion in order to advocate for a version of social constructionism over another.[1]

In this section, I will examine some of the main characterizations of social constructionism discussed by Mallon, as well as Mallon's claim that traits that are socially constructed in those ways do or do not travel. This will help us examine whether different social constructionist theories of race can or cannot do justice to the intuition that race does not travel, that is, it varies from context to context.

First of all, it is important to distinguish between two versions of the idea that race does not travel.[2] The first version is an ontological claim, according to which someone's race can change if they move to a different time or place. The second version is merely a conceptual or linguistic claim, according to which our concepts or terms about race change from context to context. This second claim seems very plausible, but this is just a claim about our concepts of race, not about the property of race itself. It could be argued that the concepts of race in the US and in Brazil, for instance, are different (e.g. it might refer to ancestry in the US whereas it might refer to superficial properties such as skin color in Brazil). David Ludwig (2018) also argues that the concept of race in contemporary Germany is probably an empty concept, that is, a concept that is not satisfied by anything, since it purports to refer to properties that do not actually exist according to contemporary biology. But he acknowledges that the concept of race in the US, for instance, might be different. It is important to distinguish this claim about concepts of race from the ontological claim that is at issue in this chapter, namely, the claim that someone's race itself changes from context to context. As we will see, this kind of ontological claim typically requires that being a member of a race requires bearing a certain relation to certain cultural or social practices, so that if the subjects travel to contexts with different cultural or social practices, the race that they instantiate might change.

Mallon introduces a version of social constructionism about races in terms of "interactive kinds," as follows:

(INTERACTIVE KIND CONSTRUCTIONISM ABOUT RACE) In order for a person to be race R, they must be at a site where the concept of R is used to divide people since being labeled by a term expressing the concept R is causally necessary to becoming an R.

(2004, 659)

The main idea of (INTERACTIVE KIND CONSTRUCTIONISM ABOUT RACE) is that a property P is so constructed just in case the following condition holds: when it is necessary for an individual to instantiate property P that the individual is labeled by a term referring to P, where that label has to have certain causal effects on the individual. For example, we can argue that in order to become a refugee, one needs to be labeled as a refugee, and this label needs to have some causal effects on the individual. This is what being a refugee amounts to. Likewise, it could be argued that being black is a matter of being labeled "black" and suffering discrimination because of that label (as Mills 1998 argues).

It seems clear that if (INTERACTIVE KIND CONSTRUCTIONISM ABOUT RACE) is true, then race *does not* travel (to other contexts where the required label is missing). That is, someone can be race R in one context but not another, such as follows: imagine subject S is at a site where the concept of race is used to divide people, and S is labeled as belonging to race

R, and in addition this label has causal effects since the subject is discriminated because of being perceived as falling under R. What happens if subject S travels to a context where label R is not employed? Then, according to (INTERACTIVE KIND CONSTRUCTIONISM ABOUT RACE), S would no longer fall under R in that context.

However, there are two possible readings of (INTERACTIVE KIND CONSTRUCTIONISM ABOUT RACE). According to the first, the view requires that subject S grows up at the site where the concept of R is used to divide people and is labeled as R (in order to be R). But if this person travels to another place where the concept of R is not employed, S would still fall under R (since they grew up at a site where the concept of R was used to divide people, and this fact about their growing up does not change wherever they go). Nonetheless, according to this reading, races could still vary from context to context in another way, namely, there could be other times in history where an individual who was phenotypically identical to S would not count as R, if they did not grow up in a site where the concept of R was used to divide people.

According to an alternative, stronger reading, (INTERACTIVE KIND CONSTRUCTIONISM ABOUT RACE) requires that subject S is at a site where R is used to divide people at time *t*, in order for S to have race R at time *t*. Therefore, if S travels to a different place with different concepts of race, S's race will change. This version of (INTERACTIVE KIND CONSTRUCTIONISM ABOUT RACE) seems too strong, and hence less plausible. And as we have seen, we do not need this stronger reading in order to entail the view that races change from context to context.

2. Passing

Mallon argues that characterizations of social constructionism about race (or about other social kinds) should also be able to explain why some members of a racial group could "pass" as members of a different group, that is, some members of R are perceived and treated as members of another race. Some common examples include the following: black people passing as white; white people passing as black; gay people passing as straight.

Mallon argues that (INTERACTIVE KIND CONSTRUCTIONISM ABOUT RACE) cannot explain the possibility of passing, since it makes it impossible that someone is race R but is perceived and treated as a member of a different race. For this reason, it is not a very plausible version of social constructionism, even if it allows the possibility of not travelling, since allowing the possibility of passing at the same time is a very important criterion.

That is to say, according to (INTERACTIVE KIND CONSTRUCTIONISM ABOUT RACE), if subject S is not labeled as R and treated as a member of R, then S cannot be a member of R. Therefore, this theory of race cannot explain the possibility of passing, since it does not allow that someone can be treated as a member of race R even if they are not actually a member of R. This is a problem because it seems intuitive that someone can pass as a race they do not have. Therefore, if we want to find a plausible version of social constructionism that entails that races do not travel, we have to find an alternative characterization, since (INTERACTIVE KIND CONSTRUCTIONISM ABOUT RACE) is incompatible with the possibility of passing. That is, we want a version of social constructionism that is compatible with both the possibility of passing, and the claim that races do not travel.

Mallon argues that the following theory of race can explain both the possibility of passing and the no-travelling requirement at the same time:

(FOLK OBJECTIVIST INSTITUTIONAL CONSTRUCTIONISM ABOUT RACE) A person is race R iff (1) he or she is the type of person that satisfies the criteria central to the application of a folk racial concept, and (2) the person is at a site where the concept R is used to divide people.

(2004, 661)

First, this theory can explain passing because someone could satisfy the criteria central to the application of a folk racial concept, such as having certain phenotypic features or certain ancestors or certain geographical origins, even if other people are ignorant about this. And the account requires only that the person is at a site where label R is used, not that the label is applied to that person, so someone can be R without being labeled as R.

Second, this theory can explain no-travelling because being a member of R entails that one is a site where the concept of R is used to divide people, that is to say, the concept is applied to people and they are privileged or discriminated because of it. Therefore, it is possible that some people satisfy the criteria associated with the folk racial concept, unbeknownst to others, in a site where people who are believed to satisfy those criteria are labeled as R and discriminated because of it. So they belong to R, even if they "pass" as not being members of R. And race doesn't travel, on this view, because if members of R move to a site where the concept of R is not used to divide people, they are no longer members of R. Or so Mallon argues.

In my view, this account of the possibility of both passing and no-travelling is on the right track but needs some clarifications. The crucial question is the following: which concept is used to divide people that appears in condition (2)? I will examine two different answers to this question, namely, that the concept used at the site to divide people must be concept R itself, or that it must be the folk conception instead.

Could it be concept R itself? Not really, for the following reason. Folk subjects do not explicitly associate their concept of race with application conditions (1) and (2). They use a folk conception of race, which typically purports to refer to a biological property and uses phenotypic or geographical or ancestor-based criteria as a guide. This conception is the one that is used to divide people at that site. Thus this seems to be the most plausible option. Hence, condition (2) should not appeal to concept R, since the concept R is characterized by conditions (1) and (2) themselves, and it seems very unlikely that folk subjects have in mind a concept of race as complex as this. As Joshua Glasgow (2007) puts it:

> So people who are not self-avowed Institutionalists can still have institutionalized races. This is plausible, again, because while people could arguably realize that just as cocktail parties and wars are institutionalized, so is race, they don't need to have this epiphany for the ontological fact of institutionalization to hold (and to hold in the way relevant to satisfying the No Travel Constraint).

(2007, 563)

But now we face a further worry: if we do not use the concept of R itself to divide people, but rather the folk conception, how can we argue that races such as R do not travel? That is, if we analyze race as follows, then it is not so clear why the property so defined does not travel:

(FOLK OBJECTIVIST INSTITUTIONAL CONSTRUCTIONISM ABOUT RACE★) A person is race R iff (1) he or she is the type of person that satisfies the criteria central to the application of a folk racial concept, and (2) the person is at a site where the *folk conception* is used to divide people.

In particular, it could be argued that someone could satisfy the criteria that are central to the folk racial concept, and being at a site where the folk conception is used to divide people, but such that if they move to a different context, they will still satisfy the folk criteria.

My answer is the following: we can distinguish two different versions of this view, with condition (2) involving the folk conception, and still get the consequence that race so characterized does not travel. Let me elaborate.

On the first reading, having R is just a matter of having the folk features that have actually been subject to discrimination at that site, although a subject with the folk features would still be a member of R if they moved to a site where people are not discriminated in virtue of having the folk criteria. This view explains passing but not no-travelling. For what is required in order to be member of R is that one satisfy the folk criteria that are central to the folk concept. What makes this view a version of social construction is that the criteria that happen to be central to the folk concept are those features that have been the target of discrimination, such as phenotypic features, or ancestry, or geographical origin, and so on. The folk conception of race itself is socially constructed, since it focuses on arbitrary features of subjects and makes them a target of privilege or discrimination. But whether someone satisfies those arbitrary features is an objective matter that does not depend on bearing a certain relation to social or cultural practices, so races do travel on this view. (Charles Mills' version of social constructionism can be understood along these lines.)

The second reading is the following: having R at a site S is a matter of having the folk features *and* being subject to discrimination at site S in virtue of those features (that is, having R requires that people use the folk conception at that site in order to divide people). So if subject S travels to a different site where people are no longer discriminated because of satisfying the folk criteria, then they cease to have the property R. This view explains both passing and no-travelling. So, no travelling across times, places nor possible worlds.

A final worry arises: we said previously that according to many versions of social constructionism about race, the folk conception purports to refer to an underlying biological property, using folk criteria such as phenotypic features, geographical origins, or common ancestors as a fallible guide. But according to social constructionism the underlying biological property is not real. Therefore if concept R purports to refer to the underlying biological property but this does not exist, then R would be an empty concept. And then there could be no passing, since a subject could not be member of race R but treated as a member of a different race, since they couldn't be a member of race R to start with.

In response to this worry, many social constructionists have said that the concept R can come to refer to the socially constructed property of being treated in such and such way when one is perceived to have the folk features, even if the term purports to refer to an underlying biological property that actually does not exist. That is to say, racial concepts might come to refer to socially constructed properties such as the one discussed in this section, even if such concepts are associated with the belief that race is a biological property. In particular, these philosophers appeal to semantic externalism in order to argue that these beliefs associated with the concept or term do not fix the referent, so the referent might turn out to be socially constructed property after all. (See Haslanger 2006; Mallon 2017 for further discussion of this worry.)

3. Conclusion

To conclude, we can summarize the kinds of social constructionist views that can explain both the passing and the no-travelling requirements. As we have seen, many authors argue that a plausible version of social constructionism should satisfy these two criteria at the same time. We have discussed what versions of social constructionism have this consequence. In my view, there are two ways in which social constructionism could satisfy these two requirements at the same time.

These two ways correspond to the distinction between social construction of the concept-dependence variety, and social construction of the merely culture-dependence variety (without concept-dependence).[3] The distinction amounts to this: there are versions of social constructionism according to which a property P is socially constructed in the sense that in order for an individual to instantiate P, the individual must bear some relation to the employment of some concept of P, that is to say, certain concepts aiming to refer to property P itself must be employed (at that site). We have seen versions of this view earlier, such as (FOLK OBJECTIVIST INSTITUTIONAL CONSTRUCTIONISM ABOUT RACE). On the other hand, there are versions of social constructionism about property P that require something weaker than this, namely, that in order for subject S to instantiate P, they need to bear a relation to certain cultural and social practices which may require the employment of certain concepts *but not the concept of P itself.* We can apply this distinction to social constructionism about race as follows:

- *Concept-dependent* version of social constructionism about race: someone is race R iff they satisfy the folk criteria that people associate with concept R, and people use the folk concept of race in order to discriminate or privilege people at that site.
- *Culture-dependent* version of social constructionism about race: someone is race R iff they satisfy the folk criteria that are the target of discrimination and privilege, without ascribing label R itself.

As we have seen, Mallon argues that versions of social constructionism of the concept-dependent variety can satisfy both passing and the no-travel requirement at the same time. For some subjects could instantiate the folk criteria that people associate with the concept of race, unbeknownst to others. And it satisfies the no travel requirement because someone cannot satisfy R at a site unless the concept of R is used at that site. But it is useful to see that versions of social constructionism about race that do not require concept-dependence per se, but only culture-dependence, can also explain the passing and no-travel requirements at the same time. This version of social constructionism also satisfies passing for the same reason, namely, because people can instantiate the folk criteria unbeknownst to others. And it satisfies the no travel requirement because someone cannot instantiate R at a site if people are not the target of discrimination and privilege in virtue of being taken to satisfy the folk criteria, at that site. That is to say, according to the culture-dependent versions of social constructionism about race, the folk do not need to employ racial concepts per se, but there need to be certain social and cultural practices of discrimination and privileging people, in virtue of being taken to satisfy certain folk criteria, such as phenotypic features, ancestry, geographical origins, etc. Even if there are no racial terms associated to these social practices of discriminating or privileging people in virtue of their perceived phenotypic traits, ancestry, and geographical origins, it seems clear that this view about the metaphysics of race counts as social constructionism because what determines someone's race is whether someone is privileged or discriminated in virtue of being perceived as satisfying the folk criteria, and this is a social practice. Hence, the culture-dependent version of social constructionism satisfies passing because someone could satisfy the folk criteria but not being perceived as doing so. And this view also entails that race does not travel, since it entails that someone's race constitutively depends on certain social and cultural practices which may vary from context to context. This is an important result, given that culture-dependent varieties of social constructionism about race are weaker than concept-dependent varieties, but they both have the same consequence in this respect.

Therefore, we can conclude that the no travel requirement seems to be a consequence of many standard versions of social constructionism about race. Thus, according to many versions of social constructionism, someone's race is relative to their context.

Notes

1 Some recent discussions of the notion of social construction, as well as how to characterize social constructionism about different human kinds, include Stein (1992, 1999), Hacking (1999), Haslanger (2003), Mallon (2007, 2008), and Diaz-Leon (2015, 2018).
2 See Glasgow (2007) for a similar distinction.
3 See Diaz-Leon (2018) for further discussion of this distinction.

References

Diaz-Leon, E. (2015), "What Is Social Construction?" *European Journal of Philosophy* 23(4): 1137–1152.

———. (2018), "Kinds of Social Construction," in *Bloomsbury Companion to Analytic Feminism*, edited by P. Garavaso, London: Bloomsbury Academic, 103–122.

Glasgow, J. (2007), "Three Things Realist Constructionism About Race – or Anything Else – Can Do," *Journal of Social Philosophy* 38(4): 554–568.

Hacking, I. (1999), *The Social Construction of What?* Cambridge, MA: Harvard University Press.

Haslanger, S. (2003), "Social Construction: The 'Debunking' Project," in *Socializing Metaphysics*, edited by F. Schmitt, Lanham, MD: Rowman & Littlefield, 301–325.

———. (2006), "What Good Are Our Intuitions? Philosophical Analysis and Social Kinds," *Proceedings of the Aristotelian Society* 80(1): 89–118.

Ludwig, D. (2018), "How Race Travels: Relating Local and Global Ontologies of Race," *Philosophical Studies*, doi:10.1007/s11098-018-1148-x.

Mallon, R. (2004), "Passing, Travelling and Reality: Social Constructionism and the Metaphysics of Race," *Nous* 38(4): 644–673.

———. (2007), "A Field Guide to Social Construction," *Philosophy Compass* 2(1): 93–108.

———. (2008), "Naturalistic Approaches to Social Construction," *The Stanford Encyclopedia of Philosophy*, Winter 2008 edition, edited by E. N. Zalta, https://plato.stanford.edu/archives/spr2019/entries/social-construction-naturalistic/.

———. (2017), "Social Construction and Achieving Reference," *Nous* 51(1): 113–131.

Mills, C. (1998), *Blackness Visible: Essays on Race and Philosophy*, Ithaca, NY: Cornell University Press.

Root, M. (2000), "How We Divide the World," *Philosophy of Science* 67(3): 628–639.

Stein, E. (1992), "The Essentials of Constructionism and the Construction of Essentialism," in *Forms of Desire: Sexual Orientation and the Social Constructionist Controversy*, edited by E. Stein, New York: Routledge, 325–353.

———. (1999), *The Mismeasure of Desire: The Science, Theory, and Ethics of Sexual Orientation*, New York: Oxford University Press.

29

RELATIVISM IN THE PHILOSOPHY OF LAW

Torben Spaak

1. Introduction

Relativism comes in many shapes and forms, though it is common to make a distinction between cognitive and moral, or, more generally, between cognitive and evaluative, relativism. There is no specifically legal form of relativism, however. Instead, the types of relativism that find a natural home in legal philosophy are versions of moral relativism.

I begin with a few words about what I take relativism, especially moral relativism, to be (Section 2) and proceed to offer an overview of work involving relativistic arguments or theories in the field of legal philosophy (Section 3). The chapter concludes with some suggestions for future work involving relativistic arguments or theories in the field of legal philosophy (Section 4).

2. Cognitive and moral relativism

Philosophers often distinguish between cognitive and moral relativism (see, e.g., O'Grady 2002, 4; Krausz and Meiland 1982a, 1). *Cognitive* relativists maintain (i) that truth, or knowledge, or rationality, or even reality itself is relative to a certain starting point, such as a paradigm or a person's or a group's conceptual scheme, and (ii) that no such starting point is privileged as the objectively true one. Thomas Kuhn (1970, 21–22), for example, maintains that one can only say that one scientific theory is better than another within a given paradigm, that no one paradigm is objectively privileged as the one true paradigm, and that scientific change can only be explained by reference to psychological or sociological factors. Indeed, Kuhn is explicit that there is not and cannot be any paradigm-independent reality that can serve as an arbitrator between different paradigms (1962, 206). The type of relativism that I attribute to Kuhn here is usually called conceptual relativism, the idea being (i) that reality (and therefore truth and knowledge) depend on our concepts, and (ii) that there is no set of concepts that is objectively privileged as the one true set.[1]

Moral relativism comes in at least three different shapes, namely descriptive, normative, and meta-ethical relativism (Brandt 2001; Frankena 1973, 109–110). Whereas *descriptive* relativism has it that as a matter of fact different people often have different fundamental moral views in the sense that the different views are not due to factual disagreements, *normative* relativism holds that a person *ought to* act in accordance with his or his group's views. *Meta-ethical* relativism, on the other hand, is the view that moral truth or validity is always relative to a moral framework, and that no such framework is objectively privileged as the one true framework. As Gilbert Harman (1996, 3) puts it,

> moral right and wrong (good and bad, justice and injustice, virtue and vice, etc.) are always relative to a choice of moral framework. What is morally right in relation to one moral framework can be morally wrong in relation to a different moral framework. And no moral framework is objectively privileged as the one true morality.

3. Overview of work done in the area

The fact of the matter is that there are not many relativistic theories or arguments to be found in contemporary legal philosophy. While some prominent legal philosophers have been non-cognitivists (see, e.g., Ross 1959, 6–11, 274–275, 280; Olivecrona 1951, 129–130; Hart 1982, 159–160), very few have explicitly defended any form of relativism. I shall therefore consider two very prominent legal philosophers from an earlier period, who do appear to have been relativists, namely, Hans Kelsen and Gustav Radbruch. Both writers appear to espouse what I have called meta-ethical relativism.

3.1. Hans Kelsen

As is well known, Kelsen (1934, 18) takes the view that the so-called pure theory of law aims to understand law as it is, not as it ought to be; and one reason why he takes this view is that he holds that morality is inescapably subjective, that it cannot be the object of rational cognition (1934, 15–18). He returns to the question of the status of morality in the *General Theory of Law and State* (1945, 6–8). Having explained that saying a social order is just amounts to saying that all men find their happiness in it, he points out (1945, 6) that there is no rational way of ranking the interests of different people in case of conflict and adds that value judgments are "determined by emotional factors and [are], therefore, subjective in character, valid only for the judging subject and therefore relative only."

Although his talk of emotional factors could be taken to mean that Kelsen is here espousing some form of emotivism, and although the phrase "valid only for the judging subject" could be taken to mean that he is espousing some form of moral subjectivism (on moral subjectivism, see Mackie 1977, 17–18), the next quotation, in which he maintains that a system of morality is a social phenomenon and that value judgments are *relative* not to individuals, but to groups of individuals, suggests that he is best understood as a meta-ethical relativist:

> A positive system of values is not an arbitrary creation of the isolated individual, but always the result of the mutual influence the individuals exercise upon each other within a given group, be it family, tribe, class, caste, profession. Every system of values, especially a system of morals and its central idea of justice, is a social phenomenon, the product of a society, and hence different according to the nature of the society within which it arises.
>
> *(1945, 7–8)*

What is the significance of Kelsen's relativism to Kelsen's legal philosophy? First, Kelsen maintains in the second edition of *Reine Rechtslehre* (1960, 71) that it is a mistake to accept the natural law view that only a system of norms that satisfies certain minimum moral requirements can be law, because there are a number of moral systems none of which is truer or more correct than the others, and because any legal system will correspond to some moral system but not to others. Clearly, if this is so, almost any legal system will pass the test, and as a result natural law theory will be more or less devoid of content.

Secondly, Kelsen (1948) maintains that democracy presupposes relativism, in the sense that only a relativistic stance could justify giving priority to democratic principles over substantive moral considerations when it comes to legislation and political decision-making in general. The idea appears to be that since relativism means that what is right today may be wrong tomorrow (premise), those who are in the minority today, and who are therefore wrong, must have a chance to become – through the democratic procedure – the majority tomorrow and thus to be right (conclusion).

> Solely because of this possibility, which only philosophical relativism can admit – that what is right today may be wrong tomorrow – the minority must have a chance to express freely their opinion and must have full opportunity of becoming the majority. Only if it is not possible to decide in an absolute way what is right and what is wrong is it advisable to discuss the issue and, after discussion, to submit a compromise.
>
> *(1948, 913–914)*

I cannot, however, see that the premise supports the conclusion. First, on the relativist analysis, the premise is descriptive and the conclusion normative, and, as Kelsen would be the first to point out, given his theory of the basic norm, a descriptive premise cannot entail a normative conclusion. Secondly, if, as seems to be the case, the idea is that the conclusion expresses a non-relative "ought," this type of relativism is incoherent, since the theory itself clearly rules out any non-relative "ought" (on this, see Williams 1982).[2]

3.2. Gustav Radbruch

It is common in the literature to distinguish between Radbruch's pre-war and Radbruch's post-war philosophy. Radbruch's pre-war stance (1932) was that of a legal positivist and moral relativist, who held that there is no necessary connection between law and morality, that moral judgments can be true or valid only in relation to a given moral framework, and that no such framework is objectively privileged as the one true framework. His post-war position (1946, 7), on the other hand, was that law and morality are necessarily connected; and whereas laws that are intolerably unjust are flawed law and must yield to justice (the intolerability formula), laws that do not even aim at justice lack legal character (the disavowal formula) (on the two formulas, see Alexy 1999, 16). In what follows, I shall, of course, focus on Radbruch's pre-war philosophy.

Radbruch explains that legal philosophy is (what he refers to as) the evaluating view of law (1932, 49–51), and that this view of law can be understood in terms of a combination of methodical dualism and relativism. On his analysis, *methodical dualism* is the view that existence and value belong in separate spheres and that there is no way to derive an "ought" from an "is" (1932, 53), and *relativism* is the view that a value judgment can be true or correct only in light of a particular outlook on values and the world and that no such outlook on values and the world is true or correct in and of itself. He points out, however, that relativism does not mean that the individual can escape the *choice* between competing legal views that have been developed on the basis of competing starting points. For relativism (1932, 57), "is limited to presenting to him [the agent] exhaustively the ultimate presuppositions, but it leaves his decision itself to the resolution he draws from the depth of his personality – by no means, then, to his pleasure, but rather to his conscience." Radbruch concludes that the task of a legal philosophy based on relativism is to develop a system of values and standards without deciding between them (1932, 69).

What is the significance of Radbruch's relativism to Radbruch's legal philosophy? First, Radbruch holds that the concept of law is a necessary general concept that is oriented toward the *idea of law*, which is *justice*. On this analysis (1932, 73), law is "the reality the meaning of which is to serve the legal value, the idea of law." He is, however, careful to point out that justice determines only the *form*, not the content of law. The reason is that justice, which on Radbruch's analysis requires that equals be treated equally, leaves it an open question whom to consider equal and how they should be treated. To get the *content* of law, Radbruch says (1932, 91), we must add the value of *expediency* to the value of justice. But, he continues, what is legally expedient depends on what the purpose of law is, and the purpose of law may be found (i) in individual values, (ii) in collective values, or (iii) in work values (*Werkwerte*), none of which is more right than the others. He also reckons with a third element, in addition to the values of justice and expediency, namely the value of *legal certainty* (*Rechtssicherheit*). We need legal certainty, he points out (1932, 108), because there is no non-relativistic answer to the question "justice or expediency?"

Radbruch (1932, 108–109) explains that whereas (formal) justice and legal certainty are *absolute* values, expediency is only a *relative* value, and that the *ranking* of these three elements is a relative matter, too. He thus acknowledges that the three elements may sometimes conflict with one another, and he argues that in a conflict between legal certainty and justice, the judge ought to give priority to *legal certainty* (1932, 119). He reasons that since according to relativism there is no correct answer to the question, "What is right or just?," the task of the law is to determine what should be done (1932, 116–117). Note, however, that while Radbruch appears to have in mind here a non-relative "ought," the relevant "ought" must, for the reasons given in the section on Kelsen, be a relative "ought." If, however, it is a relative "ought," the conclusion will be of rather limited interest.

Finally, Radbruch maintains (1932, 48) that relativism is apt to counteract moral intolerance, because, he says,

> if no partisan view is demonstrable, each view is to be fought from the standpoint of an opposite view; yet if none is refutable either, each is to be respected even from the standpoint of the adverse view. Thus relativism teaches both determination in one's own attitude and justice toward that of another.

I believe, however, that it is a mistake to think that acceptance of moral relativism necessarily leads to, or should lead to, tolerance of other moral views. First, as Geoffrey Harrison (1982, 240) points out, the claim advanced by relativists that one moral system is "as good as" another, or, as I have put it, that no moral framework is objectively privileged as the one true framework, can be understood morally or non-morally. If we take "as good as" to be a non-moral notion, we cannot, of course, arrive at a moral conclusion (about tolerance or anything else), unless we add a moral premise; if, on the other hand, we take "as good as" to be a moral notion, we can indeed arrive at a moral conclusion (about tolerance or something else). The moral interpretation is problematic, however. For one thing, it appears to be logically as well as psychologically impossible to *believe* that one's own moral theory is not morally better than any other moral theory. As Harrison puts it, this is one moral judgment that no *participant* of a moral practice – as distinguished from an observer of a moral practice – could ever make (1982, 240):

> Could, for example, a Christian who admitted that other religious/moral positions were just as good as Christianity still be regarded as a Christian? I think not, in that

adopting a morality will necessarily involve rejecting at least some aspects of any rival doctrine which is not compatible with one's own.

I find Harrison's line of reasoning persuasive, and this leaves us with the non-moral interpretation, which is incapable of supporting a moral principle of tolerance.

Note, however, that if it is logically as well as psychologically impossible to believe that one's own moral theory is not morally better than any other moral theory (on a similar issue, see Fish 1980, 319), that is, if it is impossible to embrace a first-order moral theory and combine it with meta-ethical relativism, a moral *agent*, or if you will, a *participant* cannot be a meta-ethical relativist. And this suggests that the theory of meta-ethical relativism is, in this specific sense, self-refuting. I do not think that this difficulty undermines the theory, however. For, as I see it, the important question is not whether a moral *agent* can be a meta-ethical relativist, but whether meta-ethical relativism is a true or at least a defensible theory (on this, see also Harrison 1982, 239). I believe my view here is in keeping with Harman's view (1996, 4, 17), that (his version of) meta-ethical relativism is not a theory about what people *mean* when they make moral judgments, but a theory about the truth-conditions of moral judgments. As Harman puts it (1996, 17),

> [m]oral relativism is a thesis about how things are and a thesis about how things aren't. Moral relativism claims that there is no such thing as objectively absolute good, absolute right, or absolute justice; there is only what is good, right, or just in relation to this or that moral framework.

I conclude that the theory of meta-ethical relativism properly belongs on the level of the *observer*, not on the level of the participant.

One might, however, argue that the *combination* of meta-ethical relativism and a moral principle which provides that we should not interfere with people, unless we can justify this interference to them, would entail that we ought to tolerate their moral views. This is David Wong's approach. Chris Gowans (2015, 43) renders Wong's principle as follows: "the principle is, roughly speaking, that we should not interfere with people unless we can justify this interference to them (if they were rational and well-informed in relevant respects)." This does not show that meta-ethical relativism enjoins tolerance, however, because the justification principle is *not* part of the theory of meta-ethical relativism. Moreover, as I have said, if meta-ethical relativism is true, the tolerance principle itself can only be true or valid relative to a given moral framework, and this makes the argument considerably less interesting; if instead the idea is that the conclusion is to be framed in terms of a non-relative morality, then this type of relativism is, of course, incoherent.

4. Suggestions for future work

I shall suggest two topics for future work on relativism in the philosophy of law. First, there is the question of the correct analysis of legal statements, say, that a person has a legal right to free speech, or that the judge has a legal obligation to (or should) decide the case in accordance with the plain meaning of the statute, unless doing so would yield an absurd or manifestly unjust result, or that like cases should be treated alike, etc. Such statements come in at least three different forms, namely, (i) first-order (normative) legal statements, (ii) detached (normative) legal statements (see Raz 1979, 140–143, 1990, 170–177), and (iii) second-order (descriptive) legal statements. But, one wonders, what is the correct meta-ethical account of first-order legal statements? This is a very difficult question to answer, but I believe meta-ethical relativism is at the very least a contender. Consider, for example, Stanley Fish's theory of legal and literary interpretation. According to Fish's theory, linguistic meaning, including the existence of entities such as a novel or a statute, depends on *interpretive communities* (1980, 13–4, 322), that is, groups of people who share certain *interpretive strategies* (1980, 171–2). On Fish's analysis, there is no way to establish what is the correct interpretation of a text independently of every interpretive community. As he puts it (1980, 16), "there is no single way of reading that is correct or natural, only 'ways of reading' that are extensions of community perspectives." Whether we like it or not, he says (1989, 141–142), "interpretation is the only game in town."

We see that on Fish's analysis, a first-order legal interpretive statement will be true or false only in light of the interpretive strategies embraced by a given interpretive community, and there is no interpretive community that is privileged as the one objectively true community. I find the idea appealing, that the correct interpretation of a legal text depends on interpretive communities none of which is privileged as the objectively true interpretive community, because I believe there is no way to determine the correct interpretation of a statute, say, that is independent of the methods and techniques of legal reasoning that are accepted in the relevant jurisdiction. I have, however, argued elsewhere (Spaak 2008) that Fish is a conceptual relativist, and that his theory of interpretation is vulnerable both to the objection that relativism about truth is self-refuting and to Donald Davidson's well-known objection to conceptual relativism (Davidson 1984). There may, however, be more to say about Fish's relativism.

Secondly, there is the question of whether relativists can account for legal – or moral – *disagreement*. David Lyons (1982, 211–213) makes a distinction between agent's-group relativism, which has it that an action is right if, and only if, it is in keeping with the norms and values accepted by the agent's group, and appraiser's-group relativism, according to which an action is right if, and only if, it is in keeping with the norms and values accepted by the appraiser's group. This distinction is of interest in this context, because *appraiser's-group* relativism can give rise to a situation in which two persons, Smith and Jones, who are members of different groups with different moral standards, make what appear to be *contradictory* moral judgments about an agent's, *A*, performing an act, *X*.

Lyons explains (1982, 215) that the appraiser's-group relativist might try to account for such a situation by arguing either that moral judgments lack truth-value, or else that truth in this context is *relative* and that therefore there is no contradiction. I shall focus here on the second alternative, which I see as the obvious choice for the appraiser's-group relativist. As Lyons points out (1982, 222), an appraiser's-group relativist who chooses to argue that seemingly contradictory judgments are not really contradictory because they are true relative to different moral frameworks, will find that he is now facing a new problem, namely, that he cannot explain what seems to be obvious, that the parties are *disagreeing*. For, on this analysis, the parties – who believe they are disagreeing with each other – will really be arguing at cross-purposes, Smith arguing that the action under consideration is required (or permitted) according to the standards endorsed by *his* group and Jones arguing that the action is not required (or permitted) according to the standards endorsed by *her* group.

Lyons suggests that the appraiser's-group relativist might attempt to account for such disagreement either in terms of clashing *attitudes*, or else by limiting the scope of his relativism to a subgroup of moral judgments, say, "inner" moral judgments in Gilbert Harman's sense (1982, 223). However, Lyons objects to the former alternative that one cannot analyze conflicting *attitudes* without also analyzing conflicting *beliefs*, that such beliefs have to be analyzed relativistically along the lines of the analysis of the relevant moral judgments, and that this means that we are right back where we started from (1982, 222). And he objects to the latter alternative that the problem with Harman's attempt to limit the scope of his relativism to "inner" judgments is that Harman's idea that such statements are necessarily – by virtue of their *meaning* – connected to the agent's motivational set is arbitrary. Lyons notes that, on Harman's analysis (1982, 190–193), an "inner" moral judgment is a judgment that the agent ought to do *X*, or that it was right (or wrong) of the agent to do *X*; that Harman's idea is that such judgments are necessarily connected to the agent's motivational set, because they would make no sense unless one presupposed that the agent in question was capable of being motivated by the relevant moral considerations; and that this means that they require an agent's-group relativistic analysis. But, Lyons objects, one might equally well hold that the reason why we do not say that a person ought to do (or not do) something, or that it was wrong (or right) of him to do it, in cases where we do not believe that the agent is capable of being motivated by the relevant moral considerations, is *not*, as Harman holds, that such statements by their very *meaning* invoke the agent's motivational set (and so require an agent's-group relativistic analysis), but that it would be *pointless* to advise a person who is unlikely to care about the advice (1982, 223–224). Lyons therefore concludes that Harman's attempt to avoid the problem of disagreement by restricting his relativism to "inner" judgments, which require an agent's-group relativistic analysis of the relevant moral judgments, has failed.

In a later publication, Harman (1996, 32–44) proposes instead that meta-ethical relativists can account for moral disagreement by invoking the theory of quasi-absolutism (or quasi-realism), according to which each party to the disagreement *projects* his or her (relative) moral framework onto the world and speaks as if it were the one true moral framework, while admitting that this way of talking is only "as if" (on quasi-realism, see also Blackburn 1984, ch. 6, 1993). If Harman is right, two relativists could then argue in a meaningful way about, say, the permissibility of eating meat, because in making their competing moral statements they would be expressing their respective conflicting *attitudes* to a system of standards that permits (or prohibits) eating meat. But, Harman points out (1996, 39–41), a problem then arises about how to distinguish between quasi-absolutism and absolutism (non-relativism); and he proposes that the difference is that whereas the absolutist holds that there are objective truth conditions for the relevant statements, the quasi-absolutist holds that there are no such objective truth conditions. But is he right?

Acknowledgements

I would like to thank Krister Bykvist for discussing questions of relativism with me and Robert Carroll for checking my English. The usual caveat applies, of course: The author alone is responsible for any remaining mistakes and imperfections. Note also that several paragraphs in Section 3 on Radbruch can be found, more or less verbatim, in Spaak (2009).

Notes

1 For a penetrating discussion of Kuhn's relativism, see Doppelt (1982).
2 For a helpful discussion of Kelsen's relativism, see Vinx (2007, 134–144).

References

Alexy, R. (1999), "A Defence of Radbruch's Formula," in *Recrafting the Rule of Law: The Limits of the Legal Order*, edited by D. Dyzenhaus, Oxford: Hart Publishing, 15–39.
Blackburn, S. (1984), *Spreading the Word: Groundings in the Philosophy of Language*, Oxford: Oxford University Press.
———. (1993), *Essays in Quasi-Realism*, Oxford: Oxford University Press.

Brandt, R. (2001), "Ethical Relativism," in *Moral Relativism: A Reader*, edited by P. K. Moser and T. L. Carson, Oxford: Oxford University Press, 25–31.

Davidson, D. (1984), "On the Very Idea of a Conceptual Scheme," in his *Inquiries into Truth and Interpretation*, Oxford: Oxford University Press, 183–198.

Doppelt, G. (1982), "Kuhn's Epistemological Relativism: An Interpretation and Defense," in Krausz and Meiland (1982b), 113–146.

Fish, S. (1980), *Is There a Text in This Class?* Cambridge, MA: Harvard University Press.

———. (1989), *Doing What Comes Naturally*, Durham, NC: Duke University Press.

Frankena, W. (1973), *Ethics*, second edition, Englewood Cliffs: Prentice-Hall.

Gowans, C. (2015), "Moral Relativism," in *The Stanford Encyclopedia of Philosophy*, Summer 2015 edition, edited by E. N. Zalta, https://plato.stanford.edu/archives/sum2018/entries/moral-relativism/.

Harman, G. (1982), "Moral Relativism Defended," in Krausz and Meiland (1982), 189–204.

———. (1996), "Moral Relativism," in *Moral Relativism and Moral Objectivity*, edited by G. Harman and J. J. Thomson, Oxford: Blackwell, 3–64.

Harrison, G. (1982), "Relativism and Tolerance," in Krausz and Meiland (1982b), 229–243.

Hart, H. L. A. (1982), *Essays on Bentham*, Oxford: Oxford University Press.

Kelsen, H. (1934), *An Introduction to the Problems of Legal Theory*, translated by B. Litschewski Paulson and S. L. Paulson, Oxford: Oxford University Press, 1992. (Originally published 1934 in German under the title *Reine Rechtslehre. Einleitung in die rechtswissenschaftliche Problematik* by Franz Deudicke, Vienna.)

———. (1945), *General Theory of Law and State*, translated by A. Wedberg, Union, NJ: The Lawbook Exchange.

———. (1948), "Absolutism and Relativism in Philosophy and Politics," *The American Political Science Review* 42: 906–914.

———. (1960), *Reine Rechtslehre*, second edition, Wien: Österreichische Staatsdruckerei.

Krausz, M. and J. W. Meiland (1982a), "Introduction," in Krausz and Meiland (1982b), 1–9.

———. (eds.) (1982b), *Relativism: Cognitive and Moral*, Notre Dame and London: University of Notre Dame Press.

Kuhn, T. S. (1962), *The Structure of Scientific Revolutions*, second edition, Chicago: University of Chicago Press, 1970.

———. (1970), "Logic of Discovery or Psychology of Research?" in *Criticism and the Growth of Knowledge*, edited by I. Lakatos and A. Musgrave, Cambridge: Cambridge University Press, 1–23.

Lyons, D. (1982), "Ethical Relativism and the Problem of Incoherence," in Krausz and Meiland (1982b), 209–225.

Mackie, J. (1977), *Ethics: Inventing Right and Wrong*, London: Penguin Books.

O'Grady, P. (2002), *Relativism*, Chesham: Acumen.

Olivecrona, K. (1951), "Realism and Idealism: Some Reflections on the Cardinal Point in Legal Philosophy," *New York University Law Review* 26: 120–131.

Radbruch, G. (1932), "Legal Philosophy," in *The Legal Philosophies of Lask, Radbruch, and Dabin*, edited by E. W. Patterson, Cambridge, MA: Harvard University Press, 1950, 43–224. (This is a translation into English of Gustav Radbruch, *Rechtsphilosophie*, Third Edition, Leipzig: Quelle & Meyer, 1932).

———. (1946), "Statutory Lawlessness and Supra-Statutory Law," *Oxford Journal of Legal Studies* 26: 1–11, 2006. (Translated by B. Litschewski Paulson and S. L. Paulson. This article was first published in *Süddeutsche Juristen-Zeitung* 1: 105–108 (1946) under the title "Gesetzliches Unrecht und Übergesetzliches Recht.")

Raz, J. (1979), *The Authority of Law*, Oxford: Oxford University Press.

———. (1990), *Practical Reason and Norms*, Princeton, NJ: Princeton University Press.

Ross, A. (1959), *On Law and Justice*, Berkeley and Los Angeles: University of California Press.

Spaak, T. (2008), "Stanley Fish and the Concept of an Interpretive Community," *Ratio Juris* 21: 157–171.

———. (2009), "Meta-Ethics and Legal Theory: The Case of Gustav Radbruch," *Law and Philosophy* 28: 261–290.

Vinx, L. (2007), *Hans Kelsen's Pure Theory of Law: Legality and Legitimacy*, Oxford: Oxford University Press.

Williams, B. (1982), "An Inconsistent Form of Relativism," in Krausz and Meiland (1982), 171–174.

PART 5

Relativism in epistemology

30

SELF-REFUTATION

Steven D. Hales

1. Introduction

The chief complaint against relativism is as ancient as relativism itself: the charge that the doctrine is self-refuting. How exactly the self-refutation purportedly comes about, and whether relativists can adequately answer the objection has been a topic of sporadic interest for over two millennia. Plato first filed the grievance, sketching an argument that has been described by Robert Nozick as "the quick and standard refutation [of relativism]" (Nozick 2001, 16). Even those often considered relativists, like Richard Rorty and Hilary Putnam, have admitted the power of the peritrope.[1] Precisely how the argument is supposed to work – as well as how effective it really is – is not obvious. Here I will review some of the more prominent discussions.

Proposals of very general applicability constitute their own tribunal; they must clear the bar that they themselves have set. "All sentences are five words long" is immediately seen to be false even without consulting additional evidence. Other examples of self-reference are neither clearly true nor clearly false. Eubulides, a contemporary of Plato, was well-known for generating such paradoxes, including the famous liar.[2] "A man says he is lying. Is what he says true or false?" The liar is probably the most celebrated of the self-referential puzzles. If the man is lying, then he speaks the truth and therefore cannot be lying, but if he is speaking the truth then it follows that he is in fact lying and therefore cannot be speaking the truth. The truth value of the liar sentence is unstable. Not everyone noticed that this is a problem; in St. Paul's Letter to Titus (5:12–13) he writes, "One of themselves, a prophet of their own [Epimenides] said, 'Cretans are always liars' . . . this testimony is true." Paul didn't catch on that the truth of Epimenides's testimony also made it false.

The most famous statement of relativism from ancient times was given by Protagoras. Reportedly the first sentence of his book *Alēthia* (Truth, a book now lost) was "Man is the measure of all things: of the things which are, that they are, and of the things which are not, that they are not." Protagoras's ancient commentators, including contemporaries such as Plato and Aristotle, took him to be defending a form of global relativism, and pounced on the issue of self-reference as a way to refute him.

In Theætetus, Plato marshals several arguments against Protagoras's "man is the measure of all things."[3] One is that it leads to various counterintuitive results, particularly a great flattening of knowledge: we each turn out to be as wise as any of the gods (162c), yet we may not even be wiser than pigs or baboons (161c), no one is ignorant or makes mistakes (170c), and there are no true experts on any topic (178 et passim). At 171a Plato offers "a really exquisite conclusion," namely that Protagoras's Measure Thesis can be turned on its head. Since Protagoras claims that everyone's opinion is true, it follows that his opponents who consider the Measure Thesis to be false are right – it is false. Protagoras must concede that their opinion is as true as any other, which means he is forced to admit either that the Measure Thesis is both true and false (a reductio ad absurdum on his own view), or that his endorsement of the thesis is no better off than the rejection of it. Worse, when Protagoras acknowledges that his opponents' denial of the Measure Thesis is true, it implies that truth is not relative to the individual. Therefore, for his critics, truth is not simply objective for them, it is objective simpliciter, which directly dismisses Protagoras's relativism.

Aristotle in Metaphysics Γ gives a somewhat similar argument. He begins the chapter by positing the law of noncontradiction: it is impossible for anything at the same time to be and not to be (1006a). Noncontradiction was more than a mere logical law for Aristotle. He also thought it is doxastically impossible for anyone to believe the same thing to be and not to be (1005b25). Protagoras seems to fall afoul of these bedrock principles. Suppose we accept the position of the Measure Thesis that every opinion is a true opinion. Clearly people dispute with each other and each disputant believes the other to be wrong. It follows that each party is right; the same proposition is both true and false and each person is both right and wrong. Likewise, Protagoras's Measure Thesis itself is subject to dispute and therefore it too is both true and false. These results violate noncontradiction and Protagoras is thereby refuted (1009a5).

In the second century CE, Sextus Empiricus echoes these concerns. In Against the Logicians (2005, 60–64) he offers Protagoras a fair hearing, conceding that human beings testify from the point of view of human beings, children from the perspective of childhood, the insane from the standpoint of the insane, and so on. It would be a mistake for a person of sound mind to reject the ideas of the crazy since they are not in the right circumstances to do so, or for adults to dismiss the views of children. Be that as it may, Protagoras's proposal that every appearance is true fails because of "the turning about," i.e. self-refutation. In Sextus's words, "if every appearance is true, then even not every appearance's being true, since it takes the form of an appearance, will be true, and thus every appearance's being true will become false" (Empiricus 2005, 389–390). As a Pyrrhonian skeptic, Sextus denies that we are justified in accepting any claims as either true or false, but for different reasons than the "turning about" argument that skewers Protagorean relativism.

It is worth noting that one way to skirt the problem of self-refutation entirely is to limit the domain of the relativist's claims. Someone might be a local relativist about ethics, aesthetics, epistemic modals, knowledge attributions, or a variety of other matters without running the risks of universal relativism.[4] Take for example, Rousseau's remark that "every one gives the title of barbarism to everything that is not in use in his own country. As, indeed, we have no other level of truth and reason than the example and idea of the opinions and customs of the place wherein we live."[5] Rousseau's quotation is from an essay on cannibalism, and he is suggesting that behavior in other cultures that we find shocking is no more than the result of our own enculturation, and that we are savages from the point of view of others, just as they are to us.

Suppose Rousseau's position is MR: "the truth of any moral proposition is relative to a particular culture." Thus cannibalism is immoral in one society but not in another. MR is not itself a moral claim, but a meta-level claim about the truth conditions of object-level moral propositions. Therefore the truth value of MR is not – on the basis of MR alone – culturally relative. Likewise relativism about predicates of taste is not a matter of taste, and so on. To the extent that self-refutation is a problem, it is a problem for global relativism, which will be the focus of what follows.

The ancient critics, especially Aristotle, struggled to grasp the radical idea of relative truth, and their attempts at a self-refutation argument don't clearly distinguish between objective and relative truth. As a result they may be illicitly assuming that truth is objective to make their case. A really compelling charge of self-refutation needs to explicitly assume that truth is merely relative to a point of view, and then show how that idea is self-undermining.

To be a relativist about truth, minimally you have to think that truth is partly a function of some parameter like perspective, culture, world view, conceptual scheme, historical period, point of view, etc. For simplicity, let's just call this parameter a "perspective." There are two possible anti-relativist positions: (i) Deny that there are perspectives. If the very idea of a perspective is incoherent or useless, then truth is independent of perspective and propositions are true absolutely. (ii) Admit that there are perspectives, but insist that propositions are true (or false) in all perspectives and are true absolutely for that reason.

An analogy to the theistic debate over God's eternality helps to illustrate the preceding distinction. One idea about eternality is Augustine's and Aquinas's view that God exists outside of time. He is eternal in the sense that he is atemporal or timeless, separated from the temporal order of the world, and able to survey all of points in time at once. Positing the existence of time is irrelevant to understanding God's eternality, just as option (i) maintains that positing perspectives is irrelevant to understanding absolute truth. Another option is that God exists within time, and that he is eternal in the sense that reality has no beginning or end, and God exists at every moment of it. This is analogous to option (ii) in that eternality is existence at every moment and absolute truth is truth within every perspective.

Absolutists certainly can and have argued that there are no perspectives to which truth is indexed, and that relativism is mistaken for that reason. (Famously, Davidson (1984) takes this line). Whatever the merits of that approach, it won't help build a self-refutation argument. An absolutist about truth who rejects the existence of perspectives from the start cannot effectively mount a self-refutation argument against the relativist, since without admitting the relativist's key contention that there are perspectives to which truth is indexed, there is no chance of constructing a reductio ad absurdum. If relativism really is self-undermining, then it has to be false by its own lights; the assumption of the truth of relativism needs to imply its own falsehood.

So let us assume for the sake of argument that there are perspectives and that, like times or possible worlds, they are parameters of truth. A proposition is true at a world, at a time, and at a perspective. The position of absolutism can still be formulated even if we stipulate the existence of perspectives; it is the view that for all propositions P, P has the same truth value at each perspective.

> GLOBAL RELATIVISM (GR) Every truth is true relative to some perspectives and false relative to other perspectives.

Either GR is absolutely true (true in all perspectives) or it is relatively true (true relative to some, but not all perspectives). If GR is absolutely true, then it is true in all perspectives (by the definition of absolutism). However, GR states that every truth is true in some, but not all perspectives. Therefore if GR is true, it is not absolutely true. This seems to be the point where Aristotle stopped and, as it stands, is rather thin gruel to count as adequate self-refutation.

2. Relativism is merely relatively true

There remains the possibility that GR is true, just relatively true. What is the problem with claiming that every truth is relative to some perspectives and false relative to other perspectives, and that very fact is itself is true relative to some perspectives and false relative to other perspectives?

The relativity of relativism has appealed to various thinkers. Friedrich Nietzsche, for example, defended a form of relativism known as perspectivism. Perspectivism for Nietzsche is not one precisely defined doctrine, but a cluster of related ideas about the subjectivity of truth, anti-realist metaphysics, a bundle theory of objects, the revaluation of values and the creation of one's own virtues, and the role of varying interpretations in knowledge (Hales and Welshon 2000; Hales forthcoming). But part of it is a rejection of absolute truth. In Human, All Too Human §2, Nietzsche writes that "there are no eternal facts, nor are there any absolute truths," and doubles down on this comment in The Genealogy of Morals (III, §12), "there is only a perspective see-ing, only a perspective knowing." Nietzsche is untroubled by the possibility that his ideas about perspectival knowledge are true only relative to his perspective. "Supposing that this also is only interpretation – and you will be eager enough to make this objection? – well, so much the bet-ter" (Beyond Good and Evil §22). Is his defiance justified?

If (GR) is merely relatively true, then there are perspectives in which (GR) is true and per-spectives in which it is false. When there are points of view from which it is false, the relativist faces a kind of pragmatic challenge: what makes relativism so special that a fence-sitter would be persuaded to adopt it? After all, the relativist has just conceded that there are perspectives in which absolutism is true. Why not adopt one of those? One possible reply is that we are in certain circumstances that preclude an absolutist perspective. (GR) is a relative truth, but it is relative to me, so I had better accept it. Of course, the fact of (GR)'s truth relative to me is also relatively true, and therefore there are perspectives in which it too is false.[6] While this may be an unsatisfying dialectical eddy, it is not obvious that the relative truth of (GR) is self-referentially inconsistent. John MacFarlane is sympathetic to the "relativism is relatively true" move, writing, "it is usually conceded that there is no real contradiction in the relativist's holding that relativism is not true for everyone."[7]

The idea that (GR) is merely relatively true will fail to save relativism from self-refutation just in case admitting a perspective in which (GR) is false (and hence absolutism is true) can-not sufficiently bind absolutism. Relativism proposes to bottle truths within perspectives, but if absolutism is a kind of universal acid that dissolves any bottle in which it is put, then it cannot be safely placed on the shelf along with the other relative truths. A closer examination of the idea that even allowing (GR) to be absolutely true within a perspective will lead to the elimination of the "within a perspective" stricture will be looked at shortly.

3. Self-refutation as regress

Some interpreters of the peritrope take the real problem to be one of infinite regress.[8]

Suppose that Protagoras says

(1) "Man is the measure of all things" is true for me.

Plato then asks whether (1) is absolutely true, or just relatively true. Obviously, Protagoras isn't about to say that it is absolutely true. Instead, staying true to his own principle, he avers that

(2) (1) is true for me.

The same question then arises: is (2) absolutely true or not? Once again Protagoras makes the same move

(3) (2) is true for me.

This line of questioning and response can be repeated ad infinitum. The relativist is an elusive and slippery foe, and it is not clear what exactly the problem is with his willingness to admit that all of his assertions, at whatever meta level you wish, are simply true for him.

Putnam thinks that the problem is ultimately grounded in Wittgenstein's Private Language argument. Wittgenstein argues (Wittgenstein 1953, notably §258, §265, §268) that if I invent a word in a private language, this means that I invent a rule of application for the word. Now, how can I tell if I am using the rule correctly? Whenever it *seems* that I am using the rule correctly. There is no external check since it is my own personal language. Yet if whatever seems right is right, then there is no rule at all, only the semblance of one. Writing down the rule is no use, as I still have to count on memory to determine how to interpret the rule. Counting on my mind to confirm something else done by my mind is, as Wittgenstein says in § 265, like buying several copies of the morning newspaper to confirm that what it contains is true. He concludes that we cannot invent actual rules on our own and that language – which requires the systematic application of rules – can originate only in the public sphere.

According to Putnam, relentlessly iterating sentences as being relatively true *for me* amount to making the truth predicate part of a private language. If Wittgenstein is right, as Putnam thinks he is, then relativism is impossible. Protagoras may keep insisting that his assertions are only true for him, but he is unable to make any sense of the difference between *being right* and *thinking that he is right*, or between *genuine assertion* and *merely making noises*. The affirmation of truth is essentially public, something that holds in all perspectives, something absolute.

Paul Boghossian, on the other hand, thinks the regress problem leads to propositions with infinite conjunctions. With respect to relativizing the claim that there have been dinosaurs, he writes,

> The fact-relativist is committed to the view that the only facts there are, are infinitary facts of the form:
>
> According to a theory that we accept, there is a theory that we accept and according to this latter theory, there is a theory that we accept and . . . there have been dinosaurs.
>
> But it is absurd to propose that, in order for our utterances to have any prospect of being true, what we must mean by them are infinitary propositions that we could neither express nor understand.
>
> *(Boghossian 2006, 56)*

Boghossian's criticism here is analogous to a familiar argument in epistemology. Is all justification inferential or not? If it is not all inferential, then there are some basic justified beliefs, and we have a form of foundationalism. If it is all inferential, then we have a choice between coherentism and an infinite regress of justified propositions, each of which depends for its justification on propositions evidentially prior in the chain. One classic objection to the infinite regress idea is just Boghossian's objection to global relativism in another context, namely that finite minds such as ours cannot have an infinite number of beliefs or reasons like those required by infinitism.

Peter Klein has defended infinitism in epistemology against the finite minds objection in two ways (Klein 1999). First, it is possible for a finitely extended thing to be in an infinite number of states; even Zeno knew that an arrow was in an infinite number of infinitesmally small locations on its way to a target. So infinity alone isn't a problem. Second, grant that a conscious belief requires a non-zero amount of time to attend to or grasp, thus preventing finite human beings from entertaining an infinite number of conscious thoughts. Klein suggests that the infinite set of beliefs be *dispositional*. No one possesses an infinity of already-formed dispositional beliefs, rather, we all have 2nd-order dispositions to supply a justifying reason for any occurrent

1st-order belief. For infinitists, we are always able to devise a reason for anything we think, and if we can't then the belief is not justified. To the extent that Klein's strategies work for the structure of justification, they should also work to defuse Boghossian's complaint against the relativist. The relativist need not grasp or believe a proposition with an infinity of "there is a theory that we accept" conjuncts. The global relativist need only be prepared to dispositionally add such meta-conjuncts when called to do so.[9]

4. The logic of relativism

The final approach to the self-refutation problem is to explore the formal structure of global relativism, and develop a solution based on its logic. Ideally the logical approach would also provide an error theory that could explain why the self-refutation argument looked so promising for so long while at the same time defusing or avoiding that argument. Here is one attempt.

First, stipulate the existence of a non-zero number of perspectives. These can be taken as primitive elements in the system. Then specify that the truth of propositions is to be indexed to perspectives, analogous to the way that the truth of propositions is typically indexed to a language, a time, and a possible world. "Perspective" just becomes one more parameter. Those who think there is just one perspective, or who want to reject the idea of points of view altogether, can regard this requirement as trivially satisfied. An analogy is to those who, like Spinoza, think that every truth is necessarily true. While they have no need of possible world semantics to distinguish necessity from contingency, Spinozists could still allow that necessary truth is truth in all possible worlds; they just think that there's only one possible world, and whatever is true at that world is trivially true in all. Absolutists could allow that there are perspectives, even if they think there is only one, and insist that whatever is true is trivially true in all.

Nothing is true outside the structure of perspectives any more than there are true sentences that aren't sentences of any specific language. It remains an open question as to whether there could be a proposition that is true in all perspectives, just as there might be a proposition true in all possible worlds. Formally, a proposition is relatively true just in case it is true in some perspectives and it is absolutely true just in case it is true in all perspectives. This parallels the idea that a proposition is possibly true if it is true in some worlds and necessarily true if it is true in all worlds. Let "♦" be an operator that takes sentences and indexes them to perspectives, so that "♦Φ" is to be read as "it is relatively true (true in some perspective) that Φ." The claim that everything is relative is thus: for all Φ, ♦Φ. Further, let us introduce "■" as an "absolute" operator so that ■Φ is to be read as "it is absolutely true (true in all perspectives) that Φ."

Now let's formulate the self-refutation problem. The thesis of global relativism is everything is relative. Absolutism denies global relativism: not everything is relative. By "everything is relative," let us understand the claim that every proposition is true in some perspective and untrue in another. Absolutism is then: there is at least one proposition which has the same truth value in all perspectives. Clearly, either the thesis of relativism is true absolutely (true in all perspectives) or just relatively (true in some, but not all perspectives). Suppose that relativism is true in all perspectives. If so, then there is a proposition which has the same truth value in all perspectives – viz., the thesis of global relativism itself. Yet, if there is some proposition which has the same truth value in all perspectives, then absolutism is true. Thus if relativism is true in all perspectives, absolutism is true; equivalently, if relativism is true in all perspectives then by reductio relativism is untrue.

So far this is fairly straightforward. The more interesting work comes in assessing the idea that global relativism is merely relatively true. Here is a more contentious assumption, one that will do real work in the peritrope argument: assume that the relativist logic has an S5 structure, and particularly that S5's characteristic theorem $\blacklozenge\blacksquare\Phi \Rightarrow \blacksquare\Phi$ holds.[10] Further assume that for all propositions Φ, $\blacklozenge\Phi$. That is, every proposition is relatively true (true in some perspective). Allow Φ to be "it is absolutely untrue that everything is relative." Then the following turns out to be true: relatively, it is absolutely untrue that everything is relative. Granted the S5-like theorem that whatever is relatively absolute is absolute, then it will follow straightaway that it is absolutely untrue that everything is relative. And, by reductio, the relativist thesis is false. Absolutism really is a universal acid that cannot be harmlessly contained by a perspective.

The beauty of the preceding result is that it explains why the self-refutation argument has been so durable and yet so tricky to formulate compellingly. Without recognizing that the relativist thesis can be formulated modally, or knowing about S5 logic, there is no easy path to proving that "relativism is merely relatively true" is just as self-refuting as "relativism is absolutely true." These logical ideas are of recent vintage in the history of philosophy.[11] Another positive result is that while

(1)   Everything is relative

turns out to be false and self-undermining,

(2)   Everything true is relatively true

does not.

The difference between the two formulations can be brought out with an analogy. Compare *everything is possible* to *everything true is possibly true*. No one except the pathologically optimistic would defend the idea that *everything is possible*, but *everything true is possibly true* is so obvious as to hardly rate a comment. *Everything true is possibly true* allows the possibility that there are necessary truths that are true in all worlds and it permits that some truths are merely contingent ones that are true in some worlds but false in others. Analogously, *everything true is relatively true* is compatible with there being absolute truths that are true in all perspectives while also permitting that there are merely relative truths that are true in some perspectives and false in others. Given the logical space in which to develop their theories, honest relativists must then argue that truths about morality, aesthetics, set theory, or what have you are merely relatively true.[12]

Some critics have rejected the preceding logical analysis of the self-refutation problem on the grounds that the S5 theorem $\blacklozenge\blacksquare\Phi \Rightarrow \blacksquare\Phi$ is unmotivated.[13] It is difficult to see what such critics want. The assumption of the theorem helps provide a rigorous demonstration of the self-refutation argument while at the same time allowing for a consistent global relativism. Why S5? Because doing so does valuable philosophical work. It is like asking a carpenter, "why use a 3-inch screw?" The obvious answer is because that's the tool he needs. To demand an independent semantic reason for accepting $\blacklozenge\blacksquare\Phi \Rightarrow \blacksquare\Phi$, apart from the role it plays in sorting out relativism, is like demanding a justification for using a 3-inch screw apart from the role it plays in fastening two things together. If a 2-inch screw can do just as good a job, then one might reasonably ask "why use a 3-inch screw instead of a 2-inch screw?" By analogy the critics would then need to offer a real alternative to the S5 theorem that is as effective in explaining and defusing the peritrope for their challenge to be compelling.

The self-refutation objection to global relativism has plagued relativists from the start. It turns out that neither the structure of the argument nor its effectiveness is clear as some have thought. It is certainly not the knockdown argument that Plato thought it was. Whether the objection can be marked as solved or it is still an open question depends upon what theory of relativism is being considered. The key question is what do we want any theory of relativism to do for us;

what difficulties does it solve? Knowing that, future relativists must consider the peritrope, but there are many ways it might be accommodated.

Notes

1 "Relativism certainly is self-refuting" (Rorty 1991, 202), "Although we all know that cultural relativism is inconsistent (or say we do) I want to take the time to say again that it is inconsistent" (Putnam 1983, 236).
2 According to Diogenes Laertius (book 2, ch. 10, §108). Diogenes clearly thought that Eubulides was a pretentious lover of controversy. See Kneale and Kneale (1962, 113–114, 227–229) for discussion.
3 For an excellent close reading of both Plato and Aristotle on the self-refutation argument, see Lee (2005, esp. ch. 4).
4 See MacFarlane (2005) and Carter (2016, 58).
5 Montaigne (1877, ch. 30). Compare ch. 36 where Montaigne purports to "believe and apprehend a thousand ways of living; and, contrary to most men, more easily admit of difference than uniformity amongst us."
6 Cf. Nozick (2001, 15–16).
7 MacFarlane (2014, 30).
8 For example, Putnam (1981, 119–124) and Boghossian (2006, 54–56). Cf. MacFarlane (2014, 32–33).
9 Also see the somewhat different response to Boghossian in MacFarlane (2014, 33).
10 The S5 theorem is provable if the commensurability relation among perspectives is an equivalence relation. See the appendix in Hales (1997).
11 Cf. Hales (2006) and for a similar earlier treatment, see Hautamäki (1983).
12 This logical treatment is not the only one possible. Bennigson (1999) offers an alternative, and Ressler (2013) has the most comprehensive overview of relativist logics.
13 See Kölbel (2011, 27) and MacFarlane (2014, 30) for examples of this criticism.

References

Bennigson, T. (1999), "Is Relativism Really Self-Refuting?" *Philosophical Studies* 94(3): 211–236.
Boghossian, P. (2006), *Fear of Knowledge: Against Relativism and Constructivism*, Oxford: Oxford University Press.
Carter, J. A. (2016), *Metaepistemology and Relativism*, New York: Palgrave Macmillan.
Davidson, D. (1984), "On the Very Idea of a Conceptual Scheme," in *Inquiries Into Truth and Interpretation*, edited by D. Davidson, Oxford: Oxford University Press.
Empiricus, S. (2005), *Against the Logicians*, Cambridge: Cambridge University Press.
Hales, S. D. (1997), "A Consistent Relativism," *Mind* 106(421): 33–52.
———. (2006), *Relativism and the Foundations of Philosophy*, Cambridge, MA: MIT Press.
———. (forthcoming), "Nietzsche's Epistemic Perspectivism," in *Knowledge From a Human Point of View*, edited by M. Massimi and A.-M. Creţu, New York: Springer.
Hales, S. D. and R. Welshon (2000), *Nietzsche's Perspectivism*, Champaign, Urbana: University of Illinois Press.
Hautamäki, A. (1983), "The Logic of Viewpoints," *Studia Logica* 42(2/3): 187–196.
Klein, P. D. (1999), "Human Knowledge and the Infinite Regress of Reasons," *Philosophical Perspectives* 13: 297–325.
Kneale, W. and M. Kneale (1962), *The Development of Logic*, Oxford: Oxford University Press.
Kölbel, M. (2011), "Global Relativism and Self-Refutation," in *A Companion to Relativism*, edited by S. D. Hales, Oxford: Wiley-Blackwell, 11–30.
Lee, M.-K. (2005), *Epistemology After Protagoras: Responses to Relativism in Plato, Aristotle, and Democritus*, Oxford: Oxford University Press.
MacFarlane, J. (2005), "Making Sense of Relative Truth," *Proceedings of the Aristotelian Society* 105: 305–323.
———. (2014), *Assessment Sensitivity: Relative Truth and Its Applications*, Oxford: Oxford University Press.
Montaigne, M. de (1877), *The Essays*, London: Reeves and Turner.
Nozick, R. (2001), *Invariances: The Structure of the Objective World*, Cambridge, MA: Harvard University Press.
Putnam, H. (1981), *Reason, Truth, and History*, Cambridge: Cambridge University Press.

―――. (1983), "Why Reason Can't be Naturalized," in *Realism and Reason*, edited by H. Putnam, Cambridge: Cambridge University Press, 229–247.

Ressler, M. (2013), *The Logic of Relativism*, Increasingly Skeptical Publications.

Rorty, R. (1991), *Objectivity, Relativism, and Truth*, Cambridge: Cambridge University Press.

Wittgenstein, L. (1953), *Philosophical Investigations*, Oxford: Basil Blackwell.

31

EPISTEMIC RELATIVISM AND EPISTEMIC INTERNALISM

Duncan Pritchard

1. Epistemic relativism

Relativism comes in many forms, but the variety of relativism that is most relevant when it comes to the epistemic internalism/externalism distinction is a specifically *epistemic* relativism. In its broadest outline, this is the idea that epistemic standing is itself a relative notion, where that usually means that it is relative to a particular epistemic system.[1] That is, there is no such thing as epistemic standing *simpliciter*, but only epistemic standing relative to a particular epistemic system. Such a formulation of epistemic relativism encompasses numerous different views, of course, some of which would not be especially controversial.[2]

We can bring this point out by distinguishing between the broad categorisation of epistemic relativism just offered and a more specific formulation of the view. From the general statement of epistemic relativism, it follows that there is no way of determining epistemic standing independently of an epistemic system. It also follows that different epistemic systems might generate different epistemic verdicts as regards the same belief, or that beliefs in opposing propositions (i.e., p and not-p) can both enjoy positive epistemic standing relative to different epistemic systems.[3] There can thus be disagreements between epistemic systems in terms of what beliefs have positive epistemic standing. But what doesn't follow from the general categorisation of epistemic relativism alone is the idea that there will be *incommensurable* epistemic systems.

By epistemic incommensurability I mean the idea that there can be disagreements regarding the verdicts generated by distinct epistemic systems such that there is no epistemic way of resolving this dispute. In particular, it is the idea that distinct epistemic systems can incorporate their own norms for appraising epistemic standing such that they are effectively "closed off" from other epistemic systems, which will have their own fundamentally different epistemic norms. There is nothing in the general formulation of epistemic relativism that entails epistemic incommensurability (or excludes it for that matter), since it merely says that epistemic standing is always relative to an epistemic system. It is compatible with that articulation of epistemic relativism that distinct epistemic systems are not epistemically closed off from each other – for example, that they incorporate enough shared epistemic norms that there is always, in principle at least, an epistemic way of resolving disagreements.

I take it that where epistemic relativism is understood as particularly philosophically troubling it is usually because it is thought of as involving a commitment to epistemic incommensurability.

Call this *strong epistemic relativism*. It is strong epistemic relativism, after all, that seems especially pernicious. If we cannot resolve disagreements through epistemic means, then how else are we to resolve them? Through force, perhaps? That is clearly a problematic consequence of the view.[4] In asking how the epistemic externalism/internalism distinction relates to epistemic relativism, we should thus also consider whether it can be used to motivate strong epistemic relativism in particular, or only the more general formulation of the view that lacks the commitment to epistemic incommensurability.

Notice that epistemic relativism, on either formulation, is entirely compatible with a rejection of relativism about truth (i.e., *alethic relativism*). In particular, whether the subject's beliefs are true – along with any other factual conditions related to their epistemic situation – can be an entirely objective matter, one that is not relative (in any philosophically interesting way, at any rate) to anything else. Since knowledge entails truth, it follows that if truth isn't relative then where there is a disagreement between two agents with opposing beliefs who both think that they have knowledge, then at most only one of them can be right (since one of them must have a false belief). This point about the independence of epistemic and alethic relativism is also important because it helps us to clarify the specific way in which the epistemic externalism/internalism distinction might be especially relevant to epistemic relativism.

2. Epistemic relativism and the epistemic internalism/externalism distinction

For our purposes, we can take the epistemic internalism/externalism distinction to concern whether one's epistemic standing needs to be reflectively accessible to the subject.[5] Traditionally, epistemic internalists say that it must be, while epistemic externalists, such as reliabilists, say that it needn't be. There is plenty of nuance here, of course. For example, there are differences between epistemic internalists in terms of how this reflective access requirement is understood.[6] There are also differences between epistemic internalists and externalists in terms of what the scope of their views is. For instance, does it apply to all epistemic standings, or just to a particular epistemic standing (e.g., justification)? Relatedly, consider a core epistemic standing like knowledge. It is common here to delineate a continuum of internalist/externalist positions in this regard. A pure epistemic internalism would hold that what converts true belief into knowledge is only the satisfaction of internalist epistemic conditions. In contrast, a pure epistemic externalism would deny that there are any internalist epistemic conditions on knowledge. In between these two poles, one can find intermediate views, such as the idea that there is a significant internalist epistemic condition on knowledge, but that this is only a necessary condition, where there are other epistemic conditions on knowledge that are not cast along internalist lines (e.g., an anti-luck condition).[7]

In any case, it is only internalist epistemic conditions that are relevant to epistemic relativism, given how we have disentangled this latter thesis from alethic relativism. This is because whether one has satisfied an externalist epistemic condition, such as a reliability or an anti-luck condition, depends only on whether the relevant facts in question have obtained. Without alethic relativism in play, there will thus be an objective fact of the matter whether one has satisfied the externalist epistemic condition. Take a simple process reliabilism about justification, for example, such that a belief is justified when it is formed via a reliable belief-forming process.[8] Now imagine that there is a dispute about whether a particular belief, formed via a specific belief-forming process, is justified. Who is right in this disagreement will depend on whether this belief-forming process is in fact reliable, which means that only one of the disputants' verdicts can be correct. It is thus the facts themselves that settle a disagreement of this kind, in the sense of determining whether the agent in question genuinely has a justified belief. There is thus no straightforward way of generating a specifically epistemic relativism via appeal to epistemic externalism.

One thus needs epistemic internalism in play in order to make sense of epistemic relativism. Let's take epistemic justification as our focus in this regard (henceforth, "justification"). The following scenario is entirely compatible with epistemic internalism. One subject might have good reflectively accessible reasons for believing that p, and thereby have a justified belief that p, and another subject might have good reflectively accessible reasons for believing that not-p, and thereby have a justified belief that not-p.[9] Take the belief that the Orinoco river flows through Venezuela. The first subject might believe this is true because she read it in a respected atlas, while the second subject might believe that it's false because someone authoritative told her (falsely) that border changes mean that it no longer flows through Venezuela. Appealing to the objective facts of the matter, such as which (if any) of these beliefs is formed via a reliable belief-forming process (much less whether it is true that the Orinoco river flows through Venezuela), would not settle this issue, since these facts could obtain and yet it still be the case that both agents' beliefs are justified.

Epistemic internalism is thus compatible with the idea that two opposing beliefs can both enjoy genuine (i.e., as opposed to merely apparent) positive epistemic standing, where that disagreement cannot be resolved simply by appealing to the facts of the situation. That by itself doesn't entail epistemic relativism, of course, since there is no mention here of the idea that epistemic standing is always relative to an epistemic system (still less does it entail strong epistemic relativism). But at least we can now see how epistemic internalism creates the logical space for epistemic relativism. All we need to imagine is that these judgements about internalist epistemic standing are always relative to an epistemic system.

Consider how a disagreement of the kind just described might ordinarily pan out. As things stand, the two parties have justified beliefs in isolation from each other. Moreover, since each party is unaware of the other, then the fact of this disagreement is not something that is reflectively accessible to the subject, and hence is not epistemically relevant by epistemic internalist lights. So there is nothing essentially puzzling about both of these opposing beliefs retaining their positive epistemic standing. But suppose now that these parties become aware of the disagreement, and thus of their respective reasons in support of their opposing beliefs. Both factors can now have an influence on what the subject's reflectively accessible basis for the belief in question is, and thus on her justification for that belief. In the normal case, we would thus expect at least one of the beliefs in question to change its epistemic standing as a result.

For example, upon hearing that the other party has been authoritatively told that the Venezuelan border was recently changed, such that the Orinoco river no longer flows through it, then one might naturally doubt the veracity of the atlas on this particular point. In particular, that the atlas is in general a good source of geographical information is entirely compatible with it not reflecting such a recent border change. So unless one has some further reason to discount this testimony, then I think one would naturally regard oneself as no longer having adequate reflectively accessible grounds in support of one's belief, and hence as lacking justification for that belief. One's disputant, after all, has provided one with a defeater for one's justification, one that one cannot in turn defeat. In contrast, one's disputant would naturally still regard her belief as justified, as she hasn't been presented with a defeater for her justification. (That someone else formed the opposing belief by examining a respected atlas is entirely consistent with what she currently believes, as recent border changes will not be reflected in atlas entries.)

In light of the disagreement becoming known, and the two parties thus becoming aware of each other's reasons for holding their respective beliefs, one party thus loses her justification and the clash of epistemic support is resolved. Moreover, this is a rational resolution of this disagreement, as both parties are being suitably responsive to the epistemic norms in play. Notice too that in this case it could well be that one's belief that the Orinoco river flows through Venezuela is not only true but derived from reading a reliable atlas, but that wouldn't prevent one's (internalist) justification from being defeated. Indeed, it could even be true that one's disputant's belief is not in fact formed on a reliable basis – perhaps her informant is not authorative at all about subject maters like this, but merely a (very plausible) prankster. All that matters is that she has good reason to believe what she does, not whether those reasons are in fact grounded in the relevant facts.

There is nothing in this description of how a clash of internalist epistemic justification might be resolved that implies (or, for that matter, excludes) epistemic relativism. The issue doesn't arise because it is clear in this case that both parties have shared conceptions of what constitutes a good reflectively accessible reason. So neither party is disputing, for example, that ordinarily seeing where a river is listed as flowing in a good quality atlas is an adequate rational basis for forming a belief in this regard, such that this belief is justified. Equally, however, both parties grant that even a good quality atlas would not be an adequate rational basis for such a belief if the geography in question has recently changed. This shared conception of what constitutes a good rational basis for one's belief is what enables this clash of epistemic justification to be resolved.

Viewed from the perspective of epistemic relativism, the point would simply be that these two subjects share a common epistemic system, at least as regards the matter in hand at any rate, and that's why the dispute is so easy to resolve. If one grants, in line with epistemic relativism, that all epistemic standing is relative to an epistemic system, then the interesting cases of disagreements of this kind will be ones where two distinct epistemic systems are in conflict. Notice that even where two distinct epistemic systems are in place, it needn't follow that the dispute is rationally irresolvable. In the toy case just envisaged, for example, even if the two parties are employing distinct epistemic systems, it might nonetheless be the case that they agree on the epistemic norms that are relevant to assessing this particular disagreement (e.g., that good quality atlases offer a sound rational basis to form one's geographical beliefs). If that's right, then the disagreement could be rationally resolved in just the direct way described previously, even despite there being two distinct epistemic systems in place.

Indeed, even if the two parties diverge in terms of the epistemic norms that they endorse with regard to the subject matter in hand, so long as there is broad overlap in the epistemic norms that are part of their epistemic systems, then there is still scope for rational resolution of the disagreement at hand (though this will inevitably now be much more difficult). Suppose, for example, that the one party doesn't trust atlases at all, because she believes that there is a global conspiracy to deceive on this score (she is a "flat-Earther," say). The dispute between the two sides is now more substantive, since the one party to this dispute doesn't even grant that atlases are in general a good source of reasons for one's geographical beliefs. Nonetheless, if our conspiracy theorist shares our epistemic norms in other respects, then there might still be rational ways to resolve this disagreement. In particular, while their conspiracy theory might make them distrustful of atlases and other geographical information sources of this general type, they might nonetheless see the epistemic import of first-hand testimony. If that's right, then while appealing to the atlas might not enable a rational resolution of this dispute, bringing in a third-party who can testify to seeing the Orinoco river herself while visiting Venezuela might well suffice to persuade her to change her mind.

If we want to formulate strong epistemic relativism within the epistemic internalist framework we will thus need to go further and stipulate that the epistemic systems are not merely distinct, but that they are fundamentally different. The former just delivers the idea that there are different epistemic norms in play, but that by itself won't entail epistemic incommensurability.

This is only ensured by the epistemic norms being not only different, but fundamentally different, such that there are insufficient common epistemic norms to enable disagreements to be rationally resolved.

Imagine that one of our protagonists in our toy example doesn't just reject the epistemic authority of atlases and such like, but rejects anything that we would ordinarily consider an epistemic authority (including testimony of first-hand experience). Perhaps, say, the only epistemic authority they recognise is a religious one (scripture, testimony by religious leaders, and so on) that does not figure at all in our epistemic norms. How then would we be able to appeal to common epistemic norms? In such a scenario, even after both sides are apprised of each other's reasons, and aware of the disagreement in play, they will nonetheless continue – quite rightly, relative to their epistemic system – to regard themselves as having a justified belief, and there seems no rational process available for resolving this clash.

3. A case study: hinge epistemology

Epistemic internalism thus creates the logical space for epistemic relativism, in both its general and its more specific strong varieties. One question we might ask is why an internalist epistemic standing might be relative to an epistemic system. In particular, why wouldn't there be sufficient convergence on the epistemic norms in play, such that epistemic standing isn't relative to an epistemic system to any extent to make it philosophically interesting?

This is an important issue in the history of epistemology, since one characteristic project of epistemology has been precisely to identify universal epistemic norms. Think, for example, of the Cartesian classical foundationalist project of employing the methodology of radical doubt in order to discover indubitable foundations for our epistemic practices. In this way, we are able to identify universal epistemic foundations for our epistemic practices, and thereby establish universal epistemic norms.[10] This is also a broadly internalist epistemic project, at least to the extent that the reasoning at issue is meant to be in principle reflectively available to anyone. Such a picture runs entirely counter to the spirit of epistemic relativism, and yet does so while staying within the general contours of the epistemic internalist framework.

Many are not persuaded by such a project, however. Moreover, there are competing epistemological accounts that seem to directly lead to epistemic relativism, at least in its general form. It will be useful to consider one such proposal that is influential in the contemporary literature, in order to see how a concrete epistemology might have this implication.

The proposal that I have in mind is the kind of *hinge epistemology* that many commentators think is inspired by Wittgenstein's (1969) remarks on the nature of knowledge in *On Certainty*. Very roughly, according to hinge epistemology all rational evaluation takes place relative to certain background "hinge" commitments. These hinge commitments constitute the rational "fixed points" which determine what is to count as a good reason in favour or against a particular belief. Since these hinge commitments need to be in place in order for rational evaluation to occur, it follows that they cannot themselves be rationally evaluated (including negatively, in the sense of being found to be irrational), and hence are arationally held. Moreover, the thought is that it is not an incidental feature of our epistemic practices that they are structured like this, but that it is in the very nature of a rational evaluation that it presupposes hinge commitments. If that's right, then the classical foundational project is doomed from the outset. In particular, the very idea, key to this project, of rationally evaluating all of one's beliefs at once is simply incoherent, as that would be to undertake a rational evaluation without any hinge commitments in play. Moreover, rather than there being universal epistemic foundations at the bedrock of our epistemic practices that enjoy a privileged epistemic standing (such as being indubitable), we find instead non-universal arational hinge commitments.[11]

This last point about the hinge commitments not being universal is particularly important for our purposes. For the idea is not that the hinge commitments, while arational, are nonetheless universal and eternal. Far from it. Instead, one's hinge commitments can potentially vary from culture to culture, epoch to epoch, person to person, and so on. This is the sense in which hinge epistemology seems to lead directly to epistemic relativism, if not to strong epistemic relativism in particular. Given the role that the hinge commitments play, where we have epistemic systems that incorporate distinct hinge commitments we would thereby have distinct epistemic systems in the very sense demanded by epistemic relativism, such that epistemic standing (in this case rational standing) is relative to that particular epistemic system. Moreover, if there can be wholesale variability in one's hinge commitments, then one can also use this epistemic framework to motivate strong epistemic relativism.[12] After all, if the hinge commitments can be radically divergent, then it seems that there could be epistemic systems that embody completely different epistemic norms. Accordingly, it would appear to follow that there couldn't be rational resolution of disagreements between epistemic systems with radically divergent hinge commitments. We thus get not just epistemic relativism in its general form, but also epistemic incommensurability and, thereby, strong epistemic relativism.[13]

Notice that hinge epistemology is able to lead to epistemic relativism precisely because it is specifically concerned with an internalist epistemic standing. Moreover, this feature of the view is not sufficient to generate epistemic internalism, since one also needs the further motivation for regarding epistemic standing (in this case rational standing) as essentially relative to an epistemic system. This is what the idea that all rational evaluation is relative to one's arational hinge commitments provides. Finally, notice that the fact that hinge epistemology leads to epistemic relativism doesn't itself entail that it leads to strong epistemic relativism. For that latter claim to follow, one must also be willing to countenance a fundamental divergence in the hinge commitments in play in different epistemic systems. These points highlight general features of the relationship between epistemic internalism and epistemic relativism. In particular, that the former is necessary, but not sufficient, for the latter, and that even then it is a further question whether we are faced with epistemic relativism in its general form or the more philosophically challenging strong epistemic relativism.

Acknowledgements

I am grateful to Martin Kusch for detailed comments on an earlier version of this chapter, and to Annalisa Coliva for helpful discussions on this topic.

Notes

1 See Boghossian (2006) for an influential recent formulation of epistemic relativism of this general form. For a useful recent overview of the contemporary debate regarding epistemic relativism, see Carter (2019).
2 Arguably, for example, contextualism or contrastivist accounts of epistemic standing would entail epistemic relativism on this conception, and yet these are mainstream epistemological proposals, no more controversial than any other mainstream epistemological proposal. For some key discussions of epistemic contextualism and contrastivism, see DeRose (1995), Lewis (1996), Cohen (1999, 2000), and Schaffer (2005). For a related view about "knows," albeit one that appeals to alethic relativism, see MacFarlane (2005).

3 In order to keep matters simple, I will here take the bearer of epistemic standing to be beliefs, though of course this is not the only propositional attitude that could serve this role.

4 Keeping the general conception of epistemic relativism apart from the more specific strong epistemic relativism is also useful when it comes to clarifying what is at issue in some recent debates about epistemic relativism. To take one example, Kinzel and Kusch's (2017) recent defense of epistemic relativism is clearly concerned with the more general formulation, since they allow that the resolution of disagreements between epistemic systems, while practically difficult, is not in principle impossible. In contrast, some recent attacks on epistemic relativism, such as my own – e.g., Pritchard (2009, 2010) – are best understood as critiquing strong epistemic relativism in particular, rather than the more general conception of the view.

5 This is the *accessibilist* conception of epistemic internalism. See also *mentalism* as an account of epistemic internalism, such as Conee and Feldman (2004). Note, though, that the distinction between accessiblism and mentalism, though theoretically important, isn't that pressing for our current concerns. This is because the issue that divides these views is primarily about how best to understand the motivation for epistemic internalism. So, for example, even mentalists will usually grant that epistemic standings (in the relevant sense) need to meet a suitable reflective access requirement – their claim will just be that this requirement is met because this is needed to satisfy an overarching mentalist requirement. A further way of understanding the epistemic externalism/internalism is in terms of the *new evil demon* intuition – see Lehrer and Cohen (1983) and Cohen (1984). I offer a taxonomy of classical and non-classical forms of epistemic internalism/externalism that distinguishes them along accessibilist, mentalist and new evil demon lines in Pritchard (2011a). For a helpful recent survey of the main contours of the epistemic externalism/internalism distinction, see Pappas (2014).

6 See, for example, Alston (1988) and Bergmann (2006, *passim*).

7 For example, in terms of this taxonomy of externalist/internalist theories of knowledge, the classical tripartite account of knowledge would constitute "pure" internalism (so long as we follow orthodoxy and treat the justification condition as an internalist epistemic condition at any rate), while a process reliabilist account of knowledge (e.g., Goldman 1976) would constitute "pure" externalism, and a no-false-lemmas account of knowledge, which supplements the tripartite account with an externalist epistemic condition to deal with Gettier-style cases (e.g., Lycan 2006), would constitute a "mixed" view. Note, however, that post-Gettier the externalism/internalism debate in the theory of knowledge tends to focus on the contrast between mixed views and pure externalism, where the former is the standard-bearer for epistemic internalism.

8 Process relibialism is usually expressed in a subtler fashion than this, of course. See Goldman (1979), for an early statement of the view as applied to justification.

9 Notice that for the purposes of this chapter I will be following orthodoxy in treating these reflectively accessible reasons as being essentially non-factive. My own view, following McDowell (1995), is that there can be reflectively accessible factive reasons, and hence that epistemic internalism needn't be wedded to non-factive reasons, though it would clearly take us too far afield to defend such a claim here. See Pritchard (2012) for further discussion.

10 For an influential account of this Cartesian epistemological methodology, see Williams (1978).

11 For a more detailed overview of hinge epistemology, see Pritchard (2017b). Naturally, there are numerous distinct developments of this basic proposal. For some of the key works in this regard, see McGinn (1989), Williams (1991), Moyal-Sharrock (2004), Wright (2004), Coliva (2015), and Schönbaumsfeld (2016). I offer my own version of hinge epistemology in Pritchard (2015).

12 For example, Wittgenstein certainly seemed willing to countenance the idea of there being religious hinge commitments, which would allow there to be a fairly divergent set of hinge commitments when religious and non-religious worldviews collide. For a development of the epistemology of religious belief along hinge epistemology lines (a position I call *quasi-fideism*), see Pritchard (2011b, 2017a).

13 For further discussion of whether a Wittgensteinian hinge epistemology leads to epistemic relativism, see Haller (1995), Grayling (2001), Coliva (2010), and Kusch (2016). See also Coliva (2019). For my own take on these issues, see Pritchard (2018; cf. Pritchard 2010).

References

Alston, W. P. (1988), "An Internalist Externalism," *Synthese* 74: 265–283.
Bergmann, M. (2006), *Justification Without Awareness*, Oxford: Oxford University Press.

Boghossian, P. (2006), *Fear of Knowledge: Against Relativism and Constructivism*, Oxford: Oxford University Press.

Carter, J. A. (2019), "Epistemology and Relativism," in *Internet Encyclopedia of Philosophy*, edited by B. Dowden and J. Fieser, www.iep.utm.edu/epis-rel/.

Cohen, S. (1984), "Justification and Truth," *Philosophical Studies* 46: 279–296.

———. (1999), "Contextualism, Skepticism, and the Structure of Reasons," *Philosophical Perspectives* 13: 57–89.

———. (2000), "Contextualism and Skepticism," *Philosophical Issues* 10: 94–107.

Coliva, A. (2010), "Was Wittgenstein an Epistemic Relativist?" *Philosophical Investigations* 33: 1–23.

———. (2015), *Extended Rationality: A Hinge Epistemology*, London: Palgrave Macmillan.

———. (2019), "Relativism and Hinge Epistemology," this volume.

Conee, E. and R. Feldman (2004), *Evidentialism*, Oxford: Oxford University Press.

DeRose, K. (1995), "Solving the Skeptical Problem," *Philosophical Review* 104: 1–52.

Goldman, A. (1976), "Discrimination and Perceptual Knowledge," *Journal of Philosophy* 73: 771–791.

———. (1979), "What Is Justified Belief?" in *Justification and Knowledge*, edited by G. Pappas, Dordrecht: Reidel, 1–23.

Grayling, A. (2001), "Wittgenstein on Scepticism and Certainty," in *Wittgenstein: A Critical Reader*, edited by H-J. Glock, Oxford: Blackwell, 305–321.

Haller, R. (1995), "Was Wittgenstein a Relativist?" in *Wittgenstein: Mind and Language*, edited by R. Egidi, Dordrecht: Kluwer, 223–232.

Kinzel, K. and M. Kusch (2017), "De-idealizing Disagreement, Rethinking Relativism," *International Journal of Philosophical Studies* 26: 40–71.

Kusch, M. (2016), "Wittgenstein's *On Certainty* and Relativism," in *Analytic and Continental Philosophy: Methods and Perspectives*, edited by S. Rinofner-Kreidl and H. A. Wiltsche, Berlin: Walter de Gruyter, 29–46.

Lehrer, K. and S. Cohen (1983), "Justification, Truth, and Coherence," *Synthese* 55: 191–207.

Lewis, D. (1996), "Elusive Knowledge," *Australasian Journal of Philosophy* 74: 549–567.

Lycan, W. (2006), "On the Gettier Problem Problem," in *Epistemology Futures*, edited by S. Hetherington, Oxford: Oxford University Press, 148–168.

MacFarlane, J. (2005), "The Assessment Sensitivity of Knowledge Attributions," *Oxford Studies in Epistemology* 1: 197–233.

McDowell, J. (1995), "Knowledge and the Internal," *Philosophy and Phenomenological Research* 55: 877–893.

McGinn, M. (1989), *Sense and Certainty: A Dissolution of Scepticism*, Oxford: Blackwell.

Moyal-Sharrock, D. (2004), *Understanding Wittgenstein's On Certainty*, London: Palgrave Macmillan.

Pappas, G. (2014), "Internalist vs. Externalist Conceptions of Epistemic Justification," in *Stanford Encyclopedia of Philosophy*, Fall 2017 edition, edited by E. N. Zalta, https://plato.stanford.edu/entries/justep-intext/.

Pritchard, D. H. (2009), "Defusing Epistemic Relativism," *Synthese* 166: 397–412.

———. (2010), "Epistemic Relativism, Epistemic Incommensurability and Wittgensteinian Epistemology," in *The Blackwell Companion to Relativism*, edited by S. Hales, Oxford: Blackwell, 266–285.

———. (2011a), "Evidentialism, Internalism, Disjunctivism," in *Evidentialism and Its Discontents*, edited by T. Dougherty, Oxford: Oxford University Press, 362–392.

———. (2011b), "Wittgensteinian Quasi-Fideism," *Oxford Studies in the Philosophy of Religion* 4: 145–159.

———. (2012), *Epistemological Disjunctivism*, Oxford: Oxford University Press.

———. (2015), *Epistemic Angst: Radical Skepticism and the Groundlessness of Our Believing*, Princeton, NJ: Princeton University Press.

———. (2017a), "Faith and Reason," *Philosophy* 81: 101–118.

———. (2017b), "Wittgenstein on Hinge Commitments and Radical Scepticism," in *Blackwell Companion to Wittgenstein*, edited by H-J. Glock and J. Hyman, Oxford: Blackwell, 563–575.

———. (2018), "Wittgensteinian Hinge Epistemology and Deep Disagreement," *TOPOI*, https://doi.org/10.1007/s11245-018-9612-y.

Schaffer, J. (2005), "Contrastive Knowledge," in *Oxford Studies in Epistemology*, edited by T. Gendler and J. Hawthorne, Oxford: Oxford University Press, 235–273.

Schönbaumsfeld, G. (2016), *The Illusion of Doubt*, Oxford: Oxford University Press.

Williams, B. (1978), *Descartes: The Project of Pure Enquiry*, London: Penguin.

Williams, M. (1991), *Unnatural Doubts: Epistemological Realism and the Basis of Scepticism*, Oxford: Blackwell.

Wittgenstein, L. (1969), *On Certainty*, edited by G. E. M. Anscombe and G. H. von Wright, translated by D. Paul and G. E. M. Anscombe, Oxford: Blackwell.

Wright, C. J. G. (2004), "Warrant for Nothing (and Foundations for Free)?" *Proceedings of the Aristotelian Society* 78: 167–212.

32

RELATIVISM AND EXTERNALISM

J. Adam Carter and Robin McKenna

1. Introduction

Internalists in epistemology think that whether one possesses epistemic statuses such as knowledge or justification depends on factors that are internal to one; externalists think that whether one possesses these statuses can depend on factors that are external to one.[1] We can complicate this distinction in several ways. What it means for a factor to be "internal" or "external" to one is subject to debate. One might be an internalist about some epistemic statuses (e.g. justification) but not others (e.g. knowledge). For the purposes of this chapter we set these issues aside and focus on the relationship between externalism and epistemic relativism, which is, roughly, the view that epistemic statuses like justification and knowledge are themselves always relative to some non-trivial parameter, such as local norms or conventions.[2]

Externalism isn't straightforwardly incompatible with epistemic relativism but, as we'll see, it is very common to hold that key externalist insights block or undermine some standard arguments for epistemic relativism. Our aim in this chapter is to give a broad overview of why externalism poses a problem for standard arguments for relativism. But we also want to discuss some ways in which externalist ideas might provide support for certain forms of epistemic relativism. We start with externalist arguments against relativism. We then move on to some ways in which externalist ideas might provide support for various forms of relativism. We finish with suggestions for future work.

2. Externalist arguments against relativism

Arguments for epistemic relativism often take, as a starting point, an observation that the absolutist – the epistemic relativist's opponent – can happily concede: cultures that differ across both geography and time can (and do) differ with respect to what epistemic standards they appeal to when determining whether a given belief is justified or known.

But what follows from this? A moment's reflection reveals that not all of these standards can be (absolutely) true because some are ostensibly in conflict with one another. And *this* observation is philosophically significant. For if one thinks that there are (absolutely) correct epistemic standards, then it is incumbent upon one – at least, insofar as one wishes to avoid scepticism – to establish that one's *own* epistemic standards are the right ones.

But how to do that, exactly? At this juncture, it will be helpful to consider two famous argument strategies for relativism that utilise these observations – the *argument from circularity* and the *argument from non-neutrality*. Externalism, we will then show, offers the absolutist a straightforward way to nip each of these arguments for relativism in the bud. That said, in each case we will also show why the relativist might not find the externalist counter strategy compelling.

2.1. Externalism as a response to the argument from circularity

Establishing that one's own epistemic standards are the right ones can be difficult. This is *especially* so when one's dialectical opponent does not already accept these same standards. In fact, some argue the very *attempt* to demonstrate that one's own epistemic standards have a positive epistemic status (in comparison with alternative, competing standards) plays into the hands of the relativist. Michael Williams puts the idea nicely in this passage:

> In determining whether a belief – any belief – is justified, we always rely, implicitly or explicitly, on an epistemic framework: some standards or procedures that separate justified from unjustified convictions. But what about the claims embodied in the framework itself: are they justified? In answering this question, we inevitably apply our own epistemic framework. So, assuming that our framework is coherent and does not undermine itself, the best we can hope for is a justification that is epistemically circular, employing our epistemic framework in support of itself. Since this procedure can be followed by anyone, whatever his epistemic framework, all such frameworks, provided they are coherent, are equally defensible (or indefensible).
>
> *(2007, 3–4)*

This reasoning – call it the *argument from circularity* – offers a powerful argument for epistemic relativism because it purports to show how *all* epistemic frameworks (and thus, all the epistemic standards that make up these frameworks) are ultimately on an equal footing. None aspires to anything more than *epistemically circular* justification, including the frameworks made up of the standards that *we* think have the most going for them, epistemically.

Consider, against the background of this puzzle, what the epistemic externalist might say. In order to sharpen things a bit, let's imagine a special case of the argument from circularity that purports to show that all standards for *epistemic justification* are on an equal footing. And let's look, specifically, at what the externalist about epistemic justification can say in response. As alluded to in Section 1, the externalist about epistemic justification denies that the only[3] factors that matter for whether one is justified are *internal* factors – viz., factors that are accessible to one via reflection alone.[4]

In the face of the argument from circularity (targeting epistemic justification), the externalist about justification has a two-step reply: step one involves a disambiguation and step two involves rejecting on the basis of this disambiguation a premise of the argument from circularity. The disambiguation proceeds as follows: a *justification* for a standard for epistemic justification can be read in multiple ways. On one reading, X is a justification for epistemic standard E only if X can be adduced as a reason in favour of E. Notice how something like this reading of "justification" lies in the background of the argument Williams sketches in the preceding passage when he indicates that (by the relativist's lights) "the best we can hope for is a justification that is epistemically circular."

For the externalist, by contrast, E's being justified simply doesn't require a *justification* in this sense. Take, for example, standard process reliabilism according to which justification is entirely a matter of reliable belief production: a belief is justified iff it is reliably produced.[5] From the perspective of the reliabilist, a justification (in the sense at issue in Williams' passage) for E isn't required for E to be justified. All that is required is that certain reliability facts about E obtain. And the obtaining of these facts needn't require anyone *appealing* to any standard in order to make them true. In slogan form: facts about justification are independent from facts about the activity of justifying; the former are not grounded in the latter.

We're in a position to see now why externalism-cum-reliabilism about justification offers the anti-relativist a way to nip the argument from circularity in the bud. For the argument from circularity trades on what happens when we try to *justify* our own epistemic system. And, as we've just seen, the reliabilist is in a position to challenge the argument "upstream" by simply denying the epistemological significance of justifying for justification.

We've suggested that the relativist might not find the externalist reply to the preceding argumentative strategy persuasive. One reason why is as follows: the relativist who subscribes to the argument from circularity might insist that even if (if externalism-cum-reliabilism is true) we needn't *in ordinary practice* justify our own epistemic standards by appealing to them in order to be justified, it is incumbent upon the reliabilist qua *theorist* to *vindicate* her beliefs as justified beliefs, and her standards as justified standards. If the reliabilist qua theorist wants to do this in an adequate way, then it is not enough that the reliabilist merely *be* justified in her beliefs or for that matter in her standards, but she must be able, in addition, to provide an adequate explanation for *why* they are justified.[6] And it is at this point that it looks as though the theorist who embraces an externalist view like reliabilism will, in the course of this kind of vindicatory project, inevitably appeal to her own standards in the course of justifying them.[7]

2.2 Externalism as a response to the argument from non-neutrality

We will now turn to the *argument from non-neutrality* (e.g., Rorty 1980; Hales 2014). Perhaps the most famous example of this kind of argument strategy owes to Richard Rorty (1980), who develops the argument with reference to the historical dispute between Cardinal Bellarmine and Galileo.[8] What was principally at issue between the two disputants was the matter of the truth of *geocentrism*, the doctrine that the Earth is the geographical centre of the Universe. As Bellarmine saw things, the doctrine was true, and he believed it to be true on the basis of Scripture. Galileo, by contrast, concluded that the doctrine was false. His reasoning was that, on the basis of telescopic evidence, he could observe moons orbiting Jupiter, a phenomenon that is better explained by the heliocentric model than the geocentric model. Moreover, Galileo took the evidence he received from the telescope to not only favour the heliocentric model over the geocentric model, but *also* to indicate that Scripture was not a reliable source of evidence about the movement of celestial bodies. Bellarmine, for his part, took the authority of Scripture to indicate that Galileo's telescopic evidence must be mistaken.

As Rorty saw it, the dialectical situation we find in this kind of dispute – viz., where there is both a (i) first-order disagreement about what is so, *and* (ii) an intractable kind of meta-disagreement about what even counts as suitable evidence that would bear on whether something is so – is one we should diagnose along relativist lines: Galileo is right according to scientific standards, Bellarmine is right according to Scriptural standards, and there is no further sense in which things here can be adjudicated.[9]

There are a variety of ways one might attempt to respond to this kind of argument from an anti-relativistic perspective.[10] But perhaps the most straightforward strategy belongs to the epistemic externalist. For if the externalist (e.g., the reliabilist) is correct, then a central premise of Rorty's argument from non-neutrality is simply *undercut*. Rorty's diagnosis appeals implicitly to the idea (which he does not explicitly defend) that there can be a (non-relative) resolution of the dispute concerning the existence of the moons only if there is some kind of suitably neutral, shared epistemic standard that Bellarmine and Galileo could appeal to in order to adjudicate their dispute (something Rorty thinks there is not).[11]

The externalist is now in a position to respond: whether or not Bellarmine or Galileo is epistemically justified in believing either the first-order celestial claim at issue *or* the second-order claim about which kind of evidence is relevant to adjudicating the first-order issue, is itself *entirely orthogonal* to the matter of whether the two parties can find any common ground. If either side *in fact* has reliably formed beliefs, then these beliefs are justified, otherwise not. A broader point can be gleaned here: to the extent that considerations to do with "deep disagreements" (such as the Bellarmine/Galileo dispute) are taken to be evidence for epistemic relativism, the externalist has a principled reason to disagree. In slogan form, the factors that make you justified will continue to make you justified *even when other people think they don't* (and even if you are unable to rationally persuade them by their own lights that they do).

Does the relativist sympathetic to Rorty's non-neutrality argument have a reply here? It turns out they do, though it will ultimately be a dangerous one to rely upon. The first step in the reply is to appeal to the plausibility of a position known in the peer disagreement literature as *conciliationism*.[12] According to conciliationism, if you find that someone who you previously regarded as your epistemic peer disagrees with you about p, then you are rationally required to downgrade your confidence that p is true. To the extent that conciliationism offers a plausible way to think about the epistemic significance of disagreement *given* the absolutist assumption that at most one party to a disagreement is right, the relativist is in a position to "revive" the pro-relativist import of deep disagreements against the externalist.

Here is the idea, in outline: Given the *prevalence* of disagreement about philosophical views, including views about epistemic standards (such as the kind of view the externalist/reliabilist is advancing), conciliationism seems to lead to widespread agnosticism about epistemic standards *if* absolutism is assumed. Granted, the absolutist can avoid the agnostic result by rejecting conciliationism and accepting that each party to the disagreement can rationally hold their ground. In this way, widespread disagreement about epistemic standards (something the externalist should be willing to countenance) wouldn't imply agnosticism about those very standards. However, it follows from this view – *non-conciliationism* – that disagreeing with someone you think is just as likely to be right as you isn't something that will be epistemically significant for you *even when you both think only one of you can be right*. Forced with a choice between (i) the hard-line non-conciliationist option, (ii) wholesale agnosticism about epistemic standards, and (iii) the *denial* that at most one party to a dispute can be right, the relativist suggests the third option should look the most attractive to the (non-sceptical) externalist. For by denying that at most one party can be right, the threat of scepticism about epistemic standards is off the table for the externalist who grants that at least some epistemic peers deny externalism.

We flagged that the preceding reply is a potentially dangerous one for the relativist to rely on. Here is why. The argument relies on two points the proponent of externalism is in a position to contest. Firstly, the externalist might claim that, if they encountered someone who they previously regarded as an epistemic peer, but who denied externalism, they would rightly *no longer* view that individual as an epistemic peer. Secondly, even if the previous response is not a viable one, the argument goes through only if relativism should be thought more attractive to the externalist than should the package of absolutism, conciliationism and scepticism, or the package of absolutism and non-conciliationism. And it's far from clear that this will be the case.

3. Externalist arguments for relativism

As we have just seen, externalist views in epistemology are often taken to undercut support for relativism. But one can also argue that externalist views provide support for certain forms of relativism. In this section we will review some of these arguments. Throughout the focus will be on a particular externalist view: (process) reliabilism.

3.1. *Doxastic vs. propositional justification*

Put roughly, some subject S's belief that p is *propositionally* justified iff S has good reasons for believing that p. But S may be aware of good reasons for believing p yet not believe p on the basis of those reasons (or S may not believe p at all). Imagine Catriona is aware of good reasons for believing that it will rain later (she has read a reliable forecast), but she doesn't believe that it will rain on the basis of the forecast, but rather on the basis of superstition ("red sky at dawn, shepherd's warning"). Catriona's belief is propositionally justified, but it isn't *doxastically* justified, because it isn't believed on the basis of good reasons. The problem with Catriona's belief is that it wasn't formed in the right sort of way.

In his classic 1979 paper "What Is Justified Belief?," Alvin Goldman proposed a reliabilist theory of justification on which (put roughly) S's belief that p is (doxastically) justified iff it is formed in the right sort of way, and a belief is formed in the right sort of way iff it is produced by a reliable process. Goldman's theory treats doxastic justification (in Goldman's terms, "ex post justification") as the primary notion, and defines propositional justification (which Goldman calls "ex ante justification") in terms of it. S's belief that *p* is *doxastically* justified iff it is produced by a reliable process, and then it is propositionally justified for S at some time t iff S's total cognitive state at t is such that S could come to be doxastically justified in believing that *p*. In our earlier example, Catriona's belief that it will rain is not doxastically justified (she formed it on the basis of superstition), but it is propositionally justified, because if she were to believe on the basis of the forecast, she would be doxastically justified.

While it isn't often remarked on, this view involves an interesting, albeit mild, form of relativism about *propositional* justification.[13] On Goldman's theory, whether S has propositional justification to believe p is going to depend on a combination of their total cognitive state and what propositions they could come to be doxastically justified in believing. Two individuals could therefore be aware of precisely the same evidence,[14] yet p might be (propositionally) justified for one but not for the other. Imagine Morven, who is just like Catriona except she is psychologically incapable of trusting weather forecasts because of her deep distrust of the meteorological establishment. As a result, Morven could not become doxastically justified in believing that it will rain, because she is psychologically incapable of forming this belief in the right way. Thus, for Goldman's theory, whether S has propositional justification is relative not just to their total cognitive state, but also the intricacies of their psychological makeup. Goldman's theory is therefore tantamount to a form of *psychologism* about justification.[15]

3.2. *Reliability vs. beliefs about reliability*

There is another respect in which "What is Justified Belief?" leaves the door open for a form of relativism. There may be a difference between which processes count (by the lights of our best science) as reliable and which processes are actually reliable. We don't think that wishful thinking is reliable. But imagine that, unbeknownst to us, there is a benevolent demon who has recently decided for reasons of their own to make it so that beliefs formed through wishful thinking are almost always true. In such a scenario, wishful thinking would be a reliable way of forming beliefs, but Goldman doesn't think this would mean that beliefs formed through wishful thinking are justified:

> What we really want is an explanation of why we count, or would count, certain beliefs as justified and others as unjustified. Such an explanation must refer to our beliefs about reliability, not to the actual facts. The reason we count beliefs as justified is that they are formed by what we believe to be reliable belief-forming processes. Our beliefs about which belief-forming processes are reliable may be erroneous, but that does not affect the adequacy of the explanation. Since we believe that wishful thinking is an unreliable belief-forming process, we regard beliefs formed by wishful thinking as unjustified. What matters, then, is what we believe about wishful thinking, not what is true (in the long run) about wishful thinking.
>
> *(Goldman 1979, 101)*

For Goldman, what confers justification on a given belief is that it is produced by a process that we *believe* to be reliable. While this doesn't in itself provide support for any form of relativism, it does if we add the premise that different communities may (justifiably?) count different processes as reliable. The strength of the relativism that results will depend on the strength of this premise. The weakest version would just state that, at different points in human history, we have counted different processes as reliable (which is not to say that we haven't counted a core set of processes as reliable at all points in human history). On this version, we might get interesting results in the debate between Bellarmine and Galileo: maybe in Bellarmine's time it was justifiable to regard Scripture as reliable. A stronger version would state that different present-day communities count different processes as reliable (which again is not to say that all communities don't count a core set of processes as reliable). Either way, we get a sort of relativity of justification to what counts (and doesn't count) as reliable.

3.3. The generality problem

One central problem for reliabilism is the so-called generality problem. We will finish this section with Robert Brandom's (1998) argument that "solving" the generality problem requires acknowledging a sort of relativity.

Brandom's objection targets the claim that reliabilism is a "naturalistic" epistemology. But Brandom's version of the generality problem is meant to show that reliabilism is less naturalistic than it seems. The reliabilist holds that whether a token belief is justified depends on whether the cognitive process that produced it was sufficiently likely to produce a true belief. Consider Alvin. Alvin is looking at a barn in normal conditions and accordingly forms the belief that there's a barn in front of him. Alvin is located in fake barn county, in which there are far more fake barns than real barns. But fake barn county is located in real barn state, in which there are far more real barns than fakes, and real barn county is located in fake barn country, in which there are far more fake barns than real barns. Here's a helpful diagram (Figure 32.1):

Is Alvin's belief (that there's a barn in front of him) reliable? Brandom's point is that it depends on which reference class we evaluate for reliability relative to. If we evaluate relative to fake barn county or country, it is unreliable (if he had been standing in front of a fake, he would

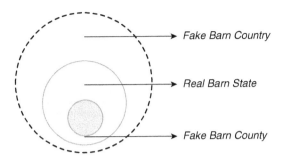

Figure 32.1 Alvin's predicament

still have believed it was a barn). But, if we evaluate relative to real barn state, it is reliable. What, then, are we to say then about Alvin? Brandom says:

> Which is the correct reference class? Is [Alvin] an objectively reliable identifier of barns or not? I submit that the facts as described do not determine an answer. Relative to each reference class there is a clear answer, but nothing in the way the world is privileges one of those reference classes, and hence picks out one of those answers.
>
> *(1998, 386)*

Brandom concludes that reliability can only be specified relative to a reference class, and there is nothing "in the world" that determines a single reference class as privileged. Thus, there is no simple fact of the matter about whether token beliefs like Alvin's are reliable. Relative to some choices of reference class (e.g. real barn state), they are; relative to others (e.g. fake barn county or country), they aren't.

The crucial thing for our purposes is that Brandom does not conclude that reliabilism should be rejected. Rather he concludes that reliabilism cannot live up to its naturalistic credentials because justification cannot be purely a function of the psychological processes that produce or preserve belief. Whether a belief is justified depends on whether it is reliable for the purposes at hand. As Michael Williams puts the point:

> Reliability itself becomes reliability for particular purposes. This is particularly evident in sophisticated forms of inquiry. In particle physics, the standard for "detecting" a particle has moved from three to five sigma, the standard in effect when the discovery of the Higgs boson was announced.... This is a very high standard, but a reasonable one given that "discoveries" at three sigma – itself a high standard – have sometimes turned out to be statistical blips. Reliability is a norm that we are not only responsible to but, in certain applications, responsible for.
>
> *(Williams 2015, 267–268)*

If reliability is a norm that we are responsible *for* then it may be that what we will require for reliability will vary from situation to situation (or community to community). Whether the form of relativism that results is benign or not is going to depend on how much (and what sort) of variation we are willing to counter. Those who push this point (like Brandom and Williams) tend to play down the degree of variation, and so the relativistic consequences. But Brandom's reflections on the generality problem leave the door open for radical forms of relativism, on which different communities are free to decide on the level of reliability they require.

4. Conclusions

We will finish by pointing to avenues for future research. Starting with externalist arguments against relativism, a recurring theme is that, while externalism promises to undercut some central arguments for relativism, the relativist has some moves at their disposal. In the case of the argument from circularity, she can insist that it is incumbent upon the reliabilist qua theorist to vindicate her beliefs as justified beliefs, and her standards as justified standards. The reliabilist may face serious difficulties in doing so, at least if the literature on the "bootstrapping" and "easy knowledge" problems is anything to go by.[16] In the case of the argument from non-neutrality, the viability of a relativist response to the reliabilist depends on the viability of conciliationism as a response to *philosophical* disagreement. So the outcome of the relativist's "encounter" with the reliabilist is going to depend on the outcome of some central epistemological debates. This suggests that relativism (and the relativist) can hardly be regarded as being off the epistemological table.

Turning to ways in which externalism might provide support for (certain forms of) relativism, we saw that externalist ideas do arguably provide support for some forms of relativism. The key question here is whether these forms of relativism are what the absolutist is really concerned to deny. We can of course define "epistemic relativism" in all sorts of ways, but the term is generally seen as denoting a view that is *threatening* to mainstream epistemology. It is an open question whether any of the forms of relativism discussed constitute such a threat. Perhaps the most interesting idea in this respect is that "we" get to determine what "reliable" amounts to. If the view is just that there is some vagueness in the idea of a reliable belief-forming process, then it is perhaps not so interesting. If the view is that the status of a belief forming process as "reliable" is subject to social negotiation, then perhaps it represents the sort of threat to epistemological orthodoxy that is worthy of the name "epistemic relativism."

Acknowledgements

Research on this chapter was assisted by funding from the ERC Advanced Grant Project "The Emergence of Relativism" (Grant No. 339382).

Notes

1 For an overview of the internalism/externalism distinction, see Pappas (2017).
2 See Baghramian and Carter (2015) for a comprehensive discussion. See also Carter (2016, Ch. 2).
3 This is not to say that the *only* things that matter for whether a belief is justified are things beyond what is reflectively accessible to one.
4 This is one common way to capture the view. But one might also be an externalist about epistemic justification because one denies a different version of epistemic internalism called "mentalism" (see Conee and Feldman 2004). Mentalist interests hold that what matters for epistemic justification are factors internal to one's mental life; captured as a supervenience thesis, the claim is that justification *supervenes* exclusively on internal factors, which are understood as mental states. An externalist who denies mentalist internalism is best understood not as denying that what matters for justification must be accessible by reflection, but rather, as denying that (in short), necessarily, mental duplicates are justificational duplicates.
5 The *locus classicus* is Goldman (1979).
6 For helpful discussion on this point, see Stroud (2008) and Sosa (2011).

7 A response to this strand of argument is developed in detail in Sosa (1997).
8 For an influential presentation and criticism of this argument, see Boghossian (2006).
9 For a more recent defence of this argumentative strategy, see Hales (2014). For critical discussion of this argument strategy, see Siegel (2011) and Carter (2016, ch. 4, forthcoming).
10 For one thing, it is unclear that the reasoning here favours relativism over scepticism. See Carter (2016, ch. 4).
11 See Siegel (2011) for an attribution of this implicit premise to Rorty.
12 For an overview of the epistemology of disagreement, see Frances and Matheson (2018).
13 What follows is based on Kornblith (forthcoming), though his discussion is more sophisticated than what we say here.
14 *Modulo* views on which it is impossible for two individuals to be aware of precisely the same evidence.
15 For more on psychologism, see Kusch (2016).
16 For both problems, see Cohen (2002).

References

Baghramian, M. and J. A. Carter (2015), "Relativism," in *The Stanford Encyclopaedia of Philosophy*, Fall 2015 edition, edited by E. N. Zalta, http://plato.stanford.edu/entries/relativism/.

Boghossian, P. (2006), *Fear of Knowledge: Against Relativism and Constructivism*, Oxford: Oxford University Press.

Brandom, R. (1998), "Insights and Blindspots of Reliabilism," *The Monist* 81(3): 371–392.

Carter, J. A. (2016), *Metaepistemology and Relativism*, London: Palgrave Macmillan.

———. (forthcoming), "Archimedean Metanorms," *Topoi*, doi:10.1007/s11245-018-9586-9.

Cohen, S. (2002), "Basic Knowledge and the Problem of Easy Knowledge," *Philosophy and Phenomenological Research* 65(2): 309–329.

Conee, E. and R. Feldman (2004), *Evidentialism: Essays in Epistemology*, New York: Oxford University Press.

Frances, B. and J. Matheson (2018), "Disagreement," in *The Stanford Encyclopedia of Philosophy*, Spring 2018 edition, edited by E. N. Zalta, https://plato.stanford.edu/archives/spr2018/entries/disagreement/.

Goldman, A. (1979), "What Is Justified Belief?" in *Justification and Knowledge*, edited by G. Pappas, Dordrecht: Springer, 89–104.

Hales, S. D. (2014), "Motivations for Relativism as a Solution to Disagreements," *Philosophy* 89(1): 63–82.

Kornblith, H. (forthcoming), "Naturalism, Psychologism, Realism" in *Social Epistemology and Relativism*, edited by N. Ashton, M. Kusch, R. McKenna and K. Sodoma. Abingdon: Routledge.

Kusch, M. (2016), "Psychologism," in *The Stanford Encyclopedia of Philosophy*, Winter 2016 edition, edited by E. N. Zalta, https://plato.stanford.edu/archives/win2015/entries/psychologism.

Pappas, G. (2017), "Internalist vs. Externalist Conceptions of Epistemic Justification," in *The Stanford Encyclopedia of Philosophy*, Fall 2017 edition, edited by E. N. Zalta, https://plato.stanford.edu/archives/fall2017/entries/justep-intext.

Rorty, R. (1980), *Philosophy and the Mirror of Nature*, Princeton, NJ: Princeton University Press.

Siegel, H. (2011), "Epistemological Relativism: Arguments Pro and Con," in *A Companion to Relativism*, edited by S. Hales, Hoboken, NJ: Wiley, 199–218.

Sosa, E. (1997), "Reflective Knowledge in the Best Circles," *The Journal of Philosophy* 94(8): 410–430.

———. (2011), *Reflective Knowledge: Apt Belief and Reflective Knowledge*, vol. II, Oxford: Oxford University Press.

Stroud, B. (2008), "Perceptual Knowledge and Epistemological Satisfaction," in *Ernest Sosa and His Critics*, edited by J. Greco, Hoboken: Wiley, 165–173.

Williams, M. (2007), "Why (Wittgensteinian) Contextualism Is Not Relativism," *Episteme* 4(1): 93–114.

———. (2015), "What's So Special About Human Knowledge?" *Episteme* 12(2): 249–268.

33

EPISTEMIC RELATIVISM AND PRAGMATIC ENCROACHMENT

Brian Kim

1. Introduction

Epistemology engages in an exploration of the epistemic realm, seeking a better understanding of epistemic concepts and norms. But, in order to provide a more synoptic view of epistemology, we might want to step back and ask, what is the relationship between the epistemic and the practical? On one extreme is the view that the epistemic realm is wholly independent of the practical realm, that our epistemic concepts are wholly independent of our practical ones. What we know, how strong our evidence is, and whether we possess epistemic virtues are all wholly independent of what is moral and what we care about. On the other extreme is the view that there is nothing which is purely epistemic. And on this view, practical considerations are always relevant when addressing epistemic concerns. Knowing, justifiably believing, having faith, and possessing epistemic virtue are all sensitive to practical considerations.

Theories that embrace pragmatic encroachment (henceforth PE) can be characterized as rejecting the former view to some degree or other. However, pragmatic encroachment resists a universally accepted characterization for a variety of reasons. First, proponents of PE represent a wide range of philosophical views and approaches. The aforementioned characterization in terms of the demarcation between the epistemic and practical realm was intended to provide as inclusive an account as possible. However, most discussions of PE are narrow in scope, focused on a specific epistemic concept, so most characterizations of PE are also fairly narrow in scope. The debate has primarily focused on knowledge, though more recent discussions have explored questions about degrees of belief and epistemic virtue.[1]

Next, when pragmatic encroachers speak more systematically about their philosophical viewpoint, they often present themselves as adopting a new philosophical methodology or approach rather than advocating a new taxonomy.[2] For example, Brian Kim and Matthew McGrath write,

> the disputes about pragmatic encroachment have helped to open up a new area of philosophy, one focused not only on the relatively narrow question of whether the orthodox approach to knowledge is correct or not, but on a range of issues concerning epistemology and its interface with practical philosophy.
>
> *(2019, 5)*

Another reason why PE resists a universally agreed upon characterization is that in order to explore the demarcation between the practical and the epistemic, we need to specify what we mean by both. And we shall see in Section 4 that this is a non-trivial task. However, for the sake of concreteness, let us begin our discussion with a standard view of "the practical" that demarcates a practical factor from a non-practical factor in terms of whether or not that factor is truth-relevant.[3] So relative to some proposition p, a factor is practical just in case it is non-truth-relevant. And a factor is non-truth-relevant just in case it does not affect the subjective or objective probability that p is true or false.

Using this account of the practical, I will offer an overview of some varieties of PE in Section 2. In Section 3, I then explore the relationship between PE and epistemic relativism. We will find that there is a minimal sense in which these varieties of PE are relativist but most pragmatist views are incompatible with a central tenet of relativism. In Section 4, I explore a radical view on which practical factors encroach on measures of epistemic strength. The view requires a wholesale revision of our understanding of the epistemic because it undermines our working distinction between the practical and the epistemic, and it embraces the view that nothing is distinctively epistemic. I motivate this view by tying it to the pursuit of a non-skeptical infallibilism about knowledge. More importantly, as it pertains to our current discussion, I show that this radical view is compatible with a robust epistemic relativism.

2. Variety of PE views

Using our initial characterization of the practical, we can work with a more precise characterization of pragmatic encroachment. Pragmatists believe that practical factors (i.e. non-truth-relevant factors) are relevant for determining whether or not certain epistemic claims are true or false. Let's start with a narrow view of the epistemic as that which concerns knowledge. So if, as is standardly assumed, knowledge is a conjunctive concept with alethic, doxastic, and justificatory components, then the epistemic concerns truth, belief, and justification (or whatever it is that makes true belief knowledge). Of course, by definition, practical factors are not relevant in determining whether or not a proposition is true. So we are left with two varieties of PE.[4]

> (BELIEF ENCROACHMENT) Practical factors are relevant in determining whether or not a subject believes that p.
> (JUSTIFICATION ENCROACHMENT) Practical factors are relevant in determining whether or not the strength of a subject's epistemic state is strong enough to meet the epistemic standards required for knowledge.

Unfortunately, this list is not comprehensive because there are a variety of epistemological questions that are not, in any obvious way, connected with questions about knowledge. For example, epistemologists are interested in the nature of faith, though whether or not one has faith is not directly relevant to whether or not one knows (Howard-Snyder 2013). Thus, in order to capture the totality of pragmatist views, we need to expand what counts as epistemic. For example, one can adopt a pragmatist view with regards to a broad range of epistemic states, such as faith, trust, and commitment. There are also pragmatist views about cognitive attitudes (e.g. degrees of belief), about epistemic virtues and vices (e.g. open-mindedness and myopia), and about epistemic standards (e.g. rationality).[5]

3. PE and relativism

Having offered an overview of the varieties of PE, we will now turn to consider the relationship between pragmatist epistemologies and epistemic relativism. I shall primarily focus on justification encroachment as it is the most widely discussed view in the PE literature. Nevertheless, most of my observations about the relationship between relativism and justification encroachment apply straightforwardly to the other pragmatic views mentioned above. In what follows, we shall consider some core tenets of and common motivations for relativism, identifying areas of agreement and disagreement with pragmatist epistemologies.

3.1. Minimal relativism

Minimally, epistemic relativism requires that the truth and falsity of epistemic claims are relative to or sensitive to factors in another domain.[6] On this minimal criteria, pragmatist views of knowledge, such as justification encroachment, are paradigmatically relativistic. And the variety of relativistic views proposed by pragmatists can be categorized them in terms of the various domains of relativization.

3.1.1. Relativization to individuals

The most common pragmatic view of knowledge is an individualistic one on which the practical situation of the subject of knowledge is relevant.[7] For example, some have proposed that the standards required for knowledge that p is directly correlated to the size of the p-related stakes governing the subject's practical context. And the size of the stakes relative to p is determined by the difference between the potential gain and loss given the truth or falsity of p in one's practical context. For example, if Bill is betting on the truth of p, and guessing correctly yields $100 and guessing incorrectly yields $0, then the size of the p-related stakes is $100. The pragmatist proposes that the larger the stakes, the higher the standards required to know. Thus, the primary type of relativism advocated by pragmatists are ones where knowing is relativize to an individual's practical situation.

Within this individualistic view, there is some debate about how to demarcate an individual's practical situation. For example, on the stakes-related view, there will be disagreement over whether we should demarcate what is at stake from a subjective point of view, in terms of what the subject personally cares about and is aware of, or from an objective point of view, in terms of what the subject ought to care about.[8]

3.1.2. Relativization to morality

While the relativization to individual viewpoints is the most common pragmatic view, other domains of relativization have been proposed. Some have argued that moral considerations are relevant (Pace 2011). For example, one may possess a decent amount of evidence for an otherwise racist belief that an Asian woman is passive and subservient. However, the morally impermissibility of this belief may deem the belief unjustified. There are, of course, many ways of viewing the interaction between moral and epistemic norms.[9] Moral and epistemic permissions and obligations can interact and conflict in a variety of ways and one must sort out how these norms fit together.

3.1.3. Relativization to communities

It may also be argued that the standards required for knowledge are relative to cultures or communities. Gerken (2019) argues that pragmatic encroachment on knowledge is incompatible with some of the distinctive features of scientific knowledge. He notes that scientific knowledge is typically backed by discursive justification, which is inter-subjectively available and replicable. As a result, scientific knowledge appears to be more stable than pragmatist views allow, and practical factors do not appear to play a prominent role.

In reply, the pragmatist could propose that when we talk about scientific knowledge, we are really talking about knowledge possessed by the scientific community. And groups can possess their own distinctive epistemic standards.[10] For example, some pragmatic accounts claim that knowledge-level justification is equivalent to the level of justification required to take a proposition for granted in inquiry (Kim 2019). Since the scientific community has a much higher standard for taking a proposition for granted in inquiry, then it may turn out that the scientific community must possess all of the distinctive features noted previously in order to know.

So we have found that (i) pragmatist views are minimally relativistic and that (ii) the variety of relativistic views in the PE literature mimics the variety found in other fields.

3.2. Incommensurability and robust relativism

While pragmatist epistemologists are minimally relativistic in virtue of relativizing the epistemic to the non-epistemic, they generally reject the relativistic view that there is an incommensurability between the norms governing different perspectives. In order to be truly relativistic, one's epistemology should entail that the epistemic norms governing one context cannot, from a neutral standpoint, be compared to any incompatible epistemic norms governing a different context. Each context possesses its own normative sovereignty, and there is no privileged perspective from which we can evaluate the norms.

One way for incommensurability to arise is when the epistemic norms themselves, such as those that determine standards of justification or knowledge, are deemed relative to different frameworks. On this view, different norms operate relative to different perspectives, and there is no way of adjudicating between the competing norms and frameworks. For example, in community A but not B, there may be an epistemic norm to consider the totality of evidence when evaluating a belief. And relativists would claim that there is no neutral perspective from which we can compare these norms.

The pragmatist theories surveyed previously do not embrace this more robust relativism. Take for example the version of justification encroachment on which knowledge-level justification is determined by, what is called, a practical adequacy standard. One version of this standard imposes the following necessary condition: "in a decision setting, one knows p only if one's actual preference and one's preference conditional on p match" Anderson and Hawthorne (2019, 108).[11] On this view, there is just a single norm that governs knowledge-level justification. There are no distinct, competing norms that arise in different perspectives. Suppose Jade and Amber possess the same evidence yet given the differences in their practical contexts, Jade knows and Amber doesn't. The practical adequacy view does not propose that there are different norms governing Jade and Amber's practical context. Rather, there is a single norm of practical adequacy, which entails that Jade's belief is practically adequate and Amber's is not because the standards required for practical adequacy are higher in Amber's practical context.

Thus, when we ask whether the higher standards governing Amber's context is better or worse than the standards governing Jade's context, we can give a straightforward answer. They are equally appropriate. And they are equally appropriate because the standards are entailed by the single demand for practical adequacy. The commensurability of standards makes these pragmatist views of knowledge incompatible with a robust relativism.

3.3. PE and the motivations for relativism

3.3.1. Disagreement

One of the primary motivations for relativism comes from the existence of pervasive and long-lasting disagreement. And pragmatist views of knowledge have been explored for similar reasons. Hawthorne (2004) developed the pragmatist account to avoid skepticism. Moreover, the resulting pragmatic view allows for both the skeptic's and dogmatist's perspective to be equally valid. Given the differences in their concerns, different standards govern the skeptic and the dogmatist. And this difference may help to understand the long-standing disagreement between the two points of view.

3.3.2. Under-determination

Another motivation for relativism, especially in the philosophy of science, comes from appeals to under-determination. And such appeals also lie at the heart of Jeremy Fantl and Matthew McGrath's argument for justification encroachment. Fantl and McGrath (2009) present their case by highlighting a central problem for fallibilist accounts of knowledge. On their view, epistemic fallibilism is the view that one can know that p even though there is a non-zero epistemic chance that not-p. This raises the obvious question, what counts as knowledge-level justification? If knowledge is compatible with an epistemic chance of being wrong, how much of a chance is acceptable?

This problem has been dubbed the "threshold problem," and both pragmatists and non-pragmatists have offered a variety of solutions.[12] Fantl and McGrath argue that the specification of knowledge-level justification is under-determined by epistemic (or truth-relevant) factors. Thus, in order to specify an appropriate and non-skeptical threshold, we must appeal to practical factors. This argument for justification encroachment is structurally identical to the arguments for pragmatic and relativistic views about theory choice in the philosophy of science. Proponents of these views have argued that the under-determination of theory choice given a set of evidence shows that others factors such as practical factors must come into play when deciding which theory ought to be accepted.

4. A radical pragmatism

While existing pragmatist epistemologies have some commonalities to epistemic relativism, their rejection of incommensurability makes them, to my mind, incompatible with a genuine relativism. So to conclude our discussion of PE and relativism, I want to identify and explore a radical type of pragmatism that could serve as a foundation for a relativistic epistemology. Given the scope of our discussion, the exploration will be limited to summarizing a reason for pursuing such an account and identifying, what I take to be, the central problem with the plausibility of the view. I end by exploring the view's connection to relativism.

4.1. Pragmatic infallibilism

As noted previously, pragmatists have argued that their view resolves the threshold problem for fallibilism. As they tell the story, a fallibilist epistemology is obscure and going pragmatist is the best and perhaps only way forward. Infallibilism lies in the other direction, but as the story typically goes, infallibilism leads to skepticism. To my mind, this narrative of the epistemological landscape reveals only half of the allures of the pragmatic turn. Just as Fantl and McGrath propose that going pragmatic offers the fallibilist the resources to address its central problem, the threshold problem, I believe that going pragmatic offers the infallibilist the resources to address its central problem, the skeptical problem. My discussion here will be limited to identifying the main difficulty in developing a plausible version of the view and sketching how we might go about addressing this difficulty. As we will see, the main difficulty is that an anti-skeptical pragmatic infallibilism is committed a radical view about measures of epistemic strength.

Let's take for granted Fantl and McGrath's view that fallibilist epistemologies are ones where knowledge that p is compatible with a non-zero epistemic chance that not-p. So infallibilism, as the denial of fallibilism, is the view that knowledge is incompatible with a non-zero epistemic chance. Alternatively, infallibilism is committed to the claim that knowledge entails epistemic certainty.

Clarifying what epistemic certainty amounts is the central task of the non-skeptical infallibilist. However, we can work with a simple characterization. Epistemic chance measures the strength of a subject's evidence for or against any given proposition. There is zero epistemic chance that not-p for some subject just in case the subject's evidence is incompatible with not-p (i.e. rules out all not-p possibilities).[13] And one is epistemically certain that p just in case there is a zero epistemic chance that not-p.

Given our aim of developing an anti-skeptical epistemology, we shall assume that any plausible infallibilism must allow for circumstances in which subjects are epistemically certain about propositions that we ordinarily presume to know. That I have hands and that the earth is very old are both ordinary propositions.[14]

On the pragmatic variant of infallibilism, whether or not one's evidence underwrites certainty may depend upon practical features of a subject's context.[15] So if we combine the anti-skeptical and pragmatic components of the view, then the type of infallibilism we seek is one in which there are contexts where one's evidence guarantees that ordinary propositions are true. This is also compatible with the possibility that there are some contexts where one's evidence does not fully eliminate skeptical alternatives. And what makes this a pragmatic view is that the difference between the former and latter contexts may be purely practical in nature.

To summarize, our anti-skeptical pragmatic infallibilism is committed to three central claims:

(1) Knowledge entails epistemic certainty.
(2) There are contexts in which one may be epistemically certain that ordinary propositions are true.
(3) Whether or not one is epistemically certain can depend upon practical features of the subject's context.

The resulting account offers an interesting twist on the pragmatic view of knowledge. Most pragmatic accounts embrace justification encroachment, where the practical encroaches on the epistemic by determining what counts as knowledge-level justification. In contrast, pragmatic infallibilism proposes that the epistemic standards for knowledge are invariant across all contexts.

Knowledge always requires epistemic certainty. Instead, what can vary from one practical context to the next is whether or not one's evidence underwrites certainty. Epistemic certainty is an upper bound on our measures of a subject's strength of evidence.[16] Thus, if practical factors encroach on epistemic certainty, then that entails that practical factors encroach on how we measure the strength of a subject's epistemic state.

4.2. PE on measures of epistemic strength

Pragmatic infallibilism has been left relatively unexplored in the recent literature. I believe that the lack of exploration is due to the fact that the view requires a commitment to pragmatic encroachment on strength of evidence.[17] And it may seem crazy to think that practical factors can affect how strong one's epistemic state is with regards to any given proposition. After all, if we fix the evidence or information that a subject possesses across any two contexts, then surely the strength of the subject's epistemic state (i.e. evidence) for or against any given proposition should also be fixed.[18] Pragmatic infallibilism denies that evidential strength supervenes on the information a subject possesses. And the rejection of this supervenience thesis is the main obstacle to its plausibility.

In order to overcome this obstacle, we need to develop an account of evidence and strength of evidence on which the former can remain fixed while the latter can vary depending upon practical features of the subject's context. Given the limited scope of this chapter, I will only try to point in the direction of how to do so.

The empirical research on choice behavior has emphasized the importance of distinguishing between how one frames a decision problem and evaluates a decision problem.[19] So if we need to develop an account of how practical factors can affect the evaluation of evidence, it is worth looking into questions about how we frame our inquiries. As it pertains to the evaluation of evidence, we can look into questions about how we demarcate what counts as evidence for or against propositions that are relevant to our decision problems. So for example, if we are betting on the outcome of a coin flip, we can ask, what questions are relevant to the evaluation of the likelihood that the next flip lands heads or not (i.e. tails)? Clearly, this proposition is practically relevant to the decision problem. And if we could ask some infallible oracle any questions which questions would be relevant to ask? Is the proportion of heads in the past flips of the coins relevant? Is the order of outcomes in the past flips relevant? Is the physical material of the coin relevant? Is the color of the coin relevant? Is the weather relevant?

Presumably, there is an infinite number of questions one could ask. This raises an important problem, how do we determine which questions are relevant?[20] Is every question and every piece of information relevant? Or are there reasons to exclude certain information from our epistemic evaluations? And if there are reasons to exclude information, then on what basis do we exclude such information?

In the ideal case, all information is relevant. We seek as much information as we can, and we consider every possibility. Unfortunately, this is practically impossible, even with computational assistance. There are obvious economic reasons why we *must* restrict what is relevant to our inquiries. We are bounded agents with resource limitations and without such restrictions, we would be overwhelmed and would be unable to proceed with deliberation.[21] Thus there appears to be an essential epistemic aspect to the framing of inquiry, which is to determine what counts as evidence for or against the propositions that are practically relevant.

How can we frame our inquiries to determine what counts as evidence for or against? Here, I simply want to raise a problem that can be used to motivate the pragmatic account. It's important to note that how one epistemically frames an inquiry will determine what count as evidence. And so presumably one cannot, in a non-circular way, use one's evidence to determine what counts as evidence. Thus, this presents an opportunity for the pragmatist to develop a new under-determination argument. Just as under-determination arguments are used to motivate pragmatic accounts of theory choice (or induction) and fallibilism, there is space to develop an under-determination argument in favor of a pragmatic account of evidential strength. Since the information one possesses under-determines what counts as evidence for or against, then practical factors could play a role in determining how strong one's epistemic state is.

4.3. PE and epistemic relativism

While my discussion of the pragmatic view of evidential strength has been incredibly cursory, I hope to have offered, in broad outlines, the path one might take to develop such a view. In the very least, I hope to have motivated the idea that how we frame our inquiries could affect what we count as evidence. And since how one frames an inquiry determines what counts as evidence, then non-epistemic factors might play a role in framing inquiry.

Pragmatist views of epistemic or evidential strength are radical. Usually when we compare different subjects, we can say which is in a stronger or weaker epistemic state with respect to any given proposition. So even if the truth of other epistemic matters is relative, we typically assume that there is some purely epistemic perspective from which comparisons can be made. Most discussions of PE have indeed assumed that our measures of epistemic strength are purely epistemic. In fact, our working distinction between the practical and the epistemic depends upon the existence of a purely epistemic measure. After all, a non-practical (i.e. truth-relevant) factor relative to p is a factor that affects the probability that p. However, if the probability that p is sensitive to practical factors, then these so-called non-practical factors can be sensitive to practical factors.[22] Thus, we might be left with the extreme view on which nothing is purely epistemic.[23]

This extreme view should be of interest to the relativist because it is compatible with (though does not entail) the incommensurability of epistemic norms and standards. Justification encroachment rejected incommensurability because on that view, a single context-sensitive norm determined what counted as knowledge-level justification in every local context. Of course, we could have a similar view about our measures of epistemic strength. On one version of our radical view, once we fix the relevant features of the practical context, we have thereby determined what information counts as evidence for or against. So there may be some general way of measuring the strength of a subject's epistemic state that is simply sensitive to practical factors. But on this view, we would be able to compare and contrast the appropriateness of different measures in different local contexts.[24]

There is, however, an alternative that is grounded in the epistemological perspective advocated by C. S. Peirce. Peirce proposed that belief is the removal of doubt but that only genuine doubt is relevant (Peirce 1877). And genuine doubts require the appropriate attitude. They do not arise simply because one has written down or asked a question. Accordingly, one of the central aims of epistemology for Peirce is to determine the norms or methods governing the resolution of genuine doubt. Furthermore, from what I can tell, Peirce did not believe that there were questions that ought or ought not to be asked. He thought it was merely a descriptive matter of fact as to which questions were asked and what doubts one possessed. If we combine this view about question-asking with our pragmatic account, then we have the start of a relativistic epistemology. For if we have no norms that determine which questions are genuine, and, as a result, have no norms governing what doubts or assumptions may or may not frame our inquiries, then there isn't a normative perspective from which we can evaluate the appropriateness of the assumptions and questions governing different inquiries.

There is simply the fact of the matter of what questions we actually ask and what assumptions we actually make. This Peircian version of the pragmatic account of epistemic strength embraces the incommensurability thesis, and I believe it a view worth exploring for both the pragmatist and the relativist.

Notes

1 All the major discussions of pragmatic encroachment – Hawthorne (2004), Stanley (2005), and Fantl and McGrath (2009) – have focused on knowledge. Armendt (2010, 2014, 2019) and Ganson (2019) discuss PE on belief and degrees of belief. Baril (2013) discusses PE on epistemic value.
2 Nolfi (2019) offers an example of this programmatic point of view.
3 This view is found in DeRose (2009, 24, fn. 25).
4 We could have included in this list contextualist views on which practical factors are relevant for determining the truth or falsity of knowledge attributions and denials. However, the inclusion of contextualism as a variety of PE is controversial since it is often presented as a competitor to pragmatic views (see DeRose 2009). In either case, since the relation between contextualism and relativism is covered elsewhere in this volume, I shall set it aside.
5 For an attempt at a more comprehensive taxonomy of pragmatist views, see Kim (2017).
6 Haack (1996) describes the family of relativistic views in this way and suggests that relativistic views can be categorized by the domain of relativization.
7 Hawthorne (2004), Stanley (2005) or Fantl and McGrath (2009) all argue for this type of view.
8 Baril (2019) and Kim (2019) both argue that we need externalist versions of pragmatic epistemologies.
9 Basu and Schroeder (2019) discuss a variety of these interactions.
10 This social view of knowledge-level justification can be derived from the individualistic view of knowledge-level justification. After all, scientific knowledge is just the knowledge possessed by the scientific community. And on the individualistic view, if the subject is a community, then the community can determine the standards required for a true belief to count as knowledge.
11 See Kim (2019) for an alternative version of the practical adequacy requirement.
12 Brown (2014) discusses the threshold problem, and Hannon (2017) offers a purist solution to the problem.
13 I take this way of talking from Lewis (1996).
14 If we accept certain closure principles, our desired infallibilism should also entail that there are circumstances in which we can be epistemically certain that radical skeptical alternatives do not obtain.
15 My discussion will limit our discussion of the subject's practical context to a subject's deliberative context.
16 This does not entail that one's epistemic position is maximally strong. There are many dimensions to the assessment of one's epistemic position. See Gardenfors and Sahlin (1998, ch. 1) for discussion.
17 The recent discussion in Weatherson (forthcoming) is the exception. There is also a critical discussion of this view in Comesaña (2013).
18 As we will see, I will be making a distinction between evidence and counting as evidence. And for that reason, I think it is useful to use the term "information" instead of "evidence" since I think the latter is ambiguous between possessing information and possessing evidence for or against relevant proposition.
19 The well-known discussion in Tversky and Kahneman (1981) of framing effects and discussion following the discovery of preference reversals Lichtenstein and Slovic (2006) have both pointed to the importance of framing in understanding choice behavior.
20 This is different from the questions of restricting which questions we think are worthwhile to ask. Relevance does not entail informational value.
21 Kim (2014) argues against all-things-considered rationality and argues that any psychologically and normatively realistic theory of instrumental rationality must incorporate how we restrict what counts as relevant to our decision problems.
22 This is assuming that the probability that p corresponds in some way to strength of evidence.
23 One may be a pluralist about measures of epistemic strength, and if one is a pluralist, there may be some measures that are purely epistemic.
24 For example, the Cartesian view that distinguishes moral and metaphysical certainty could be interpreted as a view of this kind.

References

Anderson, C. and J. Hawthorne (2019), "Pragmatic Encroachment and Closure," in Kim and McGrath (2019), New York: Routledge, 107–115.

Armendt, B. (2010), "Stakes and Beliefs," *Philosophical Studies* 147(1): 71–87.

———. (2014), "Pragmatic Interests and Imprecise Belief," *Philosophy of Science* 80(5): 758–768.

———. (2019), "Deliberation and Pragmatic Belief," in Kim and McGrath (2019), New York: Routledge, 170–180.

Baril, A. (2013), "Pragmatic Encroachment in Accounts of Epistemic Excellence," *Synthese* 190(17): 3929–3952.

———. (2019), "Pragmatic Encroachment and Practical Reasons," in Kim and McGrath (2019), New York: Routledge, 56–68.

Basu, R. and M. Schroeder (2019), "Doxastic Wronging," in Kim and McGrath (2019), New York: Routledge, 181–205.

Brown, J. (2014), "Impurism, Practical Reasoning, and the Threshold Problem," *Noûs* 47(1): 179–192.

Comesaña, J. (2013), "Epistemic Pragmatism: An Argument Against Moderation," *Res Philosophica* 90(2): 237–260.

DeRose, K. (2009), *The Case for Contextualism,* Oxford: Oxford University Press.

Fantl, J. and M. McGrath (2009), *Knowledge in an Uncertain World,* Oxford: Oxford University Press.

Ganson, D. (2019), "Great Expectations: Belief and the Case for Pragmatic Encroachment," in Kim and McGrath (2019), New York: Routledge, 10–34.

Gardenfors, P. and N. E. Sahlin (1998), *Decision, Probability, and Utility: Selected Readings,* New York: Cambridge University Press.

Gerken, M. (2019), "Pragmatic Encroachment on Scientific Knowledge?" in Kim and McGrath (2019), New York: Routledge, 116–140.

Haack, S. (1996), "Reflections on Relativism: From Momentous Tautology to Seductive Contradiction," *Philosophical Perspectives* 10: 297–315.

Hannon, M. (2017), "A Solution to Knowledge's Threshold Problem," *Philosophical Studies* 174(3): 607–629.

Hawthorne, J. (2004), *Knowledge and Lotteries,* Oxford: Oxford University Press.

Howard-Snyder, D. (2013), "Propositional Faith: What It Is and What It Is Not," *American Philosophical Quarterly* 50(4): 357–372.

Kim, B. (2014), The Locality and Globality of Instrumental Rationality: The Normative Significance of Preference Reversals," *Synthese* 191(18): 4353–4376.

———. (2017), "Pragmatic Encroachment in Epistemology," *Philosophy Compass* 12(5): e12415.

———. (2019), "An Externalist Decision Theory for a Pragmatic Epistemology," in Kim and McGrath (2019), New York: Routledge, 69–99.

Kim, B. and M. McGrath (eds.) (2019), *Pragmatic Encroachment in Epistemology,* New York: Routledge.

Lewis, D. (1996), "Elusive Knowledge," *Australasian Journal of Philosophy* 74(4): 549–567.

Lichtenstein, S. and P. Slovic (2006), *The Construction of Preference,* Cambridge: Cambridge University Press.

Nolfi, K. (2019), "Another Kind of Pragmatic Encroachment," in Kim & McGrath (2019), New York: Routledge.

Pace, M. (2011), "The Epistemic Value of Moral Considerations: Justification, Moral Encroachment, and James' 'Will to Believe,'" *Noûs* 45(2): 239–268.

Peirce, C. S. (1877), "The Fixation of Belief," *Popular Science Monthly* 12(1): 1–15.

Stanley, J. (2005), *Knowledge and Practical Interests,* Oxford: Clarendon Press.

Tversky, A. and D. Kahneman (1981), "The Framing of Decisions and the Psychology of Choice," *Science* 211(4481): 453–458.

Weatherson, B. (forthcoming), "Interests, Evidence and Games," *Episteme* 15(3): 329–344.

34

RELATIVISM AND HINGE EPISTEMOLOGY

Annalisa Coliva

1. Introduction

Hinge epistemology is a trend in contemporary epistemology, which takes its lead from Wittgenstein's remarks in *On Certainty* (1969). Among hinge epistemologists we can list Peter Strawson (1985), Michael Williams (1991),[1] Crispin Wright (1985, 2004, 2014), Danièle Moyal-Sharrock (2005), Martin Kusch (2013, 2016a, 2016b, 2017), Genia Schönbaumsfeld (2017), Duncan Pritchard (2016) and Annalisa Coliva (2010a, 2015).[2] Other contemporary epistemologists' work bears on the status of "hinges," but their views are more in keeping with G. E. Moore's. Chief among them are James Pryor (2000, 2004), Christopher Peacocke (2004), Ernest Sosa (2013) and Ralph Wedgwood (2011, 2013). For them, hinges can be known either a posteriori or a priori. Hinge epistemologists, in contrast, deny that claim, or substantially reinterpret it, by maintaining that hinges enjoy a kind of non-evidential and non-inferential justification. Insofar as a posteriori knowledge depends on empirical evidence and a priori knowledge is taken to depend on inference, hinges would thus not be known in any traditional sense.

2. Wittgenstein's on certainty and relativism

Different developments of Wittgensteinian themes often depend on dissimilar readings of *On Certainty*. On "framework" readings of *On Certainty* (McGinn 1989; Wright 1985; Moyal-Sharrock 2005; Coliva 2010a; Pritchard 2016; Schönbaumsfeld 2017), hinges are not like ordinary empirical propositions, but more like rules.[3] According to epistemic readings of *On Certainty* (Morawetz 1978; Pritchard 2001, 2005, 2011a; Williams 2004a, 2004b; Wright 2004; Kusch 2016b), hinges may be non-evidentially justified and might even become the object of knowledge, if knowledge extends to propositions for which we do not possess evidential justifications.[4] According to naturalistic readings (Strawson 1985, but see also Stroll 1994; Moyal-Sharrock 2005), hinges are propositions we believe because of our upbringing within a community that endorses them and are thus "second nature" to us. Finally, according to therapeutic readings (drawing on Conant 1998, see Crary 2005; Maddy 2017), *On Certainty* contains no theory of hinges at all. Rather, its aim is to cure us from the kind of intellectual cramp that makes us think we could sensibly doubt them.

These different interpretations of *On Certainty* have led to different positions about the allegedly relativistic implications of that work. Clearly, if there is nothing substantive to be said about the status of hinges, they might vary for different people, but this would not by itself lend support to any relativistic view. Naturalistic readings could be more hospitable to relativism, inasmuch as "second nature" is a culturally inherited trait. Insofar as different cultures may pass on different hinges and yet may be considered to be on a par, a form of relativism would become more readily available. Yet, it is notable that no supporter of a naturalistic reading of *On Certainty* has proposed a relativistic interpretation of it. This is largely because they have focused on universal hinges like "There are physical objects," "The Earth has existed for a very long time," etc. Yet, in principle, and in keeping with the garden variety of hinges proposed in *On Certainty*, naturalistic readings could make room for relativism. Epistemic readings, in contrast, would be the least hospitable to relativistic interpretations. For if hinges are at least minimally true and non-evidentially justified, if there were a clash between them, it would be epistemically decidable. It is thus unsurprising that most scholars, who have supported an epistemic reading of *On Certainty*, have also argued against a relativistic interpretation of it.[5] Lastly, framework readings could, in principle, be hospitable to a relativistic interpretation of *On Certainty*. For, if hinges are similar to rules rather than to propositions, and would be neither true nor false, or knowable or unknowable, there would be no way to epistemically resolve the issue. Yet, most contemporary supporters of a framework reading have not coupled it with relativism.

While none of the aforementioned contemporary practitioners of hinge epistemology has endorsed relativism, with the exception of Kusch, epistemic relativism, if not itself a theme in *On Certainty*, is at least an open issue. A whole generation of earlier interpreters has indeed read *On Certainty*, and Wittgenstein's later works more generally, as supporting it.[6] Several passages could be read as hinting at it, even if, as the numerous divergent interpretations mentioned so far show, they are not conclusive.[7] The difficulty has to do with exegetical issues, but also with the very characterization of relativism used to establish whether Wittgenstein's remarks would fall within it.

While a precise characterization of relativism would require no less than a book,[8] we can at least distinguish two forms of relativism: one based on the ideas of equal validity and faultless disagreement, and one centered on the idea of an unbridgeable distance between subjects who would hold different and incompatible hinges. The former is certainly more fashionable today and is a "stronger" notion of relativism than the latter.[9] Key to it is the idea that relativism arises when people hold beliefs with contradictory propositional contents, which, however, are considered to be equally valid both by them and by third parties to their dispute, such that none of them is considered to have made a mistake in forming them. Examples are usually drawn from taste discourse, with pairs of beliefs like "Sushi is tasty" and "Sushi is not tasty." In such cases, people who disagree and hold contradictory propositions, may recognize that their opponent is equally entitled to their view, while persisting in the disagreement.

This kind of relativism does not seem to be readily applicable to *On Certainty*. First, according to framework readings, hinges are not at all like "Sushi is (or isn't) tasty." Their role is not descriptive but normative, in that they determine the conditions of evidential significance – what may be counted as evidence for what – and, in some cases, the conditions of semantic significance.[10] Second, on framework and most epistemic readings, hinges are not the content of an attitude of belief, if that is supposed to entail having evidence in favor of the proposition believed. According to one version of the framework reading, hinges, unlike "Sushi is (isn't) tasty," are not even propositions (Moyal-Sharrock 2005); and, according to Pritchard (2016), they are not the content of any propositional attitude – let it be belief, acceptance, or trust. Thus, on several prominent interpretations of *On Certainty*, there would be no room for genuine disagreement about hinges, let alone a faultless one. Nor does one get a sense that Wittgenstein is concerned with the idea that opposite parties to a debate would recognize the legitimacy of conflicting views; indeed, they would call each other "heretic" and "fool" (OC 608–612). The issue of whether their observers would be more even-handed is moot. The problem depends on how to read those passages where Wittgenstein does recoil from that idea (OC 108–111, 286). In particular, whether they should be taken at face value or, as often happens with the later Wittgenstein, as claims he is bringing up merely for dialectical purposes. Conversely, the exegetical crux concerns the interpretation of those passages in which he does seem to suggest even-handedness, without ever actually stating it (OC 92, 238, 336, 608–612). Yet, overall, there is little room for this first kind of relativism in *On Certainty*.

The second kind of relativism, which, following Bernard Williams (1974–1975), we may call "relativism of distance," does not frame relativism as centering on the idea of (faultless) disagreement. Rather, the key notion is that of "alternatives," to borrow Carol Rovane's terminology (2013). Accordingly, there may be areas of discourse, such as morals, in which more than one system of values is legitimate. On certain views, like Williams', these different systems of values happen to be factually insulated from one another, because of temporal factors. Rovane, in contrast, considers the different moral codes a contemporary woman from rural India and an American woman from NYC may have. She concedes that while having different values they may engage in a meaningful conversation which would disclose to them their deep differences (2013, 38–58). Yet, Rovane also holds that the two women's beliefs would stand in no logical relation whatsoever. Due to their deep differences regarding what is good or bad, they would end up assigning different meanings to the corresponding terms, such that their respective claims "ϕ-ing is good" and "ϕ-ing is not good" would neither be compatible nor incompatible, because "good" would not mean the same in these two sentences. This view is highly problematical, however, for at least the following two reasons. First, because it would depend on a holistic conception of meaning which is difficult to defend. Often parties to a moral debate do agree on some central uses of words such as "good" and "bad." Secondly, if by "good" they mean different things, their claims are not incompatible. For, if A says "The bank is near," while B says it isn't, but they mean two different things by "bank" (i.e. the bank of the river ($bank_r$) as opposed to the financial institution ($bank_{fi}$)), then their claims are compatible. Hence, once the relevant disambiguation has occurred, subjects should agree, *ceteris paribus*, that $bank_r$ is near, while $bank_{fi}$ is not. In particular, they should each endorse both beliefs together. Thus, on Rovane's rendition of relativism of distance, it would be difficult to understand why opposite parties could not embrace together what *prima facie* seemed to be remote alternatives. In fact, she seems drawn to the notion of incommensurability Donald Davidson (1973) warned us against long ago. Yet, it is not clear that she has succeeded in making it any more palatable. Furthermore, there is little in *On Certainty*, which resonates with the idea that people might hold incompatible hinges and thus end up assigning different meanings to the terms used to express them. On the one hand, the examples Wittgenstein uses are such that people seem to understand each other perfectly well when they use "Moon," "rain," "Earth," and so on. On the other hand, he sometimes suggests that doubting of a hinge, like, for instance, "Here is my hand" (in a context where one holds up one's hand in good lighting conditions) would cast doubt on what one means by "hand." Yet, he does not seem to be drawn to conceptual relativity. Rather, he is concerned that doing so would annihilate meaning altogether. In other writings, he actually recommends revising the interpretation of their words to re-establish accord.[11] In *On Certainty*, in contrast, he insists on the role that that proposition plays for us and makes it clear that by using it we would be giving a piece of instruction with respect to our basic categories (OC 36), which one has to accept to produce intelligible sentences at all.

We are thus left with a relativism of distance similar to Bernard Williams'. When applied to knowledge and justification, and once embedded in a Wittgensteinian framework, it amounts to the claim that any epistemic judgement depends on a set of hinges, which may vary across time and space (factual relativism), or that could conceivably be (or have been) different (virtual relativism). Yet, any adjudication between different sets of hinges would always depend on embracing the same or different ones, and would thus never be objective – that is, independent of any hinges, or "unhinged." Hence, one could never rationally go over to a different epistemic system, with its characteristic hinges, but could only convert to it.

Undoubtedly, some passages in *On Certainty* may be taken to instantiate this kind of relativism of distance (cf. OC 92, 238, 336, 608–612). Still, several other passages are much more guarded and seem to point in a different direction. The most telling one is (OC 108):

> "But is there then no objective truth? Isn't it true, or false, that someone has been on the moon?" If we are thinking within our system, then it is certain that no one has ever been on the moon. Not merely is nothing of the sort ever seriously reported to us by reasonable people, but our whole system of physics forbids us to believe it. For this demands answers to the questions "How did he overcome the force of gravity?" "How could he live without an atmosphere?" and a thousand others which could not be answered. But suppose that instead of all these answers we met the reply: "We don't know how one gets to the moon, but those who get there know at once that they are there; and even you can't explain everything." We should feel ourselves intellectually very distant from someone who said this.

According to Wittgenstein, we would need to be given evidence compatible with a set of beliefs about physics. Should that prove impossible, we would feel very distant from them. Yet, there is no parity claim being made here[12] via the suggestion that, when we say that our system of physics is better than theirs – and Wittgenstein does say that (OC 286) – we are only projecting our preference for it, while in fact the two systems are epistemically on a par.[13]

Furthermore, relativism of distance would predict (or prescribe) a shrug of shoulders or recourse to coercive and completely a-rational means of conversion from one set of hinges to the other. Yet, Wittgenstein elaborates the consequences of this encounter differently. That is, he would ask perfectly rational questions. If answered, one would have good reasons to abandon one's hinge. Key to this dynamics is the initial move of calling the hinge in question and, by so doing, demoting it from that role, to re-insert it into the category of empirical hypotheses and propositions, for or against which empirical evidence may be provided. Thus, at least in this case, we see how Wittgenstein is not recommending the kind of indifferent or squarely coercive attitude relativism of distance would command.[14]

It is true that he holds that reasons and giving grounds come to an end (OC 110, 204), with the suggestion that, in such a predicament, one could not be moved by evidence to go over to a different system of beliefs with its characteristic hinges. Yet, even then, he doesn't seem to lean towards a relativism of distance, which, to repeat, would predict either indifference or recourse to totally a-rational means of persuasion. Rather, he suggests that we might be moved to embrace this alternative epistemic system if it had certain explanatory virtues, like simplicity and the possibility of extending explanations across different domains (OC 92). Surely, all this would fall short of epistemic rationality, if that meant having evidence and reasons that would corroborate the likely truth of hinges. Yet, it would not be like coercion or indifference either. On a wider notion of rationality, this form of inference to the best explanation would count as a rational procedure after all. For these and further reasons I explored at length in Coliva (2010a, 2010b), I still think that Wittgenstein was not an epistemic relativist, not even a relativist of distance.[15]

3. Contemporary hinge epistemology and relativism

Looking at contemporary versions of hinge epistemology, most hinge epistemologists today are not in favor of relativism. A key reason is that they focus on general and universal hinges like "There is an external world," "Our sense organs are mostly reliable," "We are not victims of lucid and sustained dreams," "There are other minds," "What has constantly happened in the past will keep repeating in the future," or "We cannot be massively mistaken in our inquiries." Thus, irrespective of how exactly they conceive of their nature and epistemic status, contemporary hinge epistemologists agree that there is no room for meaningful dissent about hinges, especially if we are interested in understanding the possibility of human knowledge. Given that aim, it is difficult to find people or communities that actually embrace hinges incompatible with the ones just mentioned. That is, none of these hinges seems to be subject to the kind of cultural variability we could find with respect to hinges like "Nobody has ever been on the Moon," "The Earth is flat," "I could make it rain," or "The Earth has come to existence with my birth" (both held by a king of a "primitive" tribe), and the like.[16]

Explanations of why hinges do not give rise to relativism vary. On an epistemic reading such as Wright's (2004, 2014), we would have various kinds of rational entitlement to them, and we should further consider them as propositions to which we bear an attitude of trust. According to my own reading (Coliva 2015, 2017, 2019), they would be constitutive of epistemic rationality, and we should consider them genuine propositions, which are the object of an attitude of acceptance. According to Pritchard's interpretation (2016), they would make the acquisition of knowledge possible in the first place, while we should consider them the content of visceral commitments. Nevertheless, the common theme running along these accounts is that hinges would be essential and universal components of human inquiry, for that inquiry would always depend on an amalgam of perceptual, inductive, testimonial and reasoning methods.

For similar reasons, if one were to extend hinge epistemology to the realm of morals and religion, the hinges one would have to recognize as constitutive of these domains would have to be general ones, which all parties to possible moral or religious disputes would embrace. Something like "unmotivated harm towards innocent creatures is wrong" or "God exists" (Pritchard 2000) would probably do, while "God is triune" or "good is maximized utility" would not. This would still leave room for disagreement in the application conditions of those hinges, for there may be a dispute about what counts as unmotivated harm or as an innocent creature, but that should not cast doubt on the legitimacy and universality of the relevant hinge. Furthermore, there could be meaningful disagreement about the beliefs made possible by the appropriate hinges in each of these domains (e.g. whether God is triune, or whether eating meat is morally permissible). Clearly, with "God exists," we could have doubters and dissenters, but, in that case, they would refuse (or refuse to engage) in religious discourse and/or practices.[17] Conversely, insofar as one were to engage in that kind of discourse or practice, and do so in more than a purely hypothetical or notional way, one would have to endorse that hinge.

Notice, however, that despite contemporary hinge epistemologists' willingness to preserve the idea that hinges are truth-apt and propositional, their nature and role makes them such that, even if, *per impossibile*, we clashed about them, we would not have the normative trappings of disagreement. In particular, there would be nothing like offering evidence for or against them, at least in the following sense: that evidence could not rationally prove or disprove them because it would itself depend on taking them for granted. Thus, their willingness to recognize hinges' propositionality and truth-aptness would not re-open the way to a kind of relativism based on disagreement intuitions.

In closing, let us consider the objection that, after all, even when we are dealing with the kind of general propositions contemporary hinge epistemologists are concerned with, it would still be possible to have different and incompatible ones.[18] In an idealist or phenomenalist scenario, "there is an external world" would not be taken for granted. As remarked in Coliva (2015), however, phenomenalists could not make sense of the content of our perceptual experiences, which is as of objects with certain perceptual properties.[19] Moreover, the view that in perception we just have a bundle of sense data, which need to be grouped together through the exercise of concepts to give rise to objects of perception, is definitely discredited.[20] Indeed, nowadays it is the consensus view that we can perceive objects well before exercising concepts. Thus, this alternative, while conceivable, is not plausible as a reconstruction of human perception.

One may then wonder whether its intelligibility is all that matters in order to bring grist to the relativist mill. Notice, however, that one of the main motivations behind relativism is that, at least in some areas of discourse, it seems *prima facie* descriptively adequate. In the case of taste, it seems to capture the fact that when people disagree about the tastiness of a given food, none of them seems to have made a mistake, and they may agree to disagree. Moreover, when it comes to moral matters, we are all familiar with the sense of deep distance that encountering very different moral codes gives rise to. Thus, the appeal of relativism goes hand in hand with its *prima facie* capability to make sense of important aspects of human experience. If all we can claim for it is that it would be instantiated in some conceivable scenario, which, however, is at odds with human experience and what we know about it, then, while still a possibility, it would be supremely toothless. So, why bother?

Notes

1   Williams rejects the attribution, but Williams (1991) sparked an interest in Wittgenstein's *On Certainty* within mainstream epistemology.
2   See also Coliva and Moyal-Sharrock (2016) for essays devoted to this trend.
3   Moyal-Sharrock (2005) radicalizes the claim by denying that they are propositions *at all* and by maintaining that, as such, hinges are manifested only in action.
4   Morawetz (1978) and Pritchard (2001, 2011a) maintain the possibility that hinges be known, if knowledge is conceived, externalistically, as not requiring being able to offer rational corroboration. This interpretation finds little support in *On Certainty* and Pritchard has abandoned his earlier views, which, at any rate, were not presented as an exegesis of Wittgenstein but as a possible development of his ideas.
5   A notable exception is Kusch (2016b), but it is not clear how he would square epistemicism about hinges with relativism.
6   See Phillips (1977), Rorty (1979), Lukes (1982), Haller (1995), Hintikka and Hintikka (1986), Gullvåg (1988), Grayling (2001), Vasiliou (2004). Glock (2007) and Hacker (1996) attribute to Wittgenstein a view that comes close to conceptual relativism, Winch (1964) has been instrumental in defending an epistemic relativistic reading of Wittgenstein.
7   Other earlier anti-relativist interpretations are Bambrough (1991), Blackburn (2004), Luckhardt (1981), Marconi (1987), O'Grady (2004), Rhees (2003), Schulte (1988), Williams (1974–5), and von Wright (1982).
8   See Baghramian and Coliva (2019, chapter 2).
9   See Kölbel (2002) and MacFarlane (2014). MacFarlane is critical of the idea of faultless disagreement. Boghossian (2006) claims that equal validity is a key feature of relativism.
10 See OC 369, cf. also 114, 126, 370, 456.

11 This is indeed what he invites us to do in the celebrated passages about odd wood-sellers in Wittgenstein (1978, §150).

12 *Contra* Kusch (2013; cf. also 2016a). Kusch prefers to talk of "symmetry" rather than parity. Yet symmetry consists in the prohibition to rank epistemic systems or practices, and this comes close to a parity claim.

13 See also Williams (2007, 109).

14 Bernard Williams (1974–1975) is skeptical that relativism of distance would be applicable in the scientific domain (cf. also Williams 2007; Rovane 2013; Kusch 2016a, 2017).

15 Two further characteristic moves Wittgenstein makes and that sit badly with relativism are (i) "containment" and (ii) "translation revision." (i) consists in allowing for the possibility that different systems of justification with their characteristic hinges might be applicable to *different* domains; and (ii) consists in revising translations (including homophonic ones), when persisting in the old ones would force us to conclude that the other party holds totally irrational beliefs (from our own point of view). (ii) is similar to Quine's and Davidson's recourse to the principle of charity as a guiding maxim of radical translation/interpretation.

16 Kusch (2016a) allows that when it comes to these general hinges there is universal agreement, but he contends that there are many more hinges, which are subject to cultural variability. Yet, as we have seen in our discussion of Wittgensteinian hinges, it is not clear that a wider set of hinges would automatically translate into a more hospitable attitude towards epistemic relativism. Williams (1991, 2007) allows for a wide variety of hinges, endorses "hinge contextualism," and argues that it is not a form of epistemic relativism. According to most contemporary hinge epistemologists, these differences would be more doxastic than about hinges properly so regarded.

17 The absence of epistemic evidence for God's existence (either a posteriori or a priori), together with the fact that positing its existence would not be compatible with the best explanations we have of physical phenomena, inclines atheists to treat it as a hypothesis that has no evidence in its favor. Indeed, even the Roman Catholic Church, in *Fides et Ratio*, defends God's existence on purely existentialist grounds. For it claims that it is only by assuming it that our lives would have a meaning. The considerations Catholicism marshals in favor of God's existence are not epistemic in nature and yet can move people to embrace that hinge. This might count as a form of relativism of distance (cf. Kusch 2011), but I will have to defer its discussion to another occasion. Pritchard (2017) extends hinge epistemology to the religious domain. He thinks of hinges, including religious ones, as basic a-rational commitments. Being a-rational, Pritchard argues, there can be no evidence for or against them. Yet, it seems difficult to reconcile this reading with the kind of epistemic anti-relativism Pritchard (2011b) defends. In particular, that a theist and an atheist do have *other* hinges in common and can thus converse, seems irrelevant *vis-à-vis* their different attitudes with respect to the hinge "God exists." It is thus unclear that their disagreement could be resolved by utilizing a common epistemic ground.

18 Ashton (2019) presses this point against me. I have responded in Coliva (2019). Here we look at case (i) in which people would use the same methods and embrace different hinges. The other possibilities are (ii) different methods, same hinges; and (iii) different methods and different hinges. I discuss (i) in Coliva (2015, ch. 4, 2019), and (ii) in Coliva (2019).

19 Cf. Burge (2010).

20 For a more extended discussion, see Coliva (2015, 2016, 2019).

References

Ashton, N. (2019), "Extended Rationality and Epistemic Relativism," in *Non-Evidential Epistemology*, edited by N. Pedersen and L. Moretti, Leiden: Brill.

Baghramian, M. and A. Coliva (2019), *Relativism*, London and New York: Routledge.

Bambrough, R. (1991), "Fools and Heretics," in *Wittgenstein: Centenary Essays*, edited by A. Phillips Griffiths, Cambridge: Cambridge University Press, 239–250.

Blackburn, S. (2004), "Relativism and the Abolition of the Other," *International Journal of Philosophical Studies* 12: 245–258.

Boghossian, P. (2006), *Fear of Knowledge*, Oxford: Oxford University Press.

Burge, T. (2010), *Origins of Objectivity*, Oxford: Oxford University Press.

Coliva, A. (2010a), *Moore and Wittgenstein: Scepticism, Certainty and Common Sense*, London: Palgrave Macmillan.

———. (2010b), "Was Wittgenstein an Epistemic Relativist?" *Philosophical Investigations* 33(1): 1–23.

———. (2015), *Extended Rationality: A Hinge Epistemology*, London: Palgrave Macmillan.

———. (2016), "How to Perceive Reasons," *Episteme* 13(1): 77–88.

———. (2017), "Replies to Commentators," *International Journal for the Study of Skepticism* 7(4): 281–295.

———. (2019), "Hinges, Radical Skepticism, Relativism and Alethic Pluralism," in *Non-Evidential Episte-mology*, edited by N. Pedersen and L. Moretti, Leiden: Brill.

Coliva, A. and D. Moyal-Sharrock (eds.) (2016), *Hinge Epistemology*, Leiden: Brill.

Conant, J. (1998), "Wittgenstein on Meaning and Use," *Philosophical Investigations* 21: 222–250.

Crary, A. (2005), "Wittgenstein and Ethics: A Discussion with Reference to *On Certainty*," in *Readings of Wittgenstein's On Certainty*, edited by D. Moyal-Sharrock and W. H. Brenner, London: Palgrave Macmillan, 275–301.

Davidson, D. (1973), "On the Very Idea of a Conceptual Scheme," reprinted in *Inquiries into Truth and Interpretation*, Oxford: Oxford University Press, 1984, 183–198.

Glock, H. J. (2007), "Relativism, Commensurability and Translatability," *Ratio* 20: 377–402.

Grayling, A. (2001), "Wittgenstein on Skepticism and Certainty," in *Wittgenstein: A Critical Reader*, edited by H. J. Glock, Oxford: Blackwell, 305–321.

Gullvåg, I. (1988), "Remarks on Wittgenstein in *Über Gewissheit* and a Norwegian Discussion," *Inquiry* 31: 371–385.

Hacker, P. (1996), "On Davidson's Idea of a Conceptual Scheme," *Philosophical Quarterly* 46: 289–307.

Haller, R. (1995), "Was Wittgenstein a Relativist?" in *Wittgenstein: Mind and Language*, edited by R. Egidi, Dordrecht: Kluwer, 223–231.

Hintikka M. B. and J. Hintikka (1986), *Investigating Wittgenstein*, Oxford: Blackwell.

Kölbel, M. (2002), *Truth Without Objectivity*, London, Routledge.

Kusch, M. (2011), "Disagreement and Picture in Wittgenstein's 'Lectures on Religious Belief,'" in *Publications of the Austrian Ludwig Wittgenstein Society – New Series 17*, Berlin and Boston: De Gruyter, 35–58.

———. (2013), "Annalisa Coliva on Wittgenstein and Epistemic Relativism," *Philosophia* 41: 37–49.

———. (2016a), "Wittgenstein's *On Certainty* and Relativism," in *Analytic and Continental Philosophy: Methods and Perspectives*, edited by S. Rinofner-Kreidl and H. A. Wiltsche, Berlin and Boston: De Gruyter, 29–46.

———. (2016b), "Wittgenstein on Mathematics and Certainty," in *Hinge Epistemology*, edited by A. Coliva and D. Moyal-Sharrock, Leiden: Brill, 48–71.

———. (2017), "When Paul Met Ludwig: Wittgensteinian Comments on Boghossian's Antirelativism," in *Realism – Relativism – Constructivism: Proceedings of the 38th International Wittgenstein Symposium in Kirchberg*, edited by K. Neges, J. Mitterer, S. Kletzl and C. Kanzian, Berlin and Boston: De Gruyter, 203–214.

Luckhardt, G. C. (1981), "Wittgenstein and Ethical Relativism," in *Sprache, Logik und Philosophie, Akten des 4. Internationalen Wittgenstein Symposium 1979*, edited by R. Haller and W. Grassl, Wien: Hölder-Pichler-Tempsky, 316–320.

Lukes, S. (1982), "Relativism in Its Place," in *Rationality and Relativism*, edited by M. Hollis and S. Lukes, Oxford: Blackwell, 261–305.

MacFarlane, J. (2014), *Assessment Sensitivity and Relative Truth*, Oxford: Oxford University Press.

Maddy, P. (2017), *What Do Philosophers Do?* Oxford: Oxford University Press.

Marconi, D. (1987), *L'eredità di Wittgenstein*, Roma-Bari: Laterza.

McGinn, M. (1989), *Sense and Certainty: A Dissolution of Scepticism*, Oxford: Blackwell.

Morawetz, T. (1978), *Wittgenstein and Knowledge: The Importance of On Certainty*, Atlantic Highlands, NJ: Humanities Press.

Moyal-Sharrock, D. (2005), *Understanding Wittgenstein's On Certainty*, London: Palgrave Macmillan.

O'Grady, P. (2004), "Wittgenstein and Relativism," *International Journal of Philosophical Studies* 12: 315–337.

Peacocke, C. (2004), *The Realm of Reason*, Oxford: Clarendon Press.

Phillips, D. (1977), *Wittgenstein and Scientific Knowledge*, London: Palgrave Macmillan.

Pritchard, D. (2000), "Is 'God Exists' a 'Hinge Proposition' of Religious Belief?" *International Journal for Philosophy of Religion* 47(3): 129–140.

———. (2001), "Radical Scepticism, Epistemological Externalism, and 'Hinge' Propositions," in *Wittgenstein-Jahrbuch 2001/2002*, edited by D. Salehi, Berlin: Peter Lang, 97–122.

———. (2005), "Wittgenstein's *On Certainty* and Contemporary Anti-Scepticism," in *Readings of Wittgenstein's On Certainty*, edited by D. Moyal-Sharrock and W. H. Brenner, London: Palgrave Macmillan, 189–224.

———. (2011a), "Wittgenstein on Skepticism," in *The Oxford Handbook of Wittgenstein*, edited by M. McGinn and O. Kuusela, Oxford: Oxford University Press, 521–547.

———. (2011b), "Epistemic Relativism, Epistemic Incommensurability and Wittgensteinian Epistemology," in *A Companion to Relativism*, edited by S. Hales, Oxford: Blackwell, 266–285.

———. (2016), *Epistemic Angst: Radical Skepticism and the Groundlessness of Our Believing*, Princeton, NJ: Princeton University Press.

———. (2017), "Faith and Reason," *Philosophy* 81: 101–118.

Pryor, J. (2000), "The Skeptic and the Dogmatist," *Noûs* 34: 517–549.

———. (2004), "What's Wrong with Moore's Argument?" *Philosophical Issues* 14: 349–378.

Rhees, R. (2003), *Wittgenstein's On Certainty: There – Like Our Life*, Oxford: Blackwell.

Rorty, R. (1979), *Philosophy and the Mirror of Nature*, Princeton, NJ: Princeton University Press.

Rovane, C. (2013), *The Metaphysics and Ethics of Relativism*, Boston: Harvard University Press.

Schönbaumsfeld, G. (2017), *The Illusion of Doubt*, Oxford: Oxford University Press.

Schulte, J. (1988), "World-Picture and Mythology," *Inquiry* 31: 323–334.

Sosa, E. (2013), "Intuitions and Foundations: The Relevance of Moore and Wittgenstein," in *The A Priori in Philosophy*, edited by A. Casullo and J. C. Thurow, Oxford: Oxford University Press, 186–204.

Strawson, P. (1985), *Naturalism and Skepticism: Some Varieties*, London: Methuen.

Stroll, A. (1994), *Moore and Wittgenstein on Certainty*, New York: Oxford University Press.

Vasiliou, I. (2004), "Wittgenstein, Religious Belief and *On Certainty*," in *Wittgenstein and the Philosophy of Religion*, edited by R. L. Arrington and M. Addis, London: Routledge, 29–50.

von Wright, G. (1982), *Wittgenstein*, Oxford: Blackwell.

Wedgwood, R. (2011), "Primitively Rational Belief-Forming Processes," in *Reasons for Belief*, edited by A. Reisner and A. Steglich-Petersen, Cambridge: Cambridge University Press, 180–200.

———. (2013), "A Priori Bootstrapping," in *The A Priori in Philosophy*, edited by A. Casullo and J. C. Thurow, Oxford: Oxford University Press, 226–248.

Williams, B. (1974–1975), "The Truth in Relativism," *Proceedings of the Aristotelian Society* 75: 215–228.

Williams, M. (1991), *Unnatural Doubts*, Cambridge, MA: Blackwell.

———. (2004a), "Wittgenstein's Refutation of Idealism," in *Wittgenstein and Scepticism*, edited by D. MacManus, London and New York: Routledge, 76–96.

———. (2004b), "Wittgenstein, Truth and Certainty," in *Wittgenstein's Lasting Significance*, edited by M. Kölbel and B. Weiss, London and New York: Routledge, 249–284.

———. (2007), "Why (Wittgensteinian) Contextualism Is Not Relativism," *Episteme*: 93–114.

Winch, P. (1964), "Understanding a Primitive Society," reprinted in *Ethics and Action*, London: Routledge, 1972, 8–49.

Wittgenstein, L. (1969), *On Certainty*, Oxford: Blackwell.

———. (1978), *Remarks on the Foundations of Mathematics*, Oxford: Blackwell.

Wright, C. (1985), "Facts and Certainty," *Proceedings of the British Academy* 71: 429–472.

———. (2004), "Warrant for Nothing (and Foundations for Free?)," *Aristotelian Society Supplement* 78(1): 167–212.

———. (2014), "On Epistemic Entitlement (II): Welfare State Epistemology," in *Scepticism and Perceptual Justification*, edited by D. Dodd and E. Zardini, Oxford: Oxford University Press, 213–247.

35

RELATIVISING EPISTEMIC ADVANTAGE

Natalie Ashton

1. Epistemic relativism

First I'll outline the definition of epistemic relativism that I will be using throughout this chapter. By "*epistemic* relativism" I mean relativism about justification of beliefs, as opposed to relativism about the property of truth (alethic relativism), the truth value of propositions (semantic relativism), or relativism about any other domain (such as relativism about morality or aesthetics). By "epistemic *relativism*" I mean accounts of justification as dependent on an epistemic system or practice. The term "relativism" is sometimes used loosely by epistemologists who want to indicate that a view renders justification as arbitrary, unimportant, or non-existent. Often this loose usage also carries negative connotations – a view on which justification turns out to be arbitrary doesn't do justice to our intuitions or practices and so is problematic. My usage of the term "relativism" is intended to be merely descriptive, and not to carry such connotations. However I will evaluate this view of relativism in Section 6.

On my definition relativism has three components, which I take from Martin Kusch:[1]

(DEPENDENCE) A belief has an epistemic status (as epistemically justified or unjustified) only relative to an epistemic system or practice.

(PLURALITY) There are, have been, or could be, more than one such epistemic system or practice.

(SYMMETRY) Epistemic systems and practices must not be ranked.

(Kusch 2016, 33–34)

The third element is a placeholder which can be filled out in various ways. Kusch lists four (2016, 34–35), two of which are relevant for this chapter. The first is

(NON-NEUTRALITY) There is no neutral way of evaluating different systems and practices.

(2016, 34)

This follows from the rejection of absolutism (the idea that justification is independent of time, place, culture, and so on) embodied by Dependence. If justification is system-dependent, then the justification for any evaluation or ranking of a set of systems will be dependent too.

The second way of spelling out Symmetry is

(EQUALITY) All systems and practices are equally correct.

(2016, 34)

This is a stronger claim than Non-neutrality. Rather than simply denying the possibility of neutral rankings, it offers its own ranking: it says that all systems are equally good. This presumes a neutral standpoint from which such a claim can be made, meaning it requires absolutism and so denies Dependence. Equality can't, consistently, be incorporated into relativist views. Charitable discussions of relativism (as I intend this one to be) will therefore characterise relativism as based on Non-neutrality unless there is evidence to suggest that the relevant authors intended otherwise.

2. Feminist standpoint theory

Feminist Standpoint Theory is one of the three main branches of feminist epistemology. Feminist epistemologies explore the influence of social factors (such as gender and race) on knowledge, via justification. The idea that these social factors have such an effect on justification is often called the

(SITUATED-KNOWLEDGE THESIS) Differences in social factors create epistemic differences (e.g. in the kinds of things that inquirers are justified in believing).

In addition to standpoint theory, the other main branches of feminist epistemology are feminist empiricism (e.g. Anderson 1995; Longino 1997) and feminist postmodernism (e.g. Haraway 1988). There are various similarities and differences between these, at both the level of the branches themselves, and also at the level of specific accounts within these branches.[2] However, in this chapter I will only focus on views which can be clearly described as standpoint theories.

Feminist standpoint theories combine the situated-knowledge thesis with two further claims. The first is the

(STANDPOINT THESIS) Justification depends on "socially situated" perspectives.

According to this idea, subjects have different "social locations," or different statuses as socially oppressed or socially privileged. For example black women occupy very different social locations to white men. And these different social locations come with different experiences, which have the potential to enable different epistemic perspectives.

This idea has its roots in Marxist historical materialism. On György Lukács' (1971) interpretation, Marx argued that the different social locations of the bourgeoisie and of the proletariat lead them to have different perspectives on economic exchange and the social relations that hold between the two groups. From the perspective of the proletariat, the oppressive nature of these social relations is, or can be made, visible, whilst from the perspective of the bourgeoisie the oppressive nature of these social relations is obscured. Feminist standpoint theorists focus on social locations that are determined by gender oppression, and by multiple intersecting dimensions of oppression (e.g. oppression based on both race and gender).

The second thesis distinctive of standpoint theory is the "epistemic-advantage" (sometimes "inversion") thesis:

(EPISTEMIC-ADVANTAGE THESIS) The social oppression that socially disadvantaged groups experience can bring them epistemic benefits.

The idea is that that subjects who are socially oppressed have distinct experiences, and through critically reflecting on these can turn their perspective into a "standpoint" – an epistemically privileged perspective from which the nature of relevant social relations is visible. Subjects who aren't oppressed don't have these experiences, and as a result are less likely to achieve a standpoint.

Standpoint theorists have made several important caveats about this thesis.[3] First, the epistemic-advantage thesis does not presuppose essential categories. There needn't be any properties which all members of an epistemically advantaged group share (Hartsock 1997; Smith 1997; Wylie 2003). As Fricker puts it, standpoints don't depend on oppressed people (or even particular subsets of oppressed people, such as Latina women or gay men) being the same; they only require that their experiences are similar in certain ways (Fricker 1999, 201).

Second, possessing epistemic advantage is neither a necessary nor sufficient condition on membership of a particular group. Not all oppressed people have it, and some non-oppressed people do (Medina 2012).

Third, epistemic advantage is not automatic, but must be achieved through critical reflection. And this achievement is collaborative; it is not the work of an individual (Fricker 1999, 202–203; Medina 2012; Wylie 2003).

Fourth, the advantage can be restricted in scope. Standpoint theorists are usually clear that the argument for advantage will need to be made on a domain-by-domain basis, and that it is easiest to find in domains of knowledge which involve social relations (e.g. Harding 1991, 46; Wylie 2003, 37). Miranda Fricker suggests that knowledge of "the social world" should be "fragmented" even further into knowledge of "relevant areas of the social world" (Fricker 1999, 203).

All standpoint theories have these two theses in common. Beyond this there are differences, particularly between different accounts of the epistemic-advantage thesis, but I won't discuss these in this chapter. Instead, I will focus on the relationship between standpoint theory and epistemic relativism. In the next section I'll discuss the relationship between relativism and one particular version of standpoint theory, but first I want to make two comments on the "standard" understandings of this relationship.

The first thing to say is that it's widely reported in the literature that standpoint theory's critics draw a connection between the standpoint thesis and relativism. I have struggled to find defences of this claim in print – these critics aren't, as far as I've found, named or cited. (Perhaps the standpoint theorists mentioning them were referring to worries raised in personal conversation or by referees.) But whatever the source, there is general acknowledgement of a suspected connection between the standpoint thesis and relativism.

We can make a good guess about why some might think this connection holds. The standpoint thesis clearly incorporates Dependency and Plurality, as it says that justification depends on a perspective, and that there are multiple perspectives. So standpoint theory has two of the three components that views need to have in order to count as relativism.[4]

The second thing to note is that, in response to this, standpoint theorists standardly cite the epistemic-advantage thesis. This move is made frequently in print (most recently and explicitly in Tanesini forthcoming), although discussion of it is usually brief. But again we can identify the thinking behind it using our triadic definition of relativism: the epistemic advantage thesis seems to claim that standpoints can be ranked, and therefore contradicts Symmetry, the third component required for relativism.

My investigation into the relationship between standpoint theory and relativism turns on the claim that epistemic-advantage renders standpoint theory and Symmetry incompatible. In the next section I'll argue that one recent account of epistemic advantage is compatible with Symmetry – or at least with Symmetry based on Non-neutrality, rather than Equality. This account of epistemic advantage is not capable of showing that standpoint theory isn't relativism.

I'll discuss the relationship between relativism and other standpoint theories in Section 5. I'll argue that non-neutrality follows from the situated knowledge thesis, and so any account of epistemic advantage will either be compatible with non-neutral symmetry, or will contradict the first central thesis of standpoint theory. This means that standpoint theorists have a decision to make: either they accept relativism, or they must radically rethink, or even abandon, their view.

3. Medina on epistemic advantage

The epistemic-advantage thesis has been cashed out in different ways by different authors, for example, in terms of the nature of the work that socially oppressed groups tend to undertake (Hartsock 1983); the ability that oppressed groups have to identify constitutive values (Harding 1991); and the opportunity that oppressed groups have to compare multiple perspectives (Collins 1986). The account most relevant to this chapter is one defended by José Medina (2012), which grounds epistemic advantage in epistemic character, and specifically in terms of the development of epistemic virtues and vices.

On this view, the experiences people have influence their epistemic character. Since people who are socially oppressed and people who are socially privileged tend, in general, to have different experiences, they will tend, in general, to develop different epistemic characters. These epistemic characters influence their ability to respond to something called *epistemic friction*, which is key to achieving epistemic advantage. In the remainder of this section I'll explain the notion of epistemic friction, and the process of character development it's important to.

Like physical friction, epistemic friction is a jarring, but productive "force" which occurs when two or more objects come into contact (2012, ch. 1). When you rub your hands together there is resistance, which produces heat. In the epistemic case the "objects" that come into contact are different perspectives. Contact between perspectives can be jarring, because subjects often find it challenging to be confronted by alternative values and beliefs, but it can also be productive, because if we respond to it appropriately we can learn about both perspectives.

What does it mean to respond *appropriately* to epistemic friction? Medina offers two guiding principles. The first is the principle of

(ACKNOWLEDGEMENT & ENGAGEMENT) "all the cognitive forces [epistemic perspectives] we encounter must be acknowledged and [. . .] in some way engaged."
(Medina 2012, 50)

Medina is aware that engaging with some perspectives is extremely difficult, and that there may be cases in which "only a negative mode of engagement is possible or epistemically beneficial." I understand this as a way of anticipating the worry that some perspectives – such as explicitly and aggressively bigoted ones – have very few epistemic benefits to offer, and the potential to inflict many epistemic harms. In these cases, dismissal might be a valid form of (negative) engagement, and one which still requires initial acknowledgement. What Medina wants to rule out is ignoring, or remaining completely oblivious of, perspectives.

The second principle is the principle of

(EQUILIBRIUM) it is important to aim for "equilibrium in the interplay of [different perspectives], without some forces overpowering others, without some cognitive influences becoming unchecked and unbalanced."

(Medina 2012, 50)

I'll discuss this principle, about how to weight different perspectives, in Section 4.

I'll now return to Medina's discussion of epistemic-character development, and sketch out his understanding of the source of epistemic advantage.

As I said previously, on Medina's view the epistemic advantage oppressed people have, or can have, is grounded in their epistemic character. This epistemic character is influenced by their experiences, and in turn influences subject's abilities to respond to epistemic friction.

Table 35.1 summarises Medina's detailed discussion. The first column shows different sets of experiences, which Medina says are characteristic of privilege (top row) and oppression (bottom row) respectively. The second column shows character traits that these experiences tend to result in, and the final column contains the effects these traits have on subjects' abilities to respond to epistemic friction.

According to Medina, socially privileged people are (often) educated to dominate – they are taught to see themselves as authoritative, and others (especially those from marginalised groups) as less credible, and less worthy of respect (or even note). He illustrates this idea with accounts of education of slave owners' children in the American South, but intends it to generalise to other, less blatant, situations of oppression (2012, 31–32).

Medina says these experiences lead to the development of traits like epistemic arrogance and closed-mindedness, and make it difficult to respond to epistemic friction. Privileged subjects are less likely to notice or engage with friction, because they rarely need to do so (and often not doing so is required to keep them in power). And when they do notice it, they will struggle to balance the different views and sources of friction according to the principle of equilibrium, because they're not used to doing so.

In contrast, oppressed people are educated to be deferential, and to acknowledge and assign importance to the views of others (in particular to the dominant group, on whose approval their survival depends). Here Medina talks about slave mentality, as well as other

Table 35.1 Differential epistemic character development

	Experiences	*Character traits*	*Effects*
Socially privileged	– Educated to dominate – See self as authoritative – See marginalised groups as less credible	– Epistemic arrogance – Closed-mindedness	– Less likely to notice friction – Less likely to engage with friction – Difficulty achieving equilibrium
Socially oppressed	– Educated to defer – See other's views as important	– Epistemic humility – Open-mindedness	– Likely to notice friction – Likely to engage with friction – Experienced at achieving equilibrium

contexts of oppression (2012, 40–43). He says this leads oppressed people to develop traits like epistemic humility and open-mindedness, which give them an advantage when it comes to epistemic friction. They've already developed the kind of character which can notice and engage with multiple sources of friction, and are well practiced at keeping them in equilibrium. This is the basis of epistemic advantage on Medina's account: socially oppressed people have an epistemic advantage in how they tend to be disposed to respond to epistemic friction.

There are plenty of criticisms that could be made of this view. We could question whether oppressed groups and dominant groups really have the experiences Medina attributes to them. We might also question whether these experiences always, or even often, result in the virtues and vices he attributes to them. But Medina's account of epistemic advantage isn't my goal in this chapter. In the next section I'll consider whether Medina's understanding of epistemic advantage could (if plausible) block Symmetry, and therefore relativism.

4. Advantage and symmetry

In this section I'll show that the standpoint theorist's standard response to relativism isn't available to Medina, because on his account epistemic advantage is compatible with Symmetry. In the next section I'll argue that this is true of all standpoint theories.

My reason for beginning with Medina is that his view, on first blush, seems very amenable to relativism. His account of epistemic advantage is based on differing responses to epistemic friction, with the advantageous responses being guided by the principle of acknowledgement and engagement and the principle of equilibrium. Equilibrium says to strive to balance the influence of different systems and perspectives, which sounds like saying systems shouldn't be ranked. This would be an endorsement of symmetry rather than a contradiction of it.

Medina is aware that equilibrium sounds relativistic, but denies that it is. He says:

> the principle of epistemic equilibrium was not a relativistic principle that demanded giving equal weight to all perspectives. Rather, it was the desideratum of searching for equilibrium on the interplay of cognitive forces, without some forces overpowering others, without some cognitive influences becoming unchecked and unbalanced.
>
> *(Medina 2012, 195)*

This response shows that equilibrium is (or at least could be) incompatible with one interpretation of Symmetry, namely Equality. It says that systems can be weighted differently, leaving open the possibility that some could be more correct than others, contrary to Equality.

But this only tells us part of the story. In Section 1 we saw a second interpretation of Symmetry – Non-neutrality – which only denies that systems and practices can be evaluated *independently of a system*. Medina's response doesn't engage with this interpretation, which seems to be compatible with his account of epistemic advantage. Achieving the equilibrium Medina advocates requires a standpoint – a particular, socially located epistemic perspective – which is exactly what Non-neutrality demands.

Whilst Medina shows that his view is incompatible with relativism based on Equality, he doesn't address its (in)compatibility with relativism based on Non-neutrality. In fact, his view appears to incorporate Non-neutrality, and so it looks like Medina's standpoint theory is relativist.

5. Generalising

Other standpoint theorists might claim that Medina's view is an anomaly. They don't advocate Equilibrium, so if that's what introduces relativism their views are unaffected. However this response is too quick.

Medina's account makes standpoint theory's commitment to non-neutrality vivid because his version of epistemic advantage emphasises equilibrium, whilst most versions emphasise the opposite. But this commitment isn't unique to Medina or his account of epistemic advantage, because his account of epistemic advantage isn't the problem. The commitment to non-neutral Symmetry comes from the standpoint thesis.

Recall that Medina couldn't fully separate his account of epistemic advantage from Symmetry because achieving equilibrium (and therefore epistemic advantage) required a standpoint. This dependence on standpoints for justification isn't a quirk of Medina's epistemic advantage. It follows from the standpoint thesis. If justification depends on socially situated perspectives, then so does justification about standpoints and how they are ranked. Any view involving the standpoint thesis will involve non-neutral symmetry, so all standpoint theories are committed to non-neutrality, to symmetry, and to relativism.

6. Relativism as unproblematic

One of the standpoint theorist's central theses, the standpoint thesis, commits them to relativism based on non-neutral Symmetry. It says that justification is *dependent* on systems or perspectives, that there are a *plurality* of these perspectives, and (because it says justification is system-dependent) it's also committed to the idea that there is *no neutral*, perspective-independent way to rank these systems.

Standpoint theorists have a decision to make: either they accept relativism or they must radically rethink, or abandon, a central part of their view. In this final section I argue that a radical rethink is unnecessary. The main worries standpoint theorists have about relativism don't apply to relativism based on non-neutral symmetry, and so accepting relativism is their best option.

Sandra Harding (1991) offers the most thorough and extended discussion of standpoint theorists' problems with relativism. I'll respond to her two main worries.

Her first worry is that relativism collapses into "weak objectivity," her name for the guiding principle of mainstream science. She says this principle advocates removing all social and political values from science, but fails, instead leaving behind sexist ones. So it would be a problem if relativism were equivalent to weak objectivity.

Here's Harding on relativism and weak objectivity:

> Many thinkers have pointed out that judgemental relativism [which Harding equates with the claim that there are no rational or scientific grounds for evaluating various epistemic systems] is internally related to objectivism. For example, science historian Donna Haraway argues that judgemental relativism is the other side of the very same coin from "the God trick" required by what I have called weak objectivity. To insist that no judgements at all of cognitive adequacy can legitimately be made amounts to the same thing as to insist that knowledge can be produced only from "no place at all": that is, by someone who can be every place at once.
>
> *(1991, 152)*

The crucial comparison is in the final sentence. Harding says the "relativist" view that there are no legitimate judgements comparing epistemic systems is the same as the claim that knowledge can be produced from no place at all (i.e. that knowledge and justification can be value free, as on weak objectivity).

This claim is false, for two reasons. First, the two views she describes are not equivalent. The weak objectivist claim that knowledge can be produced from no place at all is a positive claim about the existence of justification in a social vacuum. Whilst Harding thinks this claim is incoherent and best understood as showing that no justification is possible, it's important to separate this from the negative, sceptical claim about the impossibility of justification. The claim that there are no legitimate judgements about systems is sceptical; the claim that there are judgements of a particular, value-free kind is not.

Second, relativists needn't make the claim Harding attributes to them. Whilst Equality-based relativism incorporates the denial of any legitimate judgements about epistemic systems, non-neutral relativism – the kind I've argued is present in standpoint theories – does not. Non-neutral relativism only justifies the claim that there are no *system-independent* judgements about systems. So relativism doesn't collapse back into conservative weak objectivity.

I think Harding's second worry is the real motivator behind resistance to relativism: relativism has negative connotations. She identifies several which might raise problems for standpoint theorists. The first two are that relativism is seen as "weak" compared to objectivity, and that it is coded as feminine (Harding 1995, 340). I take her point here to be strategic – it's not that these connotations run counter to feminist goals, or that they make a view less likely to be true, but they might mean a view is taken less seriously by other scholars.[5]

If this is her point then I see where she is coming from, but I don't think these connotations should trouble us too much. For one thing, *feminist* standpoint theory already runs the risk of invoking connotations of femininity and weakness given its title. For another, these connotations are (at least somewhat) less negative nowadays than they used to be.

However, the third connotation presents a greater challenge. The epistemic-advantage thesis is intended to show the importance and validity of marginalised standpoints, and that claims made from these standpoints should be taken more seriously. But according to Harding, relativism is often used in service of precisely the opposite goal; it is used by dominant groups to undermine the claims of the oppressed, saying that they are *merely* relatively justified (Harding 1991, 151).

This leaves me with some important questions to answer if I'm going to conclude that a relativist standpoint theory is unproblematic. First, how can a relativist understanding of epistemic advantage avoid reinforcing oppression? And second, how can it support the aim of tackling oppression?

The key to answering the first question lies in recognising that the relativity of a claim shouldn't be seen to diminish its credibility, or its validity. If you have an absolutist view of justification then the "mere" relativised claims will, of course, seem inferior. But this isn't the view that relativists have – they think that *all* claims are relative, and so dismissals of claims of the oppressed on the basis of relativity are only as legitimate as dismissals of claims of the dominant on this basis. If I'm right, and this relativist view follows from the standpoint thesis, then standpoint theorists don't need to worry about reinforcing oppression either. Dominant viewpoints don't come out as superior on this view.

What about the second question? How can a relativised epistemic advantage support the aims of tackling oppression? I'm unsure whether my answer to this question will satisfy standpoint theorists, although I think that it should. Remember that a non-neutral relativism (which is the kind of relativism I claim is present in standpoint theory) can allow for evaluative judgements about epistemic systems. So a standpoint theory that is relativist in this sense is compatible with the claim that some perspectives lead to better, or more accurate, claims than others – whether that claim is based on a story about virtues and vices, as with Medina's view, or on some other account. Of course this claim has to be relativised to a non-neutral perspective. It's not an absolute advantage. But I think that this is all that someone who truly understands and endorses the standpoint thesis should want to say anyway.

7. Conclusion

In this chapter I have explored the connection between feminist standpoint theory and relativism. I demonstrated that the current debate, which focuses on the epistemic advantage thesis and equality, is missing its target, and suggested that instead we shift our attention to the standpoint thesis and non-neutrality. I argued that when we do this, we quickly see that standpoint theory is committed to relativism at its very core, and so standpoint theorists have a decision to make – between relativism and a radical rethink of their view. Finally, I argued that the relativist route is a lot smoother than standpoint theorists tend to think, and so standpoint theorists should embrace the relativism within their view.

Acknowledgements

Research on this chapter was assisted by funding from the ERC Advanced Grant Project "The Emergence of Relativism" (Grant No. 339382).

Notes

1 Kusch's definition incorporates two further essential (and four non-essential) elements. All of these are important for understanding how relativism has been presented by different authors, but I have streamlined my definition for simplicity. In particular, I haven't included Kusch's "essential" *exclusiveness* and *notational confrontation*, because I think these follow from pluralism: if these differences weren't present then the multiple frameworks would collapse into one, and so pluralism would not be present either. Cf. Williams (2007) and Coliva (2015) who use similar triadic definitions of relativism.
2 Intemann (2010) has argued that some versions of standpoint theory and feminist empiricism have more in common with each other than with other accounts from their own branches.
3 Not all of these have been expressed explicitly by all standpoint theorists, though I think all standpoint theorists would (and should) endorse them.
4 I'm unsure whether this connection was drawn on the basis of the presence of these two components because it was not known that the third component was required, or because it was presumed that standpoint theory did meet the third conclusion – perhaps standpoint theory was interpreted as arguing that justification relative to a feminist standpoint is "just as good" as (and so symmetrical to) justification relative to other standpoints.
5 Although Harding criticises this principle of objectivity (and objectivity and relativism are often taken to be opposites) she does not describe herself as a relativist. Rather she sees herself as criticising both (weak) objectivity and relativism, and instead offering a new and improved principle of objectivity which she calls "strong objectivity" (1991, ch. 6).

References

Anderson, E. (1995), "Feminist Epistemology: An Interpretation and a Defense," *Hypatia* 10(3): 50–84.
Coliva, A. (2015), *Extended Rationality: A Hinge Epistemology*, London: Palgrave Macmillan.

Collins, P. H. (1986), "Learning from the Outsider Within: The Sociological Significance of Black Feminist Thought," *Social Problems* 33(6): 14–32.

Fricker, M. (1999), "Epistemic Oppression and Epistemic Privilege," *Canadian Journal of Philosophy* 29(1): 191–210.

Haraway, D. (1988), "Situated Knowledges: The Science Question in Feminism and the Privilege of Partial Perspective," *Feminist Studies* 14(3): 575–599.

Harding, S. (1991), *Whose Science? Whose Knowledge? Thinking from Women's Lives*, New York: Cornell University Press.

———. (1995), "'Strong Objectivity': A Response to the New Objectivity Question," *Synthese* 104(3): 331–349.

Hartsock, N. C. M. (1983), "The Feminist Standpoint: Developing the Ground for a Specifically Feminist Historical Materialism," in *Discovering Reality*, edited by S. Harding and M. B. Hintikka, Dordrecht: Springer, 283–310.

———. (1997), "Comment on Hekman's 'Truth and Method: Feminist Standpoint Theory Revisited': Truth or Justice?" *Signs* 22(2): 367–374.

Intemann, K. (2010), "25 Years of Feminist Empiricism and Standpoint Theory: Where Are We Now?" *Hypatia* 25(4): 778–796.

Kusch, M. (2016), "Wittgenstein's *On Certainty* and Relativism," in *Analytic and Continental Philosophy: Methods and Perspectives: Proceedings of the 37th International Wittgenstein Symposium*, edited by H. A. Wiltsche and S. Rinofner-Kreidl, Berlin and Boston: De Gruyter, 29–46.

Longino, H. E. (1997), "Feminist Epistemology as a Local Epistemology: Helen E. Longino," *Aristotelian Society Supplementary Volume* 71(1): 19–36.

Lukács, G. (1971), *History and Class Consciousness: Studies in Marxist Dialectics*, Cambridge, MA: MIT Press.

Medina, J. (2012), *The Epistemology of Resistance: Gender and Racial Oppression, Epistemic Injustice, and the Social Imagination*, Oxford: Oxford University Press.

Smith, D. E. (1997), "Comment on Hekman's 'Truth and Method: Feminist Standpoint Theory Revisited,'" *Signs* 22(2): 392–398.

Tanesini, A. (forthcoming), "Standpoint Theory Then and Now."

Williams, M. (2007), "Why (Wittgensteinian) Contextualism Is Not Relativism," *Episteme* 4(1): 93–114.

Wylie, A. (2003), "Why Standpoint Matters," in *Science and Other Cultures: Issues in Philosophies of Science and Technology*, edited by Robert Figueroa and Sandra G. Harding, New York and London: Routledge, 26–48.

PART 6

Relativism in metaphysics

36

ONTOLOGICAL RELATIVITY

David J. Stump

1. Introduction

The general idea of ontological relativity is that we cannot tell, ultimately, what we are talking about. We might say that the metaphysics of the world is unknown and that we describe the world in language using categories that are human constructions. The thesis of ontological relativity is that there are alternative constructions, descriptions, or models that work equally well for describing the world. Thus ontological relativity is based on a form of nominalism. It would also be possible to put ontological relativity in terms of metaphysics. Realists claim that we accurately describe the world as it actually is. A metaphysical version of ontological relativity would include the idea that there are multiple worlds, each relative to a conceptual scheme or to a language.

Many philosophers have advocated forms of ontological relativity without either endorsing it fully or even putting their theses in these terms. The classic text on ontological relativity is Quine's article of that name (Quine 1968). Quine does not set out to make a metaphysical claim, rather, his position follows from philosophy of language, and in particular from an analysis of reference and from Quine's thesis of the inscrutability of reference. Indeed, as we will see, Quine's relativism is quite limited. Ontological relativity comes up in translation, Quine says, but he rejects as absurd the idea that it applies to our own language. More radical views, such as for example, what is called the Saphir-Whorf Thesis in linguistics have been considered. The relativity of languages is "a thesis according to which the global structure of every language exercises a differential influence on the thought of the speaker, on the way in which he conceives reality, and the manner in which he behaves in front of it" (Rossi-Landi 1973, 1). Our descriptions of reality are tied in a very strong way to the language that we speak and, furthermore, there are radical differences in how languages conceive of the world.

Thus the claim is that ontology is relative to a language and/or to a conceptual scheme. Anyone who argues that there are frameworks (schemes, paradigms, styles, etc.) will hold some form of ontological relativity. Quine explicitly uses the language of conceptual schemes in setting up his argument for ontological relativity, as we see in the article that precedes the article, "Ontological Relativity," itself when reprinted in book form (Quine 1957, 1969, 25). Some may even hold that there are multiple worlds, following from the fact that frameworks posit alternative elements of reality, and thus take a metaphysical stance. So, for example, Thomas Kuhn held that paradigms come with ontological commitments, and he was not shy about talking about scientists who change paradigms living in different worlds (Kuhn 1962). Others, including Quine, reject metaphysical claims, and hold that there are merely alternative descriptions, not alternative worlds. This chapter will survey some of the philosophical views that approach the idea of ontological relativity from the viewpoint of the philosophy of language and will also consider structuralism in the philosophy of mathematics and the philosophy of science.

2. Carnap's tolerance

Late in his career, Carnap adopts a view that could be seen as a form of ontological relativity. According to Carnap, natural science is to be represented in a formal language. In *The Logical Syntax of Language* (Carnap 1937), he distinguishes between L-rules and P-rules, that is, he divides a scientific language into a logical part and a physical part. The logical part is analytic and conventional, while the physical part is synthetic and empirical. Many different languages are possible and the choice between them is practical, rather than cognitive. Carnap calls the choice of language an external question, guided by the suitability of a language for a given purpose. There is no right or wrong language in an objective sense but rather, only right or wrong relative to our purposes. Carnap enunciated this in his famous principle of tolerance:

> *in logic, there are no morals.* Everyone is at liberty to build up his own logic, i.e. his own form of language, as he wishes. All that is required of him is that, if he wishes to discuss it, he must state his methods clearly, and give syntactical rules instead of philosophical arguments.
>
> *(1937, 52, emphasis in the original)*

A change of language for Carnap amounts to a conceptual scientific revolution, with the change in language being a formal equivalent of a change in conceptual scheme or paradigm. The choice of a language will imply a choice of ontology as well. Even in his early work, when pressed by Neurath, Carnap came to the view that it makes no difference if your logical system is based on a physicalist language or one of phenomena (Uebel 2007). Thus the world could be thought of as either consisting of physical objects or of sense data, but the choice between these two alternatives is a metaphysical question, which Carnap sees as meaningless. Both systems could accurately describe the world. For Carnap, there is no point in asking what the world really is, as long as we have an objective description of reality in our chosen language or logic.

3. Quine's ontological relativity

Quine's views on ontological relativity are rightly considered to be difficult to pin down (Romanos 1983, 42; (Kirk 2004). Quine rejects Carnap's principle of tolerance or choice of language (Hylton 2004, 133) but he has a different argument for ontological relativity. As Peter Hylton points out, Quine's key point is that "the relation between language and the world by which our language come to have empirical meaning and empirical content is not a relation between names and objects but rather a relation between sentences and sensory stimulations" (Hylton 2004, 135).

As the famous example of the anthropologist given the task of translating the native term "gavagai" shows, to translate the term as "rabbit," the world must be assumed to contain rabbits as objects. We could just as easily translate the term as "rabbit stage" or an "undetached rabbit part," and no amount of pointing and querying would distinguish between these alternatives

(Quine 1969, 32–33). An ontology is being assumed during the task of translation. Quine is adamant that the only criterion we have to determine the proper translation is the observable behavior of the speaker of the other language. Furthermore, we can never get beyond human categories expressed in language when we describe the world. Nevertheless, Quine accepts what current science says about the world:

> The very notion of object, or of one and many, is indeed as parochially human as the parts of speech; to ask what reality is really like, however, apart from human categories, is self-stultifying. It is like asking how long the Nile really is, apart from parochial matters of miles or meters. Positivists were right in branding such metaphysics as meaningless. But early positivists were wrong if and when they concluded that the world is not really composed of atoms or whatever. The world is as natural science says it is, insofar as natural science is right; and our judgment as to whether it is right, tentative always, is answerable to the experimental testing of predictions.
>
> *(Quine 1992, 9)*

The acceptance of what our home language says we are talking about is a major concession and even seems puzzling to many, given Quine's other views. For example, Donald Davidson argues that Quine cannot avoid relativity in our home language given his premises, so although he agrees with Quine on many points, he does not see how ontological relativity can stand (Malpas 1989, 161). Davidson rejects ontological relativity (Davidson 1979), which is perhaps not surprising, given that his anti-relativist rhetoric is even stronger than Quine's (Davidson 1973).

However, acceptance of our home language at face value follows from Quine's rejection of metaphysics. Thomas Ricketts connects Quine's acceptance of the ontology of our home language to his naturalism and argues that it is completely consistent and expected, given those views (Ricketts 2011). Current science is the best that we can do, and while acknowledging its fallibility, Quine embraces it wholeheartedly:

> Within the home language, reference is best seen (I now hold) as unproblematic but trivial, on a par with Tarski's truth paradigm. Thus "London" denotes London (whatever *that* is) and "rabbit" denotes rabbits (whatever *they* are). Inscrutability of reference emerges only in translation.
>
> *(Quine 1986, 460)*

Thus, Quine rejects the view that ontological relativity applies to our own language, but we are still unable to answer the metaphysical question of what we are talking about. All we can do is say the obvious, when I say "rabbit" I mean rabbit. We can only assume that we are being understood. Quine can be nonchalant about these issues given his rejection of metaphysics. These are simply not real issues for Quine, any more than they were for Carnap.

In an article defending ontological relativity, Quine connects his thesis quite strongly to the underdetermination of scientific theories by their evidence (Quine 1970). The underdetermination thesis in science is analogous to the case of translation of language, given that no amount of evidence will distinguish between some alternatives:

> Theory can still vary though all possible observations be fixed. Physical theories can be at odds with each other and yet compatible with all possible data even in the broadest sense. In a word, they can be logically incompatible and empirically equivalent.
>
> *(Quine 1970, 179)*

Quine makes a strong case for underdetermination simply from the fact that the evidence is empirical and arrived at inductively. There will always be further evidence in the future that is relevant to the theory, so the current evidence that we have is always incomplete. Moreover, there may be past evidence that we missed.

4. Goodman's worlds

Nelson Goodman is famous for several topics in philosophy, but his most controversial stance is undoubtedly his irrealism and claim for there being multiple actual worlds. Goodman rejects the idea that there are simply alternative descriptions of a single world as not providing a robust enough account of the conflict between descriptions.

> Consider, to begin with, the statements "The sun always moves" and "The sun never moves" which, though equally true, are at odds with each other. Shall we say, then, that they describe different worlds, and indeed that there are as many different worlds as there are such mutually exclusive truths? Rather, we are inclined to regard the two strings of words not as complete statements with truth-values of their own but as elliptical for some such statements as "Under frame of reference A, the sun always moves" and "Under frame of reference B, the sun never moves" – statements that may both be true of the same world. Frames of reference, though, seem to belong less to what is described than to systems of description: and each of the two statements relates what is described to such a system. If I ask about the world, you can offer to tell me how it is under one or more frames of reference; but if I insist that you tell me how it is apart from all frames, what can you say? We are confined to ways of describing whatever is described. Our universe, so to speak, consists of these ways rather than of a world or of worlds.
>
> *(Goodman 1978, 2–3; also see 111–113)*

Goodman insists that we describe the world, not a version of the world. Our truths must be true of a world, not merely of a description of the world. Given that descriptions of the world conflict, we are led immediately to the view that there are multiple actual worlds. Furthermore, Goodman argues that there are simply no grounds to believe that all descriptions will reduce to a single account, such as that given in physics. Given that he is talking about all kinds of descriptions of the world, it is wildly implausible that all of them would reduce to any one account. Goodman asks, "How do you go about reducing Constable's or James Joyce's world-view to physics?" (Goodman 1978, 5). Reduction of one area of science to another can be a valuable exercise, he holds, but it cannot be assumed that it is always possible to reduce everything to a single account. Thus, Goodman embraces a metaphysical thesis of ontological relativity. There are actual multiple worlds, he thinks, one for each conflicting account of the world.

5. Putnam's internal realism

Hilary Putnam begins with an argument that it is impossible for mental states to fix reference. His famous example is a thought experiment in which there is a liquid on a twin earth that looks like water, but has a different chemical composition (Putnam 1975, 139). Twin earth people would have similar mental states to those on earth, but they would be referring to a different substance. Putnam's example has spawned a huge literature which is still not settled and which goes far beyond the scope of what can be covered in this chapter (Pessin et al. 1996; Yli-Vakkuri and Hawthorne 2018). I will simply note that some philosophers reject ontological relativity by claiming that it is possible to fix reference with mental states.

Having dismissed fixing reference with mental states, Putnam extends Quine's argument for ontological relativity by arguing that "no view which only fixes the truth-values of whole sentences can fix reference, even if it specifies truth-values for sentences in every possible world" (Putnam 1981, 33). Putnam uses the example of the sentence "A cat is on a mat" and shows that it does not exclude the possibility that "cat" refers to cherries and "mat" refers to trees. He gives a fairly technical proof in an appendix, but the basis of his argument is that formal systems are ontologically relative. As the famous remark attributed to Hilbert has it, in geometry we could just as well be talking about tables, chairs, and beer mugs.[1] We only know the relations between objects as set out in the axioms, not what the primitive terms stand for, given that they are undefined in a formal system. The issue of what mathematical objects really are is the subject of metaphysics, not mathematics. We will see in the next section that this has spawned a vast literature on structuralism.

Putnam's solution to the problem of reference is to reject metaphysical realism and turn instead to what he calls the internalist perspective: "it is characteristic of this view to hold that *what objects does the world consist of?* is a question that only makes sense to ask *within* a theory or description" (Putnam 1981, 49). There is, therefore, the possibility of alternative true descriptions of the world, or ontological relativity, but we are never in a position to formulate global skepticism or relativism. We are always captives of our own theory and cannot make sense of global skepticism or relativism, which is why Putnam rejects the possibility that we are brains in a vat (Putnam 1981, ch. 1).

6. Structuralisms

Structuralism, the view that our theories only tell us the relations of objects, not what the objects themselves are has gained wide acceptance in the philosophy of mathematics (Hellman 2001; Resnik 1997; Shapiro 2000). As I mentioned previously, in formal systems, primitive terms are undefined, so the axioms and theorems can apply to many different objects. Michael Resnik, a philosopher of mathematics quite influenced by Quine, puts it this way:

> Structuralists hold that since mathematics studies structures, its theories aim to characterize their objects up to isomorphism, at most. The incompleteness of mathematical objects is an inevitable consequence of this aim, which structuralists urge us to accept with equanimity.
>
> *(Resnik 1996, 83)*

Resnik's main contribution to discussions in the philosophy of mathematics is a disquotational account of reference and truth, which he proposes in order to make sense of the incompleteness of mathematical objects. Of particular significance is the point that an immanent account of reference eliminates one of the strongest arguments against mathematical realism. A transcendental approach seems to require a causal theory of reference, and since mathematical objects are causally inert, it seems impossible to be a transcendent realist in mathematics. Resnik defends Quinean holism, arguing that mathematics is not separable from the rest of science and that justification of mathematics, once we move away from local justifications within various branches of mathematics, ultimately involves the role that mathematics plays in science as a whole.

Perhaps the best argument for structuralism in mathematics is that it fits mathematical practice (McLarty 2008). Philosophers of mathematics don't need to give an account of mathematical objects that goes beyond the account given in mathematics. Ontological relativity works well in mathematics, so philosophers of mathematics should accept it too.

In the philosophy of science, structuralism was originally presented by John Worrall (inspired by Poincaré) as a middle way between scientific realism and anti-realism, and hence he labeled it structural realism (Worrall 1989). There are three tenants of scientific realism (Chakravartty 2011, 157).[2] First there is a claim that the world exists independently of what we think of it. Second, there is a semantic commitment to take theories literally. Third, there is a claim that we have as much knowledge of theoretical or unobservable entities as we do of observable ones. The debate between scientific realists and instrumentalists is relevant to ontological relativity because instrumentalists can be seen as saying that what really exists on the unobservable scale of reality does not matter, so long as our theories make accurate predictions of observable phenomena. Other issues are very relevant to ontological relativity as well, particularly the claim of whether there are natural kinds or not. As mentioned earlier, ontological relativity is based on nominalism, the view that there are no natural kinds and that therefore there are alternative descriptions of reality. There may well be real and objective properties of objects, but how objects are classified and how groups are labeled is a matter of convention.

Structural realism has splintered into epistemological and ontological versions, although all maintain the idea that relations are more primary than objects:

> there is a general convergence on the idea that a central role is to be played by relational aspects over object-based aspects of ontology – whether this reflects a fundamental fact about the world (the ontological thesis) – as opposed to our engagement with it (the epistemological thesis) – introduces still further divisions. Still further divisions within these divisions are created in the task of establishing what is actually meant by each of these theses.
>
> *(Landry and Rickles 2012, v)*

Should a structural realism leave ontology open (an epistemological thesis that we just do not know what the ultimate nature of reality is) or should it say that entities do not exist: all that there is in the world are the structures that are described in our best scientific theories. In his defense of metaphysical structural realism, James Ladyman argues that:

> Traditional realism should be replaced by an account that allows for a global relation between models and the world, which can support the predictive success of theories, but which does not supervene on the successful reference of theoretical terms to individual entities, or the truth of sentences involving them.
>
> *(Ladyman 1998, 422)*

This view has been taken up most notably by Stephen French, who advocates for an eliminativist version of ontic structural realism (EOSR), in which he responds to the challenges of underdetermination by holding that structures exist, while entities do not: "withdraw your realist commitment from particles-as-objects and invest it in the underlying structure" (French 2016, 189). For French, this does not imply that we cannot talk about objects at all, rather, he claims that we can "eliminate objects from our metaphysics, while retaining talk of them via certain metaphysical devices" (French 2016, 190). The debate between the epistemological and ontological versions of structural realism is ongoing and an area of current research.

7. Conclusion

Ontological relativity is a claim about how we describe the world and could include metaphysical claims about alternative worlds. However, we have seen that many philosophers, including Quine, couch it in terms of language, while rejecting metaphysics altogether. The reluctance to embrace metaphysical claims softens the relativism of the claim of ontological relativity, but the idea that we do not know ultimately what we are talking about still remains. To say we know what we are talking about is to make a metaphysical claim – the world really is as I describe it. To reject metaphysics is simply to give up on that claim, thus Quine's nonchalance about what "rabbits" really refers to and Carnap's claim that it does not matter what logic we choose. There is no fact of the matter about reference on this view. Neither Quine nor Carnap would want to say that we create a world through our descriptions, nor that there are alternative worlds. Nelson Goodman seems to embrace such a metaphysical view, as we saw previously. Structural realists in the philosophy of science might hold the view that there are no individual objects, rather just relations. Alternatively, they may hold that we can never know what the objects are, but we can know their relations. In either case, we must give up the realist idea that our theories actually depict the world as it actually is.

Notes

1 The remark occurs in Blumenthal's biography of Hilbert (Blumenthal 1935, 402–403) and was popularized by Constance Reid (Reid 1970).
2 Psillos also has three tenants of scientific realism, but they are slightly different (Psillos 1999).

References

Blumenthal, O. (1935), "Lebensgeschichte," in *David Hilbert Gesammelte Abhandlungen Bd. 3.* New York: Chelsea, 1965, 388–429.
Carnap, R. (1937), *The Logical Syntax of Language,* translated by A. Smeaton, London: Kegan Paul, Trench, Trubner, & Co.
Chakravartty, A. (2011), "Scientific Realism and Ontological Relativity," *Monist* 94(2): 157–180.
Davidson, D. (1973), "On the Very Idea of a Conceptual Scheme," *Proceedings and Addresses of the American Philosophical Association* 47(1973–1974): 5–20.
———. (1979), "The Inscrutability of Reference," *Southwestern Journal of Philosophy* 10: 7–20.
French, S. (2016), "Response to My Critics," *Metascience* 25(2): 189–196.
Goodman, N. (1978), *Ways of Worldmaking,* Indianapolis: Hackett Publishing.
Hellman, G. (2001), "Three Varieties of Mathematical Structuralism," *Philosophia Mathematica* 39: 184–211.
Hylton, P. (2004), "Quine on Reference and Ontology," in *The Cambridge Companion to Quine,* edited by R. F. Gibson, Cambridge: Cambridge University Press, 115–150.
Kirk, R. (2004), "Indeterminacy of Translation," in *The Cambridge Companion to Quine,* edited by R. F. Gibson, Cambridge: Cambridge University Press, 151–180.
Kuhn, T. S. (1962), *The Structure of Scientific Revolutions,* Chicago: University of Chicago Press.
Ladyman, J. (1998), "What Is Structural Realism?" *Studies in History and Philosophy of Science* 39(3): 409–424.
Landry, E. M. and D. P. Rickles (eds.) (2012), *Structural Realism: Structure, Object, and Causality,* Dordrecht: Springer.
Malpas, J. E. (1989), "Ontological Relativity in Quine and Davidson," *Grazer Philosophische Studien* 36: 157–178.
McLarty, C. (2008). "What Structuralism Achieves." In *The Philosophy of Mathematical Practice.* P. Mancosu. Oxford: Oxford University Press, 354–369.
Pessin, A., H. Putnam and S. Goldberg (1996), *The Twin Earth Chronicles: Twenty Years of Reflection on Hilary Putnam's "The Meaning of Meaning,"* Armonk: M. E. Sharpe.
Psillos, S. (1999), *Scientific Realism: How Science Tracks Truth,* London and New York: Routledge.

Putnam, H. (1975), "The Meaning of 'Meaning,'" in *Language, Mind and Knowledge*, edited by K. Gunderson, Minneapolis, MN: University of Minnesota Press, 131–193.

———. (1981), *Reason, Truth and History*, Cambridge: Cambridge University Press.

Quine, W. V. O. (1957), "Speaking of Objects," *Proceedings and Addresses of the American Philosophical Association* 31(1957–1958): 5–22.

———. (1968), "Ontological Relativity, the Dewey Lectures 1968," *Journal of Philosophy* 65: 185–212.

———. (1969), *Ontological Relativity and Other Essays*, New York: Columbia University Press.

———. (1970), "On the Reasons for Indeterminacy of Translation," *Journal of Philosophy* 67: 178–183.

———. (1986), "Reply to Paul A. Roth," in *The Philosophy of W. V. Quine*, edited by L. E. Hahn and P. A. Schil, La Salle, IL: Open Court, 459–461.

———. (1992), "Structure and Nature," *Journal of Philosophy* 89(1): 5–9.

Reid, C. (1970), *Hilbert*, London, Berlin and New York: Springer.

Resnik, M. D. (1996), "Structural Relativity," *Philosophia Mathematica* 4: 89–99.

———. (1997), *Mathematics as a Science of Patterns*, Oxford: Oxford University Press.

Ricketts, T. (2011), "Roots of Ontological Relativity," *American Philosophical Quarterly* 48(3): 287–300.

Romanos, G. D. (1983), *Quine and Analytical Philosophy: The Language of Language*, Cambridge, MA: MIT Press.

Rossi-Landi, F. (1973), *Ideologies of Linguistic Relativity*, The Hague: Mouton.

Shapiro, S. (2000), *Philosophy of Mathematics: Structure and Ontology*, New York and Oxford: Oxford University Press.

Uebel, T. (2007), "Carnap and the Vienna Circle: Rational Reconstruction Refined," in *The Cambridge Companion to Carnap*, edited by M. Friedman and R. Creath, Cambridge: Cambridge University Press, 153–175.

Worrall, J. (1989), "Structural Realism: The Best of Both Worlds?" *Dialectica* 43: 99–124.

Yli-Vakkuri, J. and J. Hawthorne (2018), *Narrow Content*, Oxford: Oxford University Press.

37

QUANTIFIER VARIANCE

Eli Hirsch and Jared Warren

1. What is quantifier variance?

Different possible languages have different concepts of what "exists" or what "there is." Despite this key difference between them, some of these languages are equally good as tools for describing reality. These are two of the central claims of *quantifier variance*, a highly influential deflationary view in contemporary metaontology. Variance is rooted in widely accepted metasemantic principles, yet it remains controversial, since it deflates the pretensions of philosophical ontology. Additionally, variance is very widely misunderstood. A proper understanding starts with the metasemantic background leading to variance, in both of its principle forms.

Language and Meaning. Linguistic meaning is determined by actual and possible language use. Taken baldly, this is almost a truism. But more plausibly, this slogan about use expresses a commitment to *charity* in interpretation. Minimally, charity ties meaning to use by ruling out interpretations that see those we are interpreting as uttering falsehoods inexplicably.

Say that an *interpretation* of language L assigns coarse-grained truth conditions to sentences of L relative to each possible context of utterance. Coarse-grained truth conditions can be modeled as sets of possible worlds (Stalnaker 1984); and these functions from contexts of utterance to coarse truth conditions are called "characters" (Kaplan 1989). A charity-based metasemantics assigns L the interpretation that, when all is said-and-done, when every disposition to correct and revise is accounted for, makes the best sense of the linguistic behavior of L-speakers by making their considered utterances come out true in actual and possible circumstances, *ceteris paribus*.

Charity-based approaches are *top down* – they explain the meanings of subsentential expressions in terms of the meanings of whole sentences. Once again, this respects the dictum that meaning is use, as the meanings of subsentential expressions will be fully explained in terms of their usage in the language. By contrast, *bottom up* theorists start with the meanings of subsentential expressions and then go on to explain the meanings of sentences in terms of them. To top down theorists, this is mysterious – what magic could attach meaning to a subsentential expression apart from its use?

Quantifier Expressions. What makes an expression in a given language a "quantifier" on a charity-based, top down approach is that it is *used as* a quantifier. The inferential role of a quantifier expression is its defining feature – an expression in a given language is an existential quantifier, for instance, if it plays the inferential role of an existential quantifier in that language. No doubt the inferential role of "there is" or "exists" in natural language is more complex than the role of "∃" in formal logical languages, but the formal-syntactic role of "∃" provides a tidy approximation of the informal inferential role of "exists" or "there is" in English. The expression "there is" is an existential quantifier, in English, roughly because for name "*a*" and predicate "*F*," from "*a* is *F*," "there is an *F*" follows; and if a non-"*a*" claim follows from "*a* is *F*," with no auxiliary assumptions made about "*a*," then that same thing also follows from "there is an *F*." Expressions that obey this role unrestrictedly, for all names and predicates that could be introduced into the language, express the language's unrestricted concept of existence.

Combining a top down charity approach to meaning with the inferential role priority account of what it is to be a "quantifier" results in a deflationary metasemantics for quantificational claims. This metasemantics entails *modest quantifier variance*. Call a language with quantifier expressions a "quantifier" language. All human languages are quantifier languages, and maybe all possible languages are too; our terminology is neutral on this. Modest variance says that there are many *distinct* quantifier languages – quantifier languages where translating one language's quantifier into the other's results in massive failures of charity. This follows almost immediately from top down charity and our account of quantifiers. Obviously there are many possible patterns of language use, distinct from each other but each with expressions playing the role that "there is" plays in English (of course, related expressions like "refers" and "object" will likewise vary in meaning between distinct quantifier languages, and variation in the meaning of the referential apparatus of the language will induce variation in the meaning of ordinary predicates like "red" and "on" as well).

Quantifier variance is often associated with a deflationary view of philosophical ontology, so we must stress that modest variance is not necessarily hostile to ontology. In fact, several prominent contemporary ontologists, including Cian Dorr (2005) and Theodore Sider (2009), count as modest variantists by our reckoning. The anti-ontological arguments associated with quantifier variance rely on a stronger form, one that builds upon modest variance. This stronger form must now be explained.

Equivalent Descriptions. Languages are, among other things, tools for describing the world. And like most tools, languages can be better, worse, or equal for the task at hand. When two languages are equal for any possible descriptive task, we say that the languages are *equivalent*, and that, informally, despite any differences between them, that they describe the very same facts or states of affairs. The top down charity approach allows for a simple account of language-wide equivalence – two languages are equivalent just in case they can express all and only the same characters, the same functions from contexts to coarse-grained truth conditions. When languages are equivalent in this sense, then for any sentence in the language of one, there is a sentence in the language of the other that is true in all and only the same possible circumstances.

This provides a very natural sense in which speakers of each language can express all and only the same facts. There is no state of the world, considered in a coarse-grained sense, that speakers of one language but not the other are sensitive to. Of course, each language may well describe these states of the world in apparently incompatible ways, using their own idiosyncratic notions. But what could it mean to say that one of these ways of describing things was closer to the truth than the other, when they are both *literally true*, in their home language? It is difficult to see how one of a pair of equivalent languages could be a better description of reality than another. Of course, there are fine-grained notions of equivalence that distinguish between character-wise equivalent languages, but when considered merely as tools for describing reality, this doesn't seem to matter.

Metaphysical Merit. Accepting that equivalent languages are of equal metaphysical merit, along with modest variance, leads to *immodest* or *strong quantifier variance*. Strong variantists endorse an egalitarian version of the pluralism about quantifier languages endorsed by modest variantists. Strong quantifier variance applies to quantifier languages the general thought that truth-conditionally equivalent languages are equally good, *metaphysically speaking*. It claims that when two quantifier languages are equivalent, there is no use asking which of them is metaphysically better or which better reflects objective reality. Nothing else about the metaphysical ordering of languages is implied; but for strong variantists, if there is a single, metaphysically special language, it can only be because that language can express truth conditions inexpressible in any other language. Of course, while strong variance is *metaphysically* egalitarian, it is not egalitarian in any stronger sense. Variantists can allow that there are often important practical reasons for preferring one quantifier language over an equivalent language.

Strong quantifier variance takes the quantifier language pluralism of modest variance and adds to it an account of the metaphysical merit of languages in terms of their truth-conditional equivalence. Variance, both modest and strong, is interesting in itself, but it can also be applied in philosophically fruitful ways.

2. Applications of quantifier variance

Here we explain five of quantifier variance's most important philosophical applications.

Merely Verbal Disputes in Ontology (Putnam 1987, 2004; Hirsch 2008a, 2009). According to modest variance, for many ontological disputes, there are possible languages associated with each side in the dispute that make the typical assertions of that side come out true. So, there is a possible language N in which the standard assertions of mereological nihilists come out true, and a possible language U in which the standard assertions of mereological universalists come out true. There is no metasemantic glue sticking words to meanings, so if philosophers depart drastically enough from standard usage, and refuse to coordinate or defer to other speakers, we should attempt to interpret them on their own terms. When we do this, charity supports the claim that nihilists are speaking N and universalists are speaking U. Since these philosophers are speaking different languages, their dispute over whether there are chairs is *merely verbal*: they each speak the truth in their own language and thus talk past each other.

They could attempt to reinstate their dispute by touting the superiority of their own language, whether N or U, over the other, metaphysically speaking. But for standard ontological disputes, including this one, the relevant languages are truth-conditionally equivalent. So, according to strong variance, N and U will also be of equal metaphysical merit. This means that there is nothing factual distinguishing these two languages – each theorist speaks the truth in their own language and each competing language is an equivalent description of the very same facts. There is no reason to prefer one language over the other, metaphysically speaking. This provides a second sense in which the dispute is "merely verbal."

Ordinary Ontology (Hirsch 2003, 2005). English speakers innocent of philosophy reject the distinctive claims of mereological nihilists and mereological universalists alike. But English-speaking ontologists seem untroubled by this, apparently thinking that since there is only one thing for "exists" to mean, ordinary English claims about material objects cannot be charitably interpreted. According to modest quantifier variance, this is false – there is a quantifier language in which the material object claims made by English speakers come out true. Charity demands that we interpret English speakers as speaking this language, so that, in English, "there are turkeys" and "there are trout" are both true, while "there are trout-turkeys" is false. Since this natural language concept of existence does not perfectly correspond to any of the standard positions in debates about material objects, if participants in these debates are speaking English, they are often speaking falsely. Strong variantists will add that, since it is plausible that English is equivalent to N, U, and the other languages of metaphysicians, English itself is a perfectly legiti-mate quantifier language, metaphysically speaking.

Vagueness about Existence (Hirsch 1999, 2002). It is widely believed that existence claims can-not be vague. This is because vagueness is usually thought to be a matter of *semantic indecision* – there are various possible things we could mean by a term like "bald," but our usage of the term doesn't decide between all of them, so the term is vague. Given this picture of vagueness, some philosophers (Lewis 1986) have argued that since there is only one possible thing that could be meant by "exists," existence claims cannot possibly be vague. But once quantifier variance is accepted, we can easily see how there could be semantic indecision and vagueness over "exists": our usage could fail to decide between various assignments of truth conditions to sentences containing "exists," while on each assignment "exists" continues to play the same formal-syntac-tic role and thus remains a quantifier.

Mathematical Freedom (Berry 2015; Warren forthcoming). Mathematicians freely introduce theories about new kinds of mathematical entities. If a mathematician introduces a new kind, the *F*s, it would be inappropriate within mathematics to object that no evidence had been given that *F*s exist (provided at least that the assumption of *F*s is consistent). On bottom up views of the metasemantics of quantifiers this procedure is either epistemically reckless or wholly inexplica-ble. However, as was perhaps first recognized by Carnap (1934), quantifier variance makes sense of mathematical freedom – mathematicians are introducing new ways of using sentences con-taining "there are" and "exists." Charity to use demands that we interpret them as speaking truly, if we can. And as quantifier variantists, we can. In this way, variance rationalizes standard math-ematical practice by making explicable its ontological freedom. Other approaches to metaontol-ogy are forced to criticize the internal norms of mathematics on purely philosophical grounds.

Paradoxes and Indefinite Extensibility (Warren 2017). Naïve set theory is beset by paradoxes, most famously Russell's paradox concerning the set of all non-self-membered sets. One response to these paradoxes, inaugurated by Russell himself, sees set-theoretic quantifiers like "all sets" as being *indefinitely extensible*. The idea being that when you attempt to quantify over all and only the sets you somehow, someway, end up being able to talk about another set that was not in the original collection. This idea has long been puzzling, since clearly we aren't *creating* a new set when we run through the reasoning of the Russell paradox! But quantifier variance makes sense of extensibility by seeing the Russell reasoning as creating not a new set, but rather a new and slightly expanded concept of "all sets," based on a slight change in the usage of sentences containing "all." In this fashion, variance provides an all purpose strategy for demystifying the hitherto puzzling paradoxes of set theory and absolute generality.

3. Misunderstandings and objections

There are many ways to *misunderstand* quantifier variance, most of them witnessed in the lit-erature. For the sake of clarity, let us make fully explicit some of what is merely implicit in the foregoing.

Quantifier Variance is not anti-realism. Ontological anti-realists think that, in some fashion or other, objects depend for their existence on human practices. But while it is undeniable that some objects depend upon humans (governments, national borders, thoughts), it seems equally undeniable that other objects do not (stars, rocks, numbers). Nothing in quantifier variance conflicts with this bit of good sense. In fact, while variance says much about the nature of our *concept* of existence, it says *nothing at all* about the nature of the things that exist (Hirsch 2002). To reason from quantifier variance to ontological anti-realism is just as confused as reasoning from the human invention of the concept of "planet" to the human invention of planets.

Quantifier Variance is not verificationism. Verificationists think that claims are only meaningful if they have clear verification conditions. This is usually understood as entailing that disputes are substantive just in case they can be settled, in principle. The logical positivists of legend used verificationism as a club with which to bash metaphysics, and some (Hawthorne 2009, for example) have worried that variantists are wielding the same club for the same purpose. But this is mistaken; strong variantists think disputes are metaphysically insubstantial when each side's language is equivalent to the other. This criterion only entails verificationism if "equivalent to" is understood to mean "has the same verification conditions as," but as we have seen, this is not how quantifier variantists understand "equivalence" (Putnam 1983; Hirsch 2011, 2016; Warren 2015, appendix). Unlike verificationism, the charity-based metasemantics behind quantifier variance allows that many substantive disputes – some disputes about the past, for example – may forever be impossible to settle.

Quantifier Variance does not venerate quantification. Some think that variantists must mean something *special* by a "quantifier," beyond obeying a particular formal-syntactic role. A persistent version of this confusion says that variantists must explain different quantifier meanings in terms of differing *domains of quantification* (Finn and Bueno 2018). Obviously this is a nonstarter – speakers of N cannot and do not admit that U's quantifiers range over a domain containing composites (charitable critics of variance, such as Sider (2009), recognize this). There are also more subtle ways to read something special into "quantifier" variance (Dorr 2014). But whether expressed simply or subtly, the thought is wrong. Those who think that something more than formal-syntactic role is required for an expression is to count as a "real" quantifier should interpret our claims as concerning only *quantifier-like* expressions – expressions that have the same formal-syntactic role as our quantifiers. Everything that is of interest to variantists can easily be said using this alternative vocabulary, though we think that talking in terms of quantifier *variance* rather than quantifier *elimination* is much more natural and suggestive (Hirsch 2011, introduction).

Even with these misunderstandings avoided, quantifier variance and its applications have been challenged in the literature. Three of these challenges warrant some discussion.

The Collapse Argument (Hale and Wright 2009; Dorr 2014; Rossberg unpublished). Imagine that we quantifier variantists are speaking N and considering U. As variantists, we admit that the sentence "Bugs is a bunny" is, while false in N, true in U. But (it is claimed) admitting that it is true in U that Bugs is a bunny, entails, by the inferential role of our N-quantifier, that *something* (in our N-sense of "something") is a bunny. Which – since bunnies are composite objects – contradicts the assumption that we were speaking N. It seems that N is not a possible language and modest variance is false.

This argument, based on one given in Harris (1982), has tempted many critics of variance. But it is based on a confusion: admitting, in N, that "Bugs is a bunny" is true in U, is not tantamount to admitting, in N, that Bugs is a bunny. To think otherwise is to confuse use and mention. In a language that calls sharks "dogs" but is otherwise exactly like English, "dogs live in the water" is true. We can all see that this does not imply, *in English*, that dogs live in the water, but the mistake made by this brainless argument seems to be the very mistake made by the collapse argument. Discussing the language U, within N, is very different than having, *within N*, the true sentence, "Bugs is a bunny." Intra-language versions of collapse, though valid, don't threaten quantifier variance, while inter-language versions threaten quantifier variance, but are fallacious (Warren 2015). The fallacy has only escaped notice because the argument is typically presented formally, in either a natural deduction system or a mathematically powerful metatheory. But a fallacy is still a fallacy, no matter how many technicalities are piled on top of it.

The Tarskian Argument (Hawthorne 2006; Eklund 2009). Imagine again that we variantists are speaking N and considering U. As variantists we should be able to freely admit that U is a possible language, in some general sense (we are not presently concerned with its psychological possibility). But according to many influential approaches in the philosophy of language and linguistics, claiming a natural language possible requires the ability to formulate a Tarski (1933) style semantic theory for that language. And this seems impossible for the "smaller" language to do in the cases of interest to variantists. For example, a Tarski-style treatment of U, within N, would explain the truth of the U-sentence "Bugs is a bunny" by saying that the referent of "Bugs" has the property expressed by "is a bunny". But there is, according to N-speakers, no suitable referent for "Bugs," since "is a bunny" has empty extension, and so a Tarski-style approach seems impossible. If this is right, then N-speakers cannot admit that U is a possible language, contradicting modest variance.

In response, some (Dorr 2005; Sider 2007) have advocated that variantists reject the standard Tarskian approach to semantics, at least as a constraint on admitting a language possible. We are sympathetic to this suggestion, but have elsewhere shown that it is not strictly required (Hirsch and Warren forthcoming). Through devious uses of the resources of set theory, N-speakers can give a perfectly standard – though complicated – Tarskian semantics for U without going beyond the resources of N. This completely undermines this particular version of the Tarskian argument, and it is plausible that similar results also hold for all other cases of interest.

Heavyweight Ontology (Sider 2001, introduction, 2009, 2011; Fine 2001, 2009). Following Quine (1948), quantifier variantists see existential quantification over Fs as expressing *ontological commitment* to Fs. This is still widely but not universally accepted in philosophy. Recently many metaphysicians have claimed that even if existential quantifiers carry some kind of "lightweight" ontological commitment, what ontologists really care about is "heavyweight" ontological commitment, which is not carried by standard quantifiers alone. There is disagreement over the particulars, but most of these heavyweight ontologists think ontological commitment is carried by some sort of special metaphysical primitive such as "in reality" or the like – for example by saying that ontological commitment to Fs is expressed by the claim that, in reality, there are Fs. With this type of move made, there no longer seems to be any reason for ontologists to worry about quantifier variance or deflationary arguments based upon it.

Despite the recent popularity of this strategy, it is difficult to see how it, alone, could salvage substantive ontology. Either sentences containing "in reality" (or the like) have clear truth conditions in the language of heavyweight ontology, or they do not. If they do, then the situation is not importantly different than it was with the quantifiers – we are able to charitably interpret each disagreeing heavyweight ontologist so that they speak the truth in their own language and thus all talk past each other (Hirsch 2008b). On the other hand, if these sentences do not have clear truth conditions, then ontological claims, questions, and disputes are problematic *for that very reason* (Warren 2016). In neither case has substantive ontology been rehabilitated. The devil is in the details, but this general situation makes us highly skeptical about the prospect of rehabilitating substantive ontology simply by moving away from quantifiers as the source of serious ontological commitment.

4. Future directions

Here we indicate four directions for future quantifier-variance-related work. Of course, our list does not exhaust the possibilities.

*Strange Languages.*Variantists think that there are possible languages that have distinct concepts of "what exists." But are there any limits on how intuitively *strange* such alternative languages can be? Some quantifier languages are bizarre, such as Hirsch's Contacti language, where the identity of an "object" over time is partly determined by its contact relations to other things, so that if two "objects" come into contact with each other they exchange all of their properties including their spatial locations and material composition (see Hirsch 1993 for details). Can we really conceive of any beings speaking Contacti as their mother tongue? Can we conceive of beings using a language like this *at all*? Contacti is a describable language (we have just described it), and what's more it seems that it is truth-conditionally equivalent to our language, so for anything we can say, they can say something to the same effect. It is not clear, therefore, what the nature of the intuitive difficulty is in imagining speakers of such a language. Nor is it clear whether the intuitive difficulty is a real difficulty. But variantists should acknowledge the insistent intuitions about such cases and attempt to account for them, in some fashion. There may be plausible metasemantic principles that exclude certain describable quantifier meanings while admitting others. The matter calls for further investigation, with quantifier variance kept firmly in mind.

Human Limits. The question we have just been sketching concerns the possibility of beings, whether human or not, speaking languages with wildly different concepts of existence than ours. A related question is whether *humans* could be raised to speak such languages, as their mother tongue. This may be principally an empirical question, informed by matters of psychology and neuroscience, but there is much in it that is grist for philosophical reflection. Many of the alternative languages are truth-conditionally equivalent and so express the same facts as our own, but are there deep connections, perhaps even *a priori* connections, between how our concept of existence articulates the facts and how human patterns of attention and learning operate? A start on these questions has been provided by Hirsch (1978, 1997, unpublished), but further work in this area would be valuable.

Hyperintensionality and Meaning. The metasemantics of quantifier variance is avowedly *intensional* – it makes no direct appeal to differences in meaning between necessarily equivalent sentences or expressions. Some critics (Hawthorne 2009) have seen this as an objection to quantifier variance. We disagree (see Hirsch 2016), but think that the connection between so-called *hyperintensionality* and quantifier variance needs further exploration, along a number of dimensions. One of these dimensions concerns cross-language belief ascriptions. Suppose that in the presence of a brown dog and a green turtle a speaker of U asserts the true U-sentence, "There is something here that is partly brown and partly green." Assuming that this sentence is false in our language, it does not seem that we have any sentence in our language that is *synonymous* with this true U-sentence. It may follow that English speakers cannot capture the fine-grained hyperintensional content of the belief expressed by the speaker of the U-language. We do not think this as a major problem for variance, but both the general and specific issues deserve further clarification.

Beyond Philosophical Ontology. Naturally enough, the original applications of quantifier variance were aimed at demystifying the ontological disputes engaged in by philosophers. But philosophers don't have a monopoly here – existence claims and questions are woven into nearly every aspect of our approach to the world. Because of this, quantifier variance can be applied to many areas of human discourse, potentially resolving various puzzles and eliminating confusions. As noted previously, this has been done in mathematics, leading to satisfying accounts of both mathematical freedom and the paradoxes of set theory. It is worth investigating what results when quantifier variance is applied to discourse about and within fiction, debates about species and natural kinds in biology, discussions of social-biological kinds like race and gender, theoretical posits in science, the posits of folk psychology and cognitive science, and beyond. Assuming that quantifier variance is the correct approach to our concept of existence, it will be applicable to existential claims in all of these areas and, as has already been the case with mathematics, some of the applications may be philosophically illuminating.

References

Berry, S. (2015), "Chalmers, Quantifier Variance, and Mathematicians' Freedom," in *Quantifiers, Quantifiers, and Quantifiers: Themes in Logic, Metaphysics, and Language*, edited by A. Torza, Cham: Springer, 191–219.

Carnap, R. (1934), *The Logical Syntax of Language*, London: Routledge & Kegan Paul.

Dorr, C. (2005), "What We Disagree About When We Disagree About Ontology," in *Fictionalism in Metaphysics*, Oxford: Oxford University Press, 234–286.

———. (2014), "Quantifier Variance and the Collapse Theorems," *The Monist* 97: 503–570.

Eklund, M. (2009), "Carnap and Ontological Pluralism," in *Metametaphysics: New Essays on the Foundations of Ontology*, edited by D. Chalmers, D. Manley and R. Wasserman, Oxford: Oxford University Press, 130–156.

Fine, K. (2001), "The Question of Realism," *Philosophers' Imprint* 1: 1–30.

———. (2009), "The Question of Ontology," in *Metametaphysics: New Essays on the Foundations of Ontology*, edited by D. Chalmers, D. Manley and R. Wasserman, Oxford: Oxford University Press, 157–177.

Finn, S. and O. Bueno (2018), "Quantifier Variance Dissolved," *Royal Institute of Philosophy Supplement* 82: 289–307.

Hale, B. and C. Wright (2009), "The Metaontology of Abstraction," in *Metametaphysics: New Essays on the Foundations of Ontology*, edited by D. Chalmers, D. Manley and R. Wasserman, Oxford: Oxford University Press, 178–212.

Harris, J. H. (1982), "What's So Logical About the 'Logical' Axioms?" *Studia Logica* 41: 159–171.

Hawthorne, J. (2006), "Plenitude, Convention, and Ontology," in his *Metaphysical Essays*, Oxford: Oxford University Press, 53–69.

———. (2009), "Superficialism in Ontology," in *Metametaphysics: New Essays on the Foundations of Ontology*, edited by D. Chalmers, D. Manley and R. Wasserman, Oxford: Oxford University Press, 213–230.

Hirsch, E. (1978), "A Sense of Unity," *Journal of Philosophy* 74: 470–494.

———. (1993), *Dividing Reality*, Oxford: Oxford University Press.

———. (1997), "Basic Objects: A Reply to Xu," *Mind and Language* 12: 406–412.

———. (1999), "The Vagueness of Identity," *Philosophical Topics* 26(1/2): 139–158.

———. (2002), "Quantifier Variance and Realism," *Philosophical Issues* 12(1): 51–73.

———. (2003), "Against Revisionary Ontology," *Philosophical Topics* 30(1): 103–127.

———. (2005), "Physical-Object Ontology, Verbal Disputes, and Common Sense," *Philosophy and Phenomenological Research* 70(1): 67–97.

———. (2008a), "Ontological Arguments: Interpretive Charity and Quantifier Variance," in *Contemporary Debates in Metaphysics*, edited by T. Sider, J. Hawthorne and D. Zimmerman, Malden, MA: Blackwell, 367–381.

———. (2008b), "Language, Ontology, and Structure," *Noûs* 42(3): 509–528.

———. (2009), "Ontology and Alternative Languages," in *Metametaphysics: New Essays on the Foundations of Ontology*, edited by D. Chalmers, D. Manley and R. Wasserman, Oxford: Oxford University Press, 231–259.

———. (2011), *Quantifier Variance and Realism: Essays in Meta-Ontology*, Oxford: Oxford University Press.

———. (2016), "Three Degrees of Carnapian Tolerance," in *Ontology After Carnap*, edited by S. Blatti and S. Lapointe, Oxford: Oxford University Press, 105–121.

———. (unpublished), "Ontological Behavior in Infants (and Adults)."

Hirsch, E. and J. Warren (forthcoming), "Quantifier Variance and the Demand for a Semantics," *Philosophy and Phenomenological Research*.

Kaplan, D. (1989), "Demonstratives," in *Themes from Kaplan*, edited by J. Almog, J. Perry and H. Wettstein, Oxford: Oxford University Press, 481–563.

Lewis, D. (1986), *On the Plurality of Worlds*, Oxford: Blackwell.

Putnam, H. (1983), "Equivalence," in his *Realism and Reason: Philosophical Papers*, vol. 3, Cambridge: Cambridge University Press, 26–45.

———. (1987), "Truth and Convention: On Davidson's Refutation of Conceptual Relativism," *Dialectica* 41(1–2): 69–77.

———. (2004), *Ethics Without Ontology*, Cambridge, MA: Harvard University Press.

Quine, W.V. (1948), "On What There Is," *Review of Metaphysics* 2(1): 21–38.

Rossberg, M. (unpublished), "On the Logic of Quantifier Variance."

Sider, T. (2001), *Four-Dimensionalism: An Ontology of Persistence and Time*, Oxford: Oxford University Press.

———. (2007), "NeoFregeanism and Quantifier Variance," *The Aristotelian Society Supplementary* 81: 201–232.

———. (2009), "Ontological Realism," in *Metametaphysics: New Essays on the Foundations of Ontology*, edited by D. Chalmers, D. Manley and R. Wasserman, Oxford: Oxford University Press, 384–423.

———. (2011), *Writing the Book of the World*, Oxford: Oxford University Press.

Stalnaker, R. (1984), *Inquiry*, Cambridge, MA: MIT Press.

Tarski, A. (1933), "Pojecie Prawdy w Jezykach Nauk Dedukcyjnych. Prace Towarzystwa Naukowego Warszawskiego," *Wyzdial III matematycznofizycznych* 34.

Warren, J. (2015), "Quantifier Variance and the Collapse Argument," *Philosophical Quarterly* 65(259): 241–253.

———. (2016), "Internal and External Questions Revisited," *Journal of Philosophy* 113(4): 177–209.

———. (2017), "Quantifier Variance and Indefinite Extensibility," *Philosophical Review* 126(1): 81–122.

———. (forthcoming), *Shadows of Syntax*, New York: Oxford University Press.

38

METAPHYSICAL ANTI-REALISM

Vera Flocke

1. Introduction

Metaphysical anti-realism is a large and heterogeneous cluster of views that do not share a common thesis but only share a certain family resemblance. Views as different as mathematical nominalism[1] – the view that numbers do not exist; ontological relativism[2] – the view that what exists depends on (something like) a perspective; and modal conventionalism[3] – the view that modal facts are conventional – all are versions of metaphysical anti-realism. As the latter two examples suggest, *relativist* ideas play a starring role in many versions of metaphysical anti-realism. But what does it mean for the existence of something to "depend on" a perspective, or for a modal fact to "depend on" a convention? We can distinguish between various dependence relations, giving rise to an array of drastically different forms of metaphysical anti-realism. This chapter offers a guided tour. I develop a systematic distinction between various forms of metaphysical anti-realism with a focus on the role of relativist ideas in this landscape.

I pursue two goals through this discussion. The first goal is to provide a *recipe book*. Suppose you are inclined towards an anti-realist view with regard to numbers. Which views are available to you? This chapter distinguishes between a few options. I will not be able to discuss all the options; but I will make a start. The second goal is to *increase intelligibility*. Sometimes anti-realist views are phrased in esoteric-sounding terms, as when Sosa (1999, 133) says that "In order for [a constituted] sort of entity to exist relative to a conceptual scheme, that conceptual scheme must recognize its constituent form as an appropriate way for a distinctive sort of entity to be constituted." What is a conceptual scheme, and how can it recognize anything?[4] I won't answer *these* questions; but I aim to show that a few versions of metaphysical anti-realism are intelligible and coherent.

My discussion will proceed as follows. In Section 2, I discuss the difference between metaphysical realism and anti-realism. In Section 3, I discuss the most straightforward versions of metaphysical anti-realism: views according to which certain entities do not exist. In Section 4, I discuss versions of metaphysical anti-realism that bring in relativist ideas. This includes views according to which certain entities – even though they exist – are not real, and views according to which certain facts – even though they exist – are not absolute. In Section 5, I conclude with a brief summary.

2. Realism vs. anti-realism

The question of how to define metaphysical anti-realism is a thorny question, but I think to a large extend terminological. "Realism" is a technical term. Various philosophers have, and legitimately can, define what they mean by this term differently. To be useful, a definition should not be entirely arbitrary, however, but relate to established uses of the term in meaningful ways, and make these established uses explicit, clearer, and more precise.[5]

I will here assume the following working definition:

(METAPHYSICAL ANTI-REALISM) is a cluster of views according to which entities of some kind X do not exist, or if they exist, are not real or (in the case of facts) absolute.

This definition does not define metaphysical anti-realism by a single thesis but by a schema. The schema is "Xs do not exist or, if Xs exist, they are not real or (in the case of facts) absolute." We can obtain versions of metaphysical anti-realism by replacing the letter "X" in this schema with the name of a kind of entities, in a way that observes certain side constraints. In particular, "X" must be the name of a kind of entity whose metaphysical status is under debate in metaphysics. For example, nominalists assert that *numbers do not exist*, and modal conventionalists assert that *modal facts are not absolute* (since they depend on conventions). Or at least this is one way of developing the view.

The side constraint, according to which instances of the schema can only be derived by replacing "X" with the name of a kind of entity whose metaphysical status is under debate in metaphysics, is vague; but it helps to distinguish metaphysical anti-realism from other forms of anti-realism. Metaphysical anti-realism is distinguished by its domain. Other forms of anti-realism include moral anti-realism – the view that moral facts are not absolute – and aesthetic anti-realism – the view that aesthetic facts are not absolute. These other forms of anti-realism also have a metaphysical core, and the distinction between the different versions of anti-realism may not always be clean-cut. An alternative approach would drop the side-constraint and treat metaphysical anti-realism as the most general form of anti-realism. But that's not the approach I pursue here.

We can distinguish between various versions of metaphysical anti-realism. For one, versions of metaphysical anti-realism differ with regard to the sort of entities which they concern. Some anti-realist views primarily concern *objects*; for instance, they may concern the metaphysical status of numbers, of universals, or of composite objects (that have parts), and so on. Other anti-realist views primarily concern *facts*; for instance, on certain antirealist views, there is no absolute fact with regard to the existence of numbers. These two approaches – objectual and fact-oriented – are crucially different.[6] For example, arguing that there is no fact of the matter with regard to the existence of numbers appears inconsistent with arguing that numbers do not exist. Furthermore, versions of metaphysical anti-realism also differ in what they say about the metaphysical status of the entities that they concern. For example, some anti-realists simply deny that Fs exist; other anti-realists allow that Fs exist, but deny that Fs are real (see Fine 2009). Similarly for anti-realist views about facts. Some anti-realists simply deny the existence of certain facts; others allow that the relevant facts hold, but argue (moreover) that they are not absolute.

Given that metaphysical anti-realism is such a broad class of views, everyone is an anti-realist about some entity or another. For example, most people think that witches and ghosts do not exist; they are anti-realists about witches and ghosts. Saying that someone is a metaphysical anti-realist without qualification therefore is nearly without content.

Metaphysical anti-realism is not the same as metaphysical deflationism. Metaphysical deflationists, such as Hirsch (2011) and Thomasson (2015) think that metaphysical inquiry is pointless or misguided. Deflationist views are often called "anti-realist," perhaps because deflationists in a sense are "against" metaphysics, and therefore "anti"-metaphysical.[7] But metaphysical antirealists propose metaphysical views. For example, the thesis that numbers do not exist is a metaphysical thesis about the fundamental structure of reality. There is at least a tension between defending a metaphysical view and arguing that metaphysical inquiry is pointless.[8] It is therefore important to clearly distinguish deflationism from anti-realism.[9]

Why this definition of metaphysical anti-realism, and not another? The literature contains a range of alternative definitions of "metaphysical anti-realism." Some philosophers think that metaphysical anti-realism equates to idealism (Smith 1999; Khlentzos 2016). Other philosophers embrace a disjunctive definition, according to which metaphysical anti-realists *either* deny that certain entities (Fs, say) exist *or* else argue that Fs exist mind dependently (Miller 2016). Yet other philosophers suggest that, according to anti-realists, certain entities exist but are not "real" (Fine 2009). A fourth approach construes metaphysical anti-realism as a semantic thesis which rejects the correspondence theory of truth (see Dummett 1982; Millikan 1986).[10] I prefer my definition over these alternatives because it groups views together those views that have important commonalities, and distinguishes these views from others that are importantly different. The first two alternatives make a strong connection between anti-realism and idealism; but that choice seems arbitrary. Some idealists simply propose a view about the nature of the physical universe – it is fundamentally mental. This view does not appear anti-realist in any way.[11] The third and fourth alternatives are not sufficiently general. They describe different ways of developing a version of metaphysical anti-realism, but do not offer a general definition of metaphysical anti-realism.

3. Anti-realist views that deny the existence of certain entities

The most straightforward versions of metaphysical anti-realism are views that deny the existence of objects of some sort F. Views in this group include nominalism about *universals* – the view that universals do not exist – and nominalist views about *abstract entities* – views on which numbers, propositions, sets and the like do not exist.[12] The opponents of nominalism about universals are *realists* about universals, and the opponents of nominalism about abstract objects are *Platonists*.[13]

Nominalists deny the existence of certain *objects*. We can distinguish between different versions of metaphysical anti-realism with regard to the kind of entities that they concern. Other metaphysical anti-realists deny the existence of certain *relations*, certain *kinds*, or certain *facts*, and so on. For example, Wilson (2014) argues that there is no capital-G grounding relation,[14] and Glasgow (2009) denies that race is a biological kind. These views are versions of anti-realism about the capital-G grounding relation and about race as a biological kind, respectively.[15] Furthermore, we can develop views on which facts of a certain sort do not exist in at least two different ways.

First, according to some meta-ethical expressivists, normative sentences do not express propositions but express non-cognitive mental states (see Ayer 1936, 110). On this view, normative facts do not exist.[16] It is hard to see, however, how one might develop an analogous expressivist view about a metaphysical domain. A possible view is that "numbers exist" does not express a proposition but instead expresses approval of numbers; but this view has little going for it.[17] Second, another way for arguing that facts of a certain sort do not exist is by rejecting the principle of bivalence – according to which every proposition is either true or false (but never both) – for a certain area of discourse. Dummett (1982) pursues this approach to anti-realism, and argues that the rejection of bivalence for a particular area of discourse is sufficient for being an anti-realist about that area.[18] For example, one way of being an anti-realist about the past is by rejecting bivalence for statements about the past (see Dummett 1968–1969); and one way of being an anti-realist about numbers is by rejecting bivalence for mathematical statements (see Dummett 1977).

Dummett argues, however, that even though a rejection of bivalence is *sufficient* for anti-realism, it is not *necessary*. Realists about a certain area of discourse, according to Dummett's conception, endorse a thesis with two components (see Dummett 1982, 56–57). On his view, a realist about mathematical statements thinks that mathematical sentences in some way relate to mathematical reality, are either true or false in virtue of how they relate to mathematical reality, where (moreover) reference to mathematical entities plays an important role for determining the truth-value of mathematical sentences. This account yields that there are two main ways of being an anti-realist about mathematics: either by denying the principle of bivalence for mathematical sentences, as intuitionists do, or by denying that mathematical objects (such as numbers) as the referents of mathematical sentences make mathematical sentences true (when they are true).[19] This latter version of anti-realism is quite close to a more "objectual" version of anti-realism that simply denies the existence of certain objects (i.e., numbers).

4. The role of relativist ideas

Proponents of metaphysical anti-realism appeal to relativist ideas in various ways and for a variety of reasons. I will discuss a few examples, first focusing on anti-realist views that concern *objects* and then focusing on anti-realist views that concern *facts*.

I will discuss two examples of versions of metaphysical anti-realism that concern objects and that appeal to relativist ideas. First an example from social metaphysics. Anti-realists about race have reasons to acknowledge the existence of racial groups. Some of these reasons are political: a fight against racism makes sense only if race-based discrimination persists, which in turn presupposes the existence of racial groups. But how can an antirealist about race acknowledge the existence of races?

Glasgow (2009) responds to this challenge by arguing that race is not a biological kind but socially constructed. On his view, races understood as biological kinds do not exist, which makes Glasgow an anti-realist about race as a biological kind. But Glasgow nevertheless thinks that races exist and are real; they are just socially constructed. Glasgow (2009, 5) puts the idea as follows: "just as journalists or doctors are real but socially constructed kinds of people, so racial kinds of people are real but socially constructed. Racial groups are real groups that have been created by our social practices, rather than by some biological process."[20] Glasgow here appeals to social construction as a metaphysical dependence relation in order to explain the nature of races. The upshot is that, since races are not biological kinds but socially constructed, it is "up to us" how to construct them, or to altogether give up on the racial categorizations of people.

I now turn to an example from mathematical ontology. Nominalists deny the truth of propositions that may appear obvious and unassailable. For example, five is a number, which trivially entails that numbers exist. Which sort of evidence could possibly override the conclusion of this straightforward argument? Many philosophers think that *no* evidence could possibly have this effect, which results in a powerful objection to nominalism.[21] One strategy that nominalists may employ in response to this objection is to grant that numbers *exist*, and to argue that numbers nevertheless are not genuine constituents of *reality* (Fine 2009). This move gives up the letter of nominalism, but retains the spirit. The distinction between things that are and things that are not real then is often drawn with appeal to broadly relativist ideas. For example, Fine (2001, 27) proposes to distinguish between realism and anti-realism as two different views about *grounding* facts. Grounding is a metaphysical dependence relation. If a fact A is grounded in another fact B, then A holds "in virtue of" B; B provides a metaphysical explanation for A. Fine proposes that, if the existence of numbers is grounded in other facts, then numbers are not genuine constituents of reality, since only ungrounded facts are genuine constituents of reality.[22]

I will now discuss views on which certain *facts* – even though they exist – hold only relatively to something else. For example, consider modal conventionalism. This is the view that what's necessary or possible depends on conventions for re-identifying objects in various possible worlds (Sidelle 2002). For example, water has the molecular structure H_2O; and we accept conventions for the cross-world identification of water according to which whatever shares water's molecular structure is identical to water. The fact that water has the molecular structure H_2O together with these conventions makes it necessary that water is H_2O – or so modal conventionalists argue. One could flesh out this view by arguing that propositions about what's necessary and possible are true or false only relative to a convention. Sidelle (2002) does not put his view this way; but it is an option.

A view on which certain facts – even though they exist – hold only relatively to something else can be developed in various ways. For instance, one could draw use the notion of grounding: ungrounded facts are in a sense absolute; but grounded facts hold only relative to their grounds. I will here discuss two further options: first, truth relativism, and then a certain modal conception.

Truth relativists, such as MacFarlane (2014), think that the truth-value of some propositions depends on the context from which they are assessed. Some propositions are true as assessed from context c_1 but false as assessed from context c_2. For example, relativists about matters of taste think that the proposition that cilantro is tasty is true or false only relative to a context of assessment which supplies a standard of taste. If speaker A does not like cilantro and is located at context c_1, then the proposition that cilantro is tasty is false as assessed from c_1. But if speaker B likes cilantro and is located at context c_2, then the very same proposition is true as assessed from c_2.

Metaphysical anti-realists who want to propose a version of truth relativism need to answer one main question: What is the new nonstandard parameter whose values are supplied by contexts of assessment? For truth relativists about predicates of personal taste the answer is obvious: the new parameter is a standard of taste. But what about other sorts of metaphysical relativism, such as perhaps relativism about modal facts?[23]

Einheuser (2006) develops a generic template that can help metaphysical anti-realists to answer this question. Einheuser suggests that each world is composed of a perspective independent *substratum* and a perspective dependent *carving*. One of the substrata is the unique *actual* substratum. The actual substratum is compatible, however, with several carvings – none of which is uniquely actual. The semantic values of declarative sentences are not ordinary propositions but sets of pairs $<s, c>$, where s is a substratum and c a carving. Suppose now that the actual substratum $s_@$ is compatible with two carvings, c_1 and c_2. A set A of ordered pairs $<s, c>$ may be such that $<s_@, c_1> \in A$ but $<s_@, c_2> \notin A$. In that case, A is true at $<s_@, c_1>$ but false at $<s_@, c_2>$. Next, one just needs to define "true" as a two-place predicate that applies to semantic contents only relative to contexts of assessment; in such a way that A is true relative to contexts of assessments that supply carving c_1 and false relative to contexts of assessment that supply carving c_2.

To make use of Einheuser's template, metaphysical relativists have to explain which sorts of entities make up the substratum, and how one may impose a carving on a substratum. Some versions of metaphysical anti-realism have a straightforward explanation.[24] Consider, for instance, Sidelle's (2002) modal conventionalism. We can explicitly develop this view as a form of relativism by arguing that non-modal facts provide a sort of perspective-independent substratum onto which conventions for cross-world identifications then impose a perspective-dependent modal structure. One may then argue that the semantic values of sentences concerning what's possible and what's necessary are sets of substratum-carving pairs that are true or false only relative to criteria for cross-world identifications.

Drawing a distinction between a perspective independent substratum and a perspective dependent carving is more difficult, however, for other versions of metaphysical anti-realism. For example, consider ontological relativism, which is the view that what exists is in some way relative. Chalmers (2009), a proponent of ontological anti-realism, suggests to represent an ontology by a "furnishing function" that maps worlds to domains of quantification. Given the notion of a furnishing function, one might argue that the semantic contents of quantified sentences are sets of pairs $<w, f>$, where w is a world and f is a furnishing function, and transfer Einheuser's template to ontology in this way. However, it is not clear what a world minus a domain of quantification is. Once one subtracts a domain from a world, there seems nothing left that is recognizably a world. It hence seems that the distinction between worlds and domains is not well-defined, and it is unclear what the parameter f is supposed to represent.

An alternative to truth relativism is given by a certain modal conception (see Flocke forthcoming). On both the relativist and the modal approach, the truth of some propositions depends on (something like) a perspective; but the two approaches explain this sort of dependence in different ways. Truth relativists defines "true" in a way that relativizes truth to a context of assessment. The modal approach, on the other hand, construes perspective-dependence as a modal contingency. For example, consider a mereological universalist who thinks that any two things compose a third. On the modal approach, this universalist is seen as adopting a standpoint at which any two things compose a third. This sort of standpoint dependence can be described using modal operators: a proposition that is true at all standpoint is in a sense necessary; and a proposition that is true at only some but not all standpoints is in a sense contingent. The modal conception does not rest on a distinction between a substratum and a carving. In comparison to relativism, the modal conception is therefore better suited to provide for a sense in which the answers to existence questions may be standpoint-dependent. But a lot of work remains to be done to fully develop the modal approach. For example, what kind of modality is perspective-dependence? Is it a more epistemic or more metaphysical modality? These and other questions require answers.

5. Conclusion

My work in this chapter has both a problem-solving and a problem-finding component. The main problem I wanted to solve is to develop a systematic distinction of the different ways in which one can be a metaphysical anti-realist. To this end, I have first distinguished between views according to the type of entity which they concern. Some versions of metaphysical anti-realism deny the existence of certain objects; others deny the existence of kinds, relations or facts. I then discussed the role of relativistic ideas in this landscape. Some metaphysical anti-realists grant that certain objects – numbers, say – exist, but maintain that they exist only relatively to something else; perhaps because their existence is grounded in the existence of other things. Other anti-realists grant that certain facts – modal facts, say – exist, but maintain that they hold only relatively to something; perhaps something like conventions for re-identifying objects in other possible worlds.

Through solving this problem I also found a couple new difficulties. For one, I have merely distinguished between various ways of being an anti-realist. But the various approaches need not be equally good! One important question for future research concerns an evaluation of the comparative advantages and disadvantages of the various approaches. Second, I have merely sketched the different approaches to anti-realism at a very high level of abstraction. Another important line of future inquiry concerns the details. Consider, for instance, the view that mathematical propositions have a merely relative truth-value. How could you spell out the details of this view?

Acknowledgements

I would like to thank the following people for helpful suggestions that influenced my work on this chapter: Daniel Durante Pereira Alves, Robin Dembroff, Sally Haslanger, Robert Michels, Panu Raatikainen, Agustín Rayo, Brad Skow, Jack Spencer, Amie Thomasson and Alessandro Torza. Thanks to David Chalmers for helpful comments on the penultimate version.

Notes

1 See Field (1980) for a defense of mathematical nominalism.
2 See Quine (1968), Davidson (1973–74) and Sosa (1999) for discussions.
3 See Sidelle (1989) for a defense of modal conventionalism.
4 Davidson (1973–74, 5) discusses these questions.
5 Using Carnap's (1950a, 3) terminology, we can also say that a useful definition of "metaphysical antirealism" should provide an explication of the vague and imprecise notion that is already in use.
6 A third anti-realist approach is property-oriented. See Spencer (2016) for a relevant discussion of the different ways in which an object can have a property only "relative to" something else.
7 For example, Biggs and Wilson (2016) argue that, according to Carnap (1950b), metaphysical claims are meaningless, and they call this view "metaphysical anti-realism."
8 Hirsch (2011) combines a deflationary view with the defense of a specific metaphysical viewpoint. Sider (2011) argues that this "combination of views" is "just more metaphysics."
9 Is Hirsch's (2011) "quantifier variance" view a form of metaphysical anti-realism? I think that there is a specific sense in which quantifier variance is an anti-realist view. Hirsch denies that there is a metaphysically distinguished quantificational structure. He is at least an anti-realist about this sort of structure.
10 I am simplifying a bit. Smith (1999) and Khlentzos (2016) speak about metaphysical anti-realism, Miller (2016) discusses anti-realism in general, and Fine (2009) discusses ontological anti-realism. I think it is fair to interpret all these authors as embracing specific accounts of metaphysical anti-realism, since metaphysical anti-realism is a specific version of anti-realism, and ontological anti-realism is a specific version of metaphysical anti-realism.
11 I will not discuss idealism in any detail since other authors have discussed the topic in much detail recently. See Hofweber (forthcoming) and Chalmers (forthcoming).
12 To give a few more radical examples, Dasgupta (2009) and Azzouni (2017) both argue that, fundamentally, no objects exist at all.
13 See Rodriguez-Pereyra (2015) for a useful overview and references.
14 Wilson (2014) thinks that there are various metaphysical dependence relations, such as type or token identity, functional realization, classical mereological parthood, and so on, but not a unique in-virtue-of relation that could do the explanatory work which metaphysicians often want the Grounding relation to do.
15 Wilson (2014) also supports her view by means of indispensability arguments. She argues that a capital-G grounding relation is not needed to explain anything, and concludes that we should not believe in its existence.
16 There are many different versions of meta-ethical expressivism; see Schroeder (2009) for a discussion.
17 Flocke (forthcoming) develops an expressivist view about ontology, but on Flocke's view, the semantic values of quantified sentences are propositions.
18 Fine (2001, 6) argues against Dummett that a rejection of bivalence is not sufficient for being an anti-realist. For more detailed discussions, see Edgington (1980–81) and Millikan (1986).
19 Rayo's (2017) view illustrates the latter form of anti-realism.

20 Haslanger (1995) proposes a similar view about gender, according to which genders are socially constructed but nevertheless exist and are real.

21 See, for example, Thomasson's (2015) "easy" arguments for the existence of numbers.

22 Analogous conceptions play an important role in various branches of philosophy of science, such as philosophy of biology (see Brigandt and Love 2017). According to so-called reductionists, each biological system (such as an organism) is constituted by nothing but (or reduces to) molecules and their interactions. These reductionists embrace a form of anti-realism about biological facts.

23 Sidelle (2002) argues that modal facts depend on conventions, which is a view that could be spelled out as a form of relativism about modal facts.

24 Gibbard's (2003) norm-expressivism applies Einheuser's template. In Gibbard's view, the semantic values of declarative sentences are sets of pairs <w, p>, where w is a possible world and p is a "hyperplan." A hyperplan is a function that represents a norm, and that maps each occasion for action in a world to a set of actions that are permitted on that occasion.

References

Ayer, A. (1936), *Language, Truth and Logic*, London: Penguin Books.

Azzouni, J. (2017), *Ontology Without Borders*, Oxford: Oxford University Press.

Biggs, S. and J. Wilson (2016), "Carnap, the Necessary A Posteriori, and Metaphysical Anti-Realism," in *Ontology After Carnap*, edited by S. Blatti and S. Lapointe, Oxford: Oxford University Press, 81–104.

Brigandt, I. and A. Love (2017), "Reductionism in Biology," in *The Stanford Encyclopedia of Philosophy*, Spring 2017 edition, edited by E. N. Zalta, https://plato.stanford.edu/archives/spr2017/entries/reduction-biology/.

Carnap, R. (1950a), *Logical Foundations of Probability*, Chicago: University of Chicago Press.

———. (1950b), "Empiricism, Semantics, and Ontology," in his *Meaning and Necessity*, Chicago: University of Chicago Press, 1956, 205–221.

Chalmers, D. (2009), "Ontological Anti-Realism," in *Metametaphysics: New Essays on the Foundations of Ontology*, edited D. J. Chalmers, D. Manley and R. Wasserman, Oxford: Oxford University Press, 77–129.

———. (forthcoming), "Idealism and the Mind-Body Problem," in *The Routledge Handbook of Panpsychism*, edited by W. Seager, London: Routledge.

Dasgupta, S. (2009), "Individuals: An Essay in Revisionary Metaphysics," *Philosophical Studies* 145: 35–67.

Davidson, D. (1973–1974), "On the Very Idea of a Conceptual Scheme," *Proceedings and Addresses of the American Philosophical Association* 47: 5–20.

Dummett, M. (1968–1969), "The Reality of the Past," *Proceedings of the Aristotelian Society* 69: 239–258.

———. (1977), *Elements of Intuitionism*, Oxford: Oxford University Press.

———. (1982), "Realism," *Synthese* 52: 55–112.

Edgington, D. (1980–1981), "Meaning, Bivalence and Realism," *Proceedings of the Aristotelian Society* 81: 153–173.

Einheuser, I. (2006), "Counterconventional Conditionals," *Philosophical Studies* 127: 459–482.

Field, H. (1980), *Science Without Numbers*, Princeton, NJ: Princeton University Press.

Fine, K. (2001), "The Question of Realism," *Philosophers' Imprint* 1: 1–30.

———. (2009), "The Question of Ontology," in *Metametaphysics: New Essays on the Foundations of Ontology*, edited by D. Chalmers, R. Wasserman and D. Manley, Oxford: Oxford University Press, 156–177.

Flocke, V. (forthcoming), "Ontological Expressivism," in *The Language of Ontology*, edited by J. Miller, Oxford: Oxford University Press.

Gibbard, A. (2003), *Thinking How to Live*, Boston: Harvard University Press.

Glasgow, J. (2009), *A Theory of Race*, London: Routledge.

Haslanger, S. (1995), "Ontology and Social Construction," *Philosophical Topics* 23: 95–125.

Hirsch, E. (2011), *Quantifier Variance and Realism: Essays in Metaontology*, Oxford: Oxford University Press.

Hofweber, T. (forthcoming), "Idealism and the Harmony of Thought and Reality," *Mind*.

Khlentzos, D. (2016), "Challenges to Metaphysical Realism," in *The Stanford Encyclopedia of Philosophy*, Winter 2016 edition, edited by E. N. Zalta, https://plato.stanford.edu/archives/win2016/entries/realism-sem-challenge/.

Lewis, D. (1968), "Counterpart Theory and Quantified Modal Logic," *The Journal of Philosophy* 65: 113–126.

MacFarlane, J. (2014), *Assessment Sensitivity: Relative Truth and Its Applications*, Oxford: Oxford University Press.

Miller, A. (2016), "Realism," in *The Stanford Encyclopedia of Philosophy*, Winter 2016 edition, edited by E. N. Zalta, https://plato.stanford.edu/archives/win2016/entries/realism/.

Millikan, R. G. (1986), "Metaphysical Anti-Realism?" *Mind* 95: 417–431.

Quine, W. (1968), "Ontological Relativity," *Journal of Philosophy* 65: 185–212.

Rayo, A. (2017), "The World Is the Totality of Facts, Not of Things," *Philosophical Issues* 27: 251–278.

Rodriguez-Pereyra, G. (2015), "Nominalism in Metaphysics," in *The Stanford Encyclopedia of Philosophy*, Summer 2015 edition, edited by E. N. Zalta, http://plato.stanford.edu/archives/sum2015/entries/nominalism-metaphysics/.

Schroeder, M. (2009), "Hybrid Expressivism: Virtues and Vices," *Ethics* 119: 257–309.

Sidelle, A. (1989), *Necessity, Essence and Individuation*, Ithaca, NY: Cornell University Press.

———. (2002), "Is There a True Metaphysics of Material Objects?" *Philosophical Issues: Realism and Relativism* 12: 118–145.

Sider, T. (2011), *Writing the Book of the World*, Oxford: Oxford University Press.

Smith, D. C. (1999), "Metaphysical Antirealism and Objective Truth: It Metaphysical Anti-Realism Self-Refuting?" *The Southern Journal of Philosophy* 37: 293–313.

Sosa, E. (1999), "Existential Relativity," *Midwest Studies in Philosophy* 23: 132–143.

Spencer, J. (2016), "Relativity and Degrees of Relationality," *Philosophy and Phenomenological Research* 92: 432–459.

Thomasson, A. L. (2015), *Ontology Made Easy*, Oxford: Oxford University Press.

van Inwagen, P. (1995), *Material Beings*, Ithaca, NY: Cornell University Press.

Wilson, J. (2014), "No Work for a Theory of Grounding," *Inquiry* 57: 535–579.

39

ONTOLOGICAL RELATIVISM AND THE STATUS OF ONTOLOGICAL DISPUTES

Delia Belleri

1. Introduction

Ontology today is accepted as a central branch of analytic philosophy, with authors discussing both long-standing, traditional problems (such as the existence of properties and universals) and issues that have become pressing only in the last few decades (such as the existence of genders and races, or the ontological status of virtual reality). Yet, not everyone is willing to take ontology at face value: philosophers with deflationary proclivities sometimes insist that ontological disputes are, at bottom, only disputes about words. This might lead them to reinterpret controversies about the existence of certain entities as covert disagreements about what one should mean with the word "exist" (or germane ontological terms like "object"). In recent terminology, this implies recasting ontological disputes as "metalinguistic negotiations" (Plunkett and Sundell 2013).

The purpose of this chapter is to explore the idea that ontological disputes are metalinguistic negotiations and to connect this idea to the main approaches that divide contemporary practitioners of ontology: realism and antirealism. In what follows, I look at three orders of reasons why one might want to link ontological disputes to the phenomenon of metalinguistic negotiation (Section 4) and I examine how positions along the realism-antirealism spectrum ("plain realism," "relativistic realism" and "relativistic antirealism") fare in the context of the idea of metalinguistic negotiation (Section 5).

2. What is at stake in ontological disputes?

The classical view is that ontological disputes are disagreements about the *existence* of putative entities. Think of the ancient controversy regarding properties, like the property of being red. In ordinary discourse, we call different things red: for instance, roses, tomatoes, or sweaters. We seem comfortable with saying that red objects have something in common, to wit, the property of being red. And yet, does this property *exist* as a separate, mind-independent object? Realism (or platonism) answers "yes." The opposite position, while granting that there are meaningful uses of the word "property," holds that properties do not exist. It is known as "nominalism" (for an overview of this debate, see Edwards 2014).

In order to understand what is at stake in such disputes, it is vital to get clear about what existence amounts to. One famous option, traceable to Willard van Orman Quine (1948), is to tie existence to the possibility of *quantifying over* certain objects. This implies redescribing the dispute just introduced in the following way. The realist maintains that we should *quantify over* properties, that is, regard properties as contained in the domain of objects that can substitute variables in sentences of the predicate calculus, thereby making them true. The nominalist denies that we can quantify over such entities. Note that, even if we establish that properties should be in our domain of quantification, we have not thereby established whether they are abstract or concrete, actual or possible, real or fictional, and so on. At this stage, this is an advantage: tying existence to quantification makes existence neutral with respect to possession of any of the aforementioned features, thus allowing us to – at least temporarily – ignore them for the purposes of this discussion.

Relying on this interpretation of ontological disputes, we can move to the next stage, that is, examining the idea that such debates might be metalinguistic negotiations.

3. Ontological disputes as metalinguistic negotiations

Here is a different way to couch the dispute canvassed earlier. Instead of being about whether properties exist, the preceding ontological dispute is about what the word "exist" should mean, and in particular, on whether the meaning of "exist" should be such that, *in light of this meaning*, properties could (or should) be in our domain of quantification.

This position amounts to saying that what, on the surface, looks like a disagreement over the existence of certain objects is really a normative metalinguistic dispute about what we should mean with the word "exist." This is what Plunkett and Sundell (2013) and Plunkett (2015) call a "metalinguistic negotiation." What would follow from such view?

A *prima facie* consequence of reconceiving ontological disputes as metalinguistic negotiations might be that of "deflating" them, that is, downplaying their theoretical significance by showing that they do not concern any matter of substance (namely, existence itself), but just concern terminological decisions. Of course, an immediate reaction to such contention could be: "Matters of terminological decision may be of ontological substance too!" After all, an ontologist might be convinced that choosing one description rather than another could help to more effectively "carve nature at its joints." Following authors like David Lewis, David Armstrong or Ted Sider, they might see ontological advantages in fixing the meaning of "exist" one way or another. Taken in this way, reconceiving an ontological dispute as a metalinguistic negotiation would not remove the ontological substance from the dispute, but rather show that questions of linguistic choice can be of ontological substance as well.

It follows from these considerations that the project of connecting ontological disputes with the idea of metalinguistic negotiation can be of interest for both theorists whose goal is to deflate ontology, and for theorists who regard it as a worthwhile philosophical endeavour.

4. Three connections, three claims

There are different options when it comes to forging the connection between the phenomenon of ontological disputes and the idea of metalinguistic negotiation. One could claim that these disputes *must be*, *are*, or simply *could be* metalinguistic negotiations. The present section explores the scope of, and motivation for, these three ways of identifying the connection.

(1) *The "must be" claim.* Let us first focus on the claim that ontological disputes *must be* – or cannot be anything but – metalinguistic negotiations. Following Plunkett and Sundell, this claim should be interpreted as stating that *the only sensible interpretation* of such exchanges is in terms

of metalinguistic negotiation, and not (yet) that these disputes must be "turned into" explicitly metalinguistic disagreements.

Why would one endorse such a claim? One strong reason may be that if these disputes were not interpreted as metalinguistic negotiations they would be semantically defective. This train of thought is highly reminiscent of the arguments against metaphysics advanced by Logical Positivism and can, arguably, find its premises the in the work of Rudolf Carnap. In *The Logical Syntax of Language* (1937), Carnap holds that the task of philosophy is to provide the logic of science. In his view, logic is nothing but a system of symbols, or a syntax. As a consequence, doing philosophy is, or should be, more of a language-analysis or language-engineering business than a substantive inquiry into the world. Philosophers have tended to conflate these two endeavours and have embarked on endless philosophical disputes "in the material mode," that is, using certain terms, not realizing that all they could really do was dispute "in the formal mode," that is, at the metalinguistic level. For instance, they debated whether numbers are objects, not realizing that all they could do was argue whether "number" is an object-word or not (Carnap 1937, 298–301).

Carnap's remarks suggest that the only way in which an ontological dispute can be philosophically respectable is by being interpreted as metalinguistic. If the only reasonable option is a metalinguistic *normative* dispute, then in a sense ontological disputes *must* be interpreted as metalinguistic negotiations, on pain of nonsense.

The foregoing also suggests that the "must be" claim seems to sit particularly well with a deflationary agenda with regard to ontological disputes. Although a non-deflationist theorist could in principle also endorse the "must be" claim, there seems to be no strong reasons for them to do so: if one thought that ontological disputes at the object-level are fine, it's not clear why they would urge to systematically re-interpret them as metalinguistic exchanges.

(2) *The "are" claim.* Let us now consider the thesis that ontological disputes *are*, as a matter of fact, metalinguistic negotiations (I understand this as a thesis wholly independent of the must-claim just considered). This would be a descriptive claim, but since it is not at all self-evident, evidence should be provided to substantiate it. David Plunkett (2015, 847, 856) lists the following as possible indicators that a dispute is metalinguistic: (i) evidence that an exchange describable as a dispute is going on; (ii) evidence that a genuine disagreement obtains between the parties; (iii) evidence that the disputants mean different things with certain key terms – for instance, patterns of expression-usage; (iv) evidence that the disagreement is normative – for instance, persistence of disagreement despite agreement on the actual meaning of a key term.

With regard to the ontological dispute reviewed previously, (i) and (ii) seem uncontroversial. (iii) and (iv) seem more like open questions. For instance, focusing on (iii), one may ask: how can the mere observation that the platonist and the nominalist use "exist" differently conclusively support the idea that they *mean different things* by "exist"? Why could divergence in usage not be evidence that they *hold different beliefs* as to what exists, in one and the same sense of "exist"? Similarly, moving to (iv), it may be asked: what if there is no determinate meaning to agree about? Or, what if the disputants insist that their disagreement has no normative element? In short, producing sufficient evidence that an ontological dispute is a metalinguistic negotiation is going to be a challenging task. My suspicion is that (i)–(iv) will generally underdetermine the desired diagnosis.

(3) *The "could be" claim.* In contrast with the conclusion just drawn, it seems almost indubitable that ontological disputes *could be* metalinguistic negotiations (and hence, that they could be interpreted as such). Why would one be attracted to such a modest possibility claim?

Suppose one were confronted with the thesis, propounded in several works by Eli Hirsch (2010), that theorists engaged in certain object-level ontological debates are just "talking past each other." If the dispute could have easily been a metalinguistic negotiation, Hirsch's contention may be downplayed and the practice of ontological disputes rehabilitated (Belleri 2017). Similarly, suppose one were confronted with the point that, in an ontological dispute, at least one of the contenders is somehow flouting the grammar of certain key terms, and hence is uttering an obvious falsehood (given the extant linguistic rules). If the dispute were a metalinguistic negotiation, the party who seems to speak ungrammatically would really be advocating a linguistic change (Thomasson 2016). Thus, the "could be" claim may have a strategic role, helping a theorist avoid untoward implications like the ones just mentioned. This also suggests that the "could be" claim may be of interest for the friend of ontological disputes, or for someone who at the very least cares about vindicating their rationality. I suspect, however, that the "could be" claim would not allow one to accomplish much more than these defensive moves.

So far, I have reviewed three ways of forging the connection between ontological disputes and the idea of metalinguistic negotiation. Whether or not someone embraces them, these options are clearly "out there," and much can be learned by even just laying them out.

5. Metalinguistic ontological disputes and the realism–antirealism spectrum

This section is devoted to the following question: Which preconditions should a metaphysical theory satisfy in order for it to be "hospitable" to, and benefit from, the idea of metalinguistic negotiation? I will argue that one central precondition is some form of *semantic equivocalism*, that is, a form of pluralism about the semantics of "exist." I will show this by considering the interplay between equivocalism and metalinguistic negotiation in three broad theoretical settings, which I will refer to as "plain realism," "relativistic realism" and "relativistic antirealism."

(1) *Plain realism.* Plain realism could be summarized as the conjunction of two theses:

(1) the world exists independently of our language and mind;
(2) language can describe how the world is independently of us.

The plain realism I have in mind here is a global position, about reality as a whole, and not about specific portions of it (e.g. material objects, numbers, normative facts). Such a global position is usually held as a very powerful default assumption which can at best be abductively justified (Devitt 1991; Sider 2011), or justified by rejecting its opposite – some form of global antirealism (Alston 2001; Boghossian 2006).

The role of language in such a setting is to "limn the structure of reality," as it were: words and sentences are supposed to match, *to some degree*, the objects and relations existing in reality. In recent times, proponents of plain realism have also promoted the idea that certain linguistic expressions are better suited than others to describe how the world is at its most *fundamental* level. David Lewis (1983, 1986) and David Armstrong (1978) adopt the distinction between predicates denoting "natural" properties and predicates denoting "non-natural" properties. A property is natural when it captures a highly explanatorily relevant resemblance: for instance, being negatively charged, or having atomic number 79. Similarly, Ted Sider argues that since our most successful theories employ existential quantification, the existential quantifier carves reality at its joints (2011, 97). In such a picture, linguistic expressions, language-fragments or even whole languages, as well as the descriptions that can be formulated by means of them, can be assessed and ranked according to how well they perform the task of "carving the world at its joints." It is therefore apparent that, at least according to authors like Lewis, Armstrong and Sider, linguistic choices *can* have an ontological import.

(2) *Relativistic realism*. Plain realism can be questioned by attacking thesis (2): that language can describe how the world is independently of us. Consider the following argument. Suppose we have two different but empirically equivalent descriptions of some facts. We could say that these are equally good renditions of the same portion of the world. However, the latter statement invites the question: how is the world like, independently of those descriptions? (Putnam 1987, 19, 33ff.; Button 2013, 201–203). A third way of describing the world would be needed. And yet, how could one know that this third way of describing the world – or any other way, for that matter – captures how the world "really" is? There seems to be no way of answering this question. This casts doubt on thesis (2): for any description, there is no way of knowing whether it is allowing us to describe how the world is independently of us. And thus there is no way of knowing that language in general *can* do so.

Furthermore, it is reasonable to think that our conceptual and linguistic system serves (among other things) the function of organizing in the best way possible our experience, cognition and human interaction. This, however, makes it difficult, from within the language, to distinguish the "mind-dependent" components of a certain description from the "objective" ones (see Putnam 1990, 28). Again it is difficult to know, for any description, whether it is allowing us to describe things as they really are; this casts even more doubt on thesis (2).

Questioning our *access* to the "real" description of the world does not mean, however, giving up on the idea that *there is* a mind-independent world "out there." To see this, imagine Putnam's (1987, 18–20) mini-world, containing three dots. Using our ordinary conceptual scheme, we would say there are three objects; using the conceptual scheme of classical mereology, we would say there are seven. This does not imply, however, that the two schemes *create* the facts; we can continue to assume that there are scheme-independent facts. All we need to acknowledge is that there is no way of describing them other than by adopting a conceptual scheme. That there are three objects is thus a mind-independent fact, *relative to our ordinary conceptual scheme*; equally, that there are seven objects is also a mind-independent fact, *relative to the scheme of classical mereology*.

The upshot of these considerations is relativistic realism; it can be summarized as the conjunction of the following theses:

(1)   the world exists independently of our language and mind;
(2)   we can describe how the world is only relative to a conceptual or linguistic framework;
(3)   it makes no sense to ask whether any description captures the way the world is independently of our language and mind.

In the context of relativistic realism, the role played by language is more tied to pragmatic considerations than to representational concerns. The language we speak, and the theories we formulate in it, are developed as devices for organizing experience and guiding practical tasks. We find this insight already in Quine's work: "We adopt . . . the simplest conceptual scheme into which the disordered fragments of raw experience can be fitted and arranged" (1948, 35–36; cf. also 1951, 44). Putnam follows suit: "there are only the various points of view of actual persons reflecting various interests and purposes that their descriptions and theories subserve" (1981, 50). As a result of this characterization, linguistic expressions, language-fragments or even whole languages are judged and ranked according to pragmatic criteria. These include their expediency in aiding description, prediction and explanation of experience, or the organization of our social life. How accurately they represent the world cannot be subject to assessment. Relativistic realism thus departs from plain realism in its stressing the pragmatic genesis of conceptual/linguistic schemes, and the theories that these allow to formulate.

(3) *Relativistic antirealism.* The last position, relativistic antirealism, differs from relativistic realism in being skeptical regarding the notion of "mind-independent world."

Nelson Goodman argues that it would be misguided to trace all the different descriptions (or "versions") produced by humankind to one single object, identified as "The World." This is because it would be impossible to state how this object is like, independently of a description; and, absent a privileged description, we would have to accept The World as an ineffable, and therefore for all we know structureless entity: "a world without kinds or order or motion or rest or pattern – a world not worth fighting for or against" (Goodman 1978, 20). Instead of positing one single world, we should feel free to admit as many "worlds" as there are descriptions. The realist's monistic world is therefore dissolved into a plurality of world-versions.

A different route is taken by Richard Rorty, who questions realism by attacking some of its core concepts. Rorty (1979a, 1979b) advocates abandoning the idea of truth as correspondence, or of knowledge as mirroring reality; this implies rejecting that something like the world, or reality, is represented by our thoughts and statements. By rejecting a representational approach to truth and knowledge, Rorty makes the very notion of the world redundant, and therefore theoretically disposable. In his view, a true belief is not a belief that accurately represents the world, but rather a belief that is shared, or agreed upon within our community.

Once the notion of mind-independent world is discarded, what is left is a plurality of descriptions or "versions" that we produce in accordance with our needs and purposes. This picture can be relativistic in two ways: first, whether a description or "version" is good, is relative to a social activity or practice; in Goodman's terms, whether a version is good depends on how it "fits" with the rest of the practice(s); for Rorty, a description is good ("true") insofar as it's agreed upon within a community. And second, there is no way of declaring one version absolutely superior to the others.

The resulting account is relativistic antirealism; it can be summarized as follows:

(1) the notion of a mind-independent world is devoid of sense;
(2) there are only descriptions suited to certain purposes;
(3) it makes no sense to ask whether any description captures the way the world is independently of our language and mind.

In the relativistic-antirealist framework, language plays again a pragmatic role. Languages and linguistic descriptions develop first and foremost to suit a range of human purposes. Goodman (1956, 1978, 128) famously argues that it is our practices, and what we deem as adequate descriptions in light of them, that affect the development of our categories – for example, the category "green" as opposed to "grue." Rorty too ties language development to human practices: "We speak a language which includes the word 'giraffe' because it suits our purposes to do so. The same goes for words like 'organ,' 'cell,' 'atom.' . . . All the descriptions we give of things are descriptions suited to our purposes" (Rorty 1996, xxvi). Consequently, it is entirely possible for linguistic tools – be they single expressions, language fragments or whole languages – to be evaluated and ranked based on criteria of practical convenience and expediency.

(4) *Metalinguistic negotiation across the realism-antirealism spectrum.* Let us now go back to the platonist-nominalist dispute about properties. The platonist claims "There are properties." The nominalist denies this. What is the difference, within each of the views just examined, between arguing that abstract objects exist and arguing that "exist" should have this-or-that meaning?

I predict that answers across the realism-antirealism spectrum will vary according to what attitude is maintained about the semantics of the key term under discussion: in this case, the example focusses on "exist." The contention is that "univocalist" attitudes regarding the semantics of "exist" (or other germane terms) will correlate with little or no excitement about the metalinguistic negotiation hypothesis; conversely, pluralistic or "equivocalist" attitudes about the semantics of this term (and similar ones) will come with a more favorable reception of the same hypothesis.

A univocalist attitude towards the semantics of "exist" consists in regarding this expression as non-ambiguous as well as non-context-sensitive, thus having a single meaning or sense. Van Inwagen (2009, 482ff.) advocates univocalism while pursuing a plain realist project, while Thomasson (2015) combines univocalism about existential quantification with a deflationary approach closer to relativistic realism. Now, suppose one combines the thesis whereby "exist" is semantically univocal with a Tarski-style semantics, and compares the following sentences: the sentence "Properties exist (at *t*)," which might be used in an object-level ontological dispute; and the sentence "Properties should fall under the extension of 'exist' (at *t*), given its univocal meaning," which would be typically associated with a metalinguistic negotiation. Under the previous assumptions, these sentences are going to be *extensionally equivalent*. Engaging in a metalinguistic negotiation would merely imply a detour through the metalanguage that would in no way advance the discussion. The univocalist about "exist" should not be attracted to this option.

In a number of special cases, it might seem that the extensional equivalence between the object-language statements and the metalinguistic statements is lost, thus making the metalinguistic option sensible by the univocalist's lights. In the first special case, one of the two parties in the object-level dispute might be using "exist" in a way that deviates from its actual semantics, attempting to establish a new usage which could, eventually, determine a new meaning. In the second special case, the party who denies the existence-claim might be trying to ban the term "exist," to replace it with a semantically deviant one, say "schmexist." In the third special case, the party who denies the existence-claim might be trying to communicate that the term "exist" should be banned, but propose no replacement.

What should the univocalist say about these cases? If being a univocalist means holding that there are not even *possible* alternative meanings for "exist" in possible alternative versions of English, then such a univocalist should say that the first and second special cases are impossible: using "exist" (or the existential quantifier) in a deviant way just results in nonsense, and introducing "schmexist" either does not allow one to express a different concept than that expressed by "exist" or results, once again, in nonsense. As to the third special case, the univocalist might try to show that attempts to ban "exist" could only have to do with whether the *word itself* is acceptable, not with whether it has a meaning or extension (for in a trivial sense, "exist" applies to everything). This would imply that the metalinguistic negotiation would be addressing a different problem than the ontological one.

An "equivocalist" attitude towards the semantics of "exist" consists in regarding this expression as susceptible of several interpretations. One could believe that "exist" is multiply ambiguous: Sider (2011, §9.3), an advocate of plain realism, holds that the word "exist" could, in different contexts, express different concepts, although only one – the one that carves at the joints – is suitable for ontological inquiry. Alternatively, one could believe that the meaning of the word "exist" is stable in the actual version of English, but that there are alternative possible meanings in the vicinity of the actual one, belonging to alternative versions of English: Hirsch (2010, 70–71) calls this position "Quantifier Variance," and develops it together with an approach inspired by relativistic realism. Finally, one could theorize that the English word "exist" has a fixed but indeterminate meaning (Kaplanian character), to be specified in different ways in different contexts: Alan Sidelle outlines such a view (2002, 141) by coupling it with an approach close to relativistic antirealism (2002, 134).

In light of this semantic plurality, there may be a point, for the purposes of an ontological dispute like the platonist-nominalist dispute, in distinguishing between what two theorists can establish by debating at the object-level and what two theorists can establish by debating at the metalinguistic level. At the object-level, two ontologists can only discuss whether the Fs exist in one sense of "exist." This might be a worthwhile project, but it need not be the only project two ontologists could pursue in this setting. Another project may be that of moving at the metalinguistic level and advocating different interpretations of "exist." Thus, the platonist could advocate an interpretation of "exist" liberal enough to allow that properties (conceived as *abstracta*) exist, while the nominalist would advocate a more restrictive interpretation.

To see how the two projects dramatically differ, think of the role assigned to language in plain realism, relativistic realism and relativistic antirealism respectively. In the setting of plain realism, the platonist and nominalist can debate over which interpretation of "exist" is more joint-carving, or which one allows us to chart the most fundamental aspects of reality (they could support these claims by using abductive considerations or by appealing to empirical arguments). By contrast, if the platonist and the nominalist were debating at the object level, they could only use one notion of "exist," no matter how joint-carving, and they could only establish whether Fs exist in that particular sense. In the framework of relativistic realism and antirealism, the two theorists can compare the platonist and nominalist interpretations of "exist" according to their practical usefulness for, say, the purposes of scientific inquiry or common-sense discourse. If they were debating at the object level, all they could accomplish would be to settle whether Fs exist in one specific sense of "exist."

The conclusion of this section is therefore the following: no matter the approach taken along the realism-antirealism spectrum, a palpable difference between arguing at the meta-level and arguing at the object-level becomes apparent under equivocalist assumptions concerning the semantics of "exist." As long as the equivocalist believes that different interpretations of "exist" can compete in point of, say, joint-carvingness (in the plain realism setting) or pragmatic convenience (in the relativistic realism and antirealism settings), then it is to be expected that they will find the metalinguistic negotiation option compatible with, if not congenial to, their views.

6. Conclusion

The purpose of this chapter was to introduce the reader to the question whether ontological disputes are metalinguistic negotiations, and examine the interplay between this idea and three major approaches along the realism-antirealism spectrum. The result of the present discussion is that theorists who have equivocalist sympathies with regard to the language of ontology's semantics are more likely to be hospitable to, and benefit from, the idea that ontological disputes are metalinguistic negotiations.

References

Alston, W. (2001), *A Sensible Metaphysical Realism*, Milwaukee: Marquette University Press.
Armstrong, D. M. (1978), *A Theory of Universals*, vols. 1–2, Cambridge: Cambridge University Press.
Belleri, D. (2017), "Verbalism and Metalinguistic Negotiation in Ontological Disputes," *Philosophical Studies* 174(9): 2211–2226.
Boghossian, P. (2006), *Fear of Knowledge: Against Relativism and Constructivism*, Oxford: Oxford University Press.

Button, T. (2013), *The Limits of Realism*, Oxford: Oxford University Press.

Carnap, R. (1937), *Logical Syntax of Language*, London: Routledge, 2001.

Devitt, M. (1991), *Realism and Truth*, Oxford: Blackwell.

Edwards, D. (2014), *Properties*, Oxford: Polity Press.

Goodman, N. (1956), *Fact, Fiction, and Forecast*, Cambridge, MA: Harvard University Press.

———. (1978), *Ways of Worldmaking*, Brighton: Harvester Press.

Hirsch, E. (2010), *Quantifier Variance and Realism: Essays in Metaontology*, Oxford: Oxford University Press.

Lewis, D. K. (1983), "New Work for a Theory of Universals," *Australasian Journal of Philosophy* 61(4): 343–377.

———. (1986), *On the Plurality of Worlds*, London: Wiley-Blackwell.

Plunkett, D. (2015), "Which Concepts Should We Use? Metalinguistic Negotiations and the Methodology of Philosophy," *Inquiry: An Interdisciplinary Journal of Philosophy* 58(7–8): 828–874.

Plunkett, D. and T. Sundell (2013), "Disagreement and the Semantics of Normative and Evaluative Terms," *Philosophers' Imprint* 13(23), 1–37.

Putnam, H. (1981), *Reason, Truth and History*, Cambridge: Cambridge University Press.

———. (1987), *The Many Faces of Realism*, Chicago: Open Court.

———. (1990), *Realism with a Human Face*, Cambridge, MA: Harvard University Press.

Quine, W. V. O. (1948), "On What There Is," *Review of Metaphysics* 2(1): 21–38.

———. (1951), "Two Dogmas of Empiricism," *Philosophical Review* 60(1): 20–43.

Rorty, R. (1979a), "Pragmatism, Relativism, and Irrationalism," *Proceedings and Addresses of the American Philosophical Association* 53(6): 717–738.

———. (1979b), *Philosophy and the Mirror of Nature*, Princeton, NJ: Princeton University Press.

———. (1996), *Philosophy and Social Hope*, London: Penguin Books.

Sidelle, A. (2002), "Is There a True Metaphysics of Material Objects?" *Noûs* 36(1): 118–145.

Sider, T. (2011), *Writing the Book of the World*, Oxford: Oxford University Press.

Thomasson, A. L. (2015), *Ontology Made Easy*, New York: Oxford University Press.

———. (2016), "Metaphysical Disputes and Metalinguistic Negotiation," *Analytic Philosophy* 57(4): 1–28.

van Inwagen, P. (2009), "Being, Existence and Ontological Commitment," in *Metametaphysics: New Essays on the Foundations of Ontology*, edited by D. Chalmers et al., Oxford: Oxford University Press.

PART 7

Relativism in philosophy of science

40

THE RELATIVISTIC LEGACY OF KUHN AND FEYERABEND

Howard Sankey

1. Introduction

Thomas S. Kuhn (1922–1996) and Paul K. Feyerabend (1924–1994) were central figures in the historical turn that took place in the philosophy of science in the latter half of the twentieth century. In a break with positivist orthodoxy, advocates of the historical approach sought to understand the sciences in terms of the developmental processes which underlie scientific change. With growing recognition of the extent to which science is subject to change, the suggestion that the sciences may be approached in relativistic terms came increasingly to the fore. In no small part, the work of Feyerabend and Kuhn was responsible for this trend.

The importance of the history of science emerges clearly in Kuhn's famous book, *The Structure of Scientific Revolutions*. (First published in 1962, reference will here be made to the 4th edition of *Structure*, published in 2012.) Kuhn sought to bring about a shift in our "image of science" by analysing the processes of scientific change revealed by the historical study of the sciences (Kuhn 2012, 1). He proposed a model of scientific change, on which consensus on paradigm emerges from disunified beginnings. A paradigm is a set of theoretical beliefs and exemplary scientific achievements that form the basis for an enduring tradition of research in a field of science. Scientists work within a paradigm during "normal science." They devote themselves to solving "puzzles" which arise in applying the paradigm to the world. Normal scientific practice is occasionally confronted by crisis-inducing "anomalies" which raise doubts about the reigning paradigm. Alternative candidates for paradigm are proposed. Debate ensues between advocates of the current paradigm and proponents of the paradigm candidates. Paradigm debate is characterized by lack of complete communication due to the incommensurability of the paradigm and its competitors (Kuhn 2012, 147–149). As a result, the choice between the paradigm and candidates for paradigm is unable to be resolved in a fully rational manner. On those occasions where the original paradigm is rejected in favour of one of the candidates for paradigm, a scientific revolution occurs. Following a revolution, normal science resumes under the auspices of the new paradigm which has taken over as the basis for research in the field.

Unlike Kuhn, Feyerabend did not propose a model of scientific theory-change. Like Kuhn, Feyerabend argued that incommensurability plays a role in scientific change. He argued against the logical empiricist idea that scientific change involves reduction of the laws of an earlier scientific theory to the laws of a later theory. The reason that such reductive relations do not obtain is due to semantic differences between the vocabulary employed by the theories, which Feyerabend described as a form of incommensurability. Apart from such semantic considerations, Feyerabend also made controversial suggestions in relation to scientific method. In his best-known book *Against Method* (1975) Feyerabend argued that all rules of scientific method have been violated at some point in the history of science. His point was not simply the descriptive one that in actual practice a scientist may fail to employ the rules of method. The point had normative force, as well. In many such cases scientists are in fact justified in violating the rules of method. Feyerabend formulated his view about the violability of the rules of method in terms of the position of "epistemological anarchism," which he famously expressed with the slogan "anything goes."

The idea that theories or paradigms may be incommensurable for semantic reasons is found in both authors. In Section 2, I will discuss the considerations that give rise to the claim that theories or paradigms may be incommensurable in a semantic sense, as well as the implications of such incommensurability with respect to relativism about truth. Apart from semantic considerations, Kuhn and Feyerabend endorsed views about the methodology of science that have potential relativistic implications. For Kuhn, incommensurability involves variation of standards between paradigms, which may lead to the relativity of epistemic rationality to paradigm. For Feyerabend, there are no universally binding standards applicable throughout the history of the sciences. As with Kuhn, the rejection of a universal methodology for science suggests a relativistic view of epistemic rationality. I will discuss the relativistic implications of methodological variation in Section 3. In Section 4, I will indicate how relativistic themes from Kuhn and Feyerabend have influenced subsequent work.

2. Semantic incommensurability

In *Structure*, Kuhn uses the term "incommensurability" to characterize methodological, perceptual and semantic differences between paradigms. In work after *Structure*, Kuhn restricted use of the term to semantic relations between theories or paradigms. By contrast with Kuhn, Feyerabend consistently employed the term to describe semantic relationships between theories. In this section, I will focus on semantic incommensurability and its relativistic implications with respect to truth.

In *Structure*, debate between supporters of an established paradigm and proponents of a paradigm candidate is characterized by partial failure of communication (cf. 2012, 148). This is due to difference in conceptual apparatus between the paradigms, which results in semantic variation between the vocabulary that the paradigms employ. Kuhn placed special emphasis on changes in classification which take place in the transition between paradigms (e.g. 2012, 199). After a revolution, entities which were previously assigned to the same class may be assigned to different classes. Entities previously in different classes may be assigned to the same class. Because of such classificatory change, terms which are retained throughout the transition undergo change of meaning, which may affect both their sense and their reference. New terminology may also be introduced which differs in meaning from terminology employed by the previous paradigm.

In describing the semantic shift that takes place during change of paradigm, Kuhn spoke increasingly in terms of translation. At first, he drew a parallel between semantic incommensurability and Quine's indeterminacy of translation (cf. 1970, 268ff.). However, he later realized the parallel was incomplete. Rather than translation being indeterminate, Kuhn argued for failure of exact translation between the vocabulary employed by theories. As his thinking evolved, he came to downplay failure of communication. He recognized that speakers of mutually untranslatable languages may learn each other's language and thereby understand each other. In his mature view, incommensurability is a local phenomenon (see, e.g., 2000, 36ff.). It consists in the inability to exactly translate between sets of holistically interdefined terms within the special vocabulary of competing theories. Because the translation failure is localized, it takes place against a background of semantically stable vocabulary that is shared between the theories. This semantic common ground provides a basis for comparison of at least some of the content of the theories.

While Kuhn employed the notion of incommensurability in describing relations between paradigms, Feyerabend's use of the notion emerged in the course of his critical analysis of the logical empiricist account of inter-theory reduction. Logical empiricists understood reduction as a deductive relationship in which the laws of an earlier theory are subsumed under the laws of the later theory that replaces it. Feyerabend pointed out that such a deductive account of reduction implies that a condition of meaning invariance must apply to the vocabulary of successive theories. For if the vocabulary employed by the theories did not possess the same meaning, the laws of the reduced theory could not be logical consequences of the laws of the reducing theory.

Against the empiricist account of reduction, Feyerabend argued that the condition of meaning invariance is violated in actual science due to conceptual change that occurs in the transition between theories. He rejected the empiricist idea of a theory-neutral observation language, arguing that neither experience nor pragmatic conditions of use fix the meaning of observational vocabulary (1981a, 21–29). Instead, the meaning of observational vocabulary depends on the theory that applies to the relevant domain of observable phenomena. Because of this, the meaning of observational terms may vary with change in the theory that explains the observable phenomena to which the terms apply (1981a, 31). As for theoretical terms, Feyerabend argued that their meaning depends upon the theoretical context in which they are employed. As a result, it may be impossible to define the terms of one theory in the context of another theory (1981b, 66–67). The dependence of meaning on theoretical context entails that there may be meaning variance in the transition between theories. Given meaning variance, the deductive consequences of the theories form disjoint classes, so that the logical relations required for deductive subsumption fail to obtain. Feyerabend concludes that "incommensurable theories may not possess any comparable consequences, observational or otherwise" (1981b, 93).

The semantic notion of incommensurability is not as such an inherently relativistic notion. However, it can be used to support a variety of relativistic views. In the next section, I will indicate how semantic incommensurability may combine with variation of methodological standards to produce a form of epistemic relativism. For now, I shall restrict myself to relativism with respect to truth. The semantic variation between successive paradigms or theories which takes place, according to Kuhn and Feyerabend, provides a basis for relativism about truth. I will first describe a weak form of relativism about truth before sketching a stronger form of the view.

Semantic variation may give rise to the inability to translate from the vocabulary of one theory into the vocabulary of another theory with which it is incommensurable. If translation fails between theories, a true proposition asserted using the vocabulary of one theory may be unable to be expressed using the vocabulary of the other theory owing to the failure of translation. The inability to express a true assertion of one theory in the vocabulary of another theory yields a sense in which truth is relative to theory. Truth is relative to theory in the sense that a true proposition that may be asserted using the vocabulary of one theory is not able to be formulated in terms of the vocabulary of the other theory. Thus, the ability to express a specific true proposition becomes relative to theory. Such relativism does not make truth relative to theory in the sense that a proposition may be true in one theory while its negation is true in the context of another theory. Rather, truth is relative in the weaker sense that a truth assertible in one theory is not able to be expressed in the other theory. It is the ability to express truth, rather than truth itself, that is relative to theory. Such a form of relativism differs from what is usually meant when truth is said to be relative. But by avoiding the claim that one and the same proposition is true in one context and false in the other, it avoids the traditional objection of incoherence that is usually levelled against relativism about truth. (For the incoherence objection, see Siegel 1987, ch. 1.)

I will now describe a stronger form of relativism about truth that makes use of the notion of semantic incommensurability. In characterizing the profound effects of scientific revolution, Kuhn sometimes spoke as if the world varies with paradigm: e.g. "when paradigms change, the world itself changes with them" (2012, 111); "the proponents of competing paradigms practice their trades in different worlds" (2012, 149). Such remarks may be read as hyperbole. Kuhn was merely employing the world-change metaphor to emphasize the dramatic impact of paradigm change. But some commentators interpret remarks such as these in anti-realist terms. For example, Paul Hoyningen-Huene has proposed a neo-Kantian interpretation of Kuhn on which the phenomenal world of the scientist varies with paradigm, though the world-in-itself remains fixed (1993, 32–35). Such a neo-Kantian interpretation of Kuhn has ontological implications which may serve as the basis for a relativistic conception of truth. Paradigms are not competing theories about the same phenomenal world. Incommensurable paradigms are located within their own distinct phenomenal worlds. The entities to which the terms used by scientists refer vary with paradigm because phenomenal world varies with paradigm. This allows for a more substantive sense in which truth is relative. For the states of affairs that make propositions true vary with phenomenal world. As such, a proposition that is true relative to the phenomenal world of one paradigm may be false relative to the phenomenal world of another. Truth is relative to phenomenal world. The prospects for this form of relativism about truth will depend on whether the neo-Kantian position itself is ultimately defensible, something which those of a realist persuasion are inclined to doubt. (For discussion of truth in a Kuhnian phenomenal world, see Devlin 2015.)

I will now briefly present two important lines of criticism that emerged in response to the semantic incommensurability thesis. The first is the translational response due primarily to Donald Davidson. The second is the referential response associated with the causal theory of reference of Saul Kripke and Hilary Putnam.

In his famous paper, "On the Very Idea of a Conceptual Scheme" (1984), Davidson sought to undermine the conceptual relativist idea that there may be alternative legitimate conceptual schemes. One of his main targets is the idea that different conceptual systems may be incommensurable, and, thereby fail to be inter-translatable. Davidson questions whether the idea even makes sense. In the first place, it is not clear what might constitute evidence for the existence of an untranslatable language. Failure to translate purportedly linguistic material might be taken as evidence that the material is not in fact linguistic rather than as evidence that it is a language that is unable to be translated. If an attempt is made to establish the untranslatability of a language by providing an example of an untranslatable term of the language, the very fact of providing the example of the untranslatable term in the language into which it purportedly cannot be translated belies the claim of untranslatability. Even the assumption that one might understand terms in an untranslatable language suggests translation is possible after all, since one has been able to arrive at an understanding of the allegedly untranslatable terms. (For a closely related argument, see Putnam 1981, 114–119.)

Davidson subjects the idea of an untranslatable language to serious criticism. But it is possible to weaken the force of his objections. Especially if one understands semantic incommensurability as a local phenomenon, it may be possible to address the concerns about evidence of untranslatability. If translation failure is restricted to a narrowly circumscribed set of interdefined terms within the special vocabulary of a theory, such translation failure occurs within the context of a background natural language as well as surrounding scientific vocabulary. As such, there is no need to determine whether a completely untranslatable language is indeed linguistic, since the failure of translation is between special vocabularies within a language (see Sankey 1990). It is also possible to remove the problem of how to understand an untranslatable language without being able to translate it. Both Kuhn and Feyerabend distinguish learning a new language from translation into one's own language (Kuhn 2000, 43–47; Feyerabend 1987, 266). Given the distinction, failure to translate does not entail failure to understand what is said in vocabulary that is untranslatable into the terms of one's own theory (see Sankey 1991).

The referential response emerged in Israel Scheffler's book, *Science and Subjectivity* (1967). Scheffler employed Frege's distinction between sense and reference to argue that meaning variant theories may be comparable for content provided that the terms used by the theories have the same reference. Kuhn and Feyerabend did not, however, restrict the claim of meaning variance to variation of sense. They held that there is variation in reference as well. In this context, the claim by Kripke (1980) and Putnam (1975) that reference is determined by causal relation between speaker and reality rather than by description took on relevance to the problem of incommensurability. For, if the reference of a theoretical term is fixed upon the original introduction of a term by a scientist, reference may remain stable throughout subsequent change of theory. If reference remains stable in theory-change, then terms of a later theory may refer to the same things as terms of earlier theories. Such shared reference would provide semantic common ground based on which the content of competing or successive theories may be compared. But, as the subsequent literature has demonstrated, it is not possible to enforce complete stability of reference across theory-change, since descriptive elements play a role in reference-determination for theoretical terms and post-introductory use may alter reference (e.g., Devitt 1979; Fine 1975; Nola 1980).

3. Methodological variation

Traditional theorists of scientific method sought to identify a scientific method employed throughout the sciences. By contrast, advocates of the historical approach to the philosophy of science emphasized variation in the methods employed in the sciences. Rejection of a single scientific method in favour of a variety of methods provides a further source for relativism in respect of the sciences.

In this section, I will consider the implications of Kuhn's and Feyerabend's methodological views with respect to relativism and the rationality of science. Though Kuhn initially took incommensurability to involve difference in methodological standards, he later restricted the notion of incommensurability to semantic relations between theories. Moreover, Feyerabend consistently limited his use of the notion to the semantic realm. For this reason, I will frame the discussion in terms of methodological variation rather than in terms of methodological incommensurability.

At one point in *Structure*, Kuhn refers to "commitments without which no man is a scientist" (2012, 42). These include a concern to understand the world in a precise manner using empirical means. But Kuhn makes no attempt in *Structure* to identify or articulate a single universal method for the sciences. Instead, he emphasizes standards that apply to scientific research within the context of a paradigm. The puzzle-solving activity of normal science is governed by methodological standards which are based on the established paradigm. The standards take the form of rules of puzzle-solving adequacy. Like the puzzles themselves, the rules derive from the paradigm. The rules include the basic laws of the paradigm, procedures for the use of instrumentation and the conduct of experiment, as well as metaphysical commitments about the nature and structure of the world (2012, 40–41). Because rules of puzzle-solving adequacy derive from the paradigm, they are subject to variation with change of paradigm.

The rules of puzzle-solving adequacy provide a basis for the evaluation of puzzle-solving activity within normal science. Precisely because of their paradigm-dependence, the rules can play no role in the choice between competing paradigms (2012, 94). In the absence of neutral standards for the comparative appraisal of paradigms, the question arises of how choice between paradigms may be made on a rational basis. The matter is further complicated by the point that Kuhn took observation to be theory-dependent, which means that no appeal may be made to a neutral domain of theory-independent empirical facts (2012, 111–134). Bearing in mind the communication breakdown produced by semantic incommensurability, the choice must be made without appeal to neutral standards or facts, and without an understanding of the opposing view. Given this, it is hard to see how the decision to adopt one paradigm over another may be made on a rational basis. It is no wonder that Kuhn tended to describe the choice between paradigms in quasi-religious terms as a "conversion experience" (2012, 150).

The combination of paradigm-dependent standards with lack of a rational basis for paradigm choice is a recipe for relativism. For while there may be no rational ground for the adoption of a paradigm, the rules of puzzle-solving adequacy provide justification for normal scientific activity within the paradigm. In the context of an accepted paradigm, the rules constitute standards of rationality based on which beliefs and actions of scientists may be justified. Because the standards of rationality depend upon and vary with paradigm, the rationality of scientists is relative to paradigm. Given the paradigm-dependence of standards and absence of a neutral basis for paradigm appraisal, Kuhn's account of the rationality of science in *Structure* was widely regarded as a relativist one. (For authoritative coverage, see Siegel 1987, Part II.)

To Kuhn's surprise, *Structure* encountered a somewhat hostile philosophical reception. Philosophers were critical of the relativism to which Kuhn's account of science appeared to lead. Seeking to defend the objectivity of science against Kuhn, philosophers emphasized the need for independent criteria of paradigm-appraisal. According to Israel Scheffler, Kuhn simply failed to distinguish between standards that are internal to a paradigm and "second-order criteria" that are external to paradigms and are employed to evaluate them (1967, 84–85). Philosophers have made a range of attempts to identify independent criteria that might be used in the comparative appraisal of paradigms. One influential example is that of the progressiveness of research programmes which Imre Lakatos proposed as part of his own model of theory-change (Lakatos 1970). Another example is problem-solving effectiveness, which Larry Laudan introduced in the context of his account of scientific research traditions (Laudan 1977).

In response to the criticism, Kuhn undertook to clarify his position in a postscript which appeared in the second and later editions of *Structure* (see, especially, 2012, 184–185), as well as in his paper "Objectivity, Value Judgment and Theory Choice" (1977). Contrary to his apparent denial of extra-paradigmatic standards, Kuhn held that there are independent criteria of theory-choice. The criteria function as values which guide theory-choice, rather than as rules which dictate the choice. Kuhn lists as examples the values of accuracy, simplicity, consistency, breadth and fruitfulness. While choice of theory or paradigm is informed by a stable set of shared values, the guidance provided by the values allows for divergent outcomes between scientists. Taken individually, the values are open to alternative interpretations. One and the same value may be understood differently and applied in different ways to the same cases. Taken collectively, there is scope for potential conflict between the values. Different theories may satisfy different values. One theory may display a higher level of accuracy, while a competitor is simpler or has greater breadth. The system of shared values may fail to uniquely pick out a specific theory from among a set of competing theories.

The position that emerges from Kuhn's clarification is not one that denies an objective basis for scientific theory-choice. The shared scientific values constitute objective criteria of theory appraisal (1977, 336–338). But while the values have an objective status, appeal to the shared values may fail to deliver a unique outcome. As Kuhn expressed the point, there is no "algorithm of theory-choice" which leads all scientists to the same decision (2012, 198). Rather than deny the rationality of theory-choice, Kuhn's view is that scientists may disagree on a rational basis. Scientists may adopt opposing theories even though their choice of theory is rationally grounded in objective criteria of theory-appraisal.

Kuhn's idea that the criteria of theory-choice are values rather than rules is similar to the view of method defended by Feyerabend. In *Against Method*, Feyerabend argued that all rules of scientific method have been violated at some point in the history of science. His point was not just the descriptive point that the rules of method are violated in the actual practice of science. It was the normative point that on some occasions violation of the rules of method may be justified. In order for science to progress, there are circumstances in which it is necessary to violate a rule of method. In effect, Feyerabend's point is that all rules have exceptions. There are no rules which are binding in all circumstances. Feyerabend's conception of violable rules closely parallels Kuhn's view that criteria of theory-choice are values that guide rather than rules that dictate choice.

In *Against Method*, Feyerabend provocatively claimed that there is only one methodological principle that may be supported in all circumstances. This is the principle "anything goes" (1975, 28). In line with this principle, Feyerabend described his theory of scientific method as "epistemological anarchism." Understandably, Feyerabend's view was widely taken to be an extreme form of relativism on which there are no methodological principles which apply to scientific inquiry. But, as later emerged, this was not what he intended (e.g. 1978, 188). The slogan "anything goes" was intended as a rhetorical response to a theorist of method who insists that there must be a single, universally binding method that applies throughout the sciences in all stages of the history of science. Given the violability of the rules of method, the only principle that may be supported for all circumstances in the history of science is "anything goes." But Feyerabend's point was not the extreme relativist claim that there are no methodological principles at all. It was an exaggerated way of expressing the point that all rules of scientific method may be justifiably violated in some circumstances.

4. Conclusion

More than half a century has passed since the original publication of *Structure* and more than four decades since that of *Against Method*. The work of Kuhn and Feyerabend prompted lively discussion of the nature of scientific theory-change and the rationality of scientific theory-choice. Alternative models of scientific theory-change were proposed. Theories of scientific rationality were developed which take account of variation of methodological standards. An extensive literature explored the implications of the theory of reference with respect to semantic incommensurability and conceptual change.

Debate about wholesale scientific change has faded into the past. Contemporary philosophers of science tend to focus on detailed analysis of specific areas of the sciences rather than to make broad generalizations about the nature of theory-change intended to apply to all sciences throughout the history of science. But while the problem of scientific theory-change is a problem from an earlier phase in the history of the philosophy of science, some of the ideas of Kuhn and Feyerabend have become commonplace. Few would now maintain a strong thesis of incommensurability. But the idea that conceptual and semantic change occurs in the development of the sciences is widely accepted. Some may still hanker for a general theory of scientific method. But the thought that there may be variation of at least some of the methodological rules or procedures employed in the practice of science no longer appears overly controversial.

The relativistic challenge to the rationality of science spawned several divergent tendencies. In the philosophy of science, one tendency was to develop a theory of scientific method on the basis of which to resist the idea that rational theory acceptance is relative to shifting standards. An important example of this is the development by Larry Laudan of the theory of normative naturalism. According to normative naturalism, rules of scientific method may be empirically evaluated by historical investigation of the past use of such rules to achieve scientific aims (e.g. Laudan 1987). But not all who were influenced by Kuhn and Feyerabend sought to defeat relativism. The advocates of the strong programme in the sociology of science explicitly endorsed relativism. Indeed, it has recently been suggested by Bojana Mladenović that Kuhn dramatically changed his epistemological views late in his career because he recognized that his original position was open to relativistic appropriation by the sociology of science which he opposed (2017, 62–70).

The figures of Kuhn and Feyerabend continue to exert an influence on the philosophy of science. The dramatic impact of their most controversial claims has lessened with time. Their moderate claims have been incorporated into the mainstream of the philosophy of science while extravagant claims have dissolved.

References

Davidson, D. (1984), "On the Very Idea of a Conceptual Scheme," in his *Inquiries into Truth and Interpretation*, Oxford: Oxford University Press, 183–198.

Devitt, M. (1979), "Against Incommensurability," *Australasian Journal of Philosophy* 57: 29–50.

Devlin, W. J. (2015), "An Analysis of Truth in Kuhn's Philosophical Enterprise," in *Kuhn's Structure of Scientific Revolutions – 50 Years on*, edited by W. J. Devlin and A. Bokulich, Switzerland: Springer, 153–166.

Feyerabend, P. K. (1975), *Against Method*, London: New Left Books.

———. (1978), *Science in a Free Society*, London: New Left Books.

———. (1981a), "An Attempt at a Realistic Interpretation of Experience," in Feyerabend (1981), 17–36.

———. (1981b), "Explanation, Reduction and Empiricism," in Feyerabend (1981), 44–96.

———. (1981), *Realism, Rationalism and Scientific Method: Philosophical Papers*, vol. I, Cambridge: Cambridge University Press.

———. (1987), *Farewell to Reason*, London: Verso.

Fine, A. (1975), "How to Compare Theories: Reference and Change," *Nous* 9: 17–32.

Hoyningen-Huene, P. (1993), *Reconstructing Scientific Revolutions: Thomas S. Kuhn's Philosophy of Science*, Chicago: University of Chicago Press.

Kripke, S. (1980), *Naming and Necessity*, Oxford: Blackwell.

Kuhn, T. S. (2012), *The Structure of Scientific Revolutions*, fourth edition, Chicago: University of Chicago Press, 1962.

———. (1970), "Reflections on my Critics," in *Criticism and the Growth of Knowledge*, edited by I. Lakatos and A. E. Musgrave, Cambridge: Cambridge University Press, 231–278.

———. (1977), "Objectivity, Value Judgment and Theory Choice," in his *The Essential Tension*, Chicago: University of Chicago Press, 320–339.

———. (2000), "Commensurability, Comparability, Communicability," in his *The Road Since Structure*, edited by J. Conant and J. Haugeland, Chicago: University of Chicago Press, 33–57.

Lakatos, I. (1970), "Falsification and the Methodology of Scientific Research Programmes," in *Criticism and the Growth of Knowledge*, edited by I. Lakatos and A. E. Musgrave, Cambridge: Cambridge University Press, 91–196.

Laudan, L. (1977), *Progress and Its Problems*, London: Routledge and Kegan Paul.

———. (1987), "Progress or Rationality? Prospects for Normative Naturalism," *American Philosophical Quarterly* 24: 19–31.

Mladenović, B. (2017), *Kuhn's Legacy: Epistemology, Metaphilosophy, and Pragmatism*, New York: Columbia University Press.

Nola, R. (1980), "Fixing the Reference of Theoretical Terms," *Philosophy of Science* 47: 505–531.

Putnam, H. (1975), "Explanation and Reference," in his *Mind, Language and Reality: Philosophical Papers*, vol. 2, Cambridge: Cambridge University Press, 196–214.

———. (1981), *Reason, Truth and History*, Cambridge: Cambridge University Press.

Sankey, H. (1990), "In Defence of Untranslatability," *Australasian Journal of Philosophy* 68: 1–21.

———. (1991), "Incommensurability, Translation and Understanding," *Philosophical Quarterly* 41: 414–426.

Scheffler, I. (1967), *Science and Subjectivity*, Indianapolis: Bobbs-Merrill.

Siegel, H. (1987), *Relativism Refuted*, Dordrecht: Reidel.

41

RELATIVISM AND ANTINOMIANISM

David Bloor

> The conditions which logic sets for rational thinking . . . are not to be understood
> as possessing some absolute metaphysical validity or as resting on the will of God.
>
> *(Rudolf Carnap 1937, 117)*

1. Introduction

Carnap was telling his audience that logic is not immutable. There can be no unique logical system, no unchanging definition of logical validity, and no fixed delimitation of logical terms. Carnap made this point by contrasting logical constraints with the will of God – traditionally understood as immutable. His contrast involved a resort to metaphor, but the metaphor works because it calls upon salient strands in a shared culture. Cultures provide exemplars which guide the thinking of their members. Thus Carnap's audience may have agreed or disagreed with his rejection of the "absolute metaphysical validity" of logic but they knew what he meant because he anchored the language of absolutism in a familiar, theological context.

I want to exploit those same cultural resources in order to understand the current debate about relativism. Relativism and anti-relativism both need to be understood in the context of traditional, theological disputes. I shall argue that an awareness of these precedents will expose the strengths of relativism and the weakness of the current, anti-relativist case. Before developing the historical argument, however, it is appropriate to look at one well-known, modern definition of relativism.

2. A puzzling definition

The American philosopher Paul Boghossian shares with many of his colleagues the conviction that relativism is a danger to science and culture. Allegedly, it undermines the respect for objective evidence. As Boghossian understands them, relativists say that science is merely one way of viewing the world and stands on a par with more traditional, myth-based cosmologies. This stance opens the door to all the worst, irrational tendencies in our culture. Relativists, says Boghossian, challenge "the principles by which society ought to be organized" (2007, 5).

Boghossian proposes a two-part definition of relativism. In the first part he says that relativists reject any claim to possess absolute knowledge. In the second part he declares that relativists subscribe to what he calls the "Equal Validity Thesis" (2007, 84–85). The Equal Validity Thesis can be illustrated by the claim, mentioned previously, that mythical accounts of the origin of human beings are just as valid as an account based on Darwinian evolution. Generalised to cover all knowledge, a commitment to this thesis expresses an unqualified irrationalism. Let "A" stand for absolutism, and "EVT" stand for the Equal Validity Thesis, then:

Relativism★ = (not-A) & (EVT).

The asterisk is to remind us that, as yet, it is an open question whether the definition is a good one. But, good or bad, there is nothing idiosyncratic about it, though others use different terminology. Instead of speaking of equal validity some say informally that, for the relativist, "anything goes," while others, more formally, refer to this as "antinomianism" (Cohen 1953, 434).

Notice that the rejection of absolutism is presented as a necessary, but not a sufficient, condition for the attribution of relativism. This immediately creates a puzzle. It would have been simpler to discard the second part, focus on essentials, and define relativism as the rejection of absolutism. The rejection of absolutism would then be both a necessary and a sufficient condition for relativism. Why does Boghossian complicate matters by adding on the Equal Validity Thesis?

I shall now look at the antecedents of this definition. In some respects they bear a remarkable similarity to the modern formulation although, in others, there are important differences. Both the similarities and the differences shed light on the puzzling character of Boghossian's definition.

3. Against enthusiasm

John Locke, the philosopher of the Glorious Revolution of 1688, was well aware of the capacity of religion to fan the flames of civil war and the discontent of the lower orders (Macpherson 1962). It was for this reason that he added a denunciation of "enthusiasm" to his *Essay Concerning Human Understanding* (Locke 1689, Bk. IV, chap. ixx). His targets were the radical, protestant sectaries. Locke saw a threat to society in their unbridled individualism and antinomian subjectivity. The sectaries believed they were in direct contact with the Deity. They recognised no church or legal institutions as mediators of this relationship and rejected all criteria external to the individual mind that could be used to evaluate the claims of personal revelation. Thus, warned Locke:

> Every conceit that thoroughly warms our fancies must pass for an inspiration, if there be nothing but the strength of our persuasions, whereby to judge of our persuasions. If reason must not examine their truth by something extrinsical to the persuasions themselves, inspiration and delusion, truth and falsehood, will have the same measure, and will not be possible to be distinguished.
>
> *(1689, Bk. IV, chap. ixx)*

The tone of Locke's denunciation resonates with Boghossian's polemic. Both identify a threat from groups who have, allegedly, discarded the constraints that sustain an orderly culture. Locke's passage even invoked the Equal Validity Thesis. For the enthusiasts, truths and falsehoods are said to have "the same measure" and are accorded equal validity. Locke called for "extrinsical" constraints on the subjectivity of the enthusiasts and declared that these constraints were furnished by the exercise of "reason." But there were shortcomings in Locke's polemic. He too believed in revelation and all he could say was that his revelations were more reasonable than their revelations. Today's anti-relativists have a rhetorically easier task. They can point to the well-entrenched institution of professional science to furnish the desired "extrinsical" criteria. But, whatever Locke's rhetorical shortcomings, it is plausible to say: for "enthusiast" read "relativist."

Obviously, we cannot exactly equate enthusiasts and relativists because the targets of Locke's criticism believed in God: they were absolutists. It is the common identification of an antinomian threat that carries the real analogy between the historical and modern cases. This may explain the role of the Equal Validity Thesis in Boghossian's definition. Boghossian declares that the very principles of social order are threatened; but the announced threat does not sound very acute if it resides in nothing but a metaphysical error. Something more substantial is needed if public opinion is to be aroused. The charge of traducing science, the central institution of our society, the modern equivalent of religion, makes a more convincing headline. This may explain why the attractive simplicity of the first part of the definition was willingly compromised by the addition of the clumsy second part.

Locke's polemic also reminds us that an absolutist position can be given expression in a number of different social forms. It is an historical platitude that there is more than one form of Christianity, and more than one form of religious community. Some of these communities appear to embrace antinomianism and some of them do not. But, if there can be different forms of absolutism, surely the same is true for relativism. Should we not take it as equally platitudinous that there is more than one form of relativism – some with and some without a commitment to the Equal Validity Thesis? But if that is true then Boghossian's definition of relativism is wrong. He has not defined relativism as such: he has merely defined a special case of relativism.

4. Pragmatism *versus* nihilism

I now want to look at another argument put forward by Boghossian. He calls it a "semantic" argument. It has a particular bearing on moral relativism although it also applies to cognitive relativism. The idea is that relativists must be beset by some form of indifference or incapacity when they try to make judgements or to act. Boghossian describes the situation by referring to the "nihilism" of any thorough-going relativism. Relativists, he says, have eliminated all genuine moral discourse. Allegedly, for the relativist, all that can be asserted is that some behaviour, call it B, is condemned from the standpoint of a certain moral code, call it M_1. Such an assertion, says Boghossian, is a mere expression of a sociological fact. In itself it carries no moral force. Suppose that B is to be praised, rather than condemned, according to another moral code M_2. That is just another sociological fact. There is no logical inconsistency between the two factual claims and no moral confrontation is signalised.

Suppose that a self-confessed relativist tried to meet this claim by insisting that, on the contrary, he or she was fully capable of voicing moral disapproval of, say, slavery. Boghossian would reply that, to him, this stance does not "look like relativism" (Boghossian 2011, 63). Moral advocacy, he says, doesn't "amount to relativism of a recognisable and intuitive kind." It looks and sounds wrong to Boghossian because, in the imagined confrontation, the relativist is adopting an "intolerant, objectivist view of the subject matter at issue." All such moral assertiveness is, "anti-relativist/objectivist in spirit" (Boghossian 2011, 63–64). Boghossian's intuitions are all built into the second clause of his definition of relativism. He therefore insists that a consistent relativist would have to make the following admission.

> I know that my claim is true relative to my standards. I acknowledge that your standards are just as correct as mine.... Hence each of us counts as having spoken correctly because of having said something that is true relative to standards that are no more correct than one another.
>
> *(2011, 66)*

Boghossian is right to distinguish between the propositions of morality and those of the sociology of morality. Nevertheless there is something badly wrong with the words he is putting into the relativist's mouth. This is because there is an entire dimension missing from Boghossian's analysis. His "semantics" needs to be supplemented by a "pragmatics." A pragmatic analysis would draw into the analysis the context of the utterance and include the factors determining its credibility, the interests that are operative, the differentiated roles adopted by the speaker, and the division of cognitive labour. Making allowance for these factors the relativist in the quotation should be saying: I acknowledge that my standards are no more *absolute* than yours, rather than saying they are no more correct than yours. There is a need to distinguish between the practice and application of a moral code, that is, local attributions of correctness, and the ultimate status and justification of the practice. The simple exercise of a moral judgement on a particular occasion should not be confused with claims about, or concern with, its ultimate ground or origin. The charge of nihilism is based on precisely this confusion.

It is usual to exercise routines of evaluation with little regard for their ultimate status. Conversely, there can be shifts of opinion on the reflective level which leave no trace in day-to-day practice. When a person is confronted by the challenge that, if they had been born in another age, their values would have been different, the typical and correct response is: *So what?* The absence of absolute grounds of correctness does not mean that judgements cannot be made authentically, and sustained pragmatically, as locally accepted conventions. And to say that conventions are "local" does not mean that the subject matter to which they are applied has to be local. We all confidently apply our moral intuitions to cases that are remote in space and time, for instance, we condemn the slavery of past times and distant places. David Hume, the Edinburgh historian and sociologist, had a deep appreciation of the pragmatic dimension of knowledge. He made the point that needs to be made when he said: "Nature, by an absolute and uncontrollable necessity has determin'd us to judge as well as to breathe and feel" (Hume, 1739, Bk. I, Pt. IV, sect. i). 'Absolute' here refers to the overwhelming force of a law of nature.

5. Images in the soul

Hume, by any reasonable definition, was a relativist. He did not use the label but he pioneered the practice. Metaphysical absolutes played no role in Hume's theory of knowledge. Knowledge comes from finite experience and the psychological mechanisms of association and constructive imagination. It is not sustained by divinely gifted reason or revelation. There is no non-circular justification for any of our reasoning. The answer to the problem of finding a justification for inductive inference is the realisation that there is no justification. It is a natural propensity. The basis of both cognitive and social order, he argued, resides in the unaided resources of human nature and nothing else.

In Book III of his *Treatise of Human Nature* Hume demonstrated how social interaction leads from the "natural" to the "artificial" virtues – that is, to the virtues that do not spring from innate psychological tendencies but which are the artefacts of social interaction. He studied the origin of conventions and made brilliant contributions to the understanding of collective action, the nature of public goods, the free-rider problem, and what game-theorists today call "salient solutions" to problems of co-ordination. In the course of this argument he specifically illustrated the social construction of the respect for the property of others, respect for the magistrate, and the obligations to honour contracts and keep promises.

Thomas Reid of Glasgow, Hume's critic and contemporary, was the equivalent of today's anti-relativist philosophers. He unhesitatingly imputed the Equal Validity Thesis to Hume, complaining that he "leaves no ground to believe any one thing rather than its contrary" (Reid 1863, vol. 1, 95). Furthermore, insisted Reid:

> The obligation of *Contracts* and *Promises* is a matter so sacred, and of such consequence to human society, that speculations which have a tendency to weaken that obligation, and to perplex men's notions on a subject so plain and so important, ought to meet with the disapprobation of all honest men.
>
> *(1863, vol. 2, 663)*

In place of Hume's historical and sociological analysis of moral phenomena Reid substituted the assumption that humans have a direct moral intuition that furnishes them with all they need to know. The notion of right and wrong, said Reid, "is discerned, not by any train of reasoning, but by an immediate perception" (1863, vol. 2, 666). How does this immediate perception work? Reid's answer was theological. God has an infinite power to discern truth and perceive the distinction between right and wrong. Individual human beings are created in his image with a faint copy of that power embodied in their intuitive moral faculties: It is the prerogative of man that he has, "the image of God in his soul" (1863, vol. 2, 652).

Hume's political links were numerous and fascinating (Forbes 1975). But, in his scholarly work, he was laying the foundation for a scientific sociology of knowledge.[1] Hume was seeking an empirical understanding of human culture, hence the bold subtitle to the *Treatise* which was: "an attempt to introduce the experimental method of reasoning into moral subjects." Reid had a shrewd mind, but his intellectual endeavours were not driven by a corresponding empirical curiosity or a concern with causal explanation. Where Hume was laying the groundwork for the self-understanding of a secular society, Reid was still articulating the ideology of a society dominated by pre-scientific forms of self-awareness. While Hume was showing how individuals generated social order through competent interaction, Reid was invoking an image of God in the soul. It would never have occurred to Reid, as it certainly occurred to Hume, that this image was no more than the image of a society whose workings were not understood.[2]

6. Science versus sophistry

To appreciate the explosive quality of this culture clash – that is, the clash between Hume's nascent social science and Reid's conservative metaphysics – it is worth recalling the threat with which Hume ended his *Enquiry Concerning Human Understanding* (Hume 1748). Referring to the empirical principles informing his own approach, Hume said:

> When we run over libraries, persuaded of these principles, what havoc must we make?
> If we take in our hand any volume; of divinity or school metaphysics, for instance; let

us ask, *Does it contain any abstract reasoning concerning quantity or number?* No. *Does it contain any experimental reasoning concerning matter of fact and existence?* No. Commit it then to the flames: for it can contain nothing but sophistry and illusion.

(Hume 1748, XII, iii)

The Hume-Reid debate was not just a conflict between two philosophers but, rather, a conflict between science and philosophy.

It is important to appreciate that the cultural fault-line was between an empirical, curiosity-driven approach to knowledge and morality, and a non-empirical approach where the over-riding concern was justificatory. This same tension is central to the present-day conflict over relativism. Seeing this fact prefigured in the Hume-Reid debate can help us to appreciate the full scope of what is at stake today. Present-day anti-relativist philosophers, like their forefathers, present the debate over relativism in a false light. Reid presented Hume as attacking morality. He wasn't: he was attacking Reid's philosophical account of morality. Today, relativist historians and sociologists of science are represented by their critics as attacking science. They aren't: they are attacking the accounts of science given by philosophers of science. The relativists are saying that knowledge and morality are to be understood, as Hume understood them, as natural phenomena.

7. The mystery of incarnation

Boghossian identifies himself as an absolutist, but he is well aware of the profound problems of making absolutism intelligible. How do we know when we are confronted with an absolute truth in science or morality? How can an absolute truth be absolute and yet lodged in the human brain and made incarnate in human institutions? Don't the two conditions of existence exclude one another? Reid could only offer obscure talk of an image of God in the soul. Shockingly, Boghossian admits he has no answers at all, of any kind, to offer to the epistemological and ontological problems he so clearly acknowledges. He merely asserts that he is "very confident that there are solutions to these problems."[3] The historical evidence suggests that his confidence is misplaced. The history of theology can once again offer guidance. The problem of how to understand the incarnation of absolute truth in human cognition is a version of the traditional problem of incarnation in Christian theology.

How is the divinity of God related to human existence? How does the divine presence manifest itself in human history? The link is made through the figure of Christ. God becomes incarnate in Christ. How is that incarnation to be understood? Is Christ a real human being or is Christ really identical with God? Theologians insist that both requirements must be met but realise that neither option alone will do the required job. The Council of Chalcedon in AD 451 met the problem by decreeing that Christ is both one and the same thing as God *and* a real human being, with no modification or diminution to either status. Such things surpass human understanding and must be accepted on faith (Camelot 2003). But this solution to the problem of incarnation is no solution at all. It is a dogmatic declaration having the opaque propositional form of "P & not-P." Despite two thousand years of effort, Christian theologians have failed to discover anything more illuminating to say on this question. Perhaps they are happy to keep it as an inexplicable mystery. It is unlikely that Professor Boghossian will do any better when he eventually tries to explain his particular, epistemological version of the problem of incarnation.

The situation is extraordinary. Boghossian criticises relativists for their alleged rejection of scientific standards of evidence. When rhetorically convenient he invokes the authority of science and poses as the friend and defender of science. Yet it transpires that this spokesman of science is basing his defence of science on impenetrable, quasi-theological premises. Scientific naturalism is subverted by the assumption that science rests on a supernatural foundation. He berates relativists for deserting rationality and then offers implicit contradictions. All the developments in science that have led to the increasing distance between religious modes of thought and professional scientific practice are, in the end, treated as nothing. It is as if the professionalisation of science over the past 150 years had never taken place.[4]

8. Is there a third way?

Perhaps there is a simple way for other anti-relativist philosophers to avoid the intractable problems that beset Boghossian's absolutist argument. Why not reject both absolutism and relativism? Surely, these are "extreme" positions, so there must be a middle ground. Why not construct a third way between absolutism and relativism?

To demonstrate that there is a third way calls for a clear example that meets the required specifications. Simply switching terminology to avoid mention of the words "absolute" or "relative" is not enough. For example, advocates of the third way might stress the "objective" nature of knowledge, but they also need to prove that their "objectivity" is not a disguised or unacknowledged expression of relativism or absolutism. Superficially plausible candidates turn out to be exercises in evasion, as when Popper's declared "evolutionary" approach to objective knowledge turns out to be nothing but a fig-leaf for absolutism (Popper 1972). It may seem that Boghossian's proposed definition of relativism yields easy access to the third way. On this definition, surely, all that is required is (i) the rejection of absolutism and (ii) the rejection of the Equal Validity Thesis. But this is not the third way that is being sought. Hume's work falls into this category and exposes the error. All that is on offer under this rubric is relativism – in disguise.

To show the pragmatic impossibility of a middle ground I shall make one final appeal to the theological tradition. Previously I have likened anti-relativists to religious believers, now I shall liken relativists to religious unbelievers. I shall argue that relativism is analogous to atheism. In order to trace the consequences of this comparison I shall start by asking: How is atheism to be defined? The standard definition is that an atheist denies the existence of God. This can be written:

Atheism = not-God.

The rejection of God's existence is a necessary and sufficient condition for the attribution of atheism. On the basis of the analogy between atheism and relativism these facts point to the adequacy of the simple, intuitive definition of relativism. Thus:

Relativism = not-Absolutism.

On this definition, relativism and absolutism constitute a dichotomy. All forms of absolutism demand belief in the possession of at least one instance of absolute knowledge. All forms of relativism demand belief in the impossibility of possessing even a single instance of absolute knowledge. These two positions are mutually exclusive and jointly exhaustive. It follows that one cannot reject both. One must choose one or the other, just as a reflective person has to decide whether they think that God exists or that God does not exist. There is no third way.

9. Rival agendas

If absolutism and relativism form a dichotomy then what is the status of the dichotomy itself? Is it absolute or relative? Obviously, a relativist must treat the dichotomy as relative. Boghossian might declare that invoking dichotomies does not sound like relativism, and many anti-relativists would share that assumption, but they are wrong. A dichotomy is an ordering device, collectively constructed and pragmatically maintained by its users on the basis of its perceived utility. A dichotomy is relative to the interests of its users, for example the interests that sustain science as a central institution of society distinct from the church.

Pragmatically, definitions represent the attempt to set an agenda. Thus the standard definition of atheism poses the question of religious belief in terms of a dichotomy. It carries the message that one thing, and only one thing, is in question: the truth or falsity of a certain belief. The behavioural consequences of the belief are set aside. Philosophers such as Reid want to frame the issue differently. They believe atheism is a threat to society so they want to keep the alleged consequences of religious unbelief on the agenda. They want to add an antinomian clause to the definition. However, history tells us that the alleged consequences are not intrinsic to the phenomenon of unbelief. An ordered, secular society is possible because it is a reality. The analogy between the case of atheism and the case of relativism is exact. A Boghossian-style definition of relativism, just like a Reid-style definition of atheism, is no more than an ungrounded attempt to shift the debate from what is essential and constant to what is contingent and variable.

If definitions are understood in this pragmatic way is the dichotomy really compelling? Why cannot an advocate of the third way simply decide to ignore the dichotomy on the grounds that it is "merely" relative? Such a posture is fantasy and presupposes an unrealistic freedom of action – indeed, a freedom of action, and a degree of individualism, that would ultimately spell the death of shared knowledge. The dichotomy between absolutism and relativism is too salient to ignore and too entrenched to discard. As Carnap said, logical constraints on rational thought are real, even though they are not absolute.

10. Conclusion

The current debate over relativism is a re-run of the traditional debate over the truth of religion. Relativists are the unbelievers; anti-relativists are the believers. The belief in absolute truth is a belief in God, and anti-relativism is a belief in God in disguise. All of the old theological themes and tropes re-surface in one form or another and the big choices remain the same. You must decide whether you are an absolutist or a relativist: there is no third way. Undeniably, there are anti-scientific, antinomian, forms of relativism. Some of these are a disgrace, but philosophers exaggerate their significance. Such tabloid-press relativists are cherished by those eager to benefit from a convenient moral panic. They are prized by philosophers in search of employment, who can then pose (however implausibly) as defenders of science and guardians of culture. The real enemies of rationality are absolutists who leave a trail of infantilising obscurity in their wake. A true concern for the principles by which society ought to be organised demands the rejection of this sophistry.[5]

The interesting and important forms of relativism are expressions of scientific naturalism and give rise to empirical investigations in fields such as the sociology of knowledge, the history of science, anthropology and experimental psychology.[6] This is the true Humean tradition in which conclusions are based on the experimental method, that is, the empirical study of causes. For scientific relativists, as for Hume, the relativities that they investigate are causalities. Causal relations are the subject matter of empirical science and scholarship, not the non-empirical discipline of philosophy. Anti-relativist philosophers have been dismissive of the relevant empirical work. They have correctly identified studies in, say, the history and sociology of science, as relativist, but they have then mischaracterised what that relativism consists in. Boghossian's bad definition of relativism is a representative example. There is nothing sophisticated about the level at which the anti-relativist case falters. The errors are low level. They are the errors of evasion, misrepresentation and over-generalisation – the predictable errors of the polemicist and ideologue.[7]

Notes

1 For an overview of Hume's work as a contribution to political science, but with no mention of the sociology of knowledge, see Whelan (1982). Whelan says Hume was not a relativist but Whelan is using a Boghossian-style definition, e.g. (1982, 217).
2 Hume said the failure to understand the social processes underlying promising leads us to *"feign* a new act of the mind, which we call the *willing* an obligation" (Hume 1739, III, II, v). This idea is generalised in Durkheim (1961).
3 Boghossian (2017, 310). The reference to nihilism is on (2017, 308).
4 For a historical analysis, see Turner (1978).
5 The civilising necessity of scepticism (i.e. relativism) was the message of Hume's *History of England* (Phillipson 1989).
6 For an introduction to the literature, see Shapin (1982), Barnes and Bloor (1982), and Bloor (2011).
7 Sociologists of knowledge are equally curious about the credibility of true and false beliefs. On p. 3 of his 2007 book, Boghossian imputes to them an acceptance of the Equal Validity Thesis. He confuses equal curiosity with equal validity.

References

Barnes, B and D. Bloor (1982), "Relativism, Rationalism and the Sociology of Knowledge," in *Rationality and Relativism*, edited by M. Hollis and S. Lukes, Oxford: Blackwell, 21–47.
Bloor, D. (2011), "Relativism and the Sociology of Scientific Knowledge," in *A Companion to Relativism*, edited by S. D. Hales, Oxford: Wiley-Blackwell, 433–455.
Boghossian, P. (2007), *Fear of Knowledge: Against Relativism and Constructivism*, Oxford: Clarendon Press.
———. (2011), "Three Kinds of Relativism," in *A Companion to Relativism*, edited by S.D Hales, Oxford: Wiley-Blackwell, 53–69.
———. (2017), "Relativism About Morality," in *Realism – Relativism – Constructivism*, edited by C. Kanzian, S. Kletzl, J. Mitterer and K. Neges, Berlin: De Gruyter, 301–312.
Camelot, P.T. (2003), "Council of Chalcedon," in *The New Catholic Encyclopedia*, vol. III, Detroit: Thomson Gale, 363–366.
Carnap, R. (1937), "Logic," in *Factors Determining Human Behavior*, Harvard Tercentenary Publications, Cambridge, MA: Harvard University Press, 107–118.
Cohen, M. R. (1953), *Reason and Nature: The Meaning of Scientific Method*, New York: Free Press of Glencoe.
Durkheim, E. (1961), *The Elementary Forms of the Religious Life*, translated by J. W. Swain, New York: Collier.
Forbes, D. (1975), *Hume's Philosophical Politics*, Cambridge: Cambridge University Press.
Hume, D. (1739), *A Treatise of Human Nature*, edited by L. A. Selby-Bigge, Oxford: Clarendon Press, 1960.
———. (1748), *Enquiry Concerning Human Understanding*, edited by L. A. Selby-Bigge, Oxford: Clarendon Press, 1902.
Locke, J. (1689), *An Essay Concerning Human Understanding*, edited by A. S. Pringle-Pattison, Oxford: Clarendon Press, 1926.
Macpherson, C. B. (1962), *The Political Theory of Possessive Individualism: Hobbes to Locke*, Oxford: Clarendon Press.
Phillipson, N. (1989), *Hume*, London: Weidenfeld and Nicolson.
Popper, K. (1972), *Objective Knowledge: An Evolutionary Approach*, Oxford: Clarendon Press.

Reid, T. (1863), *The Works of Thomas Reid*, edited by Sir William Hamilton, Edinburgh: MacLachlan and Stewart.

Shapin, S. (1982), "History of Science and its Sociological Reconstructions," *History of Science* 20: 157–211.

Turner, F. (1978), "The Victorian Conflict Between Science and Religion: A Professional Dimension," *Isis* 69: 356–376.

Whelan, F. G. (1982), *Order and Artifice in Hume's Political Philosophy*, Princeton: Princeton University Press.

42

RELATIVISM, PERSPECTIVISM AND PLURALISM

Hasok Chang

1. Introduction

Perspectivism and pluralism are currently prominent, along with relativism, among philosophical positions that call for a more relaxed view concerning the nature of knowledge. "Relaxed" is not a philosophical term, of course, but the very vagueness of that word is intended to indicate the imprecision and ambiguity with which these philosophical positions have been understood, which has also added to the uncertainty and the complexity of the relationships between them. In this chapter I explore and clarify the relations between relativism and the other relaxed views on the nature of knowledge. It is not my aim here to attempt new definitions of these positions. However, I hope that the comparisons offered in this chapter will help us sharpen our understanding of each position.

Among analytic philosophers and various advocates of scientific realism, relativism tends to be a term of abuse. It is widely seen, especially by those who have not considered the issue carefully, as an irresponsible position that renounces all judgement and regards what anyone at all says or believes as equally valid or acceptable. I was myself guilty of accepting something close to this crude characterisation of relativism, in trying to defend my advocacy of pluralism against the charge that it was "just relativism" (Chang 2012, 261).

It is helpful to start by taking relativism literally as the opposite of absolutism. Relativists deny that any standards or statements are valid absolutely, i.e., without regard to circumstances or conditions. Other relaxed epistemological positions share this anti-absolutism with relativism, but I think relativism is the only one that defines itself primarily by this opposition to absolutism. Richard J. Bernstein (1983, 8–13) took objectivism rather than absolutism as the chief doctrine in opposition to relativism, because he thought that absolutism was no longer such a live and pertinent option. But a kind of absolutism that incorporates fallibilism is still a widespread view: there *is* absolute truth, which *would be* attainable by employing absolutely correct methods, though we human inquirers are incapable of knowing or applying such methods, so our actual knowledge will always harbor uncertainties.

Let us now consider relativism in more specific terms: what are its chief characteristics? Here I refer to what Martin Kusch (2016, 106–107) has proposed as the "standard model of (epistemic) relativism," along with an updated version that he wrote a few years later (Kusch 2019).

He does not intend the "standard model" in an absolute way, but merely as a list that "seeks to capture some currently popular characterization of variants of the position."

(1) (DEPENDENCE) A belief is justified "only relative to an epistemic system or practice (=SP)."
(2) (PLURALITY) "There are, have been, or could be, more than one such [SP]."
(3) (EXCLUSIVENESS) "SPs are exclusive of one another."
(4) (NOTIONAL CONFRONTATION) This feature is rendered as "Conversion" (*Bekehrung*) in the 2019 version: a change from one SP to another cannot be forced by rational arguments (as it could be in a "real" confrontation). Note that in the 2019 version, instead of SP Kusch speaks of "a system of epistemic principles or a bundle of precedents."
(5) (SYMMETRY) "[SPs] must not be ranked."

There are some other items on Kusch's 2016 list (contingency, groundlessness, underdetermination and tolerance), but they are already deemed "not essential" there, and dropped in the 2019 list. Meanwhile the 2019 list adds "semantic relativity" and "metaphysical implications," which I think are consequences of relativism as defined by the other features. It also adds "Non-absolutism," but I would take that as an overall summary of relativism, or even a synonym for it, rather than a specific characteristic of it.

2. Perspectivism and relativism

With this characterisation of relativism in mind, I now want to move on to the task of drawing comparisons with perspectivism and relativism. Ronald Giere, in his well-known book on perspectivism, declares that "a *scientific* perspectivism does not degenerate into a silly relativism" (2006, 13; emphasis original). But how does perspectivism relate to relativism that is *not* silly? Giere has done more than anyone else to bring perspectivism to the attention of philosophers of science, but he does not lay down a definitive account of it in his definitive book on the subject. What he gives us most of all is a set of analogies (2006, 13–14): "the idea of viewing objects or scenes from different places" provides an "auspicious point of departure," and his "prototype for a scientific perspectivism will be *color vision*": "colors are real enough, but … their reality is perspectival." And then, "the considerations suggesting that color vision is perspectival can be extended to human perception more generally," with a further "extension to scientific observation" involving instruments. The most "controversial" and interesting extension is to scientific theorising: according to Giere, "the grand principles objectivists cite as universal laws of nature are better understood as defining highly generalized models that characterize a theoretical perspective." Throughout this exposition, the meaning of "perspective" remains deeply metaphorical.

Michela Massimi has done a great deal recently to develop and apply perspectivism in the philosophy of science, consciously building on Giere's work. Her take on perspectivism is perhaps less metaphorical than Giere's (Massimi 2018, 164; emphases original): "a family of positions that in different ways place emphasis on our *scientific knowledge being situated*." According to her, there are two main ways of being situated:

(1) Our scientific knowledge is *historically situated*, that is, it is the inevitable product of the historical period to which those scientific representations, modelling practices, data gathering, and scientific theories belong. And/Or (2) Our scientific knowledge

is *culturally situated*, that is, it is the inevitable product of the prevailing cultural tradition in which those scientific representations, modelling practices, data gathering, and scientific theories were formulated.

In addition to Massimi's distinction of different ways of being situated, it may be useful to distinguish three separate layers of perspectivality, which I laid out in a recent work comparing perspectivism and pragmatism (Chang 2019). (i) The same content can be expressed in different ways, in different languages, or using different expressions, that are *not* incommensurable with each other. The different expressions will typically have different connotations embodying divergent expectations and facilitating divergent courses of action. For example, the Newtonian, Lagrangian and Hamiltonian formulations of classical mechanics are formally equivalent to each other, but with significantly differing affordances in problem-solving and further theorising. (ii) Different perspectives can highlight different aspects of a given object, and also blind us to other aspects. This sense of perspectivism is consonant with quite a literal reading of "perspective." If we look at a three-dimensional object in the normal way, we will only see a two-dimensional picture whose content depends on the direction of gaze. Following Giere, we can generalise and extend this thought in a metaphorical way. (iii) Going deeper, one can argue that the relation between our knowledge and the world cannot be spelled out in an objectivist way. Any phenomena that we can discuss are already expressed in terms of concepts, and we choose from different conceptual frameworks that are liable to be incommensurable with each other. Only unspeakable Kantian *noumena* are guaranteed to be shared between incommensurable frameworks. Each perspective offers knowledge that answers to reality in its own way, but not in a way that is straightforwardly comparable to other ways.

Having given a brief characterisation of perspectivism and some pertinent distinctions within it, let us consider how exactly perspectivism and relativism relate to each other in their various aspects and strands, by taking in turn the key aspects of relativism laid out in Kusch's standard model of epistemic relativism.

(1) Perspectivism would seem to exhibit clearly the first feature of the Kusch model of relativism, namely dependence on an epistemic system or practice. Dependence is about justification, and perspectivism as characterised previously may not concern itself with justification, but Massimi (2018, 171) does state explicitly: "Truth-conditions for scientific knowledge claims vary in interesting ways depending on the context in which they are uttered and used."

(2) Plurality is implied quite strongly in the central perspectivist metaphor: if there is an object of a finite size in physical space, there are multiple vantage points from which to view it. However, in the more literal aspects of Giere's description of perspectivism, there seems to be no requirement of plurality of perspectives. Massimi's characterisation of perspectivism in terms of situatedness makes it obvious that plurality is present but only as a matter of contingent fact: historical situatedness is plural because science has in fact existed long enough for there to have been various stages in its history, and cultural situatedness is plural because science has in fact developed in a variety of cultural settings. In this regard perspectivism is very similar to relativism as Kusch defines it, as I will elaborate further in the next section.

(3) Exclusiveness is also implied in the perspectivist metaphor, to a degree: taking one perspective would seem to exclude a different perspective. But this is not completely obvious. And what exactly does exclusiveness mean, anyway? Kusch (2016, 107) notes that there are two possibilities: roughly speaking, logical incompatibility and incommensurability. It seems to me that different perspectives are unlikely to contradict each other directly, but they could easily be incommensurable with each other. In my schema of three layers of perspectivality, the first layer

indicates no exclusiveness between perspectives. The second layer does indicate an exclusiveness in terms of what can be noticed at a given time (we cannot look at the bright side of the moon while we're looking at the dark side), though there is no contradiction or any serious incommensurability involved. But the third layer of perspectivality strongly implies the possibility of semantic incommensurability, if not also methodological incommensurability.

(4) Notional confrontation, or Conversion, might be implied in a sense in perspectivism, but that is not certain. It may seem that changing someone's perspective would take something like conversion as in Thomas Kuhn's account of scientific revolutions, especially if we take perspectivism as relating to what I designated as the third layer of perspectivality. However, it is not clear that different perspectives are always subjects of confrontation and choice. Especially with the first two layers of perspectivality, there is no real need to choose between perspectives; one would do well to maintain all the different perspectives together, even if they cannot be actively taken all at once.

(5) Concerning Symmetry, with the first two layers of perspectivality, it may well be that it is more advantageous or more insightful to use one expression rather than another, or highlight one set of aspects of an object domain rather than another. In such cases it would make sense to rank perspectives in relation to one another. Most perspectivists would approve of such rankings, which seem to go against the relativist prohibition ("must not be ranked"). However, the rankings in question here would be a local and contingent matter, themselves within some larger-scale perspective. If that larger-scale perspective is not ranked in relation to other larger-scale perspectives, then that may not trouble the relativist so much, since the relativist can freely grant that epistemic judgements can and should be made *within* an SP. With the third layer of perspectivality, it seems that the ranking of different perspectives becomes quite impossible, in line with the second sense of Symmetry distinguished by Kusch (2016, 107): "There is no neutral way of evaluating different *SPs*." Or the situation could be more fitting for Kusch's fourth sense of Symmetry (2016, 108): "For a reflective person the question of appraisal of… *SPs* does not arise."

The relationship between relativism and perspectivism is complex and multifaceted. On the whole, the first two layers of perspectivality would not imply any particularly strong form of relativism that satisfies all or most of its tenets. But if perspectivism takes seriously the last layer of perspectivality, then it would become quite indistinguishable from a certain form of epistemic relativism.

3. Pluralism and relativism

Before entering into a comparison between relativism and pluralism, some clarificatory remarks about the meaning of pluralism are in order. There are many different things meant by pluralism even if we restrict ourselves to epistemology and the philosophy of science, without entering the realms of ethics, aesthetics or political philosophy. I will try to show how different parts of the pluralist landscape may connect and compare with relativism. There are two important distinctions within pluralism concerning science. The first distinction is between descriptive and normative pluralism (Chang 2012, 269). Descriptive pluralism notes that there is in fact a good deal of plurality in science, perhaps more than one might have thought, but makes no normative judgement about whether that plurality is good or bad. Normative pluralism holds plurality to be a good thing, however much actual plurality there is in present science. It is important to note that descriptive pluralism and normative pluralism do not imply each other, even though they often go together. In fact, a perceived lack of actual plurality can be a very strong motivation for the normative advocacy of plurality.

Jack Meiland and Michael Krausz open their collection on relativism thus (1982, 1): "Relativism begins with the observation of diversity." As implied quite strongly in that statement, there is certainly a close relationship between relativism and pluralism. However, I want to argue that there is a key difference between them, in that relativism does not in fact rule out monism. Note how carefully Kusch (2016, 107) phrases plurality, the second tenet in his standard model of relativism: "There are, have been, or could be" more than one epistemic system or practice. This formulation allows the possibility of having only one system or practice, and expresses no normative judgement that there ought to be more than one. Going back to Meiland and Krausz's statement: an "observation" of existing diversity does not imply that diversity is inevitable, nor that it should be promoted.

Relativism is not reliant on descriptive pluralism. Most epistemic relativists observe that there are different systems of practice, as a matter of contingent fact; however, relativism does not rely on that fact of plurality, and has no commitment to the existence of that plurality. (From here on I will speak in terms of "systems of practice" in a given domain, a rather general unit of analysis that I adopted in Chang (2012, 15–18); a "system of practice" is not to be confused with a "system or practice" (SP) that Kusch speaks of, though there are sufficient similarities between these notions.) With only one system in place some of the key tenets of relativism, such as items 3–5 in the preceding list from Kusch, would become empty; however, the very idea of non-absoluteness and the tenet of Dependence (item 1 in the preceding list) would still coherently remain as the core of relativism.

Are relativists (or perspectivists) committed to normative pluralism? The answer here is complex. As noted previously, Kusch's tenet of Plurality is in itself purely descriptive. At first glance the tenet of Symmetry, in forbidding the ranking of alternatives, would seem to push clearly towards normative pluralism: if there *are* multiple systems in existence and they cannot be ranked, then we should not make an arbitrary choice to keep only one of them. But if there is in fact only one system in place (as Kuhn argued concerning the "normal" state of science), then relativism has no reason to object to this *status quo*, while it is the essence of normative pluralism to advocate the creation of other systems in such a situation. Perspectivism is quite similar to relativism in this regard: if there is only one perspective around, perspectivism does not demand that more perspectives should be created. Giere (2016, 139) says, very similarly to Kusch: "at any given time, there may well be different, complementary perspectives on the same subject matter. So, it is a short step from perspectivism to pluralism." But it is nevertheless a distinct step, and an important one. Coming back to relativism, there is a subtler sense in which the tenet of Symmetry is actually contrary to pluralism. In any real-life situation, it is impossible to maintain an unlimited number of systems in place, due to the limitations of our material and mental resources. Therefore, any responsible normative pluralism should be committed to making a selection of a fairly small and manageable number of systems to retain and develop. This *requires* some kind of ranking of alternatives, contrary to the dictates of Symmetry.

The second key distinction within pluralism is between "tolerant pluralism" and "interactive pluralism," each of which can be taken in a normative or descriptive way (see Chang 2012, 270–284 for further details). Tolerant pluralism is about allowing the simultaneous co-existence of different systems of practice (or plurality in some more specific types of elements). Interactive pluralism goes a step further, by seeking the additional knowledge and successful practices that may result from various types of interactions between different systems of practice. A leading example of interactive pluralism is Sandra Mitchell's (2003) "integrative pluralism," according to which complex real-life objects of study require an *ad hoc* integration of various theories or models to suit each situation. I have myself advocated interactive pluralism, of the normative variety; the full benefits of pluralism, in science as in general society, only come from the productive interactions of different systems, in addition to their mere co-existence.

Interactive pluralism presupposes tolerant pluralism, as the interaction of systems obviously requires the existence of different systems. On the other hand, tolerant pluralism does not imply interactive pluralism. Relativists, to the extent that they are pluralists, tend to focus on tolerant pluralism but not on interactive pluralism. In Kusch's standard model of relativism the tenets of Exclusiveness and Notional confrontation/Conversion seem to presume that the different systems do not interact with each other. On closer reading they do not strictly rule out inter-system interactions, but interactions tend not to be a focus of relativists.

4. Realism and judgement

I would now like to draw a comparison among relativism, perspectivism and pluralism in relation to how they each bear on two crucial issues in the philosophy of science. The first issue is realism. Ultimately I want to argue that all three positions can accommodate a kind of realism, but first of all there is a tension to be defused within perspectivism. The metaphor of "perspective" introduces an intuitive pull towards an objectivist or metaphysical realism, which is actually rejected by perspectivists. Talking about various "perspectives" does not make sense unless we are talking about different perspectives on *the same thing*. But what is that presumed "same thing"? So perhaps perspectivists are stuck with the same dilemma that Kant faced in postulating the *Ding an sich* but denying that one could say anything about it – then why postulate it at all? The very seductive yet deeply misleading aspect of the perspective metaphor is that a three-dimensional object we are perspectivally studying exists "out there" in itself, already well-formed independently of our cognition and action. This image works intuitively against the third layer of perspectivism just outlined, and makes the second layer the most comfortable resting-place for perspectivists, because that layer allows the possibility that we can build a true picture of the object by unifying enough well-placed perspectival pictures of it, just as we can construct a 3-D image of an organ in a CT scan based on various 2-D cross-sections taken with X-rays. Or more generally, we can hope for the kind of "foliated pluralism" advocated by Stéphanie Ruphy (2016), according to which ontology is enriched by the coherent addition of new perspectives on to existing ones. But this sort of resting-place is ultimately detrimental to perspectivism, because it is perfectly compatible with the standard sort of scientific or metaphysical realism of the "ready-made world." If that is what perspectivism comes down to, it becomes difficult to see why we need perspectivism as a novel and separate position.

Now we can revisit Giere's objection to relativism with some profit. In his commentary on "Feyerabend's perspectivism" he suspects that Feyerabend wants to relativise reality itself, as well as our conception of reality. That would constitute "a deep relativism in his views that goes beyond perspectivism." He adds: "Relativizing *actual* as opposed to merely *supposed* physical existence to culture . . . is too radical" (Giere 2016, 140; emphasis original). Giere (2016, 137–138) makes it clear that he does not wish to enter into the discussion of a very strong form of perspectivism, which holds that "there are no perspective[-]transcendent facts that could be the object of perspective[-]transcendent knowledge." But, again, I think Giere is stepping into the Kantian dilemma: while he doesn't want to deny that there are "perspective-transcendent facts," he denies that we can know anything about such facts. According to Giere's perspectivism (or, perspectival realism), "claims made from within a perspective are nevertheless *intended* to be genuinely about the world, and thus 'realistic,' even though not fully precise or complete." He stresses the contrast between perspectivism and an objectivist (or metaphysical) realism;

according to the latter, "claims about the world are *intended* to be perspective[-]transcendent, or at least framed in a uniquely correct perspective" (Giere 2016, 138, emphases original).

Giere is at best walking a fine line here. I think a more straightforward solution would be to accept the third layer of perspectivality mentioned previously, and accept that perspectivism is not so different from epistemic relativism. If we accept the third layer of perspectivality, the very meanings of any concepts or terms we use can only be contextually fixed; therefore, there cannot be any knowledge-claims at all that are perspective-transcendent, not to mention their truth-conditions. But in that case, can this perspectivism/relativism still be realist? My answer is yes –no less realist than perspectivists want to be, in any case. If "perspectival realism [is] not a doctrine about what *is real*, but about what *is taken to be real* from within a perspective" (Giere 2016, 140, emphases original), then there is no reason for perspectivists to be so afraid of relativism. Now, we may agree with the common notion endorsed by Massimi (2018, 171) that "there are perspective-independent worldly states of affairs that ultimately make our scientific knowledge claims true or false"; however, I remain with Kant in insisting that such states of affairs, if they exist, are not expressible. All we can ever talk about are conceptualised objects, not things-in-themselves.

I have argued elsewhere that there is a kind of realism that pluralists can embrace (Chang 2018). Each good system of practice offers a true account of its proper object domain that is worth preserving and developing. Truth here needs to be taken in a pragmatist sense (as articulated in a preliminary way in Chang 2017). This kind of realism can be maintained in relativism as well, and also in strong perspectivism with the metaphysical–objectivist metaphor suppressed.

Epistemic judgement is the second major issue that I would like to consider in relation to relativism, perspectivism and pluralism. Now, no one will deny that people make epistemic judgements. Dependence, the first tenet in Kusch's standard model of relativism, is a useful reminder: there *is* judgement of the epistemic status of beliefs, though of course relatively to a system or practice. Meiland and Krausz (1982, 2–3) make a similar point when they make a distinction between skepticism and relativism: "The relativist observes that the absence of criteria for ascertaining a single objective truth does not mean that there are no criteria of truth at all." Similarly Ian Hacking (1982, 66) speaks of an "anarcho-rationalism" that he learned from Paul Feyerabend, which is "tolerance for other people combined with the discipline of one's own standards of truth and reason." But none of that is likely to placate the standard realist determined to push back on relativism, pluralism and even perspectivism. The difficult and crucial question is what we can say, from these various positions, about the desirability and feasibility of the judgement of epistemic merit between competing systems of practice.

It may seem that relativism has a clear answer to this question, in the tenet of Symmetry: "Epistemic systems and practices must not be ranked." But the situation is not so straightforward, as one can see from Kusch's distinction between at least four possible meanings of Symmetry. The second one is noteworthy: "There is no neutral way of evaluating different *SPs*." (This condition is termed "Non-neutrality" by Kusch, but in a way it is more like "neutrality," as opposed to "non-neutrality" meaning "bias.") This formulation recognises the reality that those working in any given SP do routinely make appraisals of the goodness of other SPs, from their own perspective. And as Kusch acknowledges: "it does not preclude the possibility that some SPs agree on the standards by which their overall success should be judged." He quickly adds that such "local agreement" does not justify "the hope for a global or universal agreement." However, as it happens with international treaties, literally global systems (such as the SI system of physical measurement units) may actually be based on a broad-as-can-be version of local agreement.

Other systems are based on a general principle of local neutrality (as with refereeing in international sporting events).

On this issue of cross-system judgement, perspectivism has a similar stance to relativism. Giere says (2016, 141): "Perspectival realism. . ., by itself, does not provide any means for judging one perspective superior to another. Yet, it also does not imply that all perspectives are epistemologically equal." Claiming that Feyerabend "found himself left with no possibility of finding any principles for judging one perspective superior to another," Giere advocates what he calls "relative explainability": for example, Newtonian gravitational theory cannot explain why general relativity works, while general relativity can explain why in low-curvature spacetime Newtonian theory works out well enough. Whether this principle is generally applicable in making cross-perspective judgements is doubtful, but it is a persuasive guideline at least for an important class of cases.

How the pluralist handles this question is an interesting and complex one. As the case of Feyerabend shows, pluralists – as well as anyone – can point out that some enterprises are more successful than others, when each one is judged by its own standards. Even Popperian falsification comes down to this kind of judgement, once the theory-ladenness of observation is recognised. Feyerabend is a good example of a staunch pluralist who did not renounce the desirability or feasibility of cross-system judgements. As Kusch points out, despite his "ontological relativism" criticised by Giere, Feyerabend was critical of what he called "relativism" in a more epistemological context: "not all approaches to 'reality' are successful" and success is "a matter of empirical record, not of philosophical definitions" (quoted in Kusch 2016, 111). But does this imply that there is a measure by which we can compare the degree of success of one system judged by its own standards, and the degree of success of another system judged by its own standards? Does such a comparison require system-transcendent standards? Are there any? If not, how can we make the ranking of systems that is necessary for the practice of responsible pluralism, as discussed in the last section? This may seem like a devastating objection to pluralism, but at least it is not a novel one. It is readily admitted that interactive pluralism requires some commonalities for communication, and those are bound to embody judgements, too. It is neither required by pluralism, nor commonly found in actual practice, that different systems of practice are completely disjointed from each other. Enough basis for comparisons may be given by some broadly shared elements such as basic arithmetic, widely spoken everyday languages, agreed-upon measurement standards, common material features of life on earth, and shared human desires and values.

References

Bernstein, R. J. (1983), *Beyond Objectivism and Relativism: Science, Hermeneutics, and Praxis*, Oxford: Basil Blackwell.

Chang, H. (2012), *Is Water H_2O? Evidence, Realism and Pluralism*, Dordrecht: Springer.

———. (2017), "Operational Coherence as the Source of Truth," *Proceedings of the Aristotelian Society* 117: 103–122.

———. (2018), "Is Pluralism Compatible with Scientific Realism?" in *The Routledge Handbook of Scientific Realism*, edited by J. Saatsi, Abingdon: Routledge, 176–186.

———. (2019), "Pragmatism, Perspectivism and the Historicity of Science," in *Perspectivism: Scientific Challenges and Methodological Prospects*, edited by M. Massimi and C. D. McCoy, Abingdon: Routledge, 10–27.

Giere, R. N. (2006), *Scientific Perspectivism*, Chicago: University of Chicago Press.

———. (2016), "Feyerabend's Perspectivism," *Studies in History and Philosophy of Science* 57: 137–141.

Hacking, I. (1982), "Language, Truth, and Reason," in Meiland and Krausz (1982), 48–66.

Kusch, M. (2016), "Relativism in Feyerabend's Later Writings," *Studies in History and Philosophy of Science* 57: 106–113.

———. (2019), "Epistemischer Relativismus," in *Handbuch Erkenntnistheorie*, edited by M. Grajner and G. Melchior, Stuttgart: Metzler, in press.

Massimi, M. (2018), "Perspectivism," in *The Routledge Handbook of Scientific Realism*, edited by J. Saatsi, London and New York: Routledge, 164–175.

Meiland, J. W. and M. Krausz (eds.) (1982), *Relativism: Cognitive and Moral*, Notre Dame: Notre Dame University Press.

Mitchell, S. D. (2003), *Biological Complexity and Integrative Pluralism*, Cambridge: Cambridge University Press.

Ruphy, S. (2016), *Scientific Pluralism Reconsidered: A New Approach to the (Dis)unity of Science*, Pittsburgh: University of Pittsburgh Press.

43

RELATIVISM AND SCIENTIFIC REALISM

Stathis Psillos and Jamie Shaw

1. Introduction

Richard Rorty famously said: "'Relativism' is the view that every belief on a certain topic, or perhaps on any topic, is as good as every other. No one holds this view" (1982, 166). What is relativism, then? In this chapter we won't try to offer a general answer to this question. Indeed, the sheer variety of relativisms abound would require a momentous effort to enumerate in detail (see Sankey 1997, ch. 1; Baghramian 2004). The same can be said of realism, for which there may be as many realisms as there are realists. For this chapter, there are two core commitments of scientific realism, one metaphysical, the other epistemic. The first is that the world the science investigates has a definite mind-independent structure; the second is that empirically successful scientific theories are (approximately) true of the world; hence, (most of) the entities posited by them inhabit the world (cf. Psillos 1999). We take it that these two commitments are both minimal and robust.

In this chapter, we shall focus our attention on two particular instances of relativism, with important implications about science: the Strong Programme (SP) in the Sociology of Scientific Knowledge (SSK) and the Incommensurability Thesis associated with Thomas Kuhn and Paul Feyerabend. We argue that these kinds of relativism are incompatible with realism.

2. The argument from the Strong Programme

Let us start with a broad outline of the kind of relativism advocated by SP. The key feature of SP is captured by the "equivalence postulate," viz., that, as Barry Barnes and David Bloor put it, "all beliefs are on a par with one another with respect to the causes of their credibility" (1982, 22). This is not taken to mean that all beliefs are equally true or false. Rather their point is that "regardless of truth and falsity the fact of [all beliefs'] credibility is to be seen as equally problematic" (1982, 23). In other words, in offering a causal explanation of why a certain belief is held (is being taken as credible), the truth-value of the belief should be bracketed. It shouldn't play any role in the explanation of the credibility of a belief. Fitting with the facts is dismissed as not being a causal relationship "that bodies of belief bear to their referents" (Bloor 1999a, 89).

What aligns SP with relativism is the further thought that "knowledge" and "justification" are always and invariably tied to local circumstances, cultures, and communities. Incidentally, the

SP-ers reserve "the word 'knowledge' for what is collectively endorsed, leaving the individual and idiosyncratic to count as mere belief" (Bloor 1991, 5). In other words, truth drops out of the concept of knowledge and justification is replaced by "being confidently held to and lived by" (ibid). The relativist crux then is that there is no distinction between being-taken-to-be-justified (by a community) and being-justified. Whatever justification-conferring properties are taken by a community to confer justification on a belief are the "right" properties. As Barnes and Bloor put it: "For the relativist there is no sense attached to the idea that some standards or belief are really rational as distinct from merely locally accepted as such" (1982, 27).

It is within this relativist framework that SP sets out to investigate "the specific, local causes" of the "credibility" of beliefs. Notably, this investigation is guided by two methodological principles:

> (IMPARTIALITY TENET) (IT) It would be impartial with respect to truth and falsity, rationality or irrationality, success or failure. Both sides of these dichotomies will require explanation.
> (SYMMETRY TENET) (ST) It would be symmetrical in its style of explanation. The same types of cause would explain, say, true and false beliefs.
>
> *(Bloor 1991, 7)*

SP presents itself as an empirical-naturalistic theory concerning the causes of scientific "knowledge." As such, it aims to describe the social and natural causes of scientific beliefs. As Michael Friedman has noted: "Rather than articulating the structure of what *ought to* be believed, SSK simply describes and explains what is *in fact* believed" (1998, 243). Bloor makes the same point when he says that "The aim isn't to explain nature, but to explain shared beliefs about nature" (1999a, 87). But then, Friedman is right to wonder whether we could simply acknowledge the "fundamental divergence in aim and methods" between SSK and traditional philosophy and "leave it at that" (1998, 244).

The same worry could arise when it comes to scientific realism. The *explanandum* of scientific realism, *qua* philosophical theory of science, is the success of science, whereas the *explananda* of SP (and SSK), *qua* scientific theory in sociology, are the beliefs of scientists. Given the difference in the *explananda*, a difference in the *expanans* should be expected. Consider the following analogy. One might want to explain how and why (i.e., what is the best explanation of) the fact that hammers are typically successful in driving nails into walls. Part of the *explanans* will be the properties of hammers and walls and the details and the epistemic credentials of the hammer-theory. But one might want to explain why, say, smiths believe that hammers can be employed in driving nails in the wall. Part of the *explanans* (the full of the *explanans* for SP-ers) might well be various natural and social factors, e.g., the "technical competences handed down from generation to generation" (Barnes and Bloor 1982, 23). There is no incompatibility, here. SP and SR can live happily ever after. But we cannot leave it at that, the reason being that SP-ers endorse a broader philosophical agenda, which is incompatible with scientific realism.

Bloor and other SP-ers claim to be committed to scientific realism. Bloor says: "We take for granted that trees and rocks, as well as electrons and bacilli, have long been stable items amongst the furniture of the universe" (Bloor 1999a, 86). Nick Tosh (2007) has interpreted this as implying that SSK adopts a realist view of science. But as Jeff Kochan (2010) has rightly pointed out Bloor refers to how the world is viewed by common sense (as consisting of "objects of nature" independent "from our ideas about them") and not to a commitment of SSK. The credibility of this kind of belief is the *topic* of study of SSK as opposed to a *resource* to be relied upon in the study of this topic. A commitment to the reality of the world as the topic of first-order scientific beliefs is suspended and bracketed in the causal explanation of these beliefs. Bloor says explicitly that "[t]he point of the symmetry postulate is to enjoin sociologists to draw back from making first-order judgements. The point is to make such judgements the objects of enquiry" (Bloor 1999a, 102). The tension between SSK and SR is now obvious. SR takes scientific theories as its topic and claims that the best explanation of their successes is that they are true of the world. Then it relies on these theories as resources for the theoretical understanding of reality. By contrast SSK simply takes for granted these theories and makes its topic of inquiry the causes of scientists' belief in them, bracketing the issue of how these theories fit with the world.

Martin Kusch (2018, 268) has noted that SSK adopts a "straightforward and bold realism about the social world". But is this right? In his classic (1995), Barnes introduces the idea of a "realist mode of speech" which allows us to distinguish between appearances and reality. This mode of speech is a "cultural universal." It amounts to all-embracing system by means of which sense is made of the world. But he is clear that this "realist strategy" is a mode of speech. In fact, given that "the underlying 'reality' referred to is invisible and inaccessible and cannot itself govern what is said about it" (1995, 115), it drops out of the picture. The "articulation of existing knowledge in a realist mode of speech makes its future entirely open-ended"; it is not constrained by this "reality"; nor does it capture more about it. The skeptical origin of this point is obvious. But its significance for SSK should be obvious too: a realist mode of speech does not imply a *commitment* to realism. What is more, for SSK the belief in a mind-independent world is itself a belief that needs causal explanation and should not imply any first-order commitments to a mind-independent world (cf. Bloor 1999a, 109). In case the reader still doubts that SSK is anti-scientific-realism, Bloor reminds us that "independent reality" is constituted by the conceptual tools used to study it, pretty much like "the talk (about money) and the thing talked about (the money 'itself') are one and the same" (Bloor 1999a, 109).

Now, suppose we grant that SSK is consistent with some conception of mind-independent noumenal world. It would still be far away from the realist view that the world has an objective natural-kind structure. In fact, it is constitutive of SP that there is no privileged and objective classification of worldly things into natural kinds: the world simply has no objective natural kind structure. Bloor starts from the fact that classifications (which are "projected" onto nature) change over time and concludes that "(I)t is best to think of theoretical systems as a whole having utility and embodying an overall adaptation to reality, where reality is rich enough to permit numerous possible adaptations and numerous possible descriptions and classifications" (1999a, 94). For him, a successful classification of things into kinds "doesn't mean some uniquely direct or successful reference has been achieved" (ibid). Now, one does not have to be a realist to think that there are natural classifications and that science, as it grows, approaches a natural classification of reality. Pierre Duhem (1906) is a case in point. Realists, typically, think of nature as having natural joints, which science aims at, and has succeeded in, uncovering. But there is no obstacle for a realist to accept the view that nature permits different classifications. In fact, if classification (kind-membership) is a matter of circumscribing similarity classes of resembling particulars, then what kinds are circumscribed depends on whether similarity is a fully objective relation or not. A realist will, typically, admit that there are objective similarities and differences among worldly particulars, but it is fully consistent with realism that as we move from elementary particles, where natural kinds are fully objective, to higher-level entities, the natural kinds are fixed jointly by nature and us. In any case, realism entails that some classifications are better than others, based on explanatory and predictive considerations. For instance, René Descartes posited vortices for the circulation of the planets and Isaac Newton showed that, given natural principles about motion, the vortices were fictions. Instead, he correctly identified the masses of the objects as being relevant to their gravitational behaviour. Moreover, realism would not accept the idea that there are two or more mutually inconsistent and equally correct classifications of worldly stuff, especially if we think of the case of two or more "total theories" of the world. Although perspectival approaches might be inevitable in our conceptual interaction with the world, for a realist there must be a fact-of-the-matter as to which is the correct one – especially when incompatible perspectives are in play.

Where SP-ers look for symmetries in the explanation of beliefs, scientists "work hard to establish asymmetry," as Bloor admits (1999a, 101). They, typically, look for evidence that tips the balance in favor of one theory over the other. Bloor agrees that it would be an indictment for SSK if it ignored "the possible role of sensory input in 'tipping the balance'" (1999a, 102). But his reaction is puzzling. He reiterates the symmetry principle and adds that there are only locally credible reasons, and not absolute proofs, that one scientific theory is better than another. Note that talk of "absolute proof" in this context is very misleading. For it is a fact of life that there is no absolute proof of any scientific theory which goes, even minimally, beyond the phenomena. But to equate the differential bearing of evidence on theories with "locally credible reasons" requires adopting a strong skeptical attitude according to which the evidence for a theory is either conclusive or no evidence at all. (For more on this issue, see Tim Lewens 2005, 574ff.).

Perhaps we are unfair to SP-ers. Kusch has noted that Barnes and Bloor accept that "any differences in the sampling of experience, and any differential exposure to reality must be allowed for" (1982, 35). But let's take a look at an oft-cited passage:

> If we believe, as most of us do believe, that Millikan got it basically right, it will follow that we also believe that electrons, as part of the world Millikan described, did play a causal role in making him believe in, and talk about, electrons. But then we have to remember that (on such a scenario) electrons will also have played their part in making sure that Millikan's contemporary and opponent, Felix Ehrenhaft, didn't believe in electrons. Once we realise this, then there is a sense in which the electron 'itself' drops out of the story because it is a common factor behind two different responses, and it is the cause of the difference that interests us.
>
> *(Bloor 1999a, 93)*

In what sense does the electron "itself" drop out of the story? Part of the Millikan-Ehrenhaft controversy was about the value of the smallest electric charge found in nature. Robert A. Millikan argued that the smallest electric charge was the charge of an electron and claimed that his famous oil-drop experiment supported this conclusion. Though initially Felix Ehrenhaft claimed to have found the "elementary quantum of electricity," in 1910 he announced that he had identified charges much smaller than the charge of the electron, which he called "sub-electrons." A long and occasionally bitter debate ensued which ended with Millikan winning the argument. The story is long and intricate and is told in detail by Gerald Holton (1978). It is hard to make sense of this debate if the electron "itself" drops out of the picture. If we drop the electron "itself" we fail to understand why the debate had to do with the prospects of the atomic theory of matter. We fail to take into account for what Holton has called "suspension of disbelief" which guided Millikan acceptance of the electronic charge. We fail to account for the prevailing of the theory of non-radiating electronic orbits and a lot more. In the end, we fail to explain the success of Millikan's experiments. Could we just say that Millikan "observed something he attributed to, and explained by, a postulated entity he called 'an electron'" (Bloor

1999a, 93)? Well, we could if we had in mind a causal theory of reference fixing of theoretical terms. But Bloor has something else in mind: "In this way we might be less tempted to think that nature has an automatic tendency to generate those particular verbal descriptions or responses" (ibid). It's clearly absurd to posit such a tendency. What is not absurd is to think that nature reveals itself to us causally in a way that it can be correctly captured by theoretical terms and descriptions.

The key point here is made by Kochan (2010, 132): "Bloor cautions sociologists and historians not to assume that the external world automatically determines the success of our descriptions of it." This "methodologically neutrality" is not as innocuous as it might seem. It renders irrelevant to the explanation of the disagreement that the evidence supported more one theory than another. For a realist, at least in many typical cases, getting worldly facts right is what determines the success of a theoretical description and the failure of another.

An important element of SP is the claim that scientific knowledge involves a conventional element, which SP-ers take to imply that a significant part of scientific knowledge (which, as we have seen, amounts to entrenched collective beliefs) could be different from what it is, given that alternative conventions could have been chosen. The possibility of alternative conventions is taken by Bloor to be one of the "basic principles of relativism" (1999a, 105). In support of this principle, Bloor mounts a historical argument. Roughly put, it is this. The history of science shows that there have been "alternative ways of understanding the data." This evidence can support the "inductive generalization" that despite current consensus on certain theories (i.e., ways of understanding the data), they too "will have alternatives." Now, to realist ears, this sounds too close to being a version of the pessimistic induction (or the argument from unconceived alternatives), the difference being that in the SSK version it is not the falsity of current theories that is concluded, but rather the dispensability of current theories as the best explanations of the available evidence. For this inductive argument to be cogent it must be the case that as science grows there is no substantial change in how scientific theories map out the world. In other words, it must be assumed that current theories are not better supported by the evidence than the past ones. Bloor is aware of this and upfront. He states that the inductive generalization is cogent "unless someone can produce remarkable and cogent reasons for thinking that qualitative changes have suddenly taken place in the nature of knowledge" (Bloor 1999b, 105). But that's precisely what the realist claims about current science and the arguments produced against the pessimistic induction have driven this point home (see Psillos 2018 for an overview).

The bottom line is this. Scientific realism is a view of science which, based on versions of the so-called No Miracles argument, claims that the truth of scientific theories (approximate truth, or partial truth, but still truth, viz., getting the world by and large right), is the best explanation of their empirical success and that the reason for being optimistic about science is precisely that past theories build on the successes of their predecessors. That's precisely the view that SSK denies.

3. The argument from incommensurability

There are two kinds of incommensurability that would undermine realism and suggest a robust relativism:

(1) Ontological incommensurability
(2) Methodological incommensurability

(1) posits the existence of genuinely distinct theoretical alternatives. However, it isn't enough, for realists, that there *are* distinct alternatives; these alternatives must be *equally successful*. Since "success" is a function of methodological appraisals, (2) is necessary to show that there are successful distinct alternatives. If (1) and (2) are correct, then relativism follows. Thus, (1) and (2) are necessary for realism to be undermined by an argument from incommensurability.

For (1) to be true, two (or more) theories, T_1 and T_2, must posit different pictures of reality. Obviously, many entities we thought existed we now think do not (e.g., phlogiston, witches, absolute velocities, etc.) and scientific progress often involves discovering new entities. That ontologies "shift" in this sense is uncontroversial. Ontological incommensurability requires that entities in T_1 *cannot be formulated* within T_2. While this takes different forms in Kuhn and Feyerabend at different points of their career, the common claim is that it is impossible to express a factual claim of T_1 coherently in T_2. In Feyerabend's earlier career, he defends the pragmatic theory of meaning (see Kuby 2018). Here, utterances gain their meaning by participating within a theoretical framework. When pressed on incommensurability by his critics, Kuhn defends a quasi-Quinean holism approach where radical translatability is impossible given the inscrutability of reference caused by paradigm shifts (see Kuhn 1962, 1982; Sankey 1993, 765, fn. 10). In both cases, we see of incommensurability depends on some version of *holism*. This leads to the basic issue of translation, as put by Martin Carrier:

> Translation requires the coordination of a linguistic item with another one taken from a different theoretical framework but possessing the same meaning. The understanding underlying the entire discussion of the incommensurability thesis is that translation needs to be precise (clumsy paraphrases don't suffice) and to provide a one-one correlation between expressions.
>
> *(2001, 72)*

There are two semantic issues here. One is reference. The other is the inferential roles a term may play. Consider Carrier's illustrative example of translating "phlogiston" into the oxygen theory. On the one hand, the reference is maintained across the two theories: in most cases in which partisans of the phlogiston theory thought it legitimate to apply the predicate "phlogiston escape," adherents of the oxygen theory would speak of "oxygen bonding" (2001, 75). However, the inferential role of the phlogiston and oxygen differ. High proportions of phlogiston are responsible for the "oily fatty" nature of a substance which allows for inferences about the combustibility of resin. No such inference is afforded by translating "high proportion of phlogiston" with "capacity to combine with large amounts of oxygen." As will be seen, these semantics are at odds with realist semantics.

A special case of ontological incommensurability includes the "Kant on wheels" view. This is a popular reading of Kuhn and Feyerabend. This view was first famously articulated by Paul Hoyningen-Huene (1993) and has since become quite popular (see Donhauser and Shaw 2018 and the citations therein). In short, the view is as follows:

> While Kant conceived [of] categorization, whereby our experience is molded, as being immutable, Kuhn, on this interpretation, regards this function of the mind as subject to change. An individual's categorizations are, in part, the product of his or her immersion in a paradigm-governed tradition.
>
> *(Bird 2002, 454)*

On this interpretation, our ability to conceive the world is limited by *a priori* constraints. While Kant thought that these constraints were necessary categories of the understanding and the forms of pure intuition, Kuhn thought that they were relative to a paradigm. While a realist can admit that what can be meaningfully said is dependent on theoretical assumptions, what makes the Kant-on-wheels approach anti-realist is (i) the plurality of the phenomenal worlds, since each paradigm is constitutive of its own world; and (ii) the commitment to the view that the noumenal world has no inherent natural kind structure.

Some insist on an even stronger reading of Kuhn as an "extravagant idealist" (Scheffler 1967) due to a literal reading of Kuhn's claim that scientists inhabit "different worlds." This reading is defended and elaborated by Carl R. Kordig:

> It must follow that before Lexell there were more stars in the world of the professional astronomer. But if this is true one can no longer say that these astronomers before Lexell were mistaken about the number of stars. . . . This is because from Kuhn's viewpoint there really were more stars in their world. It is not just that they believed there were more stars in their world. According to Kuhn there really was this number of stars.
>
> *(Kordig 1971, 18)*

The contrast between this interpretation of Kuhn and realism should be obvious. The realist maintains, against the Kantian constructivist, that (i) there *is* a mind-independent reality that science seeks to describe; and (ii) successful scientific theories approximately reflect the nature of this reality.

To be clear, ontological incommensurability does not commit one to being a Kantian constructivist or an idealist. Feyerabend, for instance, defended realism (though of an importantly different kind from the one considered here) and ontological incommensurability simultaneously. However, it should be clear that any proponent of ontological incommensurability cannot be a realist in our current sense.

As mentioned, it is not enough to have theoretically distinct alternatives. These alternatives must both be *successful*. Kuhn famously argued that the methodological appraisal of scientific theories comes part and parcel with a particular paradigm. As a result, the standards of success are *internal* to a paradigm. As Harvey Siegel put it: "Since there are no paradigm-neutral criteria of evaluation, paradigm debate can rely on no objective criteria of evaluation of paradigms" (1980, 361). Others, e.g., Kordig (1971), have pointed out that there may be standards that are external to paradigms which can allow for cross-paradigm comparison; for instance, "the sharing of second-order criteria by enthusiasts of rival paradigms" (1971, 106).

This point is of particular relevance to the debate about realism. Realism is justified by the No Miracles argument, which uses *empirical success* to appraise theories. Kuhn considers standards that scientists *happen* to have held. The fact that scientists might have held different standards is beside the point; for realists, the abductive argument from predictive success to approximate truth holds regardless. The No Miracles argument fixes what constitutes "success."

Does this mean that (ii) is irrelevant to the realist debate? Not quite. Consider the following claim from Feyerabend. To compare predictive success, we must compare their "content-classes" (i.e., the list of facts that can either confirm or falsify a theory). Hence, to compare T_1 and T_2, we must be able to formulate an overlapping set of facts which one can be said to predict more accurately than the other. However, such a comparison cannot be made since it involves using inconsistent theories. We arrive at the following conclusion:

> Incommensurable theories, then, can be *refuted* [or confirmed] by reference to their own respective kinds of experience; i.e., by discovering the *internal contradictions* from which they are suffering. . . . Their *contents* cannot be compared. Nor is it possible to make a judgment of *verisimilitude* except within the confines of a particular theory.
>
> *(1975, 284)*

We are now in a position to rephrase the problem of methodological incommensurability. The problem is not that the *methods themselves* are internal to the theory, but that the method of comparison used by realists *cannot be applied* to incommensurable theories.

To this charge realists have replied by adopting the causal theory of reference of theoretical terms and predicates, which was developed by Hilary Putnam. Here is a nice description by Ian Hacking:

> Putnam saved us from such questions [about holism and incommensurability] by inventing a referential model of meaning. . . . Our initial guesses may be jejune or inept, and not every naming of an invisible thing or stuff pans out. But when it does, and we frame better and better ideas, then Putnam says that, although the stereotype changes, we refer to the same kind of thing or stuff all along. We and Dalton alike spoke about the same stuff when we spoke of (inorganic) acids. J. J. Thomson, H. A. Lorentz, Bohr and Millikan were, with their different theories and observations, speculating about the same kind of thing, the electron.
>
> *(1982, 75)*

The gist is that while theoretical claims about a posited entity may change, they merely represent differing attempts to describe the same feature of reality. As such, reference is retained across theory-change, though the inferential roles of the terms appearing in various theories may differ. Hence, while the inferential roles afforded by a particular theoretical posit are important for describing the correct features of reality, changes in inferential roles does not mean a change in reference. Furthermore, a realist can maintain that those features of reality that are being referred to across theory-change do have the properties attributed to them by successful theories. As such, against the incommensurabilitist, the realist maintains that successful theories, though radically distinct, can be ontologically continuous.

4. Concluding remarks

Despite the potential plausibility and insight of relativism and realism, they cannot be held simultaneously. While we have not argued for the superiority of either approach in this chapter, we hope to have provided a better sense of what a commitment to realism or relativism would entail.

References

Baghramian, M. (2004), *Relativism*, New York: Routledge.
Barnes, B. (1995), *The Elements of Social Theory*, London: UCL Press.

Barnes, B. and D. Bloor (1982), "Relativism, Rationalism and the Sociology of Knowledge," in *Rationality and Relativism*, edited by M. Hollis and S. Lukes, Cambridge, MA: MIT Press.

Bird, A. (2002), "Kuhn's Wrong Turning," *Studies in History and Philosophy of Science* 33: 443–463.

Bloor, D. (1991), *Knowledge and Social Imagery*, second edition, Chicago: University of Chicago Press.

———. (1999a), "Reply to Bruno Latour," *Studies in History and Philosophy of Science* 30: 131–138.

———. (1999b), "Anti-Latour." *Studies in History and Philosophy of Science Part A*, 30(1): 81–112.

Carrier, M. (2001), "Changing Laws and Shifting Concepts: On the Nature and Impact of Incommensurability," in *Incommensurability and Related Matters*, edited by P. Hoyningen-Huene and H. Sankey, Dordrecht: Kluwer, 65–90.

Donhauser, J. and J. Shaw (2018), "Knowledge Transfer in Theoretical Ecology: Implications for Incommensurability, Voluntarism, and Pluralism," *Studies in History and Philosophy of Science Part A*, doi:10.1016/j.shpsa.2018.06.011.

Duhem, P. (1906), *The Aim and Structure of Physical Theory*, Princeton, NJ: Princeton University Press.

Feyerabend, P. (1975), *Against Method*, London: Verso Books.

Friedman, M. (1998), "On the Sociology of Scientific Knowledge and its Philosophical Agenda," *Studies in History and Philosophy of Science Part A* 29(2): 239–271.

Hacking, I. (1982), "Experimentation and Scientific Realism," *Philosophical Topics* 13: 71–87.

Holton, G. (1978), "Subelectrons, Presuppositions, and the Millikan-Ehrenhaft Dispute," in *Historical Studies in the Physical Sciences*, edited by R. McCormmach et al., Baltimore: Johns Hopkins University Press, 161–224.

Hoyningen-Huene, P. (1993), *Reconstructing Scientific Revolutions: Thomas S. Kuhn's Philosophy of Science*, Chicago: University of Chicago Press.

Kochan, J. (2010), "Contrastive Explanation and the 'Strong Programme' in the Sociology of Scientific Knowledge," *Social Studies of Science* 40(1): 127–144.

Kordig, C. (1971), *The Justification of Scientific Change*, Dordrecht, Holland: D. Reidel Publishing Company.

Kuby, D. (2018), "Carnap, Feyerabend, and the Pragmatic Theory of Observation," *HOPOS: The Journal of the International Society for the History of Philosophy of Science* 8(2): 432–470.

Kuhn, T. S. (1962), *The Structure of Scientific Revolutions*, Chicago: University of Chicago Press.

———. (1982), "Commensurability, Comparability, Communicability," *PSA: Proceedings of the Biennial Meeting of the Philosophy of Science Association* 1982(2): 669–688.

Kusch, M. (2018), "Scientific Realism and Social Epistemology," in *The Routledge Handbook of Scientific Realism*, edited by J. Saatsi, London: Routledge, 261–275.

Lewens, T. (2005), "Realism and the Strong Program," *The British Journal for the Philosophy of Science* 56(3): 559–577.

Psillos, S. (1999), *Scientific Realism: How Science Tracks Truth*, London: Routledge.

———. (2018), "Realism and Theory Change in Science," in *The Stanford Encyclopedia of Philosophy*, Summer 2018 edition, edited by E. N. Zalta, https://plato.stanford.edu/archives/sum2018/entries/realism-theory-change/.

Rorty, R. (1982), *Consequences of Pragmatism*, Minneapolis, MN: University of Minnesota Press.

Sankey, H. (1993), "Kuhn's Changing Concept of Incommensurability," *The British Journal for the Philosophy of Science* 44(4): 759–774.

———. (1997), *Rationality, Relativism and Incommensurability*, London: Routledge.

Scheffler, I. (1967), *Science and Subjectivity*, New York: Bobbs-Merri.

Siegel, H. (1980), "Review: Objectivity, Rationality, Incommensurability, and More," *The British Journal for the Philosophy of Science* 31(4): 359–375.

Tosh, N. (2007), "Science, Truth and History, Part II. Metaphysical Bolt-Holes for the Sociology of Scientific Knowledge?" *Studies in History and Philosophy of Science Part A* 38(1): 185–209.

44

RELATIVISM IN THE SOCIAL SCIENCES

Stephen Turner

1. Introduction

Relativism, in some sense, is the basic phenomenon of the social sciences. Political institutions, religions, customs, kinship systems, normative ideas, and legal orders vary between societies and across historical periods. The concept of "relativism" appears very early in the history of social science as a way to talk about these differences. There are, however, five more or less distinct senses of relativism that figure in the history of the social sciences. The first is descriptive relativism: the recording of differences without evaluation or explanation. The next two kinds are both in some sense explanatory. The first involves the notion that there is something that the object of explanation, such as knowledge, is relative *to* something. Comte, for example, writes that "all our knowledge is necessarily relative, on the one hand, to the medium, in as far as it is capable of acting on us, and on the other to the organism, in as far as it is susceptible of that action" (1855, v. 3, 367). The "medium" Comte has in mind is society. The second kind is social evolutionism. A third, more familiar, sense emphasizes the arbitrariness of differences, and is part of what has been called "the Standard Social Science Model." The fourth kind might be called methodological relativism: it deals with the pluralism of methodological approaches, which is typically accounted for by reference to one of the other forms of relativism. The fifth is associated with social constructivism and science studies, involves epistemic relativism, and is in a sense the most radical of these forms of relativism.

Relativism in the third sense is typically the source of controversy, because it is supposed to warrant the thesis of the normative equivalence of all moral systems, and thus to undermine the idea of ethical truth. Relativism in the second sense has more often, but not always, been aligned with a non-relativist or universalistic account of the explainer. Initially, and for most of the nineteenth century, the explainer was evolution. The larger story of the development of relativism in social science is the story of the decline of evolutionary accounts, and therefore of the moral assurance that came from claiming to be at the pinnacle of evolution. Without this fixed explainer, cultural differences, interpreted in terms of neo-Kantianism, came to be seen as arbitrary. Methodological relativism arose from historicism allied to neo-Kantianism, and was applied initially to the historical or social sciences and later to ideology. From there it was extended to epistemic contexts, with many of the same issues at play: unilineal evolution, holism, and the questioning of the notion of progress and of a teleology for the sciences.

2. Explanatory structures and devices

Although there are "relativistic" comments in antiquity, there is little there that approximates the social science explanations of the nineteenth century that are the source of social science relativism. The difference is in explanatory structure: relativism in the social sciences is concerned with explanation, and uses a distinctive array of explanatory approaches. We can get a rough account of the differences in explanatory structures by looking at four key questions:

(1) Is the explanatory structure symmetric between the explananda in the class to be explained, or asymmetric, meaning that some things – scientific truth, enlightened belief, fundamental moral intuitions, the end state of historical development – are not explained in the same way or not explained at all, as explained by Charles Taylor in terms of inertia:

> The Principle of Inertia does not single out any particular direction in which bodies "naturally" tend to move or any constellation which they tend to move towards. And thus it can be said to be neutral between the different states of any system of which it may be invoked to explain the behavior (that is, where the theory of which it is one of the foundations is invoked). But this cannot be said of a principle of asymmetry, whose function is precisely to distinguish a privileged state or result (Taylor 1964, 23).

Teleological explanations, such as those of Aristotelian physics, are paradigmatic cases of asymmetric explanations. But asymmetries often are not apparent. A typical asymmetric explanation is Lecky's account of the rise of Western rationalism as the elimination of superstition: superstition is explained; the rationalism that is left after the defeat of superstition is not explained (Lecky 1865–66).

(2) What is the explananda? The explananda might include such things as culture, the *Zeitgeist*, the mentality of an epoch, paradigms or conceptual schemes, objective mind, ideologies, *Weltanschauungen*, traditions, and so forth.

(3) What is the explainer, and is there an explainer? What is it that accounts for the differences between explananda? These might vary between such things as class interests, the material conditions of cultures, the technology of an epoch, race psychologies, and so forth. Or there may be no explainer: the explananda may be taken to be themselves foundational and incapable of further explanation.

(4) What are the mechanisms of explanation? If one says that one of the explananda is "socially determined," or the product of history, how are the explainers connected to the explananda?

We can ask these questions of each of the accounts provided in social science, and then ask the question "in what sense is the account relativistic?"

3. The Saint-Simonians, Comte, and Marx

The intellectual dynamic that produces social science relativism is a response to the universalism of the enlightenment, and its evolutionism. Enlightenment thinkers, such as Turgot (1751) and Condorcet (1793), regarded intellectual development, which they construed as the progressive improvement of and wider distribution of knowledge, as the driving force of progress. The hallmarks of this approach were their non-relative concept of knowledge and superstition. These are therefore examples of asymmetric explanation. But they are also accounts of difference. Typically they appealed to stage notions: Turgot foreshadowed Comte's law of the three stages.

In both cases the primary explainer was science and its development. Science developed through an innate, asymmetric tendency. The asymmetry was a result of the favored end state of fully developed science, so these were in effect error theories about past and outdated beliefs. The mechanisms of error were taken to be cognitive, such as credulity and errors resulting from natural but mistaken inferences, such as reasoning falsely by analogy, but reinforced by social processes, such as the prestige of priests and the authority of the church or merely by the social ambience of superstition. The development in science in turn explained social development non-asymmetrically, one as a consequence of the other. This order was inverted in historical materialism, another stage theory.

The Saint-Simonians provided what was essentially a descriptive account, but it added some novel and ultimately very influential correctives to this kind of reasoning, based on a new explainer, a philosophical anthropology in which a kind of need for faith plays a role. This was a lesson of the failures of the French revolution. The key argument was a novel periodization of history into successive "organic" and "revolutionary" phases. In the organic phases, there was a strong coherence between official belief or ideology and economic, social and political organization, such that religion validated social hierarchies and social hierarchies cohered with the state of technology and economic organization of the society in question. This coherence did not last: elements changed, and the moral ideals, for example, became irrelevant to behavior, hypocritical, and even served to undermine good behavior (The Saint-Simonians 1971, 293–298).

In Saint-Simon there seems to be a telic, asymmetric tendency to the organic society, as well as an inevitable tendency to decay. Technology or science is treated in each as a largely autonomous explainer or force to which the rest of society, its hierarchies and ideologies, need to accommodate, and this accommodation seems also to be a kind of telic force, and accommodation of an asymmetric end point. In Marx, following Saint-Simon, the driving force of change is the conflict between means of production and the social organization of production, which becomes retrograde in the course of the development of a particular technological epoch.

Sorting out the answer to the list of four questions in these cases turns out to be virtually impossible, but the attempt is nevertheless revealing. They each rely on asymmetries, particularly teleological pulling forces, such as order and progress, and on an underlying philosophical anthropology driving toward a state of stability, despite explicit denials of relying on such a doctrine. But the relativistic aspects are clearer. The thought of a given epoch is not only relative to the period, but indirectly (and with some degrees of freedom) determined by the economic or technological conditions of production: this is made explicit in Marx's theory of the Überbau or superstructure (Marx and Engels 1848; Turner 2019).

This gets us a standard form of relativism: the social determination of knowledge. The explainer is technological development and its human mode of work, which in turn "determines" or influences social institutions, which influence ideas, which in turn serve to validate institutions. From Marx through the Marxian tradition, these ideas are taken to be associated with class interests: the dominant ideas are those which support the dominance of the dominant class. The oppressed class has increasingly divergent interests, and develops resistance and eventually a counter-ideology to this class, pointing to the next historical stage. This account, in the truncated form of a reductive analysis of ideas as concealed articulations of class interests, becomes a standard exoteric version of the "sociology of knowledge," though few developed versions of this field actually make this simplified claim.

4. Anthropological relativism

Descriptive relativism dates from antiquity, including travelers' stories and the recording of legal codes of European tribes by the Romans, and tales of other cultures formed a major genre prior to the Enlightenment. But with fictional traveller's books like *Rameau's Nephew* by Diderot (1805), we get to something like a relativist challenge to contemporary morality, prior to any systematic theorizing about these differences. These challenges multiply in the nineteenth century. Initial "scientific" anthropology individuated institutions, particularly kinship institutions, and used them as a gauge of the level of development they correlated – rather imperfectly – with (cf. Tylor 1889).

These stages, given a more elaborate form by Lewis Henry Morgan (1877), has proven to be enduring. The current distinction, between hunter-gatherer, tribal, and state societies, reproduces it. These were relativist explanatory accounts, in the sense that the contents of social life were relativized to stages. But they were also asymmetric accounts, in the sense that the notion of progress was unexplained, and taken as given, in a different way than it was in Lecky, because the object of explanation was the society as a whole, which was characterized by a set of customs with roughly similar properties which had a coherent relation to one another. This was different than Lecky-style listings of superstitions. It also contemplated a wider range of societal options at a given stage of development.

The explanatory object was given conflicting definitions, but they converged on a general notion of culture. Tylor defined the object of explanation as follows: "Culture or civilization is that complex whole which includes knowledge, belief, art, morals, law, custom and any other capabilities and habits acquired by man as a member of society" (Tylor 1871, 1). This combined two dominant ideas: holism about culture and the basis of culture in the explanatory fact of "society" and membership in society. The further determinant of development, for Tylor, was religion, whose forms corresponded to the three stages: savagery to animatism, barbarism to polytheism, and civilization to monotheism.

Independent beliefs, as distinct from those organized into wholes, were the source of a different form of social science relativism. James G. Frazer, in his classic *The Golden Bough* (1890), was not a holist: he identified discrete myths and story forms in many cultures and compared them without respect to context. This produced its own form of descriptive relativism, shared with writers like Robertson Smith, which undermined Christian claims to exclusive validity by showing the presence of similar tropes in primitive Semitic tribes. Vilfredo Pareto did the same thing by comparing Christian baptism to lustral rites (Pareto 1935, §863). This form of relativization – relativization *to* primitive beliefs – was the most potent of cultural weapons, when combined with evolutionism.

Human psychology itself became the subject of relativistic and anti-relativistic interpretations. One argument was that the customs of primitive groups were the result of degeneration: a line that preserved an asymmetric form of explanation in which developed societies did not need to be explained. This was generally rejected by anthropologists. One view, championed and later abandoned by Lucien Levy-Bruhl, was that primitives had a different psychology, oriented to "mystical participation" (1975) rather than causal reasoning. There was a similar claim to the effect that primitive magic was based on false causality, replaced by religion, which was based on uncertainty and prayer, which in turn was replaced by science, with its true correlations. But Frazer, unlike Levy-Bruhl, accounted for progress by a teleological conception of the tendency of thought to perfect itself (Harris 1968, 205). Sapir and Whorf argued for a kind of linguistic determination of psychology that led to a radical form of cognitive relativism (Whorf 1940).

5. Neo-Kantianism and culturalism

Paralleling these (mostly English language) developments was a complex confluence of ideas that result in what Tooby and Cosmides call the standard social science model, which is intrinsically relativistic, because it holds that culture is free to vary in any direction with respect to any trait (1992, 24), and is almost completely determinative of cognitive processes. Leaving aside the controversy over whether this is a straw man argument, there is some truth in it. Consider Margaret Mead's image of culture (1928, 13) as a selection of features from a basket. This emphasizes the arbitrary basis of culture.

The arbitrariness of the basis of culture is a claim that follows from the internal logic of neo-Kantianism, which formed an important influence on sociology and anthropology, including through Mead's teacher, Franz Boas. The fact that Boas, as a young geographer living among Eskimos, took along and read a copy of Kant is almost too good to be true, but it captures the issue. The neo-Kantians, notably Hermann Cohen, were the source of a particular model of analysis based on Kant but generalized into the "transcendental method." This method worked on coherent intellectual systems (the original example was physics, then law), and purported to identify the a priori logical or conceptual conditions for these systems: their logically necessary presuppositions. This was itself a form of relativism: the presuppositions were relativized to the organized body of concepts that made up the system. But to be "necessary" logical conditions meant to entail concepts in the system, and therefore the results of this presuppositional analysis had to be unique. This proved to be a fatal feature in the neo-Kantian account of physics, when, during the Einsteinian revolution, it was discovered that the same physical theory could be articulated in, and thus "presuppose," different and incompatible mathematical systems. Moreover, the transcendental conditions were foundational: there was nothing logically prior to the concepts it identified.

The foundational character of "presuppositions" was what made this method attractive and generalizable. It could be and was applied to culture, and the "historical a priori" applied it to historical epochs, considered as intellectual unities. But the original form of the neo-Kantian transcendental method, following Kant, was resolutely anti-psychological: the presuppositions were supposed to underpin a conceptual order, not mental processes. In its application to culture and history, however, it was necessarily something possessed by some individuals and not others. This led to the commonplace notion of "shared presuppositions" as the basis for cultures, perspectives, thought communities, and so forth. There were many variations on these ideas, including some that could be construed as forms of "social determination." Thus Durkheim and Mauss, in a text on classification, argued that primitive classification systems of natural phenomena reproduced the social classifications of primitive societies (Durkheim 1903).

The general idea that tacit assumptions, described in a variety of ways, were the determinants of opinion, became a cliché. The claim is expressed by Carl Becker, in his study of the Enlightenment, where he called this the climate of opinion:

> Whether arguments command assent or not depends less upon the logic that conveys them than upon the climate of opinion in which they are sustained. What renders Dante's argument or St. Thomas' definition meaningless to us is not bad logic or want of intelligence, but the medieval climate of opinion – those instinctively held preconceptions in the broad sense, that Weltanschauung or world pattern – which imposed upon Dante and St. Thomas a peculiar use of the intelligence and a special type of

logic. To understand why we cannot easily follow Dante or St. Thomas it is necessary to understand (as well as may be) the nature of this climate of opinion.

(Becker 1932, 5)

This was relativistic in the sense that opinion was relative to the "instinctively held preconceptions" of a time and group, and that comprehension and agreement depended on the sharing of these preconceptions. By the 1980s, writers like Clifford Geertz would speak of culture as "what the mind full of presuppositions concludes" (1983, 84). Throughout the spread of this idea there was little attention to its philosophical origins. It simply became a conventional way of speaking about culture and the variant applications of culture and related concepts.

Versions of this basic model were tied to the sociology of knowledge in the twentieth century, most famously in Karl Mannheim's *Ideology and Utopia* ([1929] 1936). Mannheim renovated the Marxian idea of the social determination of knowledge, but his account of this relation was incoherent, and was subject to a comprehensive critique by Robert K. Merton (1968, 543–562). Subsequent Marxian accounts of the theory of superstructure and of ideology, for example by Althusser (1984), extended the notion of ideology to all thought, but were similarly vague on the mechanisms by which the thought was constrained and produced by, and thus relative to, political interests (initially unconscious mechanisms, later "interpellation," the system of social interaction itself).

6. Methodological relativism

Methodological relativism had a more direct source in neo-Kantianism, as a consequence of a sustained controversy over the status of historical knowledge, which was taken to be knowledge of individuals and individual events, and the contrast between it and knowledge in the physical sciences, which was taken to be knowledge of principles. Dilthey extended the problematic to the perspectives of the historical actor, and thus, reflexively, to the historian, whose perspective was shaped by both their historical conditions and the conventions of historical thinking of their time (Mul 2004, 311).

The most important interpretation of these issues in neo-Kantian terms was Max Weber's "Objectivity in Social Science and Social Policy" (1904), which combined an account of the pluralism of disciplinary approaches with an account of how problems in the historical sciences became defined, or salient, as a result of historical changes. In this text he used the notion of presupposition in the manner that became conventional in the social sciences, but applied it consistently to the social sciences themselves.

As "presuppositions," the foundational basis of the different sciences could be compared, but not ranked, nor was a higher synthesis possible. Methodological diversity was irreducible. Moreover, what was "meaningful" was meaningful as a result of the constitutive presuppositions that made them meaningful. These were understood as at least partly valuative or value laden, and this included *Weltanschauungen* or world views, the common sense of a period. Weber took values to be a matter of ungroundable decision, so this picture was rooted in radical value-relativism, and he argued that value-related topics, such as history and economics, necessarily constituted their objects of study on the basis of these ultimately ungroundable and choices. His picture of "science," was thus relativistic in two major directions: in relation to historically changing values and in terms of disciplinary approaches. But Weber exempted basic logic and calculation from these relativistic considerations, meaning that the analysis of facts constituted by values could and should nevertheless be value-free.

Although this picture was too complex to be taken up as a dogma, its elements, including the fact-value distinction and the idea that the factual part of social science should be value-free, and the relativistic view or values, came to be part of conventional social science, though they were subject to various kinds of (mostly misinformed) criticism.

7. Epistemic relativism

Despite exceptions, including Marxists such as Bukharin, who wrote that "Any science, whatever it be, grows out of the demands of society or its classes" (Bukharin 1921, 1), science was not treated as subject to relativistic explanation. The emphasis of earlier relativism was on ideology, the diversity of morals, customs, and so on. A crucial change occurred as a result of Evans-Pritchard's study of the Azande, an African tribe with a coherent ideology of witchcraft that seemed immune to refutation on its own terms. This example was repeatedly discussed in the philosophical literature, though as a model for accounting for pseudosciences such as Marxism and psychoanalysis. In the 1960s the Azande became the subject of a major Oxford debate (Lukes 2000) which in turn became the basis for two major collections of essays, *Rationality* (Wilson 1970) and *Rationality and Relativism* (Hollis and Lukes 1982).

The issue in these books derived from a claim by Peter Winch. Winch had argued that societies were constituted by and could only be understood through their own concepts (1958). But he also raised the question of whether different societies could have different concepts of rationality. This in turn raised questions about translatability, the possible need for a common rational "bridge" between cultures as a condition of understanding, and questions about the nature of concepts themselves. This discussion was eventually cut off by the recognition of the significance of Donald Davidson's "The Very Idea of a Conceptual Scheme," which denied that the notion of radically different conceptual schemes was intelligible, on the ground that we could not know there was one if we encountered it (Davidson 1974).

A second source was Thomas Kuhn's *The Structure of Scientific Revolutions* (1970). Kuhn's book was subject to a vast amount of interpretation, some of which reached back to prior relativistic perspectives, notably that of Karl Mannheim's sociology of knowledge, which had exempted science (Mannheim 1929, 244; cf. 1952, 193). A movement called the "Strong Programme in the Sociology of Science" focused on scientific knowledge as a subject of explanation, explicitly rejecting this exemption (Barnes et al. 1996).

A basic methodological rule adopted by this movement was explanatory "symmetry," meaning that "error" and scientific truth needed to be explained in the same way. The core idea was simple: scientific beliefs changed, but the world it explained did not, so the differences needed to be explained, and could not be explained by reference to "facts" that had not changed. It was further argued that there was no reason to think that normal psychological processes applied to scientific beliefs in the same way as other beliefs.

With respect to scientific truth, this explanatory program and its successors did not go beyond Kuhn's relativism. But because it presented itself as scientific, it raised issues of reflexivity and self-exemption, which were extensively debated (Ashmore 1989; Pickering 1992). One alternative view was that symmetry could be treated as a methodological precept only, immunizing it from considerations of reflexive consistency. The "Strong Programme" was also misleadingly characterized as asserting that scientific facts were determined by "social forces," which would make it into a different kind of relativism, in which the "relative" part would be to some external kind of social fact, such as class interests. This was not the argument: rather the claim was that epistemic norms, norms of inference, for example, needed to be treated as part of a determinate social context and social activity and thus explained as are other social norms, rather than being treated as having universal, ahistorical, validity (Turner 2003).

References

Althusser, L. (1984), *Essays on Ideology*, London: Verso.

Ashmore, M. (1989), *The Reflexive Thesis: Wrighting Sociology of Scientific Knowledge*, Chicago: University of Chicago Press.

Barnes, B., D. Bloor and J. Henry (1996), *Scientific Knowledge: A Sociological Analysis*, Chicago: University of Chicago Press.

Becker, C. (1932), *The Heavenly City of the Eighteenth-Century Philosophers*, New Haven: Yale University Press, 1959.

Bukharin, N. I. (1921), *Historical Materialism: A System of Sociology*, Ann Arbor, MI: The University of Michigan Press, 1969.

Comte, A. (1855), *The Positive Philosophy of Auguste Comte, Freely Translated and Condensed by Harriett Martineau*, New York: C. Blanchard.

Condorcet, Marquis de (1793), "Fragment on the New Atlantis, or Combined Efforts of the Human Species for the Advancement of Science," in *Condorcet: Selected Writings*, edited by K. M. Baker, Indianapolis: Bobbs-Merrill, 283–300, 1976.

Davidson, D. (1974), "The Very Idea of a Conceptual Scheme," in *Inquiries into Truth and Interpretation*, edited by D. Davidson, Oxford: Clarendon Press, 183–198, 1984.

Diderot, D. (1805), *Rameau's Nephew*, translated by I. C. Johnston, 2002, http://gutenberg.net.au/ebooks07/0700101h.html.

Durkheim, É. (1903), *Primitive Classification*, edited by R. Needham, translated by E. Sagarin, Chicago: University of Chicago Press, 1963.

Frazer, J. G. (1890), *The Golden Bough*, New York: MacMillan and Co.

Geertz, C. (1983), *Local Knowledge: Further Essays in Interpretive Anthropology*, New York: Basic Books.

Harris, M. (1968), *The Rise of Anthropological Theory: A History of Theories of Culture*, New York: Columbia University Press.

Hollis, M. and S. Lukes (eds.) (1982), *Rationality and Relativism*, Cambridge, MA: MIT Press.

Kuhn, T. (1970), *The Structure of Scientific Revolutions*, third edition, Chicago: University of Chicago Press, 1996.

Lecky, W. E. H. (1865–66), *History of the Rise and Influence of the Spirit of Rationalism in Europe*, 2 vols., London: Longman's Green & Co.

Lévy-Bruhl, L. (1975), *The Notebooks on Primitive Mentality*, translated by P. Riviere, Oxford: Basil Blackwell.

Lukes, S. (2000), "Different Cultures, Different Rationalities?" *History of the Human Sciences* 13(1): 3–18.

Mannheim, K. (1929), *Ideology and Utopia: An Introduction to the Sociology of Knowledge*, translated by L. Wirth and E. Shils, New York: Harcourt, Brace & World, 1936.

———. (1952), "Competition as a Cultural Phenomenon," in *Essays on the Sociology of Knowledge*, edited by K. Mannheim, London: Routledge & Kegan Paul, 191–229.

Marx, K. and F. Engels (1848), *The Communist Manifesto*, London; New York: Verso, 1998.

Mead, M. (1928), *Coming of Age in Samoa: A Psychological Study of Primitive Youth for Western Civilization*, New York: W. Morrow.

Merton, R. (1968), "Karl Mannheim and the Sociology of Knowledge," in *Social Theory and Social Structure*, enlarged edition, edited by R. Merton, New York: The Free Press, 543–562.

Morgan, L. H. (1877), *Ancient Society or Researches in the Lines of Human Progress from Savagery, Through Barbarism to Civilization*, New York: Henry Holt & Co.

Mul, J. de (2004), *The Tragedy of Finitude: Dilthey's Hermeneutics of Life*, translated by T. Burrett, New Haven and London: Yale University Press.

Pareto, V. (1935), *The Mind and Society*, New York: Harcourt Brace.

Pickering, A. (ed.) (1992), *Science as Practice and Culture*, Chicago: University of Chicago Press.

The Saint-Simonians (1971), "On Moral Education," in *French Utopias: An Anthology of Ideal Societies*, edited by F. E. Manuel and P. M. Fritizie, New York: Schocken Books, 285–292.

Taylor, C. (1964), *The Explanation of Behavior*, New York: Humanities Press.

Tooby, J. and L. Cosmides (1992), "The Psychological Foundations of Culture," in *The Adapted Mind: Evolutionary Psychology and the Generation of Culture*, edited by L. Barkow, L. Cosmides, and J. Tooby, Cambridge: Oxford University Press, 119–136.

Turgot, A-R. J. ([1751] 1973), "Plan for Two Discourses on Universal History," in *Turgot on Progress, Sociology and Economics: A Philosophical Review of the Successive Advances of the Human Mind on Universal History Reflections on the Formation and the Distribution of Wealth*, edited and translated by R. L. Meek, Cambridge: Cambridge University Press, 63–118.

Turner, S. (2003), "The Third Science War," *Social Studies of Science* 33(4): 581–611.

———. (2019), "The Philosophical Origins of Classical Sociology of Knowledge," in *The Routledge Handbook of Social Epistemology*, edited by M. Fricker, P. J. Graham, D. Henderson, N. Pedersen and J. Wyatt, London: Routledge, 31–39.

Tylor, E. B. (1871 [1958]), *Primitive Culture*, 2 vols., New York: Harper Torchbook.

———. (1889), "On a Method of Investigating the Development of Institutions, Applied to Laws of Marriage and Descent," *Journal of the Royal Anthropological Institute* 18: 245–269.

Weber, M. (1904), "'Objectivity' in Social Science and Social Policy," in *Max Weber: Collected Methodological Writings*, edited by H. H. Bruun and S. Whimster, translated by H. H. Bruun, London and New York: Routledge, 100–138.

Whorf, B. L. (1940), "Science and Linguistics," *Technology Review* 42(6): 229, 231, 247–248.

Wilson, B. R. (ed.) (1970), *Rationality*, New York and London: Harper & Row.

Winch, P. (1958), *The Idea of a Social Science: And Its Relation to Philosophy*, London: Routledge & Kegan Paul.

45

RELATIVISM IN THE PHILOSOPHY OF ANTHROPOLOGY

Inkeri Koskinen

1. Introduction

A chapter on relativism in the philosophy of anthropology should cover two quite different perspectives on its topic. On the one hand, it must deal with arguments presented by philosophers, and on the other, with ideas and practices developed by anthropologists. Sometimes the two perspectives complement each other. Both anthropologists and philosophers have taken part in some of the same debates about relativism; philosophers' arguments have influenced anthropological theory and ethnographic research practices; and ideas developed in anthropology have inspired philosophical discussions. On other occasions, however, there has been little communication across disciplinary boundaries, and sometimes the communication has been ridden with misinterpretations.

I will mostly focus on arguments presented in the tradition of analytic philosophy, and on ideas and practices developed in social and cultural anthropology. For the past few decades, however, connections between anthropological theory and philosophy have largely happened through continental thinkers. I will explore the rather fragmented interactions between the mainstream of professional philosophy and this literature. Within these limits, this chapter covers roughly a century's worth of discussions and debates.

2. Relativism in anthropology

Popular conceptions of anthropology often take cultural relativism to be its *sine qua non*. But though it is still a necessary part of any undergraduate anthropology curriculum, in the writings of contemporary anthropologists it is hard to find the thoroughgoing cultural relativism endorsed in the American cultural anthropology in the 1940s and 50s.

Cultural relativism became an important idea in anthropology in a specific social and intellectual context. It was developed as a part of the criticism directed against the evolutionary views of nineteenth-century anthropologists such as James Frazer and Edward Tylor. Franz Boas and his students, following similar ideas presented in Europe, rejected as ethnocentric and racist the way in which the evolutionists classified cultures on a scale ranging from primitive to modern (Westermarck 1932; Boas 1940; Benedict 1934; Herskovits 1955). According to these critics, anthropologists should be wary of using their own cultural norms when evaluating the cultures they study. In other words, from the well-documented relativity of e.g. moral judgements and epistemic practices, they proceeded to the relativistic claim that anthropologists could or should not move beyond this relativity.

The multifaceted and often unclear idea of cultural relativism was formulated most radically in the writings of Boas' students, particularly Melville Herskovits, who emphasised "the validity of every set of norms for the people whose lives are guided by them" (1948, 76), and formulated the most cited definition of cultural relativism: "Judgements are based on experience, and experience is interpreted by each individual in terms of his own enculturation" (1955, 15).

Cultural relativism was met with immediate criticism (e.g. Williams 1947; Kluckhohn 1955). The most common worry was that it would lead to moral nihilism. After Herskovits, it is hard to find anyone who endorsed a full-blown form of it. For instance, Clifford Geertz, in his interpretivist reading of cultural relativism, defends it to a certain degree, but also notes its shortcomings:

> The truth of the doctrine of cultural (or historical – it is the same thing) relativism is that we can never apprehend another people's or another period's imagination neatly, as though it were our own. The falsity of it is that we can therefore never genuinely apprehend it at all.
>
> *(Geertz 1983, 44; see also Geertz 1984)*

Cultural relativism is linked to several other, sometimes conflicting forms of relativism. Moreover, at least one form of relativism that rejects the idea of culture has recently been defended in the anthropological literature.

Let us loosely characterise relativism as the claim that the propositions of a certain domain (such as knowledge, ethics, or rationality) are true or false only relative to an underlying set of standards, of which there can be several. Moreover, such sets of standards must not be ranked, and one cannot choose between different sets on the basis of a neutral rational comparison (Kusch 2016). The often-noted ambiguity of cultural relativism stems from its definition only mentioning the framework encompassing a set of standards: culture. It is not clear what domains exactly it is supposed to cover. When also the domain is named, cultural relativism gets disintegrated. We can distinguish several forms of relativism where culture is often named as the overarching framework to which propositions of some domain are relative.

Here I will focus on the following forms of relativism:

(1) Conceptual relativism, according to which conceptual frameworks are or can be incommensurable, and they shape human thought so thoroughly that statements made in two different conceptual frameworks cannot be meaningfully compared.
(2) Relativism about rationality, according to which judgements of rationality are relative to a framework, and there is no neutral criterion of rationality.
(3) Moral relativism, according to which moral judgements are relative to some framework, and there is no neutral criterion for adjudicating between conflicting moral judgements made in different domains.
(4) Epistemic relativism, according to which there are several knowledge systems, and there is no neutral criterion for adjudicating between conflicting claims made in different systems.
(5) And finally, ontological relativism, according to which there are many different ontologies in which different kinds of objects emerge, and there is no one privileged ontology. As we shall see, the proponents of this form of relativism (if it indeed is a form of relativism) usually reject the notion of culture, thus breaking with cultural relativism altogether.

Before focusing on discussions and debates around these forms of relativism, it is however necessary to note that in anthropology, questions about relativism arise first and foremost in the context of ethnographic practice.

3. Methodological relativism in ethnography

Even though no single, well-defined relativistic stance can be claimed to be generally accepted in anthropology, a form of methodological relativism is very common. Ethnographers generally avoid the appraisal of their informants' knowledge claims and moral views. Even a harsh critic of cultural relativism such as Ian Jarvie agrees that if cultural relativism is interpreted purely as a methodological approach, a crucial factor of which is the suspension of judgement and censure, it "is co-terminus with good anthropology" (2006, 582).

Many misunderstandings between philosophers and anthropologists originate from different approaches to questions about relativism. Philosophers are typically interested in justifying philosophical stances. But in modern anthropology questions about relativism surfaced for practical and methodological reasons. With the rejection of evolutionary hierarchies, anthropologists had to accommodate the anti-racist conviction that there was but a single humanity, with the radical diversity evident for any ethnographer (Boas 1940; Haines 2007; Theunissen 2017). And methodological relativism offers a solution to a number of issues that arise in ethnographic practice: one must avoid ethnocentric bias, show respect to the informants, and refrain from hasty interpretations, as they could lead to misunderstandings. It is a practice rather than a stance, a way to encounter the observed radical differences without diminishing them.

The suspension of judgement is, however, limited in scope. Anthropologists typically treat the views of their colleagues very differently from those of their informants: colleagues do face criticism. Talal Asad has expressed this practical difference aptly when reflecting on a disagreement with Ernst Gellner: "In taking up a critical stance toward his text I am contesting what he says, not translating it, and the radical difference between these two activities is precisely what I insist on" (1986, 156).

4. Conceptual relativism and cultural translation

Foreign people and their exotic mores have throughout the times inspired relativistic musings, and for the past century, fictional anthropologists and linguists attempting to understand alien cultures and languages have often appeared in philosophical thought experiments. One of the best-known ones was presented by Willard van Orman Quine in 1960. He describes a linguist who attempts to understand the language of a hitherto not contacted people, and calls the task that of *radical translation*. The example of the linguist in an unlikely situation is used to make a philosophical point about the indeterminacy of translation: the linguist can never be sure of having correctly translated the utterances of the people he studies, as many incompatible conceptual schemes can account for their verbal behaviour. As Quine (1960) notes, the doctrine of indeterminacy of translation will sound familiar to readers acquainted with Ludwig Wittgenstein's remarks on meaning. Wittgenstein argued that meaning cannot be private; that it is constituted by the public use of words; and that we cannot understand individual sentences of a language without understanding the language as a whole (Wittgenstein 1953, 1958).

The idea of languages where each sentence has meaning only as a part of a whole framework, together with the idea that we have no way of assessing whether we have succeeded in translating between two languages, can be interpreted in a way that leads to conceptual relativism.

A linguistic and cultural variant of such relativism would claim that conceptual frameworks are tied to natural languages, that natural languages are integral parts of human cultures, and that statements made in two different cultures cannot be meaningfully compared. And a form of relativism following similar lines, though emphasising also a link between language and world view, had already been developed in anthropology and linguistics. It is often ascribed to Boas' student Edward Sapir (1929), and his student Benjamin Whorf (1956; see also Hoijer 1954), and referred to as the Sapir-Whorf Hypothesis, even though they never formulated any thesis together. Sapir stressed the differences between different languages and argued that language influences human thought. Whorf continued by suggesting that the differences between different languages lead to drastically different ways of thinking and perceiving: "the world is presented in a kaleidoscopic flux of impressions which has to be organized by our minds – and this means largely by the linguistic systems in our minds" (Whorf 1956, 213).

Later, philosophical ideas about conceptual schemes and translation became influential in interpretative anthropology where ethnography was often understood as cultural translation. Ethnographers of course could not accept a total lack of translatability between different cultures. However, as Geertz (1973; see also Risjord 2007) realised, combined with hermeneutical views of understanding, Wittgensteinian ideas could be used to justify established ethnographic practices. And thus Geertz stressed that culture is public and observable, not private. Even if the ethnographer's language differs from the language used in the culture the ethnographer studies, the former language could be expanded in a way that made it possible to translate expressions that originally gained their meaning in the latter.

Also philosophers have repeatedly rejected strong forms of conceptual relativism. The best-known argument against conceptual relativism was presented by Donald Davidson (1974). He claimed that the idea of alternative, incommensurable conceptual schemes is untenable: for the schemes to be truly incommensurable, we must fail to translate between them. But if translation fails, we have no reason to claim that what we are attempting to translate is in fact a *language*, not just random noise. It follows that to identify something as language, we have to be able to translate it. Davidson further argues that if we are faced with the task of radical translation, we have no other option than to adopt the principle of charity: we must assume that the speakers whose words we attempt to translate are rational and share some beliefs with us. This is because we cannot assign meanings to utterances without knowing the speaker's beliefs, nor can we learn the beliefs without understanding what the utterances mean. According to Davidson, the only way out of the impasse is to assume a common rationality and some shared beliefs. (Davidson 1974; see also Henderson 1987; Risjord 2000.)

5. Rationality and relativism

The so-called rationality and relativism debates were sparked by Wittgensteinian ideas about language and rationality, and especially by Peter Winch's famous article "Understanding a Primitive Society" (1964; see also Winch 1958). Both Wittgenstein and Winch maintained that we must question not only the applicability of our familiar concepts, but also our norms of rationality, when evaluating alien cultures. Wittgenstein (1967) found fault with James Frazer for treating religious practices as mistakes, whereas Winch turned his critical eye towards Evans-Pritchard's (1937) famous ethnography of the South-Sudanese Azande.

Winch pays particular attention to a set of Zande beliefs that Evans-Prichard claimed were formally inconsistent. The clearest case of apparent irrationality he describes is related to Zande witchcraft. The Azande held that one can identify a witch through the use of a poison oracle, and that the substance that causes witchcraft is inherited from fathers to sons and from mothers to daughters. In other words, a positive result from the poison oracle would implicate a whole paternal or maternal line of descent. And if the number of positive results were high enough, they should prove that all living Azande were witches. But according to Evans-Prichard "Azande see the sense of the argument but they do not accept its conclusions" (1937, 24).

Winch argued that it is wrong to assume that the Azande were doing something comparable to scientific explanation and confirmation of hypotheses: "Oracular revelations are not treated as hypotheses and, since their sense derives from the way they are treated in their context, they therefore are not hypotheses" (Winch 1964, 312). The apparent irrationality of the Zande beliefs is a result of Evans-Prichard using our familiar standards of rationality when assessing Zande thought (Winch 1964; see also Winch 1958).

Winch's views provoked objections, leading to a lengthy controversy. Its key papers have been published in two collections (Wilson 1970; Hollis and Lukes 1982). The debate was understood to be between two positions: relativism versus a position that treats rationality as universal. Many philosophers, anthropologists, and sociologists agreed that there is something of a shared core rationality; though there was no agreement on the nature and central features of such rationality. The debate remained closely connected to philosophical questions about translation, as the central problem of apparent irrationality was thought to arise when interpretation fails. Some appealed to arguments in the proximity of Davidson's (1974) "principle of charity": if we are to understand foreign cultures, we must attribute some shared principles of rationality to them.

Some anthropologists, particularly cognitive anthropologists (e.g. Sperber 1982), took part in the debate, arguing against Winch's views. But many of the philosophers engaged in the debate mainly used the threat of relativism as a foil in their defences of different conceptions of rationality, which may not have been conducive to much engagement with anthropology. And the anthropological reaction to the whole controversy was at times fairly critical towards all parties. This was because many anthropologists found the philosophers' general disregard for empirical facts disconcerting. For instance, according to Ernest Gellner (1968) Evans-Pritchard once pointed out that both Winch and Winch's critic, Alasdair MacIntyre, wrongly referred to cattle in Azande culture. The Azande do not have cattle.

6. Moral relativism and human rights

In anthropology discussions about cultural relativism have focused much more on moral issues. The debate about cultural relativism understood as moral relativism started in 1947 when the American Anthropological Association (AAA) published a critical statement on the UN's Universal Declaration of Human Rights. The statement's main author was Herskovits. Some of the key points of the criticism were that the declaration privileged individual rights; that it ended up disregarding colonial cultural oppression; and that respect for cultural differences demanded the acceptance of different moral codes:

> Standards and values are relative to the culture from which they derive so that any attempt to formulate postulates that grow out of the beliefs or moral codes of one culture must to that extent detract from the applicability of any Declaration of Human Rights to mankind as a whole.
>
> *(AAA 1947, 542)*

The anthropological community's reaction to the association's statement has often been described as embarrassment. The reason for embarrassment, however, has undergone changes over the years. In their initial comments H. G. Barnett (1948) and Julian H. Steward (1948) did question the moral viability of the statement's overarching tolerance, but were more concerned with its lack of scientific rigor. Steward (1948, 351) concluded that "the statement is a value judgment any way it is taken," and thus jeopardised the association's reputation as a scientific organisation. Later, especially when some human-rights violators had invoked the kind of moral relativism the statement endorsed, the focus shifted to worries about moral nihilism (Geertz 1984; Engle 2001).

Compared with the original declaration, the current human-rights discourse is less dominated by Western legal thought, and worries about cultural rights resembling some of the ones expressed in the 1947 statement have been brought up by thinkers from developing countries, as well as by representatives of Indigenous peoples. Over the years many anthropologists have lamented that the 1947 statement prevented anthropological involvement in discussions about human rights. Many attempts have been made to formulate positions that would make anthropological viewpoints taken more into account in the human-rights discourse. Finally in 1999, the AAA adopted a declaration on anthropology and human rights, which the association's Human Rights Committee has called a "complete turnaround" from the stance taken in the 1947 statement (AAA 1999; Engle 2001; Goodale 2006; Brown 2008).

7. Epistemic relativism in the postcolonial critique of anthropology

Postcolonial theory is often ambivalent in its treatment of relativism. From the literature it is easy to find both statements that sound fiercely relativistic, and statements that are resolutely antirelativist. These latter positions have often been overlooked by the critics of overtly political or "postmodern" theorising. For instance Paul Boghossian claims that in many fields in the humanities and the social sciences it is by now orthodoxy to claim that there are "many radically different, yet 'equally valid' ways of knowing the world, with science being just one of them" (Boghossian 2006, 2). Some of his examples point towards postcolonial theory, and indeed, notions such as "alternative epistemologies" and "different ways of knowing" abound in postcolonial literature. They are sometimes interpreted as expressions of postmodern epistemic relativism (mainly called so by its critics). This Nietzschean and Foucauldian form of relativism takes knowledge to be always perspectival, partial and tied to power structures: it is the dominant social group or culture that gets to set its own criteria for assessing knowledge claims as universally valid (Baghramian 2004, 79–88; Foucault 1977/1980).

Postcolonialism is overtly political. Already in early key works (Fanon 1952; Deloria 1969; Said 1978), a part of the political criticism was directed towards Western science, and particularly anthropology, which was claimed to have served colonial rule. But while science is criticised, the focus in the multifaceted postcolonial literature is usually on social injustices colouring every aspect of the lives of the oppressed, the demand for liberation, and the development of new, more socially just research methods. Not on the formulation or defence of some relativistic stance. And indeed, rather than on relativism, the countercriticism in anthropology has focused on cultural essentialism and the reintroduction of the romantic notion of "the native" (Kuper 2003; McGhee 2008).

In fact, the political aims of postcolonial theory can lead to antirelativism. Influenced by critical theory and feminist philosophy, some postcolonial thinkers embrace Marxist forms of universalism, and some reject epistemic relativism in a way that echoes arguments presented in standpoint theory (see Figueroa and Harding 2003; Harding 2011). For instance, Boaventura de Sousa Santos argues in his influential book *Epistemologies of the South* (2016) that "epistemicide" is an integral part of systematic colonial oppression. According to him, scientific knowledge is limited in ways that prevent it from grasping "the inexhaustible diversity of the world" (2016, 108). He argues that different epistemologies, particularly ones endorsed in the emancipatory movements of the "global South," should complement scientific knowledge. But he rejects epistemic relativism as an unsuitable position for anyone striving for social emancipation: "If all the different kinds of knowledge are equally valid as knowledge, every project of social transformation is equally valid or, likewise, equally invalid" (Santos 2016, 190).

So the claim that there are multiple epistemologies or knowledge systems can be interpreted as pluralistic rather than relativistic (see Chang, this volume). However, such postcolonial pluralism does not always extend to science. As Arun Agrawal (1995) has noted, mistrust in science has led many postcolonial critics to treat science as a single, coherent knowledge system, opposed by the multitude of oppressed knowledge systems. Nevertheless, the aim of many postcolonial scholars is to integrate science more closely with Southern, Indigenous etc. knowledge systems, and to develop decolonised research methods. Still, power asymmetries are often taken to impede such integration, as the marginalised knowledge systems can easily get either misrepresented, or as in e.g. ethnomedicine, commercially exploited (Smith 1999; Denzin et al. 2008; Ludwig 2016).

The postcolonial critique of science and the development of postcolonial methodology have engendered new forms of research, such as Indigenous activist research. In anthropology, they have altered ethnographic research practices. Co-research and participatory projects, where anthropologists collaborate with representatives of the people they study, have become more common. As I have argued elsewhere (Koskinen 2014), these developments challenge methodological relativism in anthropology. When informants become co-researchers in participatory projects, suspension of judgement is no longer an either epistemically or ethically viable attitude towards their views.

8. Ontological relativism and anthropology without culture

The central role that the concept of culture has long held in anthropology has been questioned in many ways during the past fifty years. The influences of Marxism and postcolonialism, and in 1986 the publication of *Writing Culture* (Clifford and Marcus 1986), a seminal collection of essays addressing the rhetorical techniques ethnographers use to establish their epistemic authority, added to an already existing uneasiness. Ethnographies depicted unified cultures, thus hiding conflicts from view. Moreover, as Roy Wagner (1975) had noted, if concepts gained their meaning in their cultural contexts, then surely this applied also to the concept of culture. And indeed, for instance Marilyn Strathern (1980) has claimed that the Hagen do not have anything resembling our distinction between the invariant nature on the one hand and culture as human elaboration upon it on the other. Ethnographers had to consider the possibility that the concept of culture might not be in any meaningful way applicable to the people they study, and that using it might be misleading (Kuper 1999; Risjord 2007).

The recent "ontological turn" in anthropology has added new arguments to the arsenal of the anthropologists who wish to do away with the concept of culture. Not only do the ontological anthropologists prefer to use local concepts instead of overarching ones – such as culture – but they have questioned the whole representationalist idea embedded in the notion of culture. According to these critics, both cognitive and interpretative anthropological theories agree that there is one world, which different cultures represent in diverse ways. This dualist position is what the ontological turn rejects. Anthropology is not translation between worldviews or cultures (Henare et al. 2006 Sivado 2015; Holbraad and Pedersen 2017; Heywood 2017).

In practice, the aim is "to take things encountered in the field as they present themselves, rather than immediately assuming that they signify, represent, or stand for something else" (Henare et al. 2006, 2). And instead of focusing on speech and human interactions, ontological anthropology pays attention to the different objects that emerge in different human-nonhuman interactions. Many of its proponents reject the idea of matter and ideas as distinct categories, often citing Gilles Deleuze's and Bruno Latour's ideas. This leads to talk about plural ontologies.

It is not entirely clear whether the ontological turn entails ontological relativism. As Martin Paleček and Mark Risjord (2012) note, ontological anthropologists hold that different objects emerge in different networks of interaction, and no ontology is privileged over others. However, the view can also be interpreted as perspectivism, or as David Ludwig (2018) has argued, pluralism. But if it is interpreted as a form of relativism, then from a philosophical point of view it is an interesting one. Paleček and Risjord claim that it is immune to the Davidsonian critique, which rests largely on the principle of charity and the identification of incommensurability with a failure of translation. The ontological anthropologists explicitly reject the idea of anthropology as translation, as well as any representationalist distinction between scheme and content. So if the view is interpreted as a form of relativism, it could spur new philosophical debates about relativism in anthropology.

References

Agrawal, A. (1995), "Dismantling the Divide between Indigenous and Western knowledge," *Development and Change* 26: 413–439.

American Anthropological Association [AAA] (1947), "Statement on Human Rights," *American Anthropologist* 49: 539–543.

———. (1999), "Declaration on Anthropology and Human Rights," https://www.americananthro.org/ConnectWithAAA/Content.aspx?ItemNumber=1880 (Accessed September 26, 2019).

Asad, T. (1986), "The Concept of Cultural Translation in British Social Anthropology," in *Writing Culture*, edited by J. Clifford and G. Marcus, Berkeley: University of California Press, 141–164.

Baghramian, M. (2004), *Relativism*, New York and London: Routledge.

Barnett, H. G. (1948), "On Science and Human Rights," *American Anthropologist* 50: 352–355.

Benedict, R. (1934), *Patterns of Culture*, Boston and New York: Houghton Mifflin.

Boas, F. (1940), *Race, Language, and Culture*, New York: Macmillan.

Boghossian, P. (2006), *Fear of Knowledge: Against Relativism and Constructivism*, Oxford: Clarendon Press.

Brown, M. F. (2008), "Cultural Relativism 2.0," *Current Anthropology* 49: 363–383.

Clifford, J. and G. Marcus (eds.) (1986), *Writing Culture*, Berkeley: University of California Press.

Davidson, D. (1974), "On the Very Idea of a Conceptual Scheme," *Proceedings and Addresses of the American Philosophical Association* 47: 5–20.

Deloria, V. Jr. (1969), *Custer Died for Your Sins: An Indian Manifesto*, New York: Palgrave Macmillan.

Denzin, N. K., Y. S. Lincoln and L. T. Smith (2008), "Introduction: Critical Methodologies and Indigenous Inquiry," in *Handbook of Critical and Indigenous Methodologies*, edited by N. K. Denzon, Y. S. Lincoln and L. Tuhiwai Smith, Los Angeles: Sage, 1–20.

Engle, K. (2001), "From Skepticism to Embrace: Human Rights and the American Anthropological Association from 1947–1999," *Human Rights Quarterly* 23: 536–559.

Evans-Pritchard, E. (1937), *Witchcraft, Oracles and Magic Among the Azande*, Oxford: Clarendon Press.

Fanon, F. (1952), *Peau noire, masques blancs*, Paris: Seuil.

Figueroa, R. and S. Harding (eds.) (2003), *Science and Other Cultures: Issues in Philosophies of Science and Technology*, New York and London: Routledge.

Foucault, M. (1977/1980), "Truth and Power. Interview with A. Fontana and P. Pasquino," in *Power/Knowledge: Selected Interviews and Other Writings 1972–1977*, edited by C. Gordon, translated by C. Gordon, L. Marshall, J. Mepham and K. Soper, New York: Pantheon Books, 109–133.

Geertz, C. (1973), *The Interpretation of Cultures: Selected Essays*, New York: Basic Books.

———. (1983), *Local Knowledge: Further Essays in Interpretive Anthropology,* New York: Basic Books.

———. (1984), "Anti Anti-Relativism," *American Anthropologist* 86(2): 263–278.

Gellner, E. (1968), "The Entry of the Philosophers," *Times Literary Supplement* 4: 347–349.

Goodale, M. (2006), "Toward a Critical Anthropology of Human Rights," *Current Anthropology* 47: 485–511.

Haines, V. (2007), "Evolutionary Explanations," in *Philosophy of Anthropology and Sociology*, edited by S. P. Turner and M. W. Risjord, Amsterdam and Boston: Elsevier, 249–310.

Harding, S. (2011), "Other Cultures' Sciences," in *The Postcolonial Science and Technology Studies Reader*, edited by S. Harding, Durham and London: Duke University Press, 151–158.

Henare, A., M. Holbraad and S. Wastell (eds.) (2006), *Thinking Through Things: Theorising Artefacts Ethnographically,* London: Routledge.

Henderson, D. K. (1987), "The Principle of Charity and the Problem of Irrationality," *Synthese* 73: 225–252.

Herskovits, M. J. (1948), *Man and His Works: The Science of Cultural Anthropology*, New York: Alfred A. Knopf.

———. (1955), *Cultural Anthropology*, New York: Knopf.

Heywood, P. (2017), "The Ontological Turn," *The Cambridge Encyclopedia of Anthropology*, doi:10.29164/17ontology.

Hoijer, H. (1954), "The Sapir Whorf Hypothesis," in *Language in Culture*, edited by H. Hoijer, Chicago: University of Chicago Press, 92–105.

Holbraad, M. and M. A. Pedersen (2017), *The Ontological Turn: An Anthropological Exposition,* Cambridge: Cambridge University Press.

Hollis, M. and S. Lukes (eds.) (1982), *Rationality and Relativism*, Oxford: Basil Blackwell.

Jarvie, I. (2006), "Relativism and Historicism," in *Philosophy of Anthropology and Sociology*, edited by S. P. Turner and M. W. Risjord, Amsterdam and Boston: Elsevier/North-Holland, 553–589.

Kluckhohn, C. (1955), "Ethical Relativity: Sic et Non," *Journal of Philosophy* 52: 663–677.

Koskinen, I. (2014), "Critical Subjects: Participatory Research Needs to Make Room for Debate," *Philosophy of the Social Sciences* 44: 707–732.

Kuper, A. (1999), *Culture: The Anthropologists' Account*, Cambridge, MA: Harvard University Press.

———. (2003), "The Return of the Native," *Current Anthropology* 44: 389–402.

Kusch, M. (2016), "Relativism in Feyerabend's Later Writings," *Studies in History and Philosophy of Science* 57: 106–113.

Ludwig, D. (2016), "Overlapping Ontologies and Indigenous Knowledge: From Integration to Ontological Self-determination," *Studies in History and Philosophy of Science Part A* 59: 36–45.

———. (2018), "Revamping the Metaphysics of Ethnobiological Classification," *Current Anthropology* 59: 415–423.

McGhee, R. (2008), "Aboriginalism and The Problems of Indigenous Archaeology," *American Antiquity* 73: 579–597.

Paleček, M. and M. Risjord (2012), "Relativism and the Ontological Turn within Anthropology," *Philosophy of the Social Sciences* 43: 3–23.

Quine, W. van O. (1960), *Word and Object*, Cambridge, MA: MIT Press.

Risjord, M. W. (2000), *Woodcutters and Witchcraft: Rationality and Interpretive Change in the Social Sciences*, New York: SUNY Press.

———. (2007), "Ethnography and Culture," in *Philosophy of Anthropology and Sociology*, edited by S. P. Turner and M. W. Risjord, Amsterdam and Boston: Elsevier, 399–428.

Said, E. (1978), *Orientalism*, New York: Pantheon Books.

Santos, B. de S. (2016), *Epistemologies of the South: Justice Against Epistemicide*, London and New York: Routledge, Taylor & Francis Group.

Sapir, E. (1929), "The Status of Linguistic as a Science," *Language* 5: 209–212.

Sivado, A. (2015), "The Shape of Things to Come? Reflections on the Ontological Turn in Anthropology," *Philosophy of the Social Sciences* 45: 83–99.

Sperber, D. (1982), "Apparently Irrational Beliefs," in *Rationality and Relativism*, edited by M. Hollis and S. Lukes, Oxford: Basil Blackwell, 149–180.

Steward, J. H. (1948), "Comments on the Statement on Human Rights," *American Anthropologist* 50: 351–352.

Strathern, M. (1980), "No Nature; No Culture," in *Nature, Culture, Gender*, edited by C. MacCormack and M. Strathern, Cambridge: Cambridge University Press, 174–222.

Theunissen, M. (2017), "Rationality, Naturalism, and Critique in the Philosophy of Social Science," PhD thesis, New York: The New School University.

Tuhiwai Smith, L. (1999), *Decolonizing Methodologies: Research and Indigenous Peoples*, London and New York: Zed Books; Dunedin, New Zealand: University of Otago Press.

Wagner, R. (1975), *The Invention of Culture*, Englewood Cliffs, NJ: Prentice-Hall.

Westermarck, E. (1932), *Ethical Relativity*, New York: Littlefield, Adams & Company.

Whorf, B. (1956), "Science and Linguistics," in *Language, Thought, and Reality: Selected Writing of Benjamin Lee Whorf*, edited by J. B. Carrol, Cambridge, MA: MIT Press, 207–219.

Williams, E. (1947), "Anthropology for the Common Man," *American Anthropologist* 49: 84–90.

Wilson, B. R. (ed.) (1970), *Rationality*, Oxford: Blackwell.

Winch, P. (1958), *The Idea of a Social Science and Its Relation to Philosophy*, London: Routledge & Kegan Paul.

———. (1964), "Understanding a Primitive Society," *American Philosophical Quarterly* 1: 307–324.

Wittgenstein, L. (1953), *Philosophische Untersuchungen: Philosophical Investigations*, translated by G. E. M. Anscombe, Oxford: Blackwell.

———. (1958), *The Blue and Brown Books*, Oxford: Blackwell.

———. (1967), "Bemerkungen über Frazers, *The Golden Bough*," *Synthese* 17: 233–253.

46

RELATIVISM IN LOGIC AND MATHEMATICS

Florian Steinberger

1. Introduction

While truths in other domains of inquiry and other human affairs may perhaps be thought to be relative to particular standards of assessment determined by the individual characteristics, cultural background, moral or epistemic framework, or even biological makeup of the assessor, logic and mathematics have traditionally been understood as the ultimate touchstone for objectivity. If there is a sturdy foundation that undergirds all of our intellectual pursuits, it is in the domain of logic and mathematics that we should seek it, or so many would maintain. In this vein, arguing against a form of relativism about logic – psychologism, according to which the laws of logic are relative to one's physiology or psychological constitution – that Frege memorably described the laws of logic as the "boundary stones set in an eternal foundation, which our thought can overflow but never dislodge" (Frege 1893, xv). Some take logic to set forth binding norms for thought regardless of its subject matter. Others take (not necessarily incompatibly) logic, much like other sciences, to concern itself with reality, albeit having greater (perhaps universal) scope and therefore operating at a greater level of abstraction than other sciences. Either way, many maintain with Frege that there must be one True Logic, which is not relative to any set of further facts. Similarly, mathematics, among all human intellectual endeavors, tends to be the paragon of objectivity and universality.

As we will see, however, there are a significant number of relativist positions concerning both logic and mathematics. Before discussing these, we would do well to get clearer about what, for the purposes of this chapter, we mean by "relativism."

Relativism is a family of views that has been variously articulated. A common strand among most versions is that relativism about some phenomenon X means that the correct account of X will depend on a further set of facts Y. Sometimes X is said to be the dependent variable, which is relative to Y, the independent variable. I am a relativist about rules of etiquette if I maintain that a rule of etiquette can be said to be correct only relative to, for example, a culture or group. A semantic version of relativism has it that corresponding claims such as "Belching after a meal is polite" cannot be said to be true or false absolutely. Such claims are really elliptical, acquiring a truth-value only once we supply the value of the hidden parameter, thereby making explicit the independent variable. Thus, properly understood, the claim comes to "Belching after a meal is polite-relative-to-local-customs Y," where its truth-value may vary with different values of Y.[1]

Typically, as we noted, the independent variable –Y– is taken to range over individual standards, cultures, moral or epistemic frameworks and so on. However, it will behoove us to follow Cook (2010) and Shapiro (2014) in adopting a broader definition including, for example, contexts of use and so treating cases of lexical ambiguity as forms of relativism. So, "bank" may designate a financial institution relative to one context of use, and a river bank relative to another context.

The philosophy of logic has focused primarily on the question of pluralism and monism. Logical pluralists maintain that there is more than one correct, legitimate or admissible logic; monists denies this. Pluralism and relativism about logic are interestingly related, but they are not identical views. Relativism implies pluralism only when the independent variable Y can take on more than one value. Conversely, there could be forms of pluralism that are not rooted in an underlying relativism (see Cook 2010). Nevertheless, the two often coincide. The more generous definition adopted here leads us to classify as "relativist" certain positions often labeled "pluralist" in the literature (see Russell 2014 for a helpful survey). Nothing of substance hinges on this.

In the following I survey various forms of relativism, first about logic and then about mathematics, showing how the two can be intertwined.

2. Relativism about logic

Now that we have a working definition of "relativism," "logic" would seem to be next on the list of terms requiring elucidation. The trouble is that the question of *what logic is* is itself a matter of controversy, which, according to some (e.g. Shapiro 2014), is bound up with a certain kind of relativism. On this view, the very existence of a multitude of different conceptions of logic and the difficulty of adjudicating between them, calls for a broadly relativistic attitude. Rather than regarding these different conceptions as rivals, they can be seen as equally legitimate ways of rendering precise the (as the proponents of this view see it) multi-faceted central concepts of logical consequence and validity.[2]

Let us consider two examples. Some invoke a primitive notion of modality, maintaining that C is a logical consequence of a set of truth-bearers $A_1, \ldots, A_n$ if and only if it is not possible for all of $A_1, \ldots, A_n$ to be true while C is false. By contrast, the influential, austere Tarskian view does away with any primitive modality: C is a logical consequence of $A_1, \ldots, A_n$ just in case C is true under all re-interpretations of the non-logical vocabulary that make all of $A_1, \ldots, A_n$ true. Several alternative candidates have been proposed (see Shapiro (1998) for a helpful survey). While many philosophers of logic consider these disputes to be substantive in nature and continue to fight their corner, some reject the idea that there is a monolithic notion of logical consequence at all and hence that there could be a unique correct account of logical consequence. Different accounts may provide equally legitimate "sharpenings" of the cluster concepts "logical consequence" and "validity." As Tarski points out with respect to the concept of logical consequence:

> Any attempt to bring into harmony all possible vague, sometimes contradictory, tendencies which are connected with the use of this concept, is certainly doomed to failure. We must reconcile ourselves from the start to the fact that every precise definition of this concept will show arbitrary features to a greater or lesser degree.
>
> *(Tarski 1936, 409)*

The pre-existing underlying use of the concept to be codified often fails to be univocal, but rather comprises numerous, often incompatible strands. We should therefore not expect there to be a single formal concept capable of capturing every aspect of the concept, but several. This relativistic approach might be thought to apply equally to deductive conceptions of logical consequence according to which C is a logical consequence of $A_1, \ldots, A_n$ just in case there is a gap-free proof via primitively legitimate deductive steps leading from $A_1, \ldots, A_n$ to C. Consequently, the (apparent) clash between those who regard model-theoretic approaches to logical consequence (such as the Tarskian notion) as primary and those who give a deductive, proof-theoretic conception of logical consequence pride of place is not a genuine conflict at all, but rather is simply a further example of how the polysemy of "logical consequence" can give rise to distinct "precisifications" foregrounding different features of the informal notion.

In light of this polysemy, Shapiro asks "why think that one of these is to be dubbed as the One True relation of logical consequence?" He continues:

> there are different, mutually incompatible articulations or sharpenings of the intuitive notion or notions of logical consequence. At least some of the supposedly rival explications of logical consequence are (more or less) accurate theories of different aspects of consequence, different members of the cluster.
>
> *(Shapiro 2014, 24)*

This, according to our broad definition, can be seen as a form of relativism about logical consequence.

A related but distinct point familiar from Carnap's notion of explication (Carnap 1950) concerning the process of formalizing any concept is that the theoretical utility of the concept's formal correlate must often be traded off against the degree of match with the informal target concept. The formal concept's theoretical virtues of simplicity, its ability to be integrated into a body of background theory and so on, often come at the expense of being a less faithful codification of the informal concept. This dynamic might be taken to be at work in the debate between advocates of relevance logics and their opponents. According to the relevance logician, the classical consequence relation has the wrong extension and so fails to align with our ordinary conception of what (logically) follows from what. According to classical logic, for instance, anything follows from an inconsistent set of premises, and a logical truth follows from any set of premises despite the fact that, in neither case, does there have to be any connection between the premises and the conclusion in terms of content. The classical logician maintains that the trade-off is justified: while the classical consequence relation may at points deviate from our pre-theoretic intuitions, the benefits of the resulting simple and elegant meta-theory outweigh these modest costs. The relativist take is that these are two distinct, but equally viable ways of explicating the notion of logical consequence: one, the relevantist explication, prioritizes fit with prior practice, while the classical approach weights theoretical virtues such as simplicity and systematicity more highly. Both accounts are correct relative to the relevant weightings of the desiderata, but there is no reason to assume that there is a uniquely correct balance to be struck.

A further possible dependent variable may be the choice of logical vocabulary. Familiar conceptions of logical consequence and validity presuppose a distinction between logical vocabulary – the vocabulary that is constitutive of the logical form of the premises and the conclusion – and the non-logical vocabulary, which is not integral to logical form. Different choices of logical vocabulary may generate extensionally different relations of logical consequence. Let L_1 and L_2 be two first-order languages. Assume, that identity, "=", is a logical constant according to L_1, but not according to L_2, where it is relegated to a run-of-the-mill non-logical two-place predicate. It follows that "Mark Twain is a writer; Mark Twain = Samuel Clemens; therefore, Samuel Clemens is a writer" is valid relative to L_1, but not relative to L_2, where "=" is open to re-interpretation.

Whether this form of relativism is warranted turns on whether one takes there to be a defensible principled distinction between the logical and the non-logical vocabulary of one's language (Hacking 1979; Sher 1991; Tarski 1986), or whether one regards it as a choice of convenience, constrained, perhaps, by one's theoretical goals. On the latter view (Tarski 1936; Varzi 2002), the correct account of the relation of logical consequence is relative to how one draws the logical/non-logical distinction.

We saw earlier that relativism about logical consequence might be thought to arise from the fact that we might mean different things by "logical consequence," "valid" and its cognates, and that it might stem from different ways of demarcating logical from non-logical vocabulary. But even once we agree upon a particular conception of logical consequence – the model-theoretic conception, say – and a particular set of logical constants, it might be thought that a residual flexibility in the meaning of "logical consequence" subsists that is sufficient to engender a different form of relativism. In the literature this type of relativism has been dubbed "meaning-variance pluralism" (see also Hjortland 2012). This type of relativism can again be motivated as an antidote to what some might regard to be futile disputes; in this case disputes over which logic is the correct logic (see e.g. Carnap 1937). Advocates of (purportedly) rival logics – classical and intuitionistic logic, say – need not be in conflict after all. There is no conflict because the disputants attach different meanings to the terms involved. Meaning-variance can take multiple forms depending on where the difference in meaning is located. Take the claim that the argument form '~~A; therefore, A' is valid. According to classical logic, the meta-logical statement is true, whereas within intuitionistic logic it is not. The source of variance might be located in the meaning of "valid," or in the meanings of the logical constants, or in both.

I have elsewhere (Steinberger forthcoming) called meaning-variance stemming from a difference in the meaning of "valid" *structural* meaning-variance, and meaning-variance based on differences in the meanings of some of the logical connectives *operational* meaning-variance.[3] It is important to set aside a crude version of the structural view. According to the crude version, "valid," in the mouth of the advocate of classical logic, really means "valid-in-classical logic," whereas it means "valid-in-intuitionistic logic" in the intuitionist's mouth. But this misses the point. Of course, no one – classical or intuitionistic logician alike – has ever disputed the claim that the law of double negation elimination is valid in classical but not in intuitionistic logic. The real question, many would maintain, is which of the senses of "valid" (if any) adequately captures genuine validity. To this it might be retorted that there is no rigorous system-independent concept of validity; all we have are system-immanent standards of validity. But there are less radical and, arguably, more interesting forms of relativism.

Such a more sophisticated brand of structural meaning-variance has been advanced by J. C. Beall and Greg Restall (Beall and Restall 2006, henceforth B&R). According to their account, there is a core concept of validity which can be characterized via a set of jointly sufficient and individually necessary conditions – necessary truth-preservation, formality and normativity – and via the so-called

(GENERALIZED TARSKI THESIS): An argument with premises $A_1, \ldots, A_n$ and conclusion C is valid$_x$ if and only if, in every case$_x$ in which all of the members of $A_1, \ldots, A_n$ are true, so is C.

Pluralism arises from the fact that the core concept of validity can be elaborated in a specified range of equally legitimate ways depending on how we interpret "valid" and "case," both of which co-vary depending on the value of the parameter x.[4] Depending on how we understand "case" in our definition – e.g. as Tarskian models (classical logic), stages (intuitionistic logic), situations (relevant logic), etc. – we arrive at different concepts of validity and logical consequence.

B&R's pluralism has received considerable attention and criticism.[5] I here want to develop only one objection, which has come to be known as the "collapse argument."[6] One formulation is this: if logic is normative, it presumably tells us something about how we ought to reason. But if B&R are right there are distinct equally correct consequence relations making incompatible pronouncements about how I ought to reason. Suppose $\models_1$ and $\models_2$ are two such relations and that A is known to be true. Assume $A \models_1 C$ but not $A \models_2 C$. Plausibly, given that both logics are correct and necessarily truth-preserving, I ought to believe C (assuming, as we will, that C is a proposition we care to know). Our broader epistemic norms seem to dictate that we follow the stronger of the two consequence relations.[7] More generally, this line of thought threatens to collapse all of the permissible logics generated by the Generalized Tarski Thesis into the strongest logic. Caret (2017) defends B&R by invoking a form of contextualism. Blake-Turner and Russell (forthcoming) and Russell (2017) do so by denying that logic is normative in the relevant sense.

Operational meaning variance locates the difference of meaning in (all or some of) the logical connectives. On this view, the classical logician's claim might be understood as "$\sim_C \sim_C A$; therefore, A is valid" (where the subscript indicates that the negation is to be understood classically (C) or intuitionistically (I)), whereas the intuitionist's claim might be thought of as "$\sim_I \sim_I$ A; therefore, A is invalid." Again, there is no disagreement except, perhaps, over the correct use of the logical connectives.[8] The view is often associated with Quine's (1970) famous quip that "change of logic is change of subject." The thought is that in denying the validity of the classical law, the intuitionist attaches different meanings to the logical connectives, which, according to Quine, is tantamount to a change of subject matter. However, it is not clear how finely the meanings of the logical constants are to be individuated. Does any difference in a constant's deductive behavior translate into a difference of meaning? Or only certain differences? See Field (2009b) and Hjortland (2012) for critical discussions of the underlying conception of the meanings of the logical connectives. Operational meaning variance, like the possible third view – hybrid meaning variance – where the difference resides both in the meanings of "valid" and the logical operators, is difficult to adjudicate absent a robust account of the meanings of the logical constants.

Hartry Field (2009b) has argued against many of the aforementioned forms of logical relativity and instead advances a different form of relativism of his own. Field's point of departure is his contention that "validity" is not definable in terms of necessary truth-preservation.[9] "Validity" must instead be treated as a primitive. Grasping its meaning, however, requires an appreciation of its conceptual role, which, in turn, is characterized by the normative constraints validity imposes on our doxastic attitudes. Field now couples his normative account of validity with his non-factualism about the normative (Field 2009c). There is, for him, no intelligible sense in which any one set of norms can be said to be uniquely correct. Saying that there is no correct set of logical norms is not to say that all logical norms are equally good – some can be better than others. This is because on Field's view, as a species of epistemic norms, logical norms are selected with a view to promoting our epistemic goals. Logical norms can thus be assessed based on how effectively they achieve this objective. All the same, the picture points to two possible sources of logical pluralism: (i) logical pluralism could be a result of pluralism about epistemic goals;

(ii) even if we agree on the epistemic goals we wish to further, it may be indeterminate which set of norms is most conducive to those goals. We have no reason to assume there to be a unique system that best optimizes for our often competing constraints.

A characteristic feature of logic as a discipline in the eyes of many is its universality: the fundamental laws of logic, unlike the laws of other sciences, apply unrestrictedly to any domain of inquiry whatsoever. Superficially at least, this assumption seems highly dubious. After all, there are any number of logics that are legitimate objects of study and which are fruitfully applied to different domains. For instance, classical propositional logic is used to model electric circuits, the Lambek calculus naturally models phrase structure grammars. Does the application of logics specifically tailored to their domains not tell against the conception of logic as universal? It does not. The universalist about logic can concede that logics may be usefully applied in different domains. She maintains, however, that there is, over and above questions of local applicability, a core or "canonical" (Priest 2006, 196) application of logic. It is in relation to this core application that the thesis of universalism must be understood. But what exactly does the canonical application of logic amount to? According to Priest, logic's central application is to deductive reasoning. It consists in determining "what follows from what – what premises support what conclusion – and why" (Priest 2006, 196). Philosophers may disagree over the nature of the core application. It is central to the notion of logic's universality, though, that there should be such a core application.

That said, even assuming that it is possible to make sense of the notion of a core application of logic, one might still deny that there should be one logic that best responds to the demands of all the different domains of inquiry. Perhaps the world is simply irreducibly compartmentalized into domains such that, even with respect to logic's core application, different domains – the domains of very large and very small objects, the domain of mathematics, and so forth – call for different logics. This thought gives rise to a domain-specific relativism. Whether or not a particular argument form is valid depends on the nature of the domain of discourse in question. Different domains are associated with different logics, which may yield differing validity verdicts (see e.g. Da Costa 1997). Another route to domain-specific relativism is via alethic pluralism. According to this line of thought different domains of discourse are governed by different truth properties. Given the close connection between certain conceptions of logical consequence and truth, different truth properties may induce different logics (see Pedersen (2014), Lynch (2009) and Pedersen and Wright (2013) for a useful overview and Priest (2006) and Steinberger (forthcoming) for further discussion).

3. Relativism in mathematics

So much for logic, but can there be such a thing as relativism about mathematics? At first blush this again might seem hard to imagine. Whatever one's position with respect to the metaphysical status of mathematical objects and truths, the objectivity of mathematics would seem to be a datum that any account worth its salt must accommodate. That there are infinitely many prime numbers, that the square-root of the number two is not expressible as a ratio between two natural numbers and so on – such theorems cannot be said to be true for some people or cultures or at some times, but not others. It is assumed that any mathematical truth amenable to being proven, can, in principle, at least, be rigorously proved. But given precise definitions of the fundamental notions, axioms codifying our basic assumptions concerning the subject matter, and clearly stated logical rules specifying the permissible deductive transitions, mathematicians can hardly reach different conclusions – the conclusions seem inescapable. Hence, if mathematical results can be said to be relative, the relativity would seem to have to reside here, at the level of the axioms and the logical rules governing our mathematical practice.

The advent of non-Euclidian geometries may serve as an illustration of relativity to the axioms. Since times Euclidian, geometers sought to show that the famous Fifth Postulate in Euclid's Elements follows logically from the remaining axioms and postulates. They attempted to do so via a *reductio ad absurdum*: they negated the Fifth Postulate, hoping thereby to generate a contradiction. In the nineteenth-century it transpired that the resulting axioms – the core axioms along with the negations of the Fifth Postulate – proved to be consistent. New systems of geometry, hyperbolic and elliptic geometry, were born and, in time, earned their keep. However, they did so only through an arduous intellectual process, the end result of which is well described by Alberto Coffa:

> [T]his had all the appearance of being the first time that a community of scientists had agreed to accept in a not-merely-provisory way all the members of a set of mutually inconsistent theories about a certain domain.
>
> *(Coffa 1986, 8)*

Many philosophers and mathematicians eventually took the lesson to be that geometries must be divorced from Kantian spatial intuition and indeed from physical space altogether. A geometry is to be regarded not as a theory of physical space at all, but rather as an axiomatic theory that characterizes certain abstract structures (which may, of course, *qua* piece of applied mathematics, be applied to space). The epitome of this formal axiomatic treatment of geometry was David Hilbert's *Grundlagen der Geometrie* (1899), which is aptly summarized by his protégé, Paul Bernays.

> A main feature of Hilbert's axiomatization of geometry is that the axiomatic method is presented and practiced in the spirit of the abstract conception of mathematics that arose at the end of the nineteenth century and which has generally been adopted in modern mathematics. It consists in abstracting from the intuitive meaning of the terms . . . and in understanding the assertions (theorems) of the axiomatized theory in a hypothetical sense, that is, as holding true for any interpretation . . . for which the axioms are satisfied. Thus, an axiom system is regarded not as a system of statements about a subject matter but as a system of conditions for what might be called a relational structure. . . . [On] this conception of axiomatics, . . . it is insisted that in reasoning we should rely only on those properties of a figure that either are explicitly assumed or follow logically from the assumptions and axioms.
>
> *(Bernays 1967, 497)*

This view of mathematical theories as the deductive closure of a set of axioms conceived of not as primitive truths, but as fruitful assumptions, opens the door to certain kinds of relativism. For instance, that the sum of the interior angles of a triangle is 180° is true relative to Euclidian geometry, but false relative to non-Euclidian geometries. Thus, the truth of a mathematical claim may be taken to be relative to a set of axioms. So long as different sets of axioms are not taken to be vying for the exclusive claim to truth with respect to a particular domain, but rather as equally legitimate alternatives, we find ourselves with a form of axiom-based relativism.

But our description of the axiomatic approach makes it plain that there is another possible form of mathematical relativism, as a direct consequence of logical relativism. The theorems provable in our theory are a function of the axioms but also of the logical rules that govern our theory. Hence, a plurality of logics gives rise to a plurality of different mathematical theories based on these different logics. For instance, to return to our previous example, we can formulate a formal theory of arithmetic on the basic of classical logic (Peano Arithmetic) or on the basis of intuitionistic logic (Heyting Arithmetic). The resulting relativism is harmless because the latter theory is simply a sub-theory of the former. But Shapiro (2014) considers more interesting examples where a seemingly viable theory, when formulated on the basis of intuitionistic logic, is sunk into inconsistency when giving a classical formulation of it. One such example is Heyting Arithmetic augmented by an intuitionistic version of Church's thesis. What are we to make of such cases? Shapiro recommends we return to the Hilbertian idea according to which consistency is the sole requirement for theory acceptance, there are "no further metaphysical hoop[s] the proposed theory must jump through" (Shapiro 2014, 67). In his famous exchange with Frege, Hilbert writes:

> if the arbitrarily given axioms do not contradict each other with all their consequences, then they are true and the things defined by them exist. This is for me the criterion of truth and existence.
>
> *(Letter from Hilbert to Frege of December 29th, 1899,*
> *in Frege (1980, 39))*

Since consistency is a matter of the underlying logic, the existence of the structures and objects postulated by one's theories would seem to be relative to the logic in question. Generally, the weaker the logic, the more theories turn out to be consistent. However, Shapiro extends the Hilbertian approach allowing also for inconsistent but non-trivial theories based on non-explosive logics.

4. Future work

Relativism about logic and mathematics is not itself a semantic doctrine, but it poses important questions about meaning. These issues have been investigated thoroughly in the philosophy of language and some (Caret 2017; Shapiro 2014) have explored to what extent the competing semantic views – indexical and non-indexical contextualism, and assessment-sensitive relativism among others – can be usefully applied in the domain of logic and mathematics and if so, whether this domain might constitute an interesting testing ground for the competing semantic theories. More work is to be done here, however. A further thread of our discussion that merits further investigation is the question of which kinds (if any) of relativisms about logic are compatible with a view that accords logic a normative role in reasoning. An important component of such research, I suspect, will be to gain a clearer understanding of the precise sense in which logic might be said to be normative.

Notes

1 See Baghramian and Carter (2018), Wright (2008). For relativism in this sense applied to logic and mathematics, see Cook (2010) and Shapiro (2014).
2 Standardly, validity and logical consequence are intimately related in that an argument with premises $A_1, \ldots, A_n$ and conclusion C is valid if any only if C is a logical consequence of $A_1, \ldots, A_n$ (i.e. if $A_1, \ldots, A_n \vDash C$).

3 The terminology is inspired by Gentzen-Prawitz-style proof theory, in which inference rules are divided into those that feature specific logical operators (operational rules); and those that codify general constraints on the deducibility relation (structural rules).

4 B&R explicitly reject the label "relativism" (2006, 88), but that is because they operate with a more restricted definition of "relativism."

5 See for instance Bueno and Shalkowski (2009) and Priest (2006) for sustained criticisms.

6 The collapse argument originates with Priest (2001). See also Ferrari and Moruzzi (2017), Keefe (2014), Read (2006), Stei (2017) and Steinberger (forthcoming).

7 For discussion of cases where the consequence relations cannot be straightforwardly ordered by inclusion, see Keefe (2014) and Steinberger (forthcoming).

8 Operational meaning-variance only gives rise to pluralism on the assumption that the alternative meanings are equally legitimate. This is by no means obvious. For example, according to the semantic anti-realist tradition (Dummett 1991; Prawitz 1977; Tennant 1987), meaning-theoretic considerations reveal the classical meanings of the logical constants to be defective, thus favouring weaker constructive logics.

9 See Field (2009a, 2009b, 2015).

References

Baghramian M. and J. Carter (2018), "Relativism," in *The Stanford Encyclopedia of Philosophy*, Winter 2018 edition, edited by E. N. Zalta, https://plato.stanford.edu/archives/win2018/entries/relativism/.

Beall, J. C. and G. Restall (2006), *Logical Pluralism*, Oxford: Oxford University Press.

Bernays, P. (1967), "Hilbert, David," in *The Encyclopedia of Philosophy*, vol. 3, edited by P. Edwards, New York: Palgrave Macmillan, 496–504.

Blake-Turner, C. and G. Russell (forthcoming), "Logical Pluralism Without the Normativity," *Synthese*.

Bueno, O. and S. Shalkowski (2009), "Modalism and Logical Pluralism," *Mind* 118: 295–321.

Caret, C. (2017), "The Collapse of Logical Pluralism Has Been Greatly Exaggerated," *Erkenntnis* 82: 739–760.

Carnap, R. (1937), *The Logical Syntax of Language*, London: Routledge.

———. (1950), *Logical Foundations of Probability*, Chicago: University of Chicago Press.

Coffa, A. (1986), "From Geometry to Tolerance: Sources of Conventionalism in Nineteenth-Century Geometry," in *From Quarks to Quasars: Philosophical Problems of Modern Physics*, edited by R. Colodny, Pittsburgh: University of Pittsburgh Press, 3–70.

Cook, R. (2010), "Let a Thousand Flowers Bloom: A Tour of Logical Pluralism," *Philosophy Compass* 5(6): 492–504.

Costa, N. da (1997), *Logique classique et non classique: Essaie sur le fondement de la logique*, Paris: Masson.

Dummett, M. (1991), *The Logical Basis of Metaphysics*, Cambridge, MA: Harvard University Press.

Ferrari, F. and S. Moruzzi (2017), "Logical Pluralism, Indeterminacy and the Normativity of Logic," *Inquiry*: 1–24.

Field, H. (2009a), "What Is the Normative Role of Logic?" *Proceedings of the Aristotelian Society* 83: 251–268.

———. (2009b), "Pluralism in Logic," *Review of Symbolic Logic* 2: 342–359.

———. (2009c), "Epistemology Without Metaphysics," *Philosophical Studies* 143: 249–290.

———. (2015), "What Is Logical Validity?" in *Foundations of Logical Consequence*, edited by C. Caret and O. Hjortland, Oxford: Oxford University Press, 33–70.

Frege, G. (1893), *Grundgesetze der Arithmetik*, Paderborn: Mentis, 2009.

———. (1980), *Philosophical and Mathematical Correspondence*, Oxford: Basil Blackwell.

Hacking, I. (1979), "What Is Logic?" *Journal of Philosophy* 76: 285–319.

Hilbert, D. (1899), "Grundlagen der Geometrie," in *Festschrift zur Feier der Enthüllung des Gauss-Weber-Denkmals in Göttingen*, Leipzig: Teubner, 1–92.

Hjortland, O. (2012), "Logical Pluralism, Meaning-Variance, and Verbal Disputes," *Australasian Journal of Philosophy* 91: 355–373.

Keefe, R. (2014), "What Logical Pluralism Cannot Be," *Synthese* 191: 1375–1390.

Lynch, M. (2009), *Truth as One and Many*, Oxford: Oxford University Press.

Pedersen, N. J. L. L. (2014), "Pluralism × 3: Truth, Logic, Metaphysics," *Erkenntnis* 79: 259–277.

Pedersen, N. J. L. L. and C. Wright (2013), "Pluralist Theories of Truth," in *Stanford Encyclopedia of Philosophy*, Spring 2013 edition, edited by E. N. Zalta, https://plato.stanford.edu/archives/spr2013/entries/truth-pluralist/.

Prawitz, D. (1977), "Meaning and Proofs: On the Conflict Between Classical and Intuitionistic Logic," *Theoria* 43: 2–40.

Priest, G. (2001), "Logic: One or Many?" in *Logical Consequences: Rival Approaches*, edited by J. Woods and B. Brown, Oxford: Hermes Scientific Publishers, 23–38.

———. (2006), *Doubt Truth to Be a Liar*, Oxford: Oxford University Press.

Quine, W. V. O. (1970), *Philosophy of Logic*, Oxford: Oxford University Press.

Read, S. (2006), "Monism: The One True Logic," in *A Logical Approach to Philosophy: Essays in Honour of Graham Solomon*, edited by D. DeVidi and T. Kenyon, Berlin: Springer, 193–209.

———. (2014), "Logical Pluralism," in *Stanford Encyclopedia of Philosophy*, Spring 2014 edition, edited by E. Zalta, https://plato.stanford.edu/archives/spr2014/entries/logical-pluralism/.

———. (2017), "Logic isn't Normative," *Inquiry*: 1–18.

Shapiro, S. (1998), "Logical Consequence: Models and Modality," in *The Philosophy of Mathematics Today*, edited by M. Schirn, Oxford: Oxford University Press, 131–156.

———. (2014), *Varieties of Logic*, Oxford: Oxford University Press.

Sher, G. (1991), *The Bounds of Logic: A Generalized Viewpoint*, Cambridge, MA: MIT Press.

Stei, E. (2017), "Rivalry, Normativity, and the Collapse of Logical Pluralism," *Inquiry*: 1–22.

Steinberger, F. (forthcoming), "Logical Pluralism and Logical Normativity," *Philosophers' Imprint*.

Tarski, A. (1936), "On the Concept of Logical Consequence," in *Logic, Semantics, Metamathematics*, second edition, edited by J. Corcoran, Indianapolis: Hackett, 1983, 409–420.

———. (1986), "What Are Logical Notions?" *History and Philosophy of Logic* 7: 143–154. (Transcript of a 1966 talk, edited by J. Corcoran).

Tennant, N. (1987), *Anti-Realism and Logic*, Oxford: Oxford University Press.

Varzi, A. (2002), "On Logical Relativity," *Philosophical Issues* 12: 197–219.

Wright, C. (2008), "Relativism About Truth Itself: Haphazard Thoughts About the Very Idea," in *Relative Truth*, edited by M. García-Carpintero and M. Kölbel, Oxford: Oxford University Press, 157–186.

47

LOGIC AND THE PSYCHOLOGY OF REASONING[1]

Catarina Dutilh Novaes

1. Introduction

According to a familiar slogan, logic is the discipline concerned with the fundamental principles for *inference and reasoning*. As such, it is understood to have close connections with human mental processes, insofar as inference and reasoning are viewed as phenomena pertaining first and foremost to the realm of human thought. This view has illustrious proponents: Kant famously stated that (general) logic deals with the "absolutely necessary rules of thought without which there can be no employment whatsoever of the understanding" (KrV, A52/B76) (Kant, 1781). Similarly, for Frege, the laws of logic "are the most general laws, which prescribe universally the way in which one ought to think if one is to think at all" (Frege 1884, 12).

Prima facie, such statements can be understood in at least two ways:

(1) Descriptive claim: as a matter of fact, humans *do* reason following the canons of logic.
(2) Normative claim: logic comprises the principles of reasoning that humans *ought* to follow, even if they may de facto deviate from these principles.

In the 19th and 20th centuries, many philosophers and psychologists of reasoning initially took logic to be a descriptively adequate model for human reasoning. This is in effect an empirical claim, which accordingly has been systematically studied by the empirical sciences of the mind such as psychology and cognitive science since the 20th century. This extensive body of experimental research revealed significant, systematic discrepancies between the reasoning performance of participants in experiments and the canons of traditional logic as codified in Aristotelian syllogistic and classical deductive logic (Evans 2002).

Currently, the consensus among psychologists of reasoning is that classical, deductive logic is not at all an adequate descriptive model for human reasoning. The shift from these earlier views to the current status quo has been described as an authentic Kuhnian paradigm shift in the psychology of reasoning (Elqayam 2018). But this conclusion gives rise to a host of new questions, such as: if traditional logic is not an adequate description of human reasoning, is there an alternative theoretical framework that is more descriptively adequate? Some prominent candidates are Bayesian probability calculus (Oaksford and Chater 1991) and non-monotonic logics (Stenning and van Lambalgen 2008). Another crucial question is whether traditional logic remains *normatively* adequate despite these descriptive discrepancies, which would mean that humans make systematic and frequent mistakes when reasoning. However, it has been argued that deductive logic is also thoroughly inadequate even as a *normative* model for reasoning (Harman 1986; Dutilh Novaes 2015). But then, what is logic actually *about*, if it is neither descriptively nor normatively adequate as an account of human reasoning?

In this chapter, I start with a brief account of the historical developments leading to the view that traditional logic has (descriptive and/or normative) import for thought. I then survey some well-known empirical results showing that there are significant discrepancies between how humans de facto reason and traditional logical systems such as syllogistic and propositional logic. I close by considering some theoretical options still available for the relation between logic and the psychology of reasoning.

2. Historical background

The strong association between logic and reasoning is a relatively recent phenomenon if we consider the history of philosophy as a whole. In the European tradition up to the early modern period, the chief application of logical theories was to teach students how to perform well in *debates and disputations*, and to theorize on the logical properties of what follows from what insofar as this is an essential component of such argumentative practices.[2]

The tight connection between traditional logic and debating practices dates back to the classical Hellenistic period. Intellectual activity then was quintessentially a dialogical affair, as registered in Plato's dialogues. These dialogical practices, usually described as *dialectic*, constituted the background for the emergence of Aristotle's logic (Duncombe and Dutilh Novaes 2016). Indeed, two of Aristotle's logical texts, the *Topics* and the *Sophistical Refutations*, are explicitly about dialectical practices. Importantly, even Aristotle's most abstract logical text, the *Prior Analytics*, where the venerable syllogistic system was first introduced, is filled with dialectical vocabulary and references to debating practices.

The prominence of dialectic continued through late antiquity. By the Latin medieval period, the focus on debating became even more pronounced, when it was institutionalized with the emergence of what is known as "scholastic disputation" (Novikoff 2013). This tradition reached its pinnacle in the 14th century with the birth and expansion of universities, where it became a primary teaching method. The influence of disputations went well beyond universities, expanding towards multiple spheres of cultural life.

Scholastic logic was then severely criticized by Renaissance authors such as Lorenzo Valla, who deplored its lack of applicability. For Valla, syllogistic was an artificial type of reasoning, useless for orators on account of being too far removed from natural ways of speaking and arguing (Nauta 2009). The condemnation of scholastic logic as useless was later picked up on by Descartes. Speaking of how the education of a young pupil should proceed, in *Principles of Philosophy* (1644) he bemoans that scholastic logic is no more than "ways of expounding to others what one already knows" (Descartes 1644, 186). Descartes is correct in observing that scholastic logic, in particular the theory of syllogistic, is not a logic of discovery; its chief purpose is justification and exposition.

Instead, Descartes recommends "the kind of logic which teaches us to direct our reason with a view to discovering the truths of which we are ignorant" (Descartes 1644, 186). Indeed, in this period, the very term "logic" came to be used for something other than what the Greek authors and the scholastics had meant, i.e. as closely related to dialectic. By contrast, early modern authors emphasize the role of novelty and individual discovery, as exemplified by the influential textbook *Port-Royal Logic* (1662) – essentially, the logical version of Cartesianism, based on Descartes's conception of mental operations and the primacy of thought over language. In this vein, in the modern period, a number of philosophers came to see the nature of logic in terms of the faculties of mind, and this became the predominant view. It culminated in Kant, for whom logic pertained above all to the structure of thought as such and the operations of the mind, such as in his interpretation of Aristotelian categories.

It is Kant who systematized and consolidated the close association between logic and thinking/ reasoning, which was to remain pervasive and influential throughout the 19th and 20th centuries (MacFarlane 2000). He selectively absorbed the notions of "judgment," "form," and "categories" as found in the logical textbooks of the time, and put them to use so as to describe the very conditions of possibility for our thinking and perceiving. The concept of judgment, for example, traditionally used to refer to linguistic claims made by speakers in the public sphere, is transformed into the mental act involved in the apperception of objects. Thus, with Kant, logic concerns exclusively the inner mental activities of the lonesome thinking subject.

This (broadly speaking) Kantian view provided the background for the tradition in experimental psychology investigating reasoning to emerge in the 20th century. In the early days of this tradition (in the first half of the twentieth century), the fact that participants' performance often deviated from the normative responses (as defined by the canons of deductive logic) had already been noticed (Dutilh Novaes 2012, ch. 4). But this was not sufficient to overturn the pervasive association between logic and thinking/reasoning. The work of Piaget and collaborators, most notably Inhelder, became very influential in this tradition. Some relevant aspects of Piaget's conception of human cognition are aptly described by P. Johnson-Laird, an influential psychologist of reasoning:

> How do people reason? The view that I learned at my mother's knee was that they rely on logic. . . . Jean Piaget and his colleagues argued that the construction of a formal logic in the mind was the last great step in children's intellectual development, and that it occurred at about the age of twelve.
>
> *(Johnson-Laird 2008, 206)*

Piaget's notion of "formal operations" entails the idea of reasoning proceeding by schematic substitution with different content: (mature, adult) reasoning is rule-based, and the "abstract" rules in question were the rules of "logic" as traditionally construed (syllogistic, classical logic). (Notice, however, that Piaget did not claim that we were born with these rules in our heads, ready to use as it were; the rule-based system of reasoning had to mature, and in principle education might play a fundamental role.)

The picture of human reasoning as proceeding by means of instantiations of rules and schematic substitution with different content began to be challenged by a series of experimental findings in the late 1960s suggesting that humans do not really reason on the basis of logical rules, or in any case not the specific rules pertaining to deductive logic. Of course, the question was then whether human reasoning is not rule-based at all, or whether it is ruled-based, but just by different (still to be discovered) rules.

3. Experimental work on reasoning

Three features are usually associated with logical reasoning: it is formal, i.e. it proceeds by instantiations of schematic rules; the truth (or believability) of premises and conclusions is not relevant for logical validity; and it is monotonic, i.e. if an inference from A and B to C is valid, then any arbitrary additional premise D will not invalidate the inference (so the inference from A, B, and D to C is equally valid). Empirical research has shown that these three principles are systematically violated by human reasoners, and these findings challenge the claim that traditional, deductive logic is an adequate descriptive model for human reasoning.

3.1. Content effects

The first severe blow to the Piagetian paradigm was inflicted by the results of Wason's famous "selection task" experiment in the 1960s (Wason 1968). In this experiment, participants are shown four cards, for example [A], [B], [4], and [7], and told that each card has a number on one side and a letter on the other side. They are then asked to turn exactly the cards they must turn in order to verify or falsify the following conditional: "If a card has an even number on one side, then it has a vowel on the other side."

Originally, the assumption was that participants would (should!) interpret the "if . . . then . . ." clause as having the logical properties of the material conditional; in that case, the conditional statement would be falsified by cards with an even number on one side and a consonant on the other side. From this point of view, one should turn the cards that show either an even number or a consonant to see what is on the other side. Similarly, this conditional is, according to classical logic, equivalent to its contrapositive, namely "If a card has a consonant on one side, then it has an odd number on the other side." Thus, it is clear that cards showing consonants should also be turned (that is, if the conditional is interpreted as a material implication). Indeed, the response predicted by the assumption that participants would be using classical logic (competently) to solve this task is that they would turn [4] and [B].

But this is not what the great majority of participants do: in fact, less than 10% of the participants turn [4] and [B] (or equivalent cards in other formulations). A significant number of participants turn only [4], many others turn [4] and [A]. (These results have been replicated hundreds of times since Wason's initial experiment.) Clearly, they are not using (classical) logic to solve the task, either because they do not interpret the conditional as a material implication (as argued in (Oaksford and Chater 1994)), or simply because they do not follow the rules that determine the meaning of the material implication according to classical logic. Hence, these results seem inconsistent with the view that adults reason on the basis of "formal" rules (formal in the sense of schematic (Dutilh Novaes 2011), which can then be instantiated by different occurrences of the "gaps") or Piagetian "formal operations."

Participants do realize that they must turn the cards that fit the description in the antecedent (virtually all of them turn [4]), but they fail to realize that they must also turn the cards that do *not* fit the description in the consequent. Wason and collaborators then embarked on a series of similar experiments, manipulating different elements of the task in attempts to elicit a higher proportion of normative responses (i.e. the responses dictated by classical logic), but in first instance with no success.

As if this was not a sufficient blow to the Piagetian conception of reasoning, the experiment where participants finally performed much closer to the normative response was the one where, instead of using "abstract" material (letters and numbers), cards with actual content were used (Wason and Shapiro 1971). The conditional used was "Every time I go to Manchester I travel by train," and the cards used were [Manchester], [Leeds], [train], [car]. Unlike in the abstract version of the task, participants were much more likely to realize they must also turn [car] to see if it does or does not have [Manchester] on the other side (10 out of 16 gave the "correct" answer).

These results have been replicated several times, with different kinds of contents, and it is clear that the use of material with content often (though not always) has a significant facilitating effect. What exactly is the nature of this facilitating effect is still an open question, and different views have been proposed.

The "Piagetian" view of human reasoning predicts that there should not be a difference in reasoning performance with either abstract or concrete material; it might even predict that performance would be better in abstract tasks, as this would avoid the additional step of actually interpreting the formal rules presumably guiding reasoning with specific content. And yet, these were not at all the results that emerged from these experiments. It is clear that the view of human reasoning as being governed by the abstract rules of (classical) logic which are then interpreted in specific cases (essentially the procedure of instantiating logical schemata in particular cases/domains) is inconsistent with these findings. Indeed, (untrained) human agents typically reason 'better' when they reason with content rather than with schematic letters and symbols.

3.2. Belief effects

The discrepancy between these experimental results and the deductive normative responses to the different reasoning tasks made it patent that, as a matter of fact, humans do not typically reason according to the Piagetian "formal operations" model (Dutilh Novaes 2012, ch. 4). Instead, humans systematically bring background beliefs to bear when reasoning, thus operating on the basis of content rather than logical structure/form. This general tendency towards projecting one's beliefs into a problem has been described as a "fundamental computational bias" by cognitive psychologist K. Stanovich; it is

> [t]he tendency to automatically bring prior knowledge[3] to bear when solving problems. That prior knowledge is implicated in performance on this problem even when the person is explicitly told to ignore the real-world believability of the conclusion illustrates that this tendency toward contextualizing problems with prior knowledge is so ubiquitous that it cannot easily be turned off – hence its characterization here as a fundamental computational bias.
>
> *(Stanovich 2003, 292–293)*

One of the most extensively studied phenomena in the psychology of reasoning literature since the 1980s, which is a specific manifestation of this tendency, is known as *belief bias*. Belief bias is the tendency that reasoners exhibit to let the (un)believability of the conclusion influence their judgment of the (in)validity of an argument. It has been described as "perhaps the best known and most widely accepted notion of inferential error to come out of the literature on human reasoning" (Evans 1989, 41). Belief bias is closely related to a number of other empirically observed cognitive phenomena which all point in the same direction: human reasoners typically seek to maintain the beliefs they already hold, and conversely to reject contradictory incoming information (a tendency also known as *confirmation bias* (Nickerson 1998)).

Belief bias is typically studied using fully formulated categorical syllogisms; participants are then asked to evaluate whether the conclusion follows from the given premises. The different syllogisms used are such that the validity and believability of conclusions are manipulated, thus defining four classes of arguments: (1) deductively valid arguments with a believable conclusion; (2) deductive valid arguments with an unbelievable conclusion; (3) deductively invalid arguments with a believable conclusion; and (4) deductive invalid arguments with an unbelievable conclusion.[4] The systematic combination of argument validity and conclusion believability produces problems in which validity and believability are either in opposition (known as conflict items, classes 2 and 3) or problems in which validity and believability are congruent (known as no-conflict items, classes 1 and 4); indeed, what is under investigation is the presumed conflict between "logic and belief" (as stated in the title of the seminal study (Evans et al. 1983).

A very robust pattern of results in these experiments is that, typically, "belief trumps logic" for conflict items: participants generally endorse invalid arguments with believable conclusions, and often deem valid arguments with unbelievable conclusions as invalid. In (Evans et al. 1983) the results were as follows: 71% of "wrong" answers for invalid arguments with believable conclusions, and 44% of "wrong" answers for valid arguments with unbelievable conclusions (see table; % of validity endorsements).

	Believable conclusions	*Unbelievable conclusions*
Valid	89	56
Invalid	71	10

These results have been replicated numerous times. Indeed, researchers agree unanimously on the pervasiveness of the belief bias phenomenon, but disagree on the underlying mechanisms (Ball and Thompson 2018). According to most of these accounts, what reasoners are in fact doing when asked to evaluate the (in)validity of deductive arguments is constructing plausible models of the premises and evaluating the conclusion on this first model that comes to mind (Evans et al. 1999; Dutilh Novaes and Veluwenkamp 2017). But this is not at all what they should be doing as far as the deductive canons are concerned, given that deductive validity requires that in *all* the situations in which the premises are true, the conclusion also be true (necessary truth-preservation). In practice, reasoners do not consider all models of the premises (for validity), and they do not specifically look for counterexamples to invalidate arguments (for invalidity). Instead, they produce a plausible model of the premises with additional background belief, and then check whether the conclusion looks plausible (rather than following necessarily) in this model.

3.3. Non-monotonicity

These observations lead us directly to one of the most fundamental aspects of mismatch between the canons of logical/deductive reasoning and how human agents indeed seem to reason: the issue of monotonicity v. non-monotonicity. As is well known, one of the key characteristics of deductive validity is monotonicity (also known as indefeasibility): if C follows deductively from a set of sentences K, then it follows deductively from any set K' resulting from the addition of new sentences to K. So deductive validity is construed as indifferent to the addition of extra information; if it holds, it holds come what may. Conversely, to show that a given inference is deductively invalid, it is sufficient to provide one single counterexample, one single situation (no matter how far-fetched it is) where the premises are the case and the conclusion is not the case.

The defeasibility and non-monotonicity of "ordinary" human reasoning has been stressed by Oaksford and Chater for decades (e.g. 2002), and more recently by Stenning and van Lambalgen (2008). They disagree however on how best to capture the defeasible nature of human reasoning:

Oaksford and Chater defend a Bayesian conception of human reasoning, whereas Stenning and van Lambalgen favor the perspective of non-monotonic logics.

One famous experiment whose results suggest that human reasoning has a strong non-monotonic component is Byrne's "suppression task" (1989). Byrne herself did not have non-monotonicity or defeasibility in mind when designing the experiment or interpreting the results, but her results are readily interpreted from the point of view of these concepts.

In this experiment, participants were first presented with the following premises: "If she has an essay to write, she will study late in the library. She has an essay to write." In this case roughly 90% of participants drew the conclusion "She will study late in the library." This seems like good news for logicist views of human reasoning: the great majority of participants are able to perform a simple instance of modus ponens. However, when another premise was added, namely "If the library is open, she will study late in the library," from the very same group of participants now only 60% drew the conclusion "She will study late in the library"; so 30% of them *retracted* their original conclusion. Prima facie, this seems like a breach of monotonicity: if the argument from the original premises to the conclusion was valid, then it should have remained valid even with the addition of an extra premise. Equally puzzling is the fact that when a different premise was added to the original pair of premises, namely "If she has a text-book to read, she will study late in the library," then 94% of the participants drew the original conclusion. Why is it that the first additional premise defeats the argument for 30% of the participants, while the second additional premise, seemingly "identical" in "logical form" (they are both conditional sentences) does not?

Byrne's results do suggest something important about human reasoning, namely the fact that we typically view arguments as *defeasible*. The effect is so subtle that even the addition of a conditional rather than a categorical premise may have the effect of compelling us to retract a given conclusion. In the aforementioned case, the premise that caused the "suppression effect" is the conditional "If the library is open, she will study late in the library," not the categorical "The library is closed." What does the additional conditional premise add to the participant's reasoning process? Well, it seems to add the possibility that the library *might* be closed, something that had not been under consideration so far; and if the library is indeed closed, then, no matter how many essays to write she has, she will not study late in the (closed) library. In the second case, however, the addition of the premise "If she has a textbook to read, she will study late in the library" does not bring in any new elements that might prevent the conclusion from coming about. Thus, clearly not just any added premise may lead a participant to revise her previously drawn conclusion; this will only happen when the added premise brings in a new element that may affect the realization of the situation described in the conclusion. This goes to show that, when drawing modus ponens in the simple case (with just two premises), participants seem to take the conditional in question to be a defeasible rule robust to exceptions, not as the material conditional, which in turn means that the whole argument, while "valid" in some sense of "valid" (perhaps something like "sufficiently reliable"), is not deductively valid because it can be defeated by additional information.

Stenning and van Lambalgen (2008) model these reasoning processes in terms of "closed world reasoning," a framework originally developed within artificial intelligence, and which they put to use to describe actual reasoning patterns of human reasoners. The key idea in this framework is that there is typically an implicit premise in instances of reasoning, namely that in the absence of positive information to the contrary, nothing "funny" is going on and the reasoner can assume normal conditions (in the preceding example, that the library will be open).[5] However, incoming information may then defeat the inference if it indicates that the reasoner may not assume normal conditions. This framework thus offers a compelling (but by no means the only) account of the non-monotonicity of human reasoning.

4. One logic, many logics, or no logic for reasoning?

Now that the descriptive, empirical inadequacy of deductive logical systems (syllogistic, classical logic) for human reasoning has been thoroughly established (Evans 2002; Elqayam 2018), the theoretical options still available seem to be:[6]

> (DESCRIPTIVE "LOGICAL" MONISM) some alternative theoretical framework (Bayesian probabilities, non-monotonic logics) may still offer a unified, empirically adequate account of human reasoning (Oaksford and Chater 2018; Stenning and van Lambalgen 2008[7]). The conventional view was only wrong in identifying *which* formal system best captures human reasoning, but not in maintaining that there should be a "logic" that captures the entirety of human reasoning.
>
> (DESCRIPTIVE LOGICAL PLURALISM) humans in fact rely on a variety of "logics" to reason, depending on the specific contexts and characteristics of the tasks in question. So there is not one unique formal/theoretical framework that adequately captures human reasoning in its multi-faceted, contextual nature, but different logical systems can aptly describe how humans reason in different circumstances.
>
> *(Stenning and Varga 2018)*

> (DESCRIPTIVE LOGICAL NIHILISM) as a matter of fact, human reasoning is so multi-faceted and context-dependent that no logical/theoretical framework can do justice to its complexities. No logic or comparable formal system (including Bayesian probabilities) can be rightly viewed as adequate accounts of human reasoning, and researchers should focus on purely descriptive, piecemeal approaches to reasoning.
>
> *(Elqayam and Evans 2011)*

Of these three positions, descriptive pluralism has immediate implications for issues pertaining to relativism. Indeed, some recent influential versions of logical pluralism present themselves explicitly as relativistic (Shapiro 2015). But which parameters determine which logic is used (and ought to be used) in different situations? Is there exactly one adequate logic for each situation? Or is it the case that even within one and the same context, different logics may be adequate (either descriptively or normatively, or both)? These difficult questions remain.

Notes

1 Research supported by ERC-Consolidator grant 771074 for the project 'The Social Epistemology of Argumentation'.
2 In other logical traditions such as in India and China, close connections between logic and argumentation/debate were also pervasive.
3 Stanovich then clarifies that "belief" may be a more appropriate term than "knowledge," especially given that the philosophical meaning of "knowledge" involves factivity. He introduces the term "belief projection" for this phenomenon.
4 Some studies also use arguments with "neutral" conclusions with respect to believability, for example arguments with made-up words such as "Jamtops" and "opprobine" (Sá et al. 1999). These are referred to as "neutral items."

5 But see (Dutilh Novaes and Veluwenkamp 2017) on the limits of non-monotonic logics as descriptive accounts of human reasoning.
6 These positions have natural counterparts in current discussions in the philosophy of logic on monism, pluralism, and nihilism (Russell 2018), which however take place at the normative level rather than the descriptive level adopted here.
7 In fairness, Stenning and van Lambalgen (2008) do not defend a purely monolithic conception of reasoning, in particular as they distinguish between reasoning to an interpretation and reasoning from an interpretation.

References

Ball, L. and V. Thompson (2018), "Belief Bias and Reasoning," in *International Handbook of Thinking and Reasoning*, edited by L. Ball and V. Thompson, New York: Routledge, 16–36.

Byrne, R. M. (1989), "Suppressing Valid Inferences with Conditionals," *Cognition* 31: 61–83.

Descartes, R. (1644), "Principles of Philosophy," in his *The Philosophical Writings of Descartes*, edited and translated by J. Cottingham, R. Stoothoff and D. Murdoch, Cambridge: Cambridge University Press, 1985, 177–292.

Duncombe, M. and C. Dutilh Novaes (2016), "Dialectic and Logic in Aristotle and His Tradition," *History and Philosophy of Logic* 37: 1–8.

Dutilh Novaes, C. (2011), "The Different Ways in Which Logic Is (said to be) Formal," *History and Philosophy of Logic* 32: 303–332.

———. (2012), *Formal Languages in Logic – a Philosophical and Cognitive Analysis*, Cambridge: Cambridge University Press.

———. (2015), "A Dialogical, Multi-Agent Account of the Normativity of Logic," *Dialectica* 69(4): 587–609.

Dutilh Novaes, C. and H. Veluwenkamp (2017), "Reasoning Biases, Non-Monotonic Logics, and Belief Revision," *Theoria* 83: 29–52.

Elqayam, S. (2018), "The New Paradigm in Psychology of Reasoning," in *Routledge International Handbook of Thinking and Reasoning*, edited by L. Ball and V. Thomson New York: Routledge, 130–150.

Elqayam, S. and J. Evans (2011), "Subtracting Ought from Is: Descriptivism Versus Normativism in the Study of the Human Thinking," *Behavioral and Brain Sciences* 34: 233–248.

Evans, J. S. (1989), *Bias in Human Reasoning: Causes and Consequences*, Hillsdale, NJ: Erlbaum.

———. (2002), "Logic and Human Reasoning: An Assessment of the Deduction Paradigm," *Psychological Bulletin* 128: 978–996.

Evans, J. S., J. L. Barston and P. Pollard (1983), "On the Conflict Between Logic and Belief in Syllogistic Reasoning," *Memory & Cognition* 11: 295–306.

Evans, J. S., S. J. Handley, C. N. Harper and P. N. Johnson-Laird (1999), "Reasoning About Necessity and Possibility: A Test of the Mental Model Theory of Deduction," *Journal of Experimental Psychology: Learning, Memory, & Cognition* 25: 1495–1513.

Frege, G. (1884), *Basic Laws of Arithmetic: Exposition of the System*, translated and edited by M. Furth, Berkeley: University of California Press, 1967.

Harman, G. (1986), *Change in View*, Cambridge, MA: MIT Press.

Johnson-Laird, P. (2008), "Mental Models and Deductive Reasoning," in *Reasoning: Studies in Human Inference and Its Foundations*, edited by L. A. Rips, Cambridge: Cambridge University Press, 206–222.

Kant, I. (1781), *Critique of Pure Reason*, translated and edited by P. Guyer and A. Wood, Cambridge: Cambridge University Press, 1998.

MacFarlane, J. (2000), "What Does It Mean to Say That Logic Is Formal?" PhD dissertation, Pittsburgh, PA: University of Pittsburgh.

Nauta, L. (2009), *In Defense of Common Sense: Lorenzo Valla's Humanist Critique of Scholastic Philosophy*, Cambridge, MA: Harvard University Press.

Nickerson, R. S. (1998), "Confirmation Bias: A Ubiquitous Phenomenon in Many Guises," *Review of General Psychology* 2(2): 175–220.

Novikoff, A. (2013), *The Medieval Culture of Disputation: Pedagogy, Practice, and Performance*, Philadelphia, PA: University of Pennsylvania Press.

Oaksford, M. and N. Chater (1991), "Against Logicist Cognitive Science," *Mind & Language* 6: 1–38.

———. (1994), "A Rational Analysis of the Selection Task as Optimal Data Selection," *Psychological Review* 101: 608–631.

———. (2002), "Commonsense Reasoning, Logic and Human Rationality," in *Common Sense, Reasoning and Rationality*, edited by R. Elio, Oxford: Oxford University Press, 174–214.

———. (2018), "Probabilities and Bayesian Rationality," in *International Handbook of Thinking and Reasoning*, edited by L. Ball and V. Thomson, London: Routlegde, 415–433.

Russell, G. (2018), "Logical Nihilism: Could There Be No Logic?" *Philosophical Issues* 28: 308–324.

Sá, W., R. F. West and K. E. Stanovich (1999), "The Domain Specificity and Generality of Belief Bias: Searching for a Generalizable Critical Thinking Skill," *Journal of Educational Psychology* 91: 497–510.

Shapiro, S. (2015), *Varieties of Logic*, New York: Oxford University Press.

Stanovich, K. E. (2003), "The Fundamental Computational Biases of Human Cognition: Heuristics That (Sometimes) Impair Decision Making and Problem Solving," in *The Psychology of Problem Solving*, edited by J. Davidson and R. J. Sternberg, Cambridge: Cambridge University Press, 291–342.

Stenning, K. and M. van Lambalgen (2008), *Human Reasoning and Cognitive Science*, Cambridge, MA: MIT Press.

Stenning, K. and A. Varga (2018), "Several Logics for the Many Things That People Do in Reasoning," in *International Handbook of Thinking and Reasoning*, edited by L. Ball and V. Thomson, London: Routledge, 523–541.

Wason, P. C. (1968), "Reasoning About a Rule," *Quarterly Journal of Experimental Psychology* 20: 273–281.

Wason, P. C. and D. Shapiro (1971), "Natural and Contrived Experience in a Reasoning Problem," *Quarterly Journal of Experimental Psychology* 23: 63–71.

PART 8

Relativism in philosophy of language and mind

48

CONCEPTUAL SCHEMES

Drew Khlentzos

1. Introduction

Conceptual schemes encode core assumptions about the natural and human world that inform thought. History attests to schemes once unquestioningly accepted, now rejected. We no longer believe, as Shakespeare did, that the air is full of spirits, or, as Aristotle, that the heavens are perfect and immutable, nor, with Aquinas, that a place of eternal torment located at the centre of the Earth awaits the wicked after death.

Yet, philosophers are divided over conceptual schemes: where Carnap, Putnam and Hirsch embrace their plurality, Donald Davidson rejects their very intelligibility. This chapter examines whether there can be radically different conceptual schemes we can accept *alongside* any we do accept. We are not asking: "Can we accept radically different *theories* (about space, time, identity, material constitution, etc.) alongside those we do accept?" Our question is whether thinkers in general could accept radically different assumptions about what a certain domain is like.

If so, then some form of Conceptual Pluralism would appear to hold. It may well be that we cannot accept that core assumptions $A_1, A_2, A_3 \ldots$ hold and another set of incompatible core assumptions $A_1', A_2', A_3' \ldots$ also holds in the very same context. The fact that we settled on $A_1, A_2, A_3 \ldots$ when we could just as well have chosen $A_1', A_2', A_3' \ldots$ however, might show our own scheme is not obligatory. Two theses necessary for conceptual pluralism, then, are that our own scheme is not rationally obligatory and that it is not unique.

We shall not attempt to draw conclusions about pluralism from "first-order" ontological disputes. We wish to know instead whether those who lack any theory about X could be persuaded that their approach to solving X problems is not rationally obligatory. We shall look at human reasoning for an answer. Here, subjects typically have no theories about why certain conclusions they think follow do follow from their information. Is there an alternative to logic in reasoning? We review an influential theory that aims to supply one. Yet, we cannot ignore Carnap's approach to conceptual pluralism. Eli Hirsch provides a "Carnapian" argument for pluralism that we discuss before broaching the question of reasoning. But we begin at a natural point: Davidson's argument against conceptual schemes.

2. Davidson on conceptual schemes

Davidson (1973) is a critique of conceptual relativism that starts by identifying conceptual schemes with sets of inter-translatable languages. Davidson uses Quinean methodology for translating unknown languages into one's own to prove that there could not be a language even partially untranslatable into ours. So, radically incommensurable conceptual schemes are impossible and conceptual relativism is false. From this Davidson infers that we must reject the notion of a conceptual scheme altogether:

> Given the underlying methodology of interpretation, we could not be in a position to judge that others had concepts or beliefs radically different from our own.
>
> *(Davidson 1973, 20)*

This invites the reply "so much the worse for that methodology if others *do* have beliefs radically different from our own." But Davidson thinks radical difference is not possible as we must interpret "others" to be "right in most matters." This sounds bewildering. Why should the beliefs we hold be determined by an external interpreter? Must the linguist who makes first contact with an isolated tribe believe epilepsy is caused by evil spirits in order for them to credit her with the power of speech?

Davidson rejects the idea of a conceptual scheme because he thinks it rests on an indefensible dualism between scheme and content. Yet the proscribed "dualism" isn't needed to explain why relativism is false. Davidson has already told us why: nothing counts as a language unless it is translatable into our own.

In fact, we have no difficulty imagining how an interpreter might recognise she'd come across an untranslatable language. An artificial intelligence, a more advanced "AlphaZero," say, could readily develop a language computationally too complex for any human intelligence to master – a disturbing possibility now all too real.

Davidson's maxims for discounting untranslatable languages and counting others right therefore appear too strong. Following Quine, Davidson claims that all a radical interpreter has to go on are a speaker's verbal dispositions (crucially, assent/dissent dispositions):

> there can be no more to the communicative content of words than is conveyed by verbal behavior…"meaning is use" quoth Wittgenstein. The idea is obvious, but its full force is still mostly unappreciated or misappropriated.
>
> *(Davidson 1999, 80)*

Indeed so – this idea has profound implications for core semantic notions such as entailment. Quine informs us that:

> [W]e … learn that an alternation is implied by its components with the very learning of the word "or."
>
> *(Quine 1974, 80)*

Commenting on this passage, Alan Berger notes that a reader might assume Quine has in mind the standard view that:

> [I]mplication is a deductive relation that holds among sentences regardless of whether a speaker recognizes the corresponding inference as valid or even draws the inference.
>
> *(Berger 1980, 272)*

But Quine regards the standard view as unintelligible. How could it be otherwise if the full "communicative content" of "implies" is "conveyed by verbal behaviour"? Quine is adverting to a certain correlation between speech dispositions: "whenever we assent to a disjunct 'P,' we assent to the disjunction 'P v Q'" (Berger 1980, 271).

Davidson follows Quine here. Indeed, if the meaning of functional words: OR, AND, NOT etc. cannot be tied to assent/dissent dispositions radical interpretation is doomed.

Quine posits the preceding generalisation as a putative law governing disjunction. Is there any evidence speakers of English (or other natural languages) assent to a disjunction P ∨ Q whenever they assent to a disjunct P? No. Speakers typically assent to P ∨ Q when they do not know which of P or Q is true. Whether speakers endorse the inference P ∴ P ∨ Q depends on the content of Q.[1] If they know Q is false, they will likely dissent from the inference. The evidence for the general tendency to dissent from basic inferences, depending on the content of their premises, is overwhelming.

In one recent study, university students were given inferences involving minimal pairs of sentences (De Neys 2012). While they accepted:

(NC) All vehicles have wheels. Bikes are vehicles ∴ Bikes have wheels.

The vast majority rejected:

(C) All vehicles have wheels. Boats are vehicles. ∴ Boats have wheels.

The first syllogism, (NC), is a "No-conflict" syllogism; the second (C) a "Conflict" syllogism. Conflict syllogisms cue a heuristic response to reject conclusions that are manifestly false. A similar heuristic response prompts subjects to reject sentences containing alternatives they know to be false. Moreover, Quine's behavioural ersatz for inference is palpably inadequate, as Berger points out: even if we were inclined to assent to "The Earth is round" whenever we assented to "There have been black dogs," we'd never think one implied the other (Berger 1980, 272).

Davidson's argument against conceptual schemes rests on behaviouristic semantic and psychological assumptions he inherited from Quine. Examining only speakers' verbal dispositions, blithely assuming most of what they hold true is true, the radical interpreter is blind to the neuro-computational machinery grounding a speaker's semantic competence. There is an alternative perspective, one, ironically, that Quine himself adverted to when he wrote, chiding Carnap's "reconstruction" of science:

> Better to discover how science is in fact developed and learned than to fabricate a fictitious structure to a similar effect.
>
> *(Quine 1969, 78)*

3. The Neo-Carnapian path to pluralism

Carnap, the father of conceptual pluralism, argued that there is no unique logic of science, offering a plurality of formalised "linguistic frameworks" to choose from. Scientific questions could only be answered within a framework, he averred (Carnap 1950). After Quine's influential critique of Carnap's program (Quine 1966, 1969), interest in pluralism subsided. Putnam then re-energised the debate with his notion of "conceptual relativity." I focus on Eli Hirsch's recent argument for pluralism (Hirsch 2011).

Hirsch thinks certain meta-ontological disputes, e.g. about material composition, persistence through time, personal identity, are purely verbal – a deflationary construal of them is required.[2] While his argument for pluralism does not presuppose deflationism, it assumes the quantifier "there exists" varies in meaning across languages: *Quantifier Variance*. Hirsch follows Putnam's lead:

> the symbol "∃x," and its ordinary language counterparts, the expressions "there are," "there exist" . . . do not have a single absolutely precise use but a whole family of uses.
> *(Putnam 2004, 37)*

Putnam motivates "conceptual relativity" through examples, as in this version of one: Consider a world w consisting of the letters a, b, c. How many objects are there in w? Audrey, unschooled in metaphysics, is joined by her friends Neil, a mereological nihilist, and Maxi, a mereological maximalist. Audrey and Neil say there are three objects. Maxi disagrees. She says the right answer is seven: three individuals, three pairs and one triple. Who is right? Neil and Maxi are both right, says Putnam. Hence, pluralism is true.

There's a better answer: "How many *objects* are there in w?" is an ill-posed question. Unless a kind of object is first specified in the given context, there can be no counting. "Object" in English is a context-dependent functional expression, on a par with "thing," "item," "entity," "part," "component" etc. Audrey takes "object" to mean LETTER; Neil and Maxi take it to mean ONTOLOGICAL ENTITY: let's use "thing" for this. Ontologically naïve Audrey rightly says there are three letters. Maxi says there are seven things. Repudiating mereological objects, recognising only individuals, Neil says there are three things. Whose tally of things is correct, Maxi's or Neil's?

Arguably, neither. If we're intent on tallying *all* the things there are in w, then we shouldn't overlook the fact that there are six ordered pairs and six ordered triples, of letters.

Still, we can use Putnam's example to illustrate Hirsch's idea of a purely verbal dispute: Even though Neil and Maxi disagree about the number of things world w contains, Hirsch contends each can see that the other articulates a truth "in their own language." Thus, Maxi speaks a "Maximalist" language just like English except for the fact that "object" means MEREOLOGICAL (M-) OBJECT. Call this language L_M. Neil speaks "Nihilist" English, L_N, wherein "object" means INDIVIDUAL. Hirsch thinks Neil should rate "There are seven objects" true in L_M as "object" in L_M means M-OBJECT and Maxi should grant "There are three objects" is true in L_N as "object" in L_N means INDIVIDUAL. Their disagreement is purely verbal – they are using the same term "object" in different ways.

This is puzzling. The L_M sentence "w contains seven objects" is true in L_M if and only if there are seven mereological objects in w. But if Nihilism is true, as Neil believes, there are *no* mereological objects. So, Neil must judge "w has seven objects" *false* in L_M not true. In fact, Hirsch thinks it is not so much "object" as "there exists" which differs in meaning between L_M and L_N: this is Quantifier Variance. Does QV deliver a different verdict?

Perhaps "there exists" in L_M means THEORY M ENTAILS THERE ARE, where M is Mereological Maximalism, "there exists" in L_N correspondingly means THEORY N (Nihilism) ENTAILS THERE ARE. This is Carnapian Internalism: ontological questions can only be formulated within a linguistic framework. In asking "Do composite objects exist?" prior to choosing a framework, we ask a meaningless "external" question. If we choose L_M the answer is "yes," if L_N the answer is "no." If that is what QV amounts to, then, trivially, Neil's and Maxi's divergent answers express truths in their own language. Yet for Hirsch, unlike Carnap, not all meta-ontological disputes are "purely verbal," so this is not his view.

Like Davidson, Hirsch thinks the Principle of Charity plays a crucial role in interpretation. Protagonists of rival meta-ontological theories take them to express necessary, *a priori* truths.

Hirsch invokes Quantifier Variance to explain the puzzle of conflicting intuitions about necessity and apriority; in particular, how these can be at odds with common sense. His argument calls to mind a famous earlier one about alternative logics, due to Quine:

> Here, evidently, is the deviant logician's predicament: when he tries to deny the doctrine, he only changes the subject.
>
> *(Quine 1970, 81)*

Thus, it is a basic, common-sense truth that material objects exist and endure through time. Yet the "deviant" ontologist appears to deny such commonplaces as *tables exist* (Nihilism) or *people endure through time* (Perdurantism); even to assert rank falsehoods such as COWGAROOS[3] *exist* (Maximalism). However, our dispositions to use terms like "table," "person," and "exist" cannot conflict with our basic beliefs. So, the "deviant" ontologist's apparent denial that material objects exist, or are composed of parts, merely reflects *different* learned dispositions to use words like "table," "part," "exist" in contexts far removed from those in which we acquired them. But our dispositions to use words thus and so *constitute* their meanings. Whence, attempting to deny the common-sense doctrines, the deviant ontologist merely changes the subject.

The argument rests on Wittgenstein's thesis that meaning is wholly determined by use. Hirsch may or may not wish to endorse this thesis, but it does provide a principled reason to accept his verdict of meaning variance across rival theories for "there exists." If $\exists_E$, $\exists_M$, $\exists_N$ represent English THERE EXISTS, Maximalist THERE EXISTS and Nihilist THERE EXISTS, respectively, then by QV, $\exists_E \neq \exists_M \neq \exists_N$. Just as Quine argues that it is not the Classical Law of Excluded Middle $p \vee \sim p$ which Intuitionists deny but its Constructive analogue $p \vee \neg p$, so the QV proponent argues that it is not "$\exists_E x.\text{Table}(x)$" the Nihilist denies but "$\exists_N x.\text{Table}(x)$." The term "Table" for Nihilists refers to simples arranged table-wise and "$\exists_N x$" means: THERE IS A SIMPLE x. So, "$\exists_N x.\text{Table}(x)$" says THERE IS A SIMPLE X THAT IS ARRANGED TABLE-WISE.

Yet it is not just Nihilists who deny $\exists_N x.\text{Table}(x)$ – Maximalists, indeed, everyone else, do so as well. On the other hand, since Nihilists agree the plural quantification "$\exists_N xx.\text{Table}(xx)$" is true, as there are simples arranged table-wise, they no longer have cause to deny that there are tables. Interpreting Maxi's "There are tables" utterance *charitably*, Neil should hear her as claiming no more than he himself claims by asserting $\exists_N xx.\text{Table}(xx)$.

Hirsch's desired outcome for protagonists in a purely verbal meta-ontological debate has them realising they are talking past each other and then recognising the other articulates truths in their own language. The cost is the very intelligibility of the debate. To an ideally rational, charitable interpreter, Nihilism would seem incomprehensible – everyone agrees $\exists_N x.\text{Table}(x)$ is false and everyone agrees $\exists_N xx.\text{Table}(xx)$ is true.

What exactly is the Nihilist then denying? Presumably, that there are composite objects, e.g. that $\exists_E y[y = \text{Comp}(\iota xx.\text{Table}(xx))]$, where "Comp" is the Composition operator. The Maximalist asserts there are composites, e.g. that $\exists_M y[y = \text{Comp}(\iota xx.\text{Table}(xx))]$. But once again, s/he also agrees with the Nihilist in judging $\exists_N y[y = \text{Comp}(\iota xx.\text{Table}(xx))]$ false, since it clearly isn't true that there is a simple that is a composite of simples. Quine's words thus recur, ever to haunt the aspiring variantist: *In denying the doctrine, the Nihilist merely changes the subject.*

4. Reason knows no (formal) bounds?

The Carnapian path to Conceptual Pluralism is a path only the cognoscenti can discern, invisible to the philosophical naïf. Yet the doctrine of Pluralism betrays no gnostic origin. Freedom from conventional concepts is a promise to all who believe, not just the sage. Is there no common road all can traverse? Could we, perhaps, have reasoned differently? To take the most radical possibility, could we have entirely eschewed any use of logic? Philip Johnson-Laird contends we actually do, using "mental models."

Mental models, MMs, are quasi-iconic representations of propositions that recognise grammatical but ignore logical, form. MMs warrant heuristic rather than logical responses. Suppose you hear *Ed played soccer or some game* then learn *Ed didn't in fact play any game*. The conclusion *Ed played soccer* is one that: "No one in her right mind (apart from a logician) would draw" (Johnson-Laird 2008, 214). Despite the fact that "the inference is absurd" (2008, 214) it is the logical response. But mental models for the argument S OR G, NOT-G ∴ S block the construction of the possibility that NOT-G and S both hold (2008, 214). Standard logic rules the conclusion S is entailed by S OR G and NOT-G, but since S OR G is equivalent to G, no one who learned NOT-G was true could continue to believe S OR G true also.

Johnson-Laird contends that exceptions can be found for every formal inference rule since:

> The validity of inferences can be decided only on a case-by-case basis: valid entailments are valid not by virtue of form, but in virtue of content.
>
> *(Johnson-Laird 2010, 124)*

His reason is that while logical operators in natural language all have a "single underlying meaning," "a mechanism of *modulation* can transform this meaning into an indefinite number of different sorts of interpretation" (2010, 121). So, the reason one cannot infer *Ed played soccer* from *Ed played soccer or some game* together with *Ed didn't play any game*, is, apparently, that the meaning of the word "or" has been modulated by knowledge that soccer is a game.

Is our reasoning mental model based? Is modulation real, logical form chimerical? If Johnson-Laird is right, all our reasoning is heuristic: the human mind doesn't detect logical form. This position is undermined if some grammaticality judgments concern logical form. Such is the claim for expressions known as Negative Polarity Items (NPIs): ANY, EVER, EITHER, AT ALL, CARE LESS ... To determine whether a sentence containing an NPI is grammatical, we first need to check whether the NPI occurs in a *downward-entailing* context, it is claimed. DE contexts license NPIs. Thus, whereas (2) is grammatical, (1) is ungrammatical:

(1) James ate *any cakes.
(2) James never ate any cakes.

NEVER is a downward-entailing operator: it validates inferences from sets to subsets. If JAMES NEVER ATE ANY CAKES, then JAMES NEVER ATE ANY CHOCOLATE CAKES.

Not everyone is convinced of the NPI-DE link. Timothy Williamson denies NPIs are only licensed in DE contexts. He invites us to consider the following two sentences (Williamson 2007, 108):

(3) Exactly four people in the room were of any help.
(4) Few people in the room were of any help.

Williamson contends "[(3)] is acceptable provided that in the context it is taken to imply [(4)], but not generally otherwise" (2007, 108). He then infers:

> Thus, the phenomenon involves a significant pragmatic element: which contexts are suitable for "any" cannot be determined on purely logico-linguistic grounds.
>
> *(2007, 108)*

Williamson further maintains that:

> If a speaker has deviant views as to which contexts are D.E. but uses "any" in just those contexts that she treats as D.E., we might find her deviant use of "any" inappropriate without regarding her as linguistically incompetent precisely because the deviation in use is explained by logical rather than linguistic unorthodoxy.
>
> *(2007, 108)*

Williamson thinks speakers do not need to recognise downward entailment to classify sentences as grammatical. Yet his own example appears to show the opposite – a non-DE operator EXACTLY FOUR licenses a specific occurrence of an NPI ANY *only when* that operator in the context comprises FEW, inheriting any NPI-licensing power it has from its context-determined connection with a DE operator. So, any speaker who did not realise that EXACTLY FOUR comprised a DE expression FEW in the context could therefore be predicted to reject (3) as ill-formed.

Williamson suggests a competent speaker might "treat" a non-D.E. like EXACTLY FOUR as D.E. How could this be? D.E. operators license inferences from sets to subsets, as (6) from (5):

(5)   Few boys or girls at our school speak Farsi.
(6)   ∴ Few girls at our school speak Farsi.

Yet EXACTLY FOUR fails this crucial diagnostic even when it comprises FEW:

(7)   Exactly four boys or girls at our school speak Farsi.
(8)   ∴ # Exactly four girls at our school speak Farsi.

One who thought (8) followed from (7) couldn't understand what "exactly four" means.[4] Suppose, though, that it's *not* their connection with D.E. that grounds our grammaticality judgments for NPIs so the case for MM-based purely heuristic reasoning isn't undermined. We might still wonder whether a MM-account of validity and logical meanings is correct. Does MMT classify only genuine entailments as valid? Are the meanings of operators like OR and IF subject to variation? Consider the following proposition:

EITHER TODAY IS NOT MONDAY OR TOMORROW IS WEDNESDAY.

This proposition will be true every day of the week except Monday. The mental models for it are these:

Proposition:	MENTAL MODELS	
NOT-M OR W	Not-M	Not-W
	~~M~~	~~W~~
	Not-M	W

Modulation rules out the mental model on line 2 since it cannot be that today is Monday and tomorrow will be Wednesday. But now consider models for a different proposition:

IF TODAY IS MONDAY, THEN TOMORROW IS WEDNESDAY.

Proposition:	MENTAL MODELS	
IF M THEN W	~~M~~	~~W~~
	Not-M	W
	Not-M	Not-W

Once more, Modulation rules out the possibility that today is Monday and tomorrow is Wednesday. However, the remaining mental models are exactly those for the disjunction:

EITHER TODAY IS NOT MONDAY OR TOMORROW IS WEDNESDAY.

The inference from (I) EITHER TODAY IS NOT MONDAY OR TOMORROW IS WEDNESDAY, to: (II) IF TODAY IS MONDAY THEN TOMORROW IS WEDNESDAY is thereby valid according to MMT since IF M THEN W holds in all the models where the premise NOT-M OR W holds.[5] But while (I) can be true, the conclusion (II) is manifestly false. Hence, MMT misclassifies some patently invalid inferences as valid.

What of Modulation? Modulation can eliminate possibilities from our models, as in the soccer argument. What then happens to operator meanings when there are *no* possibilities? Suppose Enzo is learning English and is confused about English number words. He opines: "Either three plus three is seven or four plus four is nine." How would you correct him? It won't do to simply say: "No, three plus three is six not seven." For then Enzo could still believe four plus four is nine. For the same reason you cannot just tell him: "No, four plus four is eight, not nine." You have to point out that three plus three is six *and* four plus four is eight.

But why should you have to do this?

If MMT is correct, the word OR in the proposition 3 + 3 IS 7 OR 4 + 4 IS 9 has been divested of its normal content since there are NO possibilities that correspond to it. But far from being divested of content, OR appears to retain its inclusive disjunction meaning, represented by the operator "∨" in standard logic, for which de Morgan's law holds:

$$\sim(A \vee B)| = \sim A \ \& \sim B$$

This interpretation of negated disjunction is no mere quirk of English. For the same law holds in every known language. Examples (10–13) are translations of (9) JOHN DIDN'T SEE TED ORDER PASTA OR SUSHI, into (10) Mandarin-Chinese, (11) Japanese, (12) Dutch and (13) Russian. The interpretation is the same in each language, (M):[6]

(M) John did not see Ted order pasta *and* John did not see Ted order sushi.
 (9) John did NOT see Ted order pasta OR sushi.
(10) Yuehan MEI kanjian Ted dian yidalimianshi HUOZHE shousi.
(11) John-wa Ted-ga sushi KA pasuta-o tanomu-no-o mi-NAKAT-ta.
(12) John zag Ted NIET pasta OF sushi bestellen.
(13) Dzhon NE videl/uvidel chto/kak Borja zakazal/zakazyval pastu ILI sushi.

Along with De Morgan's Law: ~(A OR B) ⊨ ~A AND ~B, speakers across languages also endorse the same distinctive entailments for OR and EVERY. For example, speakers of English, Mandarin-Chinese or Japanese hear the following English sentence or its translation into Mandarin or Japanese as a conjunction of two universally quantified sentences:

(14$_E$) English: Everyone who ordered pasta or sushi became ill [meaning indicated by (∧)]
(∧) Everyone who ordered pasta became ill <u>and</u> everyone who ordered sushi became ill.
(14$_{MC}$) Chinese: Mei-gedian-le yidalimianshi HUOZHE shousi de ren DOU bing-le.
Translation: Everyone who ordered pasta or sushi became ill, meaning (∧).
(14$_J$) Japanese: Pasuta KA sushi o tanonda zenin/MIN'NA ga byooki ni natta.
Translation: Everyone who ordered pasta or sushi became ill, meaning (∧):

A very different pattern results when disjunction occurs in the predicate phrase, PP, of a universally quantified sentence in these three languages, however. Here, a conjunctive meaning is unavailable, the common meaning being given by (∨):

(15$_E$) English: Everyone who became ill ordered pasta OR sushi.
(∨) Each person who became ill either ordered pasta <u>or</u> that person ordered sushi.
(15$_{MC}$) Chinese: Mei-ge sheng bing de ren DOU dian-guo yidalimianshi HUOZHE shouse.
(15$_J$) Japanese: Byooki ni natta zenin/MIN'NA ga pasuta KA sushi o tanonda.

No morphological or distributional cues distinguish OR in the PP from OR in the NP. The difference between (14$_{E/MC/J}$) and (15$_{E/MC/J}$) is their logical form − if the disjunction occurs in the restrictor of the universal quantifier, the NP, it receives a conjunctive interpretation, if in the nuclear scope, PP, the interpretation is disjunctive:

(1) Every [. . . OR . . .]$_{NP}$ [. . . . .]$_{PP}$ = Conjunctive Interpretation
(2) Every [. . . . .]$_{NP}$ [. . . OR . . .]$_{PP}$ = Disjunctive Interpretation
(3) Every [P |≠ P ∨ Q]$_{NP}$ [P ⊨ P ∨ Q]$_{PP}$

Why should there be a marked difference in the behaviour of OR depending on whether it occurs in the NP or PP of a quantified sentence? How does it come about that OR can have a "conjunctive" interpretation at all, given its meaning? There is an answer to this question.
    According to the semantics of ∀ and of ∨ as explicated in Generalised Quantifier Theory:

(1) ⟦ [∀x: Ax ∨ Bx] Cx⟧ is true if and only if ({x: A(x)} ∪ {x: B(x)}) ⊂ {x: C(x)}.

In contrast:

(2) ⟦ [∀x: Ax] Bx ∨ Cx⟧ is true if and only if {x: A(x)} ⊂ ({x: B(x)} ∪ (x: C(x)})

But (1) entails ⟦ [∀x: Ax ∨ Bx] Cx⟧ is true if and only if:

(3) {x: A(x)} ⊂ {x: C(x)} AND {x: B(x)} ⊂ {x: C(x)}

In contrast, (2) entails ⟦ [∀x: Ax] Bx ∨ Cx⟧ is true if and only if:

(4) {x: A(x)} ⊂ (x: B(x)} OR {x: A(x)} ⊂ (x: C(x))

Drew Khlentzos

We have an explanation of some otherwise puzzling facts about OR as it appears in the NP and PP of universally quantified sentences in the meaning of the quantifier EVERY. No one unable to detect the downward-entailing clauses in these constructions could be competent in using them. Could a speaker hold a heterodox theory about *which* clauses are the DE ones without compromising her competence, as Williamson suggests? Suppose Diva believes the PPs of universally quantified sentences are DE rather than their Restrictors. Suppose also that Ivy's father Kirk, who takes Ivy to playgroup each week, informs Diva that *every parent at Ivy's playgroup is either pregnant or a father*. Diva immediately infers Kirk is pregnant. If she tells Kirk this, shouldn't he harbour at least *some* suspicion Diva doesn't know what "pregnant" means?

We wondered whether there could not be a conceptual scheme for ordinary reasoning that was genuinely different from our own – a scheme, using mental model theory, for instance, that eschewed logic entirely, privileging heuristic over logical judgments for conflict inferences. The short answer appears to be "No." Not only would our reasoning suffer irreparable harm if we misclassified valid inferences (like the soccer game) as invalid or invalid inferences (like the Monday/Wednesday one) as valid, semantic processing would be undermined.

In conclusion, *inaccessible* conceptual schemes clearly are possible, but accessible schemes alternative to our own appear uncertain. Hirsch's quantifier variance argument actually threatened to *undermine* the variant meta-ontological schemes he considers – schemes, moreover, that are *not* accessible to ordinary folk. The prospect of a heuristic alternative to logical reasoning appeared elusive. So, while inaccessible alternative schemes are possible, it is an open question whether there are generally accessible alternative schemes.

Acknowledgements

Thanks to Stephen Crain, Denis Robinson and, especially, Georgina Pullar for their advice and assistance.

Notes

1 And other factors, especially scalar implicatures. See, for instance, Rips (1994).
2 "I claim that the dispute between Endurantists and Perdurantists is verbal . . . each party ought to agree the other party speaks a truth in his own language" (Hirsch 2011, 229).
3 Where "cowgaroos" are defined to be cows above the neck, kangaroos below.
4 (7) would "entail" exactly eight children speak Farsi.
5 "A conclusion is valid provided it holds in every possibility in which the premises hold" (Johnson-Laird 2010, 121).
6 The expressions for negation and disjunction are in small capitals in each of the examples.

References

Berger, A. (1980), "Quine on 'Alternative Logics' and Verdict Tables," *Journal of Philosophy* 77(5): 259–277.
Carnap, R. (1950), "Empiricism, Semantics and Ontology," *Revue Internationale de Philosophie* 4: 20–40.
Davidson, D. (1973), "On the Very Idea of a Conceptual Scheme," *Proceedings and Addresses of the American Philosophical Association* 47: 5–20.
———. (1999), "Reply to W.V. Quine," in *The Philosophy of Donald Davidson*, edited by W. Hahn, Chicago: Open Court, 80–86.
De Neys, W. (2012), "Bias and Conflict: A Case for Logical Intuitions," *Perspectives on Psychological Science* 7(1): 28–38.
Hirsch, E. (2011), *Quantifier Variance and Realism: Essays in Metaontology*, Oxford: Oxford University Press.
Johnson-Laird, P. (2008), "Mental Models and Deductive Reasoning," in *Reasoning*, edited by J. Adler and L. Rips, Cambridge: Cambridge University Press, 206–222.

———. (2010), "The Truth About Conditionals," in *The Science of Reason*, edited by K. Mantelow, D. Over and S. Elqayam, Hove: Taylor and Francis, 119–143.

Putnam, H. (2004), *Ethics Without Ontology*, Cambridge, MA: Harvard University Press.

Quine, W. V. O. (1966), "On Carnap's Views on Ontology," in his *The Ways of Paradox and Other Essays*, New York: Random House, 126–134.

———. (1969), "Epistemology Naturalized," in his *Ontological Relativity and Other Essays*, New York: Columbia University Press, 69–90.

———. (1970), *Philosophy of Logic*, Englewood Cliffs: Prentice-Hall.

———. (1974), *The Roots of Reference*, La Salle: Open Court.

Rips, L. (1994), *The Psychology of Proof: Deductive Reasoning in Human Thinking,* Cambridge, MA: MIT Press.

Williamson, T. (2007), *The Philosophy of Philosophy*, Oxford: Blackwell.

49

SEMANTICS AND
METAPHYSICS OF TRUTH

Manuel García-Carpintero

1. Introduction

Given his perceptive discussion of some of the topics that will be covered in this chapter, Devitt (2001) might complain that its title betrays a confusion – one which, according to him, is easily overlooked in debates on the relations between two popular theories of truth, deflationism and the correspondence theory. By *an account of truth* we might have in mind a theory of the truth-*property*, or rather one about the truth-*concept* – equivalently, both for Devitt and for me henceforth, of the truth-*predicate*. Deflationism, Devitt points out, focuses on the latter, because, being fundamentally a skeptical view about the metaphysics of truth, it has little to say on the former: perhaps only that its nature is exhausted by what can be gleaned from what it says about the predicate.[1] Just the opposite is the case on the correspondence view: it is essentially an account of the metaphysics of truth, and has very little to say on the predicate – say, that, like most basic monadic predicates, it signifies a (relational) property, and thereby has an extension, dependent on the correspondence features that the view more substantively articulates. Discussions of the semantics of truth focus on the concept, while discussions of its metaphysics primarily tackle the property. Contributions in recent years have taken note of Devitt's distinction. Eklund (2017) provides a good discussion of problems that deflationists encounter in trying to move from their positive claims about the truth predicate, to their distinctive negative claims about the property; further references to recent literature can be found there.

An additional oxymoronic feature of our title is that it would fit an encyclopedic discussion of everything philosophical that there is to say about truth. Of course, the reader cannot expect this – merely some related material that impinges on the topic of this book, relativism. Fortunately, truth is a subject for most of whose different edges we have excellent and fully up-to-date introductory presentations, in the *Stanford Encyclopedia* and similar undertakings: the relevant entries and chapters referenced in Glanzberg (2018a) and included in Glanzberg (2018b) will offer first-rate help to interested readers.[2] What will mainly be found in this chapter is a discussion of a distinction that, questionably in my view, both Devitt (2001, 606) and Glanzberg (2018a, §6.1) take to be irrelevant to the topic of the semantics and metaphysics of truth, namely, whether *propositions* or *representational states/vehicles* are primary truth-bearers.

Against Devitt and Glanzberg, I'll defend a contrary view which also has strong supporters in the literature; for, as Jackson (2006, 50) puts it, "the reason for favoring a kind of correspondence theory for truth of beliefs and sentences does not carry over to truth for propositions." In the next section I'll present recent debates relevant to our question, which show, I believe, that it is advisable to keep apart issues concerning propositions from issues concerning the representational vehicles to which they are ascribed. In these terms, and along lines that Jackson also (2006, 58) suggests, I'll indicate how "truth substantivists" might deal with an issue that rightly troubles Devitt (2001, 602), "How come so many people innocent of the correspondence theory nonetheless believe instances of the equivalence schema? A deflationist might reasonably claim that there is a strong argument for deflationism here." As I have argued elsewhere (see references at the end), the distinction between the truth of propositions and that of the acts/vehicles expressing them impinges on relativism-related debates concerning the semantics and metaphysics of truth. Among them are the appraisal of recent truth-relativist proposals, indeterminacy and the question of whether truth might be conventional.

2. Propositions, propositional vehicles and unity problems

As we saw previously, Jackson contends that the decision regarding truth-bearers is significant for debates between truth substantivists like correspondence theorists, and deflationists; Field (1992, 322) had famously made the point in his critical review of Horwich's (1990), "on most conceptions of propositions, the question of what it is for a proposition to be true is of little interest, . . . what is of interest are the issues of what it is for an utterance or a mental attitude to be true (or, to express a truth or represent a truth)" (cf. also Lewis 2001, 602–603).

A deflationary view regarding propositions and their features, like what it takes for them to be true, can be predicated on a general attitude about the philosophical topics that are theoretical posits of the discipline – such as *propositions, a priori justification* or *possible worlds*.[3] This attitude might be motivated by Yablo's (2014a) "quizzical" standpoint: the notion that there is no reasonable way that controversies among proponents of *prima facie* conflicting substantive ontologies for such topics can be adjudicated; or equally by the view that such entities are posits of an "easy ontology" (Thomasson 2015) and, as such, lack a "hidden nature" worth investigating and debating about. Deflationism might be grounded instead on a view specific to the case of propositions, like the measure-theoretic perspective on their explanatory role promoted by Davidson, Perry, Stalnaker and others.[4] Like numbers vis-à-vis quantitative properties, on this view propositions are just convenient resources used to represent through their relations the semantic relations among attitudes and acts in their "job description" as contents of attitudes like belief or speech acts like sayings, referents of that-clauses, and so on.[5] The second motivation is of course consistent with the first.

In the past decade, several philosophers, including Gaskin (2008), King (2007), Soames (2015) and Hanks (2015), have nonetheless advanced different substantive theories of propositions, to deal with issues those authors have raised, close to a concern with a long pedigree in philosophy, the so-called *problem of the unity of propositions*:[6] what is it that holds the constituents of a proposition together? In fact, as it has been pointed out,[7] there are several problems that have been discussed under that label. Eklund (2019) calls the "representation problem" the more specific *unity of the proposition* question that has been prominently discussed in the recent literature. This is the alleged problem of explaining the representational properties of propositions, in particular their truth-aptness and their having truth-conditions. It is related to the more traditional concern: "we may reasonably expect that any account of the unifying relations of propositions will help us to understand their truth-aptness" (Liebesman 2015, 550).

King (2019) motivates a substantivist standpoint by contending that unity problems concern properties that

> seem to call out for further explanation and whose possession seems as though it should be grounded in the possession of 'more basic' properties. It may be hard to give a criterion for being such a property, but properties like *being alive, believing that snow is white*, and *being morally good* seem to be examples of such properties.
>
> *(2019, 1346)*

Against this, philosophers like Bealer (1998) and Merricks (2015) advance a "quietist" or "primitivist" view, opposing the project of providing full-blooded accounts. As Lewis (1983, 352) protests in a similar context discussing the related *Third Man*-like regresses:

> Not every *account* is an *analysis*! A system that takes certain Moorean facts as primitive, as unanalysed, cannot be accused of failing to make a place for them. It neither shirks the compulsory question nor answers it by denial. It does give an account.

A deflationary attitude has also been advanced for the representation problem:

> the truth-value aptness of a proposition is at least a good candidate for belonging to the fundamental essence of a thing. But if this is so, the only legitimate reply . . . may consist not in a direct answer, but instead in a rejection of the question.
>
> *(Schnieder 2010, 300)*

On an influential account – still the default in contemporary semantics – propositions are sets of worlds, and these in turn are understood by Stalnaker (1976) as given by *ways* or *properties* the world might have. My own deflationary take on propositions views them as just such properties.[8] As properties of truth-makers, propositions might be finer-grained than the sets of worlds at which they are instantiated, and thus not identical to them, thereby dodging well-known difficulties with the identification of propositions with such sets. They can also be taken as properties of entities smaller than worlds, as in the "truthmaker semantics" of Yablo (2014b) and Fine (2017). This gives additional support to the deflationary standpoint on the representation problem, for no substantive explanation should be expected for why properties apply to things: this is just what properties do, by their very nature.[9]

Finally, the deflationary standpoint can be justified as well by the spurious problems substantivist views generate. A case in point of the latter is King's (2007, 2019) "interpretativist" view. He takes propositions to be facts constituted by interpretations of syntactic structures given by the "Language Module." Alas, this makes propositions too fine-grained to adequately serve the goals in their job-description (Collins 2014; Keller 2019). Soames' (2015) and Hanks' (2015) "act-theoretic" view, on which propositions are types of cognitive acts, is seriously challenged by the "Frege-Geach point" that propositions can occur and be truth-conditionally operative without being the contents of any cognitive act, as in the condition for a conditional bet or command (Jespersen 2012; Hom and Schwartz 2013; Reiland 2013). Finally, substantivist views either are prone to create the regress that traditional accounts of unity so easily engendered (Pickel 2015), or wallow in it (Gaskin 2008), thus imperiling understanding.

Unlike primitivists like Bealer (1998) and Merricks (2015), I insist that we should couple the deflationary view on propositions with robust, explanatory answers to concerns in the vicinity of those confronted by substantive proposals. Thus, Liebesman (2015) provides some illumination on the traditional unity concerns by offering an account of the representational vehicles – how predicates work – instead of one of the propositions ascribed to them. This is also the stance on truth promoted in the quotations from Field and Jackson: while deflationism is perfectly fine when it comes to the truth of propositions, a substantive account might be needed for that of representational vehicles, sentences or sentences in context, and assertive acts deploying them. Azzouni (2018, §4) outlines the main reason for this:[10] truth-deflationism makes unavailable adequate, illuminating explanations for why truth applies to assertive acts and their representational vehicles.

Field (1972) influentially defended this view for referential relations linking referential items in representational vehicles and worldly items. Taylor (2017) provides a version of this "missing explanation" argument against deflationary accounts of reference based on indeterminacy issues. In contrast, it is not difficult to explain why core instances of the disquotational schema for propositions (*the proposition that snow is white is true iff snow is white*) feel true, and are so if we put aside problematic cases which I cannot discuss here: the proposition referred to on the left-hand side is just the one expressed on the right-hand side (cp. Jackson 2006, 58). We do need an account that properly subsumes instances of the disquotational schema under some adequate generalization (Hill 2002; Künne 2003; Azzouni 2018).[11] Given it, nothing more needs to be explained about the truth of propositions.[12]

MacFarlane (2014, 47–48) questions whether it makes intuitive sense to predicate truth of acts or their vehicles: "there is something a bit odd about calling utterances or assertions, in the 'act' sense, true or false at all. We characterize actions as correct or incorrect, but not as true or false" (2005, 322). But, firstly, as Zardini (2008, 546) and Hanks (2015, 67–69) note, we may use adverbs like "truly" for acts, the way we use "frankly": a statement is truly made just in case the stated proposition is true.[13] Besides, as Austin (1950, 119) pointed out, it is also intuitively odd to predicate truth of propositions in the philosophers' sense. MacFarlane is in fact led to solecisms that seem to me to be as awkward as the ones that he objects to: he (2014, 78) needs to appeal to a distinction between the "context of use" and the "context of assessment" *of propositions*, contexts being in his view concrete historical situations. Given that propositions are abstract entities, this only makes sense derivatively: the context of a proposition is that of an act in which it is expressed, or an act expressing it appraised. This justifies, I think, a direct discussion of the truth-related normative features of the acts.[14]

I'll conclude with a brutally simplified summary of my own reasons for objecting to the truth-deflationist claim that there is nothing to explain about truth beyond its disquotational features. First, truth as a predicate of propositions signifies a non-normative property that just classifies them into two classes, the true and the untrue ones, thereby effecting a corresponding division of all representational acts of which they are contents (García-Carpintero 2011, 2012). This includes, say, imaginings, promises or directives, assuming, as I do, the traditional view that all such acts have propositional contents. However, as Dummett (1959/1978) pointed out, truth as a predicate of acts only constitutes a norm for some of them: those in the "assertive family" with "thetic," word-to-world direction of fit (Green 2017) such as judgments, assertions, guesses or presuppositions. Now, although the declarative mood can be conventionally used to perform different specific acts in the assertive family, some of them – "flat-out" assertions (Williamson 1996, 246), in my view, *tellings* – play a fundamental role. They are constituted by an epistemic, truth-entailing norm, and, as such, they have a crucial metasemantic task: semantic value is to be assigned to lexical items in a way that properly explains – in my view along teleological lines – how such a factive epistemic norm has come to be in force for them in our communities.[15]

Considerations of this kind provide support for externalist, causal-historical accounts for the truth-conditional contribution of expressions like names, predicates and natural kind terms, and thereby for core claims of correspondence theories of truth. But they do it in a restricted, nuanced way. Consider an utterance of (the Russian counterpart of) "Bezukhov and Kutuzov looked at Borodino from a hill" by Tolstoy as part of *War and Peace*. If Predelli (1997) is right, this is a straightforward true assertion whose circumstances of evaluation have been contextually shifted from the actual world to a fictional one. To properly evaluate it there, our metasemantics still needs to ascribe a referent to the fictional name "Bezukhov"; following Predelli (2002) and others, we might take this to be an exotic entity, such as an abstract character created by Tolstoy in producing the novel. Our metasemantics might even require us to ascribe the same kind of referents to real names occurring in such fictional contexts, like "Borodino" and "Kutuzov" in our example. Now, I am just offering this for illustrative purposes; I myself prefer a fictionalist stance towards such utterances. But even so, we need to ascribe contents to the fictional acts to which such names make contributions; and the outlined proposal at the very least offers a good (fictional) model for them (García-Carpintero 2018b, forthcoming-a). Hence, given the account sketched, the way in which these representational acts are truly made substantially differs from that of claims about the instantiation of natural properties by physical objects.[16]

Issues concerning truth-aptness and robust representationality offer a second illustration. On the traditional view of propositions embraced previously, they provide "fulfilment" conditions (Ludwig 1997) for all sorts of speech acts, including those we don't intuitively take to be truth-apt, like promises or directives. Given this, the explanation canvassed previously for the intuitive correctness of core instances of the disquotational schema extends to cases such as *a request that* p *is fulfilled iff* p, and so on; and we can also obtain for them adequate generalizations of the deflationary kinds envisaged by Hill (2002) or Künne (2003). The fact that we only feel entirely confident applying the intuitive notion of truth to some of those acts (intuitively, those in the *assertive family*, *sayings*) is one more reason for questioning MacFarlane's claim that the intuitive truth predicate applies only to philosophers' propositions, discussed previously. But there is a stronger point to be made here, because the feeling of inapt application extends also to acts literally performed with declarative sentences, including cases in which their deflationary fulfilment conditions are satisfied.

Thus, in joint work with Teresa Marques we have provided a "hybrid" view to account for impressions of disagreement about taste predicates and the behavior of slurs (Marques and García-Carpintero 2014, forthcoming). On such a view, the Spanish slur "polaco" has a descriptive meaning, equivalent to *Catalan*; but it also triggers an expressive presupposition, conveying the appropriateness of despising Catalans on certain grounds. Consider an assertion made with "I never invite *polacos* home." On the outlined view, its descriptive truth-condition might well be fulfilled. However, we feel reluctant to classify such assertions as *true*. A substantive account might explain this. We might say that only representational acts with the *thetic*, word-to-world direction of fit are straightforwardly classified as true or false. Acts that have only the "telic," world-to-word direction of fit are not to be classified in this way. And mixed acts with both directions of fit (even if vis-à-vis different contents) require at the very least disambiguation, which might account for our reluctance. All of this should of course be elaborated; but I am very doubtful that deflationists can provide comparable explanations.[17]

In previous work on relativism-related issues, I have shown that taking truth to be fundamentally a property of representational acts is crucial in order to properly confront them. Based on this assumption I have discussed recent truth-relativist proposals, highlighting in such terms the philosophically important differences between non-indexical contextualism and assessment relativism (cf. Ferrari's and Marques' contributions, and García-Carpintero 2008, 2013a,

2013b, 2016); the different issues raised by indeterminacy in general and vagueness in particular (García-Carpintero 2007, 2008, 2010, 2013b); and whether truth by convention is intelligible (García-Carpintero & Pérez Otero 2009). I here lack the space to go further into these issues, which I think confirm the main point I have made.

Acknowledgements

Financial support was provided by the DGI, Spanish Government, research project FFI2016–80588-R, and through the award "ICREA Academia" for excellence in research, 2013, funded by the Generalitat de Catalunya. This work received helpful comments from Filippo Ferrari, Dan López de Sa, Teresa Marques and from the editor. Thanks also to Michael Maudsley for the grammatical revision.

Notes

1 This is the standard move for deflationism – to take the truth-property to be "read-off" the truth-concept, cf. Taylor (2019, §1); but see Eklund (2017, §3) for concerns.
2 See also Wyatt and Lynch (2016), who emphasize the pluralist tendencies in recent literature, endorsed later; Kölbel (2013) has a good discussion of whether such pluralism should apply to the truth concept, or only to the property, and relevant methodological considerations.
3 In contrast to *knowledge* or *moral value*, topics which predate the discipline.
4 Cf. Green (1999), Matthews (2007, 2011), Field (2016) and Ball (2018).
5 See King's "What Role Do Propositions Play in Our Theories?" in King et al. (2014).
6 Cf. Davidson's (2005, 76–97) presentation of the early history of the debate.
7 Cf. King (2009), Schnieder (2010), Eklund (2019) and Collins (2018).
8 Cf. Speaks' "Propositions are Properties of Everything or Nothing" in King et al. (2014), Richard (2013) and Pautz (2016). I don't take the identification of propositions with properties of truth-makers to be substantive – an ascription to them after all of a "hidden nature" – because the deflationary attitude extends to them. They could be understood along the lines of resemblance nominalism: propositional properties might ultimately be equivalence classes of representational vehicles, as in traditional deflationary views – cf. Grzankowski and Buchanan (2018), Field (2016) and Sainsbury (2018) for related views. Understood as properties, contents/what is said might be more or less fine-grained for different explanatory purposes; cp. Moore's (1999) view that they are "individuatively vague."
9 Ostertag (2013, 519) reports that Stalnaker himself pointed this out. The properties that I will take propositions to be correspond to the *states of affairs* that on Matthews's (2007, 153) presentation of the measure-theoretic account are *representatives* of the attitudes.
10 I argued for related points in earlier work, García-Carpintero (1996), (1998), (1999) and (2011). McGee (2016), Moore (2018) and Taylor (2019) offer recent variations on similar themes.
11 Simmons (2018, 1012–1019) develops a compelling criticism of Horwich's (1998) "minimal" articulation of a deflationary theory for propositions; García-Carpintero (1998, 59–60, fn. 3), makes a similar criticism for the earlier presentation in the first edition of the book.
12 The "language-immanence" of this account also causes trouble for the negative claim of truth-deflationism that there is nothing to explain about truth beyond what the disquotational schema properly generalizes affords, as McGee (2016) and Moore (2018) point out.
13 This needs qualification on account of indeterminacy and other issues I cannot go into here (García-Carpintero 2007, 2010, 2013b).
14 Cf. Soames (2015, 25ff.) for related points, and Schnieder (2006) for a compelling critical discussion of philosophical arguments of MacFarlane's style.
15 García-Carpintero (2012, 2018a, 1125, forthcoming-b); cp. Williamson (2007, 264), and McGlynn (2012). The view outlined here is developed in García-Carpintero (forthcoming-c).
16 The outlined proposal is thus close to *pluralist* views on truth such as Horgan's (2001); see also Barnard and Horgan (2013), Sher (2016).
17 Cf. Taylor (2019, §2.4) and cp. Hill (2016, 3173–3176).

References

Austin, J. (1950), "Truth," *Proceedings of the Aristotelian Society, Supplementary Volume* 24: 111–128; reprinted in his *Philosophical Papers*, edited by J. O. Urmson and G. J. Warnock, Oxford: Oxford University Press, 1961, 117–133. [Page reference to the reprinted version.]

Azzouni, J. (2018), "Deflationist Truth," in *The Oxford Handbook of Truth*, edited by M. Glanzberg, Oxford: Oxford University Press, 477–502.

Ball, D. (2018), "Semantics as Measurement," in *The Science of Meaning*, edited by B. Rabern and D. Ball, Oxford: Oxford University Press, 381–410.

Barnard, R. and T. Horgan (2013), "The Synthetic Unity of Truth," in *Truth and Pluralism*, edited by N. Pedersen and C. Wright, Oxford: Oxford University Press, 180–196.

Bealer, G. (1998), "Propositions," *Mind* 107(425): 1–32.

Collins, J. (2014), "Cutting It (Too) Fine," *Philosophical Studies* 169(2): 143–172.

———. (2018), "The Redundancy of the Act," *Synthese* 195: 3519–3545.

Davidson, D. (2005), *Truth and Predication*, Cambridge, MA: Harvard University Press.

Devitt, M. (2001), "The Metaphysics of Truth," in *The Nature of Truth: From the Classic to the Contemporary*, edited by M. P. Lynch, Cambridge, MA: MIT Press, 579–611.

Dummett, M. (1959/1978), "Truth," in his *Truth and Other Enigmas*, Cambridge, MA: Harvard University Press, 1–24.

Eklund, M. (2017), "What Is Deflationism About Truth?" *Synthese*, doi:10.1007/s11229-017-1557-y.

———. (2019), "Regress, Unity, Facts, and Propositions," *Synthese* 196: 1225–1247.

Field, H. (1972), "Tarski's Theory of Truth," *Journal of Philosophy*, LXIX: 347–375.

———. (1992), "Critical Notice of Horwich's *Truth*," *Philosophy of Science* 59: 321–330.

———. (2016), "Egocentric Content," *Noûs* 51(3): 521–546.

Fine, K. (2017), "Truthmaker Semantics," in *A Companion to the Philosophy of Language*, second edition, edited by B. Hale, C. Wright and A. Miller, Oxford: Blackwell, 556–577.

García-Carpintero, M. (1996), "What Is a Tarskian Theory of Truth?" *Philosophical Studies* 82: 1–32.

———. (1998), "A Paradox of Truth-Minimalism," in *Truth in Perspective*, edited by C. Martínez, U. Rivas and L. Villegas-Forero, Aldershott: Ashgate, 37–63.

———. (1999), "The Explanatory Value of Truth Theories Embodying the Semantic Conception," in *Truth and Its Nature (if any)*, edited by J. Peregrin, Dordrecht: Kluwer, 129–148.

———. (2007), "Bivalence and What Is Said," *Dialectica* 61: 167–190.

———. (2008), "Relativism, Vagueness and What Is Said," in *Relative Truth*, edited by M. García-Carpintero and M. Kölbel, Oxford: Oxford University Press, 129–154.

———. (2010), "Supervaluationism and the Report of Vague Contents," in *Cuts and Clouds: Essays in the Nature and Logic of Vagueness*, edited by S. Moruzzi and R. Dietz, Oxford: Oxford University Press, 345–359.

———. (2011), "Truth-Bearers and Modesty," *Grazer Philosophische Studien* 82: 49–75.

———. (2012), "Foundational Semantics, I & II," *Philosophy Compass* 7(6): 397–421.

———. (2013a), "Critical Study: Relativism and Monadic Truth," *Philosophical Quarterly* 63: 597–602.

———. (2013b), "Relativism, the Open Future, and Propositional Truth," in *Around the Tree*, edited by F. Correia and A. Iacona, Berlin: Springer, 1–27.

———. (2016), "Mark Richard's *Truth and Truth-Bearers*," *Notre Dame Philosophical Reviews*, http://ndpr.nd.edu/news/69604-truth-and-truth-bearers-meaning-in-context-volume-ii/.

———. (2018a), "The Mill-Frege Theory of Proper Names," *Mind* 127(508): 1107–1168.

———. (2018b), "Co-Identification and Fictional Names," *Philosophy and Phenomenological Research*, doi:10.1111/phpr.12552.

———. (forthcoming-a), "Singular Reference in Fictional Discourse?" *Disputatio*.

———. (forthcoming-b), *Tell Me What You Know*, Oxford: Oxford University Press.

———. (forthcoming-c), "Reference-fixing and Presuppositions", S. Biggs & H. Geirsson (eds.), *Routledge Handbook on Linguistic Reference*.

García-Carpintero, M. and M. Pérez Otero (2009), "The Conventional and the Analytic", *Philosophy and Phenomenological Research* 78: 239–274.

Gaskin, R. (2008), *The Unity of the Proposition*, Oxford: Oxford University Press.

Glanzberg, M. (2018a), "Truth," in *The Stanford Encyclopedia of Philosophy*, Fall 2018 edition, edited by E. N. Zalta, https://plato.stanford.edu/archives/fall2018/entries/truth/.

———. (ed.) (2018b), *The Oxford Handbook of Truth*, Oxford: Oxford University Press.

Green, M. (1999), "Attitude Ascription's Affinity to Measurement," *International Journal of Philosophical Studies* 7: 323–348.

———. (2017), "Assertion," *Oxford Handbooks Online*, doi:10.1093/oxfordhb/9780199935314.013.8.

Grzankowski, A. and R. Buchanan (2018), "Propositions on the Cheap," *Philosophical Studies*, doi:10.1007/s11098-018-1168-6.

Hanks, P. (2015), *Propositional Content*, Oxford: Oxford University Press.

Hill, C. (2002), *Thought and World*, Cambridge: Cambridge University Press.

———. (2016), "Deflationism: The Best Thing Since Pizza and Quite Possibly Better," *Philosophical Studies* 173: 3169–3180.

Hom, C. and J. Schwartz (2013), "Unity and the Frege-Geach Problem," *Philosophical Studies* 163: 15–24.

Horgan, T. (2001), "Contextual Semantics and Metaphysical Realism: Truth as Indirect Correspondence," in *The Nature of Truth: From the Classic to the Contemporary*, edited by M. P. Lynch, Cambridge, MA: MIT Press, 67–95.

Horwich, P. (1990/1998), *Truth*, second edition, Oxford: Clarendon Press.

Jackson, F. (2006), "Representation, Truth and Realism," *The Monist* 89: 50–62.

Jespersen, B. (2012), "Recent Work on Structured Meaning and Propositional Unity," *Philosophy Compass* 7: 620–630.

Keller, L. (2019), "What Propositional Structure Could Not Be," *Synthese* 196: 1529–1553.

King, J. (2007), *The Nature and Structure of Content*, Oxford: Oxford University Press.

———. (2009), "Questions of Unity," *Proceedings of the Aristotelian Society* 109(3): 257–277.

———. (2019), "On propositions and Fineness of Grain (again!)," *Synthese* 196: 1343–1367.

King, J., S. Soames and J. Speaks (2014), *New Thinking About Propositions*, Oxford: Oxford University Press.

Kölbel, M. (2013), "Should We Be Pluralists About Truth?" in *Truth and Pluralism*, edited by N. Pedersen and C. Wright, Oxford: Oxford University Press, 278–297.

Künne, W. (2003), *Conceptions of Truth*, Oxford: Oxford University Press.

Lewis, D. (1983), "New Work for a Theory of Universals," *Australasian Journal of Philosophy* 61: 343–377.

———. (2001), "Truthmaking and Difference-Making," *Noûs* 35: 602–615.

Liebesman, D. (2015), "Predication as Ascription," *Mind* 124: 517–569.

Ludwig, K. (1997), "The Truth About Moods," *Protosociology* 10: 19–66.

MacFarlane, J. (2005), "Making Sense of Relative Truth," *Proceedings of the Aristotelian Society* 105: 321–339.

———. (2014), *Assessment Sensitivity: Relative Truth and Its Applications*, Oxford: Oxford University Press.

Marques, R. and M. García-Carpintero (2014), "Disagreement About Taste: Commonality Presuppositions and Coordination," *Australasian Journal of Philosophy* 72(4): 701–723, doi:10.1080/00048402.2014.922592.

———. (forthcoming), "Really Expressive Presuppositions and How to Block Them," *Grazer Philosophische Studien*.

Matthews, R. J. (2007), *The Measurement of Mind*, Oxford: Oxford University Press.

———. (2011), "Measurement-Theoretic Accounts of Propositional Attitudes," *Philosophy Compass* 6: 828–841.

McGee, V. (2016), "Thought, Thoughts, and Deflationism," *Philosophical Studies* 173: 3153–3168.

McGlynn, A. (2012), "Interpretation and Knowledge Maximization," *Philosophical Studies* 160: 391–405.

Merricks, T. (2015), *Propositions*, Oxford: Oxford University Press.

Moore, G. S. (2018), "Theorizing About Truth Outside One's Own Language," *Philosophical Studies*, https://doi.org/10.1007/s11098-018-1211-7.

Moore, J. (1999), "Propositions Without Identity," *Noûs* 33(1): 1–29.

Ostertag, G. (2013), "Two Aspects of Propositional Unity," *Canadian Journal of Philosophy* 43(5–6): 518–533.

Pautz, A. (2016), "Propositions and Properties," *Philosophy and Phenomenological Research* 93(2): 478–486.

Pickel, B. (2015), "Are Propositions Essentially Representational?" *Pacific Philosophical Quarterly* 97(2): 470–489.

Predelli, S. (1997), "Talk About Fiction," *Erkenntnis* 46: 69–77.

———. (2002), "'Holmes' and Holmes – A Millian Analysis of Names from Fiction," *Dialectica* 56(3): 261–279.

Reiland, I. (2013), "Propositional Attitudes and Mental Acts," *Thought* 1: 239–241.

Richard, M. (2013), "What Are Propositions?" *Canadian Journal of Philosophy* 43(5–6): 702–719.

Sainsbury, M. (2018), *Thinking About Things*, Oxford: Oxford University Press.

Schnieder, B. (2006), "'By Leibniz Law': Remarks on a Fallacy," *Philosophical Quarterly* 56: 39–54.

———. (2010), "Propositions United," *Dialectica* 64(2): 289–301.

Sher, G. (2016), *Epistemic Friction*, Oxford: Oxford University Press.

Simmons, K. (2018), "Three Questions for Minimalism," *Synthese* 195: 1011–1034.

Soames, S. (2015), *Rethinking Language, Mind, and Meaning*, Princeton, NJ: Princeton University Press.

Stalnaker, R. (1976), "Possible Worlds," *Nôus* 10(1): 65–75.

Taylor, D. E. (2017), "Deflationism and Referential Indeterminacy," *Philosophical Review* 126(21): 43–79.

———. (2019), "Deflationism, Creeping Minimalism, and Explanations of Content," *Philosophy and Phenomenological Research*, doi:10.1111/phpr.12572.

Thomasson, A. L. (2015), *Ontology Made Easy*, Oxford: Oxford University Press.

Williamson, T. (1996), "Knowing and Asserting," *Philosophical Review* 105: 489–523; reprinted and revised in his *Knowledge and Its Limits*, New York: Oxford University Press, 2000. [Page reference to the reprinted and revised version.]

———. (2007), *The Philosophy of Philosophy*, Oxford: Blackwell.

Wyatt, J. and M. Lynch (2016), "From One to Many: Recent Work on Truth," *American Philosophical Quarterly* 53(4): 323–340.

Yablo, S. (2014a), "Carnap's Paradox and Easy Ontology," *Journal of Philosophy* 111(9–10): 470–501.

———. (2014b), *Aboutness*, Princeton, NJ: Princeton University Press.

Zardini, E. (2008), "Truth and What Is Said," *Philosophical Perspectives* 22: 171–205.

50

ASSESSMENT RELATIVISM

Filippo Ferrari

1. Introduction

Assessment relativism (henceforth AR) is a type of truth relativism[1] that has been developed by John MacFarlane in a series of works,[2] culminated in his 2014 book *Assessment Sensitivity: Relative Truth and Its Applications*.

Relativism about truth is the thesis that (some) truths are true merely relatively. This view is mainly motivated by the attempt of making sense of the possibility of disputes where none of the competing opinions seems less legitimate, or *less true*, than the others. This phenomenon is known under the label "faultless disagreement" and, roughly put, it concerns situations where one party accepts while the other rejects that things are so-and-so but neither of them is, not even in principle, off-track and guilty of any mistake.[3]

To get a proper grip on AR and to distinguish it from other versions of truth relativism, three questions are particularly relevant: (i) which truths are relative and which aren't? – Section 2; (ii) what is truth relativism and what are the bearers of (relative) truth? – Section 3; (iii) in what sense is truth relative according to AR? – Sections 4–6. Section 7 discusses two challenges to AR.

2. The scope of assessment relativism

AR is a *local* thesis targeting a restricted range of truths. Its exact range of application is ultimately an empirical question, having to do with whether and to what extent a given language – e.g. English – contains expressions which are apt for a relativistically engineered notion of truth. MacFarlane takes "tasty," "knows," "tomorrow," "might," and "ought" to be paradigmatic examples of expressions which are apt for AR. As a litmus test for detecting the presence of AR, MacFarlane suggests to look not only at the phenomenon of faultless disagreement but also at that of retraction – a speech act targeting a previously made assertion which speakers perform by saying things like "I retract that," "I no longer stand by that." In this chapter, I will leave the question whether retraction has an adequate empirical support aside – see, for instance, Knobe and Yalcin (2014), Marques (2018), and Wright (2007) – focusing on introducing, explaining, and assessing the theoretical framework of AR in relation to one specific application – matters of taste.

3. What is truth relativism?

To understand the mechanics of AR, we need first to understand the general thesis of relative truth. A natural starting point is to ask what are the bearers of (relative) truth. In this regard, a distinction between utterances, sentences, and propositions has to be made. Utterances are speech acts. They typically involve the use of a language and the intentional acts of speakers at a certain time and place. Sentences are linguistic expressions. They constitute the smallest unity of meaning that can be assessed for truth and falsity. Propositions are non-linguistic items that are taken to be the content of declarative uses of sentences. Clearly, two speakers can say the same thing by uttering different sentences, whether in the same or different languages. That *same thing* is the proposition expressed by those sentences. Sarah in uttering the English sentence "John is a philosopher" expresses the proposition that John is a philosopher; Elisabetta in uttering the Italian sentence "John è un filosofo" expresses the very same proposition expressed by the sentence uttered by Sarah.

So, what are the primary bearers of truth? It sounds infelicitous to say that a speech act of uttering – e.g. Sarah's *uttering* "John is a philosopher" – is true or false. Strictly speaking, an act of uttering might be correct or incorrect, appropriate or inappropriate, but it is not true or false.[4] Thus, we are left with either sentences or propositions to play the role of primary truth bearers. MacFarlane opts for propositions. One reason for this, as we will see, is that it's easier to capture the kind of relativity at the core of AR (MacFarlane 2014, ch. 3).

Let's now introduce a standard piece of semantic machinery, the so-called Kaplanian two-dimensional semantics. In his seminal paper "Demonstratives" (Kaplan 1989), operating within the framework of compositional semantics,[5] Kaplan distinguishes two aspects of meaning: character and content. The character of a linguistic expression – a sentence or a sub-sentential expression – reflects semantic rules governing how the content of that expression may vary from one context of use to the next. The sentence "I am a philosopher" contains the indexical expression "I" which is governed by the following rule: "I" refers to the user (speaker or writer). Such a sentence expresses different propositions in different contexts, depending on who's using it. As used by MacFarlane, it expresses the proposition that MacFarlane is a philosopher, while as used by Merkel it expresses the proposition that Merkel is a philosopher. These two propositions might have different truth values: in fact, in the actual world, the former is true while the latter is false. Formally, the character of a linguistic expression is a function from contexts to contents – where a context, in its abstract sense, is a possible occasion of use of an expression that can be individuated as a sequence of parameters including an agent, a world, a time, a location. The second aspect of the meaning of a linguistic expression is the content. If the expression in question is a sentence, the content will be a proposition, whereas if it is a sub-sentential expression – e.g. a name or a predicate – the content will be an object, an individual, or a property. Formally, the content is represented by a function from circumstances of evaluation to an appropriate extension. By "circumstances of evaluation" Kaplan means

> [B]oth actual and counterfactual situations with respect to which it is appropriate to ask for the extensions of a given well-formed expression. A circumstance will usually include a possible state or history of the world, a time, and perhaps other features as well.
>
> *(Kaplan 1989, 502)*

The sentence "snow is white" expresses the same propositions in all contexts, but whether it is true will depend on the world at which it is evaluated: in the actual world the proposition expressed is true, but in a possible world where snow isn't white, it would be false.

What other information beside possible worlds should be included in the circumstances is "a matter of language engineering" (Kaplan 1989, 504). If we think, like temporalists do, that the sentence "Filippo is writing" expresses a temporally neutral content – i.e. a proposition which doesn't contain any information about the time at which Filippo is writing – then we might include a time-parameter in the circumstances in order to assess whether the proposition expressed by such a sentence is true or false. If we consider as the relevant value of the time parameter in the circumstances Wednesday, 26–12–2018 at 9am, then the proposition expressed is true, whereas if we consider Wednesday, 26–12–2018 at 3am, the proposition is false.

It is important to bear in mind that the use of a sentence always occurs in a context which contains information about the time and the world at which the sentence is used by an agent. However, if we believe that modally and temporally neutral contents should be allowed for, such information provided by the context does not leak in the proposition. Nevertheless, the context at which the sentence is used – let's call it, following Kaplan, *the context of use* – provides the default values of the relevant parameters (i.e. time and world in our example) in the circumstances of evaluation. Thus, if Filippo is uttering the sentence "Filippo is writing" in the actual world on December the 26th, 2018, at 9am, the context in which such a sentence is used by Filippo will contain information about the world and the time at which it is used. If we then want to assess whether the proposition expressed by that sentence – i.e. the proposition that Filippo is writing – is true or false in relation to the context in which it is used, we fill the world and time values in the circumstances of evaluation with the information provided by the context of use. In this sense we can talk of *the circumstances of the context of use*.

The context of use thus plays two distinct roles, as MacFarlane puts it: a content-determining role – which proposition is expressed depends on features of the context of use – and a circumstances-determining role, selecting the circumstances of evaluation that are relevant to assess the truth of the proposition. To these two roles of the context of use correspond two ways in which a sentence can be context-sensitive: it can be use-indexical in that it expresses different propositions at different contexts of use (e.g. "I am a philosopher"); or, it can be use-sensitive in that it expresses the same proposition at all contexts of use but its truth depends on features of the circumstance of the context of use (e.g. "Filippo is writing" is, according to temporalists, use-sensitive but not use-indexical).

As we can now appreciate, that sentence truth is relative because the sentence is use-indexical, is a mundane kind of relativity. A more interesting relativity occurs once the proposition is fixed for all contexts, and yet its truth value might vary from one context to the next. If temporalism is true, my utterance of "Filippo is writing" as used by me on 26–12–2018 at 11am expresses the proposition that Filippo is writing which is true as evaluated at the circumstance of the context of use but false when evaluated at different circumstances of evaluation (e.g., on 26–12–2018 at 3am).

4. Non-indexical-contextualism as truth relativism

The idea of use-sensitivity without use-indexicality was adapted by some relativists to model the intuitive talk of relative truth in some domains – e.g., that of taste (see Egan 2007; Kölbel 2002; Recanati 2008; Richard 2008). The thought is to take Kaplan's framework and to enrich the circumstances of evaluation with a taste parameter tracking the gustatory sensibility of the speaker. According to this model – known as "Non-Indexical-Contextualism" (NIC for short) – an utterance of a sentence containing a taste predicate, e.g. "oysters are tasty," expresses the same proposition in all contexts of use, i.e., the proposition that oysters are tasty, but its truth-value varies from one context of use to the next in tandem with variation in the agent's gustatory sensibility. Thus "oysters are tasty" would invariably express the proposition that oysters are tasty which is true as used by me but false as used by my brother. In short, according to NIC the truth of taste propositions is relative to the gustatory sensibility operating at the context of use. In this way NIC effectively accounts for faultless disagreement. If, in claiming "these oysters are tasty," I endorse the proposition that these oysters are tasty while my brother, in claiming "nah, these oysters are not tasty," endorses the proposition that these oysters aren't tasty, we would disagree but each of us would say something true relative to our respective contexts of use. Because there's nothing wrong in endorsing a true proposition, there's a clear sense in which our disagreement is faultless.

Whether the truth relativity exemplified by NIC deserves the honorific title of "genuine truth relativism" is open to discussion. MacFarlane thinks that since NIC preserves an element of absolutism it doesn't count as a thoroughgoing form of relativism. Once we take the proposition used by an agent and we pair it with the context of use we get a once-and-for-all truth value. This is sometimes expressed, perhaps misleadingly, by saying that within NIC utterance truth is absolute (MacFarlane 2014, 107).[6] Take the sentence "oysters are tasty" as uttered by me. Given that there is one privileged context of use – that of my utterance – there's a privileged default value for the taste parameters in the circumstances of evaluation. In this sense, the question whether the proposition expressed by the sentence "oysters are tasty" as used by me is true has a unique and absolute answer.

5. Assessment relativism

If we want to cross "the philosophically interesting line between truth absolutism and truth relativism [we need to] relativize truth not just to a context of use ... but also to a context of assessment" (MacFarlane 2014, 60). Thus, what makes a view about truth a genuinely relativistic one is not the mere addition of a special parameter in the circumstances of evaluation – e.g., a taste parameter – but rather the addition of a new context besides the context of use – i.e., a context of assessment. Such a context, which is structurally analogous to the context of use, is a possible situation in which a use of a sentence might be assessed. Its primary function is to provide the value for the special parameters.

This gives the semantic machinery an interesting twist. In fact, AR can be defined as the thesis that the truth of certain propositions – i.e., those that are expressed by a use of sentences containing assessment-sensitive expressions – is relative not only to aspects of the context of use but also to aspects of a context of assessment. Focusing on taste propositions, MacFarlane characterizes assessment-relative truth as follows:

> (REL) A proposition p is true as used at the context of use (c_1) and assessed from a context of assessment (c_2) if and only if p is true at $<w_{c1}, g_{c2}>$, where w_{c1} is the world of c_1 and g_{c2} is the taste of the agent of c_2 (the assessor).
>
> *(MacFarlane 2014, 105)*

While the job of c_1 is that of determining the proposition expressed, that of c_2 is to provide the values for the gustatory parameter (g) in the circumstances of evaluation. To illustrate: I'm at the restaurant with my brother and I claim "oysters are tasty." Here the context of use (c_1) is the context in which I utter the sentence "oysters are tasty." It determines the proposition expressed – namely the proposition that oysters are tasty – and it fixes the default value of the world parameter in the circumstances of evaluation (wc$_1$) – i.e., the actual world. Since there is no use-indexical expression, such a content remains invariant across contexts. The role of a context of assessment (c$_2$) is that of providing the value for the taste parameter in the circumstances of evaluation (gc$_2$) which determines whether the extension of "tasty" includes the oysters (making the proposition true) or not (making it false). Since I like the oysters while my brother dislikes them, the proposition that oysters are tasty is true as assessed from my context of assessment but false as assessed from my brother's context of assessment. In this way, AR can account for the possibility of faultless disagreement.[7]

(REL) shows the formal difference between AR and NIC. While the two views agree that an utterance of "oysters are tasty" invariantly expresses a single proposition, they disagree about what gustatory standard is relevant for assessing the truth of the proposition expressed. According to NIC the default value for the relevant standard is provided by the context of use whereas according to AR it is provided by a context of assessment. At first sight, this might look an insignificant difference. On a closer look, though, it has important consequences since it frees AR from the residual element of absolutism that affects NIC. Given that there's no privileged context of assessment which sets the default value of the taste parameter in the circumstances of evaluation, in AR utterance truth isn't absolute. Even keeping fixed the context of use and the proposition expressed, we don't get a once and for all truth value.[8] Moreover, relativising truth to both the context of use and a context of assessment has important normative consequences, to which we now turn.

6. The normative profile of assessment relativism

Dummett argued that to have a proper grasp of what truth is, we need not merely to know under what circumstances propositions are true, but also to understand the connection between their truth and the proprieties of their use (Dummett 1959). Following Dummett, MacFarlane discusses at length the practical difference that his theory makes with respect to the making and retracting of assertion. Starting with the idea that truth is the core – but not necessarily the sole – normative notion of assertion, MacFarlane subscribes to the following reflexive truth norm:

> (RTN) An agent is permitted to assert that p at context c$_1$ only if p is true as used at c$_1$ and assessed from c$_1$.
>
> *(MacFarlane 2014, 103)*

RTN is a bridge principle linking semantics and pragmatics. It says that the truth of an assertion made at c$_1$ and assessed from c$_1$ is a necessary condition for the permissibility of the asserting at c$_1$. However, if RTN were the only normative principle there would be no practical difference between AR and NIC since the assessment context plays no distinctive role there: whenever we make an assertion in a context, we are also assessing it from that same context. NIC and relativism turn out to be normatively equivalent theories according to RTN. This means that RTN doesn't give us the full story concerning the normative significance of AR.

It is only when we turn to a different conversational phenomenon – namely, that of retracting – that the operational difference between NIC and AR is appreciable. By "retraction" MacFarlane means the speech act one performs in saying "I take that back" or "I retract that." The target of retraction is a previously made but unretracted speech act (e.g., an assertion). As any other speech act, retraction has its distinctive normative profile which MacFarlane characterizes as follows (Retraction Norm):

(RN) An agent in context c_2 is required to retract an (unretracted) assertion of p made at c_1 if p is not true as used at c_1 and assessed from c_2.

(MacFarlane 2014, 108)

RN is a prescriptive norm – it says that the untruth of a proposition as used at c_1 and assessed from c_2 is sufficient to require retracting a previous assertion of it. The effect of retracting is that of rendering the previous assertion null and void – to disavow the assertoric commitments undertaken in the original assertion.

With both RN and RTN on board, we can fully appreciate the practical difference between NIC and AR. Because NIC does not distinguish operationally between a context of assertion and a context of assessment, when it is equipped with a retraction norm it predicts that a subject ought to retract a previously unretracted assertion just in case that assertion was impermissible – i.e., it expressed a false proposition in the context in which it was performed. By contrast, AR predicts that a subject is required to retract an assertion whenever the proposition it expresses is false from her current context of assessment regardless of whether such an assertion was deemed permissible by RTN. To illustrate: suppose that in c_1 I like oysters and I assert "oysters are tasty" but in c_2 I change my tastes and I no longer enjoy oysters. AR – but not NIC – predicts that at c_2 I am required to retract my assertion of "oysters are tasty" made at c_1 since the proposition it expressed is assessed as false in c_2, even though that assertion was permissible in c_1 according to RTN.

7. Assessing assessment relativism

AR is an ingenious proposal which sets a milestone in the history of relativism. The view is, however, not immune to criticisms. Among the many objections that have been raised against AR,[9] I'll just mention two issues in relation to retraction and faultless disagreement.

Recall, an act of retraction targets a previously made speech act – e.g., an act of asserting – not its content. However, according to RN, what triggers the requirement to retract are characteristics of the asserted proposition, namely its untruth relative to the context of assessment where the retraction takes place. By retracting, an agent undoes the normative changes effected by her assertion in the original context with the result that she "is no longer obliged to respond to challenges to the assertion. . ., and that others are no longer entitled to rely on [her] authority for the accuracy of this assertion" MacFarlane (2014, 108). However, in retracting an agent is not required to admit fault, and for two reasons: first, given that RN states that the untruth of a proposition is only a sufficient condition for the obligatoriness of retracting, nothing precludes an agent to retract an assertion which is true as assessed from her context of assessment. Second, even when a subject is required to retract, the retracted assertion might have complied with RTN when it was made in the original context of assertion. In both cases, an attribution of fault is inappropriate.

Do we need AR and the specific sense of retraction that MacFarlane has in mind to obtain this normative effect? I'm inclined to agree with Marques (2018) and Raffman (2016) that we do not. Let's consider an example concerning matters of taste: until the age of twenty I didn't like the taste of mustard and, as I recall now, I've asserted several times "mustard is disgusting." Then my taste sensibility changed, and I started liking mustard. During a dinner at which my mother has prepared mustard, noticing, with surprise, that I'm the first to grab some mustard, she claims: "Didn't you think mustard is disgusting?" To that I reply: "Well, I've changed my mind, and I no longer stand by my previous assertion." Let's label this speech act "withdrawing." Everybody at dinner would take my withdrawing to signal my intention of distancing myself from my previous assertions that mustard is disgusting and in fact to withholding from the assertoric commitments engendered by my previous assertions. This, as expected, does not involve an admission of fault for having asserted in the past that mustard is disgusting: since I disliked mustard, I consider my previous assertions as perfectly reasonable.

What I did with my withdrawing is thus: (i) to deny that mustard is disgusting, and (ii) to manifest my disposition to withhold from the assertoric commitments associated with my previous assertions. It thus seems that my withdrawing would achieve the normative revisions that characterize an act of retraction without, however, requiring AR. An advocate of NIC, for instance, could easily account for the normative effect of my withdrawing. The challenge for an advocate of AR is thus to individuate some normative effects distinctive of retraction that are not already carried out by an act of withdrawing.[10]

The second issue concerns what I take to be an ingredient, and important, part of the folk conception of faultless disagreement, namely the idea that the faultlessness of certain kinds of disagreement can be appreciated and coherently expressed not just from the abstract point of view of the formal framework of AR, but also from within a committed perspective taking part to the dispute. Wright calls this extra ingredient "parity," and according to him it is "meant to be implicated by faultlessness – conveyed in the acknowledgment that your opinion is just as good as mine" (Wright 2012, 439). For this reason, intuitions about parity are arguably explanatorily prior to intuitions about faultlessness: it is because disputants think that their opinions are roughly on a par that they think that the disagreement between them is faultless. A fully satisfactory account of the phenomenon of faultless disagreement must include an account of the parity ingredient.

Can AR account for it? In its current form, it does not seem that it can. The gist of the argument is the following. Suppose I disagree with John about whether oysters are tasty: I claim they are tasty while John claims they are not. My assessment context is one in which the proposition that oysters are tasty is true: relatively to that context the proposition that oysters aren't tasty is assessed as false. But that's exactly what John has asserted. So how can I judge from within my context of assessment that John's assertion is on a par with mine if what I've claimed is true while what John has claimed is false – as assessed from my context of assessment? (cf. MacFarlane 2014, 106–107). If an attribution of falsity carries some normative weight – as it should do (cf. Wright 1992, ch. 1) – it seems that I cannot but assess John's claim to be *not* on a par with mine: after all it expresses a proposition I'm committed to assess as false. It is thus clear that the framework of AR, as it stands, does not allow for the possibility of expressing equal legitimacy of contrary opinions within a committed perspective. This isn't, of course, meant to be a conclusive objection to AR: it just illustrates that more work has to be done by an advocate of AR both on the normative functions of truth and falsity and on the metaphysics of taste.[11]

Acknowledgements

I'd like to thank Elke Brendel, Manuel García-Carpintero, Andrea Crepoli, Martin Kusch, Teresa Marques, Sebastiano Moruzzi, Nikolaj Jang Lee Linding Pedersen, Stefano Pugnaghi, Elisabetta Sassarini, Giorgio Volpe, and Dan Zeman for their useful feedback. Moreover, I'd like to acknowledge the support of the Deutsche Forschungsgemeinschaft (DFG – BR 1978/3–1) for sponsoring my postdoc at the University of Bonn. While working on this chapter, I have benefitted from participation in the Pluralisms Global Research Network (National Research Foundation of Korea grant no. 2013S1A2A2035514). This support is gratefully acknowledged.

Notes

1 See García-Carpintero and Kölbel (2008) for a variety of models of relative truth.
2 The view is first sketched in MacFarlane (2003).
3 Kölbel (2003) takes faultless disagreement as the main motivation for truth relativism. Rovane (2013) offers some criticisms to this strategy.
4 For a dissenting voice, see García-Carpintero (2019).
5 Compositional semantics provides rules governing the interpretation of subsentential expressions and their modes of combination in order to explain how the meaning of whole sentences is determined by the meanings of their parts.
6 MacFarlane can express this point without using the notion of utterance truth but that of accuracy instead: "an attitude or speech act occurring at c_1 is accurate, as assessed from a context c_2, just in case its content is true as used at c_1 and assessed from c_2 " (MacFarlane 2014, 127). For NIC an utterance made by Ben of "oysters are tasty" in context c_1 and an utterance of "oysters aren't tasty" made by Bob in c_2 can both be accurate as assessed from any context. AR predicts that the accuracy of Ben's utterance precludes the accuracy of Bob's utterance. Thus, utterance's accuracy is invariant in NIC but not in AR.
7 While AR and NIC agree on the relevant notion of faultlessness, they work with different notions of disagreement. See MacFarlane (2014, 132).
8 See note 6 for an elaboration of this point.
9 Just to mention a few criticisms: for objections against the application of AR to (i) epistemic modals see Ross and Schroeder (2013), von Fintel and Gillies (2008), Wright (2007); (ii) to future contingent see Moruzzi and Wright (2009), Brogaard (2008), Dietz and Murzi (2013); (iii) to knowledge ascription see Brendel (2014), Kompa (2002), and Wright (2017).
10 See Ferrari (2016b) for a criticism of MacFarlane's take on retraction.
11 Ferrari (2016c) develops some suggestions for how to improve AR with a pluralist conception of the normative function(s) of truth. Ferrari (2016a) and Ferrari (2018) give a full account of normative alethic pluralism in relation to a comparative analysis of disagreement in different domains.

References

Brendel, E. (2014), "Contextualism, Relativism, and the Semantics of Knowledge Ascriptions," *Philosophical Studies* 168: 101–117.
Brogaard, B. (2008), "Sea Battle Semantics," *The Philosophical Quarterly* 58(231): 326–335.
Dietz, R. and J. Murzi (2013), "Coming True," *Philosophical Studies* 163: 403–427.
Dummett, M. (1959), "Truth," *Proceedings of the Aristotelian Society* 59: 141–162.
Egan, A. (2007), "Epistemic Modals, Relativism and Assertion," *Philosophical Studies* 133: 1–22.
Ferrari, F. (2016a), "Disagreement About Taste and Alethic Subrogation," *Philosophical Quarterly* 66(264): 516–535.
———. (2016b), "Assessment Sensitivity," *Analysis* 76(4): 516–527.
———. (2016c), "Relativism, Faultlessness and Parity," *Argumenta* 3: 77–94.
———. (2018), "Normative Alethic Pluralism," in *Pluralisms in Truth and Logic*, edited by J. Wyatt et al., London and New York: Palgrave Macmillan, 145–168.
García-Carpintero, M. (2019), "Semantics and Metaphysics of Truth," this volume.
García-Carpintero, M. and M. Kölbel (eds.) (2008), *Relative Truth*, Oxford: Oxford University Press.
Kaplan, D. (1989), "Demonstratives: An Essay on the Semantics, Logic, Metaphysics, and Epistemology of Demonstratives," in *Themes from Kaplan*, edited by J. Almog, J. Perry and H. Wettstein, New York: Oxford University Press, 481–563.
Knobe, J. and S. Yalcin (2014), "Epistemic Modals and Context: Experimental Data," *Semantics and Pragmatics* 7: 10–11.
Kölbel, M. (2002), *Truth Without Objectivity*, London: Routledge.
———. (2003), "Faultless Disagreement," *Proceedings of the Aristotelian Society* 105: 53–73.

Kompa, N. (2002), "The Context Sensitivity of Knowledge Ascriptions," *Grazer Philosophische Studien* 64: 1–18.

MacFarlane, J. (2003), "Future Contingents and Relative Truth," *Philosophical Quarterly* 53(212): 321–336.

———. (2014), *Assessment Sensitivity: Relative Truth and Its Applications*, Oxford: Oxford University Press.

Marques, T. (2018), "Retractions," *Synthese* 195: 3335–3359.

Moruzzi, S. and C. Wright (2009), "Trumping Assessments and the Aristotelian Future", *Synthese* 166(2): 309–331.

Raffman, D. (2016), "Relativism, Retraction, and Evidence," *Philosophy and Phenomenological Research* 92(1): 171–178.

Recanati, F. (2008), "Moderate Relativism," in García-Carpintero and Kölbel (2008), 41–62.

Richard, M. (2008), *When Truth Gives Out*, Oxford: Oxford University Press.

Ross, J. and M. Schroeder (2013), "Reversibility or Disagreement," *Mind* 122: 43–84.

Rovane, C. (2013), *The Metaphysics and Ethics of Relativism*, Cambridge, MA: Harvard University Press.

von Fintel, K. and A. Gillies (2008), "CIA Leaks," *Philosophical Review* 117(1): 77–98.

Wright, C. (1992), *Truth and Objectivity*, Cambridge, MA: Harvard University Press.

———. (2007), "New Age Relativism and Epistemic Possibility: The Question of Evidence," *Philosophical Issues* 17(1): 262–283.

———. (2012), "Replies Part III: Truth, Objectivity, Realism and Relativism," in *Mind, Meaning and Knowledge: Themes from the Philosophy of Crispin Wright*, edited by A. Coliva, Oxford: Oxford University Press, 418–450.

———. (2017), "The Variability of 'Knows': An Opinionated Overview," in *Routledge Handbook of Epistemic Contextualism*, edited by J. Ichikawa, New York: Routledge, 13–31.

51

FAULTLESS DISAGREEMENT

Dan Zeman

1. Introduction: the phenomenon and the challenge

Disagreement is ubiquitous in everyday life. Sometimes it has negative effects, as when it is conducive to unresolved confrontation; sometimes it has positive effects, as when it brings about beneficial change. People disagree about many things, and in many ways, too. Consider, for example, a disagreement about the age of the Earth. While a lot of scientific knowledge is needed to answer this question, a definite answer to this question exists, even if it is really hard to arrive at it. Questions like the one about the age of the Earth concern objective matters (or "matters of fact," as they are sometimes called). When disagreeing about objective matters, at most one of the participants in the disagreement can be right. Consider, in contrast, a disagreement about the tastiness of licorice. Such questions are relatively easy to answer, because, presumably, all that it takes to arrive at an answer is to be attentive to our gustatory reaction to licorice. Questions like the one about the tastiness of licorice concern subjective matters (or "matters of inclination"). When disagreeing about subjective matters, both participants in the disagreement can be right. In the first case, disagreement involves fault; in the second, it is faultless.

This notion of faultless disagreement has played a great role in the debate over the semantics of expressions we use to talk about subjective matters – predicates of taste, aesthetic and moral terms, epistemic modals, epistemic expressions, etc. (I'm going to call all of these expressions "subjective," although this shouldn't obscure the fact that there are important differences between them, semantically and otherwise.) Thus, many authors in this debate think that the fact that a certain semantic theory can account for faultless disagreement is a point in its favor. The main theories presently involved in giving a semantic account of subjective expressions are the following: *contextualism, relativism, expressivism* and *absolutism*. According to the first, the subjective character of the expressions focused on is captured by introducing perspectives[1] in the propositions asserted by sentences containing subjective expressions in a context. According to contextualism's main rival, relativism, the subjective character of the expressions at stake is captured by introducing perspectives not in the propositions asserted, but in the "circumstances of evaluation" (Kaplan's (1989) term) the relevant propositions are evaluated against.[2] Expressivism situates the subjective character in the attitudes (positive or negative) expressed by speakers when using the expressions in question. Finally, in contrast to all these views, absolutism holds that perspectives play no role in the semantics of subjective terms, and that what seem to be matters of opinion are, in fact, matters of (perhaps unknowable) fact.

But what exactly is faultless disagreement? Here is how Kölbel (2004b), one of the first authors to extensively discuss this phenomenon in contemporary literature, defines it:

> A faultless disagreement is a situation where there is a thinker A, a thinker B, and a proposition (content of judgment) p, such that:
>
> (a) A believes (judges) that p and B believes (judges) that not-p
> (b) Neither A nor B has made a mistake (is at fault).
>
> *(2004b, 53–54)*

Two features of this definition merit stressing from the outset. First, as clause (a) shows, disagreement has been conceived of in doxastic terms (i.e., believing/judging propositions). Second, clause (b), in itself, is silent about what is the relevant sense of fault. What is intended by most authors, though, is something along the following lines: someone is at fault if they say something false. The challenge for theories of subjective expressions is to account for both the disagreement and the faultlessness intuitions that exchanges about the tastiness of licorice or those involving other subjective expressions presumably give rise to. Views for which the disagreement part is problematic (e.g., contextualism) have striven to come up with ways of construing disagreement that avoid the challenge. Views for which the faultlessness part is problematic (e.g., absolutism) have aimed to find strategies for rendering faultlessness that avoid the challenge. In what follows, I briefly present the main strategies for avoiding each problematic aspect – by investigating contextualist and expressivist (Section 2), absolutist (Section 3) and relativist (Section 4) answers to the challenge. I end by mentioning two issues that I believe will shape the future debate surrounding faultless disagreement.

2. Contextualist (and expressivist) answers

A lot of recent contextualist literature in the debate over the semantics of subjective terms has been dedicated to answering the faultless disagreement challenge. In most cases, this has amounted to contextualists providing accounts of disagreement that fend off the challenge. Given that faultlessness is not a problem for contextualism (according to it, the interlocutors assert compatible true propositions), solving the disagreement issue amounts to a full solution to the challenge from faultless disagreement.

The contextualist answers can be grouped into several categories.[3] In the first category are authors who seek to show that, when focusing on certain uses of subjective expressions, accounting for disagreement is not problematic. According to these analysts, the cases put forward in supporting the disagreement intuition in matters of opinion are too indeterminate to do the trick. As Schaffer puts it, "the case for relativism relies on a misrepresentative sample of underdeveloped cases" (2011, 211);[4] other authors (Glanzberg 2007; Stojanovic 2007; Cappelen and Hawthorne 2009) follow suit. It is then pointed out that, when properly fleshed out, the examples provided in support of the challenge lead to cases in which contextualists have no problem accounting for disagreement. For example, subjective expressions have what has been called "exocentric" uses (i.e., they are used from a different person's point of view), "collective" uses (i.e., they are used from the point of view of a group) or "generic" uses (making claims involving a whole class or kind). When two interlocutors use a subjective expression in such a way, and both have the same person, the same group or the same class or kind in mind, then disagreement can be accounted for easily, even if conceived of according to the preceding definition. In addition, some of these authors argue that, when the examples provided are fleshed out in the relevant way for the challenge to arise, either the intuition of disagreement disappears or they cannot, upon reflection, be considered disagreement. This is the case, for example, with exchanges in which both interlocutors use a subjective expression "autocentrically" – that is, from their own point of view.[5] The combination of these two aspects of the answer investigated leads to a solution to the challenge posed by faultless disagreement by rejecting the problematic cases as not leading to disagreement and by showing that other, equally viable ones, are easily accounted for.

In the second category are authors who render disagreement to be not semantic (that is, at the level of the propositions expressed), but to be pragmatic. Contextualists taking this route have a vast array of phenomena to rely on, as the label "pragmatic" has a wide range of application. One type of proposal within this category is to take disagreement in subjective matters to happen at the level of, or made possible by, *presuppositions*. Thus, in Lopez de Sa's (2007, 2008) view, a "presupposition of commonality" is needed for disagreement to arise, otherwise interlocutors end up merely talking past each other. Other presupposition-based views include Parsons' (2013), according to whom the presupposition that subjective expressions carry in a context is that at least one of the interlocutors is right, and Zakkou's (2015), according to whom the presupposition doing the work is one "of superiority," whereby each of the interlocutors thinks that their tastes are superior to the other's. All these views allow the presuppositions triggered by subjective expressions to play a role in explaining disagreement – either by making it possible or simply by understanding it as a clash of presuppositions.

Another way in which proponents of the pragmatic strategy have avoided the challenge is to take disagreement to arise at the level of implicatures, either conventional or conversational (Finlay 2005 is an early example of such a strategy in the moral domain). Yet another pragmatic answer to the challenge consist in taking disagreement to be metalinguistic. According to such an answer (Sundell 2011; Plunkett and Sundell 2013; Ludlow 2014; Plunkett 2015), disagreement can be not only about facts but also about what our words *do* or *should* mean. One example often cited is the debate about whether waterboarding is torture. Disagreement here is not about facts (all interlocutors agree about them) but about what the word "torture" does or should mean. The proposal is to treat disagreement in subjective matters as metalinguistic. Finally, another answer proposed was to understand disagreement in subjective matters as consisting in opposed conversational moves. According to one particular implementation of this view by Silk (2016), for example, the conversational moves amount to opposite proposals to establish the value of a certain contextual parameter and register it in the "conversational score." Since each interlocutor has a different proposal to do so, disagreement ensues. Both views are compatible with contextualism because, while the propositions expressed by the interlocutors don't contradict each other, there is another level at which disagreement takes place.

Finally, in the third category are authors that take disagreement in subjective matters to involve conative, rather than cognitive, attitudes (Huvenes 2012, 2014; Marques 2015, 2016; Marques and García-Carpintero 2014; López de Sa 2015; etc.). While proposals differ in their details (for example, some don't commit to a particular type of conative attitude, while others are more explicit about it), the main idea is that disagreement consist in a "clash of attitudes" – an idea taken from the expressivist model of disagreement that has been proposed long ago by, e.g., Stevenson (1944). This move is usually accompanied by the postulation of another level of content besides the proposition asserted; Buekens (2011), for example, calls it the "expressive-affective dimension"; Clapp (2015) holds that sentences containing subjective expressions, besides encoding incomplete propositions, express evaluative mental states; while Gutzmann (2016) talks about a "use-conditional" level of content that includes deontic attitudes towards what should count as the property designated by a subjective expression in a context. In this way, contextualism comes very close to hybrid expressivist theories, which are the most prominent views in the expressivist literature today (see the essays in Fletcher and Ridge 2014 for a sample). While such views are not necessarily contextualist,[6] many of its variants are. For such variants, while the propositions expressed by the interlocutors don't contradict each other, there is another level at which disagreement takes place – namely the additional one postulated.

3. Absolutist answers

Absolutists about subjective expressions hold that perspectives are, in fact, not needed for the semantic treatment of such expressions. This leads to the claim that the interlocutors assert contradictory propositions whose truth doesn't vary across contexts. In contrast with contextualists, for whom the Achilles' heel was disagreement, absolutists's problem is with faultlessness.

Like contextualists, absolutists about matters of taste have not been silent in connection with this problem. In most cases, addressing the challenge has amounted to finding a notion of faultlessness that could replace the notion envisaged by the initiators of the challenge. Thus, a general strategy adopted by the absolutist camp is to explain faultlessness in a way that shies away from taking "being at fault" as asserting something false. The most widespread strategy to do so is to take faultlessness to be epistemic. Several authors (Schafer 2011; Belleri 2011; Hills 2013) trade on the idea that we have authority over what we approve or disapprove, and that we are justified to make assertions in subjective matters based on that authority. They then go on to argue that making no mistakes in asserting propositions under such an authority is enough to vindicate a sense of faultlessness worth its salt. Other authors take different routes. Thus, Ferrari (2016) explains faultlessness by attending to the normative dimension of assertion and its relation to truth. Since in the domain of taste truth is best conceived of as devoid of any axiological and deontic aspects, this allows interlocutors to judge each other as faultless despite taking the propositions asserted as false. Alternatively, Baker and Robson (2017) argue for a "Humean absolutism," according to which faultlessness is understood as convergence with the judgment of ideal critics, even if the propositions asserted by interlocutors are, in fact, false.

An interestingly different way to respond to the challenge from faultless disagreement is Wyatt's (2018) absolutism. Wyatt's response rests on two key claims. One concerns the semantic content of the target expressions and the relation between semantic content and the content of beliefs/assertions in general; the other concerns disagreement. Thus, in relation to semantic content, Wyatt contends that the semantic content of utterances containing subjective expressions[7] are minimal propositions – that is, propositions that don't contain perspectives. However, Wyatt also wants to vindicate recent experimental studies according to which the propositions the folk assert or believe are not minimal in the preceding sense – that is, such propositions as the folk assert or believe do contain perspectives. To achieve this goal, Wyatt trades on the idea that the two – the semantic content of sentences and what is asserted/believed – come apart. His view thus postulates two levels of content.

In relation to disagreement, Wyatt takes a leaf out of the expressivist textbook and construes disagreement as a clash of conative attitudes. He is also explicit about the type of conative attitude involved: in disagreements about subjective matters, he claims, interlocutors are expressing conflicting preferences about the object the subjective expressions are applied to. Now, when it comes to faultlessness, Wyatt is roughly in line with previous absolutist views, insofar as he understands fault as a failure to comply with norms of belief/ assertion of the following form (here tailored to the specific case of taste): a subject S rationally ought to believe/assert the content < licorice's flavor is pleasing to S's tastes > if S knows that S has experienced licorice's flavor first-hand and that it is pleasing to S's tastes. Both disagreement and faultlessness are thus secured, while at the same time keeping the semantics absolutist (i.e., minimalist).

Wyatt's view is a complex view in that it combines ideas that have been employed by other responses to the challenge from faultless disagreement with a semantic thesis and the results of experimental studies. A major challenge to the view is, then, to show that all the elements employed fit together in a coherent whole. Another possible worry one might have is that the proposal amounts to a less economical one than the orthodox views. I will not develop these remarks here, but they are something to be kept in mind for future debate.

4. Relativist answers

Contextualism and absolutism have been the usual targets of the challenge from faultless disagreement, launched mostly by relativists. According to relativists (Kölbel 2004a; Kompa 2005; Lasersohn 2005, 2016; Brogaard 2008; Egan 2014; MacFarlane 2014; etc.), the propositions asserted by interlocutors in a faultless disagreement are contradictory: hence disagreement is secured. And faultlessness is secured as well, since each of the interlocutors asserts a true proposition – albeit true relative to a circumstance containing a different perspective. However, the fact that relativism might have its own problems with the challenge has not remained unnoticed. For example, an impressive number of authors (Stojanovic 2007; Moruzzi 2008; Rosenkranz 2008; Iacona 2008; Francen 2010; Zouhar 2014; Hîncu 2015; etc.) have pointed out that relativism fares no better in accounting for faultless disagreement than contextualism. The main complaint these authors voice is the following. According to relativism, disagreement requires that each proposition be evaluated with respect to the same circumstance, while faultlessness required the opposite. But these two demands cannot be met at the same time, which makes it impossible for the view to account for the phenomenon.

One way for the relativist to reply to this worry is to cling to a notion of disagreement that takes it to consist of merely contradictory propositions, without taking circumstances to play any role. This simple view of disagreement has been extensively criticized by all parties to the debate (see, e.g., MacFarlane 2007; Marques 2014). Luckily, more nuanced responses from relativists are available. For example, MacFarlane (2014) adopts a pluralistic stance towards both disagreement and faultlessness. According to MacFarlane, both terms are ambiguous, and each combination of meanings for the two terms amounts to a different notion of faultless disagreement. Thus, instead of looking for a unique notion of disagreement, MacFarlane's aim is to identify several such notions and see which of them is suited for a certain (type of) domain of discourse:

> Instead of arguing about what is "real" disagreement, then, our strategy will be to identify several varieties of disagreement. We can then ask, about each dialogue of interest, which of these kinds of disagreement can be found in it, and we can adjudicate between candidate theories of meaning by asking which theories predict the kinds of disagreement we find.
>
> *(MacFarlane 2014, 119)*

This strategy has the ecumenical effect that the parties in debate over faultless disagreement are all vindicated to a certain degree: each of them can adopt a certain notion of disagreement fit for a different specific (type of) domain. The five notions of disagreement distinguished by MacFarlane are the following:

(NONCOTENABILITY) A disagrees with B's attitude if A cannot adopt that same attitude without changing their mind. Noncotentability can be **doxastic**, if the attitude at stake is doxastic (e.g., belief) or **practical**, if the attitude at stake is noncognitive (e.g., desire, hope etc.).

(PRECLUSION OF JOINT SATISFACTION) A disagrees with B's attitude if the satisfaction of A's attitude precludes the satisfaction of B's attitude.

(PRECLUSION OF JOINT ACCURACY) A disagrees with B's attitude if the accuracy of A's attitude (in any context) precludes the accuracy of B's attitude (in that same context).[8]

(PRECLUSION OF JOINT REFLEXIVE ACCURACY) A disagrees with B's attitude if the accuracy of A's attitude (in A's context) precludes the accuracy of B's attitude (in B's context).

When it comes to subjective matters, MacFarlane claims that a specific notion of disagreement is best suited for it (at least for the domain of taste). How the pluralistic strategy and its ecumenical effect work is illustrated with clarity in the following quote:

By distinguishing varieties of disagreement, we can sharpen up the question and explain why the original question provokes such disparate answers. The question is not whether there is "genuine" disagreement about matters of taste, but rather which of the varieties of disagreement we have distinguished characterizes disagreements of taste. And the main kinds of account we have considered can be defined by the answers they give to this question. In the case we were evaluating, standard contextualism and expressivism secure only practical noncotenability; nonindexical contextualism secures doxastic noncotenability; relativism secures preclusion of joint accuracy; and [absolutism] secures preclusion of joint reflexive accuracy. Evaluating the case for relativism about predicates of taste, then, does not require settling what kinds of disagreement are "genuine," an issue that seems merely terminological. It just requires determining whether disputes of taste are characterized by preclusion of joint accuracy, for example, or just by doxastic noncotenability. And we can do this by considering the diagnostics outlined above for these varieties of disagreement.

(MacFarlane 2014, 137)

MacFarlane adopts the same pluralistic strategy with respect to faultlessness. Again different senses of it are distinguished. These senses can then be variously combined with different senses of "disagreement." This leads to several (coherent) notions of faultless disagreement, and thus to a pluralist approach to the whole phenomenon.

This strategy, although sensible, opens up several issues. MacFarlane claims that the notion of disagreement fit for subjective matters is a particular one (i.e., preclusion of joint accuracy). While this might be so, and while the data MacFarlane presents seem to mandate this conclusion (the "diagnostics" mentioned in the preceding quote), I suspect that a more experimental approach to this issue is the best way of making progress here. I don't think we can rule out at this point that several notions of disagreement are needed to characterize all the ways of disagreeing found in *one and the same domain*. If this is indeed the case, then the ecumenical strategy pursued by MacFarlane won't work. The more general point to be made is that, absent any sure proof method of deciding what disagreements can be found in what domains, simply distinguishing between several senses of disagreement pushes the problem one step back instead of solving it. A second issue is that significant differences might arise within the same type of domain of discourse. For example, both the taste and the moral domains seem to fall within subjective matters; yet, there are great differences both in the behavior of the corresponding expressions and in the way people disagree with respect to such matters.[9] The thick brush of MacFarlane's notions of disagreement falls short of accounting for such differences. Thus, while tearing apart the various notions of disagreement and faultlessness and their possible combinations is a noteworthy achievement, further experimental anchoring and finessing of the pluralist approach seems to be needed.

5. Looking ahead

This exhausts my brief characterization of the main responses to the challenge from faultless disagreement. There is, of course, the further issue of whether these answers hold water – that is, whether they amount to coherent notions of disagreement and faultlessness, and whether they adequately explain the phenomena. Pointing out the shortcomings of these notions is not my aim here. Instead, I want to end with mentioning two issues that I think are important in the next phase of the debate about the semantics of subjective expressions.

First, many papers dealing with the challenge from faultless disagreement arrive at a pessimistic conclusion about the role the phenomenon does or can play in establishing semantic theses. One challenge for those who appeal to this phenomenon is to show that it is still relevant for the debate at stake. This requires something more than merely pointing out that there is a phenomenon to be explained. What is needed instead is a careful consideration of all the possibilities to account for it and a thorough comparison of the costs and benefits of the combinations of those possibilities with the semantic views on the table. With the surfacing of so many answers to the challenge (among them the ones briefly presented earlier), this is not an easy task, but one that nevertheless has to be undertaken in order to be in a position to claim an advantage with respect to faultless disagreement.

A second open issue is this. As we have seen, construing disagreement in a different way than the initiators of the challenge from faultless disagreement (by relegating it at a pragmatic level, by conceiving it as a clash of conative attitudes etc.) is a popular strategy to account for the phenomenon. This has led to a proliferation of notions of disagreement. And this raises several questions. On the one hand, are all of these notions of disagreement relevant for the semantic debate we are concerned with? It could be the case that, while some ways of disagreeing speak to the semantic issues at stake, others don't. On the other hand, assuming that they are all relevant, what, if anything, do these notions have in common so that they all count as notions of *disagreement*, rather than of something else? This has sparked interest for a notion of "minimal" or "basic" disagreement (Belleri and Palmira 2013; Coliva and Moruzzi 2014; Palmira 2015; etc.): a notion that all participants in the debate can embrace without begging the (semantic) question against their opponents. Forging such a notion might have several advantages: it would, for example, diminish the risk for theoreticians of disagreement to talk past each other, or it would provide a unified and economical theory of disagreement. A similar proliferation, with the same questions arising, can be found with faultlessness: are all notions of faultlessness present in the literature relevant to the debate at hand, and, if so, what would a notion of "minimal fault" look like? Whether such notions exist, what is the best way to construct them, and how they relate to the more substantial notions already present in the debate, are questions still waiting for answers.

Acknowledgements

I would like to thank Filippo Ferrari, Mihai Hîncu and Manuel García-Carpintero for reading and commenting on a previous version of this chapter, and to Martin Kusch for precious editorial help and the opportunity to contribute to this volume. Work on this chapter has been supported by a Lise Meitner Senior Postdoctoral grant (M 2226-G24) from the Austrian Science Fund (FWF).

Notes

1 Or other subjective elements, such as tastes for predicates of taste, aesthetic sensibilities for aesthetic terms, moral outlooks for moral terms etc. I will remain neutral on how to best capture the subjective element involved. I take "perspectives" as an umbrella term covering all the relevant subjective elements corresponding to the expressions in question.
2 In what follows, I bypass the issue of which context is it that provides the relevant circumstance – the context of utterance ("non-indexical contextualism") or that of assessment ("relativism," in MacFarlane 2014's terms).
3 One straightforward way to circumvent this problem is to flat-out reject even the appearance of faultless disagreement altogether. This is not what most authors in this debate do. More importantly, recent experimental studies confirm the existence of folk's intuitions of faultless disagreement; see, e.g., Solt (2018).
4 A stronger criticism along such lines (with focus on case studies of disagreement in science) has recently been made by Kinzel and Kusch (2018).
5 Glanzberg (2007) appeals to sheer intuition; Stojanovic (2007) and Moltmann (2010) raise the issue of what does the disagreement amount to in such cases; Cappelen and Hawthorne (2009) argue by analogy with cases in which no intuition of disagreement is present etc. See Zeman (2017) for a more detailed presentation.
6 See, for example, Boisvert (2008) for a hybrid expressivist view ("expressive-assertivism") that takes the proposition asserted to be relativistic – that is, to not contain perspectives.
7 In fact, Wyatt only applies the view to predicates of taste; here I'm extending it to the entire sphere of subjective discourse, leaving possible complications aside.
8 Accuracy is defined by MacFarlane, in a heavily theory-laden way, as follows: "An attitude or speech act occurring at c_1 is accurate, as assessed from a context c_2, just in case its content is true as used at c_1 and assessed from c_2." (2014, 127).
9 Ferrari (2016) attempts to trace the difference to a variation in the normative profile of judgments, while Stojanovic (2019) relies on syntactic/semantic considerations. The discussion is pretty much open at this point.

References

Baker, C. and J. Robson (2017), "An Absolutist Theory of Faultless Disagreement in Aesthetics," *Pacific Philosophical Quarterly* 98: 429–448.
Belleri, D. (2011), "Relative Truth, Lost Disagreement and Invariantism on Predicates of Personal Taste," in *Proceedings of the Amsterdam Graduate Philosophy Conference "Truth, Meaning, and Normativity"*, edited by M. I. Crespo, D. Gakis and G. Weidman-Sassoon, Amsterdam: ILLC Publications, 19–30.
Belleri, D. and M. Palmira (2013), "Towards a Unified Notion of Disagreement," *Grazer Philosophische Studien* 88: 124–139.
Boisvert, D. (2008), "Expressive-Assertivism," *Pacific Philosophical Quarterly* 89: 169–203.
Brogaard, B. (2008), "Moral Contextualism and Moral Relativism," *Philosophical Quarterly* 58(232): 385–409.

Buekens, F. (2011), "Faultless Disagreement, Assertions and the Affective-Expressive Dimension of Judgments of Taste," *Philosophia* 39: 637–655.

Cappelen, H. and J. Hawthorne (2009), *Relativism and Monadic Truth*, Oxford: Oxford University Press.

Clapp, L. (2015), "A Non-Alethic Approach to Faultless Disagreement," *Dialectica* 69(4): 517–550.

Coliva, A. and S. Moruzzi (2014), "Basic Disagreement, Basic Contextualism and Basic Relativism," *Iride* XXVI: 537–554.

Egan, A. (2014), "There's Something Funny About Comedy: A Case Study in Faultless Disagreement," *Erkenntnis* 79(1): 73–100.

Ferrari, F. (2016), "Disagreement About Taste and Alethic Suberogation," *Philosophical Quarterly* 66(264): 516–535.

Finlay, S. (2005), "Value and Implicature," *Philosophers' Imprint* 5(4): 1–20.

Fletcher, G. and M. Ridge (eds.) (2014), *Having It Both Ways: Hybrid Theories and Modern Metaethics*, Oxford: Oxford University Press.

Francen, R. (2010), "No Deep Disagreement for New Relativists," *Philosophical Studies* 151(1): 19–37.

Glanzberg, M. (2007), "Context, Content, and Relativism," *Philosophical Studies* 136(1): 1–29.

Gutzmann, D. (2016), "If Expressivism Is Fun, Go for It!" in *Subjective Meaning: Alternatives to Relativism*, edited by C. Meier and J. van Wijnbergen-Huitink, Berlin and Boston: De Gruyter, 21–46.

Hills, A. (2013), "Faultless Moral Disagreement," *Ratio* XXVI: 410–427.

Hîncu, M. (2015), "Predicates of Personal Taste and Faultless Disagreement," *Logos & Episteme* 6(2): 160–185.

Huvenes, T. T. (2012), "Varieties of Disagreement and Predicates of Taste," *Australasian Journal of Philosophy* 90(1): 167–181.

———. (2014), "Disagreement Without Error," *Erkenntnis* 79(1): 143–154.

Iacona, A. (2008), "Faultless or Disagreement," in *Relative Truth*, edited by M. García-Carpintero and M. Kölbel, Oxford: Oxford University Press, 287–298.

Kaplan, D. (1989), "Demonstratives," in *Themes from Kaplan*, edited by J. Almog, J. Perry and H. Wettstein, Oxford: Oxford University Press, 481–563.

Kinzel, K. and M. Kusch (2018), "De-Idealizing Disagreement, Rethinking Relativism," *International Journal of Philosophical Studies* 26(1): 40–71.

Kölbel, M. (2004a), "Indexical Relativism vs Genuine Relativism," *International Journal of Philosophical Studies* 12(3): 297–313.

———. (2004b), "Faultless Disagreement," *Proceedings of the Aristotelian Society* 104: 53–73.

Kompa, N. (2005), "The Semantics of Knowledge Attributions," *Acta Analytica* 20(1): 16–28.

Lasersohn, P. (2005), "Context Dependence, Disagreement, and Predicates of Personal Taste," *Linguistics and Philosophy* 28(6): 643–686.

———. (2016), *Subjectivity and Perspective in Truth-Theoretic Semantics*, Oxford: Oxford University Press.

López de Sa, D. (2007), "The Many Relativisms and the Question of Disagreement," *International Journal of Philosophical Studies* 15(2): 269–279.

———. (2008), "Presuppositions of Commonality: An Indexical Relativist Account of Disagreement," in *Relative Truth*, edited by M. García-Carpintero and M. Kölbel, Oxford: Oxford University Press, 279–310.

———. (2015), "Expressing Disagreement," *Erkenntnis* 80: 153–165.

Ludlow, P. (2014), *Living Words: Meaning Underdetermination and the Dynamic Lexicon*, Oxford: Oxford University Press.

MacFarlane, J. (2007), "Relativism and Disagreement," *Philosophical Studies* 132(1): 17–31.

———. (2014), *Assessment Sensitivity: Relative Truth and Its Applications*, Oxford: Oxford University Press.

Marques, T. (2014), "Doxastic Disagreement," *Erkenntnis* 79(1): 121–142.

———. (2015), "Disagreeing in Context," *Frontiers in Psychology* 6: 1–12.

———. (2016), "Aesthetic Predicates: A Hybrid Dispositional Account," *Inquiry* 59: 723–751.

Marques, T. and M. García-Carpintero (2014), "Disagreement About Taste: Commonality Presuppositions and Coordination," *Australasian Journal of Philosophy* 92(4): 701–723.

Moltmann, F. (2010), "Relative Truth and the First Person," *Philosophical Studies* 150(2): 187–220.

Moruzzi, S. (2008), "Assertion, Belief and Disagreement: A Problem for Truth-Relativism," in *Relative Truth*, edited by M. García-Carpintero and M. Kölbel, Oxford: Oxford University Press, 207–224.

Palmira, M. (2015), "The Semantic Significance of Faultless Disagreement," *Pacific Philosophical Quarterly* 96(3): 349–371.

Parsons, J. (2013), "Presupposition, Disagreement, and Predicates of Taste," *Proceedings of the Aristotelian Society* 113: 163–173.

Plunkett, D. (2015), "Which Concepts Should We Use? Metalinguistic Negotiations and the Methodology of Philosophy," *Inquiry* 58(7–8): 828–874.

Plunkett, D. and T. Sundell (2013), "Disagreement and the Semantics of Normative and Evaluative Terms," *Philosophers' Imprint* 13(23): 1–37.

Rosenkranz, S. (2008), "Frege, Relativism and Faultless Disagreement," in *Relative Truth*, edited by M. García-Carpintero and M. Kölbel, Oxford: Oxford University Press, 225–238.

Schafer, K. (2011), "Faultless Disagreement and Aesthetic Realism," *Philosophy and Phenomenological Research* LXXXII(2): 265–286.

Schaffer, J. (2011), "Perspective in Taste Predicates and Epistemic Modals," in *Epistemic Modality*, edited by A. Egan and B. Weatherson, Oxford: Oxford University Press, 179–226.

Silk, A. (2016), *Discourse Contextualism: A Framework for Contextualist Semantics and Pragmatics*, Oxford: Oxford University Press.

Solt, S. (2018), "Multidimensionality, Subjectivity and Scales: Experimental Evidence," in *The Semantics of Gradability, Vagueness, and Scale Structure: Experimental Perspectives*, edited by E. Castroviejo, L. McNally and G. Weidman-Sassoon, Springer, 59–91.

Stevenson, C. L. (1944), *Ethics and Language*, New Haven: Yale University Press.

Stojanovic, I. (2007), "Talking About Taste: Disagreement, Implicit Arguments, and Relative Truth," *Linguistics and Philosophy* 30(6): 691–706.

———. (2019), "Disagreement About Taste vs. Disagreement About Moral Issues," *American Philosophical Quarterly* 56(1): 29–42.

Sundell, T. (2011), "Disagreements About Taste," *Philosophical Studies* 155(2): 267–288.

Wyatt, J. (2018), "Absolutely Tasty: An Examination of Predicates of Personal Taste and Faultless Disagreement," *Inquiry* 61(3): 252–280.

Zakkou, J. (2015), "Tasty Contextualism: A Superiority Approach to the Phenomenon of Faultless Disagreement," PhD thesis, Berlin: Humboldt University of Berlin.

Zeman, D. (2017), "Contextualist Answers to the Challenge from Disagreement," *Phenomenology and Mind* 12: 62–73.

Zouhar, M. (2014), "In Search of Faultless Disagreement," *Prolegomena* 13(2): 335–350.

52

RELATIVISM AND PERSPECTIVAL CONTENT

Max Kölbel

1. Introduction

> The evaluativist view is, as I've said, a kind of relativism. . . . But the term "relativism"
> has had the misfortune of being defined by its opponents.
>
> *(Field 2009, 255)*

If would-be opponents of relativism define "relativism" in a way not acceptable to relativists, then they aren't opponents of relativism, but are criticizing some other view. Still: Hartry Field's remark has the ring of truth. Why? – Is it a case of "dissuasive definition" (Harman 1975, 3), i.e. a type of (self-)deception on the part of the would-be opponents? Or is it the relativists themselves who have not articulated their view clearly, and thus failed to establish definitional sovereignty?

At least in the recent debate (on which Field is commenting) there is an explanation for the difficulties relativists seem to have in imposing their own definition. Recent relativists,[1] despite their differences, have one trait in common: a preference for perspectival contents, i.e. propositional contents the truth-value of which is not settled by the (objective) state of the world. Articulating a relativist view therefore requires clarifying the nature and theoretical role of these contents. But spelling out the nature and role of propositional contents is a challenge for anyone, quite independently of the issues relativists address. It is therefore no surprise that a lot of controversy should surround the feasibility and clear articulation of theories that involve perspectival contents.

This chapter aims to report some of the progress that has been made in this area, by providing a selective overview of what is at stake in operating with perspectival representational contents. I shall begin by introducing the idea of contents of representation and distinguishing several roles that contents are expected to play. In Section 3, I distinguish traditional and perspectival contents of representation. In Sections 4, 5 and 6, I respectively discuss how perspectival contents fare as objects of belief, as compositional semantic values and as objects of assertion.

2. Contents and their roles

Contents (also called "propositions," Fregean *Gedanken*, etc.) serve to characterize how things are represented as being by the representation whose content they are. They are the "objects" of belief, assertion, supposition, desire, etc., i.e. of "propositional attitudes" and "propositional acts." These roles interlock: I may *believe* the content that the pie is hot, and then *assert* that content to warn you. You may *believe* what I have asserted or test the *hypothesis* that that pie is hot and perhaps come to *reject* it, i.e. believe the negation of that content: the content that the pie is not hot. In this episode, the very same thing (the content that the pie is hot) figures as the object of belief, of assertion, as the information conveyed, as what is hypothesized and rejected.

A further role of contents[2] is as *compositional semantic values*. The semantic theory of a language assigns semantic values to the atomic expressions of that language (or of its "base language," see Lewis 1970) and states compositional rules that specify how the semantic values of compound expressions are determined by those of the constituents and the mode of composition. It is often thought that the semantic values such a theory assigns to complete sentences (or to sentence-context-pairs, if the language has context-sensitive expressions) are propositional contents.[3]

These roles (as objects of propositional attitudes and acts, and as semantic values) summarise some of the explanatory uses to which contents are put: Some phenomena concerning mental or linguistic representations can be explained by the contents of these representations. For example, the logical relations among contents may be used as a model for the logical relations among representations with these contents; or certain aspects of the meaning of expressions may be explained by pointing to the contents of expressions (in context). The general strategy is to explain some similarities between representations by the sameness of their content, and differences between them by differences in content. For example, if one belief is incompatible with another, and it is their contents that explain this, then any other two beliefs with the same contents will also be incompatible. Or, if a sentence's embeddability is explained by its content, then any other sentence with the same content will be predicted to be embeddable in the same way.

Since contents are merely theoretical entities that play the mentioned theoretical roles,[4] contents can be construed differently depending on one's explanatory aims. If certain features are regarded as within the range of phenomena to be explained by the proposition-assigning part of the theory alone, then it should assign different propositions whenever two representations differ with respect to those features. For example, if truth or correctness is a feature that should be explained by propositional content alone, then two representations that differ with respect to truth-value or correctness should not be assigned the same propositional content. On the other hand, if certain similarities, e.g. in psychological role, are amongst the primary explanatory targets of our content assignment, then some representations that differ as to correctness may still need to be construed as equal in content. The decision to adopt contents as objects of propositional attitudes and acts, and as semantic values, leaves a lot of room for differences in implementation, possibly depending on explanatory aims. Anyone constructing a theory assigning propositional contents to representations of some sort will inevitably have certain explanatory interests and preferences which guide his or her choice of similarities and differences to be modeled as sameness or difference of propositional content. There is thus a pragmatic element to the choice between different conceptions of propositions.[5]

3. Traditional and perspectival contents

In addition to the mentioned roles, contents have traditionally been taken to be the primary truth-bearers. This assumption is usually accompanied by the view that the truth-values of propositions vary only with how things are.[6] Since there is only one way things actually are, this amounts to saying that (actual) truth-values of propositional contents are absolute. Thus, traditional conceptions assume that contents have absolute truth values:

> (ABS) Propositional contents have absolute truth-values.

Some representations, such as beliefs, aim to represent things as they are, while other representations, such as hopes, do not. Let's call representations of the first type "belief-like."[7] On the traditional conception, any belief-like representation will therefore be representationally correct if and only if its content is true:

> (CORR) It is correct to accept (believe, assert etc) a propositional content iff it is true.

(ABS) and (CORR) entail that the representational correctness of a belief-like representation depends solely on its propositional content:

> (¬PERSP) The propositional content of a belief-like representation (together with the objective state of the world) determines whether the representation is correct.

However, some theoretical purposes and roles for contents suggest that there are contents that it may be correct for one person to believe at one time, but not correct for another person or at another time. For example, suppose we are interested in the similarity amongst all those who believe themselves to be philosophers and want to model this similarity as a sameness in propositional content, i.e. by saying that there is a content they all believe: call it "C." Then we must operate with a conception of propositional content that rejects (¬PERSP) (and therefore either (ABS) or (CORR)). For some of those who believe C do so correctly, while others don't: whether it is correct for a thinker to believe C depends on the thinker and on the time at which he or she has the belief.

Let us call any conception of propositions that rejects (¬PERSP), i.e. that allows that the correctness of belief-like representations can depend on factors other than the representation's propositional content and the objective state of the world, a "perspectival" conception, and let us call the propositions with which such a conception operates "perspectival contents." Assuming the connection between truth and the correctness of belief-like representations (CORR), perspectivalists will also reject (ABS). Thus, perspectival contents typically satisfy (PERSP) and (¬ABS):[8]

> (PERSP) The propositional content of a belief-like representation (together with the objective state of the world) does not determine whether the representation is correct.
> (¬ABS) Propositional contents do not have absolute truth-values.

Misgivings about, and problems with, perspectival contents are often associated with the expectation that contents play this or that traditional role of contents. I shall therefore present my selection of considerations about perspectival contents under three roles for propositions: as objects of belief, as compositional semantic value and as objects of assertion.

4. Contents as objects of belief

A typical worry about perspectival contents is that if the state of the world does not determine whether believing a given content is correct, then we don't know what state the world needs to be in for a belief with that content to be correct. Hence, we don't know which state the belief represents the world as being in. Consider again "C," a perspectival content that it is correct to believe for those, and only for those, who are philosophers at the time of belief. Consider also a perspectival content that it is correct to believe for those, and only for those, who have, at the time of believing, a disposition to experience pleasure from receiving a foot massage. Think of this as the content that foot massages are pleasant, and call it "F." There is no determinate way the world needs to be for beliefs with one of these contents to be correct. If *you* believe C *now*, the requirement on the world is quite different from what it would be if it was believed by someone else or at a different time. If Anna believes F, what foot massages need to be like for the belief to be correct may be quite different from what they need to be like for Ben's belief with the same content – if Anna's and Ben's dispositions are relevantly different. Since there is no determinate way the world is represented as being by these contents, they may seem unsuitable as contents of representation.

One can, of course, stipulate, when starting to talk about representational contents, that no two (actual) representations with the same content can differ as to their correctness. Frege makes such a stipulation (e.g. 1897, 147–148, 1918), and many have followed his example. But it is merely a pragmatic decision, and one that prejudices what kinds of phenomena a mere assignment of contents will be capable of explaining.

There are clearly other ways of thinking about how things are represented as being by a representation. There is a good sense in which any two people who believe it to be raining represent things in the same way. Their beliefs may concern different places and times, but the correctness of these beliefs requires the same of those places and times: namely that it rain there at the time. There is an important representational similarity between all these believers: there are certain conditions that need to hold for each such belief to be correct. There are also psychological similarities between all holders of such a belief. If there is such a representational similarity, then why not model this similarity by assigning the same representational content? Just as it is fine for Frege and his followers to have made their stipulation, there is nothing in principle wrong with treating this similarity as a sameness of representational content.

Now consider again all beliefs with content F. True, such beliefs may differ as to what foot massages need to be like for the belief to be correct. But in each case foot massages are represented as standing in the same relation to the thinker having the representation. This similarity is potentially explanatorily interesting because it accounts for similarities in behaviour. Those who operate a classic framework of contents will have to account for such similarities in a different way, for example by saying that the distinct contents are related in a certain way.

There is no principled reason to prefer a traditional, Fregean, stipulation in line with (¬PERSP). So what we find in the literature is usually just overt stipulation or dogmatic insistence. Wright declares that having contents that obey something like (¬PERSP) is constitutive of representation (Wright 1992, 91, 2008, 170–171). But others are happy to drop such a requirement (see, e.g. Moore 1997, 4–5, 10). Cappelen and Hawthorne (2009) defend (ABS) and (¬PERSP), but assume that this is the default position. Cappelen and Dever (2013) again, primarily argue against selected arguments that try to establish a need for a modification of the traditional framework of representational content in line with (PERSP).

There is, however, a conclusive argument by David Lewis (1979b) to the effect that a certain version of a perspectival content framework, namely centred contents, can do everything that the corresponding traditional framework of contents as uncentred sets of possible worlds can do. Centred contents are sets of centred worlds, i.e. sets of world-center pairs, where a center is simply a thinker and a time. Uncentred contents are simple sets of worlds. There is a small subset of the centred contents, the "boring" or "portable" ones,[9] which have exactly the same theoretically relevant properties as traditional uncentred propositions. Portable centred contents are those sets of world-center pairs where, if one world-center pair is a member, all other world-center pairs with the same world element are also a member. Lewis' result holds for *unstructured* contents, conceived of as simple sets of evaluation points. But the same holds for perspectival versions of *structured* propositions that comply with (PERSP), when compared with corresponding traditional versions that comply with (¬PERSP): there will be a portable counterpart amongst the perspectival structured propositions for each traditional structured proposition.[10]

Lewis's argument, and the extension of his argument to structured propositions, shows that generalizing from a traditional framework of contents complying with (¬PERSP) to a corresponding perspectival framework that complies with (PERSP), will not jeopardize any explanatory resources that were present before the generalization.

I want to discuss three further worries concerning perspectival propositions as the objects of belief. One is the worry that, if belief contents are construed in such a way that whether it is correct to believe them can depend on the perspective of the believer, then every such belief will be automatically correct. This worry simply assumes that everyone is infallible about, and in control of, their own perspective. But the assumption is unmotivated. For example, suppose perspectives are construed simply as Lewisian centers, i.e. thinker-time pairs: thinkers are neither in control of, nor infallibly knowledgeable about, what time it is or who they are. Consider again content F: a content that it is correct to believe for those, and only those, who have at the time of belief a disposition to experience pleasure when undergoing a foot massage. People are not in direct control of what makes them experience pleasure, nor are they infallibly knowledgeable about which procedures will cause pleasure in them. Thus, believing this content is subject to error just like belief in most traditional contents.

The second worry is that if we concede that another person's belief with a certain content is correct, this will force us to accept that content ourselves (e.g. Boghossian 2011) But if we concede that it is correct for Mia to believe C, then this doesn't commit us to believing C ourselves. For if Mia has the dispositions required for actually believing this correctly, this does not mean that we also have such dispositions (see Kölbel 2015a; Dinges 2017, §3).

The third worry is that somehow allowing belief contents to be perspectival will open the floodgates towards skepticism about objective reality and objective knowledge. However, simply stipulating that contents of representation are traditional (i.e. comply with (ABS) and (¬PERSP)) is not going to make the objectivity skeptic disappear. On the contrary, stipulating all representation to be objective will reinforce the impression that objective reality is merely an ideology (Rorty 1985). A conception of representational content that leaves room for both objective and non-objective representations provides a better starting point for arguing about objective reality.[11]

5. Contents as compositional semantic values

A compositional semantic theory of a language will assign to each sentence (in context) a semantic value. It will do this by assigning semantic values to primitive expressions and stating how the semantic values of complexes result from the semantic values of constituents and the way they are put together (compositional axioms). Given these assumptions, we can motivate the view that we need contents that do not comply with (ABS). Take, for example, the sentence

(1) The sun is shining.

What should be the semantic value when (1) is uttered on Vienna's Stephansplatz at Easter 2019? A traditional answer would have it that it would be a traditional content that is true if and only if the sun is shining on Stephansplatz at Easter 2019 (the place and time of the context of utterance). Thus, the utterance would have the very same semantic value as an utterance of

(2)   The sun is shining here now.

or of

(3)   The sun is shining on Stephansplatz at Easter 2019.

would have had if uttered in the same context. Lewis (1980) points out a problem with that view. If the utterances of (1)–(3) in that context have the same semantic value, then why does prefixing them with "Somewhere" not yield equivalent compounds?

(S1)   Somewhere the sun is shining.
(S2)   Somewhere the sun is shining here now.
(S3)   Somewhere the sun is shining on Stephansplatz at Easter 2019.

While (S1) is fine if uttered in the same context – it is true if there is a place where the sun is shining Easter 2019 – (S2) and (S3) make no sense.

If we want to treat "Somewhere," for example, as an intensional locational operator then we should treat (1) as having something like a location-unspecific proposition (a kind of perspectival content) as semantic value, i.e. a content that it is correct to believe at some places but not at others. This would explain the difference with (2) and (3), which are not in this way unspecific. Since we could repeat the argument with "Sometimes," the semantic value of (1) in the context has to be a content that is unspecific as to time *and* place.

Kaplan (1977, 503–504, fn. 28) very casually offers a similar argument (often called "operator argument"), in which he assumes, without argument, that certain expressions (like "sometimes," "always," "somewhere," etc.) are intensional operators, and then concludes that the expressions with which they compose must have intensions that are variable with respect to time, place etc.

In recent debates about relativism, the possibility of such intensional treatments has sometimes been cited by relativists as an advantage of operating with perspectival contents. Thus, for example, "for John" or "for some people" could be treated as intensional operators that can be concatenated with sentences (or perhaps predicates) like "Licorice is tasty" (or the predicate "is tasty").[12]

Defenders of traditional contents have rejected the premise that "sometimes," "somewhere" etc. are intensional operators, and offer quantificational accounts instead. They also argue against "for John," "sometimes" etc. being *sentential* operators (see Cappelen and Hawthorne 2009, 74–82).

The second point, about the syntactic category of the operators in question, has been answered: they can also be categorized as intensional predicate modifiers, thus the assumption that they are sentential is not essential to their being intensional.[13] But the first point simply says that there is an alternative account, so it does not provide any reason against the relativist's perspectival contents as semantic values.[14]

One point that is often overlooked is that by denying the assumption, in Kaplan's operator argument, that certain expressions are intensional operators, no answer has yet been given to the version of Lewis's argument that I have presented previously.[15] For that argument did not assume that "Somewhere" was an intensional operator. The premise was merely that (S1) differed in meaning from (S2) and (S3), so that, if (S1)–(S3) were the result of applying "Somewhere" respectively to (1)–(3), and if the semantic values are compositional, then (1) should not have the same semantic value as (2) and (3). This means that *independently* of whether we use an intensional or a quantificational treatment of "Somewhere," if we assume that (S1) is the result of concatenating (1) and "Somewhere," then the semantic value of (1) must be something that is variable with respect to place. On our assumptions (i.e. the claim that (S1) = "Sometimes"^(1) and compositionality) the semantic value of (1) cannot be a traditional proposition that has the same truth-value at every place. On a quantificational treatment too, it needs to be the semantic value of an open sentence containing an unbound variable.[16]

Thus, there are reasons to allow non-portable perspectival contents as compositional semantic values.[17] One reaction is to question whether the same thing needs to play all three roles. For example, one might say that the compositional semantic value of a sentence in context need not be *identified* with the assertoric content of an utterance of the sentence in that context – it is enough if the former determines the latter, perhaps together with other factors (see Rabern 2012). This brings us to the last role for contents that I want to discuss here.

6. Contents as objects of assertion

Propositional contents also have a role as objects of propositional acts. That perspectival contents can play this role has been questioned for many reasons, but mostly this has concerned the act of assertion.

One set of worries starts from the Fregean idea that to assert a content is to present it as true, or from the idea that knowledge or truth are "norms of assertion."[18] It seems clear that, to the extent to which a perspectivalist wants to accept these ideas about assertion at all, she can modify them to accommodate that the objects asserted do not have absolute truth-values.

A related area of discussion concerns what happens in successful assertoric communication. On a very simple model, preferred by many, thinker A believes a content c, asserts c in order to pass on this information to thinker B. Thinker B witnesses the assertion, trusts A and as a result comes to believe c herself. The same content is believed and subsequently asserted by one thinker, and then received and believed by another thinker. The merits of this simple picture have long been questioned in discussions of *de se* or indexical thought and language. For example, following Frege (1918), Evans (1981) and Pollock (1982) deny that the content of the source's belief and that of the recipient's acquired belief are always the same. More recently, there is a lively literature on how *de se* contents, or other perspectival contents, figure in linguistic communication.[19] One of the standard positions in this debate ("uncentering," following Kindermann 2016) is to say that while the source's belief may have a (non-portable) perspectival content, the content of a resulting assertion will not be perspectival, or will be a portable perspectival content.[20] Another standard position ("recentering") is to claim that when the source believes and asserts a non-portable perspectival content, the recipient acquires a belief with a distinct, but systematically related non-portable perspectival content.[21]

The debate about communication with perspectival contents is too large to be summarized adequately here. Instead, I shall briefly mention two frameworks for theorizing about assertion, and the role they can afford to perspectival contents: accounts of assertion and conversational updating that follow Stalnaker's (1999) and Lewis's (1979a) pioneering work, and normative accounts in the style of Brandom (1983). To start with the latter: assertion and other conversational moves are here characterized in terms of the norms they give rise to within a game-like activity of conversation. In a Brandomian account, the significance of assertion is that of undertaking an obligation to justify what one has asserted when challenged to do so, and of issuing a license to others to use what one has asserted as a premise (in their justifications). This type of account can be readily adapted to perspectival contents. The best developed account is that of John MacFarlane (2014, ch. 5),[22] who adds to norms much like Brandom's an obligation to retract what one has asserted in those "contexts of assessment" where the asserted proposition is not true. This is meant to explain under what conditions speakers have an obligation to retract. MacFarlane's account is tailor-made for perspectival contents and claims (◊ Assessment Relativism).

On the Stalnaker-Lewis model of conversational updating, at every stage of a conversation, there is a set of contents that are accepted: let's call that the "score" of the conversation (similar terms used include "presupposition set," "context set" and "common ground"). What is accepted, i.e. the score, can be changed by making conversational moves.[23] The most paradigmatic example of this is making an assertion. Making an assertion counts as a proposal to add the content asserted to the score. The proposal will come into effect if no-one challenges the assertion, e.g. by saying "no". If no challenge is made the asserted content is added to the score. Moreover, all moves in the conversation are added (as a matter of conversational record) as well as other obviously manifest information. Which moves are acceptable at any point, and the interpretation of such moves, will depend on the score.

Misgivings about perspectival contents as objects of assertion in this conversational model derive from the plausible idea that the acceptance of contents in the score is importantly related to participants' belief of these contents, and their chances of being correct if they were to believe them. In line with the simple picture of communication mentioned previously, a (portable) content would be asserted, the score would be updated by it, and other conversationalists could then simply adopt any contents from the score as suitable objects of belief. The recipient would come to believe the very same content that the source already believed and asserted.

However, the conversational model covers more than just these ideal cases of so-called serious conversations that are held with the sole purpose of information transfer. It is not unusual for an audience to fail to challenge (and thereby accept) assertions of contents they themselves do not believe. Sometimes contents get accepted that some or all participants do not believe, even if this is common knowledge amongst them. Asserters may transparently lie, while audiences may simply play along. The rules of score change remain the same.[24] Of course, a conversation sometimes results in a change of belief in its participants. However, this will not just be a matter of the audience coming to believe whatever is in the score. Rather, the nature of the conversation, the reliability of asserters and other factors will influence to what extent, if at all, a conversation changes a participant's beliefs. Thus, even if we allow perspectival contents to be asserted, and to update the conversational score, there will be a further step before this has effects on participants' acquired beliefs. "Recentering," as in the debate about *de se* communication, is always an option. Thus, there is no difficulty in principle in admitting perspectival contents as objects of assertion and updating, even in this model.[25]

Notes

1 E.g. Kölbel (2002, 2015a, 2015b), MacFarlane (2005, 2014), Lasersohn (2005), Recanati (2007), Stephenson (2007), Egan (2007, 2012), Zeman (2015), Dinges (2017), Coppock (2018).
2 Cf. Kaplan (1977/89), Lewis (1980), Rabern (2012).

3 Historically, this role has sometimes preceded the other roles, e.g. in Frege (1892).
4 Some theorists propose to reduce propositions to certain act types (see, e.g. Soames (2011), Hanks (2011)). However, even in this context, the proposed reductions answer to, and are justified by, the intended theoretical roles, not vice versa (cf. Lewis 1970, 22).
5 See Kölbel (2015a, §5) for more detail. It has often been discussed whether contents can play all these roles at once. See, e.g., Dummett (1973/81), Lewis (1980) or Rabern (2012).
6 For discussions of propositions as truth-bearers, see, e.g. Kirkham (1992) and Künne (2003).
7 Stalnaker (1984, 79) speaks of "acceptance" as a general attitude of which each belief-like attitude is a species.
8 Wright (1992, 148–157, 2001, 58) combines (Persp) with (ABS) and thus denies (CORR). On this view, a content's truth is not necessary or sufficient for it being correct to believe it. See Kölbel (1997; 2003).
9 See, e.g. Egan (2007) and Kölbel (2013).
10 See Kölbel (2014).
11 I have argued this point elsewhere, see Kölbel (in preparation). See also Coppock (2018), who makes the point that the framework of representational contents should not prejudice the question which aspects of reality are objective. She therefore operates with "outlooks" as primitive evaluation points without presupposing that these will be structured as world-center-pairs.
12 See, e.g. Kölbel (2009, 2011, 2015a, 2015b).
13 See Kölbel (2011b, 144, 2015b, 58).
14 King (2003) argues that a quantificational account of tense is confirmed by the fact that semanticists have moved from Prior's intensional paradigm to quantificational accounts nowadays. However, he does not offer principled objections againt an intensional treatment of tense either. See Rey (2016) for a detailed account.
15 Schaffer (2018) is an exception.
16 For temporal and locational modifiers, Cappelen and Hawthorne (2009) deny "uniformity," i.e. that (S1) = "Sometimes"^(1), thus allowing them to maintain compositionality.
17 Reasons that have been amply discussed both in the relativism literature and in the unarticulated constituents literature.
18 See, e.g. Frege (1892), Dummett (1973/81), Evans (1979/85), Rosenkranz (2008), Pagin (2016).
19 See, e.g. Kölbel (2002), Stalnaker (2008, 2014), Ninan (2010), Torre (2010), Moss (2012), Stojanovic (2012), Weber (2013) and the papers collected in García-Carpintero and Torre (2016).
20 E.g. Egan (2007, 2012), Stalnaker (2008).
21 E.g. Weber (2013). Another proposal is "multicentering," e.g. Ninan (2010), Torre (2010), Kindermann (2016).
22 Kölbel (2002, §§1.7 and 6) also uses a Brandom-style account of assertion to carve out a role for perspectival contents.
23 For discussion of the nature of the score, see Lewis (1979a), Kölbel (2011), Stalnaker (2014).
24 Some claim that the linguistic moves I am describing do not count as assertions. However, conversational updating effects are the same in these cases. See Kölbel (2010).
25 See Kölbel (2013) for a detailed proposal with regard to the role of unstructured perspectival contents in a Stalnaker-Lewis update model.

References

Boghossian, P. (2011), "Three Kinds of Relativism," in *A Companion to Relativism*, edited by S. D. Hales, Oxford: Blackwell, 53–69.
Brandom, R. (1983), "Asserting," *Nous* 17: 637–650.
Cappelen, H. and J. Dever (2013), *The Inessential Indexical*, Oxford: Oxford University Press.
Cappelen, H. and J. Hawthorne (2009), *Relativism: A Defence of Monadic Truth*, Oxford: Oxford University Press.
Coppock, E. (2018), "Outlook-Based Semantics," *Linguistics and Philosophy* 41: 125–164.
Dinges, A. (2017), "Relativism and Assertion," *Australasian Journal of Philosophy* 95(4): 730–740.
Dummett, M. (1973), *Frege: Philosophy of Language*, London: Duckworth, 1981.
Egan, A. (2007), "Epistemic Modals, Relativism and Assertion," *Philosophical Studies* 133: 1–22.
———. (2012), "Relativist Dispositional Theories of Value," *Southern Journal of Philosophy* 50: 557–582.

Evans, G. (1979), "Does Tense Logic Rest on a Mistake?" in his *Collected Papers*, Oxford: Clarendon Press, 1985, 341–363.

———. (1981), "Understanding Demonstratives," in *Meaning and Understanding*, edited by H. Parret and J. Bouveresse, Berlin: De Gruyter.

Field, H. (2009), "Epistemology Without Metaphysics," *Philosophical Studies* 143: 243–290.

Frege, G. (1892), "Über Sinn und Bedeutung," *Zeitschrift für Philosophie und philosophische Kritik* NF 100: 25–50.

———. (1897), "Logik," in *Nachgelassene Schriften*, edited by Hans Hermes et al., Hamburg: Felix Meiner Verlag, 1983, 137–163.

———. (1918), "Der Gedanke. Eine logische Untersuchung," *Beiträge zur Philosophie des deutschen Idealismus* 1: 58–77.

García-Carpintero, M. and S. Torre (eds.) (2016), *About Oneself: De Se Thought and Communication*, Oxford: Oxford University Press.

Hanks, P. (2011), "Structured Propositions as Types," *Mind* 120: 11–52.

Harman, G. (1975), "Moral Relativism Defended," *Philosophical Review* 84: 3–22.

Kaplan, D (1977), "Demonstratives," in *Themes from Kaplan*, edited by J. Almog, J. Perry and H. Wettstein, Oxford: Oxford University Press, 1989, 481–563.

Kindermann, D. (2016), "Varieties of Centering and De Se Communication," in García-Carpintero and Torre (2016), 307–340.

King, J. (2003), "Tense, Modality, and Semantic Values," *Philosophical Perspectives* 17: 195–245.

Kirkham, R. (1992), *Theories of Truth*, Cambridge, MA: MIT Press.

Kölbel, M. (1997), "Wright's Argument from Neutrality," *Ratio* 10: 35–47.

———. (2002), *Truth Without Objectivity*, London: Routledge.

———. (2003), "Faultless Disagreement," *Proceedings of the Aristotelian Society* 104: 53–73.

———. (2009), "The Evidence for Relativism," *Synthese* 166: 375–395.

———. (2010), "Literal Force: A Defence of Conventional Assertion". In *New Waves in Philosophy of Language*, edited by S. Sawyer, Basingstoke: Palgrave Macmillan 2010, 108–137.

———. (2011a), "Conversational Score, Assertion and Testimony," in *Assertion: New Philosophical Essays*, edited by H. Cappelen and J. Brown, Oxford: Oxford University Press, 49–77.

———. (2011b), *Objectivity, Relativism and Context Dependence*, Hagen: Fernuniversität, www.ub.edu/grc_logos/files/user126/1343744586-ObjectivityRelativismContextDependence.pdf.

———. (2013), "The Conversational Role of Centered Contents," *Inquiry* 56: 97–121.

———. (2014), "Agreement and Communication," *Erkenntnis* 79: 101–120.

———. (2015a), "Relativism 1: Representational Content," *Philosophy Compass* 10(1): 38–51.

———. (2015b), "Relativism 2: Semantic Content," *Philosophy Compass* 10(1): 52–67.

———. (in preparation), "Objectivity and Perspectival Content."

Künne, W. (2003), *Conceptions of Truth*, Oxford: Oxford University Press.

Lasersohn, P. (2005), "Context Dependence, Disagreement, and Predicates of Personal Taste," *Linguistics and Philosophy* 28: 643–686.

Lewis, D. (1970), "General Semantics," *Synthese* 22: 18–67.

———. (1979a), "Scorekeeping in a Language Game," *Journal of Philosophical Logic* 8: 339–359.

———. (1979b), "Attitudes De Dicto and De Se," *Philosophical Review* 88: 513–543.

———. (1980), "Index, Context, and Content," in *Philosophy and Grammar*, edited by S. Kanger and S. Öhman, Dordrecht: Reidel, 79–100.

MacFarlane, J. (2005), "Making Sense of Relative Truth," *Proceedings of the Aristotelian Society* 105: 321–339.

———. (2014), *Assessment Sensitivity*, Oxford: Oxford University Press.

Moore, A. (1997), *Points of View*, Oxford: Oxford University Press.

Moss, S. (2012), "Updating as Communication," *Philosophy and Phenomenological Research* 85: 225–248.

Ninan, D. (2010), "De Se Attitudes: Ascription and Communication," *Philosophy Compass* 5: 551–567.

Pagin, P. (2016), "Assertion," in *The Stanford Encyclopedia of Philosophy*, Winter 2016 edition, edited by E. Zalta, https://plato.stanford.edu/archives/win2016/ entries/assertion/.

Pollock, J. (1982), *Language and Thought*, Princeton, NJ: Princeton University Press.

Rabern, B. (2012), "Against the Identification of Assertoric Content with Compositional Value," *Synthese* 189: 75–96.

Recanati, F. (2007), *Perspectival Thought*, Oxford: Oxford University Press.

Rey, D. (2016), "In Defense of Implicit Times," PhD thesis, University of Barcelona.

Rorty, R. (1985), "Solidarity or Objectivity?" in *Post-Analytic Philosophy*, edited by J. Rajchman and C. West, New York: Columbia University Press, 3–19.

Rosenkranz, S. (2008), "Frege, Relativism and Faultless Disagreement," in *Relative Truth*, edited by Manuel García-Carpintero and Max Kölbel, Oxford: Oxford University Press, 225–237.

Schaffer, J. (2018), "Confessions of a Schmentencite: Towards an Explicit Semantics," *Inquiry*. DOI: 10.1080/0020174X.2018.1491326.

Soames, S. (2011), "Propositions," in *The Routledge Companion to Philosophy of Language*, edited by D. Graff Fara and G. Russell, London: Routledge.

Stalnaker, R. (1984), *Inquiry,* Cambridge, MA: MIT Press.

———. (1999), *Context and Content*, Oxford: Oxford University Press.

———. (2008), *Our Knowledge of the Internal World*, Oxford: Oxford University Press.

———. (2014), *Context*, Oxford: Oxford University Press.

Stephenson, T. (2007), "Judge Dependence, Epistemic Modals, and Predicates of Personal Taste," *Linguistics and Philosophy* 30: 487–525.

Stojanovic, I. (2012), "On Value Attributions: Semantics and Beyond," *Southern Journal of Philosophy* 50: 621–638.

Torre, S. (2010), "Centered Assertion," *Philosophical Studies* 150: 97–114.

Weber, C. (2013), "Centered Communication," *Philosophical Studies* 166: 205–223.

Wright, C. (1992), *Truth and Objectivity*, Cambridge, MA: Harvard University Press.

———. (2001), "On Being in a Quandary," *Mind* 110: 45–98.

———. (2008), "Relativism About Truth Itself," in *Relative Truth*, edited by Manuel García-Carpintero and Max Kölbel, Oxford: Oxford University Press, 157–185.

Zeman, D. (2015), "Relativism and Bound Predicates of Personal Taste," *Dialectica* 69: 155–183.

53

THE CASE AGAINST SEMANTIC RELATIVISM

Teresa Marques

1. Intuition-based arguments

Intuitions about retraction and disagreement play a crucial role in arguments for semantic relativism about epistemic modals, deontic modals, conditionals, knowledge attributions, or value and personal taste claims: the use of these expressions seems to be dependent on people's perspectives, either on a relevant body of information available, or a relevant standard of taste. But while objectivists seem unable to capture the perspective-dependence of claims in that range, indexical-contextualists seem unable to account for disagreement- and retraction-intuitions about the same claims, at least assuming the presumed intuitions of competent speakers. Disagreements between people with different perspectives in these areas seem to occur, as do retractions of past claims made by subjects after a change of perspective (where the perspective-independent facts remain the same). Likewise, we are owed a suitable explanation of what is going on with eavesdropper cases.

The chapters by Dan Zeman, Filippo Ferrari, and Andy Egan and Dirk Kindermann in this volume provide good introductions to different aspects of the contemporary debate on relativism, and we refer to them since they offer a good background to our discussion.[1]

Eavesdropper cases are cases of this kind (here is an example from Egan and Kindermann, this volume):

> (TASTE) F and G, who are from Australia, are having breakfast. G says to F, "mm, Vegemite toast is delicious." F replies, "that's true! Such a joy of a morning." At the next table, H and J, who are from the United States, overhear. J has never tried Vegemite, but knows that H has. J turns to H, hand poised to lift a piece of Vegemite toast to her mouth, and asks, "is that true?" H replies, "no, it's not. Vegemite toast tastes like feet."

Cases like this are supposedly problematic for contextualists. If what G says is "Vegemite toast is delicious, from my perspective," H's denial is, at best, misguided. But H seems to be perfectly within his rights to reply as he does. Absolutists do not have problems with eavesdropper cases, since whatever the sentence "Vegemite is tasty" says, it is absolutely true (or, in our view, false!). Other kinds of examples are discussed in eavesdropper cases, for instance epistemic modals.

Typically, these examples involve people with disparate opinions of what might or should be the case given the information available to them when they make their judgments.

Intuitions about seeming faultless-disagreement cases are supposedly a problem for both objectivists (or absolutists) and contextualists. Faultless disagreement can be characterized as follows:

> A faultless disagreement is a situation where there is a thinker A, a thinker B, and a proposition (content of judgment) p, such that:
>
> (a)   A believes (judges) that p and B believes (judges) that not-p
> (b)   Neither A nor B has made a mistake (is at fault).
>
> *(Kölbel 2004, 53–54)*

Consider, as an illustration, this case offered by Egan and Kindermann (also this volume):

> (TASTE★) F and G, who are from Australia, are having breakfast. G says to F, "mm, Vegemite toast is delicious." At the next table, H and J, who are from the United States, are sitting down to their own breakfast. J turns over to F and G's table, spitting out a bite of Vegemite, and says to them, "Vegemite toast is not delicious! This stuff tastes like feet."

A flat-footed reaction to have about this case: When G says "Vegemite toast is delicious" and J says "Vegemite toast is not delicious," they are disagreeing. (And they are disagreeing about whether Vegemite toast is delicious.) But nobody need be making a mistake: Both are (or on suitable fillings-in of the case, both could easily be) getting it right, "from their own perspective (Egan and Kindermann this volume).

Absolutists are supposed to have trouble explaining that nobody need be making a mistake, although they have no problem explaining the existence of disagreement. The absolutist's prediction is that "Vegemite toast is delicious" is either true or false (absolutely). Hence, one of G or F in this scenario must be making a mistake.

Contextualists are supposed to have trouble explaining this scenario because although they can explain why none of the speakers is making a mistake, they apparently can't explain what the interlocutors disagree about, since G is saying something like "Vegemite toast is delicious, from my perspective" and J is saying "Vegemite toast is not delicious, from my perspective," which can both be true and accepted as such.

Finally, intuitions about the retraction of claims in disputed domains are meant to further undermine the plausibility of contextualist or absolutist semantics for the contested terms in those domains. Consider this case from MacFarlane (2011):

> Sally and George are talking about whether Joe is in Boston. Sally carefully considers all the information she has available and concludes that there is no way to know for sure. Sally says: "Joe might be in Boston." Just then, George gets an email from Joe. The email says that Joe is in Berkeley. So George says: "No, he isn't in Boston. He is in Berkeley."

In this scenario, MacFarlane claims, it would be natural for Sally to retract her assertion that Joe might be in Boston, saying something like "I guess I was wrong; I take that back," and it would not be normal for her to say, "Fine, he can't be in Boston, but I still stand by what I said." The scenario assumes that intuitions about what it is more natural is something the contextualist would have trouble explaining since the speaker would be taking back a correct assertion of a true proposition.

Relativist semantic approaches are meant to be better equipped to handle these of cases. But there are different versions of semantic relativism, making different kinds of commitments and theoretical requirements. Ferrari (this volume) mentions a distinction between two ways of understanding semantic relativism related to the one made by Gareth Evans (1985): the "non-indexical contextualism" defended by Kölbel (2003, 2004), "NIC" henceforth, and the "assessment relativism" advocated by MacFarlane and others, "AR."

NIC is primarily a view about propositions. Traditionally, propositions are possible-world-neutral: the same proposition expressed in an actual utterance might be expressed when thought of as made under counterfactual circumstances, "in" a different world. Propositional truth is hence relative to a world, in that we need a world to evaluate the truth of a proposition. Philosophers like David Lewis (1979), have argued that we also need time-neutral and subject-neutral propositions for different reasons to account for *de se* phenomena. The truth of such propositions must thus be further relativized to times or subjects. Truth-relativists of this stripe try to address the philosophical issues they confront in the previous cases described by arguing for additional relativization of propositional truth, to judges or taste standards. However, truth as a norm for assertoric acts remains absolute on NIC, because the needed parameters are provided once and for all by the context that the act concerns.

To illustrate consider that in the preceding intuitive retraction case, the apparent natural response "I take that back" would suggest that an alternative to the standard contextualist view should be correct. For example, one alternative to the canonical Kratzerian contextualist analysis of modals[2] is that the quantification over a domain of possible worlds compatible with the information available at the context of use is not part of the content. Such an alternative view could take the shape of Egan's (2007) proposal that bare epistemic modals express centered world propositions whose truth varies with respect to different centers in the same world:

> It might be the case that P is true relative to a centered world (w, t, i) iff it's compatible with everything that's within i's epistemic reach at t in w that P.
>
> *(Egan 2007, 8)*

This view yields some predictions about the falsity of bare epistemic possibility modals (BEPs) made in the past – if the prejacent of the modal (e.g., "Joe is in Boston") is false with respect to the information available to a present hearer, then the modal is false. The view makes no strong requirement about the retrospective assessment of the correctness of that claim. It would be consistent with the view to say that it is natural for the speaker to assert "it might be the case that p" when p is compatible with the information available to her at the time of utterance, and for the speaker to take back the assertion when p is no longer compatible with information available to her at a later time.

There is however a stronger normative view about the retraction of epistemic possibility modals. The alternative view is not that it would be natural for Sally to retract, but that she ought to.

> Intuitions about retraction may be used against non-indexical contextualism. Upon learning that Joe is down the hall, Sally ought to retract her assertion that Joe might

be in Boston. Non-indexical contextualists lack the resources to explain this fact, since their account of the norm of assertion appeals only to the notion of truth relative to the context of assertion (a feature Sally's assertion has).

(MacFarlane 2014, 256)

This challenges any view according to which an assertion is correct only if it is true relative to the context of assertion. But this relies essentially on the strength of the intuition that a retraction is obligatory when the information now available entails the falsity of the prejacent of the modal.

MacFarlane seems to be relying on a point Michael Dummett (1978) made in connecting the truth of what we say and the point of our saying it, i.e., the point of assertion. Dummett's claim was that a "well-defined consequence of an assertion's proving incorrect is that the speaker must withdraw it" (1978, 20). Thus, by holding that retraction is mandatory in what he calls the assessment context, MacFarlane's AR is plausible. If it were the case that a retraction of a previous assertion is required, then that assertion must be incorrect. But if the assertion was correct (as MacFarlane puts it, "accurate") when it was made (the speaker said what was true when she spoke), and is incorrect ("inaccurate") when its retraction is required, then its correctness conditions must be themselves relative. This would be the point of assessment-relative truth: to capture this feature of some assertions whose correctness would also be assessment-relative. This is the more radical relativist view on which it is the appraisal of a "use" of a proposition (as MacFarlane puts it) that may change: while an assertion of "Vegemite is tasty" or of "Joe might be in Boston" might meet a truth norm when evaluated relative to the standard prevailing in the context in which it is made, thus being perfectly accurate, it may come to be false, and hence wrong, when evaluated from another context.[3] Evans (1985) argues that this is importantly different from NIC and more difficult to motivate, and we agree with him (cf. Marques 2014a, 2018), as we argue later.

2. Contextualist replies

There are several responses to intuition-based arguments offered to motivate truth-relativism.[4] Since the point of this chapter is to argue against semantic relativism, I shall be brief in considering those replies. Zeman's contribution to this volume offers a nice compilation of the variety of responses contextualists have offered with respect to faultless-disagreement cases. He identifies three kinds of strategies. First, there are authors that argue that the examples based on the appearance of faultless disagreement are often undetermined. The descriptions of the putative intuitive cases are said to fail to distinguish between "exocentric" and "autocentric" uses of the relevant phrases, to ignore the possibility of collective uses, or of generic uses. Once those distinctions are made available, the intuitions about the existence of disagreement should dissipate (cf. Schafer 2011; Glanzberg 2007; Stojanovic 2007; Cappelen and Hawthorne 2009).

Second, some disagreements and impressions of disagreement are said to be pragmatic. For instance, some impressions of disagreement can result from a presupposition that interlocutors are speaking from the same perspective (presupposition of commonality) and may dissipate once they realize that is not the case; or they may result from presupposing that at least one of the interlocutors are right, or that one of them is speaking from a superior perspective. Alternatively, other disagreements should rather be recognized as metalinguistic. Finally, some disagreements may concern incompatible conversational moves (cf. López de Sa 2015; Parsons 2013; Zakkou

2015; Sundell 2011; Plunkett and Sundell 2013; Ludlow 2014; Silk 2016). These pragmatic disagreements, or impressions of disagreement, are compatible with contextualism. Third, disagreements that are pragmatic in nature are sometimes claimed to be motivated by conative or non-doxastic disagreements, where those who disagree in attitude may have conflicting attitudes[5] (cf. Huvenes 2012; Marques 2015, 2016, 2019).[6]

Presumably some of these strategies are available concerning responses to the intuitive strength of eavesdropper cases. Although most explanations of the intuitions of disagreement focus on evaluative or taste predicates, some focus specifically on epistemic and deontic modal cases. For instance, Dowell (2011, 2013) persuasively argues that disagreement and eavesdropper cases are not real challenges to the standard Kratzerian contextualist semantics for modals. She suggests that when the proper constraints on contexts are satisfied, namely, when speaker's domain restricting intentions are properly made manifest (a condition she names "Publicity") the intuitions about the challenging cases – disagreement, eavesdroppers – dissipate.

Finally, several people have questioned the strength of the retraction of intuitions, which we will discuss later.

3. Limitations of relativism

Many authors have pointed out that relativists are not better positioned to explain disagreement data than contextualists are.[7] NIC aims to account for relativism by ascribing exclusionary contents to utterance contents such that "it has to be the case that at least one of them is false" (Khoo and Knobe 2016, 1), or, in other words, such that they can't be both right. Like others, we have argued that exclusionary contents are neither necessary nor sufficient for disagreement. First, they don't suffice to explain disagreement: If Lewis and his followers are right, an utterance of "I am making a mess" and one of "I am not making a mess" by another agent have exclusionary contents, but we don't have impressions of disagreement about them.

Dinges (2017a, 730; 2017b, 501) buys the exclusionary-content explanation, and, in work purporting to establish that we should adopt relativism on the abductive grounds that it is more "elegant" than rival views (2017b, 497), he confronts this argument. He submits (2017b, 502) that there are no assertions (or there shouldn't be?) of *de se* contents.[8] However, proponents of *de se* contents have provided well-developed proposals to account for their felicitous assertion even in such cases.[9] And as we have argued (cf. Marques and García-Carpintero 2014; Marques 2014a), there are felicitous assertions of what on relativist accounts are contents that are neutral with respect to standard-of-taste parameters (contradicting Dinges' claim). Two foodies with fully developed taste sensitivities, fully aware that there are serious differences in their respective rulings, might engage in the debates we are considering, "this is tasty – no, it isn't," and it might be perfectly rational for them to do so.[10]

Exclusionary contents are not necessary for disagreement, either. In previous work, we mentioned Horn's cases of "metalinguistic negation"[11] (García-Carpintero 2008; Marques 2014b, 2015), in connection with Barker's (2002, 2013) observation that one can use expressions with an indeterminate meaning to make proposals for their contextual precisification. Given this, one might use the negations to reject such a proposal. Sundell (2011) persuasively elaborates on related ideas, and there is now a huge literature developing the point.[12]

Khoo and Knobe (2016) confirm previous experimental evidence that ordinary speakers might be sensitive to the point that there might be disagreements without exclusionary contents.[13] As already mentioned in the previous section, it has also been repeatedly pointed out, and developed at length in different directions, that rejections might tap into disagreements of a practical nature, without the need for exclusionary content.[14] The purported explanation of the disagreement data that NIC offers doesn't thus appear promising.

A similar diagnosis applies to AR. Here the issue turns primarily on whether there is any good reason to accept that, as MacFarlane contends, the same assertion can be accurately made in its context, but assessed as inaccurate or incorrect from another context providing a conflicting appraisal. With respect to disagreements, specifically, MacFarlane agrees that under one understanding of disagreement, relativists are not better positioned than contextualists. In fact, we should ask, why would people disagree when they don't share a perspective? MacFarlane's (2007) response is one that any contextualist can likewise offer:

> Perhaps the point is to bring about agreement by leading our interlocutors into relevantly different contexts of assessment. If you say "skiing is fun" and I contradict you, it is not because I think that the proposition you asserted is false as assessed by you in your current situation, with the affective attitudes you now have, but because I hope to change these attitudes. Perhaps, then, the point of using controversy inducing assessment-sensitive vocabulary is to foster coordination of contexts. We have an interest in sharing standards of taste, senses of humor, and epistemic states with those around us. The reasons are different in each case. In the case of humor, we want people to appreciate our jokes, and we want them to tell jokes we appreciate. In the case of epistemic states, it is manifestly in our interest to share a picture of the world, and to learn from others when they know things that we do not.
>
> *(MacFarlane 2007, 30, emphasis added)*

Considerations about the pull towards coordination can be met without embracing relativist semantic revisions. To see this, consider a case that Dinges (2017b, 503) uses for an abductive argument for relativism in which A, whom B barely knows, offers advice to her regarding a restaurant by uttering "The food is tasty." As Dinges points out, if C then asks B about the food at the restaurant, it would be inappropriate for B to produce in her turn "The food is tasty" without having yet tried it. If, however, the topic had been all along the restaurant opening hours or food specialty, it would have been perfectly ok for B to pass on A's testimony. Dinges (2017b, 511) contends that the relativist has a simple explanation, which we rather doubt, but we lack the space to go into this here.

Be this as it may, Dinges' discussion highlights the difficulties that relativism poses to make sense of exchanges like that between A and B, and the very existence of a testimonial practice understood along the lines it promotes. We must put aside situations in which A and B are justified to take A's claim to concern a common standard; for in such cases, disagreement offers no support for relativism: at least one party makes a false claim, and whatever good standing both parties have can only be epistemic. But, if so, A's claim must be equivalent to "I like the food there" (2017b, 514). Given that a commonality assumption is not in place, that doesn't help B to make her own mind, and should discourage her adoption of the relativist proffered proposition as good enough for her own circumstances. Moreover, as can be appreciated from the possibility of pragmatic and conative conflicts, contextualists can appeal to conflicts over incompatible attitudes consistent with conversations like the described being attempts to change other people's attitudes.

MacFarlane later adopted a different approach. Instead of trying to explain the point of communicating and disagreeing over perspectival contents, he now argued for pluralism about disagreement and faultlessness:

> Instead of arguing about what is "real" disagreement, then, our strategy will be to identify several varieties of disagreement. We can then ask, about each dialogue of interest, which of these kinds of disagreement can be found in it, and we can adjudicate between candidate theories of meaning by asking which theories predict the kinds of disagreement we find.
>
> *(MacFarlane 2014, 119)*

Zeman (this volume) nicely introduces the varieties MacFarlane considers.[15] We will not go over these here. Yet, once the discrimination between different kinds of putative disagreements has been done, it is no longer clear whether speakers may plausibly be expected to have intuitions about each of those senses of disagreement.[16] Some of the more "minimal" forms MacFarlane declares to be types of disagreement (such as simple noncotenability, i.e. exclusionary contents) can of course be stipulated to be a form of disagreement, but it's unclear what independent motivation can be offered in its support.

Egan (2007) and MacFarlane (2011), among others, also argued that contextualist semantics cannot account for eavesdropper cases. But, with respect to deontic and epistemic modals, not only do contextualists have a response, as Dowell (2011; 2013) claims; competent speakers don't appear to share the intuitions that would support revisionary relativist theories. In a joint paper, Knobe and Yalcin (2014) experimentally tested eavesdropping and retraction in cases involving epistemic modals. With respect to extra-contextual assessors their experiments revealed that eavesdroppers did not tend to judge that a claim like Sally's "Joe might be in Boston" would be false when new information about Joe's whereabouts came to light, against the presumed intuitions that eavesdroppers would predominantly assess the sentence as true only if "Joe is in Boston" were consistent with the information available to them (qua assessors).

And finally, the most serious intuition-based argument against contextualism concerns retraction cases. It's the presumed requirement to retract that makes AR distinctive, and without it AR would be unmotivated. As one of us argues (Marques 2014a, 2018), it's not just that the requirement to retract doesn't seem well motivated. As experimental data suggests, speakers tend to disagree that what Sally said ("Joe might be in Boston") is false, while tending to agree that it would be appropriate for her to retract (Knobe and Yalcin 2014). This seems puzzling. However, this experiment didn't address the presumed requirement to retract that, as MacFarlane holds, motivates AR. But in new experimental work, Marques (manuscript) found that indeed participants tend to disagree that the modal is false, and that participants also tend to disagree that there is any requirement to retract when the truth of the prejacent of the "might" claim is incompatible with new available information.

There is a more a serious a priori reason against relativism. In a 1985 paper, Evans raised a challenge to the coherence of relative truth. A modification of the challenge can be given against AR. The main point is that rational sincere speakers cannot be bound by assessment-relative standards of correctness or accuracy, and that's why they are not required to retract assertions that were correctly made (true with respect to the intended context) with propositions as contents that turn out to be not true with respect to a new "assessment context."

Evans's point has an a priori character. It doesn't question the rationality of particular moves in an assertoric practice legitimized by assessment relativism, but rather the rationality of such a practice itself. The question is whether it's rational to implement a relativistic assertoric practice. MacFarlane (2014, 311–319) discusses the issue, but his rationale for justifying a practice modeled by AR only takes into consideration what we think is a negligible dimension. He offers considerations of economy: against what both NIC and contextualism mandate, it makes for some sort of efficiency not to have to keep track of the contexts relative to which claims or proposals are made, appraising them only with respect to current standards. We think that this lacks significance when weighed against the crucial reasons to have a practice of making assertions, asking questions or giving directions, such as pooling our information.

Cases like the following, which illustrate Evans's point, cast doubts on the requirement to retract past taste assertions:

> Imagine that Smith is aware that [...] her culinary opinions depend on her current dispositions towards food, etc. She knows, moreover, that certain standards are bound to change with time [...] that people in her family, when they reach 60, lose the capacity to appreciate sweets. [In the future] she will not think that crème brûlée is tasty. [...] On MacFarlane's account, she should be puzzled, in envisaging now that there will come to be situations in which she will be forced to retract her current assertion, while, from the viewpoint of her present context of assessment, those situations do not constitute at all a reason for so doing. [...] Certainly, Smith could have aimed at the context she knows she will occupy when she is 60 – but had she done so, her standard of taste at 60 would still have been a standard of taste fixed by the context of utterance.
>
> *(Marques 2014a, 370)*

Similar seeming puzzles arise if we suppose that assertors of epistemic possibility modal claims should retract their present assertions when, in the future, the prejacents of the sentences they use turn out to be incompatible with the evidence they then acquire. Von Fintel and Gillies (2008, 81 ff.) point out that not all "might" claims are retracted in the face of new evidence. In fact, in the situation Sally finds herself in, it would have been perfectly natural for her to reply: "Look, I didn't say Joe is in Boston; I said he might be. Maybe he changed his mind and canceled the trip. Sheesh!"

When we make sincere assertions, we aim to speak truly, like Smith who asserts "Crème Brûlée is tasty" fully aware that she will find herself in situations where she will assert the negation of that sentence, or Sally who asserts "Joe might be in Boston" fully aware that there are probable situations in the near future in which she will assert "Joe can't be in Boston." If truth is assessment-sensitive, there is no final answer to the question of whether these assertions were correct when they were made. Speaker's committed to the truth of what they asserted. And when they find themselves in situations where they assert the negation of the earlier claims, they will commit to the truth of the new assertion. But they won't be in a position to uphold the correctness of the earlier assertion and of that of the later one. How could they purport to commit to both?[17] There's no answer as to whether the aim of speaking truly is ever achieved. At best, speakers can aim to speak truly from a context. But, from which context? If we can give an answer to this question, then we also generate, as Evans (1985, 350) put it, a "once-and-for-all assessment of utterances, according to whether or not they meet whatever condition the answer gave."

Acknowledgements

Thanks are due to Manuel García-Carpintero and to Martin Kusch. Work funded by the European Commission for the project DIAPHORA [Grant agreement number H2020-MSCA-ITN-2015–675415]; Ministerio de Economía y Innovación for projects ABOUT OURSELVES and CONCEDIS [Grant agreements: FFI2013–47948-P; FFI2015–73767-JIN], and AGAUR de la Generalitat de Catalunya for Grup Consolidat LOGOS.

Notes

1 García-Carpintero and Kölbel (2018) collects many of the contributions that have shapped the current debate.
2 On standard Kratzerian accounts, the relevant epistemic situation is the state of information that is available in the context where the speaker uses an epistemic modal sentence like "Joe might be in Boston." An alternative sentence, "From what I know, Joe might be in Boston," makes it explicit that the modal base (set of worlds selected by the context that are meant to be compatible with Joe being in Boston) are the worlds consistent with the information available to Sally. The sentence in the example is a bare epistemic modal and gives no explicit indication of how the modal base is to be restricted. Its modal base is nonetheless also determined in context: the context of use determines a set of possible worlds compatible with the relevant information available at the context of use.
3 Wright (2016, 189) also points out the significance of the distinction between NIC and AR. In contrast, Cappelen and Hawthorne (2009) and Cappelen and Huvenes (2018) adopt the sort of view we are opposing here, contending that the most salient difference among different relativist views "is arguably that relativists have different views about what propositions are true or false relative to" (Cappelen and Huvenes 2018, 521); cf. Kölbel (2015). This take informs choices in the presentation and discussion of the issues, with confounding consequences.
4 Kölbel (2009, §3) lists additional related data.
5 The notion of conflicting attitudes was already advanced by Stevenson (1944).
6 Absolutists can also offer replies to faultless disagreement cases, as Zeman (this volume) explains (cf. Schafer 2011).
7 As is argued by Stojanovic (2007), Moruzzi (2008), Rosenkranz (2008), Iacona (2008), Francén (2010), amongst others.
8 Caso (2014) also confronts the objection, like Dinges from a perspective that is consistent with NIC even if his sympathies lie with AR. He buys the explanatory exclusion explanation of disagreement too (2014, 1322). He dismisses the objection by claiming that the relativist view is not intended to be applied to the discourses for which *de se* or *de nunc* contents have been advanced (2014, 1314, 1317). This overlooks the point of the objection, that is, that the possibility of such discourses shows that the relativist alleged *explanans* for the impression of disagreement is no good, because it might be in place without any impression of disagreement being triggered.
9 For recent versions, see the papers by Recanati (2016), Maier (2016), Weber (2016) and Kindermann (2016) in García-Carpintero and Torre (2016).
10 Dinges (2017a, 735) considers cases of this sort.
11 These are cases in which one uses expressions of disagreement to object to the appropriateness of an assertion on grounds other than the falsity of its content: "John failed some students – No, he failed all of them"; "John has a car – No, he has a Ferrari Testarossa." These fall under the various forms of pragmatic disagreement mentioned earlier.
12 Cf. for instance Plunkett and Sundell (2013), Ludlow (2014), López de Sa (2015).
13 Sarkissian et al. (2011) present evidence that also appears to suggest that previous results about presumed folk intuitions can be accounted for by the fact that relevant disparities in the judges were not made salient. (See Marques and García-Carpintero (2014, 707–708) for some reservations about the interpretation of such results.)
14 Marques and García-Carpintero (2014) and Marques (2015, 2016, 2019) appeal to a "hybrid expressivist" view.
15 Others argue for a unified notion of disagreement, e.g. Belleri and Palmira (2013).
16 For additional skepticism about some of the notions involved here, see for instance Palmira (2015), Carter (2014), Eriksson and Tiozzo (2016).
17 See also Ross and Schroeder (2013, 69–70).

References

Barker, C. (2002), "The Dynamic of Vagueness," *Linguistics and Philosophy* 25(1): 1–36.
———. (2013), "Negotiating Taste," *Inquiry* 56(2–3): 240–257.
Belleri, D. and M. Palmira (2013), "Towards a Unified Notion of Disagreement," *Grazer Philosophische Studien* 88(1): 139–159.
Cappelen, H. and J. Hawthorne (2009), *Relativism and Monadic Truth*, Oxford: Oxford University Press.

Cappelen, H. and T. Huvenes (2018), "Relative Truth," in *The Oxford Handbook of Truth*, edited by M. Glanzberg, Oxford: Oxford University Press, 517–542.

Carter, J. A. (2014), "Disagreement, Relativism and Doxastic Revision," *Erkenntnis* 79(1): 1–18.

Caso, R. (2014), "Assertion and Relative Truth," *Synthese* 191: 1309–1325.

Dinges, A. (2017a), "Relativism and Assertion," *Australasian Journal of Philosophy* 95(4): 730–740.

———. (2017b), "Relativism, Disagreement and Testimony," *Pacific Philosophical Quarterly* 98: 497–519.

Dowell, J. L. (2011), "A Flexible Contextualist Account of Epistemic Modals". *Philosophers' Imprint* 11: 1–25.

———. (2013), "Flexible Contextualism about Deontic Modals: A Puzzle about Information-Sensitivity". *Inquiry: An Interdisciplinary Journal of Philosophy* 56(2–3): 149–178.

Dummett, M. (1959/1978), "Truth," in his *Truth and Other Enigmas*, Cambridge, MA: Harvard University Press, 1–24.

Egan, A. (2007), "Epistemic Modals, Relativism and Assertion," *Philosophical Studies* 133(1): 1–22.

Egan, A. and A. Kindermann, "De Se Relativism," this volume.

Eriksson, J. and M. Tiozzo (2016), "Matters of Ambiguity: Faultless Disagreement, Relativism and Realism," *Philosophical Studies* 173(6): 1517–1536.

Evans, G. (1985), "Does Tense Logic Rest upon a Mistake?" in his *Collected Papers*, Oxford: Clarendon Press, 343–363.

Francén, R. (2010), "No Deep Disagreement for New Relativists," *Philosophical Studies* 151(1): 19–37.

García-Carpintero, M. (2008), "Relativism, Vagueness and What Is Said," in García-Carpintero and Kölbel (2008), 129–154.

García-Carpintero, M. and M. Kölbel (eds.) (2008), *Relative Truth*, Oxford: Oxford University Press.

García-Carpintero, M. and S. Torre (eds.) (2016), *About Oneself*, Oxford: Oxford University Press.

Glanzberg, M. (2007), "Context, Content, and Relativism." *Philosophical Studies* 136(1): 1–29.

Huvenes, T. (2012), "Varieties of Disagreement and Predicates of Taste," *Australasian Journal of Philosophy* 90(1): 167–181.

Iacona, A. (2008), "Faultless or Disagreement," in García-Carpintero and Kölbel (2008), 287–298.

Khoo, J. and J. Knobe (2016), "Moral Disagreement and Moral Semantics," *Noûs* 52: 109–143.

Kindermann, D. (2016), "Varieties of Centering and *De Se* Communication," in García-Carpintero and Torre (2016), 307–339.

Knobe, J. and S. Yalcin (2014), "Epistemic Modals and Context: Experimental Data," *Semantics and Pragmatics* 7(10): 1–21.

Kölbel, M. (2003), "Faultless Disagreement," *Proceedings of the Aristotelian Society* 104: 53–73.

———. (2004), "Indexical Relativism vs Genuine Relativism," *International Journal of Philosophical Studies* 12(2): 297–313.

———. (2009), "The Evidence for Relativism," *Synthese* 166: 375–395.

———. (2015), "Relativism 1: Representational Content," *Philosophical Compass* 10(1): 38–51.

Lewis, D. (1979), "Attitudes De Dicto and De Se," *Philosophical Review* 88: 513–543.

López de Sa, D. (2015), "Expressing Disagreement: A Presuppositional Indexical Contextualist Relativist Account," *Erkenntnis* 80: 153–165.

Ludlow, P. (2014), *Living Words: Meaning Underdetermination and the Dynamic Lexicon*. Oxford: Oxford University Press.

MacFarlane, J. (2007), "Relativism and Disagreement," *Philosophical Studies* 132: 17–31.

———. (2011), "Epistemic Modals Are Assessment-Sensitive," in *Epistemic Modality*, edited by B. Weatherson and A. Egan, Oxford: Oxford University Press, 108–130.

———. (2014), *Assessment Sensitivity: Relative Truth and Its Applications*, Oxford: Oxford University Press.

Maier, E. (2016), "Why My *I* Is Your *You*: On the Communication of *De Se* Attitudes," in García-Carpintero and Torre (2016), 220–245.

Marques, T. (2014a), "Relative Correctness," *Philosophical Studies* 167(2): 361–373.

———. (2014b), "Doxastic Disagreement," *Erkenntnis* 79(1): 121–142.

———. (2015), "Disagreeing in Context," *Frontiers in Psychology* 6: 1–12.

———. (2016), "Aesthetic Predicates: A Hybrid Dispositional Account," *Inquiry* 59(6): 723–751.

———. (2018), "Retractions," *Synthese* 195(8): 3335–3359.

———. (2019), "Hybrid Dispositionalism and the Law," in *Dimensions of Normativity: New Essays on Metaethics and Jurisprudence*, edited by K. Toh, D. Plunkett and J. Shapiro, New York: Oxford University Press, 263–286.

Marques, T. and M. García-Carpintero (2014), "Disagreement About Taste: Commonality Presuppositions and Coordination," *Australasian Journal of Philosophy* 72(4): 701–723.

Moruzzi, S. (2008), "Assertion, Belief and Disagreement," in García-Carpintero and Kölbel (2008), 207–224.

Palmira, M. (2015), "The Semantic Significance of Faultless Disagreement," *Pacific Philosophical Quarterly* 96(3): 349–371.

Parsons, J. (2013), "Presupposition, Disagreement, and Predicates of Taste," *Proceedings of the Aristotelian Society* 113: 163–173.

Plunkett, D. and T. Sundell (2013), "Disagreement and the Semantics of Normative and Evaluative Terms," *Philosopher's Imprint* 13: 1–37.

Recanati, F. (2016), "Indexical Thought: The Communication Problem," in García-Carpintero and Torre (2016), 141–178.

Ross, J. and M. Schroeder (2013), "Reversibility or Disagreement," *Mind* 122(485): 43–84.

Rosenkranz, S. (2008), "Frege, Relativism and Faultless Disagreement," in Relative Truth, edited by M. García-Carpintero and M. Kölbel, Oxford: Oxford University Press, 225–238.

Sarkissian, H., J. Park, D. Tien, J. Cole Wright and J. Knobe (2011), "Folk Moral Relativism," *Mind and Language* 266(4): 482–505.

Schafer, K. (2011), "Faultless Disagreement and Aesthetic Realism," *Philosophy and Phenomenological Research* 82(2): 265–286.

Silk, A. (2016), *Discourse Contextualism: A Framework for Contextualist Semantics and Pragmatics*, Oxford: Oxford University Press.

Stevenson, C. L. (1944), *Ethics and Language*, New Haven: Yale University Press.

Stojanovic, I. (2007), "Talking About Taste: Disagreement, Implicit Arguments, and Relative Truth," *Linguistics and Philosophy* 30(6): 691–706.

Sundell, T. (2011), "Disagreement About Taste," *Philosophical Studies* 155: 267–288.

von Fintel, K. and A. S. Gillies (2008), "CIA Leaks," *Philosophical Review*, 117(1): 77–98.

Weber, C. (2016), "Being at the Centre: Self-location in Thought and Language," in García-Carpintero and Torre (2016): 246–271.

Wright, C. (2016), "Assessment-Sensitivity: The Manifestation Challenge," *Philosophy and Phenomenological Research* 92: 189–196.

Zakkou, J. (2015), "Tasty Contextualism: A Superiority Approach to the Phenomenon of Faultless Disagreement," PhD thesis, Berlin: Humboldt University of Berlin.

Zeman, D., "Faultless Disagreement," this volume.

54

DE SE RELATIVISM

Andy Egan and Dirk Kindermann

1. Motivations for *de se* relativism

Consider the following two cases:

> (MIGHT) A and B are looking for the surveillance microphone. They haven't yet searched the desk. B says, "it might be in the desk." A says, "that's true – let's check there." C and D are monitoring the surveillance cameras (and microphone) in A & B's apartment. They aren't sure where the microphone is (E installed it, and is currently on break), but C was watching when E planted the microphone, and while she didn't see exactly where E put it, she saw that E never went near the desk. D, hand poised over the button that will scramble the emergency team, asks C, "is that true?" C replies, "no, it's not. No need to worry."

A flat-footed reaction to have about this case: When A said "that's true," she was attributing truth to B's assertion of "it might be in the desk," and she was correct to do so. When C said, "no, it's not" in response to D's question, "is that true?," she was attributing falsity to B's assertion of "it might be in the desk," and she was correct to do so.

> (TASTE) F and G, who are from Australia, are having breakfast. G says to F, "mm, Vegemite toast is delicious." F replies, "that's true! Such a joy of a morning." At the next table, H and J, who are from the United States, overhear. J has never tried Vegemite, but knows that H has. J turns to H, hand poised to lift a piece of Vegemite toast to her mouth, and asks, "is that true?" H replies, "no, it's not. Vegemite toast tastes like feet."

A flat-footed reaction to have about this case: When F said "that's true," she was attributing truth to G's assertion of "Vegemite toast is delicious," and she was correct to do so. When H said, "no, it's not" in response to J's question, "is that true?," she was attributing falsity to G's assertion of "Vegemite toast is delicious," and she was correct to do so.

A natural proposal for how to accommodate the flat-footed reactions: The propositions asserted with "might" claims and predicates of personal taste, are (at least in some contexts) the sorts of things that can take different truth-values relative to different assessors within the same world.

A relatively theoretically conservative way of implementing the natural proposal: the propositions people assert with "might" claims, and with predicates of personal taste, are (at least in some contexts) interesting centered-worlds propositions. That is: they're propositions that take truth-values relative to centered worlds (which we can think of as <world, time, individual> triples) rather than worlds, and that sometimes take different truth values relative to centered worlds <w,t,i> and <w,t★,i★>.

Cases like (MIGHT) and (TASTE) – in which there are two assessors within a world, one of whom seems as if they'd be correct to assess a particular utterance (or sentence in context) as true, the other of which seems as if they'd be correct to assess the same utterance as false – are one of the headline motivations for *de se* relativist views about some subject matter.

The other headline motivation is cases of (purported) faultless disagreement. This is more frequently appealed to in discussions of taste than of epistemic modals, so we'll focus on that kind of case, generating another example by tweaking the eavesdropper taste case:

> (TASTE★) F and G, who are from Australia, are having breakfast. G says to F, "mm, Vegemite toast is delicious." At the next table, H and J, who are from the United States, are sitting down to their own breakfast. J turns over to F and G's table, spitting out a bite of Vegemite, and says to them, "Vegemite toast is not delicious! This stuff tastes like feet."

A flat-footed reaction to have about this case: When G says "Vegemite toast is delicious" and J says "Vegemite toast is not delicious," they are disagreeing. (And they are disagreeing about whether Vegemite toast is delicious.) But nobody need be making a mistake: Both are (or on suitable fillings-in of the case, both could easily be) getting it right, "from their own perspective."

A natural proposal for how to accommodate the flat-footed reaction: The propositions people assert with predicates of personal taste, are (at least sometimes) the sorts of things that can take different truth-values relative to different assessors within the same world. So, J is really denying the very proposition that G is asserting (hence disagreement), but that proposition is true relative to G and false relative to J, so G is correct to accept it, and J is correct to deny it (hence faultlessness).

Here's a relatively theoretically conservative way of implementing the natural proposal: the propositions people assert with predicates of personal taste (and with epistemic modals), are (at least in some contexts) interesting centered-worlds propositions.

2. The view

De se relativism (as advocated by Egan 2007, 2010, 2012, 2014; Kindermann 2012, 2019 – other views that may be counted as versions of *de se* relativism are advocated by Brogaard 2007, 2012; Stephenson 2007a, 2007b; Lasersohn 2005, 2016, and will be discussed in Section 6) has two core commitments: a proposal about what kind of semantic contents to assign to certain sentences in context, and a proposal about the theoretical role of the assignment of one semantic content rather than another to a sentence in context.

(DE SE RELATIVISM–DSR)
    (SEMANTIC CONTENT) Certain sentences involving uses of epistemic modals and predicates of personal taste (among, perhaps, other expressions) semantically express, in context, interesting centered-worlds propositions.

(FOUNDATIONS) The semantic content of a sentence in context specifies, inter alia, the conventional uptake condition for that sentence in that context. This proposal about the theoretical role of semantic content requires a bit more explanation.

According to this proposal, a semantic theory for a language is in the business of associating declarative sentences in context with conventional *production conditions* as well as conventional *uptake conditions* or *acceptance conditions*. Production conditions are something that speakers are conventionally called upon to satisfy; uptake conditions are something that assessors are conventionally called upon to satisfy. Speakers' willingness to make an utterance tracks, inter alia, the production conditions of the utterance. Assessors' judgments of truth and falsity track (perhaps among other things – more later) their willingness to accept, or inclination to reject, an utterance's uptake condition. (We use "assessors" as a general term to cover both participants in the conversation in which the utterance takes place and eavesdroppers.)

For non-subjective declarative utterances, separating production and uptake conditions is not important. If we associate a possible-worlds proposition P with a sentence S in context c, it doesn't matter if we think of P as modeling, in the first instance, the sentence's production condition (for example, what the speaker has to believe in order to sincerely assert S in c) or its uptake condition (what the assessor has to accept in order to accept as true an utterance of S made by a speaker in c). If speakers are conventionally called upon not to say S in c unless (they believe) P, the predictable uptake effect, when assessors accept that the speaker is speaking truly, is that the assessor will come to believe P as well. If an assessor's acceptance of an assertion of S in c requires updating with P (if P models the uptake condition of S in c), then cooperative speakers who don't want to mislead will strive not to assert S in c unless (they believe) P. Either way, we predict that sincere assertors of S in c will believe that P is true, and that assessors who accept the assertion of S in c will update with P.

In the case of a *de se* relativist account of subjective utterances, such as in (TASTE) and (MIGHT), the difference between production and uptake conditions matters a great deal. If we associate S in c with an interesting centered-worlds proposition Q, it matters whether we think of Q as modeling, in the first instance, a production condition or an uptake condition. If take Q to model an uptake condition, that means that an assessor's acceptance of an assertion of S, in context of utterance c, requires the assessor to update with Q. This gives rise to the possibility that there are assertions that one assessor ought to accept, but another assessor in the same world ought to reject. That's because Q, being interestingly *de se*, could be true of one assessor and false of the other. It also gives rise to the possibility of assertions whose contents are true of the speaker but false of some assessor, and so which are properly accepted by the speaker, but properly rejected by the assessor.

If, on the other hand, we take Q to model a production condition, we don't get any of these predictions. The predictable uptake effect, for trusting assessors of an assertion whose production condition is (believing) Q, will be to accept that Q is true *of the speaker*. Importantly, it *won't* be to update with Q. And so the attribution of interestingly *de se* content to S in c won't, if we understand the attribution of content to be characterizing production conditions, predict that there will be any assertions that would be correctly accepted by one assessor and correctly rejected by another – whether Q is true of the speaker won't vary across assessors.

If we (a) take the semantically determined contents of declarative sentences to be in the business of, inter alia, modeling conventional uptake or acceptance conditions, and (b) assign interesting *de se* contents as semantically determined assertoric contents in context,[1] then we get a theory that predicts variation in truth-value judgments across assessors, as well as the possibility of faultless disagreement, in which two speakers assert sentences with incompatible acceptance conditions, each of which is true of the speaker. Thus, each party to the disagreement is correct to accept their own assertion, and correct to reject the other's.

The combination of these two proposals says that the semantic content of a sentence in context specifies, inter alia, the conventional uptake condition for that sentence in that context, and that the semantic contents (in their contexts) of many uses of epistemic modal sentences and taste sentences are interestingly *de se*. It gives us a view that's responsive to, and well-positioned to explain, the peculiar phenomena from Section 1.

2.1 *Eavesdropper truth-value attributions*

When G says "Vegemite is delicious" in (TASTE), she expresses an interesting centered-worlds proposition – at a first pass, the one that's true at <w,t,i> iff i is disposed (at t, in w) to enjoy the taste of Vegemite. (This is just a first pass. See Egan 2010.)

That's true of G, and also true of F (we suppose – after all, they're from Australia). So it's not only a good thing for G to assert but also a good thing for F to accept – accepting it would help F to locate herself more precisely in logical space. It's also a good thing for G and F to add to their stock of presuppositions in their conversation (to the stock of potential objects of belief that they both accept, take each other to accept, take each other to take each other to accept, etc. See Stalnaker 2002) – presupposing it will help them to zero in on their collective position in logical space, and on ways in which their individual positions are similar, and recognized to be similar. If we suppose that truth-value attributions track with willingness to accept and to add to conversational presuppositions, this explains why it makes sense for F to respond to G's assertion with "that's true."[2]

It's not, however, true of H, nor is it true of J. (Neither H nor J is disposed to enjoy the taste of Vegemite.) And so it would be a bad thing for either H or J to accept. Since it's false of H, she is correct to reject it, and by signaling her rejection of it, to warn J off of accepting it as well. If we suppose that truth-value attributions track willingness to accept (and inclination to reject) and to add (or reject the addition of) conversational presuppositions, this explains why it makes sense for H to respond to J's request for a verdict about the truth-value of G's utterance with "that's false." It is false for J and for H, and so H ought not to accept it (ought not to bring it about that she satisfies its conventional uptake condition), and ought to warn J off of accepting it as well.

The story about A, B, C, and D in (MIGHT) is similar: when B says, "it might be in the desk," he expresses (at a first pass – see Egan 2007) the interesting centered-worlds proposition that's true relative to <w,t,i> iff i (at t, in w) doesn't have access to evidence that rules out the microphone's being in the desk. That's true of B, and it's also true of A. So it makes sense for A to update with the content of B's utterance, and to allow it to pass in to the stock of conversational presuppositions, and to signal his willingness to accept B's utterance and update accordingly by saying "that's true." C and D's evidential situation is different – they *do* have access to evidence that rules out the microphone's being in the desk. (C saw that the microphone-planter never went near the desk, and D has access to C's evidence by being in conversation with her.) So C ought to reject updating with the content of B's utterance, and ought to warn D off of updating with it as well. She can do this by saying, "that's false" in response to D's request for a verdict on the truth-value of B's assertion.

2.2 *Faultless disagreement*

Very briefly, here is how the *de se* relativist promises to deliver faultless disagreement:

We get *disagreement* because the uptake conditions of the two assertions (of e.g. "Vegemite is tasty" and "Vegemite is not tasty") are incompatible – no assessor's acceptance state, and no conversational context, can be consistently updated with both propositions.

We get faultlessness (or at least, we get something that should satisfy our intuition of faultlessness) because (i) the propositions asserted are both true of the people asserting them, and so each party to the disagreement is correct to accept their own assertion, and (ii) each party is correct to reject the other's assertion, because (for example) G's assertion is false of J, and J's is false of G.

This is very quick – things get more complicated quickly. We take up some complications with the motivating cases in Section 5.[3]

3. *De se* philosophy of mind

We have seen the barebones of *de se* relativism and how the view makes sense of eavesdropper and disagreement cases. Let's now look at the complementary motivations for *de se* content in the philosophy of mind. The introduction of *de se* content is, in the first instance, motivated by the existence of psychological phenomena that resist characterization in terms of propositional attitudes with possible-worlds content, but that are happily characterized in terms of attitudes with *de se* content. For our purposes, we will quickly rehearse some of the most compelling examples and make a few points that will be relevant for understanding how the *de se* relativist view is meant to work.

Muhammad Ali and George Foreman, in the early months of 1974, displayed an important doxastic similarity, which gave rise to similarities in their behavioral dispositions. Each of them believed of himself that he was the greatest. And that doxastic similarity disposed them to act in similar ways – to expect victory in boxing matches, regardless of the opponent; to dispute others' claims to be the greatest; to seek out opportunities to fight for the heavyweight championship of the world, etc. But clearly this doxastic similarity wasn't a matter of both of them believing some common possible-worlds proposition. Ali didn't believe *that Foreman is the greatest*, Foreman didn't believe *that Ali is the greatest*, and while they did both believe *that somebody is the greatest*, we believe that too, and we emphatically do not share the behavioral dispositions that are symptomatic of Ali and Foreman's doxastic similarity. A centered-worlds proposition, however, will do nicely as a common object of belief. In particular, the one that's true relative to <w,t,i> iff i is the greatest (boxer) at t in w.[4]

Another famous example is Perry's (1977) bear attack case, presented here in lightly modified form: John and David are walking in the woods when John is attacked by a bear. Everyone is on the same page about what the world is like – John and David are just alike with respect to their relevant possible-worlds-y beliefs. But they act very differently – John curls up into a ball, David runs for help. If we want to explain their difference in behavior in terms of a doxastic difference – in terms of each of them doxastically appreciating and appropriately responding to the differences in their situations – it won't be in terms of a difference in possible-worlds propositions believed. But we could happily explain their difference in behavior in terms of a difference in centered-worlds propositions believed. While both John and David stand in the belief relation to the proposition that includes <w,t,i> iff John is being attacked by a bear while David looks on in w (at t), only John believes the proposition that includes <w,t,i> iff i is being attacked by a bear at t in w, and only David believes the proposition that includes <w,t,i> iff i is an unattacked witness to a bear attack at t in w.

This is all very quick, and does not pretend to be a full-dress argument for the Lewisian framework for theorizing about self-locating belief. But it is, we hope, enough to see the motivation for the introduction of centered-worlds propositions as objects of belief, and to see the theoretical itch the introduction of *de se* content in the philosophy of mind is meant to scratch.[5]

Two remarks before we move on. First, we can think of centered-worlds propositions as being, or representing, properties. For instance, the centered-worlds proposition that is true at a $<$w,t,i$>$ iff i is disposed (at t, in w) to enjoy the taste of Vegemite is true of you now just in case you now instantiate the property *being disposed to enjoy the taste of Vegemite*. We will treat centered-worlds talk and property talk interchangeably. Second, everything that is captured by possible worlds propositions can also be captured by centered-worlds propositions because, as Lewis (1979) points out, there is a 1–1 correspondence between the possible-worlds propositions and a proper subset of the centered-worlds propositions. The possible-worlds proposition that is true at w iff snow is white at w corresponds to the centered-worlds proposition that is true at $<$w,t,i$>$ iff snow is white at $<$w,t,i$>$. This centered-worlds proposition does not vary with time and individual – if it's true at any location within a world, it is true at every location in that world. We call centered-worlds propositions that do not vary across locations within worlds *boring* and those centered-worlds propositions that do vary across locations within the same world *interesting*. (This terminology is from Egan 2006.)

4. *De se* semantics and *de se* communication

A central motivating thought behind *de se* relativism is that there are some sentences, the central conventional communicative role of which is to produce *de se* beliefs (more carefully, *de se* uptake effects) in addressees and/or conversational contexts. (That's the motivating thought behind the sort of *de se* relativism Egan (2007, 2010, 2012, 2014) and Kindermann (2016, 2019) advocate. But see also Section 6.) That a sentence is governed by this sort of communicative practice would explain the patterns of acceptance and rejection, truth-value attribution, and judgements about disagreement and faultlessness that we (arguably) see around epistemic modal sentences and taste sentences. The semantic theory delivers centered-worlds propositions as semantically determined contents in context is in service of that thought.

It's important to note that we don't need a *de se* semantics in order to predict *every de se* communicative effect in conversation. And under the assumption that the central theoretical role of semantic content is to model acceptance conditions, there will be many *de se* communicative effects that we will emphatically not want to explain in terms of a *de se* semantics (but will instead explain by having *de se* contents figure in pragmatics). For example:

> English speakers can use sentences containing indexicals to communicate to their interlocutors that they have a particular *de se* belief. Muhammad Ali does this when he says, at the pre-fight press conference, "I am the greatest." Explaining this doesn't require the introduction of any semantic machinery beyond that of a standard sort of Kaplanian contextualism about "I." Given a standard view of the context-dependence of "I" (that it refers, in a context c, to the speaker of c), and a *de se*-ist philosophy of mind, we can predict that sincere English speakers will only assert "I am the greatest" if they self-attribute *being the greatest*. We can also predict that competent users of English will, in virtue of their knowledge of how the context-dependence of "I" works, realize that sincere speakers will only say "I am the greatest" if they self-attribute *being the greatest*. And that speakers will, on the basis of their linguistic competence and their presupposition that their interlocutors are similarly competent, realize that their interlocutors know this, and that their interlocutors know that they know it. And so speakers will be in a position to use their productions of "I am the greatest" to

communicate to their interlocutors, by way of the usual sort of Gricean mechanism, that they self-attribute *being the greatest*.

None of this requires any fancy semantics – it's straightforwardly predicted by an off-the-shelf Kaplanian account of "I" (Kaplan 1989), on which Ali's assertion of "I am the greatest" has as its content the possible-worlds proposition (equivalently, a boring centered-worlds proposition) that Ali is the greatest. So we don't need any semantic innovations in order to explain and predict this kind of *de se* communicative phenomenon. And if we're working in a framework in which semantic content in context serves to characterize uptake conditions, the introduction of at least one kind of semantic innovation – going for a semantics for indexicals such that sentences like "I am the greatest" wind up getting assigned interestingly *de se* content in context – will be disastrous. A theory according to which the uptake condition for "I am the greatest" is self-attributing *being the greatest* (i.e., accepting the centered-worlds proposition true at <w,t,i> iff i is the greatest at t in w) is not a good theory of indexicals, as it would have competent and trusting hearers self-attribute the property *being the greatest*. So the presence of this kind of *de se* communicative effect doesn't exert theoretical pressure to go for a *de se* semantics.[6]

The same is true for a common sort of second-personal *de se* communicative effect, by which speakers can, systematically, use sentences involving second-personal pronouns like "you" to produce *de se* beliefs in their addressees. For example, Bob can systematically use the sentence, "you are a fool" to produce self-attributions of being a fool in credulous addressees. So personal pronouns and standard context-sensitive expressions such as "here," "now," "this," and "that" (and many non-standard context-sensitive expressions) do not require the implementation of *de se* effects within one's semantic theory. They can be handled by standard pragmatics combined with a *de se* philosophy of mind.[7]

The sorts of phenomena we saw in Section 1, however, are not similarly predictable on the basis of off-the-shelf Kaplanian semantics, off-the-shelf pragmatic principles, and *de se* philosophy of mind. Those phenomena were (i) variation across assessors in the appropriateness of acceptance and rejection of a single utterance; (ii) variation across assessors of the truth-values it's appropriate for them to attribute to a single utterance; (iii) judgments about disagreement and correctness, in which we're inclined to say that (for example) G and H disagree about whether Vegemite toast is delicious, but they're both correct.

These sort of *de se* communicative phenomena are not happily explained by off-the-shelf pragmatics and an off-the shelf contextualist semantics that delivers no *de se* contents, together with a *de se* philosophy of mind. But a *de se* semantics, paired with an understanding of semantic content as characterizing uptake conditions, does promise to explain them.

5. Complicating the motivation

Things are more complicated with respect to the motivation for relativism than we let on in Section 1. For complications with faultless disagreement cases, see Chapter 51 "Faultless Disagreement" by Dan Zeman in this handbook. As regards eavesdropper cases, we suggested in Section 1 (though we were careful not to quite assert) that what we in fact observe is eavesdroppers systematically and uniformly assessing extra-conversational epistemic modal claims in the light of their own evidence, rather than the evidence that's plausibly relevant to the speaker, and systematically and uniformly assessing extra-conversational taste claims in the light of their own tastes, rather than those of (or those plausibly relevant to) the speaker; and that what we in fact observe from theorists contemplating these sorts of cases is clear, distinct and universal intuitions that those are the correct assessments for eavesdroppers to make. But that is not what we actually see.

What we actually see is a complicated pattern of ambivalence and variability in the sorts of judgments that it seems appropriate, to various theorists, for eavesdroppers to make. Some kinds of cases provoke, more strongly and/or more frequently, the sorts of judgments characteristic of the "flat-footed reaction" of Section 1. Other kinds of cases provoke, more strongly and/or more frequently, the intuition that eavesdroppers' assessments ought to be based on the evidence or the standards of the speaker. (See for instance Knobe and Yalcin 2014; Fintel and Gillies 2008, 2011; Dowell 2011.)

Some *de se* relativists have argued for their view on the basis of its ability to accommodate this sort of variability, and in particular to explain the specific patterns of variation in judgments that we observe. (See Beddor and Egan 2018 cf. also Lasersohn 2005, 2016 on autocentric vs exocentric uses of predicates of personal taste.)

6. Versions of *de se* relativism and other contemporary relativisms

De se relativism (DSR) is characterized by the commitment to (SEMANTIC CONTENT) and (FOUNDATIONS) (Section 2). Moreover, perhaps *the* distinguishing feature of *de se* relativism is its *de se* philosophy of mind, on which having an attitude like the belief that Vegemite is tasty amounts to self-attributing the property *being disposed to enjoy the taste of Vegemite*. (This is captured more precisely with the formal tools of centered-worlds propositions and the idea of attitudes as self-location, but this is an optional, if useful, feature of the view.) This allows for some variation among *de se* relativists. For instance, while Egan (2010, 2012, 2014) places central importance on the foundational role of conventional *uptake* conditions (rather than production conditions) for assertoric content, Kindermann (2012, 2019) stresses the assumption that cooperative conversation is aimed at the coordination of individual perspectives in the conversation's common ground – placing equal importance on production and acceptance conditions for constraining assertoric content. As a result, Kindermann argues that a full characterization of the communication of both *de se* attitudes expressed with personal pronouns ("I," "you") and taste attitudes requires the introduction of multi-centered possible worlds content.

Several recent relativist views (Brogaard 2007, 2012; Kölbel 2013; Lasersohn 2005, 2016; Recanati 2007; Stephenson 2007a, 2007b) are more or less close to (DSR) and may be seen as versions of *de se* relativism. Minimally, they all share with (DSR) a centered-worlds semantics (or something plausibly intertranslatable with a centered-worlds semantics) on which semantically determined assertoric contents are interesting centered-worlds propositions (cf. the first commitment of (DSR), (SEMANTIC CONTENT). This does not mean, however, that these views all share the second commitment of (DSR), (FOUNDATIONS), or a *de se* account of mental attitudes.

Another contemporary version of relativism, MacFarlane's assessment-sensitive relativism (MacFarlane 2014, see Ferrari, ch. 50), differs from *de se* relativism in a number of crucial ways. For one thing, assessment-sensitive relativism (ASR) does not feature centered-worlds propositions but relativizes propositional truth to (world, time and) information states, standards of taste, and/or whatever a particular application of the view requires. This feature is central to ASR's different predictions about the felicity of assertions, e.g. in cases involving predicates of personal taste embedded under past and future tense and modal expressions (see MacFarlane 2014, ch. 7; Kindermann 2012, ch. 6). MacFarlane's ASR also differs from (DSR) in its approach to the semantic foundations of the view. MacFarlane takes speaker-centered conditions (on assertion and retraction) to be foundational for semantic content. In particular, he sees the speech act of retraction by the speaker to be the motivating foundational piece for assessment-sensitive relativism, whereas (DSR) places central importance on uptake conditions.

Notes

1 Just what the relation is between compositional semantic values and semantically determined content is controversial. Ninan (2010) and Rabern (2012) argue convincingly that the relation mustn't uncritically be assumed to be identity. Pretty much all parties to the controversy agree that, somehow, compositional semantics *determines* some kind of semantic content, which plays some important role in a theory of communication. In the main text, we remain neutral about exactly what the underlying compositional semantic theory, and mechanism for going from compositional semantics to semantic content, looks like. See Kindermann (2019) for an implementation of *de se* relativism for predicates of taste in a modified Kaplanian two-dimensional framework (Kaplan 1989).
2 The notion of presupposition is the one championed by Robert Stalnaker (1978). See Egan (2007, 2010) and Kindermann (2012, 2019) for details of how talk about taste and epistemic modality works on Stalnaker's model of communication.
3 For many criticisms of the relativist's reliance on faultless disagreement as well as the various characterizations of the phenomenon, see Zeman in this volume.
4 This case is inspired by Perry's (1977) Hume/Heimson example, but we prefer it to the Hume/Heimson case because the beliefs involved are prosaic and not pathological.
5 In favor of special contents for first-person, or *de se*, thought, see Castañeda (1966), Chisholm (1981), Lewis (1979), and more recently Ninan (2016) and Torre (2018) (and Perry 1977, 1979 for special modes of presentation). For arguments against the need for special *de se* contents, see e.g. Millikan (1990), Magidor (2015) and Cappelen and Dever (2013).
6 For recent literature on the communication of *de se* attitudes, see the papers in García-Carpintero and Torre (2016) as well as Moss (2012), Ninan (2010), Torre (2010) and Weber (2013).
7 Egan (forthcoming) and Kindermann (2012, 2016, 2019) spell out the details of *de se* effects in communication and their place within semantics and pragmatics. See also Kölbel in this volume.

References

Beddor, B. and A. Egan (2018), "Might do Better. Flexible Relativism and the QUD," *Semantics and Pragmatics* 11(7): 1–43.
Brogaard, B. (2007), "Moral Contextualism and Moral Relativism," *The Philosophical Quarterly* 58(232): 385–409.
———. (2012), "Colour Eliminativism or Colour Relativism?" *Philosophical Papers* 41(2): 305–321.
Cappelen, H. and J. Dever (2013), *The Inessential Indexical*, Oxford: Oxford University Press.
Castañeda, H-N. (1966), "'He':A Study in the Logic of Self-Consciousness," *Ratio* 8: 130–157.
Chisholm, R. (1981), *The First Person*, Brighton:The Harvester Press.
Dowell, J. J. L. (2011), "A Flexible Contextualist Account of Epistemic Modals," *Philosophers' Imprint* 11: 1–25.
Egan, A. (2006), "Secondary Qualities and Self-Location," *Philosophy and Phenomenological Research* 72(1): 97–119.
———. (2007), "Epistemic Modals, Relativism and Assertion," *Philosophical Studies* 133: 1–22.
———. (2010), "Disputing About Taste," in *Disagreement*, edited by R. Feldman and T. A. Warfield, Oxford: Oxford University Press, 247–286.
———. (2012), "Relativist Dispositional Theories of Value," *The Southern Journal of Philosophy* 50(4): 557–582.
———. (2014), "There's Something Funny About Comedy: A Case Study in Faultless Disagreement," *Erkenntnis* 79(1): 73–100.
———. (forthcoming), "De Se Pragmatics," *Philosophical Perspectives*.
García-Carpintero, M. and S.Torre (eds.) (2016), *About Oneself: De Se Thought and Communication*, Oxford: Oxford University Press.

Kaplan, D. (1989), "Demonstratives. An Essay on the Semantics, Logic, Metaphysics, and Epistemology of Demonstratives and Other Indexicals," in *Themes from Kaplan*, J. Almog, J. Perry and H. Wettstein, Oxford: Oxford University Press, 481–563.

Kindermann, D. (2012), "Perspective in Context: Relative Truth, Knowledge, and the First Person," PhD thesis, University of St Andrews.

———. (2016), "Varieties of Centering and *De Se* Communication," in *About Oneself: De Se Thought and Communication*, edited by M. Garcia-Carpintero and S. Torre, Oxford: Oxford University Press, 307–340.

———. (2019), "Coordinating Perspectives: *De Se* Attitudes and Taste Attitudes in Communication," *Inquiry*: 1–44.

Knobe, J. and S. Yalcin (2014), "Epistemic Modals and Context: Experimental Data," *Semantics and Pragmatics* 7(10): 1–21.

Kölbel, M. (2013), "The Conversational Role of Centered Contents," *Inquiry* 56(2–3): 97–121.

Lasersohn, P. (2005), "Context Dependence, Disagreement, and Predicates of Personal Taste," *Linguistics and Philosophy* 28: 643–686.

———. (2016), *Subjectivity and Perspective in Truth-Theoretic Semantics*, Oxford: Oxford University Press.

Lewis, D. (1979), "Attitudes De Dicto and De Se," *The Philosophical Review* 88(4): 513–543.

MacFarlane, J. (2014), *Assessment-Sensitivity: Relative Truth and Its Applications*, Oxford: Oxford University Press.

Magidor, O. (2015), "The Myth of the *De Se*," *Philosophical Perspectives* 29(1): 249–283.

Millikan, R. G. (1990), "The Myth of the Essential Indexical," *Noûs* 24(5): 723–734.

Moss, S. (2012), "Updating as Communication," *Philosophy and Phenomenological Research* 85(2): 225–248.

Ninan, D. (2010), "Semantics and the Objects of Assertion," *Linguistics and Philosophy* 33(5): 355–380.

———. (2016), "What Is the Problem of *De Se* Attitudes?" in *About Oneself: De Se Thought and Communication*, edited by M. Garcia-Carpintero and S. Torre, Oxford: Oxford University Press, 86–120.

Perry, J. (1977), "Frege on Demonstratives," *The Philosophical Review* 86(4): 474–497.

———. (1979), "The Problem of the Essential Indexical," *Noûs* 31(1): 3–21.

Rabern, B. (2012), "Against the Identification of Assertoric Content with Compositional Value," *Synthese* 189(1): 75–96.

Recanati, F. (2007), *Perspectival Thought. A Plea for (Moderate) Relativism*, Oxford: Oxford University Press.

Stalnaker, R. (1978), "Assertion," in *Syntax and Semantics 9: Pragmatics*, edited by P. Cole, New York: Academic Press, 315–332.

———. (2002), "Common Ground," *Linguistics and Philosophy* 25(5): 701–721.

Stephenson, T. (2007a), "Judge Dependence, Epistemic Modals, and Predicates of Personal Taste," *Linguistics and Philosophy* 30(4): 487–525.

———. (2007b), "Toward a Theory of Subjective Meaning," PhD thesis, MIT Press.

Torre, S. (2010), "Centered Assertion," *Philosophical Studies* 150(1): 97–114.

———. (2018), "In Defense of *De Se* Content," *Philosophy and Phenomenological Research* 97(1): 172–189.

von Fintel, K. and A. S. Gillies (2008), "CIA Leaks," *The Philosophical Review* 117: 77–98.

———. (2011), "'Might' Made Right," in *Epistemic Modality*, edited by A. Egan and B. Weatherson, Oxford: Oxford University Press, 108–130.

Weber, C. (2013), "Centered Communication," *Philosophical Studies* 166(1): 205–223.

55

RELATIVISM AND EXPRESSIVISM

Bob Beddor

1. Introduction

The last couple of decades have seen an explosion of work on relativism and expressivism. However, the exact relationship between these two frameworks remains unclear. This chapter aims to shed some light on this murky state of affairs.

  Both relativism and expressivism have been put forward in response to the perceived short-comings of a contextualist semantics. This chapter starts by briefly reviewing both contextualism (Section 2) and a major source of dissatisfaction with the contextualist framework (Section 3). Next, it outlines how relativists (Section 4) and expressivists (Section 5) try to improve on contextualism. The rest of the chapter canvasses potential choice points for deciding between relativism and expressivism. I focus on their relation to truth conditional semantics (Section 6), their conceptions of belief and communication (Section 7), and their strategies for explaining disagreement (Section 8).

2. Contextualism

Contextualists about some expression e maintain that the truth-values of sentences containing e are partially dependent on features of the context of utterance. A particularly clear illustration comes from sentences containing indexicals, e.g.:

(1) I am hungry now.

The truth-value of (1) depends on who is speaking, as well as the time of utterance:

  (CONTEXTUALIST SEMANTICS FOR INDEXICALS) (1) is true, as uttered in a
  context c and evaluated at a world w, iff the speaker in c is hungry at w at the time of c.

While contextualism about indexicals is widely accepted, there is vigorous debate about which other expressions should get a contextualist treatment. Here I'll focus on two controversial cases: moral discourse and epistemic modals.

First up: moral discourse. Consider a claim such as:

(2)   Stealing is wrong.

One simple (some might say cartoonishly so) contextualist semantics holds that moral claims are about the speaker's attitudes, e.g.:

> (SPEAKER CONTEXTUALIST SEMANTICS FOR MORAL DISCOURSE) (2) is true, as uttered in a context c and evaluated at a world w, iff the speaker in c disapproves of stealing at w.

While few today would advocate this particular semantics, more sophisticated contextualist approaches to moral discourse have been defended by numerous philosophers.[1] For our purposes, this simple semantics is enough to highlight the key contextualist idea, which is that utterances of moral sentences express different propositions depending on who is speaking.

Next up: epistemic modals – i.e., uses of modal language (e.g., *might, must, possibly*) to convey some distinctly epistemic species of possibility or necessity. According to contextualism, epistemic modals communicate whether some embedded proposition is consistent with – or entailed by – some body of information determined by the context of utterance. To illustrate, consider:

(3) It might be snowing.

Contextualists maintain:

> (CONTEXTUALIST SEMANTICS FOR EPISTEMIC MODALS) (3) is true, as uttered in a context c and evaluated at a world w, iff the proposition *It is snowing* is compatible with the c-determined information at w.

In the simplest case, the c-determined information is just the speaker's. In other cases, it might be that of some speaker-inclusive group, or some contextually salient agent.[2]

This just scratches the surface. Contextualist semantics have been proposed for a wide range of expressions, including gradable adjectives, conditionals, and knowledge ascriptions. Rather than discuss these applications in detail, I now turn to a common misgiving about contextualism – a misgiving that is often used to motivate a shift to relativism or expressivism.

3. The problem of lost disagreement

A common objection to contextualism is that it has trouble accounting for disagreements.[3] To introduce this concern, it is helpful to start once again with indexicals. Suppose Aliya utters (1) (*I'm hungry now*) on June 4, 2019. Suppose Bruno overhears her. It would be very odd for Bruno to disagree with her merely on the grounds that he (Bruno) isn't hungry.

Contextualism has a nice story about why there's no disagreement here. According to contextualism, different utterances of (1) express different propositions in different contexts of utterance. When Aliya utters (1), she asserts the proposition, *Aliya is hungry on June 4, 2019.* And Bruno does not disagree with this proposition.

But while this is a mark in favor of contextualism about indexicals, some have argued that it is a liability for contextualism about other domains. Suppose Aliya asserts (2) (*Stealing is wrong*). Suppose Bruno does not disapprove of stealing. It would be natural for Bruno to disagree with Aliya's claim by saying something like:

(4a) No [/that's not true],
(4b) stealing isn't wrong.

But why is this disagreement any more genuine than in the indexical case? After all, on the Speaker Contextualist Semantics, Aliya's utterance of (2) expresses the proposition that *she* disapproves of stealing. But presumably Bruno does not disagree with *this* proposition. Following MacFarlane (2014), call this "the problem of lost disagreement."

This problem extends to other domains. Suppose that Aliya asserts (3) (*It might be snowing*). And suppose Bruno overhears Aliya's remark, having just come in from the brilliant sunshine. It would be natural for Bruno to disagree with Aliya. In doing so, Bruno need not disagree with the proposition that Aliya's information leaves open the possibility of snow.

The argument from lost disagreement is controversial. Some object that it relies on naive assumptions about the disagreement data, or about how to set the contextual parameters, or about the nature of disagreement.[4] For present purposes, I won't take a stand on the merits of this argument. What's important is that this argument has been used to motivate a shift to an alternative semantic framework, such as relativism or expressivism. Let's take relativism first.

4. Relativism

According to relativism, some sentences are *assessment-sensitive*. Even when one has fixed the context of utterance and the world of evaluation, one has not thereby fixed the truth-value of the sentence. Rather, there is room for further variation in truth-value depending on who is assessing the sentence for truth or falsity.[5]

To develop this thought, let an *assessor* be any agent who is evaluating an utterance for truth or falsity. Let a *context of assessment* be any situation where an assessor is making some such evaluation. For many purposes, it is convenient to model a context of assessment as a centered world: an ordered pair of a world w and an assessor a. Then we can give our Speaker Contextualist Semantics for Moral Expressions a relativist twist:

> (RELATIVIST SEMANTICS FOR MORAL DISCOURSE) (2) is true, as uttered in a context c and evaluated at a context of assessment $<w, a>$, iff a disapproves of stealing at w.[6]

Let's see how this applies to our moral dispute. According to relativism, when Aliya utters (2), the content of her utterance is a *centered proposition*: a set of centered worlds. What is distinctive about this sort of content is that it can be true for one person and false for another. It is true for her, since she disapproves of stealing. But it is false for Bruno, since he does not disapprove of stealing. And this, relativists claim, is why they disagree. (More on this in Section 8.)

A similar diagnosis applies to modal disputes. Relativists agree with contextualists that the truth conditions of modal utterances depend on some body of information. However, relativists claim this information is determined by the context of assessment rather than the context of utterance:

> (RELATIVIST SEMANTICS FOR EPISTEMIC MODALS) (3) is true, as uttered in a context c and evaluated at a context of assessment $<w, a>$, iff the proposition *It is snowing* is compatible with the $<w, a>$-determined information at w.[7]

On this view, when Aliya asserts (3), she asserts a proposition that is true for her (since her information leaves open the possibility of snow), but false for Bruno (since his information does not). Hence, they disagree.

5. Expressivism

Another response to the problem of lost disagreement is to go expressivist. According to expressivists about some sentence φ, the conventional function of uttering φ is to express some mental state m of the speaker. This is typically paired with the idea that the meaning of φ just is m; or, at the very least, that the semantics and pragmatics of φ cannot be understood without reference to m.[8]

To flesh this out, start with moral discourse. On a simple expressivist analysis, the meaning of (2) is some mental attitude towards stealing. What sort of attitude? Historically, expressivism has gone hand-in-hand with noncognitivism. According to noncognitivism, moral beliefs do not aim to represent the world in the same way that ordinary descriptive beliefs do. Rather, moral beliefs are conative attitudes: desires, preferences, plans, states of approval/disapproval, etc. Thus, a simple expressivist analysis of (2) might go like this:

> (SIMPLE EXPRESSIVIST ANALYSIS OF MORAL DISCOURSE) The meaning of (2) is the mental state: DISAPPROVAL OF STEALING.

How does expressivism help with the problem of lost disagreement? Typically, expressivists adopt a "mind-first" picture of disagreement: disagreements between speakers are explained in terms of disagreements between the mental states that these speakers express.

The idea that mental states can stand in disagreement relations is clearest when it comes to factual beliefs. If Aliya believes that Singapore is to the south of Kuala Lumpur, whereas Bruno believes that Singapore is to the north of KL, then their beliefs disagree. More controversially, expressivists claim that there are also "disagreements in attitude": disagreements between desire-like attitudes. As Stevenson (1994, 3) puts it:

> Suppose that two people have decided to dine together. One suggests a restaurant where there is music; another expresses his disinclination to hear music and suggests some other restaurant. . . . The disagreement springs more from divergent preferences than from divergent beliefs, and will end when they both wish to go to the same place.

Applied to our moral dispute: when Aliya utters (2) she expresses disapproval of stealing. Bruno holds a different conative attitude towards stealing: he tolerates it. This mental state disagrees with Aliya's disapproval.

Not all moral expressivists equate meanings with mental states. Gibbard (2003) takes the meaning of a moral claim to be a formal object that represents the content of a conative attitude (more precisely: the content of some combination of representational and conative attitudes). On Gibbard's view, moral claims express plans to adopt reactive attitudes, such as blame and outrage. But the content of (2) is not itself a plan, but rather a set of *world-hyperplan* pairs. (Here a "hyperplan" is a formal device representing the content of a special sort of plan. It's a plan that, for any possible situation and any possible course of action, takes a stand on whether to pursue that course of action in that situation.) This gives us:

> (GIBBARDIAN SEMANTICS FOR MORAL DISCOURSE) The meaning of (2)
> is a set of world-hyperplan pairs, e.g.:
> {<*w, h*> | *h* includes a plan to blame those who steal at *w*}

The difference between the Simple Expressivist Semantics and Gibbardian Semantics will be important when it comes to evaluating whether expressivism is compatible with truth conditional semantics (Section 5).

What about epistemic modals? Expressivists about epistemic modals claim that when someone utters a sentence such as (3), they are expressing a *credal state*. Specifically, (3) expresses a credal state that leaves open the possibility that it is snowing.

The most thorough semantic implementation of expressivism about epistemic modals is due to Yalcin (2007). Yalcin's semantics resembles Gibbardian Semantics. Like Gibbard, Yalcin adopts a modest extension of a possible worlds semantics. Rather than taking the contents of sentences to be sets of world-hyperplan pairs, Yalcin takes them to be sets of *world-information state* pairs. An information state *s* is a formal representation of a credal state. Simplifying slightly, it is a set of worlds representing live doxastic possibilities. This yields:

> (CREDAL EXPRESSIVIST SEMANTICS FOR EPISTEMIC MODALS) The
> meaning of (3) is a set of world-information state pairs, specifically:
> {<*w, s*> | *s* includes at least one possible world where it is snowing}

On this view, the semantic content of (3) can be thought of as representing a property of a credal state: roughly, the property of assigning some positive credence to worlds where it is snowing.[9]

We have now laid out some of the key ideas behind relativism and expressivism. How should we decide between these two semantic frameworks? In what follows, I consider three potential choice points.

6. Truth conditional semantics and the Frege-Geach Problem

According to semantic orthodoxy, the meaning of a sentence is its truth conditions. One advantage of this view is that it provides a constructive recipe for assigning meanings to logically complex sentences on the basis of the meanings of their parts. The truth conditions of ⌜φ *or* ψ⌝ are a function of the truth conditions of φ, together with the truth conditions of ψ. So, if meanings are truth conditions, we have a nice story about how the meaning of the disjunction is a function of the meanings of its disjuncts.

Relativism is perfectly consistent with truth conditional semantics. Of course, relativists think truth conditions sometimes depend on the context of assessment. But assessment-sensitive truth conditions are still truth conditions. As a result, relativists have no difficulty handling logically complex sentences, e.g.:

(5) Stealing is wrong or stealing is harmless.

By contrast, many have thought that expressivism is inconsistent with truth conditional semantics. To see why, recall that at least our Simple Expressivist Semantics identified meanings with mental states. Such a "psychologistic" semantics is naturally construed as an alternative to truth conditional semantics.

If this is right, Simple Expressivism cannot use the standard truth conditional strategy for explaining the meanings of logically complex sentences in terms of the meanings of their parts. This is what gives rise to the notorious Frege-Geach Problem: the problem of providing an expressivist semantics for logically complex sentences such as (5).[10] And this suggests a reason for preferring relativism to Simple Expressivism. Since relativism is consistent with truth conditional semantics, it avoids the Frege-Geach Problem.

But before we place too much weight on this argument, we should bear in mind that there are other semantic implementations of expressivism. Recall Gibbardian Semantics, which identifies the meanings of moral sentences with sets of world-hyperplan pairs. As Yalcin (2012) observes, we can convert this into a truth conditional semantics. All that's needed is to take our circumstances of evaluation to be world-hyperplan pairs. This gives us:

(GIBBARDIAN TRUTH CONDITIONS) (2) is true at some $<w, h>$ iff h includes a plan to blame those who steal at w.

Similarly, Yalcin (2007) formulates his semantics for epistemic modals in truth conditional terms, where truth is relativized to world-information state pairs.

Since our Gibbardian expressivist can assign truth conditions to moral sentences, they can use the standard truth conditional strategy for explaining the meanings of logically complex sentences. For example, they can say that (5) is true at some $<w, h>$ iff either *stealing is wrong* is true at $<w, h>$ or *stealing causes harm* is true at $<w, h>$.

From a formal perspective, Gibbardian truth conditions and relativism have much in common. Both define truth and falsity not just relative to worlds, but relative to ordered pairs of a world and something else. For Gibbard, this something else is a hyperplan. For the relativist, it's an assessor. It's natural to wonder: is this a difference without a difference? Are the two frameworks notational variants?

Let me close this section by mentioning one way of trying to locate a genuine difference. Even if both relativists and Gibbardian expressivists adopt similar formalisms, there may be important differences in how the formalism is interpreted. For Gibbard a set of world-hyperplan pairs represents the content of a "plan-laden" state of mind: a combination of planning states and representational beliefs. Schroeder (2008a, 9, 2010, 131–133) questions whether Gibbardian expressivists can provide a systematic story about which plan-laden states of mind map onto arbitrary sets of world-hyperplan pairs. An example: the content of *stealing is wrong* is a set of w, h pairs that maps onto a planning state (a plan to blame thieves). The content of *stealing causes harm* is a set of w, h pairs that maps onto a representational belief (the belief that stealing causes harm). But what about the content of their disjunction? Gibbard has a recipe for associating this sentence with a set of w, h pairs. But, the objection runs, he hasn't told us how to make intuitive sense of the state of mind that corresponds to this set.

By contrast, it is less clear that relativists face this problem. First, relativists need not say that sentences express states of mind at all. Moreover, if they do say this, it would be natural for them to hold that all declarative sentences express representational mental states. When it comes to assessment-sensitive sentences, a natural option is to conceive of the relevant representational states as *de se* beliefs. This is particularly tempting for those relativists – such as Egan (2007, 2012) and Stephenson (2007) – who follow Lewis (1979) in modeling the contents of *de se* attitudes with centered propositions.

On this view, the state of mind expressed by (2) might be the *de se* belief that one disapproves of stealing. And the state of mind expressed by (5) might be the *de se* belief that either one disapproves of stealing or one inhabits a world where stealing is harmless. This would allow relativists to avoid the burden of making sense of disjunctions of representational mental states and conative attitudes.

This difference is related to the issue of how relativists and expressivists should understand beliefs on moral/modal matters, to which we now turn.

7. Belief and communication

Consider the following belief reports:

(6) Aliya believes stealing is wrong.
(7) Aliya believes it might be snowing.

What sort of mental states do these reports ascribe to Aliya?

As standardly developed, relativism and expressivism yield different answers. Start with relativism. On the most straightforward way of developing relativism, (6) and (7) ascribe beliefs in centered propositions. As noted in Section 6, one natural option is to think of this as a sort of *de se* belief. On this interpretation, (6) says Aliya has a *de se* belief that she herself disapproves of stealing. And (7) says she has a *de se* belief that her information is compatible with the possibility of snow.

Thus developed, relativism construes moral and modal beliefs as beliefs about one's own mental states (states of disapproval or states of information). In the terminology of Yalcin (2011), this makes moral and modal beliefs into *second-order* states.

By contrast, expressivists adopt a first-order conception of the relevant beliefs. According to moral expressivists, (6) is true just in case Aliya disapproves of stealing (or plans to blame thieves, etc.). On this view, moral beliefs are not *about* conative attitudes. Rather, they *are* conative attitudes. Similarly, expressivists about epistemic modals maintain that (7) is true just in case Aliya's belief state leaves open the possibility that it's snowing. To have a modal belief, on this view, is not to have a belief *about* one's information state. Rather, it's just to *be* in a certain information state.

These two pictures of belief are naturally paired with two different pictures of communication. Suppose we follow Stalnaker (1978) in holding that the goal of making an assertion is to get one's interlocutors to believe its content. Then the relativist says that the goal of asserting some moral or modal claim is to get your audience to share your second-order, *de se* belief. For example, the goal of asserting (2) is to get your listener to believe that *they* disapprove of stealing. By contrast, the expressivist says that the goal of asserting some moral or modal claim is to get your interlocutors to share your first-order mental state.

Which of these conceptions of belief – first-order or second-order – is more plausible? Different considerations pull in different directions. We've already noted (Section 6) that the first-order conception of moral belief faces the question of how to make sense of combinations of representational and conative attitudes. Another challenge facing first-order theorists is to make sense of *degrees of belief* on moral/modal matters. What is it, to have, say, 0.7 credence that stealing is wrong, if moral belief is just a conative attitude? By contrast, the second-order theorist has a comparatively easy time here: just plug in your preferred theory of *de se* credences and you'll get a theory of moral/modal credences.

At the same time, other considerations motivate the first-order conception. Suppose the psychological tests are in: turns out you disapprove of stealing. Does this fact provide a reason for you to believe that stealing is wrong? The second-order approach says "yes." But this seems counterintuitive. The first-order conception fares better: while the test results show you *do* disapprove of stealing, they do not show you *should* disapprove of it.[11]

Another argument for the first-order conception comes from motivational internalism: the idea that moral beliefs directly motivate actions in a manner similar to desires. If moral beliefs are conative attitudes, then we have a simple explanation for how moral beliefs exert their motivational magnetism.[12] By contrast, it's less clear whether the second-order conception explains this. To see why, suppose that (6) is true. According to the second-order conception, this means Aliya believes that she disapproves of stealing. But presumably she could be mistaken about this. If she is mistaken, why should we expect her to be motivated to avoid stealing? Of course, motivational internalism is controversial. But those sympathetic to the view may regard it as counting in favor of the first-order conception.

This suggests one important avenue for future research: compare the difficulties facing the first-order conception with those afflicting the second-order conception, and see which batch of problems proves more tractable.

But would settling this issue settle the relativism/expressivism debate? While relativists often embrace a second-order conception of moral/modal belief, it's not clear that this is forced; we might be able to develop a version of relativism that delivers a first-order conception. Here's a sketch of how this might go. One way to compositionally implement our relativist semantics for *wrong* is to use a contextual parameter supplying a function from contexts of assessment to whatever actions the assessor holds in disapproval.[13] We could then propose that the attitude verb *believes* shifts this parameter to a function from contexts of assessment to whatever actions the believer holds in disapproval. This would predict that (6) is true, as uttered in a context of utterance c and evaluated at a context of assessment $<w, a>$, iff Aliya disapproves of stealing at w.[14] For a structurally similar modification to the semantics of *believes* designed to predict that (7) ascribes a first-order state, see Ninan (2018).

Taking stock: there are significant differences between the standard relativist account of moral/modal belief and the standard expressivist account. However, it would be hasty to conclude that this reveals an essential difference between the two frameworks. Relativist who opt for a "first-order" semantics for belief reports may go a long way towards closing the gap between the two views.

8. Explaining disagreement

A further choice point between relativism and expressivism concerns their accounts of disagreement. As we saw in Section 4, expressivists usually explain disagreements about some domain in terms of disagreements in the mental states expressed. When it comes to moral disputes, this will be a disagreement in conative attitudes. Call this the "disagreement in attitude strategy."

Relativists typically take a different approach. On a relativist semantics, the content of Aliya's utterance of (2) is inconsistent with the content of Bruno's utterance of (4b) (*Stealing isn't wrong*). They are inconsistent in the sense that there is no context of assessment where both are true.

(Likewise, *mutatis mutandis*, in the modal case.) A natural thought is that we can leverage this fact to explain why their assertions constitute a disagreement. Call this the "discursive disagreement strategy":

> (DISCURSIVE DISAGREEMENT) Two assertions disagree with one another iff they have inconsistent contents.[15]

Which strategy for explaining disagreement should we prefer? Unfortunately, both face difficulties. Start with the disagreement in attitude strategy. Proponents of this strategy face the question: how should we understand disagreement in attitude? Without some account, there is a worry that expressivists are simply helping themselves to a phenomenon that they should be in the business of explaining.

One option is to explain disagreement in attitude in terms of what MacFarlane (2014) calls "noncotenability": two attitudes disagree iff it's not possible to hold both at the same time. However, this account risks overgenerating disagreements among *de se* desires. Suppose Aliya wants the last slice of cake. Suppose Bruno desires not to receive the last slice. If we adopt Lewis' view of the *de se*, the content of Aliya's desire is the centered proposition: $\{<w, x> \mid x$ gets the last slice at $w\}$. (Or, to put it another way, what she desires is to have the property: *getting the last slice*.) And the content of Bruno's desire is the centered proposition: $\{<w, x> \mid x$ doesn't get the last slice at $w\}$. (Equivalently, what he desires is to have the property: *not getting the last slice*.) Given this way of thinking about *de se* desires, their desires are noncotenable. But, intuitively, there's no disagreement here.

While Aliya and Bruno's cake-related desires are not cotenable, they are still jointly satisfiable: there's a way for both Aliya and Bruno to get what they want. So perhaps we should say that two attitudes disagree iff they cannot be jointly satisfied. This is a more promising approach, but it still raises a number of questions. First, what are the satisfaction conditions of the conative attitudes that constitute moral judgment? Second, does this story generalize to handle disagreements involving epistemic modals? (Do credences even have satisfaction conditions?) If not, there is a worry that the disagreement in attitude strategy does not encompass the full range of disagreement data.

Turn next to the discursive disagreement strategy. This strategy also faces its share of challenges. For example, it faces a challenge accounting for what Beddor (2018) calls "speechless disagreements": cases where two parties disagree on some matter even though they never converse about it. (For example, Aliya believes stealing is wrong, Bruno believes it isn't, but they never talk about the matter.)

To account for such cases, it seems that relativists, much like expressivists, need to make sense of disagreements in mental states. Indeed, relativists might explore analogues of the expressivist accounts of disagreement in attitude. For example, they might propose that two beliefs disagree with one another provided they are noncotenable. However, this proposal will run into an analogous worry: it overgenerates disagreements in ordinary *de se* beliefs (e.g., Aliya believes she is hungry, and Bruno believes he isn't).

Alternatively, relativists might suggest that two beliefs disagree iff they cannot be jointly accurate, where a belief is "accurate" just in case it is true relative to the believer's context of assessment (cf. MacFarlane 2014). But then they risk undergenerating disagreements. Assume that relativists stick with the picture of moral belief as a second-order, *de se* belief (Section 6). Then Aliya's belief that stealing is wrong is accurate just in case she disapproves of stealing. And Bruno's belief that stealing isn't wrong is accurate just in case he doesn't disapprove of stealing.

Since these beliefs can be jointly accurate, the proposal under discussion predicts that they do not disagree after all. It thus proves challenging for the relativist to avoid either overgenerating or undergenerating speechless disagreements.[16]

In summary, both expressivists and relativists face a number of difficulties when it comes to explaining disagreement. This should be troubling, given that disagreement data was one of the main motivations for abandoning contextualism in the first place! An important area for further research is to explore whether these difficulties can be solved, and, if so, whether the solution works equally well in a relativist or an expressivist setting.

9. Conclusion

In this chapter we've examined different ways of developing relativism and expressivism, and considered various choice points.[17] Along the way, we've found that while there are important differences between certain ways of developing relativism and certain ways of developing expressivism, it is much harder to identify points on which all expressivists and relativists disagree.

This suggests that in order to make progress, we should go one of two routes. First, we could try to develop a more rigorous characterization of both relativism and expressivism: we could lay down necessary and sufficient conditions for both. We could then try to prove an impossibility result of the form: "No relativist view can satisfy all of the conditions for being an expressivist view."

Alternatively, we could give up the assumption that there are important questions that distinguish all forms of relativism from all forms of expressivism. Rather than asking: "What are the reasons for preferring relativism to expressivism, or *vice versa*?" we should instead ask, "What are the reasons for preferring this particular version of relativism to this particular version of expressivism, or *vice versa*?"

Notes

1 See, a.o., Dreier (1990), Silk (2016), Khoo and Knobe (2018).
2 Contextualism about epistemic modals is defended by Derose (1991), Dowell (2011), Mandelkern (forthcoming), a.o. The canonical contextualist semantics for modals more generally comes from Kratzer (1981).
3 See e.g., Lasersohn (2005), Stephenson (2007), Egan (2007), MacFarlane (2011, 2014).
4 For arguments that contextualists can make sense of the relevant disagreement data, see, a.o., Dowell (2011), Plunkett and Sundell (2013).
5 For a wide-ranging overview and defense of relativism, see MacFarlane (2014).
6 Cf. Egan (2012).
7 Relativist semantics for epistemic modals are defended in Egan et al. (2005), Egan (2007), Stephenson (2007), MacFarlane (2011, 2014), Beddor and Egan (2018).
8 Expressivism traces its roots back to the emotivism of Ayer (1936). Influential developments include Stevenson (1944), Blackburn (1993, 1998), Gibbard (1990, 2003), Schroeder (2008a).
9 This approach bears important affinities to dynamic semantics for modals (e.g., Veltman 1996). According to a dynamic semantics, the meaning of a sentence is its context change potential: its ability to make a difference in the information of speakers and listeners. Epistemic modals function as "tests" on the context (modeled as a set of worlds). In particular, $\ulcorner$ *Might* φ $\urcorner$ tests to see whether the context contains at least one world where φ holds. There is an interesting question as to whether dynamic semantics should be classified as a type of expressivism.
10 The *locus classicus* of the problem is Geach (1965). For an overview of work on the problem, see Schroeder (2008b).

11 For similar arguments in favor of a first-order conception of modal belief, see Yalcin (2011), Moss (2013).

12 See e.g., Blackburn (1998), Gibbard (2003, ch.7).

13 A bit more precisely: let *g* be a function from a centered world to the set of things the center holds in disapproval at the world. Then our relativist lexical entry for *wrong* might go like this:

$$\llbracket \text{is wrong} \rrbracket^{c,g,\langle w,a \rangle} = \lambda x. x \in g(w,a).$$

14 A toy implementation: let $g_{w,S}$ be a constant function from centered worlds to the set of things that *S* holds in disapproval at *w*. And let *dox* be a function from a centered world $<w, a>$ to the set of centered worlds compatible with what *a* believes at *w*. Then our semantics for *believes* might go like this:

$$\llbracket S \text{ believes } \phi \rrbracket^{c,g,\langle w,a \rangle} = \text{true iff } \forall \langle w',x \rangle \in dox(S,w) : \llbracket \phi \rrbracket^{c,g_{w,S},\langle w',x \rangle} = \text{true}.$$

15 See Egan (2007, 2012), Stephenson (2007). Note that Discursive Disagreement is also available to certain expressivists. For example, while Gibbard pursues the disagreement in attitude strategy, Gibbardian Semantics agrees with the relativist that (2) and (4b) have inconsistent contents: there is no $<w, h>$ where both are true.

16 See Dreier (2009) for development of these concerns. For a proposed solution, see Beddor (2018).

17 I have focused on general choice points that arise in both the moral and modal domains. Another strategy is to argue for expressivism or relativism about a particular domain using domain-specific considerations. For example, Yalcin (2007) argues for expressivism about epistemic modals on the grounds that it explains the infelicity of embedded "epistemic contradictions" (e.g., *Suppose it's raining and it might not be*). For a relativist discussion of epistemic contradictions, see MacFarlane (2014, ch. 10.5).

References

Ayer, A. J. (1936), *Language, Truth, and Logic*, Dover: New York.

Beddor, B. (2018), "Subjective Disagreement," *Noûs*, doi:10.1111/nous.12240.

Beddor, B. and A. Egan (2018), "Might Do Better: Flexible Relativism and the QUD," *Semantics and Pragmatics* 11, doi:10.3765/sp.11.7.

Blackburn, S. (1993), *Essays in Quasi-Realism*, Oxford: Oxford University Press.

———. (1998), *Ruling Passions*, Oxford: Clarendon Press.

DeRose, K. (1991), "Epistemic Possibilities," *Philosophical Review* 100: 581–605.

Dowell, J. L. (2011), "A Flexible Contextualist Account of Epistemic Modals," *Philosophers' Imprint* 11(14): 1–25.

Dreier, J. (1990), "Internalism and Speaker Relativism," *Ethics* 101(1): 6–26.

———. (2009), "Relativism (and Expressivism) and the Problem of Disagreement," *Philosophical Perspectives* 23(1): 79–110.

Egan, A. (2007), "Epistemic Modals, Relativism, and Assertion," *Philosophical Studies* 133(1): 1–22.

———. (2012), "Relativist Dispositional Theories of Value," *Southern Journal of Philosophy* 50(4): 557–582.

Egan, A., J. Hawthorne and B. Weatherson (2005), "Epistemic Modals in Context," in *Contextualism in Philosophy*, edited by G. Preyer and G. Peter, Oxford: Oxford University Press.

Geach, P. (1965), "Assertion," *Philosophical Review* 74: 449–465.

Gibbard, A. (1990), *Wise Choices, Apt Feelings*, Cambridge, MA: Harvard University Press.

———. (2003), *Thinking How to Live*, Cambridge, MA: Harvard University Press.

Khoo, J. and J. Knobe (2018), "Moral Disagreement and Moral Semantics," *Noûs* 52(1): 109–143.

Kratzer, A. (1981), "The Notional Category of Modality," in *Words, Worlds, and Contexts: New Approaches in Word Semantics*, edited by H. Eikmeyer and H. Rieser, Berlin: De Gruyter.

Lasersohn, P. (2005), "Context Dependence, Disagreement, and Predicates of Personal Taste," *Linguistics and Philosophy* 28: 643–686.

Lewis, D. (1979), "Attitudes *De Dicto* and *De Se*," *Philosophical Review* 88: 513–543.

MacFarlane, J. (2011), "Epistemic Modals Are Assessment-Sensitive," in *Epistemic Modality*, edited by A. Egan and B. Weatherson, Oxford: Oxford University Press.

———. (2014), *Assessment Sensitivity: Relative Truth and Its Applications*, Oxford: Oxford University Press.

Mandelkern, M. (forthcoming), "How to Do Things with Modals," *Mind and Language*.

Moss, S. (2013), "Epistemology Formalized," *Philosophical Review* 122(1): 1–43.

Ninan, D. (2018), "Relational Semantics and Domain Semantics for Epistemic Modals," *Journal of Philosophical Logic* 47(1): 1–16.

Plunkett, D. and T. Sundell (2013), "Disagreement and the Semantics of Normative and Evaluative Terms," *Philosophers' Imprint* 13: 1–37.

Schroeder, M. (2008a), *Being for: Evaluating the Semantic Program of Expressivism*, Oxford: Oxford University Press.

———. (2008b), "What Is the Frege-Geach Problem?" *Philosophy Compass* 10: 703–720.

———. (2010), *Noncognitivism in Ethics*, New York: Routledge.

Silk, A. (2016), Discourse *Contextualism: A Framework for Contextualist Semantics and Pragmatics*, Oxford: Oxford University Press.

Stalnaker, R. (1978), "Assertion," in *Syntax and Semantics: Pragmatics*, vol. 9, edited by P. Cole, New York: New York Academic Press, 315–332.

Stephenson, T. (2007), "Judge Dependence, Epistemic Modals, and Predicates of Personal Taste," *Linguistics and Philosophy* 30(4): 487–525.

Stevenson, C. L. (1994), *Ethics and Language*, New Haven: Yale University Press.

Veltman, F. (1996), "Defaults in Update Semantics," *Journal of Philosophical Logic* 25(3): 221–261.

Yalcin, S. (2007), "Epistemic Modals," *Mind* 116(464): 983–1026.

———. (2011), "Nonfactualism About Epistemic Modality," in *Epistemic Modality*, edited by A. Egan and B. Weatherson, Oxford: Oxford University Press.

———. (2012), "Bayesian Expressivism," *Proceedings of the Aristotelian Society* 112(2): 123–160.

PART 9

Relativism in other areas of philosophy

56

RELATIVISM IN THE PHILOSOPHY OF RELIGION

Paul O'Grady

1. Philosophy of religion

Topics in philosophy of religion – for example, the existence and nature of God, faith and reason, the problem of evil – were issues traditionally dealt with by most major philosophers of the Western philosophical tradition. However, by the mid-twentieth century the scope of this subfield had dwindled to a rather narrow defensive exercise in pleading the intelligibility of religious discourse. The success of the Logical Positivists in dominating English-language thought, especially through Ayer's *Language, Truth and Logic* (1936) had a profoundly stultifying effect on philosophy of religion. Now regarded as unacceptably narrow, self-refuting and implausible, positivism's popularity in its heyday shaped the work of analytic philosophers of religion. Following on from this, ordinary language philosophy brought the great tradition of Western speculation to a rather fruitless place. Strawson's austere reading of Kant further reduced the space for metaphysical or theological speculation. Richard Swinburne noted that while he admired the rigor of the analysts, he disliked their narrowness (Swinburne 1993, 182).

Changes in general analytic philosophy brought welcome changes to philosophy of religion. The renewed interest in metaphysical thought – whether Kripkean essentialism, Lewisian possible worlds or Plantinga's defense of modal arguments – led to a renewal of constructive work, especially a re-examination of classic theistic arguments and renewed interest in the divine attributes. Work on the epistemology of theism included sustained challenges to Cliffordian evidentialism and major players in contemporary epistemology, for example Alvin Plantinga, Nicholas Wolterstorff, William Alston and Linda Zagzebski, all devoted attention to the rationality of religious belief.

A topic that has received sustained attention is Religious Pluralism. Given the enormous diversity in religious belief, what is one to make of that (e.g. Christians think that God has a triune structure, Muslims deny this; Christians and Muslims both believe in a creator, Buddhists deny this)? Perhaps the diversity and mutual incompatibility is a sign that they're all incorrect. More positively perhaps there's a way in which they can be construed as being true, or partially true. Maybe when properly interpreted the incompatibilities are only apparent. It is clear that this is terrain where the topic of relativism has relevance, since it proposes a way of allowing competing views to co-exist in some rationally acceptable way. In this chapter I wish to explore recent ways in which philosophical work about relativism has been used in philosophy of religion.

There are at least four areas in contemporary analytic philosophy that feed directly into discussions of relativism in respect of religious belief, several of which derive from Wittgenstein. In the first instance there has been recent clarification of the very idea of relativism and how this connects to the plurality of belief systems in religion, which is the topic of Section 2. Then there is work that takes religious belief as non-truth apt and hence seems to allow wide scope for diversity, covered in Section 3. Section 4 looks at an approach, also associated with Wittgenstein (especially the epistemological work of *On Certainty* (1961)), called relativism of distance that seems to lend itself to a defence of relativism about religious belief. In Section 5 the work of John Hick is examined, who has set out the most influential theory of relativism about religious belief. Finally, recent work in virtue epistemology has application to religious belief and some commentators have identified the potential for relativism in it, explored in Section 6.

2. Clarifying relativism

While the fact of religious diversity has always been apparent, the view that conflicting positions might nevertheless be equally viable, in some sense or other, emerged clearly in the twentieth century. Three terms have provided a framework for this debate: Exclusivism, Inclusivism and Pluralism. Here I follow Robert McKim's recent painstaking analysis of them, where he notes "it is striking how varied have been the interpretations of these key terms" (McKim 2012, 12).

Exclusivism, as the name suggests, is the view that a single or specific religious tradition is correct and all others are false. However, this blunt statement needs considerable finessing. McKim distinguishes between exclusivism about truth and exclusivism about salvation. While they often go together they may also separate. A tradition might claim unique access to religious truth, but allow that members of other traditions may achieve salvation (understood as ultimate happiness, whether in heaven, nirvana etc.). Alternatively they may claim that salvation is unique to this group (the phrase "*extra ecclesia nulla salus*" – outside the church there is no salvation – is attributed to the third-century Christian bishop St Cyprian of Carthage) but allow some true beliefs to other groups. However, issues arise as to what counts as membership of a tradition, e.g. has it to be explicitly publicly expressed or could it be implicit in practice? Furthermore many religious traditions hold beliefs and practices in common, so a major objection to exclusivism is that traditions hold too much in common for blunt exclusivism to be plausible. Nevertheless underlying exclusivism lie simple logical intuitions about contradiction. If someone holds a belief that flatly contradict one's own, it makes sense to think of one of them as false. So exclusivists fit in well with anti-relativistic arguments. Recent defenders of religious exclusivism include Alvin Plantinga (2000) and Peter Van Inwagen (1995). However, the boundaries between more sophisticated statements of exclusivism and inclusivism are blurred as modifiers and caveats are added.

Inclusivism weakens the severity of the exclusivist position. It still maintains a priority for one tradition in terms of truth and salvation, but allows that other traditions may have a lesser share in these and that indeed one might learn from these other traditions. Clearly questions about degree emerge in these discussions. How many beliefs might one allow as true in another tradition and how confident might one be in the superiority of one's own tradition in respect of religious truth and virtue over others? McKim accepts that there may be dissatisfaction with the looseness of the terms of the debate – but contends there is some connection between these terms and issues that really matter to people – how much truth one has, how we might be in a better position to others in respect of truth, to what extent we judge others to have truth and to what extent we can learn from them. Hence they still provide a useful framework.

Pluralism is the view that a great number of traditions are equally successful in respect of truth and salvation. Let us assume that there is a hierarchy of beliefs in a religious tradition, some central, some peripheral. Pluralism is a claim about the core beliefs. Pluralism holds that not merely traditions which hold the same or similar or even compatible beliefs (for example Christianity, Judaism and Islam holding there is a non-spatio-temporal creator) are acceptable, but that beliefs which are prima facie incompatible (e.g. Buddhist views on Nothingness and Christian views on Trinity) are nevertheless in some sense rationally acceptable. This clearly takes us to the territory of relativism. John Hick has articulated the most developed version of this view and I shall discuss it later. But first let us turn to recent terminological developments in analytic philosophy to help orient the discussion.

The resurgence of interest in relativism with New Relativism, especially the work of Max Kölbel (2002) and John MacFarlane (2014), led to a sharpening of the terms of the debate. Both MacFarlane and Kölbel believe that a defensible form of relativism is possible and well motivated for specific subject areas. This deals with aspects of experience which are less than fully objective, for example predicates of taste or treatments of future contingent propositions. Whether it is applicable to religious belief has not been explored, but the clarification in options available is helpful for the religious discussion.

A basic distinction is that between shifts in meaning and shifts in substantive beliefs. Some disputes which seem substantive can turn out to be merely verbal, for example an argument whether the distance between A and B is 5 miles or 8 kilometres. More subtle discussions in philosophy of language show how the same sentence may express different propositions depending on context of usage – e.g. "It is raining here, now." Relative to different contexts it is perfectly permissible to affirm and deny this sentence. However, when the meaning is clearly fixed, is it possible for the very same proposition to be treated as true relative to some parameter and false relative to another? Points of view or epistemic norms are some such potential parameters and current debates revolve around the plausibility of such relativization to parameters. An important distinction is between forms of contextualism and forms of genuine relativism. With contextualism, the impasse between P and not-P is explained away as a difference between P and Q. Indexical contextualism (or Revisionism) locates the difference in the realm of meaning, non-indexical contextualism puts it into the world. In either version, there is not genuine disagreement. Relativism involves what Kölbel has popularized as Faultless Disagreement, there is an impasse which is not explained away as involving difference, but which is nevertheless philosophically permissible. As Williamson puts it – contextualism is relativism tamed (Williamson 2005, 91). Armed with these distinctions let's now turn to diversity in religious belief.

3. Wittgenstein's influence

Whether Wittgenstein can be classified as a relativist has been debated. There are strands in his thought which point that way and other strands which resist it. However, in respect of religious discourse Wittgenstein's work is frequently understood as supporting relativism.

First, it is important to get a sense of the role of religion in Wittgenstein's thought as this is frequently not appreciated by those only familiar with his work on language, mind or mathematics. He famously remarked to his friend Drury, "I am not a religious man, but I cannot help seeing every problem from a religious point of view" (Rhees 1984, 79). He was brought up a Catholic but subsequently rejected religious belief as a young man. Russell noted that

Wittgenstein was a fierce critic of religion while at Cambridge. However, the experiences of the First World War had a significant impact on him. He read Tolstoy's version of the Gospels, to the extent that he became known as "the one with the Gospels." Under the influence of these writings he disbursed himself of his share in the large family fortune. His writings include references to God's will, the mystical and various other specifically religious topics. His reading included Augustine and Kierkegaard and the Vulgate. He opens his *Philosophical Investigations* (Wittgenstein 1953) with Augustine and the militantly atheistic Vienna Circle were bemused by his religious attitude when he was invited to meet with them. At various times he thought of becoming a monk and his friends thought it proper to have prayers said at his deathbed (influenced by Wittgenstein's own stated approval for a similar move by Tolstoy). However, he was not an orthodox religious believer and didn't identify with any religious group.

While he never wrote systematically in the area, comments on religious topics recur throughout Wittgenstein's work. He consistently took religion seriously, didn't share the dismissive attitude of positivists or naturalists and explicitly dissociated himself from scientism and "progress." The remarks in his later work have had the most impact in philosophy of religion. Wittgenstein's frequent method in philosophy was to expand the range of examples and contexts used in relation to a concept to show their plasticity and new uses. The constant leitmotif of his later work is practice – concepts make sense in a context which involves human activities. How this relates to religious belief is that his focus is less on propositional "belief" and more on practices, emphasizing the importance of ritual and ceremony.

In his early work he tends to equate talk about the meaning of life with talk about God: "To believe in God means to see that life has a meaning" (Wittgenstein 1961, 74). However, in line with his distinction between showing and saying, the mystical is not grasped in the manner of facts, rather it is shown. The very fact that there is anything in existence, rather than the specific configurations of things, excited in Wittgenstein the attitude he chose to label the mystical. This contrasts sharply with a scientific attitude. Science seeks to give explanations to all things, religion acknowledges that some things cannot be explained.

His later work contains several references to the Last Judgment (see Wittgenstein 1969). He retains an element of his early attitude in holding that explicit propositional beliefs are not relevant, rather belief in the Last Judgment manifests itself in how one lives. Responding to a student who had just converted to Catholicism, Wittgenstein said that he was not impressed by anyone buying a tightrope walker's costume, rather he was interested in what they did with it.

With a focus on practice, Wittgenstein's work seems to allow for wide diversity of beliefs. He criticizes the work of James Frazer, whose *Golden Bough*, an exploration of primitive religions, had been enormously influential (Wittgenstein 1993). Frazer thinks of these religions as advancing mistaken factual claims. In contrast, Wittgenstein argues that the function of utterances in such a context is not assertoric. In a famous passage he says:

> Was Augustine in error then, when he called upon God on every page of the Confessions?
>
> But – one might say – if he was not in error, surely the Buddhist holy man was – or anyone else – whose religion gives expression to completely different views. But none of them was in error, except when he set forth a theory.
>
> *(Wittgenstein 1993, 119)*

This suggests that religious utterances are not truth-apt. And this seems similar to the view advanced by Braithwaite who sees the essence of religion as having to do with right living and the associated doctrines serve as a narrative or story which has psychological power, but is not taken as true in the same way as factual claims are (Braithwaite 1971). A powerful objection against this approach is that it seems highly revisionary of the beliefs of most religious believers. Perhaps a few sophisticates interpret such beliefs in a non-cognitive way, but most people in the pews think of them as straightforwardly true (it counts as Revisionism in Kölbel's taxonomy). With this kind of non-cognitivism, diversity is rescued at the price of content. However, if one thinks that disagreement is constitutive of relativism (i.e. preserving clashing truth-claims) this approach doesn't genuinely count as relativistic.

A recent very interesting development of the strategy is found in Howard Wettstein's *The Significance of Religious Experience* (2012). He argues for the possibility of full-blooded religious practice shorn of metaphysical interpretation, analogous to how mathematicians engage with their practices while shelving questions of the metaphysical status of numbers. He argues that the importation of Greek metaphysics into the Hebrew religious tradition was a mistake. Its natural idiom is poetic and impressionistic rather than rationalistic and apodictic. Philosophy is not irrelevant to this – it helps clarify and extend our concept – but it doesn't provide a foundation. While not being in full agreement with Wettstein, Eleonore Stump argues that attention to the literary quality of biblical texts can allow us access to affective and relational states which is lacking in more traditional propositional approaches (Stump 2010). These approaches are sensitive to the specificities of religious discourse, allow for diversity and multiplicity, but are not properly relativistic. There is, however, another use of Wittgenstein which is closer to relativism.

4. Relativism of distance

There is debate about whether *On Certainty* articulates relativism. Martin Kusch has recently defended a reading which makes sense of this disagreement and which identifies a certain kind of relativism there (Kusch 2016). He discounts a non-cognitive reading of Wittgenstein on religious belief, citing *Lectures on Religious Belief* (1967) (so also does Steven Mulhall, arguing that a crude bifurcation between cognitivism and non-cognitivism goes right against the kind of gradualist and distinction-making strategies of later Wittgenstein, see Mulhall 2015). Instead of this Kusch identifies 30 different instances where one might speak of "certainty" in the text. He categorizes these into five groups. The first are perceptual beliefs, for which evidence is overwhelming (e.g. here is a table, I am in pain). The second are mathematical propositions. The third are fundamental scientific beliefs (e.g. water boils at 100°C). Fourthly come very general beliefs which have to be assumed (e.g. the earth exists). Finally, the fifth kind are religious beliefs, for example "Jesus only had a human mother." Wittgenstein treats different kinds of certainties in different ways – hence he allows relativism in some instances but not in others.

Kusch interprets Wittgenstein as maintaining that the way one might respond to a conversation partner depends on the degree of socio-cultural distance one might have with them. There is confidence in dealing with a friend, or a teacher might have confidence in discounting erroneous beliefs in children. Even with alien cultures one might be confident in dismissing some of their views, when there is empirical refutation possible. However, with religious beliefs things are different. We might think of ourselves as unfamiliar with what the beliefs mean or with how to assess them.

This naturally leads to thoughts about incommensurability. Kai Nielsen (1982) famously interpreted the later Wittgenstein as (wrongly) defending a form of fideism, where religious believers entertain different content to unbelievers. The work of D. Z. Phillips (1965) is often read this way, sealing off religious discourse from rationalistic challenge. Against this Kusch thinks that religious beliefs can indeed be understood and challenged and hence denies incommensurability of meaning. But there are some circumstance in which a kind of relativism is possible. To substantiate this he cites a discussion in *Lectures on Religious Belief.*

Wittgenstein speaks there of someone believing in the Last Judgment and someone else not believing this. Does a contradiction ensue? Not necessarily. To explain this Kusch distinguishes "belief" from "extraordinary belief." An extraordinary belief is constituted by the attitude of the believer. It is tied to strong emotions and pictures, it guides people's life and its expression can be the culmination of a form of life. It is not falsified or verified by empirical evidence. The contradiction is averted by distinguishing these two propositional attitudes. A non-believer rejects the Last Judgment as an ordinary belief, whereas a believer, who holds it as an extraordinary belief, uses it in a way that is on a different plane. Kusch interprets Wittgenstein as defending a kind of relativism of distance about this. This is not incommensurability of meaning since a non-believer can come to terms with the grammar of the term and its nexus of conceptual connections. However, it doesn't play the embedded role in his or her life that it does for the believer. So the kind of relativism in play is where opposing propositional attitudes (asserting/denying) do not come into play on the same content, since the content is too strange or distant from one of the parties.

One possible objection to this approach is to think of cases where people have been on both sides of the disagreement. Think of the case of one of Wittgenstein's literary executors, Anthony Kenny (see Kenny 1985). He had been a Catholic Priest, and therefore treated a certain set of beliefs as being "extraordinary beliefs." However, he then came to doubt the truth of those beliefs, with philosophical critique playing an important role in that process. So he rejected many of the beliefs and therefore came to deny them, but in an ordinary way. In this case distance doesn't seem to be in play, despite the shift between extraordinary and ordinary belief and relativism is not sustained. While it may well be useful to think that the way a believer holds certain kinds of belief (say dogmas or matters of faith) is usefully distinguished from other beliefs they might hold, this doesn't seem to be a kind of relativism. It seems closer to a kind of agnosticism, where one refuses to draw a conflicting judgment about those beliefs. I shall return to this topic later in thinking about virtue epistemology, but now I shall turn to a classic statement of relativism about religious belief, John Hick's Religious Pluralism.

5. Hick's religious pluralism

John Hick has articulated the most fully developed theory of relativism about religious belief, under the label "Religious Pluralism" (Hick 1989). He starts by drawing attention to the religious ambiguity of the universe. Certain considerations (religious experience, morality, natural theology, revelation) draw one to a religious understanding of reality, whereas others (evil, naturalistic conceptions) push against this. Hick believes that one's own interpretation of this mixture and especially the phenomenon of religious experience makes it rational for one to accept a religious interpretation of reality (drawing on anti-skeptical insights from Alston and Swinburne who defend a principle of credulity with appropriate defeaters). He particularly approves Alston's defense of the epistemic value of religious experience, while critiquing Alston's non-pluralistic stance.

Given the multiplicity of conflicting religious interpretations of reality, it is not rational to hold that one's own view is unique, or the only true and justified account. Indeed if one holds the exclusivist position that one's own tradition is true and all others false, this cuts against the general default claim defended by Alston that religious experience is veridical. Hick points to the contingency of one being born into a particular tradition and the dubiety of arguing for exclusivism from within such a tradition. Since religion has a soteriological as well as a cognitive dimension, Hick's Religious Pluralism holds that each religious tradition has as much validity in respect of soteriology and truth as any other. Religious traditions help one to move from a position of being self-centred to being reality-centred and numerous examples can be cited from a wide diversity of traditions to support this. He cites two key historical figures to support his views. From Kant, he takes the distinction between phenomenon and noumenon. Hick holds that the beliefs of each religious tradition have the status of phenomenal realities, which are human constructs attempting to respond to a noumenal reality beyond them. Hick calls this reality "The Real," but argues that it is transcendent and not adequately captured by any of the phenomenal accounts. He notes that some religions have personal account of the Real (e.g. Christians), whereas others have impersonal accounts (e.g. Buddhists), but these can be construed as phenomenal, not capturing the transcendent Real. From Aquinas, Hick takes the epistemological principle that "things are in the knower according to the mode of the knower" (*Summa Theologiae* II–II.1.2). From this he claims that the theoretical accounts of religious belief are shaped by history and culture and so do not claim to be true of reality itself. Needless to say, this is not consistent with Aquinas's own views (or indeed Kant's views on religion). However, Hick's bold thesis, that "the great world faith embody different perceptions and conceptions of, and correspondingly different responses to, the Real" (Hick 1989, 240) has generated a great deal of discussion and response.

One line of criticism is to attack the ontological relativism inherent in Hick's position. The degree of difference between, say, Christianity and Buddhism means that no intrinsic predicates can be attributed to the Real which are shared between these two traditions. Self-Subsistent Being versus Nothing, Creator versus non-Creator. The notion of the Real cannot even be cited as the cause of its phenomenal manifestations, as this is too reifying for Buddhists. The Real becomes a kind of cosmic porridge out of which different religions are shaped, but without any positive reality itself. One possible response is to restrict the scope of pluralism. As Alston noted one might reject certain positions (like Advaita Vedanta's monistic system, as too much at odds with the massive evidence of experience and critical common sense to be credible (Alston 2001, 49)). But one might find greater continuity among the Abrahamic traditions. Several other lines of critique have been leveled, including the one of revisionism. However, I agree with Robert McKim's assessment where he says

> Hick is one of the religious pioneers of our time and is so much more engaging than his various detractors; he can be read for inspiration and vision, whereas his critics can be read for distinctions and sometimes for new ways to dig in one's religious heels.
>
> *(McKim 2012, 108)*

6. Virtue epistemology

The basic insight of virtue epistemology is that the epistemological value a belief has derives primarily from the person holding the belief and the dispositions they have, rather than being some abstract properties of the belief itself. Hence traditional attempts to identify these properties, such as clarity and distinctness, or immediacy, or infallibility have been replaced with discussions of the reliability, conscientiousness, thoroughness etc. of the agent. Debates exist about how to characterize these features, whether to be minimalist and naturalized about them (reliabilist virtue epistemology) or to have richer accounts, akin to Aristotelian virtues (responsibilist virtue epistemology), (see Battaly 2008). Virtue epistemologists offer analyses of traditional questions (e.g. the definition of knowledge, response to skepticism, the nature of justification), but also offer ways of rethinking the epistemological project. In particular the discussion of epistemic value has raised the question of the kind of normativity involved in epistemology and how it relates to ethics. One promising answer to this question is to think of both epistemic and moral value as part of the project of human flourishing. We do well when we acquire reliably true, useful, relevant beliefs; that's why we value them.

A long-discussed topic in moral philosophy has been the nature of practical reasoning and whether there is a governing virtue that controls and adjusts all the elements in making complex moral judgments, traditionally called prudence. A more recent development has been a growth of interest in whether there is a similar virtue operational in the theoretical sphere which deals with the contingencies and complexities of theoretical judgments, Recent work on the virtue of wisdom has attempted to clarify this. The traditional bifurcation of practical and theoretical wisdom has been challenged. In a context of an account of human flourishing, perhaps a uni-fied account can be given of both moral and epistemic judgments – how to evaluate them and decide which are better or worse (Zagzebski 1996). Wisdom is a disposition to reflect on the structure of one's own thinking and orient it to support one's flourishing.

Wisdom, as a mode of meta-reflection on one's fundamental methodological and structural commitments, allows one to evaluate these commitments in terms of how they contribute to one's wellbeing. To make such an evaluation a rich range of contextual factors need to be considered – the socio-economic situation, one's values, one's goals, the role of theoretical reflection in this cluster of elements. It seems that exactly this situation applies in the case of religious belief as well as philosophical belief. An important evaluative dimension is to assess how it contributes to flourishing. Clearly these is scope for great debate – for example how does one determine flourishing or success? Yet this approach offers a promising framework for thinking about deep intellectual diversity (see O'Grady 2018).

The Wittgensteinian insight about the relationship to theory to practice is given a vehicle in this approach, but unlike the therapeutic approach advocated by many Wittgensteinians this is compatible with constructive philosophizing. How does it relate to religion and to relativism? Since religion has a soteriological dimension in that it explicitly deals with one's wellbeing, engaging with the virtue of wisdom brings philosophy closer to the goals of religious belief. William Alston explicitly connected the diversity of philosophical methods and beliefs with the diversity of religious belief (Alston 2001, 50), noting that religious diversity has an analogue in philosophical diversity. One of the main motivations for advocating religious pluralism has been the historical contingency of the religious tradition one is socialized in. One can note structurally similar features in philosophy – whether one is analytic or continental, naturalist or transcendental has a lot to do with training, mentoring, personal inclination, even geography (see Peter Hacker's discussion of attitudes to the analytic-synthetic distinction on different sides of the Atlantic in Hacker 1996). Of course this doesn't amount to determinism, or eradicate the point of argument. However, it does point to a wider range of factors in play than simple logical validity or soundness or argument in adopting a philosophical position.

Like relativism a virtue-theoretic use of wisdom is sensitive to diversity and rejects a dog-matic approach. However, such an approach is closer to contextualism than relativism. The contextual differences (the way virtues are indexed to metaphysics or worldviews) put a greater emphasis on difference rather than conflict. And furthermore one can be committed to a robust realism in this approach as long as one is sufficiently fallibilist about one's methods. After reflec-tive equilibrium has been achieved in the balance between belief, environment and wellbeing, one can affirm one's views in a realist manner. However, if one is fallibilist and allows that one may be mistaken and open to revision, the kind of dogmatism, opposition to which motivates the relativistic position, is taken away.

Acknowledgements

Thanks to Shane Ryan for comments on this chapter.

References

Alston, W. (2001), "Responses and Discussion" in *Dialogues in the Philosophy of Religion*, edited by J. Hick, Basingstoke, London: Palgrave-Macmillan, 37–52.

Aquinas, T. (1947), *Summa Theologiae*, translated by Fathers of the English Dominican Province, New York: Benziger Bros. edition.

Ayer, A. J. (1936), *Language, Truth, and Logic*, New York: Dover Publication, 1952.

Battaly, H. (2008), "Virtue Epistemology," *Philosophy Compass* 3: 639–663.

Braithwaite, R. (1971), "An Empiricist's View of Religious Belief," in *Philosophy of Religion*, edited by B. Mitchell, Oxford: Oxford University Press, 72–91.

Hacker, P. (1996), *Wittgenstein's Place in Twentieth Century Analytic Philosophy*, Oxford: Blackwell.

Hick, J. (1989), *An Interpretation of Religion*, Basingstoke: Palgrave Macmillan.

Kenny, A. (1985), *A Path from Rome*, Oxford: Oxford University Press.

Kölbel, M. (2002), *Truth Without Objectivity*, London: Routledge.

Kusch, M. (2016), "Wittgenstein's *On Certainty* and Relativism," in *Analytic and Continental Philosophy*, edited by S. Rinofner-Kreidl, Berlin and Boston: De Gruyter, 29–46.

MacFarlane, J. (2014), *Assessment Sensitivity: Relative Truth and Its Applications*, Oxford: Oxford University Press.

McKim, R. (2012), *On Religious Diversity*, Oxford: Oxford University Press.

Mulhall, S. (2015), *The Great Riddle: Wittgenstein and Nonsense, Theology and Philosophy*, Oxford: Oxford University Press.

Nielsen, K. (1982), *An Introduction to the Philosophy of Religion*, Basingstoke: Palgrave Macmillan.

O'Grady, P. (2018), "Epistemology and Wellbeing," *European Journal of Philosophy of Religion* 10(1): 97–116.

Phillips, D. Z. (1965), *The Concept of Prayer*, London: Routledge and Kegan Paul.

Plantinga, A. (2000), *Warranted Christian Belief*, Oxford: Oxford University Press.

Rhees, R. (ed.) (1984), *Recollections of Wittgenstein*, Oxford: Oxford University Press.

Stump, E. (2010), *Wandering in Darkness: Narrative and the Problem of Suffering*, Oxford: Oxford University Press.

Swinburne, R. (1993), "The Vocation of a Natural Theologian," in *Philosophers Who Believe*, edited by K. J. Clarke, Downers Grove, IL: Intervarsity Press, 179–202.

van Inwagen, P. (1995), "Non Est Hick," in *Rationality of Belief and the Plurality of Faith*, edited by T. Senor, Ithaca: Cornell University Press, 216–241.

Wettstein, H. (2012), *The Significance of Religious Experience*, Oxford: Oxford University Press.

Williamson, T. (2005), "Knowledge, Context and Agent's Point of View," in *Contextualism in Philosophy*, edited by in G. Preyer and G. Peter, Oxford: Oxford University Press, 91–114.

Wittgenstein, L. (1953) *Philosophical Investigations*, translated by G.E.M. Anscombe, Oxford: Basil Blackwell.

———. (1961), *Notebooks 1914–1916*, edited by G. E. M. Anscombe and G. H. von Wright, Oxford: Blackwell.

———. (1967), "Lectures on Religious Belief," in *Lectures and Conversations on Aesthetics, Psychology and Religious Belief*, edited by C. Barrett, Berkeley: University of California Press.

———. (1969), *On Certainty*, edited by G. E. M. Anscombe and G. H. von Wright, Oxford: Blackwell.

———. (1993), *Philosophical Occasions*, edited by J. Klagge and A. Nordmann, Indianapolis: Hackett.

Zagzebski, L. (1996), *Virtues of the Mind*, Cambridge: Cambridge University Press.

57

RELATIVISM AND EXPERIMENTAL PHILOSOPHY

Stephen Stich, David Rose, and Edouard Machery

1. Introduction

As the chapters in this volume make abundantly clear, the term "relativism" has been applied to a wide range of theories within philosophy and without. The relativist theories that will be our focus here can be roughly characterized by their commitment to four claims.

> (R-i) Some phenomenon of philosophical importance varies across individuals, cultures, or other demographic groups like genders, speakers of particular languages, socio-economic groups, or age cohorts. Among the phenomena that have been claimed to exhibit this sort of variation are moral principles, epistemic norms, aesthetic values, important philosophical concepts and entire conceptual schemes. Views of this sort are often labeled "descriptive relativism."
>
> (R-ii) The alternatives that vary across individuals or groups are incompatible or incommensurable; one can't coherently embrace more than one at any given time.
>
> (R-iii) There is no rational, non-question begging way of adjudicating these differences and determining which alternative is better or closer to the truth.

These three claims, by themselves, might lead, and often have led, to skepticism, the view that there is no way of knowing which alternative is better, or to nihilism, the view that notions like rational assessment or truth simply have no purchase in these areas. What relativism adds is a strategy for sidestepping these skeptical and nihilistic consequences. For the relativist, there are rational assessments and/or knowable truths in these contested domains, but

> (R-iv) truth or rational assessment, in these domains, must be relativized to some parameter that is determined by features of the individuals or groups involved.

All four of these claims have been hotly debated.

The term "experimental philosophy" has also been used in a variety of ways.[1] In this chapter, our focus will be on the sort of experimental philosophy that studies intuitions or judgments made about philosophically interesting cases and the factors that influence those judgments. We'll call it "X-Phi." To date, X-Phi has little to contribute to debates over (R-ii), (R-iii) and (R-iv). But it has played an increasingly important role in debates over (R-i), descriptive relativism, and that will be our focus in Sections 2–5. These debates are important because, as a number of authors have noted, claims like (R-i) are "often used as the starting point for philosophical debates on relativism,"[2] and "most common rationales" for claims like (R-ii) and (R-iii) "would be undermined" if (R-i) is not correct.[3] So the question we will be asking in Sections 2–5 is: *How has X-Phi contributed to debates over descriptive relativism?* In Section 6, we'll consider a very different way in which X-Phi and relativism are linked. Our focus will be on the growing literature exploring the extent to which ordinary people are relativists about moral judgments.

In the burgeoning X-Phi literature, there are studies reporting demographic variation in intuitions or judgments about philosophically important matters, including moral issues, knowledge, metaphysical questions and reference. In Sections 2–5 we'll review that literature and, in each case, ask what philosophically interesting conclusions can be drawn. But there is one contentious issue we want to sidestep. Recent years have seen a growing debate on what philosophical intuitions *are*.[4] Rather than joining this debate, we will simply stipulate that the sorts of intuitions we are concerned with are spontaneous judgments made in response to real or imagined cases, where the judgments are accompanied by little or no awareness of the psychological processes that give rise to them. This stipulation is appropriate, for present purposes, because it characterizes just about all of the intuitions surveyed in the studies we will discuss.

2. X-Phi and moral intuitions

Cultural differences in moral judgments have been acknowledged by philosophers since the time of Herodotus, and they have been widely explored in cultural anthropology following the pioneering work of Westermarck (1906) and Sumner (1934).[5] But in philosophical discussions of moral relativism, the focus is on the *moral principles* people accept and internalize, rather than on their judgments about specific cases. People's judgments, it is typically assumed, are inferred (often unconsciously) from their moral principles along with relevant non-moral beliefs. So if people have different non-moral beliefs about the agents involved in a case, the likely consequences of an agent's actions, or about apparently relevant scientific, metaphysical or theological matters, they might have different intuitions or judgments without any disagreement in underlying moral principles. In his book, *Hopi Ethics*, which is sometimes cited as a precursor of contemporary experimental philosophy, Richard Brandt (1954) reports a number of cases where the judgments of his Hopi informants differed from those he assumed would be made by contemporary Americans of European ancestry. Brandt explored an impressive range of potential non-moral disagreements that might account for these moral disagreements, but found none. Inspired by Brandt, and adopting the experimental methods of contemporary social psychology, Peng et al. (n.d.) found that Chinese and American participants differed in their judgments about a case in which an innocent man would have to be framed to prevent a riot in which many people would be injured and killed. Peng and colleagues searched for non-moral disagreements that might explain the moral disagreement, but – like Brandt – they found none. Unfortunately, most of the other recent X-Phi studies reporting cultural differences in moral intuitions have done little to determine whether the differences could be explained by differences in non-moral beliefs.

Trolley cases have loomed large in recent X-Phi work on moral judgment. In a typical trolley case, a runaway trolley is about to kill five innocent people unless

(1) it is rerouted on to another track, where only one innocent person will be killed (the "switch" case)

or

(2) it is stopped by pushing a large person off a footbridge and onto the track, killing the large person (the "footbridge" case).

A number of studies found little cross-cultural difference in judgments about trolley cases: Hauser et al. (2007) report data from participants in the USA, the UK, Canada, Australia, Brazil, and India; O'Neill and Petrinovich (1998) report data from Taiwan and the USA; Moore et al. (2011) compared Americans with mainland Chinese. However, there are also a number of studies that *have* found cross-cultural differences in trolley case intuitions, including Ahlenius and Tännsjö (2012) whose participants were American, Russian, and Chinese, and Gold et al. (2014) whose participants were English and Chinese. Xiang (2014) reports a particularly striking cross-cultural difference: 83% of Tibetan Buddhist monks and 64% of lay Tibetans judged that it is permissible to push the large person off the bridge to save five others, while less than 15% of American laypeople made that judgment. Abarbanell and Hauser (2010) found that rural and mostly uneducated Tseltal speaking Mayan informants in the Tenejapa region of Chiapas, Mexico, responded to dilemmas modeled on the switch and footbridge cases in much the same way as internet users in large-scale, highly industrialized and educated societies. But, to their surprise, they found that these rural Mayans differed from internet users in not recognizing a moral difference between actions and omissions. The Mayans did not judge actions that caused harm to be morally worse than omissions like failing to warn the victim. However, in a very large unpublished study, Kneer et al. (ms.) report finding little evidence of cross-cultural variation in the relative importance of action and omission for moral judgment.

Culture is not the only demographic variable whose effect on moral judgments has been explored experimentally. Friesdorf et al. (2015) report an impressive meta-analysis of forty studies involving 6,100 participants. The studies focused on trolley-style moral dilemmas where a protagonist can cause harm to avoid a greater harm. They found that *gender* had a substantial impact, with men more likely to make the "utilitarian" judgment.[6] Other gender effects have been reported by Banerjee et al. (2010),[7] Gao and Tang (2013), and by Gold et al. (2014), who also report an effect of *age* – older participants were more likely to judge that the action was wrong in their version of the switch case. Several studies have also found that participant's scores on tests for psychopathic personality traits and for low levels of empathic concern predict "utilitarian" responses on a variety of trolley-style sacrificial dilemmas (Kahane et al. 2015; Glenn et al. 2010).[8] And in a recently published paper, Hannikainen et al. (2018) report several studies establishing a generational effect on moral judgment: millennials are more likely than people born earlier to find the "utilitarian" option permissible in sacrificial dilemmas such as the footbridge and crying baby cases.

The X-Phi research reviewed in the previous two paragraphs provides ample reason to believe that people in different demographic groups often have significantly different moral intuitions about cases and often make very different moral judgments about them. But in discussions of moral relativism, this has rarely been doubted. What has been challenged is the claim that there are cross-cultural differences in underlying moral principles that combine with a wide range of non-moral beliefs to yield judgments about cases. And the experimental literature currently available does not come close to settling this issue. With the exception of the studies by Brandt and by Peng et al., there has been little systematic effort to determine whether demographic differences in moral intuition can be traced to differences in non-moral beliefs. Moreover, the studies reporting the influence of psychopathy and levels of empathy on moral judgments are just one facet of a vast recent literature that has been investigating the many ways in which the psychological systems responsible for emotion and for a wide range of other processes may interact with underlying moral principles to produce moral judgments. As Young and Saxe (2011, 323) report, "recent evidence . . . suggests that all moral judgments reflect the complex output of numerous psychological processes [including] controlled cognition and mental-state reasoning [in addition to] emotional responding." For example, studies indicate that "moral judgments are affected by individual differences in cognitive style, and working-memory capacity." And these differences could lead people who have identical underlying moral principles to make quite different moral judgments. Young and Saxe go on to note that:

> moral judgments are affected by individual differences in reasoning about intentions. Recently, we have discovered that individual differences in moral judgments of accidents (good intent, bad outcome) are correlated with individual differences in the engagement of a cortical region dedicated to mental-state reasoning, the right temporo-parietal junction (RTPJ) (Young and Saxe 2009). Participants with a high RTPJ response weigh beliefs and intentions more heavily when judging accidental harms, assigning less blame for the unintended bad outcome; participants with a low response blame more on the basis of the outcome alone. Temporarily disrupting RTPJ activity using transcranial magnetic stimulation also resulted in more outcome-based moral judgments (Young et al. 2010).
>
> *(Young and Saxe 2011)*

These effects could lead people to make different moral judgments even if they share all relevant underlying moral principles.

The take home message, here, is that there are many different ways in which diverging moral judgments about cases might be produced even if the people offering these diverging judgments shared *all* of the same underlying moral principles. So while the X-Phi literature exploring demographic differences in moral intuition and judgment has uncovered a number of intriguing facts, it has not (yet) made a convincing case for the claim that underlying moral principles vary in different demographic groups.

3. X-Phi and epistemological intuitions

In one of the earliest and most widely cited X-Phi papers, Weinberg et al. (2001) reported substantial cultural differences in intuitions about familiar epistemological thought experiments. Their most striking finding focused on a Gettier case:

> Bob has a friend, Jill, who has driven a Buick for many years. Bob therefore thinks that Jill drives an American car. He is not aware, however, that her Buick has recently been stolen, and he is also not aware that Jill has replaced it with a Pontiac, which is a different kind of American car.

After reading the case, participants were asked:

Does Bob really know that Jill drives an American car, or does he only believe it? The two responses available were:

REALLY KNOWS

and

ONLY BELIEVES.

The participants were undergraduates at an American university with various cultural backgrounds. Seventy-four percent of participants whose cultural background was European responded that Bob only believes, but 57% of participants whose cultural background was East Asian, and 61% of participants whose cultural background was South Asian responded that Bob really knows! One reaction to findings like this is to take them as evidence for the hypothesis that different cultural groups have different concepts of knowledge (Jackson 1998, 32; Stich and Tobia 2018; Stich and Mizumoto 2018). However, the Weinberg et al. study had a number of methodological flaws, most notably quite small sample sizes, and two more recent studies failed to replicate the Weinberg et al. results (Seyedsayamdost 2015; Kim and Yuan 2015). Other studies using somewhat different methods and materials also failed to find cultural differences in Gettier intuitions (Nagel et al. 2013; Turri 2013).

Recently, two much larger, more systematic, and more methodologically sophisticated studies found evidence that a substantial majority of participants in a wide range of cultures have the intuition that protagonists in Gettier cases do not have knowledge (Machery et al. 2017a, 2017b). The Machery et al. (2017b) study, which included more than 2,000 participants in twenty-three countries, speaking seventeen languages, also found no gender difference in Gettier intuitions.

Could it be that the concept of knowledge is innate, as the linguist Anna Wierzbicka (2018) has argued? In the wake of the findings reported in Machery et al. (2017a), we cautiously hypothesized that there may indeed be an innate core folk epistemology. But we now think that conclusion was premature, since there has been very little cross-cultural exploration of epistemic intuitions in response to other sorts of hypothetical cases.[9] Moreover, in one recently published study, Waterman et al. (2018) found that knowledge judgments of Americans and Chinese are notably more sensitive than those of Indians to a salient possibility of error in "skeptical pressure" cases. Focusing on very different demographic categories, Starmans and Friedman (2014) have found that there are striking differences in intuitions about Gettier cases between philosophers and academics in other disciplines, and Machery at al. (2017b) found that a number of personality traits, including conscientiousness, neuroticism and openness to experience have a substantial effect on Gettier intuitions. All of these findings appear to pose a challenge to the innate core folk epistemology hypothesis. However, Machery et al. (2017b) also reported that reflectiveness, as measured by the Cognitive Reflection Task (Frederick 2005), correlated with the intuition that protagonists in Gettier cases do not have knowledge. This might be interpreted as evidence that people who report that Gettier cases *are* cases of knowledge are making a performance error – a judgment that does not reflect their own underlying concept of knowledge.

The bottom line is that there is still a great deal we do not know about cultural and other sorts of demographic variation in epistemic intuitions. The epistemic relativist who contends that the concept of knowledge varies across demographic groups will get at best very limited support from recent work in X-Phi.

4. X-Phi and metaphysical intuitions

Though a number of philosophers, anthropologists and linguists have maintained that people in different cultures have fundamentally different metaphysical concepts (Foucault 1972; Lévy-Bruhl 1923; Whorf 1956), there has been very little work in this area by experimental philosophers. The single exception is a study by Rose et al. (in press) that focused on the concept of artifact persistence, using an updated version of the classic "Ship of Theseus" puzzle (Rose 2015):

> John is an accomplished woodworker and sailor, whose lifelong hobby is building rowboats by hand. He built his first rowboat – which he named "Drifter" – thirty years ago. Over the years there has been wear and tear, and every single one of the original planks in that rowboat has been replaced.
>
> John – never one to throw anything out – has stored all of the original planks in his shed over the years. Last month John – realizing that he had accumulated enough old planks for a whole rowboat – took out his old plans for Drifter and assembled these old planks exactly according to his old plans. John now has two rowboats of the same design: the rowboat that resulted from gradually replacing the original planks used to build a boat thirty years ago and that now has none of its original planks, and the rowboat just built one month ago with all and only the original planks that were used thirty years ago.
>
> John has promised two of his friends – Suzy and Andy – that they can borrow Drifter for an outing. But Suzy and Andy disagree on which of the two rowboats is actually Drifter. Andy thinks that the rowboat just built a month ago is actually Drifter since it has exactly the same planks, arranged in exactly the same way as Drifter originally had. But Suzy thinks that the rowboat that resulted from gradually replacing the original planks used to build a boat thirty years ago is actually Drifter since, even though it has all new parts, this was just the result of normal maintenance.

Participants were asked whether they agreed with Suzy or Andy, and offered two options:

(1) I agree with Suzy that Drifter is the rowboat that resulted from gradually replacing the original planks used to build a boat thirty years ago and that now has none of its original planks.
(2) I agree with Andy that Drifter is the rowboat built a month ago with the planks and plans that were used thirty years ago.

Data were collected from over 2,700 participants in twenty-three countries, speaking eighteen languages. In some cultures, including the USA, Italy and China, a substantial majority of participants chose the first option. In other cultures, including Spain, Mongolia and Indonesia, about half the participants chose the first option and about half chose the second. And in the two small scale societies included in the sample, the Nasa people in a remote area of Colombia, and Bedouin in Israel, a majority of participants chose the second option. This is an intriguing finding. Though it is only one study, it suggests that there may indeed be dramatic differences in the concept of artifact persistence both between and within cultures.

The disagreement between compatibilists, who insist that free will and moral responsibility can exist even in a deterministic universe, and incompatibilists, who deny this, has been a staple of philosophical debate from Hobbes and Hume onward, and there is now a substantial X-Phi literature exploring people's intuitions about a wide variety of cases.[10] In a cross-cultural study, Sarkissian et al. (2010) found that a majority of participants in four very different cultures – Colombia, Hong Kong, India and the USA – agree that "it is possible for a person to be fully moral and responsible for their actions" in a deterministic universe. But on a very different demographic dimension, Feltz and Cokely (2009) found that judgments about free will and moral responsibility are influenced by personality. Extroverts are more likely than introverts to give compatibilist responses. Though we know of no philosophical relativists who have focused on personality differences, we are inclined to think this may be a fruitful area for relativists to explore.[11] Finally, Hannikainen et al. (ms.) report a systematic cross-cultural difference in the role of sourcehood (being the ultimate source of one's action) in ascription of responsibility and free will: Sourcehood matters much less in Asian countries than in the rest of the world, in line with Asians' tendency to explain actions by appealing to the situations of agents (an explanatory style called "situationism") instead of intrinsic properties of agents.

5. X-Phi and intuitions about reference

Appeals to a theory of reference play a surprisingly large role in philosophical arguments in metaethics, the philosophy of mind, the philosophy of science, the philosophy of race and elsewhere (Bishop and Stich 1998; Mallon et al. 2009). Thus a great deal turns on which theory of reference is correct. Since the revolutionary work of Kripke (1972/1980) and Putnam (1975), that is a question that has been hotly debated. Although many different sorts of arguments have been invoked in these debates, hypothetical cases and appeals to intuitions about those cases have unquestionably played an important role. Indeed, Mallon et al. (2009, 338) maintain that "Kripke's masterstroke was to propose a number of cases that elicited widely shared intuitions that were inconsistent with traditional descriptivist theories."[12] But in an early and widely discussed X-Phi paper, Machery et al. (2004) set out to explore whether those intuitions were shared in other cultures. One of the cases they used borrowed both the fanciful story and much of the wording from Kripke (1972/1980):

> Suppose that John has learned in college that Gödel is the man who proved an important mathematical theorem, called the incompleteness of arithmetic. John is quite good at mathematics and he can give an accurate statement of the incompleteness theorem, which he attributes to Gödel as the discoverer. But this is the only thing that he has heard about Gödel. Now suppose that Gödel was not the author of this theorem. A man called "Schmidt," whose body was found in Vienna under mysterious circumstances many years ago, actually did the work in question. His friend Gödel somehow got hold of the manuscript and claimed credit for the work, which was thereafter attributed to Gödel. Thus, he has been known as the man who proved the incompleteness of arithmetic. Most people who have heard the name 'Gödel' are like John; the claim that Gödel discovered the incompleteness theorem is the only thing they have ever heard about Gödel. When John uses the name "Gödel," is he talking about:
>
> (A) the person who really discovered the incompleteness of arithmetic? Or
> (B) the person who got hold of the manuscript and claimed credit for the work?

The participants in this experiment were students at an American university with European cultural background, and English-speaking Chinese students at a university in Hong Kong. A majority of the American participants chose (B), which Machery and colleagues took to be the appropriate response if a Kripke-style causal-historical account of reference is correct for proper names, but a majority of the Chinese participants chose (A), the response that would be appropriate according to a description theory of reference for names. In both cultural groups, there was also a substantial minority who made the opposite choice. Sytsma et al. (2015) report similar results in a study that compared the reference intuitions of American and Japanese participants, and in his 2017 book, Machery assembles a list of twelve studies using a variety of vignettes and response options, *all* of which report an Asian/Western difference in intuitions about the reference of proper names, with Westerners always being more likely than Asians to choose the causal-historical response.[13]

The Machery et al. (2004) study provoked a firestorm of controversy, with some critics arguing that the question asked in the study was ambiguous (Ludwig 2007; Deutsch 2009; Sytsma and Livengood 2011), some arguing that the study elicited the wrong intuitions (Martí 2009, 2014), some arguing that the experiment had targeted the wrong participants (Devitt 2011a, 2011b) and some proposing and conducting experiments with different designs (Devitt 2015; Devitt and Porot 2018). Machery and colleagues have responded to just about all of these arguments (Machery 2011; Machery and Stich 2012; Machery et al. 2013; Machery 2014; Machery et al. 2015).

Space does not permit discussing any of these debates in detail. But we will offer a few admittedly partisan observations. The first is that it is now very clear that there are both cross-cultural and within-culture differences in the sorts of semantic intuitions that have often been appealed to in debates between advocates of causal-historical theories of reference and advocates of descriptivist theories. That leaves philosophers concerned with reference with a pair of uncomfortable options. The first is to develop and defend some way of deciding which theory of reference is correct without appealing to these culturally variable intuitions. While a few authors, most notably Devitt and Porot (2018) have made some progress in this direction, there is a long way to go.[14] The second is to acknowledge that the sort of semantic intuitions that must be used as evidence for a theory of reference do indeed vary both between and within cultures. This would, of course, lend support to the contention that reference works in different ways in different cultures, which is a version of (R-i), the descriptive relativist contention that has been our focus in this chapter.

It might be thought that relativism about reference is an anodyne view, quite unlike relativism about moral principles, or knowledge, or basic ontological categories. Students learning French are warned about *faux amis* – words like "bras" and "location" that have very different meanings in English and French. There is nothing philosophically problematic about these cases. So one might think that differences across cultures or individuals in how the reference of terms is determined would be philosophically unproblematic. But Mallon et al. (2009) argue that this would be a mistake. Central to their argument is the observation that if intuitions about bizarre hypothetical cases (like Kripke's Gödel case) are crucial evidence about the contours of a speaker's reference relation, then reference differences, between individuals or between cultures, are typically covert and very difficult to discover. Thus it is very difficult to determine whether a pair of philosophers who seem to disagree on an ontological question actually do disagree; and it is equally difficult to determine whether philosophers who seem to agree on an ontological question are actually talking past one another.

6. Are the folk moral relativists or moral objectivists?

Our previous four sections have been concerned with the contributions X-Phi might make to assessing the claims of descriptive relativism. In this section we consider the role X-Phi can play in determining whether ordinary people ("the folk") are moral relativists or moral objectivists. Why should philosophers care about this? For many philosophers, the folk view provides an important constraint on metaethical theorizing. If the folk are objectivists about morality, then a metaethical theory that seeks to give an account of the nature of morality should also be objectivist, or, if it is not, it owes us an explanation of how the folk came to be mistaken about the nature of morality. Michael Smith provides a straightforward statement of this view.

> [W]e seem to think that moral questions have correct answers; that the correct answers are made correct by objective moral facts; that moral facts are wholly determined by circumstances and that, by engaging in moral conversation and argument, we can discover what these objective moral facts determined by the circumstances are.
>
> *(Smith 1994, 6)*

"The philosopher's task," Smith tells us, "is to make sense of a practice having these features" (Smith 1994, 5). These passages are quoted in Sarkissian (2016) who also quotes passages from Frank Jackson, Stephen Darwall, Simon Blackburn and Richard Joyce, all of whom agree that commonsense morality is objective. Other philosophers have been more than a bit skeptical about the alleged anti-relativism of folk morality. In response to Jackson's claim that the folk are committed to moral objectivism, Stich and Weinberg (2001) couldn't help "wondering whether Jackson ever talks to undergraduates," since many of *their* undergraduates claimed to be moral relativists.

Smith also provided an idea that experimental philosophers have adopted in their attempts to test the claim that folk morality is objectivist, not relativist.

> It is a platitude that our moral judgments at least purport to be objective. . . . Thus if A says "It is right to ø in circumstances C" and B says "It is not right to ø in circumstances C" then we take it that A and B disagree: that at least one of their judgments is true.
>
> *(Smith 1994, 86)*

Do the folk agree? Two early experimental explorations of the question, by Nichols (2004) and by Goodwin and Darley (2008), concluded that the answer was yes. But a closer look at the Goodwin and Darley study suggests a more complicated picture. Their participants gave very different answers about different moral issues. Beebe and Sackris (2016) and others have reported similar findings. The picture became even more complex when a study by Sarkissian et al. (2012) showed that objectivist responses declined if the disagreeing parties were said to be from different cultures or from different planets. Providing some experimental backing for Stich and Weinberg's quip, Beebe and Sackris also found that relativist responses were strongly correlated with age; moral objectivism is at its lowest during the late teenage years and early adulthood. "At first glance," Beebe (2015, 21) writes in an article reviewing these and other findings, "these data appear to be bad news for the dominant view in analytic philosophy that ordinary individuals are moral objectivists." But he goes on to argue that there are significant methodological problems in all of these studies. The bottom line, we think, is that experimental work, to date, has certainly not shown that the folk are moral relativists, but it offers little encouragement for the many philosophers who insist that the folk are moral objectivists.

Notes

1 For useful discussion, see Sytsma and Livengood (2016, chs. 1–3), and O'Neill and Machery (2014).
2 Baghramian and Carter (2017, §2).
3 Gowans (2018, §4). Gowans' focus is moral relativism. But his comment is equally true for relativism in most other domains.
4 For useful overviews of the debate, see Alexander (2012, ch. 2), Nado (2014) and Machery (2017, ch. 1).
5 For a brief, colorful review of some memorable examples, see Prinz (2007, ch. 5).
6 Following the widespread practice in the experimental literature, we'll use the term "utilitarian" (in scare quotes) for judgments that morally approve of causing the death of a small number of people to save the lives of a larger number. However, as Kahane et al. (2015) note, in many cases these judgments would not be condoned by philosophers in the Utilitarian tradition.
7 For discussion of this interpretation of Banerjee et al. (2010), see Machery (2017, 62–63).
8 A number of other studies, including Bartels and Pizarro (2011) and Gleichgerrcht and Young (2013) have reported that psychopathic personality traits and low empathy correlate with participants' "utilitarian" responses when they are asked *which option they would choose* in these sorts of dilemmas, rather than explicitly asking for a moral judgment.
9 For further discussion, see Stich (2018).
10 For useful overviews, see Björnsson and Pereboom (2016) and Chan et al. (2016).
11 For further discussion, see Hannikainen et al. (ms.).
12 Descriptivist theories of reference for names maintain that the referent of a name is the person or object that uniquely or best satisfies a description that speakers associate with the name. By contrast, causal-historical theories, of the sort proposed by Kripke, maintain that the referent of a name is the person or object that plays an appropriate role in the causal explanation of the speaker's current use of the name. Both theories have been elaborated in a variety of ways. For a useful overview, see Cumming (2016).
13 See Cova et al. (in press) for one failed replication.
14 For an argument for this conclusion, relying on Devitt's attempt to provide a detailed theory of reference in his 1981 book, *Designation*, see Machery et al. (2013, 624ff.).

References

Abarbanell, L. and M. Hauser (2010), "Mayan Morality: An Exploration of Permissible Harms," *Cognition* 115: 207–224.

Ahlenius, H. and T. Tännsjö (2012), "Chinese and Westerners Respond Differently to the Trolley Dilemmas," *Journal of Cognition and Culture* 12: 195–201.

Alexander, J. (2012), *Experimental Philosophy: An Introduction*, Cambridge: Polity Press.

Baghramian, M. and J. A. Carter (2017), "Relativism," in *The Stanford Encyclopedia of Philosophy*, Summer 2017 edition, edited by E. N. Zalta, https://plato.stanford.edu/archives/sum2017/entries/relativism/.

Banerjee, K., B. Huebner and M. Hauser (2010), "Intuitive Moral Judgments Are Robust Across Variation in Gender, Education, Politics and Religion: A Large-Scale Web-Based Study," *Journal of Cognition and Culture* 10: 253–281.

Bartels, D. and D. Pizarro (2011), "The Measure of Morals: Antisocial Personality Traits Predict Utilitarian Responses to Moral Dilemmas," *Cognition* 121: 154–161.

Beebe, J. (2015), "The Empirical Study of Folk Metaethics," *Etyka* 50: 11–28.

Beebe, J. and D. Sackris (2016), "Moral Objectivism Across the Lifespan," *Philosophical Psychology* 29: 919–929.

Bishop, M. and S. Stich (1998), "The Flight to Reference, or How Not to Make Progress in the Philosophy of Science," *Philosophy of Science* 65: 33–49.

Björnsson, G. and D. Pereboom (2016), "Traditional and Experimental Approaches to Free Will and Moral Responsibility," in *A Companion to Experimental Philosophy*, edited by J. Sytsma and W. Buckwalter, Oxford: Wiley-Blackwell, 142–157.

Brandt, R. B. (1954), *Hopi Ethics: A Theoretical Analysis*, Chicago: University of Chicago Press.

Chan, H., M. Deutsch and S. Nichols (2016), "Free Will and Experimental Philosophy," in *A Companion to Experimental Philosophy*, edited by J. Systma and W. Buckwalter, Oxford: Wiley-Blackwell, 158–171.

Cova, F., et al. (in press), "Estimating the Reproducibility of Experimental Philosophy," *Review of Philosophy and Psychology*.

Cumming, S. (2016), "Names," in *The Stanford Encyclopedia of Philosophy*, Fall 2016 edition, edited by E. N. Zalta, https://plato.stanford.edu/archives/fall2016/entries/names/.

Deutsch, M. (2009), "Experimental Philosophy and the Theory of Reference," *Mind and Language* 24: 445–466.

Devitt, M. (1981), *Designation*, New York: Columbia University Press.

———. (2011a), "Experimental Semantics," *Philosophy and Phenomenological Research* 82(2): 418–435.

———. (2011b), "Whither Experimental Semantics?" *Theoria* 72: 5–36.

———. (2015), "Testing Theories of Reference," in *Advances in Experimental Philosophy of Language*, edited by J. Haukioja, London: Bloomsbury Press, 31–63.

Devitt, M. and N. Porot (2018), "The Reference of Proper Names: Testing Usage and Intuitions," *Cognitive Science* 42: 1552–1585.

Feltz, A. and E. Cokely (2009), "Do Judgments About Freedom and Responsibility Depend on Who You Are? Personality Differences in Intuitions About Compatibilism and Incompatibilism," *Consciousness and Cognition* 18: 342–350.

Foucault, M. (1972), *The Archaeology of Knowledge*, New York: Pantheon.

Frederick, S. (2005), "Cognitive Reflection and Decision Making," *Journal of Economic Perspectives* 19: 25–42.

Friesdorf, R., P. Conway and B. Gawronski (2015), "Gender Differences in Response to Moral Dilemmas: A Process Dissociation Analysis," *Personality and Social Psychology Bulletin* 41: 696–713.

Gao, Y. and S. Tang (2013), "Psychopathic Personality and Utilitarian Moral Judgment in College Students," *Journal of Criminal Justice* 41: 349–390.

Gleichgerrcht, E. and L. Young (2013), "Low Levels of Empathic Concern Predict Utilitarian Moral Judgments," *PLoS One* 8: e60418.

Glenn, A., S. Kolevam, R. Iyer, J. Graham and P. Ditto (2010), "Moral Identity in Psychopathy," *Judgment and Decision Making* 5: 497–505.

Gold, N., A. Colman and D. Pulford (2014), "Cultural Differences in Response to Real-Life and Hypothetical Trolley Problems," *Judgment and Decision Making* 9: 65–76.

Goodwin, G. and J. Darley (2008), "The Psychology of Meta-Ethics: Exploring Objectivism," *Cognition* 106(3): 1339–1366.

Gowans, C. (2018), "Moral Relativism," in *The Stanford Encyclopedia of Philosophy*, Summer 2018 edition, edited by E. N. Zalta, https://plato.stanford.edu/archives/sum2018/entries/moral-relativism.

Hannikainen, I., E. Machery and F. Cushman (2018), "Is Utilitarian Sacrifice Becoming More Morally Permissible?" *Cognition* 170: 95–101.

Hannikainen, I., E. Machery, D. Rose, S. Stich, et al. (ms). "Sourcehood and Free Will in 21 Countries."

Hauser, M., F. Cushman, L. Young, R. Kang-Xing Jin and J. Mikhail (2007), "A Dissociation Between Moral Judgments and Justifications," *Mind and Language* 22: 1–21.

Jackson, F. (1998), *From Metaphysics to Ethics: A Defense of Conceptual Analysis*, Oxford: Oxford University Press.

Kahane, G., J. Everett, B. Earp, M. Farias and J. Savulescu (2015), "'Utilitarian' Judgments in Sacrificial Moral Dilemmas Do Not Reflect Impartial Concern for the Greater Good," *Cognition* 134: 193–209.

Kim, M. and Y. Yuan (2015), "No Cross-Cultural Differences in the Gettier Car Case Intuition: A Replication Study of Weinberg et al. 2001," *Episteme* 12: 355–361.

Kneer, M., C.Y. Olivola, Y. Kim, E. Machery, S. Stich, D. Rose, et al. (ms.), "The Action/Omission Distinction Across Cultures."

Kripke, S. (1972/1980), *Naming and Necessity*, Cambridge, MA: Harvard University Press.

Lévy-Bruhl, L. (1923), *Primitive Mentality*, London: George Allen & Unwin.

Ludwig, K. (2007), "The Epistemology of Thought Experiments: First-Person Approach vs. Third-Person Approach," *Midwest Studies in Philosophy* 31: 128–159.

Machery, E. (2011), "Variation in Intuitions About Reference and Ontological Disagreements," in *A Companion to Relativism*, edited by S. Hales, Oxford: Wiley-Blackwell, 118–136.

———. (2014), "What Is the Significance of the Demographic Variation in Semantic Intuitions?" in *Current Controversies in Experimental Philosophy*, edited by E. Machery and E. O'Neill, New York: Routledge, 3–16.

———. (2017), *Philosophy Within Its Proper Bounds*, Oxford: Oxford University Press.

Machery, E., R. Mallon, S. Nichols and S. P. Stich (2004), "Semantics, Cross-Cultural Style," *Cognition* 92(3): B1–B12.

———. (2013), "If Folk Intuitions Vary, Then What?" *Philosophy & Phenomenological Research* 86: 618–635.

Machery, E. and S. Stich (2012), "The Role of Experiment," in *The Routledge Companion to Philosophy of Language*, edited by G. Russell and D. Graff Fara, New York: Routledge, 495–512.

Machery, E., S. Stich, D. Rose, A. Chatterjee, K. Karasawa, N. Struchiner, S. Sirker, N. Usui and T. Hashimoto (2017a), "Gettier Across Cultures," *Nous* 51: 645–664.

———. (2017b), "The Gettier Intuition from South America to Asia," in *Journal of the Indian Council of Philosophical Research*, special issue on experimental philosophy edited by J. Knobe, E. Machery and S. Stich, doi:10.1007/s40961-017-0113-y.

Machery, E., J. Sytsma and M. Deutsch (2015), "Speaker's Reference and Cross-Cultural Semantics," in *On Reference*, edited by A. Bianchi, Oxford: Oxford University Press, 62–76.

Mallon, R., E. Machery, S. Nichols and S. Stich (2009), "Against Arguments from Reference," *Philosophy and Phenomenological Research* 59: 332–356.

Martí, G. (2009), "Against Semantic Multi-Culturalism," *Analysis* 69: 42–48.

———. (2014), "Reference and Experimental Semantics," in *Current Controversies in Experimental Philosophy*, edited by E. Machery and E. O'Neill, New York: Routledge, 17–26.

Moore, A., N. Lee, B. Clark and A. Conway (2011), "In Defense of the Personal/Impersonal Distinction in Moral Psychology Research: Cross-Cultural Validation of the Dual Process Model of Moral Judgment," *Judgment and Decision Making* 6: 186–195.

Nado, J. (2014), "Why Intuition?" *Philosophy & Phenomenological Research* 86: 15–41.

Nagel, J., V. San Juan and R. Mar (2013), "Lay Denial of Knowledge for Justified True Beliefs," *Cognition* 129: 652–661.

Nichols, S. (2004), "After Objectivity: An Empirical Study of Moral Judgment," *Philosophical Psychology* 17: 3–26.

O'Neill, E. and E. Machery (2014), "Experimental Philosophy: What Is It Good for?" In *Current Controversies in Experimental Philosophy*, edited by E. Machery and E. O'Neill, New York: Routledge, vii–xxix.

O'Neill, P. and L. Petrinovich (1998), "A Preliminary Cross-Cultural Study of Moral Intuitions," *Evolution and Human Behavior* 19: 349–367.

Peng, K., J. Doris, S. Nichols and S. Stich (n.d.), Unpublished data, discussed in Doris and Plakias (2008).

Prinz, J. (2007), *The Emotional Construction of Morals*, Oxford: Oxford University Press.

Putnam, H. (1975), "The Meaning of 'Meaning'," in *Language, Mind and Knowledge, Volume 7 of Minnesota Studies in the Philosophy of Science*, edited by K. Gunderson, Minneapolis: University of Minnesota Press, 131–193.

Rose, D. (2015), "Persistence Through Function Preservation," *Synthese* 192: 97–146.

Rose, D., E. Machery, S. Stich, et al. (in press), "The Ship of Theseus Puzzle," in *Oxford Studies in Experimental Philosophy*.

Sarkissian, H. (2016), "Aspects of Folk Morality: Objectivism and Relativism," in *A Companion to Experimental Philosophy*, edited by J. Sytsma and W. Buckwalter, Oxford: Wiley-Blackwell, 212–224.

Sarkissian, H., A. Chatterjee, F. de Brigard, J. Knobe, S. Nichols and S. Sirker (2010), "Is Belief in Free Will a Cultural Universal?" *Mind and Language* 25: 346–358.

Sarkissian, H., J. Park, D. Tien, J. Wright and J. Knobe (2012), "Folk Moral Relativism," *Mind and Language* 26(4): 482–515.

Seyedsayamdost, H. (2015), "On Normativity and Epistemic Intuitions: Failure of Replication," *Episteme* 12: 95–116.

Smith, M. (1994), *The Moral Problem*, Oxford: Blackwell.

Starmans, C. and O. Friedman (2014), "No, No, KNOW! Academic Disciplines Disagree About the Nature of Knowledge," Paper presented at the *Common-Sense Beliefs and Lay Theories*. Preconference at the Fifteenth Annual Society for Personality and Social Psychology, Austin, TX.

Stich, S. (2018), "Knowledge, Intuition and Culture," in *Metacognitive Diversity: An Interdisciplinary Approach*, edited by J. Proust and M. Fortier, Oxford: Oxford University Press, 381–394.

Stich, S. and M. Mizumoto (2018), "Manifesto," in *Epistemology for the Rest of the World*, edited by M. Mizumoto, S. Stich and E. McCready, New York: Oxford University Press, 1–11.

Stich, S. and K. Tobia (2018), "Intuition and Its Critics," in *The Routledge Companion to Thought Experiments*, edited by M. Stuart, Y. Fehige and J. R. Brown, Abingdon: Routledge, 369–384.

Stich, S. and J. Weinberg (2001), "Jackson's Empirical Assumptions," *Philosophy and Phenomenological Research* 62(3): 637–643.

Sumner, W. (1934), *Folkways*, Boston: Ginn.

Sytsma, J. and J. Livengood (2011), "A New Perspective Concerning Experiments on Semantic Intuitions," *Australasian Journal of Philosophy* 89: 315–332.

———. (2016), *The Theory and Practice of Experimental Philosophy*, Peterborough: Broadview Press.

Sytsma, J., J. Livengood, R. Sato and M. Oguchi (2015), "Reference in the Land of the Rising Sun: A Cross-Cultural Study on the Reference of Proper Names," *Review of Philosophy and Psychology* 6: 213–230.

Turri, J. (2013), "A Conspicuous Art: Putting Gettier to the Test," *Philosophers' Imprint* 13: 1–16.

Waterman, J., C. Gonnerman, K. Yan and J. Alexander (2018), "Knowledge, Certainty and Skepticism: A Cross-Cultural Study," in *Epistemology for the Rest of the World*, edited by M. Mizumoto, S. Stich and E. McCready, New York: Oxford University Press, 187–214.

Weinberg, J., S. Nichols and S. Stich (2001), "Normativity and Epistemic Intuitions," *Philosophical Topics* 29: 429–460.

Westermarck, E. (1906), *Origin and Development of the Moral Ideas*, 2 vols., New York: Palgrave Macmillan.

Whorf, B. (1956), *Language, Thought and Reality*, Cambridge, MA: MIT Press.

Wierzbicka, A. (2018), "I KNOW: A Human Universal," in *Epistemology for the Rest of the World*, edited by M. Mizumoto, S. Stich and E. McCready, New York: Oxford University Press, 215–250.

Xiang, L. (2014), *Would the Buddha Push the Man Off the Footbridge? Systematic Variations in the Moral Judgment and Punishment Tendencies of Han Chinese, Tibetans and Americans*, Bachelor of Arts, Harvard.

Young, L., J. Camprodon, M. Hauser, A. Pascual-Leone and R. Saxe (2010), "Disruption of the Right Temporo-Parietal Junction with Transcranial Magnetic Stimulation Reduces the Role of Beliefs in Moral Judgment," *Proceedings of the National Academy of Sciences of the USA* 107: 6753–6758.

Young, L. and R. Saxe (2009), "Innocent Intentions: A Correlation Between Forgiveness for Accidental Harm and Neural Activity," *Neuropsychologia* 47: 2065–2072.

———. (2011), "Moral Universals and Individual Differences," *Emotion Review* 3(3): 323–324.

INDEX

Milton Keynes UK
Ingram Content Group UK Ltd.
UKHW051537141024
449569UK00028B/1510